POLARIZATION PHENOMENA IN NUCLEAR PHYSICS
EIGHTH INTERNATIONAL SYMPOSIUM

AIP CONFERENCE PROCEEDINGS 339

POLARIZATION PHENOMENA IN NUCLEAR PHYSICS

EIGHTH INTERNATIONAL SYMPOSIUM
BLOOMINGTON, IN   SEPTEMBER 1994

EDITORS: EDWARD J. STEPHENSON
STEVEN. E. VIGDOR
INDIANA UNIVERSITY

American Institute of Physics   Woodbury, New York

Authorization to photocopy items for internal or personal use, beyond the free copying permitted under the 1978 U.S. Copyright Law (see statement below), is granted by the American Institute of Physics for users registered with the Copyright Clearance Center (CCC) Transactional Reporting Service, provided that the base fee of $2.00 per copy is paid directly to CCC, 27 Congress St., Salem, MA 01970. For those organizations that have been granted a photocopy license by CCC, a separate system of payment has been arranged. The fee code for users of the Transactional Reporting Service is: 0094-243X/87 $2.00.

© 1995 American Institute of Physics.

Individual readers of this volume and nonprofit libraries, acting for them, are permitted to make fair use of the material in it, such as copying an article for use in teaching or research. Permission is granted to quote from this volume in scientific work with the customary acknowledgment of the source. To reprint a figure, table, or other excerpt requires the consent of one of the original authors and notification to AIP. Republication or systematic or multiple reproduction of any material in this volume is permitted only under license from AIP. Address inquiries to Series Editor, AIP Conference Proceedings, AIP, 500 Sunnyside Boulevard, Woodbury, NY 11797-2999.

L.C. Catalog Card No. 95-77216
ISBN 1-56396-482-1
DOE CONF-9409103

Printed in the United States of America.

Contents

Preface xv
Committees xvii

I. HADRON FORM FACTORS

Spin Dependent Structure Functions of Nucleons and Nuclei 3
  A. W. Thomas
Determination of the Neutron Electric Formfactor in Quasielastic Collisions of Polarized Electrons with ^{3}He and ^{2}D 18
  H. G. Andresen, J. R. M. Annand, K. Aulenbacher, J. Becker,
  J. Blume-Werry, Th. Dombo, P. Drescher, J. E. Ducret, D. Eyl,
  H. Fischer, A. Frey, P. Grabmayr, S. Hall, P. Hartmann, T. Hehl,
  W. Heil, J. Hoffmann, J. D. Kellie, F. Klein, M. Leduc,
  M. Meierhoff, H. Möller, Ch. Nachtigall, M. Ostrick,
  E. W. Otten, R. O. Owens, S. Plützer, E. Reichert, D. Rohe,
  M. Schäfer, L. D. Schearer, H. Schmieden, K. Steffens, R. Surkau,
  and Th. Walcher
Measurement of the Proton's Axial Form Factor via Neutrino-Nucleon Scattering 32
  I. Stancu
A New Neutron Polarimeter and Measurement of G_E^n from the $^2H(\vec{e},e',\vec{n})^1H$ Reaction 47
  R. Madey, A. Lai, and T. Eden
Measuring G_E^n in Quasielastic $^3\vec{He}(\vec{e},e')$ 55
  J.-O. Hansen
The SAMPLE Experiment: Parity-Violating Electron Scattering from the Proton and Deuteron 63
  M. Pitt, J. Arrington, D. Beck, E. Beise, E. Candell, L. Cardman,
  R. Carr, G. Dodson, K. Dow, F. Duncan, M. Farkhondeh, B. Filippone,
  T. Forest, H. Gao, W. Korsch, S. Kowalski, A. Lung, R. McKeown,
  R. Mohring, B. Mueller, J. Napolitano, N. Šimičević, B. Terburg,
  and M. Witkowski
Progress at HERMES: Preparations for Measurements of Nucleon Spin Structure Functions 71
  S. F. Pate, for the HERMES Collaboration
The Deuteron Spin Structure Functions in the Bethe-Salpeter Approach and the Extraction of the Neutron Structure Function $g_1^n(x)$ 79
  A. Yu. Umnikov, L. P. Kaptari, K. Yu. Kazakov, and F. C. Khanna
Transverse Polarization in Deep Inelastic Scattering 87
  B. Vuaridel and the HELP Collaboration
RAPPORTEUR: Hadron Form Factors 95
  E. J. Beise

II. SYMMETRIES

Limit on the Electric Dipole Moment of ^{199}Hg 107
    B. R. Heckel

Search for Incomplete Parity-Violation in Leptonic and Semileptonic Weak Processes: Status and Perspectives 112
    J. Deutsch

Study of Parity and Time-Reversal Violation in Neutron-Nucleus Interactions 120
    Yi-F. Yen, J. D. Bowman, B. E. Crawford, P. P. J. Delheij, C. M. Frankle,
    K. Fukuda, C. R. Gould, A. A. Green, D. G. Haase, M. Iinuma,
    J. N. Knudson, L. Y. Lowie, A. Masaike, Y. Masuda, Y. Matsuda,
    G. E. Mitchell, S. I. Penttilä, H. Postma, N. R. Roberson, S. J. Seestrom,
    E. I. Sharapov, H. M. Shimizu, S. L. Stephenson, and V. W. Yuan

Parity Violation in P-P Scattering at TRIUMF 136
    J. Birchall, A. R. Berdoz, J. D. Bowman, J. R. Campbell, C. A. Davis,
    A. A. Green, P. W. Green, A. A. Hamian, D. C. Healey, R. Helmer,
    E. Korkmaz, L. R. Lee, C. D. P. Levy, R. E. Mischke, S. A. Page,
    W. D. Ramsay, S. D. Reitzner, G. Roy, P. W. Schmor, A. M. Sekulovich,
    J. Soukup, G. M. Stinson, T. Stocki, V. Sum, N. A. Titov, W. T. H. van Oers,
    and A. N. Zelenskii

Theoretical Overview of Parity Violation in pp Scattering 142
    M. Shmatikov

T-Violation in Neutron Spin Rotation 148
    Y. Masuda

Parity Nonconservation in ^{207}Pb 156
    M. Leuschner, B. Cain, J. E. Knott, A. Komives, W. M. Snow,
    J. J. Szymanski, A. Andalkar, I. C. Girit, M. Petrov, and J. D. Bowman

Measurement of Charge Symmetry Breaking in np Elastic Scattering at 350 MeV 162
    R. Abegg, A. R. Berdoz, J. Birchall, J. R. Campbell, C. A. Davis,
    P. P. J. Delheij, L. Gan, P. W. Green, L. G. Greeniaus, D. C. Healey,
    R. Helmer, N. Kolb, E. Korkmaz, L. Lee, C. D. P. Levy, J. Li,
    C. A. Miller, A. K. Opper, S. A. Page, H. Postma, W. D. Ramsay,
    J. Soukup, G. M. Stinson, W. T. H. van Oers, A. N. Zelenski, and J. Zhao

Overview of Charge Symmetry 172
    G. A. Miller

An Experimental Test of Parity-Even Time Reversal Invariance with MeV Neutrons 185
    P. R. Huffman, C. R. Gould, D. G. Haase, C. D. Keith, N. R. Roberson,
    M. L. Seely, and W. S. Wilburn

Test of Time-Reversal Invariance in Proton-Deuteron Scattering 191
    P. D. Eversheim, F. Hinterberger, J. Bisplinghoff, R. Jahn, J. Ernst,
    W. Kretschmer, H. Paetz gen. Schieck, and H. E. Conzett

Tests of Discrete Symmetries Using Polarization........................... 197
    B. M. K. Nefkens
RAPPORTEUR: Reflections on Symmetries at SPIN '94 202
    S. A. Page

III. FEW BODY SYSTEMS

Proton-Proton Interactions with Polarized Internal Targets in Storage
Rings... 213
    W. Haeberli
Few-Body Physics with Polarized Photons 230
    A. M. Sandorfi
The Spin Structure of ^{3}He... 245
    R. G. Milner
New Results in Nucleon-Nucleon Scattering at Low Energies................ 260
    W. Tornow
New Results in Nucleon-Nucleon Scattering at Intermediate Energies 275
    H. Spinka
Neutron-Proton Analyzing Power at 12 MeV and Charged πNN
Coupling Constant 290
    R. T. Braun, W. Tornow, D. E. González Trotter, C. R. Howell, R. Machleidt,
    C. D. Roper, F. Salinas, H. R. Setze, and R. L. Walter
A Measurement of the Spin Transfer Observable $D_{NN'}$ for p+p Elastic
Scattering at $T_p=200$ MeV 296
    S. M. Bowyer, S. W. Wissink, A. D. Bacher, T. W. Bowyer, S. Chang,
    W. Franklin, J. Liu, J. Sowinski, E. J. Stephenson, S. P. Wells, W. K. Pitts,
    and D. V. Bugg
Zero-Crossing Angle of the np Analyzing Power Below 300 MeV 302
    C. A. Davis, R. Abegg, A. R. Berdoz, J. Birchall, J. R. Campbell,
    L. Gan, P. W. Green, L. G. Greeniaus, R. Helmer, E. Korkmaz, J. Li,
    C. A. Miller, A. K. Opper, S. A. Page, W. D. Ramsay, A. M. Sekulovich,
    V. Sum, W. T. H. van Oers, and J. Zhao
Spin Observables in Neutron-Proton Elastic Scattering..................... 308
    A. Ahmidouch, J. Arnold, B. van den Brandt, M. Daum, Ph. Demierre,
    R. Drevenak, M. Finger, M. Finger, Jr., J. Franz, N. Goujon, P. Hautle,
    Z. Janout, Jr., W. Hajdas, E. Heer, R. Hess, R. Koger, J. A. Konter,
    H. Lacker, C. Lechanoine-LeLuc, F. Lehar, S. Mango, Ch. Mascarini,
    D. Rapin, E. Rössle, P. A. Schmelzbach, H. Schmitt, P. Sereni,
    M. Slunecka, R. Stachetzki, A. Teglia, and B. Vuaridel

$\Sigma^+ +p$ Scattering Experiment at KEK 314
    J. K. Ahn, B. Bassalleck, M. S. Chung, W. M. Chung, H. Enyo,
    T. Fukuda, H. Funahashi, Y. Goto, A. Higashi, M. Ieiri, M. Iinuma,
    K. Imai, Y. Itow, G. D. Kim, Y. D. Kim, J. M. Lee, A. Masaike,
    Y. Matsuda, S. Mihara, K. Okada, I. S. Park, Y. M. Park, N. Saito,
    Y. M. Shin, K. S. Sim, R. Susukita, R. Takashima, F. Takeuchi,
    P. Tlustý, S. Yamashita, S. Yokkaichi, and M. Yoshida

Cross Sections and Analyzing Powers of $p(\vec{p},\pi^+)d$ Very Near Threshold 319
    P. Heimberg, R. E. Segel, F-J. Chen, K. Ackerstaff, R. D. Bent,
    J. Blomgren, H. O. Meyer, H. Nann, B. v. Przewoski, T. Rinckel,
    A. Zhuralev, M. A. Pickar, G. Hardie, P. Pancella, E. Jacobsen,
    and J. D. Brown

Polarization Observables for the p-d Breakup Reaction and the
Nuclear Three-Body Force 325
    L. D. Knutson

Status and Future of Polarization Phenomena Investigations in
Backward Elastic Deuteron-Proton Scattering........................... 331
    I. M. Sitnik and V. P. Ladygin

Determination of the Asymptotic D- to S-State Ratio for the Triton
and ^{3}He via $(\vec{d},t)$ and $(\vec{d},^3He)$ Reactions 337
    Z. Ayer, B. Kozlowska, R. K. Das, H. J. Karwowski,
    and E. J. Ludwig

Measurement of Spin Observables in Quasielastic Scattering of
Polarized Protons from Polarized ^{3}He................................... 343
    M. A. Miller, K. Lee, A. Smith, J.-O. Hansen, C. Bloch,
    J. F. J. van den Brand, H. J. Bulten, D. DeSchepper, R. Ent,
    C. D. Goodman, W. W. Jacobs, C. E. Jones, W. Korsch, L. H. Kramer,
    M. Leuschner, W. Lorenzon, N. C. R. Makins, D. Marchlenski,
    H. O. Meyer, R. G. Milner, J. S. Neal, P. V. Pancella, S. F. Pate,
    W. K. Pitts, B. von Przewoski, T. Rinckel, G. Savopulos, J. Sowinski,
    F. Sperisen, E. R. Sugarbaker, C. Tschalär, O. Unal, T. P. Welch,
    and Z-L. Zhou

Elastic π^+ Scattering on Polarized ^{3}He.................................. 349
    M. A. Espy, S. P. Blanchard, J. E. Brash, B. Brinkmöller,
    G. R. Burleson, W. J. Cummings, B. J. Davis, D. Dehnhard,
    P. P. J. Delheij, C. M. Edwards, O. Häusser, R. Henderson,
    B. K. Jennings, M. K. Jones, B. A. Lail, J. L. Langenbrunner,
    B. Larson, W. Lorenzen, K. Maeda, C. L. Morris, B. Nelson,
    J. M. O'Donnell, M. A. Palarczyk, B. K. Park, S. I. Penttilä,
    D. R. Swenson, D. Thiessen, D. Tupa, and Q. Zhao

Polarization Transfer in p-d Scattering at 22.7 MeV 355
    W. Kretschmer and the POLAR Collaboration

RAPPORTEUR: Few Body Systems 361
    A. M. Eiró

IV. INTERMEDIATE-ENERGY HADRON-INDUCED REACTIONS

The Nuclear Spin-Isospin Response to Quasifree Nucleon Scattering 371
    T. N. Taddeucci

Spin Polarization and Weak Decays of Hypernuclei 386
    H. Ejiri

Measurement of Spin Rotation Parameters in Proton Elastic Scattering
from ^{58}Ni at E_p=300 MeV 395
    A. Tamii, H. Akimune, I. Daito, M. Fujiwara, K. Hatanaka,
    K. Hosono, T. Inomata, M. Nakamura, T. Noro, H. Sakaguchi,
    S. Toyama, M. Yamagoshi, M. Yoshimura, and M. Yosoi

Fragmentation of "Stretched" 6^- Strength in ^{28}Si$(\vec{p},\vec{p}')^{28}$Si 401
    J. Liu, E. J. Stephenson, A. D. Bacher, S. M. Bowyer, S. Chang,
    C. Olmer, S. P. Wells, S. W. Wissink, and J. Lisantti

Simultaneous Measurements of $(\vec{p},\vec{p}')$ and $(\vec{p},p'\gamma)$ Observables for
the 15.11 MeV, 1^+, T=1 State in ^{12}C at 200 MeV 407
    S. P. Wells, S. W. Wissink, A. D. Bacher, J. Beene, G. P. A. Berg,
    F. Bertrand, A. Betker, S. M. Bowyer, S. Chang, C. Foster,
    W. Franklin, M. Halbert, K. Hicks, D. Horen, J. Lisantti, J. Liu,
    P. Mueller, D. Olive, W. Schmitt, E. J. Stephenson, D. Stracener,
    and R. Varner

A Study of the Fermi (0^+) Transition in ^{14}C$(\vec{p},n)^{14}$N at 495 MeV 413
    D. A. Cooper, S. L. Delucia, B. A. Luther, J. B. McClelland,
    D. L. Prout, L. J. Rybarcyk, E. Sugarbaker, and T. N. Taddeucci

Measurement of the Polarization Transfer D_{NN} (0°) for $(\vec{p},\vec{n})$ Reactions
at 295 MeV 419
    T. Wakasa, M. B. Greenfield, K. Hatanaka, S. Ishida, N. Koori,
    H. Okamura, A. Okihana, H. Otsu, H. Sakai, N. Sakamoto,
    Y. Satou, and T. Uesaka

Isoscalar Spin Strength in ^{12}C and ^{40}Ca 425
    C. Djalali, M. Morlet, F. T. Baker, L. Bimbot, J. Guillot, C. Glashausser,
    B. N. Johnson, H. Langevin-Joliot, N. Marty, L. Rosier,
    E. Tomasi-Gustafsson, J. Van de Wiele, and A. Willis

Tensor Analyzing Power of the ^{12}C$(d,^2$He$)^{12}$B Reaction at 270 MeV 431
    H. Okamura, S. Ishida, N. Sakamoto, H. Otsu, T. Uesaka, T. Wakasa,
    Y. Satou, S. Fujita, H. Sakai, T. Niizeki, K. Katoh, T. Yamashita,
    Y. Hara, H. Ohnuma, T. Ichihara, and K. Hatanaka

Study of the ^{3,4}He$(\vec{p},n)$ Reactions at T_p=100 and 200 MeV 438
    C. M. Edwards, M. Palarczyk, L. C. Bland, B. D. Anderson,
    B. Brinkmöller, D. S. Carman, D. Dehnhard, M. A. Espy, J. L. Langenbrunner,
    R. Madey, Y. Wang, and J. W. Watson

Quasifree $(\vec{p},Np)$ Reaction Studies at 200 MeV 444
    D. S. Carman, L. C. Bland, N. Chant, T. Gu, G. M. Huber, J. Huffman,
    A. Klyachko, B. C. Markham, P. Roos, P. Schwandt, and K. Solberg

Exclusive Measurement of $s_{1/2}$ Proton Knockout Reaction 450
    T. Noro, M. Kawabata, M. Tanaka, K. Tamura, K. Hatanaka, N. Matsuoka,
    K. Takahisa, Y. Yuasa, Y. Mizuno, H. Yamazaki, K. Sagara, S. Morinobu,
    M. Nakamura, and A. Okihana

Study of Polarization in Quasi-Elastic Break-up Reaction ^{6}Li(p,2p)^{5}He
in Complete Kinematics at 1 GeV 456
    A. N. Prokofiev, N. P. Aleshin, S. L. Belostotski, Yu. V. Dotsenko,
    V. A. Efimovykh, O. Ya. Fedorov, A. A. Izotov, A. Yu. Kisselev,
    E. N. Komarov, O. V. Miklukho, V. I. Murzin, Yu. G. Naryshkin,
    D. A. Prokofiev, Yu. A. Scheglov, A. V. Shvedchikov, A. A. Zhdanov,
    and A. A. Zhgun

Scattering of Polarized Protons from Polarized Targets..................... 462
    W. G. Love

Charge Exchange Spin Observable Measurements on Pb at 795 MeV in
the Giant Resonance Region... 470
    D. Prout, E. Sugarbaker, S. Delucia, D. Cooper, B. Luther,
    C. D. Goodman, B. K. Park, J. Rapaport, L. Rybarcyk, and T. Taddeucci

Polarization Observables in $\vec{p}p \rightarrow pK\vec{Y}$ Reactions at 2.9 GeV................. 476
    J. Arvieux, F. Balestra, Y. Bedfer, R. Bertini, L. C. Bland, S. Bossolasco,
    F. Brochard, M. P. Bussa, I. V. Falomkin, L. Fava, L. Ferrero, R. Garfagnini,
    D. R. Gill, A. Grasso, W. W. Jacobs, V. I. Lyascenko, A. Maggiora,
    D. Panzieri, G. Piragino, G. B. Pontecorvo, V. Serdyuk, F. Tosello,
    V. I. Travkin, S. E. Vigdor, B. Zalikanov, and G. Zosi

RAPPORTEUR: Intermediate Energy Hadron-Induced Reactions 482
    H. Sakai

V. INTERMEDIATE-ENERGY ELECTROMAGNETIC INTERACTIONS

Spin Physics with an Intense CW Electron Beam in the 15–30 GeV
Range: the ELFE Program... 495
    J. M. Laget

The Measurement of the Target Asymmetry of the Reactions $\gamma p \rightarrow \pi^+ n$
and $\gamma p \rightarrow \pi^\circ p$ with the Bonn Frozen Spin Target and the PHOENICS-
Detector at ELSA... 505
    H. Dutz and the PHOENICS Collaboration

Photoproduction of η Mesons on the Proton............................. 515
    M. Bouché-Pillon, B. Saghai, and F. Tabakin

Photoproduction of $K^+ \Sigma^0$ on the Proton................................ 518
    J. C. David, C. Fayard, G. H. Lamot, F. Piron, and B. Saghai

Duality and $K^+ \Lambda$ Photoproduction on the Proton....................... 521
    B. Saghai and F. Tabakin

Exclusive Pion Production from ^{16}O Using Polarized Photons 524
K. Hicks, R. Finlay, J. Rapaport, R. Lindgren, V. Gladyshev, H. Baghaei,
A. Cichocki, T. Gresko, B. Norum, R. Sealock, L. Smith, S. Thornton,
A. Caracappa, S. Hoblit, O. Kistner, L. Miceli, A. Sandorfi, C. Thorn,
M. Khandaker, C. S. Whisnant, and M. Lucas

Status of the T_{20} Experiment at VEPP-3 530
S. G. Popov, S. I. Mishnev, D. M. Nikolenko, D. V. Petrov, I. A. Rachek,
A. V. Sukhanov, D. K. Toporkov, E. P. Tsentalovich, A. V. Volosov,
B. B. Wojtsekhowski, C. E. Jones, R. S. Kowalczyk, M. Poelker,
D. H. Potterveld, L. Young, R. J. Holt, R. Gilman, E. R. Kinney,
K. P. Coulter, J. A. P. Theunissen, C. W. de Jager, H. de Vries,
V. V. Nelyubin, V. V. Vikhrov, A. N. Osipov, and V. N. Stibunov

**A New Deuteron Tensor Polarimeter and a Measurement of t_{20} in e-d
Scattering at CEBAF** 545
S. Kox, C. Furget, J. S. Réal, L. Bimbot, C. Djalali, G. W. R. Edwards,
M. Garçon, C. Glashausser, B. N. R. Johnson, M. Morlet, L. Rosier,
E. Tomasi-Gustafsson, E. Voutier, A. Willis, and the CEBAF t_{20} Collaboration

Proton Knock-Out from Tensor Polarized Deuterium 551
E. Passchier, M. Ferro-Luzzi, Z.-L. Zhou, T. Botto, M. Bouwhuis,
J. F. L. van den Brand, D. Dimitroyannis, M. Doets, C. W. de Jager,
J. Konijn, D. J. J. de Lange, G. J. Nooren, N. Papadakis, I. Passchier,
P. Salle, J. J. M. Steijger, N. Vodinas, H. de Vries, C. Zegers, R. Alarcon,
S. Choi, J. Comfort, D. M. Nikolenko, S. G. Popov, I. Rachek, J. Lang,
H. Arenhövel, R. Ent, W. Leidemann, M. Bucholz, H. J. Bulten,
M. A. Miller, J. S. Neal, and O. Unal

RAPPORTEUR: Intermediate Energy Electromagnetic Interactions.......... 557
M. Garçon

VI. LOW-ENERGY NUCLEAR REACTIONS

**Cross Section and Polarization Measurements in the ^{12}C($\vec{d}$,p) and
^{12}C(d,$\vec{n}$) Reactions, and their CDCC Analysis** 563
H. Toyokawa and H. Ohnuma

Spectroscopy of ^{88}Y by Means of the ^{91}Zr($\vec{p},\alpha$)^{88}Y Reaction at 22 MeV 569
P. Guazzoni, U. Atzrott, G. Cata-Danil, G. Graw, R. Hertenberger,
D. Hofer, M. Jaskola, P. Schiemenz, G. Staudt, J. Tropilo,
E. Zanotti-Müller, and L. Zetta

**General Formulae of Cross Sections and Analyzing Powers in Low
Energy Limit—Application to ^{2}H($\vec{d},p$)^{3}H Reactions** 575
M. Tanifuji and H. Kameyama

**Inelastic Scattering of Low- and Intermediate-Energy Polarized
Protons from Various Light Nuclei** 581
A. Plavko, M. Onegin, V. Kudriashov, C. Olmer, A. Bacher, P. Schwandt,
and E. Stephenson

Singlet-State Contributions to Deuteron Elastic Scattering 587
  Y. Iseri and M. Tanifuji
A Constrained Dispersive Optical Model for the Neutron-Nucleus Interaction from −80 to +80 MeV for the Mass Region $27 \leq A \leq 32$ 593
  M. A. Al-Ohali, C. R. Howell, W. Torwnow, and R. L. Walter
Analyzing Powers for $\vec{^6Li} + {}^{12}C$ Scattering at 30 and 50 MeV 599
  P. L. Kerr, E. L. Reber, P. V. Green, K. W. Kemper, A. J. Mendez,
  K. Mohajeri, E. G. Myers, and B. G. Schmidt
Semi-Classical Analysis of Scattering of Deformed Heavy-Ions Below the Coulomb Barrier 605
  R. C. Johnson, E. J. Roberts, C. V. Sukumar, and D. M. Brink
Polarized ^6Li Studies at the Nuclear Structure Facility, Daresbury 611
  R. P. Ward, C. O. Blyth, H. D. Choi, N. M. Clarke, K. A. Connell,
  N. J. Davis, P. R. Dee, S. J. Hall, O. Karban, K. I. Pearce,
  C. N. Pinder, S. Roman, K. Rusek, D. B. Steski, and G. Tungate
Fusion of a Polarized Projectile with a Polarized Target 617
  J. A. Christley, R. C. Johnson, and I. J. Thompson
Spin Polarization of ^{23}Mg in $^{24}Mg+Au$, Cu and Al Collisions at 91 A MeV 623
  K. Matsuta, S. Fukuda, T. Izumikawa, M. Tanigaki, M. Fukuda,
  M. Nakazato, M. Mihara, T. Onishi, T. Yamaguchi, T. Miyake,
  M. Sasaki, A. Harada, T. Ohtsubo, Y. Nojiri, T. Minamisono,
  K. Yoshida, A. Ozawa, T. Kobayashi, I. Tanihata, J. R. Alonso,
  G. F. Krebs, and T. J. M. Symons
Spinflip Probability via the $^{26}Mg(^3He, t\gamma)^{26}Al$ Reaction 629
  H. Sakai, T. Aoyama, M. N. Harakeh, K. Kubota, H. Okamura,
  H. Otsu, Y. Satou, M. Tanaka, T. Uesaka, and T. Wakasa
RAPPORTEUR: Low Energy Nuclear Reactions 635
  D. Fick

VII. POLARIZED SOURCES AND TARGETS

Study of a Polarized Hydrogen Ion Source with Deuterium Plasma Ionizer 643
  A. S. Belov, G. E. Derevyankin, V. G. Dudnikov, V. S. Klenov,
  L. P. Nechaeva, Yu. V. Plohinsky, G. A. Vasil'ev, and V. P. Yakushev
Spin-Exchange Polarization Study at the TRIUMF OPPIS 650
  A. N. Zelenski, C. D. P. Levy, W. T. H. van Oers, P. W. Schmor,
  J. Welz, and G. W. Wight
A Dual-Optically-Pumped Polarized Negative Deuterium Ion Source 656
  M. Kinsho, Y. Mori, K. Ikegami, and A. Takagi
Polarized Ion Source Operation at IUCF 662
  V. Derenchuk, A. Belov, R. Brown, J. Collins, J. Sowinski,
  E. Stephenson, and M. Wedekind

The Polarized Ion Source for COSY 668
    P. D. Eversheim, R. Gebel, M. Altmeier, O. Felden, C. Heimann,
    M. Kammermann, W. Kretschmer, R. Weidmann, K. Mümmler,
    B. Aumüller, A. Glombik, H. Paetz gen. Schieck, S. Lemaitre,
    R. Reckenfelderbäumer, M. Eggert, and O. Suttorp

The HERMES-FILTEX Target Source for Polarized Hydrogen and Deuterium 674
    F. Stock and the HERMES Target Group

Polarized Internal Gas Target for Hydrogen and Deuterium at the IUCF Cooler Ring 680
    T. Wise, W. Haeberli, B. Lorentz, F. Rathmann, M. A. Ross, W. A. Dezarn,
    J. Doskow, J. G. Hardie, H. O. Meyer, R. E. Pollock, B. von Przewoski,
    T. Rinckel, F. Sperisen, and P. V. Pancella

A High Density Polarized Hydrogen Gas Target for Storage Rings 686
    K. Zapfe, B. Braun, H.-G. Gaul, M. Grieser, B. Povh, M. Rall, E. Steffens,
    F. Stock, J. Tonhäuser, C. Montag, F. Rathmann, D. Fick, and W. Haeberli

An Internal Polarized ^{3}He Target for Electron Storage Rings 692
    L. H. Kramer, D. DeSchepper, R. G. Milner, S. F. Pate, and T. Shin

Status on the Michigan-MIT Ultra-Cold Polarized Hydrogen Jet Target 698
    V. G. Luppov, B. B. Blinov, J. A. Bywater, S. Chin, V. V. Churakov,
    G. R. Court, W. A. Kaufman, D. Kleppner, A. D. Krisch, Yu. M. Melnik,
    J. B. Muldavin, T. S. Nurushev, J. S. Price, A. F. Prudkoglyad,
    R. S. Raymond, V. B. Shutov, and J. A. Stewart

High Intensity Polarized Ion Source at RCNP 702
    K. Hatanaka, K. Takahisa, H. Tamura, H. Kaneko, and I. Miura

Polarizing Stored Beams by Interaction with Polarized Electrons 708
    C. J. Horowitz and H. O. Meyer

A Possible Method to Produce a Polarized Antiproton Beam at Intermediate Energies 713
    H. Spinka, E. W. Vaandering, and J. S. Hofmann

VIII. CONCLUDING PRESENTATION

Spin Physics in the Next Decade 721
    J. M. Moss

APPENDICES

Conference Schedule and Plenary Session Talks 733
Parallel Sessions 739
SPIN'94 Participants List 755
Author Index 781

PREFACE

The Eighth International Symposium on Polarization Phenomena in Nuclear Physics was held September 15-22, 1994, on the campus of Indiana University in Bloomington, Indiana, USA. The Symposium met in conjunction with the Eleventh International Symposium on High Energy Spin Physics. Both conferences held sessions in the meeting rooms of the Indiana University Memorial Union. The Symposia were sponsored by the Indiana University Cyclotron Facility, Indiana University, the National Science Foundation, and the US Department of Energy. Additional support was made available from the International Science Foundation to underwrite the travel of visiting scientists from the countries of the former Soviet Union.

Holding these two Symposia at the same time and place recognized the strong overlap, both in scientific concerns and technical innovations, that characterizes research with spin and polarization in both nuclear and high energy physics. Both meetings were last held during the same year in 1990. At the 1990 nuclear conference it was recognized that there were benefits to bringing the meetings together, and a process was initiated that resulted in the Bloomington Symposia.

There were 260 participants registered for the Symposium on Polarization Phenomena in Nuclear Physics, and 269 registered for the Symposium on High Energy Spin Physics. A large overlap existed between the two meetings, with a total of 355 registrations.

The scientific program offered a variety of new results. Of special note since the last nuclear meeting was the progress that has occurred in the acceleration and storage of polarized beams in synchrotron rings. SLC, HERA, and LEP all report large electron polarizations, while experiments at IUCF and BNL with Siberian Snakes have demonstrated the feasibility of accelerating polarized hadron beams even in the presence of a large number of depolarizing resonances. These technical innovations will make possible a new generation of polarized beam experiments at the world's largest particle accelerators as well as open up the exploitation of the properties of storage rings for experiments with spin. Data on the spin structure functions of the nucleon were discussed extensively, with measurements from both CERN and SLAC reviewed. At the other end of the energy scale, new results were available on the measurement of parity violation using epithermal neutrons.

Plenary sessions for the two Symposia were scheduled mostly during the morning, with time shared between the two Symposia. Parallel sessions for both conferences were scheduled during the afternoon, with as many as six parallel sessions meeting at one time. There was one jointly-organized session on nucleon spin structure functions on Saturday afternoon. Toward the end of the conference, the discussions in the parallel sessions were reviewed by rapporteurs from each session.

A strong overlap in subjects for the two Symposia made it necessary to divide the topics and have the program committees for each meeting organize a part of the plenary and parallel sessions. It was agreed that the proceedings would be published in two volumes, each containing the plenary and parallel session talks organized by each committee. The nuclear organizing committee was responsible for discussions under the headings of hadron form factors, symmetries, few body systems including nucleon-nucleon scattering and resonance production, intermediate-energy hadron-induced reactions, intermediate-energy electromagnetic interactions, low-energy nuclear reactions, polarized hadron sources, and gaseous polarized targets. The nuclear organizing committee received 167 formal contributions to the program on these and other topics. These contributions were published in a separate, paper-bound volume distributed during registration, and formed the basis for the choice of parallel session speakers. A poster session was arranged for Wednesday afternoon to provide an opportunity for anyone who wished to share a result that was not included in the parallel session program.

The organizing committee for high energy physics arranged sessions around the topics of strong interactions at high energy, electroweak interactions, polarized electron sources, polarized solid targets, polarized beam acceleration and handling in both hadron and electron machines, and polarimeters for both hadrons and leptons. Speakers for the parallel sessions were taken from the list of abstracts submitted to the high energy committee. The proceedings of the joint session on nucleon spin structure functions will become part of the high energy volume.

Sunday, September 18, was set aside for an excursion to the Ohio River and Madison, Indiana. A steam-driven paddle-wheel boat was reserved for the use of participants and visitors to the symposia. Lunch was provided on the boat, along with live Dixieland band music and commentary about the history of the riverfront. Madison, Indiana played a prominent role in the settling of the Ohio River valley nearly two centuries ago, and a number of the streets still have houses dating from the early 19th century. Four of these homes were available for inspection during the tour.

The evening hours were set aside for social activities. Saturday evening contained a reception and tour of IUCF. On Monday evening, there was a concert by the Scott Chamber players, followed by refreshments. The Tuesday evening banquet included a reception with a jazz band, and an after-dinner presentation by Prof. David Baker from the Indiana University School of Music on the history of jazz. Wednesday evening, participants and guests were invited to tour the Musical Arts Center on campus. In parallel with the daytime physics sessions, there were also activities for companions that included trips to nearby points of interest.

We would like to thank the members of the program committee for their help in the selection of topics, and to the people willing to chair both plenary and parallel sessions. The committees are listed following this preface, and the names of the session chairs are included with the copies of the programs in the appendix to this volume. We also appreciate the efforts of the local organizing committee who took responsibility for many aspects of SPIN'94, including Peter Schwandt for local transportation and the companions' program, Ben Brabson for organization of the Sunday excursion, Bob Bent for social activities including the banquet, Tom Rinckel for computer setup, and Phil Thompson, Joyce Pace, and a number of workers that ran much-needed errands. We also wish to gratefully acknowledge Margie Rietel and the Indiana University Conference Bureau for attending to the numerous details of local arrangements. A special word of thanks is due to Sharon Herzel, Janet Meadows, and Sandy Smith for providing essential secretarial help with the meeting and during the months before and after. SPIN'94 would not have been possible without their dedicated support.

Edward J. Stephenson
Steven E. Vigdor
Editors

EIGHTH INTERNATIONAL SYMPOSIUM ON POLARIZATION PHENOMENA IN NUCLEAR PHYSICS

INTERNATIONAL ADVISORY COMMITTEE

J. M. Cameron (chair)	IUCF
J. Arvieux	Saturne
A. S. Belov	INR Moscow
L. S. Cardman	CEBAF
C. R. Gould	North Carolina State
R. J. Holt	Illinois
R. L. Jaffe	MIT
M. Kondo	RCNP
C. Lechanoine-Leluc	Geneva
V. M. Lobashov	Moscow
J. B. McClelland	LANL
R. D. McKeown	Caltech
W. T. H. van Oers	Manitoba
H. Ohnuma	Tokyo Inst. Tech.
J.-M. Richard	CERN
H. Sakai	Tokyo
F. D. Santos	Lisboa
M. Simonius	Zürich
J. Speth	KFA Jülich
E. Steffens	Heidelberg
K. Yagi	Tsukuba

ORGANIZING COMMITTEE

S. E. Vigdor (chair)	IUCF
T. B. Clegg	North Carolina
W. W. Jacobs	IUCF
L. D. Knutson	Wisconsin
J. T. Londergan	Indiana
S. W. Wissink	IUCF

LOCAL ARRANGEMENTS COMMITTEE

J. M. Cameron (chair)	B. B. Brabson
E. J. Stephenson (co-chair)	P. Schwandt
A. D. Bacher	J. Sowinski
R. D. Bent	

ELEVENTH INTERNATIONAL SYMPOSIUM ON HIGH ENERGY SPIN PHYSICS

INTERNATIONAL COMMITTEE

A. D. Krisch (chair)	Michigan
C. Y. Prescott (chair-elect)	SLAC
D. P. Barber	DESY
O. Chamberlain	Berkeley
E. D. Courant	Brookhaven
G. R. Court	Liverpool
A. V. Efremov	Dubna
G. Fidecaro	CERN
W. Haeberli	Wisconsin
K. J. Heller	Minnesota
V. W. Hughes	Yale
D. Kleppner	MIT
A. Masaike	Kyoto
P. W. Schmor	TRIUMF
A. N. Skrinsky	Novosibirsk
V. Soergel	Heidelberg
J. Soffer	Marseille
L. D. Soloviev	Protvino

ORGANIZING COMMITTEE

K. J. Heller (chair)	Minnesota
S. J. Brodsky	SLAC
A. W. Chao	SLAC
D. G. Crabb	Virginia
A. R. Dzierba	Indiana
R. L. Jaffe	MIT
S. Y. Lee	IUCF
D. B. Lichtenberg	Indiana
R. A. Phelps	Michigan
R. Prepost	Wisconsin
R. A. Rameika	Fermilab
J. B. Roberts	Rice
T. Roser	Brookhaven
H. M. Steiner	Berkeley
D. G. Underwood	Argonne

I. HADRON FORM FACTORS

Spin Dependent Structure Functions of Nucleons and Nuclei

A.W. Thomas
Department of Physics and Mathematical Physics
University of Adelaide
Adelaide, S.A. 5005, AUSTRALIA

Abstract

We review recent progress in the understanding of the spin structure of the nucleon. For the free nucleon the issues addressed include the status of the Bjorken and Ellis-Jaffe sum-rules and the role of the axial anomaly. We outline recent work connecting the quark models familiar from hadron spectroscopy to the spin and flavour dependence of the parton distributions. Finally we review the current understanding of nuclear spin structure functions and particularly the extraction of the neutron spin structure function from deuteron data.

1 Introduction

Measurements of the spin-dependent structure function of the nucleon, g_{1N}, continue to generate enormous interest. Not only has the precision with which the Ellis-Jaffe sum-rule is known improved, but with the addition of neutron data from SLAC and SMC the Bjorken sum-rule has also been tested [1, 2, 3]. Our aim is to provide a necessarily brief review of the situation with respect to the sum-rules. (For those interested in more detail we suggest some recent reviews [4, 5, 7, 8].) We then argue that the emphasis in future should be much more on the shape and momentum dependence of g_{1N}. There has recently been substantial progress in understanding the flavour and spin dependence of the parton distributions in terms of the quark models familiar from low energy spectroscopy and we briefly recall the main ideas.

As Nature has not seen fit to give us a free neutron target, nuclear and particle physicists must work together in order to extract g_{1n}. Some very recent progress in the treatment of deep-inelastic scattering from the deuteron is reviewed. Finally we highlight some important open questions which need experimental or theoretical attention.

4 Spin Dependent Structure Functions

2 Tests of Sum Rules

The spin structure function g_{1N} is determined by the difference in the cross section for a polarised, virtual photon to be absorbed by a nucleon with its spin anti-parallel and parallel to that of the photon

$$g_1 \propto \sigma_{\frac{1}{2}} - \sigma_{\frac{3}{2}}. \tag{1}$$

As the photon is absorbed by a spin-$\frac{1}{2}$ quark, in the former case the absorbing parton must have spin parallel to that of the nucleon ($q^\uparrow$) while in the latter case it must be anti-parallel ($q^\downarrow$). It is usual to define the net spin (helicity) carried by quarks (and anti-quarks) of a given flavour as

$$\Delta u(x, Q^2) = u^\uparrow - u^\downarrow + \bar{u}^\uparrow - \bar{u}^\downarrow, \tag{2}$$

etc. (Quarks and anti-quarks have the same charge squared and contribute with the same sign.) Thus one easily finds

$$g_{1p}(x, Q^2) = \frac{1}{2}\left\{\frac{4}{9}\Delta u(x, Q^2) + \frac{1}{9}\Delta d(x, Q^2) + \frac{1}{9}\Delta s(x, Q^2) + \ldots\right\}, \tag{3}$$

and if we define the isovector, octet and singlet flavour combinations in the usual way

$$\begin{aligned}
\Delta q_3(x, Q^2) &= \Delta u(x, Q^2) - \Delta d(x, Q^2), \\
\Delta q_8(x, Q^2) &= \Delta u(x, Q^2) + \Delta d(x, Q^2) - 2\Delta s(x, Q^2), \\
\Delta q_0(x, Q^2) &= \Delta u(x, Q^2) + \Delta d(x, Q^2) + \Delta s(x, Q^2),
\end{aligned} \tag{4}$$

we find

$$g_{1p(n)}(x, Q^2) = \pm\frac{1}{12}\Delta q_3 + \frac{1}{36}\Delta q_8 + \frac{1}{9}\Delta q_0. \tag{5}$$

Using current algebra the integrals of Δq_3 and Δq_8 can be related to the axial charges in neutron and hyperon β-decay

$$\begin{aligned}
\Delta q_3 &= \int_0^1 dx\, \Delta q_3(x, Q^2) = 1.257 \pm 0.003, \\
\Delta q_8 &= \int_0^1 dx\, \Delta q_8(x, Q^2) \stackrel{?}{=} 0.59 \pm 0.02,
\end{aligned} \tag{6}$$

where the question mark indicates some residual uncertainty over the use of flavour SU(3) symmetry. From the definition in equ.(4) it is clear that the integral of Δq_0 is just the fraction of the spin of the nucleon carried by its quarks, Σ:

$$\Sigma = \int_0^1 dx\, \Delta q_0(x, Q^2). \tag{7}$$

Within QCD there are also perturbative corrections which vanish logarithmically as $Q^2 \to \infty$, so that the integrals of the proton and neutron spin structure functions

$$S^{p(n)}(Q^2) = \int_0^1 dx g_{1p(n)}(x, Q^2), \qquad (8)$$

become [9, 10, 11]:

$$\begin{aligned} S^{p(n)}(Q^2) &= \frac{1}{12}\left(1 - \frac{\alpha_s}{\pi} - 3.58\left(\frac{\alpha_s}{\pi}\right)^2 - 20.2\left(\frac{\alpha_s}{\pi}\right)^3\right)\left(\pm\Delta q_3 + \frac{\Delta q_8}{3}\right) \\ &+ \frac{1}{9}\left(1 - \frac{\alpha_s}{3\pi} - 1.1\left(\frac{\alpha_s}{\pi}\right)^2\right)\Sigma, \end{aligned} \qquad (9)$$

(for three flavours).

The Bjorken sum-rule involves the difference of the integrals for the proton and neutron

$$\begin{aligned} S^{Bj} &= S^p - S^n = \frac{1}{6}\left(1 - \frac{\alpha_s}{\pi} - 3.6\left(\frac{\alpha_s}{\pi}\right)^2\right)\Delta q_3, \\ &= 0.185 \text{ at } Q^2 = 10\text{GeV}^2. \end{aligned} \qquad (10)$$

This particular sum-rule is extremely important as a failure would represent a failure of current algebra for the quarks [12]. The Ellis-Jaffe sum-rule [13] requires a dynamical assumption, namely that the polarisation of the strange sea of the nucleon vanishes, $\Delta s = 0$. With this reasonable assumption we find $\Sigma \simeq \Delta q_8 \simeq 60\%$ and hence

$$\begin{aligned} S^p_{EJ} &= 0.171 \pm 0.004, \ (0.161 \pm 0.004), \\ S^n_{EJ} &= -0.014 \pm 0.004, \ (-0.010 \pm 0.004) \end{aligned} \qquad (11)$$

at $10GeV^2$ ($3GeV^2$). The latest experimental information was presented in detail at this meeting [2, 3] and therefore we simply state the results.

The original EMC measurement [1] gave $S^p = 0.126 \pm 0.010 \pm 0.015$ at $10.7GeV^2$. This corresponds to Σ of only $17\% \pm 9\% \pm 14\%$ (12% to $0(\frac{\alpha_s}{\pi})$). It is clearly compatible with *none of the nucleon spin being carried by its quarks*, which was the origin of the term 'spin crisis'. This led to great interest in alternatives to the quark model, such as the Skyrme model [14]. Indeed the question has been asked whether the quark model would have been taken seriously if this had been known in the 60's – rather than the famous prediction $\mu_n/\mu_p = -2/3$. On the other hand, a cautious observer might add that the result is only 2-3 standard deviations from the Ellis-Jaffe prediction.

At this meeting new values were presented from the SMC and SLAC groups. For the proton, SMC finds $S^p = 0.136 \pm 0.011 \pm 0.011$ at $10\ GeV^2$, while a combined analysis with earlier proton data yields $S^p = 0.145 \pm 0.008 \pm 0.011$

6 Spin Dependent Structure Functions

Experiment	SMC(p)	SMC(p; all data)	SLAC(p; E143)	SLAC(n; E142)
$\langle Q^2 \rangle$ (GeV^2)	10	10	3	3
Measured	0.136 ± 0.011 ± 0.011	0.145 ± 0.008 ± 0.011	0.129 ± 0.004 ± 0.010	-0.035 ± 0.0096 ± 0.011
Corrected	0.147 ± 0.016	0.157 ± 0.014	0.166 ± 0.014	-0.042 ± 0.015
Σ	$34 \pm 14\%$	$43 \pm 13\%$	$62 \pm 13\%$	$28 \pm 14\%$

Table 1: Integrals of g_{1p} and g_{1n} with and without a correction for the non-perturbative dependence on Q^2 following refs.[4, 16].

[2]. The preliminary result reported from SLAC E143 was $S^p = 0.129 \pm 0.004 \pm 0.010$ at $3 GeV^2$, while for the deuteron and neutron they found $0.043 \pm 0.004 \pm 0.004$ and $-0.035 \pm 0.0096 \pm 0.011$, respectively [3]. These values are summarised in Table 1.

In all cases the experimental values are significantly below the theoretical expectation and it is common to conclude that the two experiments agree and that Σ is of order one third. On the other hand, we know that the integral of g_1 changes sign at $Q^2 = 0$ (to give the DHG sum-rule [15]), so that there is a dramatic variation in the sum-rule, due to non-perturbative effects, between $Q^2 = 0$ and 3 GeV^2. Vector meson dominance has been used by Ioffe and collaborators to provide a physically reasonable, smooth interpolation between the low-Q^2 and asymptotic regimes [4, 16]. At 3 and 10 GeV^2 this approach suggests a correction to the measured sum-rule for the proton (neutron) of 29% (20%) and 8% (6%) respectively. While there is, as yet, no consensus on these values [17] we find the sign and magnitude very reasonable. Applying them to the results quoted above, we find the values for the integrals and corresponding spin fractions shown in Table 1. After correction for non-perturbative effects the experiments are consistent with an average value of the spin carried by the quarks of $42 \pm 14\%$ – somewhat higher than the value of one third found without the correction. We believe that $42 \pm 14\%$ is probably the best estimate at the present time. It is roughly one standard deviation below the value of 60% implicit in the Ellis-Jaffe sum-rule. Notice that this expectation is not 100% for the same reason that g_A is not 5/3 – the lower component of the quark wave function is p-wave and the quark spin most often points opposite to the total angular momentum. This is a simple, well understood, relativistic correction. Clearly the data is no longer consistent with $\Sigma = 0$ and a 1 σ deviation is not a crisis. Nevertheless the physics is far from clear and in many ways the serious study of the problem is just beginning!

Much of the theoretical interest in the Ellis-Jaffe sum-rule, and most of the papers, have been concerned with the role of the U(1) axial anomaly in the flavour singlet piece, $\Delta q_0(x, Q^2)$. Following the work of Efremov and Teryaev and Altarelli and Ross [18] (ETAR) – see also ref.[19] – it is common to correct the naive parton model (NPM) value for $\Delta u, \Delta d$ and Δs by

$$\Delta u(x, Q^2) \rightarrow \Delta u(x, Q^2)|_{NPM} - \delta(x, Q^2),$$
$$\Delta d(x, Q^2) \rightarrow \Delta d(x, Q^2)|_{NPM} - \delta(x, Q^2),$$
$$\Delta s(x, Q^2) \rightarrow \Delta s(x, Q^2)|_{NPM} - \delta(x, Q^2), \qquad (12)$$

where

$$\delta(x, Q^2) = \frac{\alpha_s}{2\pi} \int_x^1 C^{\Delta g}\left(\frac{x}{y}, Q^2\right) \Delta g(y, Q^2) dy, \qquad (13)$$

and

$$\delta \equiv \int_0^1 \delta(x, Q^2) = \frac{\alpha_s}{2\pi} \Delta g. \qquad (14)$$

Here Δg is the fraction of the proton spin carried by gluons. In this picture

$$\Sigma \rightarrow \Sigma_{NPM} - 3\delta, \qquad (15)$$

and if (following Ellis and Jaffe) one sets Δs to zero in the NPM we find $\Delta s \neq 0$ through the anomaly. Clearly the residual discrepancy in the Ellis-Jaffe sum-rule could be fixed with $\delta \approx 5\%$, which corresponds to $\Delta g \sim 1.2$ at $10 GeV^2$, or $\Delta g \sim 1/2$ at a quark model scale ($\lesssim 1 GeV^2$). The latter agrees rather too well with the simple estimate of Δg associated with one-gluon-exchange by Brodsky and Schmidt [20]. Certainly this result is far more reasonable that the values of $\Delta g \sim 5$ required when Σ was compatible with zero!

We would like to emphasise that one wins in two ways through radiation of polarised gluons. Firstly, as we have just seen, the point-like coupling to the quarks in the gluon reduces Σ. Secondly, however, these gluons carry spin away from the quarks. This correction was shown to amount to 5-10% in Σ [21, 22] but as it involves non-perturbative QCD has generally been overlooked by the same groups who use equ.(15).

As usual with the anomaly there has been considerable controversy. The separation between $\Delta q|_{NPM}$ and δ is not gauge invariant at the operator level. Indeed this lies at the heart of the anomaly [23], which arises precisely because one cannot satisfy both chiral symmetry and gauge invariance. In the operator product expansion (OPE) there is no operator corresponding to Δg. From this a number of authors, notably Bodwin and Qiu, have argued that the first moment of $C^{\Delta g}$ should vanish [24, 25].

There has also been much debate over the infrared sensitivity of the separation of Δg_{NPM} and δ [4,5,24-27]. In the end, for the gauge usually used in the parton model ($A_+ = 0$), the correction is probably reliable. However this raises another problem, the shape of $g_1(x, Q^2)$. As emphasised in ref.[28, 29], the gluonic correction effects $g_1(x, Q^2)$ only at very small x (well below 0.1). This is well below the region where the calculations of $g_{1p}(x)$ deviate from the data. The study of the shape of $g_1(x, Q^2)$, and particularly the mechanism for

8   Spin Dependent Structure Functions

transfering strength from intermediate x to low x, is therefore a clear priority for future work. With this in mind, in the next section we present some simple ideas which help us to understand the shape of g_{1p} and g_{1n}. Before closing we should mention that not everyone is convinced that the effects of the anomaly are confined to small x, and any hint of OZI violation in the purely valence region would be most interesting indeed [30].

3   Understanding the Shape of g_{1p} and g_{1n}.

Formally the twist-two parton distributions can be written [31]

$$q^{(2)}(x,\mu^2) = \frac{m}{(2\pi)^3} \sum_n \int d\vec{p}\, |\langle n\vec{p}|\psi_+(0)|N\rangle|^2 \delta(m(1-x) - p_n^+), \quad (16)$$

with

$$p_n^+ = \sqrt{m_n^2 + \vec{p}^2} + p_z > 0. \quad (17)$$

Here $|n\vec{p}\rangle$ are a complete set of states of momentum $\vec{p}$ and rest mass m_n and equ.(16) guarantees that the distribution vanishes for $x \geq 1$. This formal expression has been used extensively to calculate parton distribution for the MIT bag model by the Adelaide group [32, 33, 34]. An important conclusion of that work is that the dominant contribution (at the quark model scale μ_-^2 – c.f. refs.[35, 36]) arises from a two-quark intermediate state, n=2. For a simple nucleon wave function, with three quarks in the 1s-state, the matrix element $\langle n=2, \vec{p}|\psi_+(0)|N\rangle$ should peak near $|\vec{p}| = 0$. Hence the maximum value of $q^{(2)}(x,\mu^2)$ should occur at $x \sim \frac{m-m_2}{m} \sim (\frac{1}{3} - \frac{1}{4})$ – from the δ-function – and the dominant contribution at large x will be the one associated with the lowest value of m_2. This was first realised by Close and Thomas who stressed the importance for both the flavour and the spin dependence of the parton distribution [37]. For example, using the standard SU(6) wave function for the proton one easily finds that a valence d-quark has only an S=1 pair as spectator, while a u-quark has a 50-50 chance of S=0 or S=1. Whatever mechanism is used to generate the N-Δ mass splitting – e.g. gluon exchange, pion exchange or instantons – gives typically $m_2(S=1) - m_2(S=0) \approx 200 MeV$. Hence the u-quarks should dominate at large x – i.e. one expects $d/u \ll 1$ as $x \to 1$, as found experimentally.

For a polarised proton (neutron) only the valence $u^\uparrow(d^\uparrow)$ is found with an $S=0$ pair and hence $u^\uparrow(d^\uparrow)$ should dominate at large x. From equ.(2) we see that g_{1p} must be positive, as must g_{1n}, at intermediate and large x. Note that all this follows from the *same physics* required in low energy spectroscopy (and a simple SU(6) wave function). It works because deep inelastic scattering involves a light-cone correlation function and the energy of the quark is as important as its momentum. Of course, a full calculation should also include the configuration mixing in the wave function, but this is not the *leading* effect. In passing we note that this same approach to calculating parton distributions leads to an unexpectedly large violation of charge symmetry for the minority valence quarks ($d^{(p)} \neq u^{(n)}$) [38, 39].

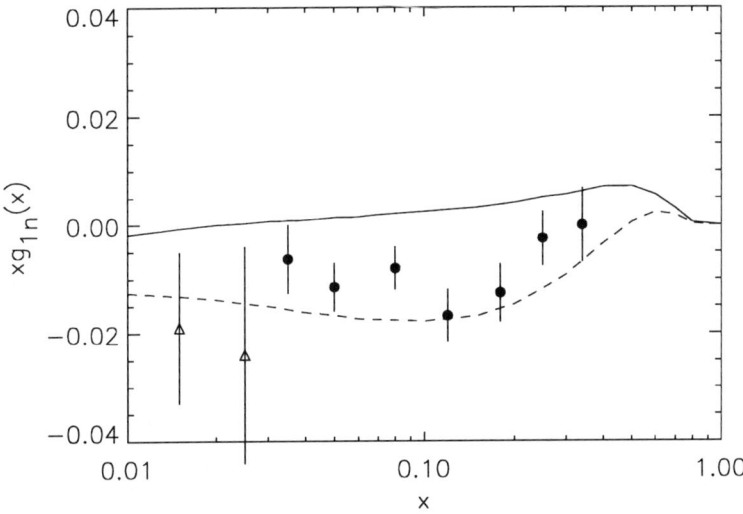

Figure 1: Data for g_{1n} from SMC and SLAC in comparison with predictions based on the bag model with (dashed) and without (solid) a phenomenological correction attributed to the axial anomaly – from ref.[40].

It is somewhat disappointing that the focus on sum-rules means that we still do not know whether g_{1n} does become positive at intermediate and large x. Although the approach just described is unambiguous, the inclusion of the correction associated with the axial anomaly could change this and it is very important to know. Figure 1 illustrates this by showing the prediction for g_{1n} from the bag model (with a bag radius R=0.6 fm) [34], with and without a phenomenological singlet term added to fit the proton data [40]. The dashed curve was then a prediction for g_{1n} published before the data. Recall that the earlier phenomenology of Carlitz-Kaur [41] and Schäfer [42] used counting rules to parameterize the effect of admixing polarised gluons in the proton wave function. As we have already noted, this complements the work of ETAR - see also refs.[21, 22]. It would be desirable to have a full microscopic calculation, including this effect and the Close-Thomas mechanism, within a model which is consistent with low energy hadronic properties.

As a final consideration in calculating the shape of the spin distributions we must also mention the role of chiral symmetry. Chiral quark models, like the cloudy bag[43], have proven very successful in dealing with low-energy hadronic properties. The charge form-factor of the neutron, G_{En}, is a famous example, where the positive core and negative tail of the corresponding charge density is a first-order effect in the perturbative dressing of the nucleon bag – through the process $n \to p_{Bag}\pi^-$[43, 44]. (Indeed, within the cloudy bag model an accurate measurement of the zero in $\rho_{ch}^n(r)$ would determine the radius within which the quarks are confined.) For the spin problem the role of the pion cloud is more complicated. For the proton the dominant, pionic correction is $p^\uparrow \to n^\downarrow \pi^+ (l_z = +1)$, while for the neutron it is $n^\uparrow \to p^\downarrow \pi^- (l_z = +1)$. Combined with the reduction in the bare nucleon contribution through the wave function renormalisation constant, Z, both of these effects tend to reduce the positive values of g_1 at large x [45]. The only complete investigation of pionic corrections has been carried out for the Ellis-Jaffe sum-rule [46], where it gives a small reduction ($O(5\%)$). This reduction would be larger were it not for an $N - \Delta$ interference term. The x-dependence of the latter is a completely new term which has never been calculated.

4 Nuclear Spin Structure Functions:

In order to study quark-parton distributions in the neutron, one must deal with either the deuteron or some heavier nucleus. In the spin-dependent case the two nuclei studied are polarised deuterium ($\vec{D}$) and ^3He. The simplest theoretical approach for deuterium is to approximate it as a free neutron and proton with polarisation $(1 - \frac{3}{2}\omega_D)$ - where ω_D is the deuteron d-state probability. At the next level of sophistication we have a convolution of the free nucleon structure function with the light-cone momentum distribution of nucleons in the nucleus [47, 48]:

$$g_{1D}(x, Q^2) = \sum_{N=n,p} \int_x \frac{dy}{y} \Delta f_{N/D}(y) g_{1N}(\frac{x}{y}, Q^2). \qquad (18)$$

Strangely the situation for polarised structure functions is better than for the unpolarised case, where data is still often analysed [49] using "smearing functions" pre-dating [50] the original EMC effect [51] - and therefore ignoring binding corrections [52, 53, 54] (see however ref.[55]). Tokarev found that for the polarised case the convolution gave a result very close to that obtained with a constant depolarisation for $x \overset{<}{\sim} 0.7$ [56] - see also Kaptari and Umnikov [57].

Reassuring as these results may be, the only way to test the reliability of the convolution is to do better. Even within the impulse approximation (IA) one may ask, for example, whether off-shell effects might spoil the convolution form. For the unpolarised case this was studied by Melnitchouk, Thomas and Schreiber [53, 54] who found that even in the Bjorken limit the Dirac structure of the nucleon involves not one but three functions

$$\hat{W}^{\mu\nu} = P_T^{\mu\nu}(\chi_0(p,q) + \not{p}\chi_1(p,q) + \not{q}\chi_2(p,q)). \tag{19}$$

Thus, whereas the free structure function involves one linear combination

$$F_2^{free} \propto m\chi_0 + m^2\chi_1 + p \cdot q\chi_2, \tag{20}$$

obtained by tracing $\hat{W}^{\mu\nu}$ with $(\not{p}+m)$, in a nucleus, where the nucleon propagator is $(\not{A}+B)$, we find

$$F_2^{bound} \propto B\chi_0 + A \cdot p\chi_1 + A \cdot q\chi_2, \tag{21}$$

which is <u>not</u> proportional to F_2^{Free}.

To study the practical significance of this breakdown, one needs a covariant description of the $\vec{D} \to \vec{N}N$ vertex as well as the off-shell nucleon structure function. For the former we can use the vertex of refs.[58]. For the latter Melnitchouk, Piller and Thomas [59] have employed a covariant version of the physics presented in sect. 3. That is, they constructed a set of phenomenological functions for the $N \to q$ (di-quark) vertices which were fitted to free nucleon data. As before, the di-quark states were either scalar or pseudovector. After considerable Dirac algebra one finds that the anti-symmetric nucleon tensor, surviving for massless quarks, in the Bjorken limit is

$$\hat{W}^{\mu\nu}_{A-S} = i\epsilon^{\mu\nu\alpha\beta}\left\{q_\alpha p_\beta \not{p}\gamma_5 G_{(p)} + q_\alpha p_\beta \not{q} G_{(q)} + q_\alpha \gamma_\beta \gamma_5 G_{(\gamma)}\right\}, \tag{22}$$

so that

$$g_{1N}^{Free} \propto p \cdot q \left[p \cdot qG_{(q)} + G_{(\gamma)}\right]. \tag{23}$$

For the deuteron case one must trace $\hat{W}^{\mu\nu}_{A-S}$ with a term involving $\gamma_5\gamma_\lambda \mathcal{A}(P,S,p)$, with the result:

$$g_{1D}(x) \propto \sum_{N=n,p} \int dy dp^2 \left\{\mathcal{A} \cdot q \left[p \cdot qG^N_{(q)} + G^N_{(\gamma)}\right] + p \cdot q\mathcal{A} \cdot pG^N_{(p)}\right\}. \tag{24}$$

The last term is not present in the free case, so that in principle the convolution must fail.

Figure 2 shows the numerical results obtained by Melnitchouk et al. [59] for the ratio of the deuteron to isoscalar nucleon spin structure function. The full calculation (solid line) is very close to the simple approximation of a constant depolarisation (dotted line) for $x \lesssim 0.7$. This agrees with the result found in earlier work using the convolution approach. Indeed we see from the dashed curve that an appropriate convolution is very close to the full off-shell

12 Spin Dependent Structure Functions

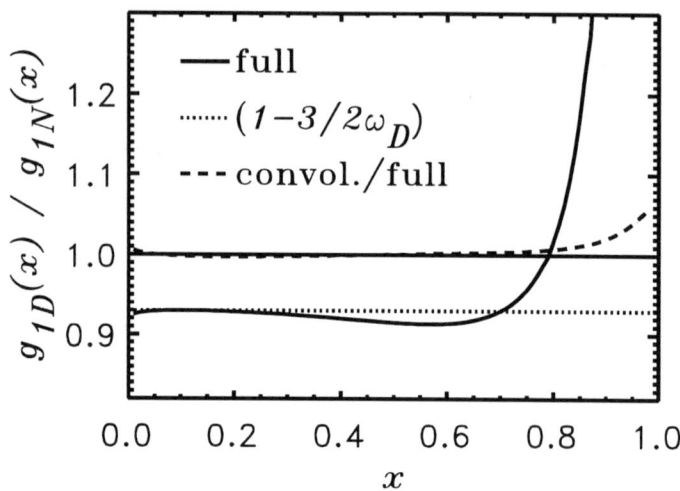

Figure 2: Ratio of deuteron and isoscalar nucleon structure functions in the full model (solid) compared with a constant depolarization factor $1 - 3/2\omega_D$ (dotted, for $\omega_D = 4.7\%$[58]). Also shown (dashed line) is the ratio of g_{1D} calculated via convolution and in the full model – from ref.[59].

calculation for all $x \lesssim 0.9$. In practice the additional term in equ.(24) is of order $(v/c)^3$ and negligible for the deuteron. In the end the major uncertainty in extracting the neutron structure function is not the failure of convolution or even the assumption of a constant depolarisation, but the old chestnut of the lack of knowledge of the deuteron d-state probability. This ranges between 4 and 7%, with modern, non-local potentials preferring $5 \pm 1\%$. At $x = 0.1$, for example, a variation of 1% in ω_D corresponds to an error of roughly 10% in g_{1n} [59]. At present this is smaller than the experimental errors but will soon be significant.

To conclude this discussion, we must note that the IA is by no means secure. For heavier nuclei it has been shown to significantly overestimate the effect of binding because deep-inelastic scattering determines the energy and momentum distribution of quarks <u>not</u> nucleons [60, 61]! This has not yet been investigated for the deuteron where the IA may not be so bad. Theoretical studies of $^3\vec{He}$ have so far not gone beyond an essentially non-relativistic convolution because of the absence of a relativistic model of the necessary vertices. In view of the deuteron results it would be surprising if there were a major problem, nevertheless it should be checked.

Finally we note that the actual structure of the bound nucleon may change. For the unpolarised case this has been shown to be a very small effect [60] within the Guichon model [62]. However, spin structure functions involve the lower components of the quark wave functions more directly and these change most in nuclear matter. It has been suggested in ref.[57], for example, that $(g_A/g_V)_{Bound}$ differs from the free value by about 5%. It is therefore an urgent matter to check this issue.

5 Concluding Remarks

As we have seen there has been remarkable progress in the experimental determination of the spin structure functions of the proton and the neutron. The Bjorken sum-rule has been verified at the 10% level (at 1σ). In view of its significance it is very important to continue to reduce this error. With respect to the Ellis-Jaffe sum-rule it is now clear that there is no longer a 'crisis' in the sense that none of the nucleon spin is carried by quarks. With a reasonable estimate of the non-perturbative corrections we find that all experiments are consistent with $\Sigma \sim 42 \pm 14\%$ compared with the expectation of about 60% of the nucleon spin carried by quarks.

However, *the end of the crisis is the beginning of some very challenging and fascinating physics*. It is now known that the Regge behaviour usually assumed in extrapolating to $x = 0$ is not appropriate to the flavour singlet component, which may have logarithmic corrections [63]. Bass and Landshoff suggest these might contribute as much as 25% to Σ [64], while Close and Roberts have suggested recently that the error might be even larger[65]. There is as yet no consensus on the size of the higher twist corrections. Both of these issues need thorough experimental study of the x and Q^2 variation of the structure functions as well as more theoretical work.

The shape of the x distribution of the neutron is still not well known. In view of the beautiful link between low energy hadron structure and parton distributions outlined in Sect. 3 it is very important to determine whether g_{1n} does become positive at intermediate and large x. Related to this is the question of whether strength is shifted from intermediate x to the region below 0.1, and if so what mechanism is responsible. The role of the axial anomaly is particularly fundamental, and can in principle be determined by comparing g_1 with a C-odd structure function like g_3 [5].

The use of light nuclear targets to extract information on the neutron seems to be well under control. We have seen that an appropriate convolution formula is very accurate when compared with a full off-shell calculation. The major problem for the deuteron is the uncertainty over its d-state probability. This may also be a problem for ^3He. Shadowing and meson exchange corrections have been investigated in detail for the unpolarised case [66] but much more must be done for the polarised case [67, 68]. There is so far no estimate of the effect on g_1 of a possible change in the internal nucleon structure.

Throughout our discussion we have followed the conventional approach of ignoring charm quarks (i.e. Δc is set to zero). Since most data points are actually above charm threshold, this needs more investigation [69, 70]. A related problem concerns the suggestion of using neutrino proton elastic scattering to measure Δs. As the neutral current couples to a non-singlet axial current it is less than obvious that such an experiment can yield information on a singlet quality like δ. Thus one may learn about $\Delta s|_{NPM}$ rather than Δs [40, 70].

The breadth of open questions in this field is enormous - ranging from issues of quark structure in nuclei, to Pomerons and the axial anomaly. It is truly an appropriate meeting ground for nuclear and particle physicists and an appropriate beginning to the first joint Spin Conference. I am grateful to the Organising Committee(s) for the invitation to present these thoughts.

Acknowledgments

It is a pleasure to thank my colleagues, particularly S. Bass, W. Melnitchouk, G. Piller, A. Schreiber and F. Steffens, who have taught me a great deal about the matters presented here. I would also like to thank A. Rawlinson and A. Shaw for their assistance in the production of this manuscript. This work was supported by the Australian Research Council.

References

[1] J. Ashman et al. (EMC), Phys. Lett. B206 (1988) 364; Nucl Phys. B328 (1990) 1.

[2] R. Windmolders (SMC), Proceedings Spin '94; D. Adams et al. (SMC), CERN PPE/94-57; B. Adeva et al., Phys. Lett. B320 (1994) 400.

[3] D. Day (E132 & E143), Proceedings Spin '94.

[4] B.L. Ioffe, preprint ITEP-61 (1994), hep-ph/9408291.

[5] S. Bass and A.W. Thomas, Prog. Part. Nucl. Phys. 33 (1994) 449; J. Phys. G9 (1993) 925.

[6] R. Windmolders, Int. J. Mod. Phys. A7, (1992) 639.

[7] J. Ellis, in Proceedings PANIC XIII, ed. A. Pascolini (World Scientific, Singapore, 1994) p.48.

[8] G. Veneziano, Okubofest lecture in "From Symmetries to Strings", ed. A. Das (World Scientific, Singapore, 1990).

[9] J. Kodaira et al., Phys. Rev. D20 (1979) 627; Nucl. Phys. 165 (1980) 129.

[10] M.A. Ahmed and G.G. Ross, Nucl. Phys. B111, (1976) 441.

[11] S.A. Larin and J.A.M. Vermaseren, Phys. Lett. B259 (1991) 345.

[12] C. Itzykson and J.-B. Zuber, "Quantum Field Theory" (McGraw-Hill,New York,1980).

[13] J. Ellis and R.L. Jaffe, Phys. Rev. D9 (1974) 1444; ibid. D10 (1974) 1669.

[14] S.J. Brodsky, J. Ellis and M. Karliner, Phys. Lett. B206 (1988) 309.

[15] S.D. Drell and A.C. Hearn, Phys. Rev. Lett. 16 (1966) 908; S.B. Gerasimov, Yad. Fiz. 2 (1965) 598.

[16] V.D. Burkert and B.L. Ioffe, Phys. Lett. B296 (1992) 223; M. Anselmino, B.L. Ioffe and E. Leader, Sov. J. Nucl. Phys. 49 (1989) 136.

[17] J. Ellis and M. Karliner, Phys. Lett. B313 (1993) 131; I.I. Balitsky et al., Phys. Lett. B242 (1990) 245; (E) B318 (1993) 648.

[18] A.V. Efremov and O.V. Teryaev, Dubna preprint JINR-E2-88-287 (1988); G. Altarelli and G.G. Ross, Phys. Lett. B212 (1988) 391.

[19] R.D. Carlitz, J.C. Collins and A.H. Mueller, Phys. Lett. B214 (1988) 229.

[20] S.J. Brodsky and I. Schmidt, Phys. Lett. B234, (1990) 144.

[21] F. Myhrer and A.W. Thomas, Phys. Rev. D38 (1988) 1633.

[22] H. Hogassen and F. Myhrer, Z. Phys. C48 (1990) 295.

[23] R.J. Crewther, Acta Physica Austriaca Suppl. XIX (1978) 47.

[24] G.T. Bodwin and J. Qiu, Phys. Rev. D41 (1990) 2755.

[25] R.L. Jaffe and A. Manohar, Nucl. Phys. B337 (1990) 509.

[26] S.D. Bass et al., J. Moscow Phys. Soc. 1 (1991) 317.

[27] L. Mankiewicz, Phys. Rev. D43 (1991) 64.

[28] S.D. Bass, N.N. Nikolaev and A.W. Thomas, ADP-90-133/T80 (1990).

[29] S.D. Bass and A.W. Thomas, Nucl. Phys. A527 (1991) 519.

[30] S.D. Bass, Z. Phys. C55 (1992) 653; ibid. C60 (1993) 343.

[31] R.L. Jaffe, Nucl. Phys. B229 (1983) 205.

[32] A.I. Signal and A.W. Thomas, Phys. Rev. D40 (1989) 2832.

[33] A.W. Schreiber, A.W. Thomas and J.T. Londergan, Phys. Rev. D42 (1990) 2653.

[34] A.W. Schreiber, A.I. Signal and A.W. Thomas, Phys. Rev. D44 (1991) 2653.

[35] R. Parisi and G. Petronzio, Phys. Lett. B62 (1976) 331.

[36] A.W. Thomas, Prog. Part. Nucl. Phys. 20 (1988) 21.

[37] F.E. Close and A.W. Thomas, Phys. Lett. B212 (1988) 227.

[38] E. Sather, Phys. Lett. B274 (1992) 433.

[39] E. Rodionov, A.W. Thomas and J.T. Londergan, Mod. Phys. Lett. A9 (1994) 1799.

[40] S.D. Bass and A.W. Thomas, Phys. Lett. B312 (1993) 345.

[41] R. Carlitz and J. Kaur, Phys. Rev. Lett. 38 (1977) 673.

[42] A. Schäfer, Phys. Lett. B208 (1988) 175.

[43] A.W. Thomas, Adv. Nucl. Phys. 13 (1984) 1; G.A.Miller, Int. Rev. Nucl. Phys. 1 (1984) 189; S. Théberge et al., Phys. Rev. D22 (1980) 2838,(E) D23 (1981) 2106.

[44] S. Théberge et al., Can. J. Phys. 60 (1982) 59.

[45] A.W. Schreiber et al., Phys. Rev. D45 (1992) 3069.

[46] A.W. Schreiber and A.W. Thomas, Phys. Lett. B215 (1988) 141.

[47] R.M. Woloshyn, Nucl. Phys. A496 (1989) 749.

[48] C. Ciofi degli Atti, E. Pace and G. Salme, Phys. Rev. C46 (1992) R1591.

[49] L.W. Whitlow et al., Phys. Lett. B282 (1992) 475.

[50] L.L. Frankfurt and M.I. Strikman, Phys. Rep. 76 (1981) 215; ibid. 160 (1988) 235.

[51] J.J. Aubert et al., Phys. Lett. B123 (1983) 275; Nucl. Phys. B293 (1987) 740.

[52] R.P. Bickerstaff and A.W. Thomas, J. Phys. G15 (1989) 1523.

[53] W. Melnitchouk, A.W. Schreiber and A.W. Thomas, Phys. Rev. D49 (1994) 1183.

[54] W. Melnitchouk, A.W. Schreiber and A.W. Thomas, Phys. Lett. B335 (1994) 11.

[55] L.P. Kaptari and A.Yu. Umnikov, Phys. Lett. B259 (1991) 155.

[56] M.V. Tokarev, Phys. Lett. B318 (1993) 559.

[57] L.P. Kaptari and A.Yu. Umnikov, Phys. Lett. B240 (1990) 203.

[58] W.W. Buck and F. Gross, Phys. Rev. D20 (1979) 2361; R.G. Arnold, C.E. Carlson and F. Gross, Phys. Rev. C21 (1980) 1426.

[59] W. Melnitchouk, G. Piller and A.W. Thomas, Adelaide preprint ADP-94-16/T157.

[60] A.W. Thomas et al., Phys. Lett. B233 (1989) 43.

[61] K. Saito, A. Michels and A.W. Thomas, Phys. Rev. C46 (1992) R2149; K. Saito and A.W. Thomas, Nucl. Phys. A547 (1994) 659.

[62] P.A.M. Guichon, Phys. Lett. B200 (1988) 235.

[63] F.E. Close and G.G. Roberts, Phys. Rev. Lett. 60 (1988) 1471.

[64] S.D. Bass and P.V. Landshoff, Cavendish preprint HEP-94/4.

[65] F.E. Close and R.G. Roberts, RAL Report RAL-94-071 (1994).

[66] N.N. Nikolaev and V.I. Zakharov, Phys. Lett. B55 (1975) 397; N.N. Nikolaev and V.R. Zoller, Z. Phys. C56 (1992) 623; B. Badalek and J. Kwiecinski, Nucl. Phys. B370 (1992) 278; W. Melnitchouk and A.W. Thomas, Phys. Rev. D47 (1993) 3783; Phys. Lett. B317 (1993) 437.

[67] L.P. Kaptari et al., Phys. Lett. B321 (1994) 271.

[68] H. Khan and P. Hoodbhoy, Phys. Lett. B298 (1993) 181.

[69] J. Ellis, M. Karliner and S. Sachrajda, Phys. Lett. B231 (1989) 497.

[70] S.D. Bass and A.W. Thomas, Phys. Lett. B293 (1992) 457.

Determination of the Neutron Electric Formfactor in Quasielastic Collisions of Polarized Electrons with ^{3}He and ^{2}D.

Collaboration A3 at MAMI: H.G. ANDRESEN[1], J. R. M. ANNAND[4], K. AULENBACHER[2], J. BECKER[2], J. BLUME-WERRY[1], TH. DOMBO[1], P. DRESCHER[2], J. E. DUCRET[1], D. EYL[1], H. FISCHER[2], A. FREY[1], P. GRABMAYR[3], S. HALL[4], P. HARTMANN[1], T. HEHL[3], W. HEIL[2], J. HOFFMANN[2], J. D. KELLIE[4], F. KLEIN[1], M. LEDUC[5], M. MEIERHOFF[2] H. MÖLLER[2], CH. NACHTIGALL[2], M. OSTRICK[1], E. W. OTTEN[2], R. O. OWENS[4], S. PLÜTZER[2], E. REICHERT[2], D. ROHE[2], M. SCHÄFER[2], L.D. SCHEARER[6], H. SCHMIEDEN[1], K. STEFFENS[1], R. SURKAU[2], TH. WALCHER[1].

[1] *Institut für Kernphysik, University of Mainz, Germany;*
[2] *Institut für Physik, University of Mainz, Germany;*
[3] *Physikalisches Institut, University of Tübingen, Germany;*
[4] *Kelvin Laboratory, Glasgow, Scotland;*
[5] *ENS, Paris, France;*
[6] *University of Missouri, Rolla, USA (deceased).*

Abstract

The determination of the neutron electric formfactor from quasielastic reactions $^3\vec{H}e(\vec{e}, e'n)$ and $D(\vec{e}, e'\vec{n})$ respectively is one of the present goals of experiments with polarized electrons at the Mainz race track microtron MAMI.

A GaAsP–photoelectron source is used at MAMI to get an 855 MeV electron beam spinpolarized to a degree of 35 % at a current of 10 μA. Polarized ^{3}He–nuclei are produced by optical pumping metastable ^{3}He. Scattered electrons are detected in coincidence with the recoil neutrons, the transverse spinpolarization of the neutrons may be analysed by neutron–proton scattering in a double wall plastic scintillator detector.

A subset of the final detector set–up has been tested successfully now by investigating the polarization transfer to the proton in reactions $H(\vec{e}, e'\vec{p})$ and $D(\vec{e}, e'\vec{p})$ and to the neutron in $D(\vec{e}, e'\vec{n})$ at a 4–momentum transfer with $-Q^2 = 8 fm^{-2}$. First data from the exclusive quasielastic collision $^3\vec{H}e(\vec{e}, e'n)$ indicate a value of the neutron electric formfactor of $G_E^n = 0.035 \pm 0.015$ at $-Q^2 = 8 fm^{-2}$.

1 Introduction.

Electron scattering from nucleons and from nuclei is a powerful method to investigate the electromagnetic structure of nuclear matter. It has provided

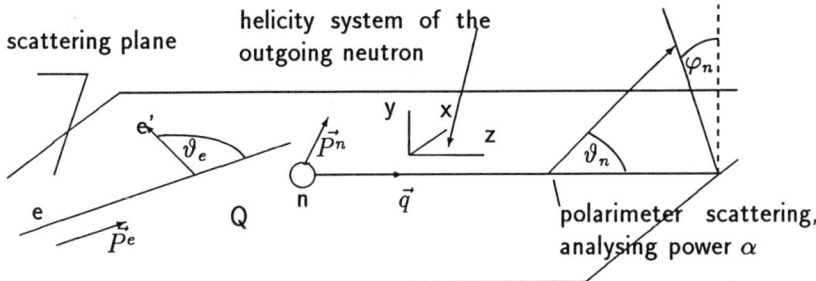

Figure 1: Kinematics of the elastic $\vec{n}$-$\vec{e}$-process. $\vec{P^e}$ = electron polarization; $\vec{P^n}$ = neutron polarization

us with a wealth of data of nuclear electromagnetic formfactors in the past. Nevertheless the neutron electric formfactor, a very fundamental structure quantity indeed, is only poorly known up today. It is small compared to the magnetic formfactor of the neutron, so the direct determination e. g. in a quasielastic neutron knockout reaction using the wellknown Rosenbluth analysis [18] would not be feasible. The most accurate data on the electric neutron formfactor at medium energies come from precise measurements of the elastic electron deuteron cross section [9, 17]. But the extraction of the neutron formfactors from the the deuteron data needs the knowledge of the deuteron wavefunctions, which are model dependent. So the analysis is model dependent, too.

Another way of determining the electric neutron formfactor exploits the spin-dependence of the elastic nucleon cross section. The kinematics in the scattering of longitudinally polarized electrons from polarized nucleons at rest is sketched in figure 1.

The elastic cross section has the form [5]:

$$\sigma = \sigma_{Mott} \cdot f_{rec}^{-1}[I_0 + (\Delta_x P_x^n + \Delta_z P_z^n)P^e] \tag{1}$$

with

$$I_0 = \frac{(G_E^n)^2 + \tau(G_M^n)^2}{1+\tau} + 2\tau(G_M^n)^2 \cdot tan^2(\vartheta_e/2) \tag{2}$$

$$\Delta_x = -2\sqrt{\frac{\tau}{1+\tau}} tan(\vartheta_e/2) \cdot G_E^n G_M^n \tag{3}$$

$$\Delta_z = -2\tau\sqrt{\frac{1}{1+\tau} + tan^2(\vartheta_e/2)} \cdot tan(\vartheta_e/2) \cdot (G_M^n)^2 \tag{4}$$

and

$$\begin{aligned}
G_E^n &= \text{electric formfactor} \\
G_M^n &= \text{magnetic formfactor} \\
\sigma_{Mott} &= \text{Mott cross section} \\
f_{rec}^{-1} &= \text{recoil correction} \\
\tau &= -\frac{Q^2}{4M^2} \\
Q &= \text{4-momentum transfer} \\
M &= \text{neutron mass} \\
P^e &= \text{electron spinpolarization} \\
P_x^n &= \text{x-component of neutron polarization} \\
P_z^n &= \text{z-component of neutron polarization}
\end{aligned}$$

The electric formfactor of the neutron G_E^n is small compared to the magnetic formfactor G_M^n. So $(G_M^n)^2$ dominates the spinindependent part of the cross section in (2) while the contribution $(G_E^n)^2$ is scarcely detectable here. But the spindependent term Δ_x in (3) is directly proportional to $G_E^n \cdot G_M^n$ and offers a good chance to determine G_E^n experimentally. Two different approaches are possible to use expression (3) for G_E^n-determination. In the first scheme longitudinally polarized electrons are scattered from neutrons that are transversely polarized with respect to the direction of 3-momentum transfer $\vec{q}$ along direction x in the helicity system of the outgoing neutron. Using equation (1) one gets

$$\sigma_x^\pm = \sigma_{Mott} \cdot f_{rec}^{-1}[I_0 \pm \Delta_x P^n P^e] \tag{5}$$

Where the $\pm$-sign stands for scattering of positive or negative helicity electrons respectively.

In experiment one determines the asymmetry:

$$A_x = \frac{\sigma_x^+ - \sigma_x^-}{\sigma_x^+ + \sigma_x^-} = \frac{\Delta_x}{I_0} \cdot P^n P^e \tag{6}$$

Inserting expressions (2) and (3) gives

$$A_x = -\frac{2\sqrt{\tau(1+\tau)}tan(\vartheta_e/2)}{(G_E^n)^2 + \tau(G_M^n)^2 + 2\tau(1+\tau)(G_M^n)^2 tan^2(\vartheta_e/2)} \cdot P^n P^e \cdot G_E^n G_M^n \tag{7}$$

which in the approximation $(G_E^n)^2 \ll (G_M^n)^2$ reduces to

$$A_x = -\frac{2\sqrt{\tau(1+\tau)}tan(\vartheta_e/2)}{\tau + 2\tau(1+\tau)tan^2(\vartheta_e/2)} \cdot P^n P^e \cdot \frac{G_E^n}{G_M^n} \tag{8}$$

So the measurement of the asymmetry A_x determines G_E^n in units of G_M^n.

Measurement of Δ_z in (1) in addition to that of Δ_x allows the determination of G_E^n/G_M^n without approximation. The cross section for scattering longitudinally polarized electrons from neutrons polarized in direction of the 3–momentum transfer $\vec{q}$ is

$$\sigma_z^{\pm} = \sigma_{Mott} \cdot f_{rec}^{-1}[I_0 \pm \Delta_z \cdot P^e P^n] \qquad (9)$$

The asymmetry now is

$$A_z = \frac{\sigma_z^+ - \sigma_z^-}{\sigma_z^+ + \sigma_z^-} = \frac{\Delta_z}{I_0} \cdot P^n P^e \qquad (10)$$

Combining (6) with (9) one arrives at

$$\frac{A_x}{A_z} = \frac{\Delta_x}{\Delta_z} = \frac{1}{\sqrt{\tau + \tau(1+\tau)tan^2(\vartheta_e/2)}} \cdot \frac{G_E^n}{G_M^n} \qquad (11)$$

or

$$\frac{G_E^n}{G_M^n} = \sqrt{\tau + \tau(1+\tau)tan^2(\vartheta_e/2)} \cdot \frac{A_x}{A_z} \qquad (12)$$

In deriving (11) and (12) respectively constant polarization values P^e, P^n have been assumed. In this case P^e and P^n drop out. So one has not to know the numerical polarization values of electron and target spin for the evaluation of G_E^n/G_M^e as long as the values do not change when switching from one asymmetry measurement to the other.

In a second scheme the spindependent parts of the cross section in (1) are determined by analysis of the spinpolarization $\vec{P}_{rec}^n$ of the recoil neutron. Using (1) one gets the x–component of recoil polarization

$$P_{rec,x}^n = \frac{\Delta_x}{I_0} \cdot P^e \qquad (13)$$

and the z–component

$$P_{rec,z}^n = -\frac{\Delta_z}{I_0} \cdot P^e \qquad (14)$$

which may be determined by a second scattering with analysing power α as indicated in figure 1 for the case of $P_{rec,x}$–determination. The asymmetry $\mathcal{A}$ measured in this scattering is given by

$$\mathcal{A} = \frac{I(\varphi_n) - I(\varphi_n + 180)}{I(\varphi_n) + I(\varphi_n + 180)} = \alpha \cdot P_{rec,x}^n \cdot cos\varphi_n \qquad (15)$$

Both schemes are used in current experiments at MAMI. The polarized neutron target, which is needed in the first scheme, is approximated by a polarized ^3He–target, which represents to a good approximation a system of two paired unpolarized protons and a single polarized neutron [3]. Investigated is the exclusive quasielastic reaction $^3\vec{H}e(\vec{e},e'n)$. In the second scheme the quasielastic reaction $^2D(\vec{e},e'\vec{n})$ is studied with polarization analysis of the recoil neutron.

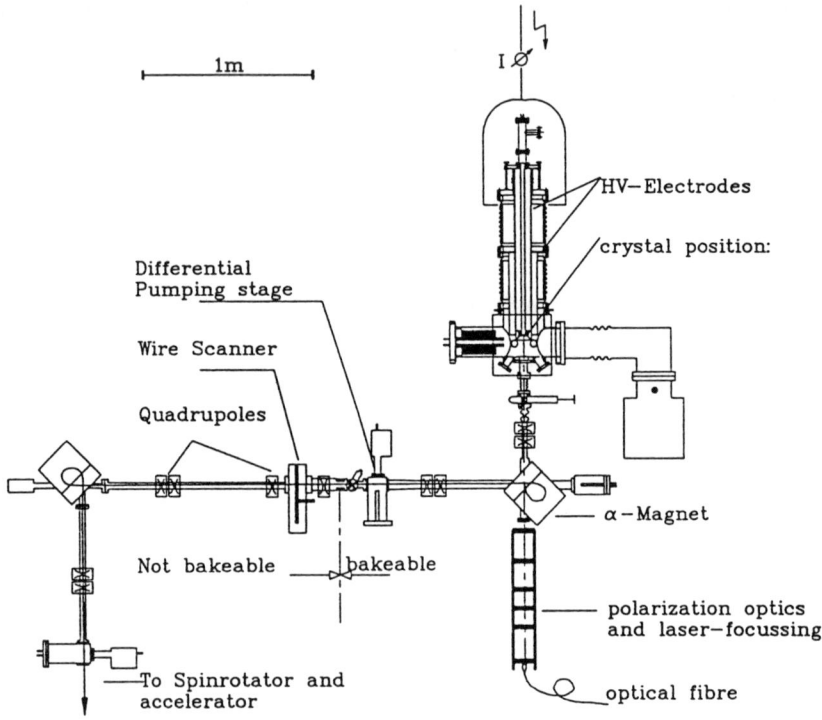

Figure 2: MAMI-source of polarized electrons

2 Spinpolarized electron beam of MAMI.

Figure 2 sketches the source of polarized electrons used at MAMI. It is based on photoelectron emission of a III–V–semiconductor cathode that is prepared for negative electron affinity (NEA) [16]. Its triode gun is equipped with a GaAsP-photocathode [10, 1]. In situ preparation of NEA is done with help of cesium channels (SAES Getters, Milano) and an NF_3-inlet attached to the gun chamber. UHV in the gun is necessary for avoiding excessive degradation of the cathode and is achieved here by use of ion– and NEG–pumps, that maintain a residual pressure of $2 \times 10^{-11} mb$. The cathode is illuminated by a circularly polarized 640 nm light beam from a laser system, which consists of a dye laser, that is pumped by an argon ion laser. The light is transported to

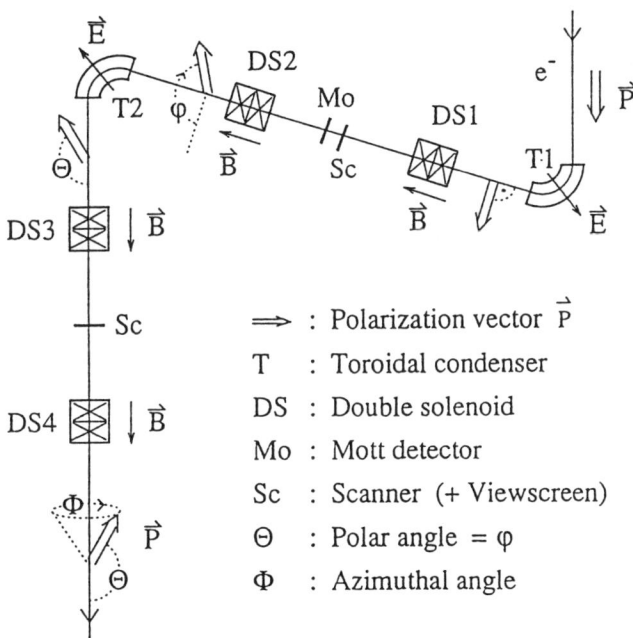

Figure 3: Spinrotator of MAMI [19]

the polarizing optics beneath the gun chamber via a mono mode fiber.

At present the MAMI source produces a beam with a degree of polarization of 35%. Emission current may be 20 μA when run continuously for 100 h. Higher currents are available for shorter times, e.g. 50 μA for 24 h. Longer operation times are achieved effectively by changing the emission spot at the cathode from time to time. The beam produced is d.c., so a greater portion of the current is lost in the chopper buncher system at injection into MAMI. Only 20% of the current emitted from the cathode is transmitted to a target at the exit of the accelerator. So the source has to deliver e.g. 50 μA polarized current for 10 μA at the target. In transverse magnetic fields the spin vector of an electron precesses faster than its momentum because of its anomalous magnetic moment [2]. The orientation of the polarization vector relative to momentum varies accordingly during beam acceleration in MAMI and the final orientation may depend severely on the energy to which the system is tuned. Therefore one has to provide means that allow adjustment of the spinorientation of the final beam. In the MAMI scheme a spin rotator is incorporated in the beamline between source and injection that is sketched in figure 3 [19]. It allows the setting of any spinorientation of the beam at the target.

3 Detector.

Figure 4 shows the detector set up used for investigating the quasifree collisions $^3\vec{H}e(\vec{e},e'n)$ and $^2D(\vec{e},e'\vec{n})$. Scattering angles of electrons are measured by a 16×16 matrix of lead glass detectors. The energy resolution is sufficient for discrimination of π–production. An imaging air–Čerenkov–detector is used to reject events that stem from electrons scattered at the entrance and exit windows of the $^3\vec{H}e$–target cell and events produced by photons from π^0–decay. Scattering angles of recoil neutrons are measured by two walls of plastic scintillators, the neutron kinetic energies are determined from the neutron time of flight.

In the case of the $^2D(\vec{e},e'\vec{n})$ experiment the transverse spinpolarization of the recoil neutrons has to be analysed (equation 13). This is done by a technique proposed by Taddeucci and coworkers [20, 13]. The neutron detector-polarimeter consists of two walls of plastic scintillators bars with two veto detectors, all of them with phototubes on either end. So both, horizontal and vertical coordinates of the neutron interaction vertices may be determined. The first wall serves as the polarimeter target (second scattering in figure 1) and exploits the analysing power of n–p–scattering in the plastic essentially. The veto detectors reject events produced by charged particles.

4 Targets

A liquid deuterium target is used in case of the $^2D(\vec{e},e'\vec{n})$ investigations. The polarized ^3He–target bases on a development started by G. K. Walters and coworkers in 1963 [4]. It produces spinpolarized ^3He–nuclei in atomic ground state ^3He via metastability exchange scattering with optically pumped and oriented metastable $(1s2s^3S_1)He$–atoms. Dense polarized ^3He–targets have been produced in recent years by application of powerful lasers in the pumping process [12, 15, 6]. Figure 5 shows the ^3He–target developed in Mainz [6]. Metastable ^3He–atoms are produced by a weak r.f. discharge in an optical pumping cell at a gas pressure of roughly 1 mb. They are polarized by absorption of cicularly polarized 1083 nm radiation from a LNA–laser. The nuclear spin polarization of the metastables is very effectively transferred to the ground state atoms in the cell via metastability exchange collisions. The scheme works only in a low pressure discharge. To get higher ^3He densities the polarized ^3He–gas is compressed by a Toepler pump [21] and fed to a target

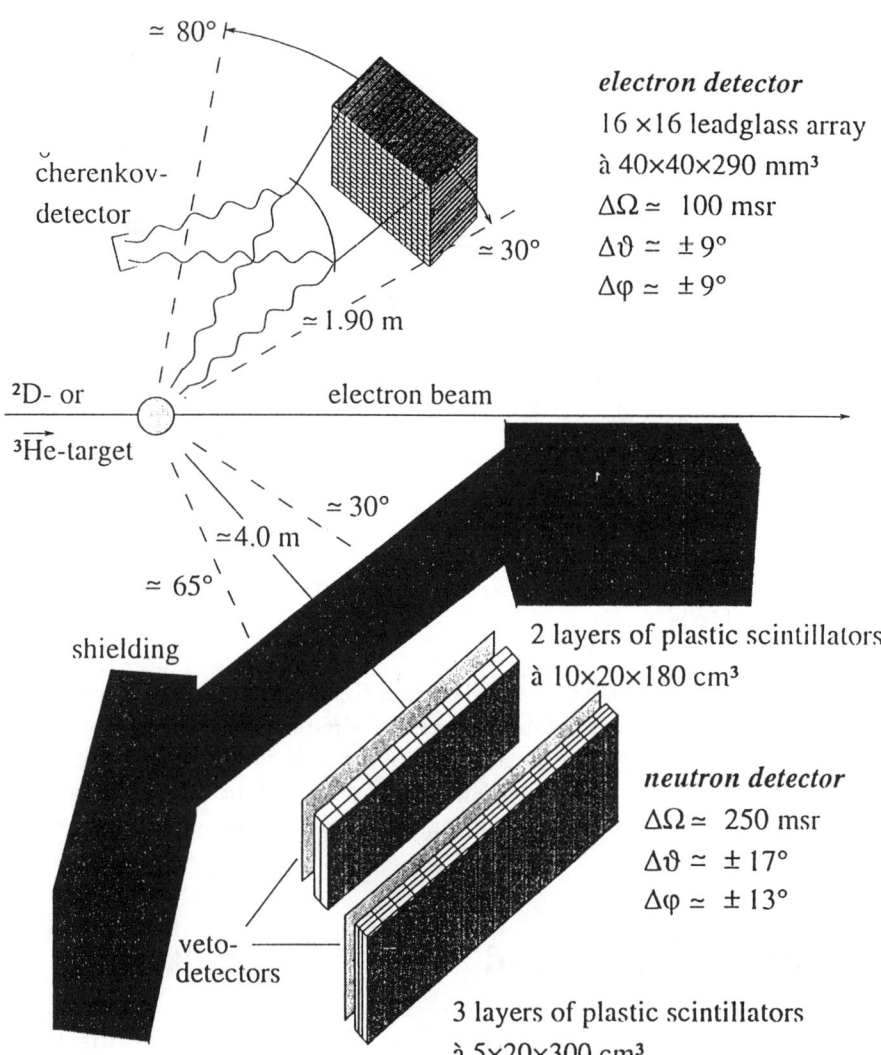

Figure 4: The MAMI–detector for investigation of quasielastic $^3\vec{H}e(\vec{e},e'n)$ and $^2D(\vec{e},e'\vec{n})$ collisions.

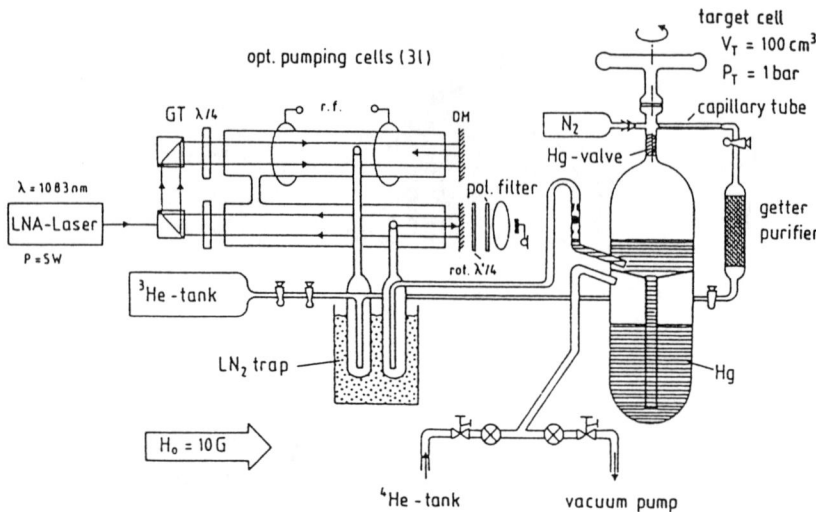

Figure 5: Polarized ^3He–target used in investigation of quasifree collisions $^3\vec{H}e(\vec{e},e'n)$ at MAMI.

cell at pressures around 1 bar. Spinrelaxation times around 50 h are achieved in the target cell. Through a capillary system some He may leak back to the optical pumping stage where its nuclear polarization is refreshed again. In equilibrium a nuclear spinpolarization of 60% is reached in the pumping cell while in the target cell at one bar the degree of polarization so far observed is just below 40%.

5 First $^2D(\vec{e},e'\vec{n})$ and $^3\vec{H}e(\vec{e},e'n)$ results

Polarization experiments at MAMI started in 1992/93 with the investigation of the polarization transfer to the proton in elastic $^1H(\vec{e},e'\vec{p})$– and quasielastic $^2D(\vec{e},e'\vec{p})$–collisions. The results are in accordance with transfer values that are calculated using proton formfactors published in literature [7, 11].

A subset of the final detector sketched in figure 4 has been used to start pilot experiments $^3\vec{H}e(\vec{e},e'n)$ and $^2D(\vec{e},e'\vec{n})$ at a 4–momentum transfer of $-Q^2 = 8fm^{-2}$. In the investigation $^2D(\vec{e},e'\vec{n})$ the neutron detector is run as a polarimeter [8]. The modularity of the plastic walls and the time difference in signal readout at the ends of the plastic bars allow full reconstruction of polar angles ϑ_n and azimuthal angles φ_n of neutrons after scattering in the first wall.

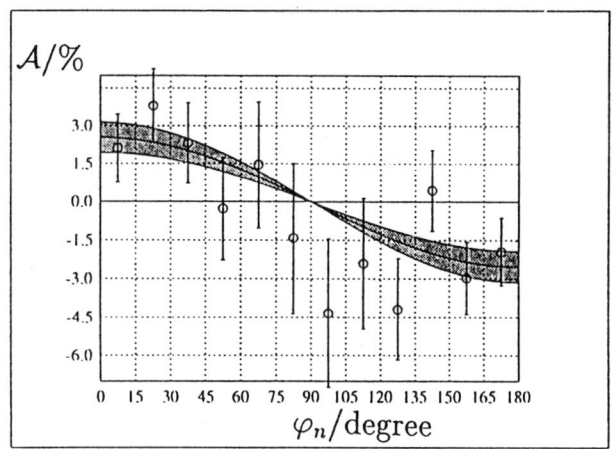

Figure 6: Asymmetry $\mathcal{A}$ detected by the neutron polarimeter as a function of azimuth angle φ_n in $^2D(\vec{e}, e'\vec{n})$.

In case of transversely polarized neutrons one expects a $\cos\varphi_n$ dependence of the scattering asymmetry $\mathcal{A}$ of equation (15) measured by the polarimeter. This is clearly observed in the experimental data of figure 6. The asymmetry $\mathcal{A}$ for events integrated over the ϑ_n-angular range $10^0 \leq \vartheta_n \leq 35^0$ is plotted as a function of azimuth φ_n. The cos–fit to the data points has an amplitude of $(2.53 \pm 0,61)\%$. The analysing power of the polarimeter is not calibrated yet. So we are not in the position to extract the G_E^n/G_M^n-ratio at present.

In the investigation $^3\vec{He}(\vec{e}, e'n)$ the neutron detector operates as a time–of–flight device only. Again only a part of the final detector set up is used in an exploratory experiment. Asymmetries Δ_x, Δ_z of equation (1) were measured using four different target spin orientations A, B, C, and D [14]. In orientations A and B the target spin is nearly transverse with respect to momentum transfer pointing in directions +x and -x of figure 1 approximately, while in orientations C and D it points essentially along +z and -z respectively. The deviations from pure transverse and longitudinal alignments of the target polarization vector are produced by the earth magnetic field that was not compensated and so adds to the guiding field H_0 of the ^3He-target sketched in figure 5. Asymmetries A_x and A_z of equations (7) and (10) respectively are evaluated from the data by taking

$$A_x = \frac{1}{2}(A_A - A_B); \qquad A_z = \frac{1}{2}(A_C - A_D) \tag{16}$$

The experimental data of table 1 give $A_x = 0.89 \pm 0.30$ and $A_z = -7.40 \pm 0.73$.

Table 1: Results of asymmetry measurements at $-Q^2 = 8 fm^{-2}$ in exclusive collisions $^3\vec{H}e(\vec{e}, e'n)$.

orientation	$\theta(deg)$	$\phi(deg)$	Asymmetry (%)		
A	88.2	-1.8	$+0.44 \pm 0.42$		
B	88.2	181.9	-1.35 ± 0.44		
			$(A_A - A_B)/2$	=	$A_x = +0.89 \pm 0.30$
C	2.1	-56.2	-7.22 ± 1.01		
D	177.7	-56.2	$+762 \pm 1.07$		
			$(A_C - A_D)/2$	=	$A_z = -7.40 \pm 0.73$
			ratio: A_x/A_z	=	$-0.120 \pm 0.042 \pm 0.01$

Using relation (11) we get

$$\frac{A_x}{A_z} = -0.120 \pm 0.042 \pm 0.01 \quad (17)$$

Using the value of the magnetic formfactor G_M^n given by the empirical dipol fit

$$G_M^n = \mu_N \cdot (1 - Q^2/18.2)^{-2} \quad (18)$$

we arrive at a final value of the electrical formfactor of the neutron

$$G_E^n(-Q^2 = 8fm^{-2}) = 0.035 \pm 0.012 \pm 0.005 \quad (19)$$

This value is in agreement with the work of Platchkov and coworkers [17]. In figure 7 the G_E^n-data of Platchkov et al are shown. They are evaluated from elastic D(e,e')-data using the Paris potential. The solid line shown is the best fit using the parametrization

$$G_E^n(Q^2) = \frac{-a\mu_N\tau}{1+b\tau} \cdot \frac{1}{(1-Q^2/18.23)^2} \quad (20)$$

with a = 1.25 ±0.13 and b = 18.3 ±3.4 . In addition corresponding curves using RSC, Argonne V14, and Nijmegen potentials are also included in figure 7. Our data point at $-Q^2 = 8fm^{-2}$ supports the analysis of Platchkov and coworkers.

6 Conclusion

Experiments with spinpolarized electrons have been started successfully at the Mainz race track microtron MAMI. First exploratory investigations of the quasielastic collisions $^2D(\vec{e}, e'\vec{n})$ and $^3\vec{H}e(\vec{e}, e'n)$ demonstrate the potentialities one gets in using spinpolarized collision partners in the study of the electromagnetic structure of nucleons and nuclei. The completion of the detector of

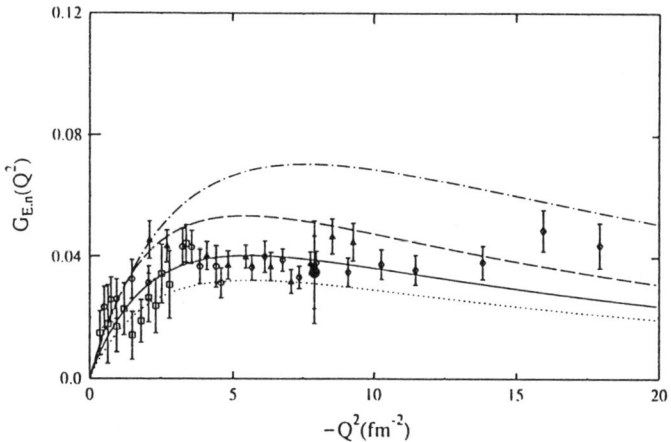

Figure 7: Comparison of the G_E^n value determined from $^3\vec{H}e(\vec{e},e'n)$ collisions (full circle) with data from [17] (open symbols). The solid curve gives the two parameter best fit of relation (20) to the data points of [17] using the Paris potential. In addition the corresponding two parameter fits using the RSC (dotted), Argonne V14 (dashed), or Nijmegen (dash–dotted) potentials are shown.

figure 4 will be finished in autumn 94. The goal then is the determination of G_E^n with an accuracy of 10% using quasifree reactions $^3\vec{H}e(\vec{e},e'n)$ as well as $^2D(\vec{e},e'\vec{n})$.

7 Acknowledgements

This work is supported by the Deutsche Forschungsgemeinschaft (SFB 201) and by the U. K. Science and Engineering Research Council.

References

[1] H. G. ANDRESEN, K. AULENBACHER, T. DOMBO, P. DRESCHER, H. EUTENEUER, H. FISCHER, D. V. HARRACH, P. HARTMANN, P. JENNEWEIN, K. H. KAISER, S. KOEBIS, H. J. KREIDEL, C. NACHTIGALL, S. PLÜTZER, E. REICHERT, K.-H. STEFFENS, AND T. WEIS, *Operating experience with the MAMI polarized electron source*, in Proceedings of the Workshop on Photocathodes for Polarized Electron Sources for Accelerarors, Stanford, September 8 – 10, 1993; SLAC–report

432, M. Chatwell, J. Clendenin, T. Maruyama, and D. Schultz, eds., Stanford, USA, 1994, SLAC, pp. 2–12.

[2] V. BARGMAN, L. MICHEL, AND V. L. TELEGDI, *Precession of the polarization of particles moving in a homogeneous electromagnetic field*, Phys. Rev. Lett., 2 (1959), pp. 435–437.

[3] B. BLANKLEIDER AND R. M. WOLOSHYN, *Quasi-elastic scattering of polarized electrons on polarized ^3He*, Phys. Rev. C, 29 (1984), pp. 538–552.

[4] F. D. COLEGROVE, L. D. SCHEARER, AND G. K. WALTERS, *Polarization of He^3 gas by optical pumping*, Phys. Rev., 132 (1963), pp. 2561–2572.

[5] T. W. DONNELLY AND A. S. RASKIN, *Considerations of polarization in inclusive electron scattering from nuclei.*, Annals of Physics, 69 (1986), pp. 247–351.

[6] G. ECKERT, W. HEIL, M. MEYERHOFF, E. W. OTTEN, R. SURKAU, W. WERNER, M. LEDUC, P. J. NACHER, AND L. D. SCHEARER, *A dense polarized ^3He target based on compression of optically pumped gas*, Nucl. Instr. Meth., A320 (1992), pp. 53–65.

[7] D. EYL, *Messung des Spintransfers in elastischer und quasielastischer Streuung polarisierter Elektronen in den Reaktionen $H(\vec{e}, e'\vec{p})$ und $D(\vec{e}, e'\vec{p})$*, PhD–thesis, Institut für Kernphysik der Joh. Gutenberg Universität Mainz, 1993.

[8] A. FREY, *Messung der Neutron-Rückstoßpolarisation in der Reaktion $D(\vec{e}, e'\vec{n})$ zur Bestimmung des elektrischen Formfaktors des Neutrons*, PhD–thesis, Institut für Kernphysik der Joh. Gutenberg Universität Mainz, 1994.

[9] S. GALSTER, H. KLEIN, J. MORITZ, K. H. SCHMIDT, D. WEGENER, AND J. BLECKWENN, *Elastic electron–deuteron scattering and the electric neutron formfactor at four momentum transfers $5fm^{-2} < q^2 < 14fm^{-2}$*, Nucl. Phys. B, 32 (1971), pp. 221–237.

[10] W. HARTMANN, D. CONRATH, W. GASTEYER, H.-J. GESSINGER, W. HEIL, H. KESSLER, L. KOCH, E. REICHERT, H. G. ANDRESEN, T. KETTNER, B. WAGNER, J. AHRENS, J. JETHWA, AND F. P. SCHÄFER, *A source of polarized electrons based on photoemission of GaAsP*, Nucl. Instr. and Meth., A286 (1990), pp. 1–8.

[11] F. KLEIN, *The electric formfactor of the neutron*, in Proceedings of the 14th International Conference on Few Body Problems in Physics, Williamsburg, Virginia, USA, May 26–31, 1994, 1994.

[12] M. LEDUC, in 7th Int. Conf. on Polarization Phenomena in Nuclear Physics, A. Boudard and Y. Terrien, eds., Coll. de Physique 51 (1990), Suppl. C6-317, 1990.

[13] R. MADEY, *Bates experiment E85-05*, tech. rep., BATES, 1988.

[14] M. MEYERHOFF, D. EYL, A. FREY, H. G. ANDRESEN, J. R. M. ANNAND, K. AULENBACHER, J. BECKER, J. BLUME-WERRY, T. DOMBO, P. DRESCHER, J. E. DUCRET, H. FISCHER, P. GRABMAYR, S. HALL, P. HARTMANN, T. HEHL, W. HEIL, J. HOFFMANN, J. D. KELLIE, F. KLEIN, M. LEDUC, H. MÖLLER, C. NACHTIGALL, M. OSTRICK, E. W. OTTEN, R. O. OWENS, S. PLÜTZER, E. REICHERT, D. ROHE, M. SCHÄFER, L. D. SCHEARER, H. SCHMIEDEN, K.-H. STEFFENS, R. SURKAU, AND T. WALCHER, *First measurement of the electric formfactor of the neutron in the exclusive quasielastic scattering of polarized electrons from polarized ^3He*, Phys. Lett. B, 327 (1994), pp. 201–207.

[15] R. G. MILNER, R. D. MCKEOWN, AND C. E. WOODWARD, Nucl. Instrum. Meth., A274 (1989), p. 56.

[16] D. T. PIERCE, F. MEIER, AND P. ZÜRCHER, *Negative electron affinity GaAs: A new source of spinpolarized electrons*, Appl. Phys. Letters, 26 (1975), p. 670.

[17] S. PLATCHKOV, A. AMROUN, S. AUFFRET, J. M. CAVEDON, P. DREUX, J. DUCLOS, B. FROIS, D. GOUTTE, H. HACHEMI, J. MARTINO, X. H. PHAU, AND I. SICK, *The deuteron $A(q^2)$ structure function and the neutron electric formfactor*, Nucl. Phys. A, 510 (1990), pp. 740–758.

[18] M. N. ROSENBLUTH, *High energy scattering of electrons on protons*, Phys. Rev., 79 (1950), pp. 615–619.

[19] K. H. STEFFENS, H. G. ANDRESEN, J. BLUME-WERRY, F. KLEIN, K. AULENBACHER, AND E. REICHERT, *A spinrotator for producing a longitudinally polarized electron beam with MAMI.*, Nucl. Instr. Meth., A325 (1993), pp. 378–383.

[20] T. N. TADDEUCCI, Nucl. Instr. Meth., A241 (1985), p. 448.

[21] R. S. TIMSIT, J. M. DANIELS, E. I. DENNIG, A. K. KIANG, AND A. D. MAY, *An experiment to compress polarized ^3He gas*, Can. J. Phys., 49 (1971), pp. 508–516.

Measurement of the Proton's Axial Form Factor via Neutrino-Nucleon Scattering

Ion Stancu

Department of Physics, University of California, Riverside, CA 92521

Abstract

Recent experimental and theoretical work indicates that strange quark pairs play a non-negligible role in determining nucleon properties. Their effects contribute to the electroweak nucleon form factors and thus can be isolated through measurements of neutral weak neutrino-nucleon scattering. In particular, the $s\bar{s}$ contribution to the nucleon spin can be determined by measuring the ratio of proton-to-neutron neutrino-induced quasielastic yield. Preliminary results from the LSND experiment are presented, experiment which, at the end of the scheduled 94-95 run period, should obtain a measurement of the strange axial form factor, $G_1^s(0)$, with an absolute accuracy of approximately ± 0.04.

INTRODUCTION

During the past decade there has been a significant amount of experimental and theoretical research that would require strange quark pairs $s\bar{s}$ to play an important role in determining nucleon properties. Measurements of opposite sign muon pairs in deep-inelastic scattering (DIS) of ν_μ and $\bar{\nu}_\mu$ from nucleons clearly reveal the presence of gluons and strange quarks within the nucleon (1), while analysis of the sigma term in pion-nucleon scattering (2) suggested that strange quarks contribute as much as one-third to the nucleon's rest mass. With the accumulation of better data and more careful theoretical analysis, the need for a significant strange quark contribution to the nucleon's mass has all but vanished (3).

The nucleon's spin also appears to be affected by the presence of strange quarks, as earlier measurements by the European Muon Collaboration (EMC) indicated. The EMC plus earlier SLAC experiments (4,5) measured the proton's spin structure function $g_1^p(x)$ over the interval $0.01 \leq x \leq 0.7$. Using reasonable extrapolation to $x = 0$ they found

$$\Gamma_1^p \equiv \int_0^1 g_1^p(x)dx = 0.126 \pm 0.010 \pm 0.015 \tag{1}$$

at $Q^2 = 10 \ GeV^2/c^2$, in disagreement with the Ellis-Jaffe sum rule, which predicts a value of 0.175 ± 0.018 (6,7) under the assumption that only u ($\overline{u}$) and d ($\overline{d}$) quarks (antiquarks) contribute to the proton spin. This difference implies that the strange quark contribution to the proton's spin must be

$$\Delta s(Q^2 = 10 GeV^2/c^2) = -0.16 \pm 0.08 \qquad (2)$$

and, when coupled with hyperon axial decay rates to fix Δu and Δd, leads to a value for the total quark contribution to the proton's spin of (7)

$$\Delta u + \Delta d + \Delta s = 0.120 \pm 0.094 \pm 0.138, \qquad (3)$$

giving rise to the much debated "spin crisis". Several polarized DIS measurements are now well underway to verify and extend the EMC results: the Spin Muon Collaboration (SMC) at CERN (8), E142 (9) and E143 at SLAC and HERMES at HERA. Recent results from the SMC proton data alone (8) yield

$$\Gamma_1^p = 0.136 \pm 0.011 \pm 0.011 \qquad (4)$$

with $g_1^p(x)$ measured over $0.003 \leq x \leq 0.7$. This raised the total quark contribution to the proton's spin to

$$\Delta u + \Delta d + \Delta s = 0.22 \pm 0.10 \pm 0.10, \qquad (5)$$

while taking into account all available data one obtains $\Delta u + \Delta d + \Delta s \approx 0.27$ with a strange quark contribution $\Delta s \approx -0.10$ (8). However, the uncertainties related to the interpretation of the data still remain: the use of $SU(3)$-symmetry in fixing Δu and Δd from hyperon decays and the extrapolation of $g_1^p(x)$ to $x = 0$, which may be more singular than assumed.

In the following, I will summarize the current status of neutrino-nucleon scattering and present a direct method of measuring the strange quark contribution to the nucleon spin using neutrino quasielastic scattering from isoscalar nuclei. This procedure has the advantage over DIS that it does not require $SU(3)$ symmetry, nor an extrapolation to $x = 0$. Preliminary results from the Liquid Scintillator Neutrino Detector (LSND) experiment at the Los Alamos Meson Physics Facility (LAMPF) will be presented, an experiment which is expected to determine the axial form factor of the proton at low momentum transfer, and hence Δs with an absolute accuracy of approximately ± 0.04.

NEUTRINO-NUCLEON SCATTERING

Following the EMC results, Kaplan and Manohar and Ellis and Karliner (10) pointed out the sensitivity of the neutrino-nucleon elastic scattering cross-section to the strange quark contribution. The charge-changing scattering

$\bar{\nu}_e + p \rightarrow e^+ + n$ is sensitive only to the isovector quark currents of the nucleon. However, neutral current scattering permits isoscalar contributions from strange or heavier quarks. Insofar these contributions couple through the axial current, they are contributions to the nucleon spin.

The formalism for describing the neutral current neutrino-nucleon scattering has already been worked out in detail and appears in several places in the literature (11,12). It is however briefly presented here in order to clarify the conventions and define the terms needed to describe the experimental results.

Consider first the coupling between the electric current of the nucleon and the photon, which can be simply expressed in terms of quark currents as

$$j_\mu^N \cdot A^\mu = <N'|\sum_i e_i \bar{q}_i \gamma_\mu q_i|N> A^\mu$$

$$= e <N'|\frac{2}{3}\bar{u}\gamma_\mu u - \frac{1}{3}\bar{d}\gamma_\mu d - \frac{1}{3}\bar{s}\gamma_\mu s|N> A^\mu, \qquad (6)$$

where $|N>$ and $|N'>$ are nucleon spinors. The summation is only over the u, d and s quarks, since the contribution from the heavier flavours is considered to be negligible (10). This can be rewritten as the following linear combination:

$$j_\mu^N \cdot A^\mu = e <N'|\frac{1}{2}(\bar{u}\gamma_\mu u - \bar{d}\gamma_\mu d) + \frac{1}{6}(\bar{u}\gamma_\mu u + \bar{d}\gamma_\mu d - 2\bar{s}\gamma_\mu s)|N> A^\mu, \qquad (7)$$

where the first term in brackets is an isovector current and the second one an isoscalar - or octet in $SU(3)_c$ parlance. However, the calculation of exact expectation values of quark currents is still beyond reach and therefore, one usually expresses the result in terms of form factors, determined experimentally:

$$j_\mu^N \cdot A^\mu = e <N'|F_1^N(Q^2)\gamma_\mu + \frac{i}{2m_p}F_2^N(Q^2)\sigma_{\mu\nu}q^\nu|N> A^\mu. \qquad (8)$$

Each of these form factors, the charge form factor $F_1^N(Q^2)$ and the magnetic form factor $F_2^N(Q^2)$ contains an isoscalar and an isovector component arising from the corresponding currents in Eq.(7) above.

Following the same approach, the neutral weak current (NWC) for the nucleon can be written as

$$j_\mu^N \cdot Z^\mu = \sqrt{\frac{G_F}{\sqrt{2}}} <N'|\sum_i [\bar{q}_i(1-\gamma_5)\gamma_\mu t_z q_i - 2e_i \sin^2\theta_W \bar{q}_i\gamma_\mu q_i]|N> Z^\mu, \qquad (9)$$

where t_z is the weak isospin, i is the summation index over the quark flavours ($i = u, d, s$), e_i is the electric charge of flavour i and θ_W is the weak mixing angle. This current can be expressed in terms of form factors as follows:

$$j_\mu^N \cdot Z^\mu = \sqrt{\frac{G_F}{\sqrt{2}}} <N'|F_1^Z(Q^2)\gamma_\mu + \frac{i}{2m_p}F_2^Z(Q^2)\sigma_{\mu\nu}q^\nu + G_1^Z(Q^2)\gamma_\mu \gamma^5|N> Z^\mu, \qquad (10)$$

with

$$F_1^Z(Q^2) \equiv \left(\frac{1}{2} - \sin^2\theta_W\right)\left[F_j^p(Q^2) - F_j^n(Q^2)\right]\tau_3$$
$$- \sin^2\theta_W \left[F_j^p(Q^2) + F_j^n(Q^2)\right] - \frac{1}{2}F_j^s(Q^2) \quad (11)$$

$$G_1^Z(Q^2) \equiv -\frac{1}{2}G_A(Q^2)\tau_3 + \frac{1}{2}G_1^s(Q^2), \quad (12)$$

where $j = 1$ or 2, $F_j^{p(n)}$ is the corresponding proton (neutron) electromagnetic form factor and $F_j^s(Q^2)$ is the strange quark contribution to the NWC vector form factor. $G_A(Q^2)$ is the nucleon isovector axial vector form factor - whose value at $Q^2 = 0$ (1.256 ± 0.003) is determined from neutron β-decay - and $G_1^s(Q^2)$ is the strange axial vector form factor. Furthermore, τ_3 in the two equations above denotes the z-component of the nuclear isospin ($\tau_3 = 1$ for the proton and $\tau_3 = -1$ for the neutron). To simplify the notation we shall drop the superscript Z on the weak form factors henceforth.

With the NWC defined above one can finally write down the expression for the neutrino-nucleon scattering which, after some manipulation, yields:

$$\frac{d\sigma}{dQ^2} = \frac{G_F^2}{2\pi} \cdot \frac{Q^2}{E_\nu^2} \left(A \pm By + Cy^2\right), \quad (13)$$

where $y = 4E_\nu/m_p - Q^2/m_p^2$ and

$$A \equiv \frac{1}{4}\left[G_1^2\left(1 + \frac{Q^2}{4m_p^2}\right) - \left(F_1^2 - \frac{Q^2}{4m_p^2}F_2^2\right)\left(1 - \frac{Q^2}{4m_p^2}\right) + \frac{Q^2}{m_p^2}F_1 F_2\right], \quad (14)$$

$$B \equiv -\frac{1}{4}G_1(F_1 + F_2), \quad (15)$$

$$C \equiv \frac{1}{16} \cdot \frac{m_p^2}{Q^2}\left(G_1^2 + F_1^2 + \frac{Q^2}{4m_p^2}F_2^2\right). \quad (16)$$

The $+$ sign in Eq.(13) above corresponds to neutrinos, whereas the $-$ sign corresponds to antineutrinos; furthermore, the explicit Q^2 of the form factors has been dropped for simplicity. At neutrino energies (E_ν) and momentum transfers (Q) that are much less than m_p - which is indeed the case for the LSND experiment - to first order in E_ν/m_p and Q/m_p the cross-section becomes:

$$\frac{d\sigma}{dQ^2} = \frac{G_F^2}{2\pi}\left\{G_1^2\left(1 + \frac{Q^2}{4E_\nu^2}\right) + F_1^2\left(1 - \frac{Q^2}{4E_\nu^2}\right)\right.$$
$$\left. - \left[\frac{1}{2}\left(G_1^2 + F_1^2\right) \pm G_1(F_1 + F_2)\right]\frac{Q^2}{E_\nu m_p}\right\}. \quad (17)$$

The Q^2 dependences of the form factors is given by:

$$F_1^p(Q^2) = \frac{0.035 + 1.054\tau - 0.956\tau/(1+5.6\tau)}{(1+\tau)(1+Q^2/M_V^2)^2} - \frac{1}{2} F_1^s(Q^2), \qquad (18)$$

$$F_2^p(Q^2) = \frac{1.019 + 0.956\tau/(1+5.6\tau)}{(1+\tau)(1+Q^2/M_V^2)^2} - \frac{1}{2} F_2^s(Q^2), \qquad (19)$$

$$G_1^p(Q^2) = \frac{-0.631}{(1+Q^2/M_A^2)^2} - \frac{1}{2} G_1^s(Q^2), \qquad (20)$$

where $M_V = 0.843\ GeV/c^2$, $M_A = 1.061 \pm 0.026\ GeV/c^2$, $\tau \equiv Q^2/4m_p^2$, while the neutron form factors are taken as in (12). It is not clear what is the proper dependence for the strange form factors F_1^s and F_2^s. It is known however that $F_1^s(Q^2=0) = 0$, and therefore, from Eq.(17) above, it becomes clear that the νp cross-section is dominated by G_1^s in the low Q^2 limit.

However, a direct measurement of the νp cross-section turns out to be rather difficult to perform in the LSND environment for at least two reasons. First, background processes such as $\nu + {}^{12}C \to \nu + p + X$, where X produces no visible signal in the detector, are easily confused with $\nu + p \to \nu + p'$. Secondly, knowledge of the absolute neutrino flux is difficult to establish to better than 12%. These difficulties can be circumvented by measuring instead the ratio of proton-to-neutron yield from an isoscalar nucleus such as ${}^{12}C$, as discussed in the following section.

QUASIELASTIC SCATTERING FROM ISOSCALAR NUCLEI

Two recent calculations by Garvey, Kolbe, Krewald and Langanke have showed that the ratio of proton-to-neutron neutrino-induced quasielastic yield is a sensitive way to determine the strange quark axial form factor of the nucleon. While the first calculation (13) employs a mean-field approximation for the neutrino-induced nucleon knockout from ${}^{12}C$, the second one (14) is based on the continuum random phase approximation and takes into account final states interactions by a finite-range particle-hole residual interaction derived from the Bonn potential. Only the results from Ref.14 will be presented here, but it should be noted that they do not differ substantially from those derived in Ref.13, which seems to indicate that the ratio is essentially independent of the adopted nuclear model.

Figure 1 shows the cross-sections for quasielastic ν- and $\bar{\nu}$- induced reactions on ${}^{12}C$ as a function of the kinetic energy of the final nucleon. The incident neutrino energy has been fixed to $E_\nu = 200\ MeV$, which is a typical value for the decay-in-flight neutrino beam available at LAMPF. The strange

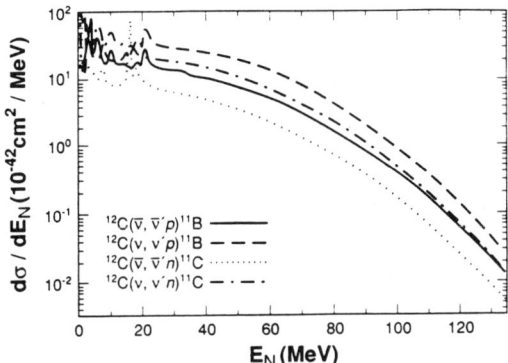

FIGURE 1. Differential cross-section as a function of nucleon energy. The incident neutrino energy was set to $E_\nu = 200\ MeV$. (From Ref.13.)

form factors are fixed at $G_1^s = -0.19$, taken from the analysis of the proton spin structure function (5), and $F_2^s = -0.22$ (15). It is apparent from the figure that the cross-section becomes essentially independent of details of nuclear structure at energies above $E_N = 30\ MeV$. The ratio of proton-to-neutron quasielastic yield for 200 MeV neutrinos on ^{12}C is illustrated in Figure 2 as a function of the nucleon energy E_N. In the absence of any strange quark contribution, the value of the calculated ratio for $E_N > 60\ MeV$ (above which contributions from free protons is negligible) is 0.88; however, with G_1^s

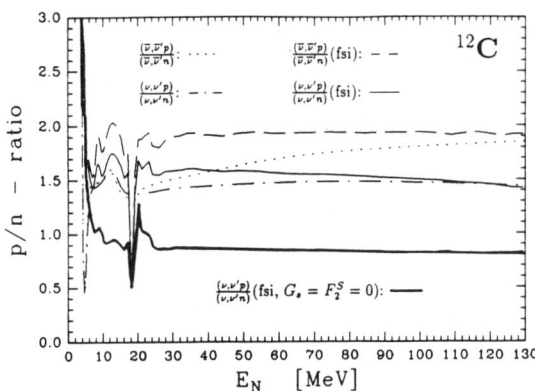

FIGURE 2. Ratios of proton-to-neutron to yields as a function of nucleon energy. The thick curve was obtained with vanishing strange form factors, the others with $G_1^s = -0.19$ and $F_2^s = -0.22$. The incident neutrino energy was set to $E_\nu = 200\ MeV$. (From Ref.13.)

= −0.19, the EMC value, the ratio becomes 1.45. In a very different calculation performed in Ref.16 at $E_\nu = 150\ MeV$, the ratio for $E_N > 36\ MeV$, the corresponding ratios are calculated to be 0.81 and 1.36, respectively.

The quantity of interest for the LSND experiment is the total nucleon cross-section - denoted by $\sigma(\nu,\nu'p)$ and $\sigma(\nu,\nu'n)$ in the following - obtained by integrating $d\sigma/dE_N$ over the energy of the emitted nucleon, after discriminating against protons that result from neutrino elastic scattering from the free protons in the LSND active volume. The dependence of the total cross-section on the strange form factors can be illustrated as follows: Noting that the quasielastic neutrino-induced knockout cross-section is dominated by the axial vector component, one roughly has

$$\sigma(\nu,\nu'N) \sim |-G_A\tau_3 + G_1^s|^2 \approx G_A^2 \left(1 \mp 2\frac{G_1^s}{G_A}\right), \qquad (21)$$

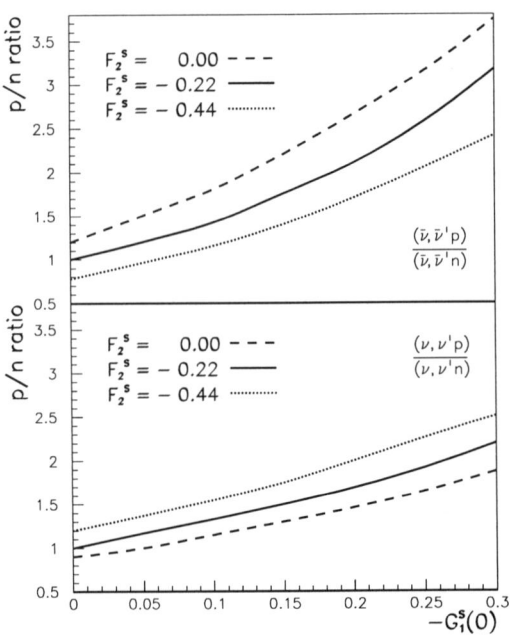

FIGURE 3. Ratio of integrated proton-to-neutron for quasielastic antineutrino- and neutrino-induced reactions on ^{12}C as a function of $G_1^s(0)$. The incident neutrino energy was set to $E_\nu = 200\ MeV$. (From Ref.13.)

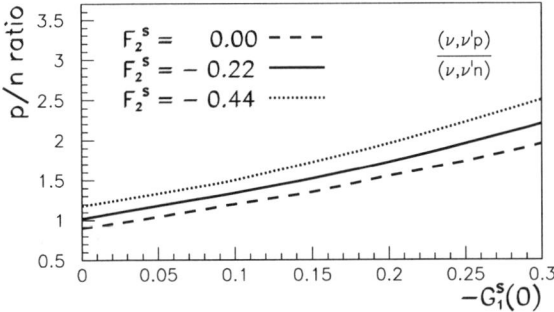

FIGURE 4. Same as Figure 3, but calculated for the LAMPF neutrino beam.

and therefore, the ratio of proton-to-neutron yield reads

$$R = \frac{\sigma(\nu, \nu' p)}{\sigma(\nu, \nu' n)} \approx 1 - 4 \frac{G_1^s}{G_A}, \qquad (22)$$

which is rather sensitive to G_1^s. As can be seen in Figure 3, the strong sensitivity of the ratio on the strange axial vector form factor $G_1^s(0)$ is confirmed as is its nearly linear dependence. However, the ratio is also dependent on the strange vector form factor F_2^s introduced by the vector-axial-vector interference term in the cross-section. Plotted in Figure 3 is the proton-to-neutron ratio as a function of $G_1^s(0)$ for three different values of $F_2^s(0)$, for both neutrino- and antineutrino-induced reactions.

Neutrino beams are not normally monoenergetic and generally consist of a mixture of neutrinos and antineutrinos. However, even after folding in the LAMPF neutrino beam profile, the dependence of the ratio on $G_1^s(0)$ shows a very similar behaviour to the one above, as illustrated in Figure 4. Therefore, the measurement of the proton-to-neutron quasielastic yield is expected to determine the strange axial vector form factors of the nucleon. An independent measurement of $F_2^s(0)$ would be desirable but, even without it, a determination of $G_1^s(0)$ to reasonable accuracy should be feasible within the accepted range of $F_2^s(0)$ values.

THE LSND EXPERIMENT

The LSND experiment has been primarily designed to search for neutrino oscillations in both $\nu_\mu \to \nu_e$ and $\bar{\nu}_\mu \to \bar{\nu}_e$ appearance channels. In addition,

LSND is able to study $\nu_{\mu(e)}C$ inclusive and exclusive processes - producing the first accurate measurement of the inclusive reaction $\nu_\mu C \to \mu^- X$ above threshold (17) - as well as quasielastic neutrino-nucleon scattering.

The neutrino source is provided by the A6 beam stop at LAMPF which produces π^+ and π^- using a 780 MeV proton beam passing through a 20 cm long water target positioned 1 m upstream of a water-cooled Cu beam dump. Essentially all of the π^- are captured before they can decay and most of the π^+ decay at rest in the target or the adjacent beam stop, to produce a monoenergetic ν_μ and a μ^+ which subsequently decays at rest to produce a ν_e and a $\bar{\nu}_\mu$ with the characteristic Michel spectrum with endpoint energy of 52.8 MeV. Less than 3% of the π^+ decay in flight over the 1 m free path to produce a ν_μ with an energy extending up to 270 MeV (Figure 5). The neutrino fluxes at any point in the detector are calculated using a beam Monte Carlo program, based on measured pion cross sections (18) and the uncertainties are estimated to be 7% and 12% for the decay-at-rest (DAR) and the decay-in-flight (DIF), respectively. Since the kinetic energy of the recoiling proton is

$$T_p = \frac{2E_\nu^2 sin^2(\theta/2)}{2E_\nu sin^2(\theta/2) + m_p} \qquad (23)$$

then its maximum kinetic energy is $T_p^{max} = 2E_\nu^2/(2E_\nu + m_p)$. For neutrinos from DAR this yields only $T_p^{max} = 5.4$ MeV which does not produce a readily observable signal in the detector; therefore only DIF neutrinos are of interest for this measurement.

The detector is a cylindrical tank, 5.7 m in diameter and 8.7 m long, with

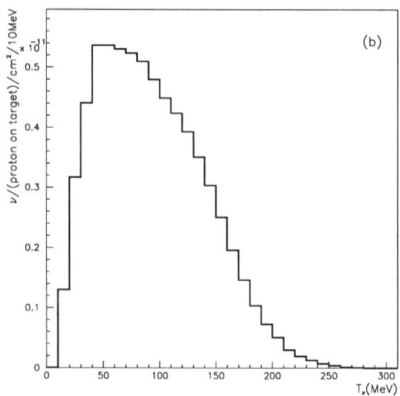

FIGURE 5. Neutrino flux at the center of the LSND detector (ν_μ from DIF).

its main axis lying horizontally, located at a mean distance of 29 m from the beam stop and an average angle of 12° relative to the incident proton direction. The inside surface supports 1220 inward looking 8-inch photomultiplier tubes, which provide 24.8% coverage. The tank is filled with 51,000 gallons (180 tons) of liquid scintillator consisting of mineral oil (CH_2) and 0.031 g/l of butyl-PBD. This mixture allows the detection of both Čerenkov and scintillation light; for relativistic particles (electrons and muons with $\beta > 0.68$) approximately 25% of the total light observed is in the Čerenkov cone (19). The detector is surrounded by an active liquid scintillator veto shield (20) with 292 5-inch phototubes, which tag cosmic muons that pass through the detector.

An elaborate data acquisition system records the phototube pulse heights and times every 100 ns and updates the detector and veto shield history over 204.8 μs. When a primary trigger occurs (i.e. > 100 detector and < 6 veto hits) the event is read out along with its past 51.2 μs activities with > 17 detector or > 5 veto hits. After a primary trigger with > 300 detector hits the threshold was lowered to 21 hit phototubes for 1 ms in order to trigger on 2.2 MeV γ from neutron capture on free protons. Furthermore, no primary triggers were allowed for 15.2 μs following each firing of the veto (> 5 veto hits) in order to reject the decay electrons following stopped cosmic muons.

Figure 6 illustrates the remarkable dynamic range of the LSND apparatus. The cosmic ray muons that escape the first level trigger of the veto shield deposit up to 2 GeV in the detector volume and dominate the high multiplicity

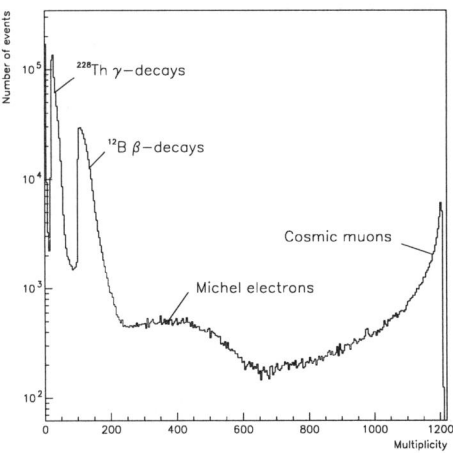

FIGURE 6. Typical event multiplicity in the LSND detector.

range of the distribution, while the γ rays recorded in the 1-ms window are as low as 0.6 MeV, produced primarily by radioactive ^{228}Th decays. The events between 200 and 600 hit phototubes are mostly Michel electrons that survive the 15.2 μs dead-time imposed by the trigger, while the peak above the primary trigger threshold (100 hits) is due to ^{12}B β-decays, produced subsequent to $\mu^- + {}^{12}C \rightarrow {}^{12}B + \nu_\mu$.

A complete event reconstruction is performed on-line: the event vertex is determined from the phototube timing information to within ~ 25 cm and, for relativistic particles, the direction is reconstructed to within ~ 15°. In addition, particle identification for particles above and below Čerenkov threshold is obtained through the fit to the Čerenkov cone and from the time distribution of the light, which is relatively later for particles below Čerenkov threshold.

Figure 7 shows the particle identification capability of the detector. Some 200 $\nu_\mu + {}^{12}C \rightarrow \mu^- + X$ beam excess events with subsequent $\mu^- \rightarrow e^- \nu_\mu \bar{\nu}_e$ have been extracted from the 1993 data sample (17), primarily by space-time correlation between the μ^- and the subsequent decay electron. As Figure 7 shows, there is good separation between the Michel electrons and the muons, which produce only scintillation light. The particle-id parameter is formed via a threefold product of the variance in an event's position, the variance in the corrected phototube timing and the event fit to the angular distribution of the observed light.

Hadron identification is performed through the particle-id parameter, which yields the same good separation from electrons as muons, while requiring no subsequent decay product consistent with a Michel electron insures misidentification as muons. Furthermore, distinction between neutrons and protons is facilitated by the detection of the 2.2 MeV γ following neutron thermalization and capture on hydrogen, $n + p \rightarrow d + \gamma$, correlated in space ($dr < 2.0\ m$) and time ($dt < 0.5\ ms$). The γ reconstruction has been verified by selecting

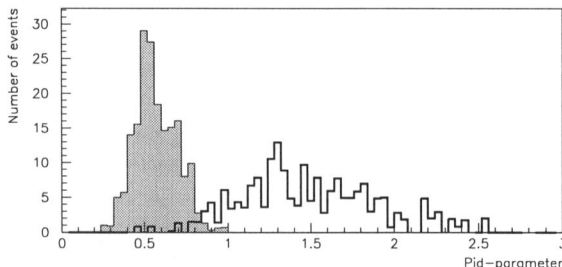

FIGURE 7. Particle-id parameters for electrons (shaded histogram) and muons from a sample of $\nu_\mu + C \rightarrow \mu^- + X$ events.

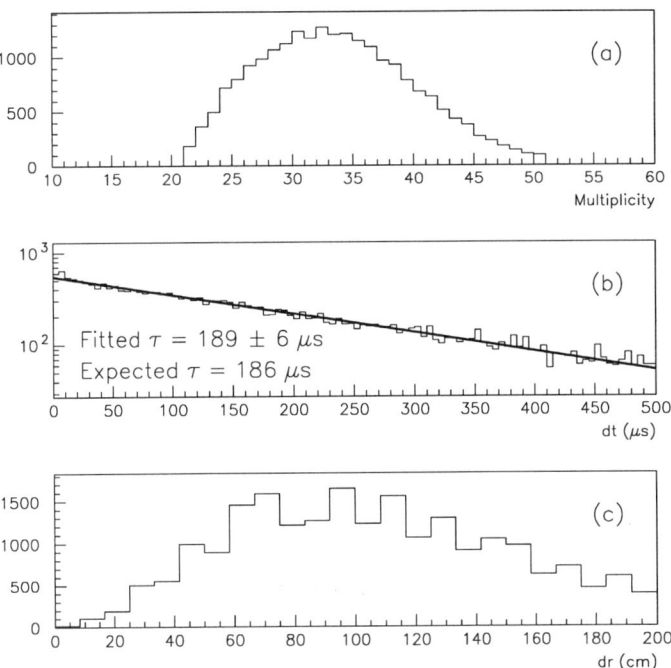

FIGURE 8. A sample of 2.2 MeV γs from cosmic neutrons: (a) γ multiplicity; (b) Time difference between the γ and the neutron; (c) Distance between the γ and the neutron reconstructed position.

cosmic ray neutrons with associated 2.2 MeV γs. The γ hit multiplicity, time difference and distance to the neutron are illustrated in Figure 8. As one can see from the figure, the average γ multiplicty is 33 hits, the time difference is 189 μs - in good agreement with the estimated neutron capture time of 186 μs - and is usually reconstructed within 2 m from the parent neutron. The efficiency for the neutron capture and for the γ to be reconstructed within 2.0 m and 0.5 ms and to have between 21 and 50 hit phototubes is 80%.

There has been only 1.5 months (September-October 1993) of operation of the detector with an intense beam (0.8 mA) from LAMPF, and therefore the statistics for a reasonable proton-to-neutron yield measurement are very limited. Figure 9 shows a preliminary sample of protons and neutrons from neutrino-induced nucleon knockout off ^{12}C extracted from the 1993 data, after

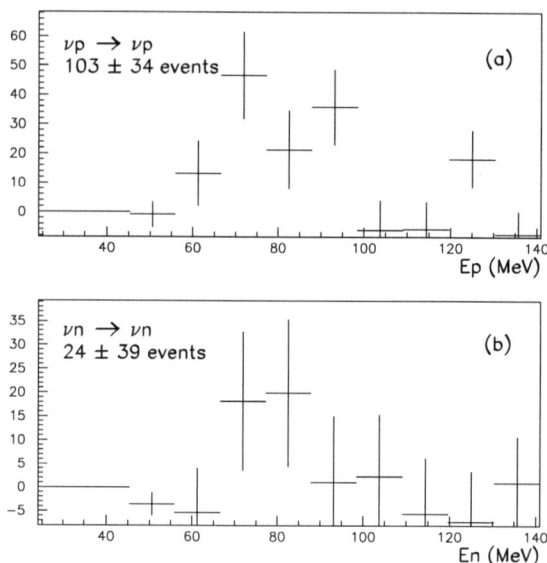

FIGURE 9. Preliminary sample of beam excess protons (a) and neutrons (b) from the 1993 LSND data.

beam-off subtraction. It should be pointed out that the selection criteria used for obtaining the above sample are still subject to change, as more detailed insight into the detector's response to hadrons is gained. The data presented here should be regarded only as an indication of the feasibility of the measurement and by no means as a definite result. After the 1994-95 run periods, with longer operation time, modified water target, longer DIF path, higher beam current (1.0 mA) and a lower threshold for the γ-window, LSND will record an order of magnitude more data which should allow for a good measurement of $G_1^s(0)$. The goal of the LSND experiment is to determine $G_1^s(0)$ to an absolute accuracy of approximately ± 0.04.

HIGH-ENERGY νN SCATTERING - E734

Before concluding, I would like to mention the recent reanalysis (21) of the currently best experiment (11) on high-energy ($E_\nu > 1\ GeV$) neutrino-nucleon elastic scattering, the AGS experiment E734, performed using the

TABLE 1. Fit results for the strange form factors $G_1^s(0) = \Delta s$, F_1^s and F_2^s and the axial-vector dipole mass M_A. (From Ref.21.)

Fit	$G_1^s(0)$	F_1^s	F_2^s	M_A	χ^2/N_{DOF}
I	0	0	0	1.086±0.015	14.12/14
II	−0.15±0.07	0	0	1.049±0.019	9.73/13
III	−0.13±0.09	0.49±0.70	−0.39±0.70	1.049±0.023	9.28/11
IV	−0.21±0.10	0.53±0.70	−0.40±0.72	1.012±0.032	8.13/11

BNL wide-band neutrino beam. In its original analysis (11), E734 used the then world average for M_A (1.032 ± 0.036 GeV/c^2) in the isovector axial form factor. However, in a subsequent publication (22), E734 reported a new value $M_A = 1.09 \pm 0.03 \pm 0.02\,GeV/c^2$, which raised the world average to its present value of $M_A = 1.061 \pm 0.026\,GeV/c^2$. Furthermore, the effects of strange vector form factors were not included in the analysis. In light of these shortcomings, the E734 data were reanalyzed, using the formalism indicated in Eqs.(13)-(16), along with a standard parametrization of the known form factors.

The new results are illustrated in Table 1 (from Ref.21), where the first three fits were performed using the current value for M_A while the last one uses the world average value for M_A prior to the E734 result. As one can easily see from the table, the data can be readily fit within the new world average for M_A with all strange form factors consistent with zero. A lower value of χ^2 can be achieved by allowing M_A to drop to 1.049 GeV/c^2 and keeping both F_1^s and F_2^s fixed to zero, whereupon $G_1^s(0) = -0.15 \pm 0.07$. The third fit allows F_1^s to float to a positive value and F_2^s to a negative value, lowering $G_1^s(0)$ to -0.13 ± 0.09 and improving slightly the χ^2.

While very interesting, E734 is not decisive in determining any of the strange form factors. The value of $G_1^s(0)$ appears to be strongly correlated to the value of M_A and thus, little progress can be made in the determination of $G_1^s(Q^2)$ unless M_A is more precisely determined. The strange vector form factors $F_1^s(Q^2)$ and $F_2^s(Q^2)$ also appear to be correlated and their sum consistent with zero. In fact, E734 does an excellent job in fixing the sum $F_1^s(Q^2) + F_2^s(Q^2)$ at $Q^2 = 0.75\,GeV^2/c^2$ to -0.01 ± 0.08.

CONCLUSIONS

Measuring the axial vector form factors in low-energy neutrino-nucleon elastic scattering is an excellent way to probe the spin structure of the nucleon. The proton-to-neutron yield ratio in neutrino-induced nucleon knockout from ^{12}C is very sensitive to $G_1^s(0) = \Delta s$, while insensitive to the details of the nuclear model, final states interactions and neutrino beam profile. Furthermore, the above method does not suffer from the inherent problems that beset the interpretation of the deep-inelastic scattering results, e.g. extrapolation to $x = 0$, higher twist corrections or $SU(3)$ flavour symmetry arguments. The early results from the 1993 LSND run are encouraging but it is yet too early to precisely predict the accuracy that $G_1^s(0)$ will be determined with after the 94-95 runs at LAMPF.

REFERENCES

1. Abramowicz, H., et al., CDHS Collaboration, *Z. Phys.* **C15**, 19-31 (1982).
   Foudas, C., et al., CCFR Collaboration, *Phys. Rev. Lett.* **64**, 1207-1210 (1990).
2. Jaffe, R.L., and Korpa, C.L., *Comments Nucl. Part. Phys.* **17**, 163-175 (1987).
3. Gasser, J., Leutwyler, H., and Sainio, M.E., *Phys. Lett.* **B253**, 252-259 (1991); *Phys. Lett.* **B253**, 260-264 (1991).
4. Ashman, J., et al., EMC Collaboration, *Phys. Lett.* **B206**, 364-370 (1988).
5. Ashman, J., et al., EMC Collaboration, *Nucl. Phys.* **B328**, 1-35 (1989).
6. Ellis, J., and Jaffe, R.L., *Phys. Rev.* **D9**, 1444-1446 (1974); *Phys. Rev.* **D10**, 1669-1670 (1974).
7. Jaffe, R.L., and Manohar, A., *Nucl. Phys.* **B337**, 509-546 (1990).
   The values of Δs and $\Delta u + \Delta d + \Delta s$ quoted here are from this paper. The slight differences arise from the F and D values extracted from hyperon decay.
8. Adams, D., et al., SMC Collaboration, *Phys. Lett.* **B329**, 399-406 (1994).
9. Anthony, P.L., et al., E142 Collaboration, *Phys. Rev. Lett.* **71**, 959-962 (1993).
10. Kaplan, D.B., and Manohar, A., *Nucl. Phys.* **B310**, 527-547 (1988);
    Ellis, J., and Karliner, M., *Phys. Lett.* **B213**, 73-80 (1988).
11. Ahrens, L.A., et al., *Phys. Rev.* **D35**, 785-809 (1987).
12. Beise, E.J., and McKeown, R.D., *Comments Nucl. Part. Phys.* **20**, 105-117 (1991).
13. Garvey, G.T., Krewald, S., Kolbe, E., and Langanke, K., *Phys. Lett.* **B289**, 249-254 (1992).
14. Garvey, G.T., Kolbe, E., Langanke, K., and Krewald, S., *Phys. Rev.* **C48**, 1919-1925 (1993).
15. Jaffe, R.L., *Phys. Lett.* **B229**, 275-279 (1989).
16. Horowitz, C.J., et al., *Phys. Rev.* **C48**, 3078-3087 (1993).
17. Albert, M., et al., LSND Collaboration, to be submitted.
18. Burman, R.L., Potter, M.E. and Smith, E.S., *NIM* **A291**, 621-633 (1990).
19. Reeder, R.A., et al., *NIM* **A334**, 353-366 (1993).
20. Napolitano, J.J., et al., *NIM* **A274**, 152-164 (1989).
21. Garvey, G.T., Louis, W.C., and White, D.H., *Phys. Rev.* **C48**, 761-765 (1993).
22. Ahrens, L.A., et. al., *Phys. Lett.* **B202**, 284-288 (1988).

A New Neutron Polarimeter and Measurement of G_E^n from the $^2H(\vec{e}, e'\vec{n})\,^1H$ Reaction

R. Madey,*† A. Lai,* T. Eden†

Kent State University, Kent, OH 44242
†*Hampton University, Hampton, VA 23668*

Abstract. In a continuing effort to improve the quality of measurements for extracting the neutron electric form factor (G_E^n) from the $^2H(\vec{e}, e'\vec{n})\,^1H$ reaction, one of us (RM) proposed a new configuration for a neutron polarimeter (NPOL) to measure G_E^n from the $^2H(\vec{e}, e'\vec{n})\,^1H$ reaction in Bates E89-04 and CEBAF E93-038. Substantially higher count rates with this new neutron polarimeter reduce the uncertainties in G_E^n. A description of the new NPOL is given here followed by an outline of the experiment with projections of uncertainties in G_E^n.

THE NEW NEUTRON POLARIMETER: A GENERAL DESCRIPTION

The configuration of the NPOL used in Bates E85-05[1] to measure G_E^n at a central squared four-momentum transfer $Q^2 = 0.255\ (\text{GeV}/c)^2$ was described previously.[2] The design of the polarimeter was based on properties of n-p scattering as a polarization analyzer. It consisted of 12 detectors: four front analyzers and two rear arrays of four detectors arranged in a V shape. The axis of each rear array was at an optimal n-p scattering angle θ from the central ray of the incident neutron flux. These detectors were contained in a large shielding enclosure with a front steel collimator. In order to shield the rear detectors from neutrons produced at the target, the height of the collimator was restricted to 25.4 cm, which in turn restricted the acceptance solid angle of the NPOL.

The new configuration for the NPOL planned for Bates E89-04 and CEBAF E93-038 is shown in Fig. 1; it consists of 20 plastic (NE102) scintillation detectors: eight front analyzers and two rear arrays **parallel** to the central ray of the incident neutron flux. Each rear array consists of two staggered layers of detectors; each layer is composed of three detectors side by side; where each detector is 101.6-cm long by 50.8-cm wide by 10.16-cm thick. The eight front scintillators are 101.6-cm long by 25.4-cm wide by 10.16-cm thick. Positioned immediately at the front and rear of the front detector array are thin (0.95-cm) plastic counters to veto charged particles. The configuration of the NPOL and its shielding is based on the criterion that a minimum thickness of one meter of iron is needed to shield the rear detectors from the direct flux of neutrons emanating from the target; however, this new configuration permits doubling the height of the front detectors, thereby doubling the acceptance solid angle of the neutron polarimeter. Doubling the neutron solid angle permits doubling the electron solid angle in a coincidence experiment. Also, this new configuration results in higher detection efficiencies than those of the old configuration while keeping comparable analyzing powers. Optimizing the configuration determines the position of the detectors in the rear array, which is characterized by the horizontal and vertical distances from the center of each array to the center point of the front detectors.

OPTIMIZING THE G_E^n EXPERIMENT

Bates E89-04 and CEBAF E93-038 were designed to measure the electric form factor G_E^n of the neutron from the $^2H\,(\vec{e},e'\vec{n})\,^1H$ reaction for Q^2 values from 0.30 to 1.5 $(\text{GeV}/c)^2$. Both the kinematics and the configuration of the NPOL are optimized to minimize the uncertainties in G_E^n.

It should be pointed out that there are two approaches to extract G_E^n from these experiments. In **Approach I** (AI hereafter), we measure the scattering asymmetry $\xi_{S'}$ related to the sideways polarization $P_{S'}$ of the neutron. With a known analyzing power $\langle A_y \rangle$ of the NPOL and a measured polarization P_L of the incident electron, we can extract the polarization transfer coefficient $D_{LS'} \equiv P_{S'}/P_L$ and thus the ratio $g(D_{LS'}) \equiv G_E^n/G_M^n$, where G_M^n is the magnetic form factor of the neutron. In **Approach II** (AII hereafter), we measure both scattering asymmetries $\xi_{S'}$ and $\xi_{L'}$ related, respectively, to the sideways and longitudinal polarizations of the neutron $P_{S'}$ and $P_{L'}$. This approach does not require knowing the NPOL analyzing power $\langle A_y \rangle$; instead, it requires a magnet to precess the longitudinal polarization $P_{L'}$ into the sideways direction. By doing the experiments with both approaches, we can obtain an internal consistency check of the results.

Optimizing the experiment means minimizing the data acquisition time T for a desired statistical uncertainty in g. Because we plan to do the experiment with both approaches, the optimization is carried out with AI that requires knowing the analyzing power of the NPOL. The data acquisition time for detecting N events with a count rate R is given by

$$T = \frac{N}{R} \propto \frac{1}{(D_{LS'}/f_1)^2 \langle \sigma_3 \rangle \langle A_y \rangle^2 \epsilon_n \Delta\theta_n (\Delta\phi_n^v)^2}, \quad (1)$$

Here, f_1 is a kinematic function, which will be given later, $\langle\sigma_3\rangle$ is the electron-neutron coincidence cross section averaged over the detector acceptances, ϵ_n is the detection efficiency of the NPOL, and $\Delta\theta_n$ and $\Delta\phi_n^v$ are the neutron horizontal and vertical angular acceptances, which are determined by the neutron flight path for a fixed size of the neutron detectors. We define two figures-of-merit (FOM), one being the kinematical FOM_k, and the other being the configurational FOM_c:

$$FOM_k \equiv (D_{LS'}/f_1)^2 \langle\sigma_3\rangle \qquad (2)$$

$$FOM_c \equiv \langle\sigma_3\rangle \langle A_y\rangle^2 \epsilon_n \Delta\theta_n (\Delta\phi_n^v)^2 . \qquad (3)$$

Then the data acquisition time can be written

$$T \propto \frac{1}{FOM_k \times FOM_c}, \qquad (4)$$

and thus the optimization of the experiment is equivalent to maximizing FOM_k and FOM_c.

OPTIMIZED KINEMATICS

The FOM_k as a function of electron scattering angle θ_e was computed with the Monte Carlo code MCEEP.[3] Figure 2 is a plot of FOM_k for $Q^2 = 1.0$ $(GeV/c)^2$. The figure shows that the electron scattering angle near the maximum FOM_k is at $\theta_e = 45.0°$ corresponding to an electron beam energy of 1.6 GeV. Table 1 lists the optimized kinematics for four Q^2 points. Because the measurements at Bates for $Q^2 = 0.30$ and 0.50 $(GeV/c)^2$ are planned to use the same neutron scattering angle, we optimized the kinematics for $Q^2 = 0.50$ $(GeV/c)^2$ and forced the neutron scattering angle for $Q^2 = 0.30$ $(GeV/c)^2$ to be the same as that of $Q^2 = 0.50$ $(GeV/c)^2$ by reducing the beam energy.

TABLE 1. Kinematic Conditions for G_E^n Experiments

Q^2 $(GeV/c)^2$	Exp. No.	E_{beam} (GeV)	θ_e (deg)	θ_n (deg)	$p_{e'}$ (MeV/c)	p_n (MeV/c)
0.30	Bates E89-04	0.586	66.4	43.1	425	572
0.50	Bates E89-04	0.880	57.5	43.1	613	757
1.0	CEBAF E93-038	1.60	45.0	41.2	1066	1133
1.5	CEBAF E93-038	2.40	36.4	40.5	1600	1464

THE NEW NEUTRON POLARIMETER: OPTIMIZED CONFIGURATIONS

The FOM_c is determined by the neutron flight path x (which determines the angular acceptances of the NPOL for neutron detectors of a given size) and the positions of the rear detectors relative to the front detectors characterized by the vertical and horizontal distances d and x_{FR}. These quantities determine the value of the analyzing power $\langle A_y \rangle$ and the detection efficiency ϵ_n. The definitions of x, d, and x_{FR} are given in Fig. 1.

To find the optimal flight path x, we computed analytically (by assuming a point-like front detector) the maxima of FOM_c as a function of x by varying d and x_{FR}. The analyzing power A_y was calculated with the code SAID.[4] Figure 3 is a plot of $(FOM_c)_{max}$ for $Q^2 = 0.50$ $(GeV/c)^2$. It shows that the optimal flight path should be at $x \simeq 450$ cm; however, in order to enhance the model insensitivity of the experiment, which was discussed elsewhere,[5] we chose $x = 500$ cm for the two Q^2 points at Bates. Similar studies show that a flight path of $x = 700$ cm is optimal for the two Q^2 points at CEBAF.

Once the flight path is fixed, we perform a simulation of the NPOL[6] to determine the values of d and x_{FR} that maximize FOM_c. The simulation takes into account the fact that the front detectors have finite sizes. Table 2 lists the values of these optimized configurational variables together with the values of $\langle A_y \rangle$ and ϵ_n.

With this new NPOL, the e-n coincidence count rates in the G_E^n measurements are typically five to six times higher than those with the old V-shaped configuration. A calibration run at $T_n = 160$ MeV [corresponding to $Q^2 = 0.30$ $(GeV/c)^2$] was performed recently at IUCF. Analysis of the calibration data is currently underway.

TABLE 2. $\langle A_y \rangle$ and ϵ_n with optimized values of x, d, and x_{FR}

Q^2 $(GeV/c)^2$	Exp. No.	x (m)	Layer	d (cm)	x_{FR} (cm)	$\langle A_y \rangle$ (%)	ϵ_n (%)
0.30	Bates E89-04	5.0	Inner Outer	69.4 84.4	222.8 260.0	42.6	0.25
0.50	Bates E89-04	5.0	Inner Outer	69.4 84.4	222.8 260.0	34.4	0.20
1.0	CEBAF E93-038	7.0	Inner Outer	51.6 66.6	223.6 251.6	24.7	0.28
1.5	CEBAF E93-038	7.0	Inner Outer	50.2 65.2	194.1 233.1	23.0	0.41

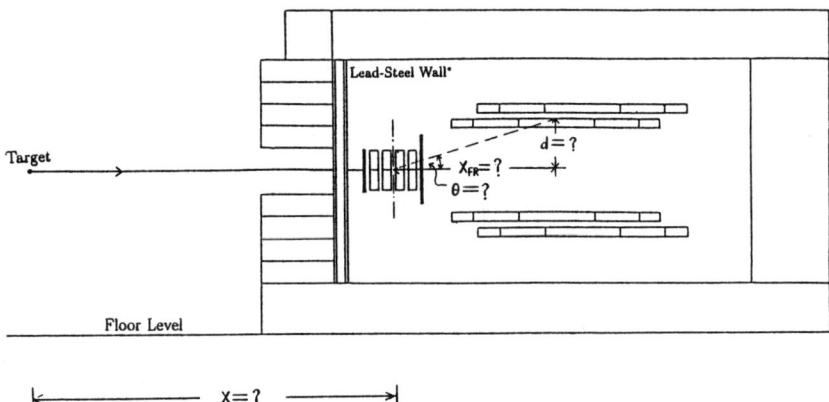

FIGURE 1. A schematic view of the NPOL. The definitions of the configuration variables x, d, and x_{FR} are given also.

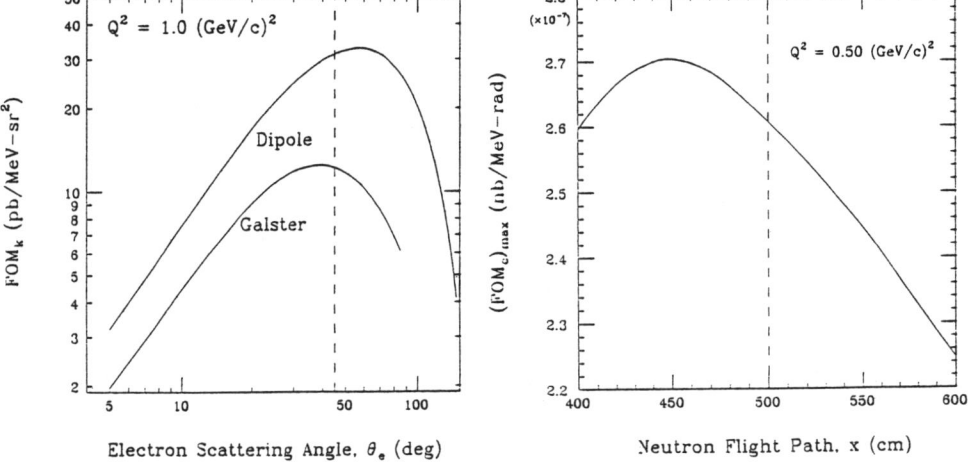

FIGURE 2. FOM_k for $Q^2 = 1.0$ $(GeV/c)^2$. The dashed line denotes the electron angle chosen for the experiment.

FIGURE 3. $(FOM_c)_{max}$ for $Q^2 = 0.50$ $(GeV/c)^2$. The dashed line denotes the flight path chosen for the experiment.

UNCERTAINTIES IN G_E^n

Measurements of G_E^n in Bates 89-04 and CEBAF E93-038 will be performed by measuring the sideways and longitudinal polarization components of the recoil neutron after quasielastic scattering of a longitudinally polarized electron from an unpolarized neutron in a liquid-deuterium (LD$_2$) target. With AI, where only $\xi_{S'}$ will be measured, we employ the technique suggested by Arnold, Carlson, and Gross,[7] which gives the ratio of the electric form factor to the magnetic form factor $g \equiv G_E^n/G_M^n$

$$g = \left[(K_1^2 - 4K_2 D_{LS'}^2)^{1/2} - K_1\right](2D_{LS'})^{-1} \quad (5)$$

where

$$K_1 = 2\sqrt{\tau(1+\tau)}\tan(\theta_e/2) \quad (\tau = Q^2/4M^2) \quad (6)$$

$$K_2 = \tau + K_1^2/2 . \quad (7)$$

With AII, where both $\xi_{S'}$ and $\xi_{L'}$ will be measured, the value of g is given by

$$g = -K_3(D_{LS'}/D_{LL'}) = -K_3(\xi_{S'}/\xi_{L'}) \quad (8)$$

where

$$K_3 = \sqrt{\tau[1+\tau\sin^2(\theta_e/2)]}\left[\cos(\theta_e/2)\right]^{-1} . \quad (9)$$

The statistical uncertainties in g with these two approaches are

$$(\Delta g/g)_{stat}^I = f_1^I(\Delta\xi_{S'}/\xi_{S'})_T \quad (10)$$

where the superscript I denotes AI, and

$$f_1^I = K_1\left(K_1^2 - 4K_2 D_{LS'}^2\right)^{-1/2} , \quad (11)$$

$$(\Delta g/g)_{stat}^{II} = f_1^{II}(\Delta\xi_{S'}/\xi_{S'})_T \quad (12)$$

where the superscript II denotes AII, and

$$f_1^{II} = 1 - g/K_3 . \quad (13)$$

The subscript T, used in the above expressions, denotes the total data acquisition time, which in AII is allocated optimally to measuring $\xi_{S'}$ and $\xi_{L'}$, respectively.

The scale uncertainty with AI arises from the uncertainties in the analyzing power of the NPOL and the beam polarization; however, the scale uncertainty arises in AII from the uncertainty in the precession angle χ:

$$(\Delta g/g)^I_{scale} = f^I_1 \left[(\Delta A_y/A_y)^2 + (\Delta P_L/P_L)^2\right]^{1/2}, \qquad (14)$$

$$(\Delta g/g)^{II}_{scale} = 1 - \cos(\Delta \chi). \qquad (15)$$

The systematic uncertainties arising from the uncertainties in the electron scattering angle are negligible in both approaches; therefore, for both approaches, the dominant uncertainty is the statistical uncertainty for data acquisition times of less than a few hundred hours.

With the optimized kinematics and configurations for the NPOL, we are able to project the uncertainties in G^n_E for both aforementioned approaches. The total relative uncertainty in G^n_E is given by

$$(\Delta G^n_E/G^n_E)_{total} = \left[(\Delta g/g)^2_{total} + (\Delta G^n_M/G^n_M)^2_{total}\right]^{1/2}. \qquad (16)$$

In the projection, we assumed $(\Delta G^n_M/G^n_M)_{total} = 0.030$, which is expected to be achievable. In Table 3, we list the projected uncertainties in G^n_E for the Galster parameterization for both approaches.

TABLE 3. Projected Uncertainties in G^n_E

Exp.	Luminosity ($\times 10^5$ nb^{-1}s^{-1})	P_L (%)	Q^2 (GeV)^2	Appr.	$\Delta G^n_E/G^n_E$ (%)	T (hr)
Bates	0.6[a]	40	0.30	AI AII	11.6 9.4	75
E89-04			0.50	AI AII	11.8 9.5	150
CEBAF	2.8[b]	49	1.0	AI AII	10.6 7.8	250
E93-038			1.5	AI AII	13.4 11.4	250
CEBAF	2.7[c]	80	1.0	AI AII	8.8 5.2	250
E93-038	2.2[d]		1.5	AI AII	10.4 8.1	250

a. With a beam current of 25 µA, and a target thickness of 5 cm.
b. With a beam current of 100 µA, and a target thickness of 5 cm.
c. With a beam current of 35 µA, and an effective target thickness of 14 cm.
d. With a beam current of 35 µA, and an effective target thickness of 11 cm.

Note that AII yields smaller uncertainties. In Fig. 4, we plot G_E^n as a function of Q^2 with projected total uncertainties in AII for the dipole, Galster, and $G_E^n = 0$ cases. The figure demonstrates clearly that precision measurements are possible and that we will be able to achieve the goal of distinguishing between different parameterizations of G_E^n.

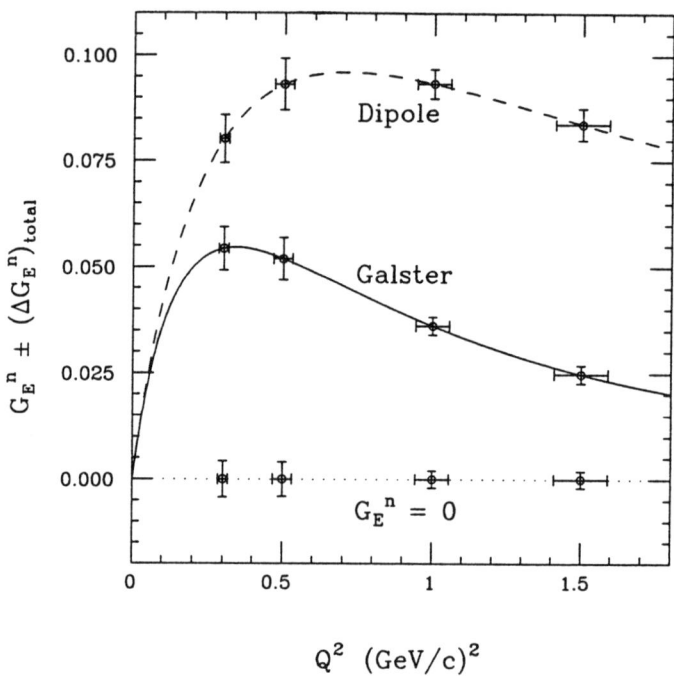

FIGURE 4. G_E^n as a function of Q^2 with projected total uncertainties

This work was supported in part by the National Science Foundation under Grant Nos. PHYS-91-07064 and HRD-91-54080

REFERENCES

1. T. Eden et al., Phys. Rev. **C50**, R1 (1994).
2. T. Eden et al., Nucl. Instrum. Meth. **A338**, 432 (1994).
3. P. Ulmer, CEBAF-TN-91-101 (1991), unpublished.
4. R. Arndt, R. Hackman, and L. Roper, Phys. Rev. **C15**, 1002 (1977).
5. R. Madey et al., Proposal for CEBAF E93-038 (1993), and updated proposal for Bates E89-04 (1994), unpublished.
6. T. Eden, A neutron polarimeter simulation code, unpublished.
7. R. Arnold, C. Carlson, and F. Gross, Phys. Rev. **C23**, 363 (1981).

Measuring G_E^n in Quasielastic $^3\vec{\mathrm{He}}(\vec{e}, e')$

J.-O. Hansen*

*Laboratory for Nuclear Science and Department of Physics,
Massachusetts Institute of Technology, Cambridge, Massachusetts 02139.*

The MIT-Bates 88-25 Collaboration

Abstract. We report a measurement of the transverse-longitudinal asymmetry $A_{TL'}$ in $^3\vec{\mathrm{He}}(\vec{e}, e')$ quasielastic scattering at $Q^2 = 0.14$ (GeV/c)^2. A straightforward extraction of the neutron electric form factor, G_E^n, is made using a PWIA model. No meaningful result for G_E^n is obtained due to large uncertainties and possible final-state interaction effects. At higher Q^2, the determination of G_E^n in $^3\vec{\mathrm{He}}(\vec{e}, e')$ appears feasible.

INTRODUCTION

Electromagnetic form factors are of fundamental importance for an understanding of the underlying structure of nucleons. Knowledge of the distribution of charge and magnetization within the neutron and proton provides constraints on nucleon models based on QCD, as well as a basis for calculations of processes involving the electromagnetic interaction with complex nuclei. The precise measurement of the neutron electric form factor, G_E^n, is a long-standing problem in medium-energy nuclear physics. Compared to the other nucleon form factors, G_E^n is small owing to the net zero electric charge of the neutron. Experimentally, one faces the difficulty that no free neutron targets exist so that almost all data on neutron form factors must be derived from nuclear scattering experiments. Only the slope of G_E^n near zero momentum transfer is known precisely from scattering of neutron beams from atomic electrons [1]. To date, most of our information about G_E^n at finite momentum transfer comes from elastic and quasielastic electron-deuteron scattering [2].

A relatively new approach to the problem of measuring G_E^n is the use of spin observables in electron scattering from polarized ^{3}He ($^3\vec{\mathrm{He}}$). In quasielastic (QE) scattering, the spin degrees of freedom introduce new response functions

---

*Present address: Physics Division, Argonne National Laboratory, Argonne, IL 60439. Electronic mail: ole@anl.gov.

into the cross section, thus providing additional information about nuclear structure which is otherwise inaccessible or difficult to measure [3]. In particular, interference terms between different multipoles may appear in the cross section which may enhance the sensitivity to small quantities. ^{3}He is an interesting nucleus for such studies because its ground state wave function is predominantly a spatially symmetric S state in which the spins of the two protons pair off and the spin of the nucleus is carried mainly by the lone neutron in the nucleus. Therefore, spin observables in QE scattering of polarized electrons from $^3\vec{\mathrm{He}}$ should be sensitive to the neutron electromagnetic form factors. Both inclusive scattering, $^3\vec{\mathrm{He}}(\vec{e},e')$ [4–7], and the neutron coincidence reaction, $^3\vec{\mathrm{He}}(\vec{e},e'n)$ [8], have been explored with the goal of extracting G_E^n.

In this talk, we present the results of a recent $^3\vec{\mathrm{He}}(\vec{e},e')$ experiment at MIT-Bates in kinematics sensitive to G_E^n at momentum transfer $Q^2 = 0.14$ (GeV/c)^2. The experiment significantly improves on the precision of two previous inclusive experiments done at similar Q^2 [6,7]. We attempt a straightforward extraction of G_E^n in the framework of a plane wave impulse approximation (PWIA) calculation and investigate the model dependence of the result. In addition, we use the PWIA model to discuss the feasibility of determining G_E^n in $^3\vec{\mathrm{He}}(\vec{e},e')$ at higher momentum transfers.

FORMALISM

For inclusive electron scattering from the spin-$\frac{1}{2}$ ^{3}He target, the unpolarized cross section can be expressed in terms of the longitudinal and transverse response functions, R_L and R_T, respectively. For both beam and target polarized, the spin-dependent contribution to the cross section is completely contained in two additional response functions, a transverse response, $R_{T'}$, and an interference between transverse and longitudinal multipoles, $R_{TL'}$. An experimentally clean signature of these polarized responses is the spin-dependent asymmetry, defined as

$$A \equiv \frac{\sigma_+ - \sigma_-}{\sigma_+ + \sigma_-}, \qquad (1)$$

where the subscript $+(-)$ refers to the helicity of the incident electrons, and σ is the differential cross section. In terms of the R_K, A can be written [3]

$$A = -\frac{\cos\theta^* v_{T'} R_{T'}(Q^2,\omega) + 2\sin\theta^* \cos\phi^* v_{TL'} R_{TL'}(Q^2,\omega)}{v_T R_T(Q^2,\omega) + v_L R_L(Q^2,\omega)}, \qquad (2)$$

where θ^* and ϕ^* are the polar and azimuthal angles defining the direction of the target spin with respect to the momentum transfer $\vec{q}$, the v_K are kinematic factors, ω is the electron energy loss, and $Q^2 \equiv |\vec{q}|^2 - \omega^2$. Hence, orienting the target spin at $\theta^* = 90°$ (0°) selects the asymmetry piece $A_{TL'}$ ($A_{T'}$) proportional to $R_{TL'}$ ($R_{T'}$).

To date, theoretical studies of inclusive QE scattering from $^3\vec{\text{He}}$ have employed the PWIA to describe the scattering mechanism, in which the $R_{TL'}$ response contains a contribution proportional to $G_M^n G_E^n$. Blankleider and Woloshyn, using closure to sum over final states, originally predicted that $A_{TL'}$ would be strongly sensitive to G_E^n at the kinematics of the present work [5]. However, recent more sophisticated analyses [9,10], which use a full spin-dependent spectral function to describe the $^3\vec{\text{He}}$ ground state, point out an inconsistency in the previous predictions for $A_{TL'}$ and show, at the kinematics of the work reported here, a dominant ($\sim 75\%$) proton contribution to $A_{TL'}$, which greatly reduces the sensitivity to G_E^n. This proton contribution arises chiefly from the presence of the 'small' S' ($\sim 1\%$) and D ($\sim 8\%$) state admixtures to the ^{3}He ground state wave function.

BATES EXPERIMENT 88-25

The experiment was carried out at the MIT-Bates Linear Accelerator Center using a 370 MeV longitudinally polarized, pulsed electron beam and a polarized ^{3}He gas target. The target was a closed double-cell system consisting of a glass optical pumping cell and a copper target cell, which was cryogenically cooled to 13 K. The metastability-exchange optical pumping technique [11,12] was employed to polarize the ^{3}He gas at a pressure of 2.2 Torr. The average beam and target polarizations were $(37.1 \pm 1.6)\%$ and $(37.1 \pm 1.7)\%$, respectively. To maximize sensitivity to $A_{TL'}$, the target spin was oriented at 42.5° with respect to the beam direction. The scattered electrons were detected in the One Hundred Inch Proton Spectrometer (OHIPS) using a detector package consisting of two crossed drift chambers, three planes of plastic scintillators, and an isobutane gas Čerenkov counter. The trigger was formed by a coincidence of the three scintillators. The spectrometer selected a scattering angle of 70.1° and had a central momentum of 285 MeV/c, corresponding to QE kinematics. The momentum resolution was $\lesssim 1.2\%$ FWHM over the momentum acceptance of 9.5%.

Charge (μA-h)	θ^*	ϕ^*	A (%)
2879	87°	0°	$1.72 \pm 0.72 \pm 0.13$
2133	93°	180°	$-1.43 \pm 0.85 \pm 0.11$
5012 (combined)			$1.60 \pm 0.55 \pm 0.12$

TABLE 1. Results of the $A_{TL'}$ asymmetry measurement averaged over the experimental energy acceptance ($72 \leq \omega \leq 99$ MeV). The uncertainties are statistical and systematic, respectively. The second row corresponds to the reversal of the target spin; hence the negative sign.

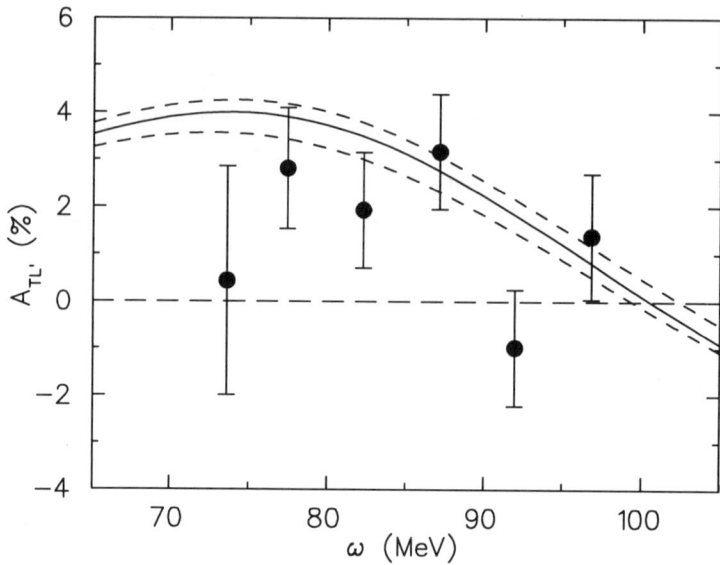

FIGURE 1. The experimental transverse-longitudinal asymmetry, $A_{TL'}$, as a function of electron energy transfer. The errors are statistical only. The dashed curves represent the upper and lower limits of our PWIA calculation (see text).

$A_{TL'}$ was extracted from the detected spin-dependent yield according to Eq. (1) as a function of the electron energy loss ω. To account for the dilution from unpolarized scattering, the result was divided by the product of the beam and target polarizations. Corrections were applied for target-related background (+5%) and radiative effects (−15%). The measured asymmetry is shown in Fig. 1. (The curves in the figure are discussed below.) Combining all the data over the experimental energy acceptance ($72 \leq \omega \leq 99$ MeV) yields a value of $A_{TL'} = 1.60 \pm 0.55$stat. $\pm\ 0.12$sys.%, as detailed in Table 1. A detailed description of experiment and analysis has been published elsewhere [13,14].

FORM FACTOR EXTRACTION

To extract G_E^n from the measured asymmetry data, predictions for $A_{TL'}(G_E^n)$ (averaged over the experimental ω acceptance) were computed [15] in PWIA using the formalism of Refs. [10,16]. G_E^n was taken to be a single constant parameter over the energy acceptance of the experiment, which is sufficiently accurate as G_E^n is expected to vary only slowly with Q^2. The dipole parameterization [18] was used for the other nucleon form factors.

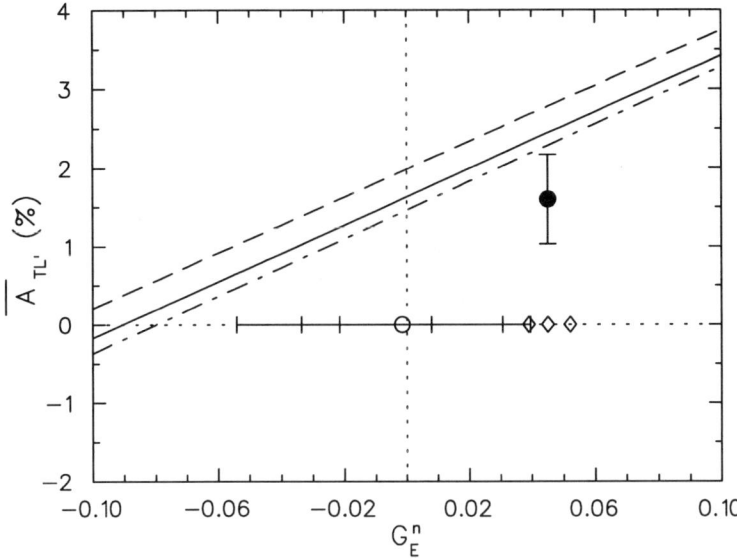

FIGURE 2. The predicted asymmetry $A_{TL'}$ in PWIA vs. G_E^n. The solid circle is the measured average asymmetry, and the open circle is the extracted G_E^n (see text).

To study the model dependence inherent in the PWIA calculation, predictions were generated for three different combinations of input parameters:

1. Paris NN potential and $CC1^{(0)}$ single-nucleon off-shell prescription [16].

2. Reid soft-core NN potential and the off-shell prescription used by the Rome group [9].

3. Bonn B NN potential and $CC1^{(0)}$ off-shell prescription.

The effect of different parameterizations of the other nucleon form factors is negligible compared to the variations arising from NN potential and off-shell prescription.

The three functions $A_{TL'}(G_E^n)$ obtained in this analysis are plotted in Fig. 2. The solid curve corresponds to case 1, the dashed curve to case 2, and the dot-dashed curve to case 3. The data point shown represents the measured asymmetry with total experimental error (arbitrarily placed at the Galster [18] value for G_E^n). The open circle on the x-axis is the extracted G_E^n value (−0.0015) obtained with calculation 1. The inner error bars represent the estimated theoretical uncertainty from different NN potentials and off-shell prescriptions (±0.015); the middle error bars the experimental error (±0.032); and the outer error bars the total uncertainty with theoretical and experimental uncertainty added linearly (± 0.047). The three open diamonds on the x-axis are the G_E^n predictions of Höhler [17], Galster [18], and Gari-Krümpelmann [19] at $Q^2 = 0.14$ (GeV/c)^2 (from left to right).

Fig. 1 shows the predictions for $A_{TL'}$ as a function of electron energy loss. Calculation 1 with Galster form factors yields the solid curve. The upper dashed curve (maximum prediction) is obtained with calculation 2 and Gari-Krümpelmann form factors, while the lower dashed curve (minimum prediction) results from calculation 3 with Höhler form factors. A more detailed discussion of the model dependence study is given in [14].

DISCUSSION

Two observations can be made: First, a precise determination of G_E^n in $^3\vec{\text{He}}(\vec{e}, e')$ appears to be very difficult at this Q^2. Even though the error on the extracted form factor is dominated by the statistical precision of the measurement, the residual theoretical uncertainty exceeds the variation in the G_E^n predictions. Therefore, in the framework of the present PWIA calculation, different form factor parameterizations cannot be distinguished at this Q^2, even if experimental uncertainties were absent. However, the theoretical uncertainty is substantially smaller than the predicted *absolute* value of G_E^n (~ 0.04–0.05). Therefore, one could determine whether G_E^n differs from zero, provided the experimental error is dramatically reduced. The achievable systematic precision is comparable to that of elastic electron-deuteron scattering [2].

Second, the extracted G_E^n value appears to be low compared to the predictions, although no meaningful conclusions can be drawn given the large uncertainties. However, it should be noted that the seemingly low value may be the result of final-state interactions (FSI) affecting $A_{TL'}$, i.e. a breakdown of PWIA, rather than an actual suppression of the form factor. The appearance of FSI at this Q^2 would not be surprising since it has been observed for the unpolarized longitudinal response, R_L, under similar conditions [21]. We discuss this issue further in Ref. [14].

In going to higher Q^2, the PWIA model-dependent uncertainties are expected to become smaller, in particular because the proton contribution to $A_{TL'}$ decreases. In addition, the ability to distinguish different G_E^n parameterizations improves because the predictions diverge. To illustrate this point, we have extended [15] the PWIA analysis beyond the present kinematics to $Q^2 = 0.7$ (GeV/c)^2, keeping the electron scattering angle fixed at $\theta = 70°$ and varying the beam energy. Fig. 3 shows the predicted $A_{TL'}$ *at the quasielastic peak* for calculation 1. Curve (a) is obtained for Galster form factors with $\eta = 5.6$, where the error band indicates the model dependence due to different NN potentials and off-shell prescriptions (as given by calculations 2 and 3). The dashed curves are the asymmetry predictions for Galster form factors with

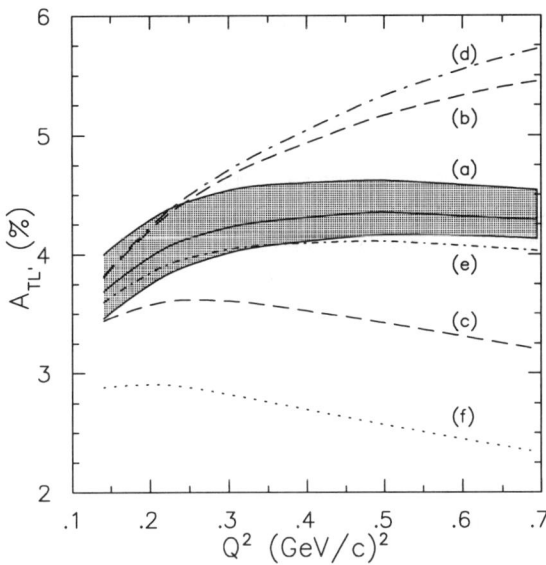

FIGURE 3. Q^2 dependence of the PWIA $A_{TL'}$ predictions (see text).

$\eta = 1.5$ (b) and $\eta = 20$ (c). These η values correspond to a variation of G_E^n by $\approx \pm 50\%$ at $Q^2 = 0.5$ $(\text{GeV}/c)^2$, which is roughly the present experimental error on G_E^n [2]. Curve (d) is obtained with Gari-Krümpelmann form factors, curve (e) with Höhler form factors, and curve (f) is the proton contribution only. For clarity, no error bands are shown around these curves. One observes that the sensitivity to form factors outgrows the model dependence for $Q^2 \gtrsim 0.4$ $(\text{GeV}/c)^2$. The effect of FSI is expected to be negligible in this region [20,21].

CONCLUSIONS

In summary, we have measured the $A_{TL'}$ asymmetry in $^3\vec{\text{He}}(\vec{e}, e')$ QE scattering at $Q^2 = 0.14$ $(\text{GeV}/c)^2$. The uncertainties in the theoretical PWIA model for this reaction have been studied in detail for the first time. The PWIA study shows a large spread of $A_{TL'}$ predictions at this Q^2, which represents a presently inescapable model uncertainty, in which the sensitivities to G_E^n, NN potential model, and off-shell prescription are comparable. The significant deviation of the experimental data from the PWIA prediction may be an indication of FSI effects, which further complicate the extraction of nucleon form factors at this Q^2. We conclude that in order to extract precise information on G_E^n from inclusive experiments using $^3\vec{\text{He}}$, it is at present necessary to work above a momentum transfer of $Q^2 \sim 0.4$ $(\text{GeV}/c)^2$.

ACKNOWLEDGEMENTS

We would like to thank the staff of the MIT-Bates Linear Accelerator Center for their efforts in making this experiment possible, M. Titko, T. W. Donnelly and R.-W. Schulze for providing the PWIA results and many fruitful discussions, and A. Stadler for calculating the ^{3}He wave function for the various potential models. This work was supported in part by the U.S. Department of Energy under cooperative agreement No. DE-FC02-94ER40818 (MIT) and W-31-109-ENG-38 (Argonne), the National Science Foundation Grant No. PHY91-15574 (Caltech) and PHY92-08119 (RPI), the Natural Sciences and Engineering Research Council of Canada (TRIUMF), and the Deutsche Forschungsgemeinschaft (DFG) under the contract Sa 247/11-2 (Hannover).

REFERENCES

[1] L. Koester, W. Nistler, and W. Waschkowski, Phys. Rev. Lett. **36**, 1021 (1976).
[2] S. Platchkov *et al.*, Nucl. Phys. **A510**, 740 (1990).
[3] T. W. Donnelly and A. S. Raskin, Ann. Phys. (N.Y.) **169**, 247 (1986).
[4] R. G. Arnold, C. E. Carlson, and F. Gross, Phys. Rev. C **23**, 363 (1981).
[5] B. Blankleider and R. M. Woloshyn, Phys. Rev. C **29**, 538 (1984).
[6] C. E. Woodward *et al.*, Phys. Rev. Lett. **65**, 698 (1990); C. E. Jones *et al.*, Phys. Rev. C **47**, 110 (1993).
[7] A. K. Thompson *et al.*, Phys. Rev. Lett. **68**, 2901 (1992).
[8] M. Meyerhoff *et al.*, Phys. Lett. **B327**, 201 (1994).
[9] C. Ciofi degli Atti, E. Pace, G. Salmè, Phys. Rev. C **46**, R1591 (1992) and *Proceedings of the VI Workshop on Perspectives in Nuclear Physics at Intermediate Energies*, ICTP, Trieste, May 1993 (World Scientific).
[10] R.-W. Schulze and P. U. Sauer, Phys. Rev. C **48**, 38 (1993).
[11] F. D. Colegrove, L. D. Schearer, G. K. Walters, Phys. Rev. **132**, 2561 (1963).
[12] R. G. Milner *et al.*, Nucl. Instr. Meth. **A274**, 56 (1989).
[13] H. Gao *et al.*, Phys. Rev. C **50**, R546 (1994); H. Gao, Ph.D. thesis, California Institute of Technology, 1994 (unpublished).
[14] J.-O. Hansen *et al.*, accepted for publication in Phys. Rev. Lett.; J.-O. Hansen, Ph.D. thesis, Massachusetts Institute of Technology, 1994 (unpublished).
[15] M. A. Titko, T. W. Donnelly, R.-W. Schulze, P. U. Sauer, A. Stadler, private communication (1994).
[16] J. A. Caballero, T. W. Donnelly, and G. I. Poulis, Nucl. Phys. **A555**, 709 (1993).
[17] G. Höhler *et al.*, Nucl. Phys. **B114**, 505 (1976).
[18] S. Galster, Nucl. Phys. **B32**, 221 (1971).
[19] M. Gari and W. Krümpelmann, Phys. Lett. **B173**, 10 (1986).
[20] J. M. Laget, Phys. Lett. **B273**, 367 (1991); Phys. Lett. **B276**, 398 (1992).
[21] E. van Meijgaard and J.A. Tjon, Phys. Rev. C **45**, 1463 (1992).

The SAMPLE Experiment: Parity-Violating Electron Scattering from the Proton and Deuteron

M. Pitt,[a] J. Arrington,[a] D. Beck,[b] E. Beise,[c] E. Candell,[e]
L. Cardman,[f] R. Carr,[a] G. Dodson,[d] K. Dow,[d] F. Duncan,[c]
M. Farkhondeh,[d] B. Filippone,[a] T. Forest,[b] H. Gao,[b]
W. Korsch,[a] S. Kowalski,[d] A. Lung,[a] R. McKeown,[a]
R. Mohring,[c] B. Mueller,[a] J. Napolitano,[e] N. Šimičević,[b]
B. Terburg,[b] and M. Witkowski,[e]

[a] *W.K. Kellogg Radiation Laboratory, Caltech, Pasadena, CA 91125*
[b] *Nuclear Physics Lab, University of Illinois, Champaign, IL 61820*
[c] *University of Maryland, College Park, MD 20742*
[d] *MIT-Bates Linear Accelerator Center, Middleton, MA 01949*
[e] *Rensselaer Polytechnic Institute, Troy, NY 12180*
[f] *CEBAF, Newport News, VA 23606*

Abstract. Recent experimental evidence on nucleon structure has provided indications that some strange quark matrix elements can be comparable to those involving up and down quarks. The SAMPLE experiment will determine the strange magnetic form factor G_M^s at $Q^2 = 0.1$ (GeV/c)^2 from a measurement of the asymmetry in the scattering of polarized electrons from the proton. The error on the extraction of G_M^s is ultimately limited by a theoretical uncertainty - the uncertain electroweak hadronic radiative correction to the axial form factor, $R_A^{T=1}$. To address this issue, the collaboration is also approved to measure the asymmetry in parity-violating quasielastic electron scattering from the deuteron. The combination of the proton and deuteron measurements will yield a value of G_M^s that is almost completely free of the uncertainty in $R_A^{T=1}$.

INTRODUCTION

The contribution of the strange quark sea to nucleon matrix elements has received considerable experimental and theoretical attention. Experimental evidence for a nonzero value of the nucleon axial vector strange matrix element $\langle N|\bar{s}\gamma_\mu\gamma_5 s|N\rangle$ has been obtained from deep inelastic lepton scattering experiments, while a nonzero value of the scalar strange matrix element $\langle N|\bar{s}s|N\rangle$ has been extracted from the sigma term in pion-nucleon scattering.[1] Little is currently known experimentally about the vector strange matrix element $\langle N|\bar{s}\gamma_\mu s|N\rangle$. It is typically written in terms of the strange electric and magnetic form factors, G_E^s and G_M^s. The SAMPLE experiment[2] will measure G_M^s at $Q^2 = 0.1$ (GeV/c)^2 using parity-violating scattering of polarized electrons from the proton at the MIT-Bates Linear Accelerator Center. Recently, the SAMPLE collaboration has received approval to also measure the parity-violating asymmetry in quasielastic scattering from the deuteron using the same apparatus[3]. This measurement is important because it provides a constraint on the electroweak radiative correction to the axial form factor, which is the dominant theoretical uncertainty in the extraction of G_M^s from the proton asymmetry.

STRANGE QUARK FORM FACTORS

The strange nucleon vector matrix element, $\langle N|\bar{s}\gamma_\mu s|N\rangle$, can be determined by making measurements with probes that couple to it; in this case we consider the photon and Z-boson. Information on the strange quark matrix elements is then extracted by making enough measurements of electroweak form factors to make a flavor decomposition.

The nucleon vector matrix elements are required to have the form

$$\langle N|\bar{s}\gamma_\mu s|N\rangle = \bar{u}_N \left(F_1^s \gamma_\mu + F_2^s \frac{i\sigma_{\mu\nu}q^\nu}{2M_N} \right) u_N, \qquad (1)$$

where F_1^s and F_2^s are the strange quark form factors. In the following discussion we will use the linear combinations of these known as the Sachs form factors: $G_E^s = F_1^s - \tau F_2^s$ and $G_M^s = F_1^s + F_2^s$, where $\tau = Q^2/(2M_N)^2$ with M_N being the nucleon mass. The vector quark form factors for the other quark flavors can be similarly defined. The electroweak form factors of the proton and neutron can be expressed in terms of the vector quark form factors using the electroweak quark couplings from the Standard Model. For our experiment the relevant form factor decompositions are:

$$\begin{aligned} G_M^{\gamma,p} &= \frac{2}{3}G_M^{u,p} - \frac{1}{3}G_M^{d,p} - \frac{1}{3}G_M^{s,p} \\ G_M^{\gamma,n} &= \frac{2}{3}G_M^{u,n} - \frac{1}{3}G_M^{d,n} - \frac{1}{3}G_M^{s,n} \\ G_M^{Z,p} &= \left(\frac{1}{4} - \frac{2}{3}\sin^2\theta_W\right)G_M^{u,p} - \left(\frac{1}{4} - \frac{1}{3}\sin^2\theta_W\right)G_M^{d,p} - \left(\frac{1}{4} - \frac{1}{3}\sin^2\theta_W\right)G_M^{s,p} \end{aligned} \qquad (2)$$

These are the conventional magnetic form factors, $G_M^{\gamma,p}$ and $G_M^{\gamma,n}$, of the proton and neutron, and the neutral weak magnetic form factor of the proton, $G_M^{Z,p}$, which SAMPLE will measure. To determine the strange quark form factor from these three measured quantities one must invoke the assumption of isospin symmetry to reduce the number of unknown quantities. In this case that means that the u quark behaves in the proton in the same way as the d quark behaves in the neutron and vice-versa. In terms of the quark form factors isospin symmetry is expressed as:

$$\begin{aligned} G_M^{u,p} &= G_M^{d,n} \\ G_M^{d,p} &= G_M^{u,n} \\ G_M^{s,p} &= G_M^{s,n}. \end{aligned} \qquad (3)$$

Using Equations (2) and (3) the neutral weak magnetic form factor of the proton can be expressed as:

$$G_M^{Z,p} = \left(\frac{1}{4} - \sin^2\theta_W\right) G_M^{\gamma,p} - \frac{1}{4}G_M^{\gamma,n} - \frac{1}{4}G_M^s \qquad (4)$$

Thus, a measurement of the neutral weak magnetic form factor, $G_M^{Z,p}$, combined with the previously measured magnetic form factors of the proton and neutron can be used to extract the strange magnetic form factor, G_M^s.

PARITY-VIOLATING ELASTIC ELECTRON SCATTERING FROM THE PROTON

The asymmetry measured in parity-violating elastic electron scattering from the proton can be used to extract the neutral weak form factor of the proton. The asymmetry is proportional to the interference between the γ and Z exchange Feynman diagrams. At tree level in the standard model the asymmetry can be written as:

$$A_p = \left[\frac{-G_F Q^2}{\sigma_p \pi \alpha \sqrt{2}}\right] \left[\varepsilon G_E^{\gamma,p} G_E^{Z,p} + \tau G_M^{\gamma,p} G_M^{Z,p} - \frac{1}{2}(1 - 4\sin^2\theta_W)\varepsilon' G_M^{\gamma,p} G_A^{Z,p}\right], \qquad (5)$$

where ε, τ, and ε' are kinematic factors. The SAMPLE experiment will detect scattered electrons at backward angles (130-170°). At the SAMPLE kinematics the last two terms in the asymmetry are dominant:

$$A \propto \frac{G_M^p G_M^{Z,p} - 0.20 G_M^p G_A^{Z,p}}{(G_M^p)^2}$$

$$\sim -\underbrace{\frac{1}{4}\frac{G_M^n}{G_M^p}}_{72\%} - \underbrace{\frac{1}{4}\frac{G_M^s}{G_M^p}}_{15\%} - \underbrace{0.20\frac{G_A^{Z,p}}{G_M^p}}_{13\%} \qquad (6)$$

To demonstrate the sensitivity to the strange magnetic form factor we have used a representative prediction of $G_M^s \sim -.30$ from Jaffe[4]. $G_A^{Z,p}$ is the neutral weak axial form factor. When one calculates beyond the tree level the axial term acquires a large and uncertain radiative correction, referred to as $R_A^{T=1}$. This theoretical uncertainty is the ultimate limit on how well one can extract G_M^s from the proton measurement. As discussed in a later section, a measurement of the deuteron asymmetry can largely eliminate this uncertainty. The expected asymmetry for the proton, assuming $G_M^s = 0$, is -7.3×10^{-6}.

THE SAMPLE EXPERIMENT

The SAMPLE experiment is currently installed on the S-beamline in the North Hall at the MIT-Bates Linear Accelerator Center. The target and detector are shown in Figure 1. The experiment is performed using a 200 MeV polarized electron beam incident on a 40 cm long circulating liquid hydrogen or deuterium target. The target is designed to handle 500 watts of deposited power (40 μA at 200 MeV). The scattered electrons are detected in a large solid angle (~ 1.4 sr) air Čerenkov detector at backward angles ($130° < \theta < 170°$). The detector consists of ten ellipsoidal mirrors which focus the Čerenkov light onto ten 8 inch photomultiplier tubes. The entire detector is enclosed in a light-tight and gas-tight skin, for the possible use of other Čerenkov gases. At the proposed beam current of 40 μA the instantaneous count rate in each phototube during a 15 μsec beam pulse will be 0.3 GHz, requiring the integration of the phototube signal over each beam burst.

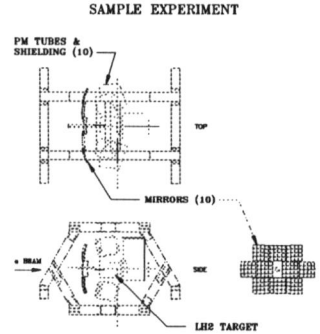

FIGURE 1. SAMPLE experimental setup.

The installation of the apparatus at Bates was completed in July 1993. Development runs since then have concentrated on target performance and detector shielding studies. As shown in Figure 2a, the target has been maintained as a subcooled liquid with up to 35 μA of beam incident. We have also studied target density fluctuations, and they appear to be at a small and manageable level. Studies of the detector have concentrated on reducing backgrounds. With some shielding additions, the observed signal is close to that expected for hydrogen elastic scattering, as Figure 2b illustrates. This will be further verified in the near future by using a NaI detector to measure the spectrum of electrons that trigger the Čerenkov counter. We have sucessfully utilized this technique with a prototype detector prior to the installation of the full experiment.

The polarized electron source at Bates uses a SLAC/CEBAF PEGGYII style diode gun[5], with photoemission from a bulk GaAs crytal to provide the $P_e \sim 40\%$ polarized electrons. The electron beam polarization is measured using a Møller polarimeter installed upstream of the experiment. Since the experiment is located on a beamline that is at an angle with respect to the accelerator the collaboration installed a Wien filter in the polarized source to prerotate the spins to compensate for the $g-2$ spin precession. This introduces the additional complication that the accelerator focussing solenoids cause the spins to precess about the beam axis. Through a series of measurements with the Møller polarimeter, we calibrated the Wien filter and accelerator solenoids so that purely longitudinally polarized beam can be delivered to the experiment. We can also adjust the Wien filter and solenoids to deliver purely transversely polarized beam at the target. This is useful because it allows us to study the response of our detector to the unwanted transverse polarization component.

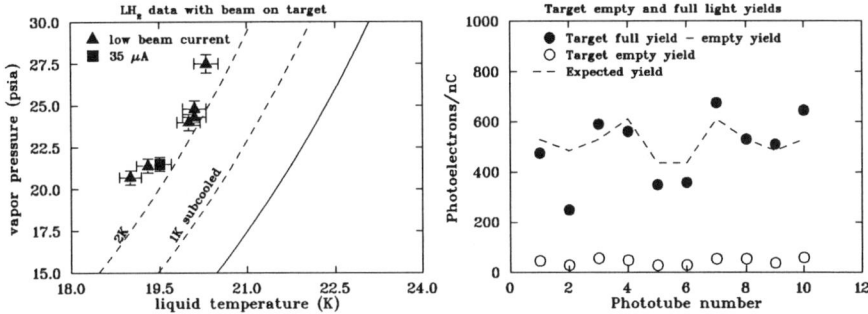

FIGURE 2. (a) Target performance. (b) Integrated phototube signals from detector.

In our experiment we will need to control the systematic uncertainty in the asymmetry at the 1×10^{-7} level. The methods necessary to achieve this have been developed previously in electron experiments on ^{9}Be at Mainz[6] and ^{12}C at MIT-Bates[7]. The main source of false asymmetries is any helicity-correlated change in beam properties, such as the energy, angle, position, and shape. The SAMPLE beamline is equipped with beam position monitors to measure the position and angle of the beam, a beam position monitor at a point of high dispersion to measure the energy, and a wire array to measure the beam shape. The response of our apparatus to changes in beam properties is studied using a computer automated system to deliberately vary steering coils and an energy vernier connected to an accelerator klystron. This allows the beam position, energy, and angle to be varied at desired times. All beam parameters are monitored during actual running to determine if they have a helicity-correlated difference. The measured asymmetry can be corrected for any beam parameters that have a non-zero helicity-correlated difference by using the measured response of the apparatus to that parameter.

The experiment on hydrogen is planned for 600 hours of production running with 40 μA of 40% polarized electrons. We expect to measure the proton asymmetry with a statistical precision of 7%. The principle experimental systematic error will be the 5% uncertainty in the determination of the beam polarization. There will also be a 6% error associated with the theoretical uncertainty in the axial hadronic radiative correction, $R_A^{T=1}$. Both of these systematic uncertainties can be dramatically reduced by making an asymmetry measurement with deuterium, as described in the next section.

PARITY-VIOLATING QUASIELASTIC ELECTRON SCATTERING FROM THE DEUTERON

The extraction of G_M^s from the proton measurement alone is ultimately limited by the uncertainty in the theoretical calculation of the electroweak radiative correction, $R_A^{T=1}$, to the isovector axial form factor, $G_A^{(3)}$, which is the dominant contribution to $G_A^{Z,p}$ in Equation 6. The theoretical difficulty in calculating these corrections is twofold: there are many possible diagrams and each one is difficult to calculate because of our lack of computational ability with low-energy QCD. The only attempt to calculate this radiative correction is that of Musolf and Holstein[8]. To illustrate the difficulty we schematically outline the approach of Musolf and Holstein. They restricted themselves to the two types of diagrams shown in Figure 3. They used low-energy effective theories of the hadronic interaction to estimate the diagrams. As shown in the figure this introduces the Desplanques, Donoghue, and Holstein[9](DDH) parity-nonconserving nucleon-meson couplings, which are very uncertain. Their result was $R_A^{T=1} = -.34$ with uncertainties of $\delta R_A^{T=1} \sim \pm 0.20$ due to Higgs and top mass variations and $\delta R_A^{T=1} \sim \pm 0.20$ due to the recommended DDH parameter range.

A measurement of the asymmetry in the quasielastic scattering of polarized electrons from deuterium can be used to eliminate reliance on this uncertain theoretical estimate of $R_A^{T=1}$. The asymmetry for the deuteron can be estimated by making the "static" approximation - assuming that the deuteron consists of a noninteracting neutron and proton at rest. In that case the quasielastic cross section can be estimated as the incoherent sum of neutron and proton cross sections, $\sigma_d = \sigma_p + \sigma_n$, and the asymmetry is

$$A_d = \frac{\sigma_p A_p + \sigma_n A_n}{\sigma_d}, \qquad (7)$$

where the expression for A_n can be written analogously to the expression for A_p in Equation 5. If we then suppress all form factors except the axial term and G_M^s we obtain

$$A_d = \left[\frac{.049}{\sigma_d}\right]\left[\frac{-G_F Q^2}{\pi\alpha\sqrt{2}}\right][1 + 0.22(1 + R_A^{T=1})G_A^{(3)} - 0.10 G_M^s]$$

$$A_p = \left[\frac{.026}{\sigma_p}\right]\left[\frac{-G_F Q^2}{\pi\alpha\sqrt{2}}\right][1 + 0.24(1 + R_A^{T=1})G_A^{(3)} - 0.61 G_M^s] \qquad (8)$$

for the deuteron and proton asymmetries. The ratio of these two asymmetries is very insensitive to the axial term:

$$\frac{A_p}{A_d} \propto 1 + 0.02(1 + R_A^{T=1})G_A^{(3)} - 0.51 G_M^s. \qquad (9)$$

So, G_M^s can be extracted from this ratio with very little systematic error contribution from the theoretical error in $R_A^{T=1}$. The beam polarization cancels in this ratio, so beam polarization systematic uncertainties will not contribute to the extraction of G_M^s.

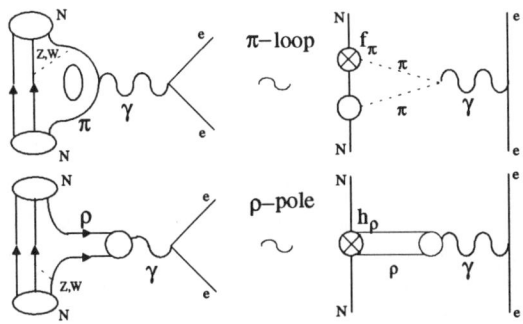

FIGURE 3. Feynman diagrams that contribute to $R_A^{T=1}$ and the approximations to them used for the Musolf-Holstein[8] calculation.

There are some new theoretical uncertainties introduced that are related to the deuteron, but they are small. Hadjimichael, Poulis, and Donnelly[10] have shown that corrections to the "static" approximation discussed above are at the 1-2% level. Since the SAMPLE detector has no energy resolution we will accept deuteron elastic scattering and threshold electrodisintegration events as well as the desired quasielastic events. It can be shown that inclusion of these processes and their asymmetries leads to only a 1% change from the pure quasielastic asymmetry.[3]

The deuterium measurement is planned for 600 hours under the same beam conditions as the proton measurement. The expected deuteron asymmetry with $G_M^s = 0$ is -9.6×10^{-6}. Combining the proton and deuteron asymmetries in the ratio discussed above we expect the following error sources for G_M^s: statistical ±0.19, beam polarization ±0.02, theoretical uncertainty in $R_A^{T=1}$ ±0.01, deuteron-related and other form factor uncertainties ±0.08. Thus, the combined measurement significantly reduces the errors arising from the theoretical uncertatinty in $R_A^{T=1}$ and the beam polarization, and it also more nearly approaches the ideal of a statistics-limited measurement. The expected $1-\sigma$ error limits are illustrated in Figure 4.

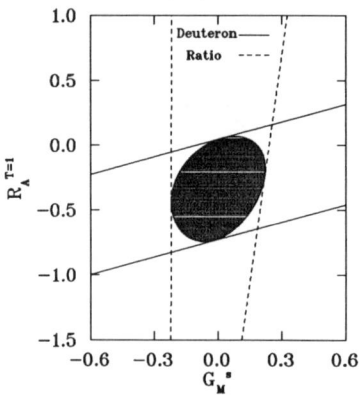

FIGURE 4. Expected $1-\sigma$ limits from the measurements of A_p/A_d and A_d.

CONCLUSION

The SAMPLE collaboration will measure the parity-violating asymmetry in the scattering of polarized electrons from both hydrogen and deuterium targets. The combined measurements will determine the strange magnetic form factor of the proton to a precision $\delta G_M^s \sim \pm 0.21$. In addition, the measurements will provide the first experimental constraints on the theoretically uncertain electroweak radiative correction to the axial form factor, $\delta R_A^{T=1} \sim \pm 0.40$.

REFERENCES

1. For a recent summary, see M.J. Musolf, T.W. Donnelly, J. Dubach, S.J. Pollock, S. Kowalski, E.J. Beise, Phys. Rep. **239**, 1 (1994).
2. MIT-Bates experiment #89-06, R.D. McKeown and D.H. Beck, contact people.
3. MIT-Bates experiment #94-11, E.J. Beise and M.L. Pitt, contact people.
4. R.L. Jaffe, Phys. Lett. **B229**, 275 (1989).
5. C.K. Sinclair, Journale de Physique (1983) C2-669.
6. W. Heil et al., Nuc. Phys. **B327**, 1 (1989).
7. P.A. Souder et al., Phys. Rev. Lett. **65**, 694 (1990).
8. M.J. Musolf and B.R. Holstein, Phys. Lett. **B242**, 461 (1990).
9. B. Desplanques, J.F. Donoghue, and B.R. Holstein, Ann. Phys. **124**, 449 (1980).
10. E. Hadjimichael, G.I. Poulis, and T.W. Donnelly, Phys. Rev. **C45**, 2666 (1992).

Progress at HERMES: Preparations for measurements of nucleon spin structure functions

S.F. Pate, for the HERMES collaboration

Laboratory for Nuclear Science,
Massachusetts Institute of Technology, Cambridge, Massachusetts 02139

Abstract: A new experiment, designed to measure spin-dependent lepton-nucleon deep inelastic scattering, is being built on the HERA electron storage ring at DESY, Hamburg. The HERMES experiment will use a windowless internal gas cell of polarized H, D, or ^{3}He nuclei as a target for the stored 27 GeV electron (or positron) beam. The pure target, coupled with a high luminosity, will enable greater statistical precision and much smaller systematic errors that has been possible thus far in previous measurements at CERN and SLAC. The construction of the experiment is underway. At the same time, a test experiment is being conducted on the HERA ring, using an unpolarized gas cell and prototypes of the detectors. This report briefly summarizes the results of the test run and the status of the construction of the full experiment.

One of the most exciting developments in hadronic physics in the last decade has been the revelation of interesting nucleon spin structure physics. The initial experiments in polarized deep inelastic muon scattering at CERN [1], followed by recent measurements at SLAC [2] and continuing efforts at CERN [3] have raised questions about the separate contributions of quarks, gluons, and the sea to the nucleon spin.

The HERMES experiment at DESY, in Hamburg, will approach the measurement of the nucleon spin structure functions in a new way, using a pure, windowless gas target in an electron storage ring (HERA). The purity and low density of the target eliminates many sources of systematic error inherent in the targets employed in the CERN and SLAC experiments, such as the dilution of the asymmetry and the energy straggling of the scattered particles. The measurement of proton properties will employ an atomic beam source (ABS) to produce a polarized hydrogen target, whereas the neutron will be studied using both a polarized deuteron target (also from the ABS), for which the subtraction of the proton data will be needed, and a separate polarized ^{3}He target, for which no subtraction of proton asymmetries is needed.

Measurements at HERMES will cover a range in Bjorken scaling variable x of $0.02 < x < 0.65$. The spin-dependent structure function $g_1(x)$ (associated with scattering of longitudinally polarized electrons and nucleons), will be measured with precision for both the proton and the neutron, allowing the determination of the Bjorken sum rule to 5-8%, a uncertainty two times smaller than is proposed in any other experiment. The $g_2(x)$ structure function, associated with transverse nucleon polarization, will be measured with good accuracy for the first time. Additional structure functions associated with tensor spin variables will also be measured. The detector scheme proposed allows for the observation of coincident hadrons, permitting the extraction of flavor dependent contributions to the structure functions. (A fuller description of the physics program at HERMES will be found in the contribution of H. Jackson.)

Naturally, there are many questions concerning backgrounds and the effect of target gas on the stored beam, to name a few, that need to be answered in order to optimize such an experiment. Of particular concern is the effect of the target gas on the beam. If the introduction of a useful amount of target gas has only a small effect on the beam, then the HERMES experiment can run simultaneously with the two collider experiments at HERA (H1 and ZEUS). On the other hand, if the effect is large, then the experiment would have to be conducted during a shorter, dedicated running period. To study these issues, a sample target cell and prototypes of the detectors have been installed in the East Hall at HERA for the 1994 operation period. Also, there have been modifications to the beam line and optics. These changes and installations include the following:

- The electron beam lattice has been modified for the final HERMES optics. This includes straightening the electron beam out over ±90 m around the HERMES interaction point and the installation of a strong bend (2.6 mrad) at ±89 m and a weak bend (0.5 mrad) at ±38 m in order to minimize the amount of synchrotron radiation hitting the target region. Furthermore, the electron and proton beam line have been separated by 72 cm horizontally and now have the same vertical height. A pair of spin rotators has been installed in front of and behind the HERMES interaction point, to enable longitudinal electron beam polarization in that region. (For a full description of electron beam polarization at HERA, see the contributions of M. Düren and M. Böge.)

- A set of collimators, upstream of the target, used to block synchotron radiation from the upstream magnets. One set of the collimators is movable — these are opened for beam injection and acceleration, and closed down after the beam is stored at 27 GeV. The other fixed collimator, downstream of the movable set and immediately upstream of the target cell, acts as an anti-scatter baffle for the upstream collimators. In this way, not even singly scattered synchrotron radiation can reach the target cell walls.

- An aluminum target cell, of elliptical cross section (9.8 mm by 28 mm), 400 mm long. Various gases (H_2, ^4He, and N_2) can be let into this cell at a controlled rate, typically in the range 10^{14}–10^{15} atoms/sec, producing target thicknesses in the range 10^{12}–10^{14} nucleons/cm^2, depending on the gas. (This prototype cell is very similar to the final ^{3}He target cell, described in the contribution of L. Kramer.) A differential pumping system, consisting of five large turbopumps installed up- and down-stream of the target, removes the target gases before they reach the rest of the HERA vacuum system.

- A silicon surface barrier detector, operated in anti-coincidence with two backing scintillator counters, to observe synchrotron light scattered from the target gas atoms. Both targets, in the full experiment, will be cryogenically cooled, and it is important to know the flux of synchrotron light on the target cell, as a high flux would heat the cell.

- A prototype of the Target Optical Monitor (TOM), placed upstream of the target cell. This device will observe light emitted from the target atoms, after excitation by the passing beam particles. When the target is polarized, these photons will also be polarized, and the TOM will serve as a target polarization monitor.

- A prototype of the forwardmost chamber, or vertex chamber, consisting of three layers of microstrip gas counter planes. This is placed immediately downstream of the scattering chamber, and is critical to proper tracking.

- A prototype of the rear scintillator hodoscope, necessary to fast triggering.

- Prototypes of the "magnet" and "backward" wire chambers. An additional small prototype of the "backward" chambers is used in the "forward" chamber position to improve tracking.

- Prototype of the transition radiation detector (one layer) only suitable for observing performance characteristics.

- A small stack of the lead glass calorimeter detectors, with a lead preshower plate and the hodoscope prototype placed immediately in front.

- Various scintillators placed so as to study backgrounds from upstream of the target, from the proton beam, and in order to allow various types of triggering that are relevant to the full experiment.

Note that no magnet is present in this test experiment.

Beam Studies

After the installation in March 1994, the first studies done were of the interaction of the collimators and gas with the circulating beam. Since the collimator apertures are very much larger than the beam size, and are only intended to block the synchrotron light, no problems were expected with them, and in fact the collimators

The HERMES Detector

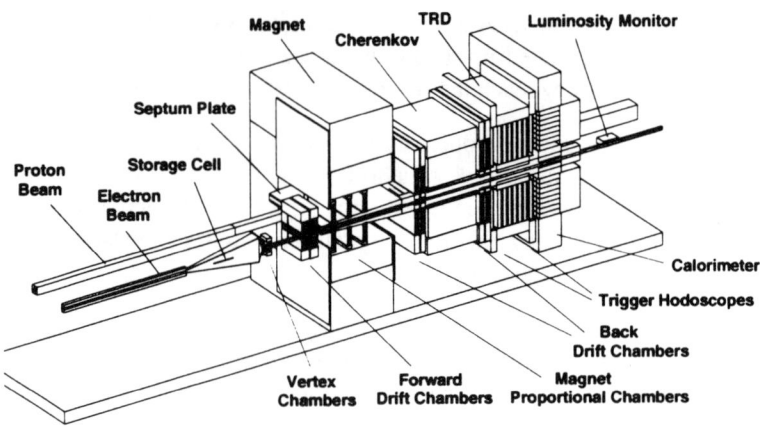

FIGURE 1. Schematic of the full HERMES detector apparatus, to be installed in the East Hall of HERA in the spring of 1995.

were successfully closed on the very first attempt and no effect was seen on the stored beam.

Initially, in the 1994 running period, electrons were used as the stored beam particle. Poor beam lifetimes, usually not better than 3-4 hours, were observed. Partly this was due to an imperfect machine vacuum, since many new components had been installed in the East section. But the observed machine vacuum was not enough to account for the poor lifetimes. Another mechanism that can contribute significantly is that of "ion-trapping" whereby positive ions in the machine vacuum are electrically attracted to the central electron orbit, increasing the number of scattering centers seen by the beam and thus reducing the lifetime. In the past, this problem has been temporarily treated by injecting positrons into the ring, travelling in the opposite direction as the electrons. The positrons repel the positive atomic ions from the central orbit. Subsequent fills of electrons would then have a longer lifetime. This year, however, it was decided to simply run with positrons as the beam for an indefinite period. This had not been done before, and much hardware and software work by the machine group was needed so that the positrons could orbit in the same direction as the electrons had. In the end, this changeover was a success, with positron lifetimes 3-4 times higher than for electrons.

Studies were to have been made of the effect of the introduction of gas into the ring, at the HERMES target location, for both electrons and positrons. One may calculate the effect of gas in the ring. At HERA energies (27 GeV) the primary mechanism for beam loss is atomic bremsstrahlung, and the method to calculate the effect of gas on the beam lifetime in a storage ring is known[4]. Unfortunately, leaks developed in the gas handling system, and these leaks have only recently been repaired. Studies will be done on the effect of gas on the beam before HERA operations come to an end in November.

Target Optical Monitor (TOM)

The prototype of the TOM consists of a combination mirror and elliptical aperture upstream of the target cell – the aperture to allow the passage of the beam, and the mirror positioned so as to reflect light coming upstream from the target. The reflected light passes through a narrow-pass filter and into a phototube. The phototube is outfitted with a shutter which is closed during injection to block out the large flux of synchrotron light that is incident when the movable collimators are open. Figure 2 summarizes the results of the work with this prototype TOM so far, in the form of a timing spectrum, where the start is the beam-crossing time. These spectra are taken without any gas in the ring. The sharp, main peak comes from prompt light emitted by the beam as it passes the east section — despite the collimator system, some light still gets to the TOM phototube. This light has a narrow time structure, and so is easily gated away in hardware. The broader peak, about 26 nsec after the prompt peak (and modulo the 96 nsec time range of the TDC), has been determined to come from light that reflects back from the first bend in the beam line downstream, about 38 meters away. The flat background present in the whole spectrum is not clearly understood at this time – it may be due to thermal heating of the movable collimators. Attempts will be made to reduce these vagrant light levels, for example by using combinations of filters and lenses.

Results of preliminary analysis of data with gas in the cell

In early running with beam, the leakage currents in the wire chambers were too high to run them effectively. What data was collected pointed to an undefined source of low energy neutrals far upstream of the target, so a 4 mm thick lead shield was put in place severals meters upstream. This reduced the leakage currents by a factor of 4, enough to allow operation at the desired voltage. It is worth noting that this sort of background will not affect all detectors in the final experiment, since the tremendous bulk of the spectrometer magnet will shield any detector behind it from neutrals, while the field will shield against low energy charged particles.

Another beam-related background that has been observed is that from the nearby proton beam. However, these events come at a very different time than electron beam events, as measured with the "HERA clock" (beam rf signal), and so are easily removed in replay. The rate of showers from proton background during injection is

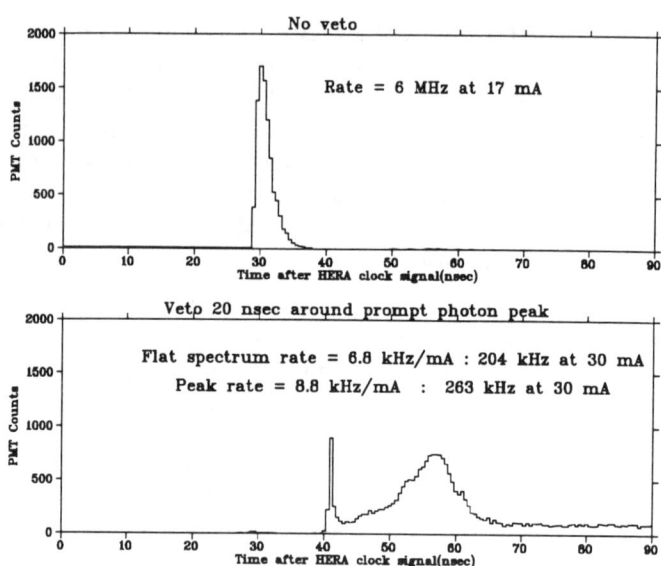

FIGURE 2. Time spectra from TOM studies, both with and without the a fast veto of the prompt peak associated with the beam crossing.

high enough to perhaps damage the calorimeter, but this aspect of the problem will be dealt with via shielding.

Some showering has been seen from the movable collimators, when they are fully closed. For the purpose of blocking synchrotron light, the collimators do not have to be fully closed, and it was found that this showering could be reduced by opening them somewhat. Some optimal position will have to be found which balances the two requirements.

Data have been taken with and without gas in the ring. With gas in, the multiplicity of particles is often rather high, and with only three chambers available (the prototype vertex chamber, the small "back" chamber, and a larger "back" chamber), tracking is only possible for low multiplicity events (a handicap which would not occur in the full experiment). Figure 3 shows the result of traceback to the calculated vertex, *i.e.*, the nearest point of approach of a track to the beam axis, of events during a run with a 27 GeV positron beam and N_2 gas let into the ring. There is a clear peak at $z = 0$ with a base of approximately 40 cm width, as expected from the target geometry. The rate of deep inelastic electron scattering events is in rough agreement with expectation. The same spectrum, taken without gas in the ring, has almost no events.

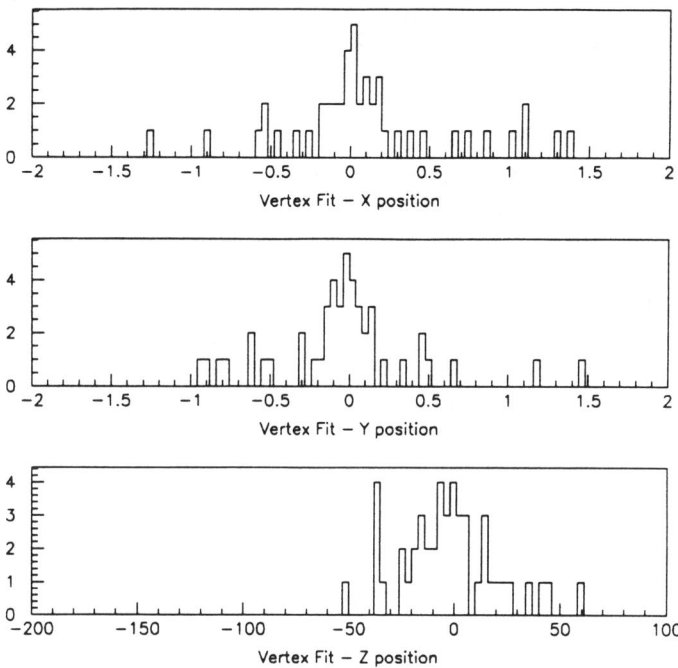

FIGURE 3. Vertex spectra, for all three Cartesian coordinates, for a positron beam run at 27 GeV with N_2 as the target gas. Distance units are centimeters.

Installation of complete apparatus for 1995 running

In parallel with the test run, work on the final detector apparatus goes on. A large platform, on rails for ease of installation into the ring, is used to support the detector and target components. At present the HERMES spectrometer magnet has been installed on this platform and the field is being mapped. The magnet was designed, using the MAFIA code, to have a field integral of 1.2 T-m, and so far the results are in very good agreement with what has been measured in the mapping. Later, the field in the target region will be mapped. For the ^{3}He target, this stray field is critical, as a field with large gradients will prevent proper polarization. A set of correction coils is being designed.

Tests and calibrations of detectors are underway at DESY and at CERN, using beams of electrons of known energy to study the responses of various components. The position resolution of some chambers and the energy calibration of the calorimeter elements are measured in this way. A phototube monitoring scheme, using a central light source whose output is distributed to all phototubes via optical fibers, is tested while these calibrations go on. This monitor will provide relative gain information during the experiment.

Detector installation will begin after the field mapping is complete, and the target installation soon after that. There are two types of targets; the atomic beam source (see the contribution of F. Stock) which can produced polarized hydrogen and deuterium, and the ^{3}He target (see the contribution of L. Kramer). A decision will be made late in September as to which target to install first. The full installation will be in place sometime in March of 1995.

1. J. Ashman et al., *Phys. Lett.* **B206**, 364 (1988) and *Nucl. Phys.* **B328**, 1 (1989).
2. D.L. Anthony et al., *Phys. Rev. Lett.* **71**, 959 (1993).
3. B. Adeva et al., *Phys. Lett.* **B302**, 533 (1993).
4. J.E. Spencer, "Storage Rings, Internal Targets, and PEP," SLAC Report 316, May 1987, p. 37.

The Deuteron Spin Structure Functions in the Bethe-Salpeter Approach and the Extraction of the Neutron Structure Function $g_1^n(x)$.

A.Yu. Umnikov, L.P. Kaptari[†], K.Yu. Kazakov[‡]
and F.C. Khanna

Theoretical Physics Institute, Physics Department, University of Alberta, Edmonton, Alberta T6G 2J1,

and TRIUMF, 4004 Wesbrook Mall, Vancouver, B.C. V6T 2A3, Canada

[†]*Bogoliubov's Laboratory of Physics, JINR, Dubna, 141980 Russia*

[‡]*Far Eastern State University, Vladivostok, 690000 Russia*

Abstract

The nuclear effects in the spin-dependent structure functions g_1^D and b_2^D are calculated in the relativistic approach based on the Bethe-Salpeter equation with a realistic meson-exchange potential. The results of calculations are compared with the non-relativistic calculations. The problem of extraction of the neutron spin structure function, g_1^n, from the deuteron data is discussed.

1. PRELIMINARIES

In this talk we present a relativistic approach to the deep inelastic lepton scattering on the deuteron based on the Bethe-Salpeter (BS) equation within a realistic meson-exchange model. The method is developed in refs. [1, 2] and now we apply it to the deep inelastic scattering with polarized particles. We calculate the leading twist spin-dependent structure functions (SF) of the deuteron, g_1^D and b_2^D [3], and we discuss the extraction of the neutron SF g_1^n from deuteron data. This investigation is partially motivated by a number of existing and forthcoming experiments on the deep inelastic scattering of leptons by deuterons (SLAC, CERN, DESY, CEBAF). A covariant theory of this process will be useful in the analysis of the experimental data.

2. THE BETHE-SALPETER AMPLITUDE FOR THE DEUTERON

An accurate description of both the NN-interaction at energies up to ~ 1 GeV, and the basic properties of the deuteron, can be provided within the meson-nucleon theory [4, 5]. The covariant description is based on the BS equation or its various approximations. We use the ladder approximation for the kernel of the BS equation [4]:

$$\Phi(p, P_D) = i\hat{S}(p_1) \cdot \hat{S}(p_2) \cdot \sum_B \int \frac{d^4 p'}{(2\pi)^4} \cdot \frac{g_B^2 \Gamma_B^{(1)} \otimes \Gamma_B^{(2)}}{(p-p')^2 - \mu_B^2} \cdot \Phi(p', P_D), \qquad (1)$$

where μ_B is the mass of meson B; Γ_B is the meson-nucleon vertex, corresponding to the meson B, $\hat{S}(p) = (\hat{p} - m)^{-1}$, m is the nucleon mass. This equation is solved, using the technique described in ref. [1].

The meson parameters, such as masses, coupling constants, cut-off parameters are taken similar to those in ref. [4], with a minor adjustment of the coupling constant of the scalar σ-meson so as to provide a numerical solution of the BS equation. All parameters are presented in Table 1, where coupling constants are shown in accordance with our definition of the meson-nucleon form-factors, $F_B(k) = (\Lambda^2 - \mu_B^2)/(\Lambda^2 - k^2)$.

The deuteron amplitude is normalized by using the conserved vector current:

$$J_\mu(0) = \langle P_D \mid \bar{\psi}\gamma_\mu\psi \mid P_D \rangle = 2P_{D\mu}. \qquad (2)$$

For the analysis of the processes with the polarized deuterons an important characteristic of the deuteron wave functions is the probability of the D-wave,

Table 1. Parameters of the model

meson B	coupling constants $g_B^2/(4\pi); [g_v/g_t]$	mass μ_B, GeV	cut-off Λ,GeV	isospin
σ	12.2	0.571	1.29	0
δ	1.6	0.961	1.29	1
π	14.5	0.139	1.29	1
η	4.5	0.549	1.29	0
ω	27.0; [0]	0.783	1.29	0
ρ	1.0; [6]	0.764	1.29	1
$m = 0.939$ GeV, $\epsilon_D = -2.225$ MeV				

$\mathcal{P}_D$, which varies in the range 3 − 6% for realistic potentials. Since the components of the BS amplitude do not have a direct probability interpretation, we consider the matrix element of the axial current on the deuteron state with the total momentum projection $M = 1$:

$$J^5_\mu(0) = \langle P_D | \bar{\psi}\gamma_5\gamma_\mu\psi | P_D\rangle_{M=1} \tag{3}$$

$$J^5_3(0) \equiv \int_0^\infty n^{BS}_{spin}(|\mathbf{p}|)|\mathbf{p}|^2 d|\mathbf{p}|, \tag{4}$$

which corresponds in the non-relativistic limit to the mean value of the spin projection:

$$\langle P_D | \sigma_3 | P_D\rangle_{M=1} = \int_0^\infty n^{n.r.}_{spin}(|\mathbf{p}|)|\mathbf{p}|^2 d|\mathbf{p}| = 1 - \frac{3}{2}\mathcal{P}_D,$$

$$n^{n.r.}_{spin}(|\mathbf{p}|) = u^2(|\mathbf{p}|) - \frac{1}{2}w^2(|\mathbf{p}|). \tag{5}$$

Our calculation of the matrix element (4) gives $J^5_3(0) = 0.922$, from which we get an approximate estimate of $\mathcal{P}^{BS}_D \approx 5.2\%$ in reasonable agreement with the estimate $\mathcal{P}^{BS}_D = 4.8\%$ from ref. [4] (compare, also, with $\mathcal{P}_D = 4.3\%$ and $\mathcal{P}_D = 5.9\%$ for Bonn and Paris potentials, respectively). Note, that accuracy of such a computations of the D−wave admixture from the (4) is of the order of the $\sim \langle p^2\rangle/m^2$, i.e. $\sim 1\%$.

The matrix elements (2) and (3) contain the explicit integration over the relative 4-momentum. To compare our results with the non-relativistic case, we define the charge density $n^{BS}_{ch.}(|\mathbf{p}|)$ in the deuteron as a primitive function in (2) and (4) prior the integration on the modulus of the 3-momentum:

$$\langle P_D | \bar{\psi}\gamma_\mu\psi | P_D\rangle \equiv \int_0^\infty n^{BS}_{ch.}(|\mathbf{p}|)|\mathbf{p}|^2 d|\mathbf{p}|, \tag{6}$$

which corresponds to the non-relativistic charge density:

$$n^{n.r.}_{ch.}(|\mathbf{p}|) = u^2(|\mathbf{p}|) + w^2(|\mathbf{p}|), \tag{7}$$

where $u(|\mathbf{p}|)$ and $w(|\mathbf{p}|)$ are $S-$ and $D-$waves components of the deuteron wave function. Similarly the spin density $n^{BS}_{spin}(|\mathbf{p}|)$ is defined from eq. (4).

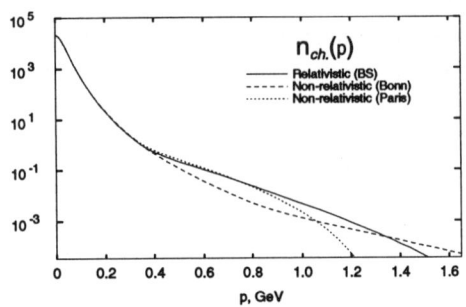

Figure 1. The charge density calculated in different models (see text).

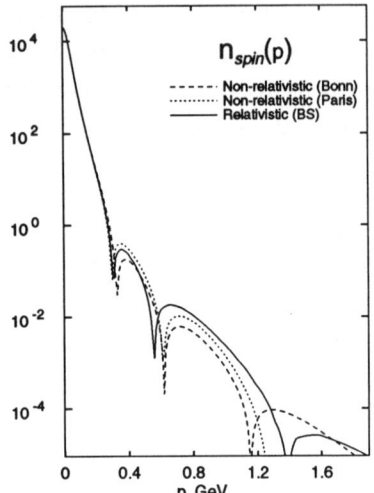

Figure 2. The spin density calculated in different models (see text).

The charge and spin densities are compared with those obtained with wave functions of the Paris [7] and Bonn [6] potentials in Fig. 1 and 2, respectively. All calculations are in a reasonable agreement in the non-relativistic region, up to $|\mathbf{p}| \sim m$.

3. THE DEEP INELASTIC SCATTERING ON THE POLARIZED DEUTERON

To calculate the structure functions (SF) of the deuteron we use the OPE method within the effective meson-nucleon theory [10, 11, 1]. This method allows us to calculate SF in terms of the Wick rotated BS amplitude [1, 2]. Neglecting possible "off-mass-shell" corrections to the deuteron SF [8, 9], we obtain the deuteron SF in the convolution form. For two leading twist spin-dependent SF of the deuteron, g_1^D and b_2^D, [3] we have:

$$g_1^D(x) = \int_0^1 f_5^{N/D}(\xi) g_1^N(x/\xi) \frac{d\xi}{\xi}, \qquad (8)$$

$$b_2^D(x) = \int_0^1 \Delta f^{N/D}(\xi) F_2^N(x/\xi) d\xi, \qquad (9)$$

where g_1^N and F_2^N are isoscalar nucleon SF. The moments of the effective distribution functions, $f_5^{N/D}$ and $\Delta f^{N/D}$ are the matrix elements of the leading twist operators on the deuteron states:

$$\mu_n(f) = \int_0^1 x^{n-1} f(x) dx, \qquad (10)$$

$$\mu_n(f_5^{N/D}) = \frac{1}{2M_D^n} \int \frac{d^4p}{(2\pi)^4} p_{1+}^{n-1} \left[\bar{\Phi}_M(p) \gamma_+^{(1)} \gamma_5^{(1)} (p_2 \gamma^{(2)} - m) \Phi_M(p) \right]\Big|_{M=1}, \qquad (11)$$

$$\mu_n(\Delta f^{N/D}) = \frac{1}{2M_D^n} \int \frac{d^4p}{(2\pi)^4} p_{1+}^{n-1} \qquad (12)$$
$$\left\{ \left[\bar{\Phi}_M(p) \gamma_+^{(1)} (p_2 \gamma^{(2)} - m) \Phi_M(p) \right]\Big|_{M=0} - [\ \cdots\ ]\Big|_{M=1} \right\},$$

where the kinematical variables in the rest frame are defined by

$$p = (p_0, \mathbf{p}), \quad P_D = (M_D, \mathbf{0}), \quad p_1 = \frac{P_D}{2} + p, \quad p_2 = \frac{P_D}{2} - p, \qquad (13)$$

where M_D is the deuteron mass, $p_+ = p_0 + p_3$, $\gamma_+ = \gamma_0 + \gamma_3$ and we use deep inelastic kinematics: $pq \approx q_0(p_0 + p_3)$.

The explicit form of the effective distribution functions, $f_5^{N/D}$ and $\Delta f^{N/D}$ is defined by the inverse Mellin transform of (11) and (12). These distributions satisfy normalization conditions:

$$\int_0^1 f_5^{N/D}(\xi) d\xi = J_3^5(0), \qquad \int_0^1 \Delta f^{N/D}(\xi) d\xi = 0. \qquad (14)$$

Using the distribution functions, $f_5^{N/D}$ and $\Delta f^{N/D}$, and realistic parametrizations of the nucleon SF, g_1^N [12] and F_2^N [11], we calculate the deuteron SF g_1^D and b_2^D. The results of calculation are presented on Figs. 3 and 4.

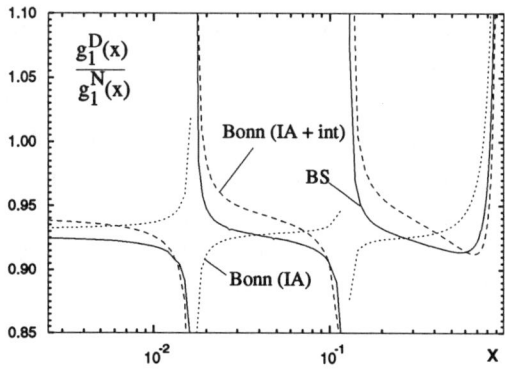

Figure 3. The ratio of the deuteron and nucleon SF, g_1^D/g_1^N

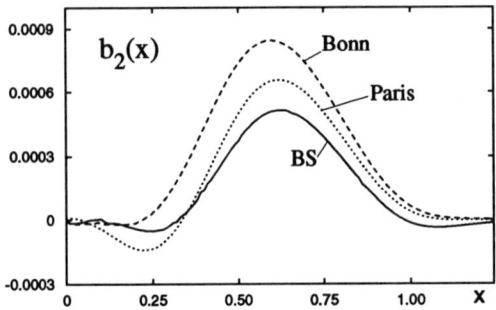

Figure 4. The deuteron SF b_2 calculated in different models (see text)

In Fig. 3 nuclear effects in g_1^D calculated within BS approach (solid curve) are compared with non-relativistic calculation [13, 14] with Bonn wave function. The non-relativistic calculation including the Fermi motion of polarized nucleons and binding effects (dashed curve) is in reasonable agreement with the relativistic calculations (solid curve). At the same time both these curves differ from non-relativistic impulse approximation (dotted), which does not include binding effects. The SF b_2^D calculated in relativistic and non-relativistic approaches is presented in Fig. 4. Similar to the case of the g_1^D, there is no special relativistic effect in b_2^D. The models are slightly distinguishable in view of different admixture of D-wave and minor variations of the nucleon momentum distributions.

4. ON THE EXTRACTION OF THE NEUTRON SF.

Mathematically the problem of extraction of the neutron SF from the deuteron data is formulated as a problem to solve the inhomogeneous integral equation (8) for the neutron SF with a model kernel $f_5^{N/D}$ and experimentally measured left hand side[1], g_1^D.

Recently we proposed a method to extract the neutron SF from the deuteron data within any model, giving deuteron SF in the form of a "convolution integral plus/minus additive corrections" [15]. The principal advantages of the method, compared with the smearing factor method, are the following. (i) Only analyticity of the SF need be assumed, (ii) the method allows us to elaborate on the spin-dependent SF, where the traditional smearing factor method does not work.

---

[1] Depending of the model, some additive corrections could be taken into account [15].

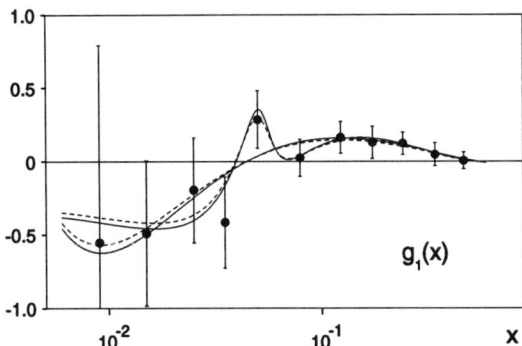

Figure 5. The SF g_1 of deuteron (dashed lines) and nucleon (solid lines) for two different parametrizations of data

As an example, we apply our method to the SMC data [16]. The resulting nucleon functions are presented in fig. 5 (solid curves). It is clear that due to a singular behavior of the ratio g_1^D/g_1^N the nucleon SF g_1^N cannot be obtained by the smearing factor method. As a more practical result we can report that even in the experimentally restricted region of x the following integral relation is valid with high accuracy:

$$\int_{0.006}^{0.6} g_2^D(x)dx \cong \left(1 - \frac{3}{2}\mathcal{P}_D\right) \cdot \int_{0.006}^{0.6} g_2^N(x)dx, \qquad (15)$$

5. CONCLUSIONS

We have presented a description of the spin-dependent deep inelastic lepton-deuteron scattering based on the Bethe-Salpeter formalism within an effective meson-nucleon theory. In particular,

1. The spinor-spinor Bethe-Salpeter equation for the deuteron is solved in the ladder approximation for a realistic meson exchange potential.

2. The leading twist spin-dependent structure functions, g_1^D and b_2^D, of the deuteron are calculated in terms of the Bethe-Salpeter amplitude.

3. Numerical results of the calculations of the deuteron structure functions g_1^D and b_2^D in the Bethe-Salpeter formalism are presented. It is found that results are in qualitative agreement with previous non-relativistic calculation.

4. A method to extract the neutron structure function, g_1^n, from the deuteron and proton data is suggested.

The reasonable quantitative agreement of the presented calculations of the deuteron SF at $x < 1$ in the non-relativistic and relativistic approaches confirms the expectation that these approaches have to give similar results within the boundaries of validity of the non-relativistic approximation. However, it does not imply that the relativistic effects in the deuteron SF are negligible in general. It only shows that in a slightly relativistic system such as the deuteron we should find *special* kinematic conditions of the experiment to display the relativistic effects. We could expect non-trivial relativistic phenomena at $x > 1$, where the precise evaluation of the SF is important for QCD analysis of the experimental data. The behavior of the nuclear SF $\sim (1 - x_N)^\gamma$ as $x_N \to 1$ may lead to errors.

References

[1] A.Yu. Umnikov and F.C. Khanna, Phys. Rev. **C48** (1994) 2311.

[2] A.Yu. Umnikov, L.P. Kaptari, K.Yu. Kazakov and F.C. Khanna, Phys. Lett. **B334** (1994) 163.

[3] P. Hoodbhoy, R.L. Jaffe and A. Manohar, Nucl. Phys. **B312** (1989) 571; R.L. Jaffe and A. Manohar, Nucl. Phys. **B321** (1989) 343.

[4] M.J. Zuilhof and J.A. Tjon, Phys. Rev. **C22** (1980) 2369.

[5] F. Gross, J.W. Van Orden and K. Holinde, Phys. Rev. **C45** (1992) 2094.

[6] R. Machleid, K. Holinde and Ch. Elster, Phys. Rep. **149** (1987) 1.

[7] M. Lacombe et al, Phys. Rev. **C21** (1980) 861.

[8] W. Melnitchouk, A.W. Schreiber and A.W. Thomas, Phys. Rev. **D49** (1994) 1183.

[9] S.A. Kulagin, G. Piller and W. Weise, Preprint TPR-94-02, Regensburg; Preprint ADP-94-1/T144, Adelaide, 1994.

[10] B.L. Birbrair, E.M. Levin and A.G. Shuvaev, Nucl. Phys. **A496** (1989) 704.

[11] L.P. Kaptari, K.Yu. Kazakov and A.Yu. Umnikov, Phys. Lett. **B293** (1992) 219; L.P. Kaptari, A.Yu. Umnikov and B. Kämpfer, Phys. Rev. **D47** (1993) 3804.

[12] A. Schäfer, Phys. Lett. **B208** (1988) 175.

[13] L.P. Kaptari, K.Yu. Kazakov, A.Yu. Umnikov and B. Kämpfer, Phys. Lett. **B321** (1994) 271; **B322** (1994) E473.

[14] L.P. Kaptari, A.Yu. Umnikov, C. Ciofi degli Atti, S. Scopetta and K.Yu. Kazakov, Preprint DFUPG 92/94, Perugia, 1994.

[15] A.Yu. Umnikov, F.C. Khanna and L.P. Kaptari, Z. Phys. **A348** (1994) 211.

[16] SM Collab., B. Adeva et al., Phys. Lett **B302** (1993) 534.

Transverse Polarization in Deep Inelastic Scattering

The HELP Collaboration[a]
presented by B. Vuaridel[b]

[a] *Brno, CERN, Ferrara, Geneva, Genova, Iowa, JINR, Lyon, Milano, Moscow, Legnaro, Praha, Trieste, Yerevan, Zilina*
[b] *DPNC Université de Genève, CH-1211 Genève 4, Switzerland*

Abstract. The semi-inclusive deep inelastic scattering, ep → e'h+X, allows to study new transverse spin distributions and spin dependent fragmentation functions which have never been measured. A polarized target in conjunction with an unpolarized lepton beam is required for this type of experiment. A possible experimental apparatus is described.

Introduction

The longitudinal spin of the quarks in the nucleons has been probed in experiments at SLAC and at CERN. So far no experiment has been devoted to the study of transverse spin distributions of the quarks in a transversally polarized nucleon. These distributions are also fundamental structure functions and are independent of the longitudinal ones. They cannot be measured in purely inclusive deep inelastic scattering (DIS) because transverse spin asymmetries are suppressed in this case. In contrast, in polarized Drell-Yan or in polarized semi-inclusive DIS, ep → e'h+X, transverse asymmetries arise at the leading twist level (1,2).

With a target polarized transversally, the unpolarized electron beam will probe the quark transverse polarization in DIS if the "struck" quark polarization is measured by some means. This "quark polarimeter" may be provided by an azimuthal dependence of the fragmentation functions into one or two leading hadrons (3). Another possibility consists in the measurement of the polarization of a leading baryon. For example, the polarization of the protons could be analyzed by rescattering or, with more ease, the polarization of "self-analyzing" baryons, like the Λ, could be measured from the angular dependence of their decay products (2). The spin dependence of the fragmentation functions may represent an important piece of hadronic physics which remains to be discovered, and their measurement is required to further probe the spin structure. In DIS the kinematics of the fragmentation is well defined by the electron initial and final momenta, and it is an ideal process to study these phenomena to enlighten the dynamics of confinement and make new tests of factorization.

Spin Distributions

The longitudinal spin distribution $\Delta_L q(x)$ or $g_1(x)$, where x is the fraction of nucleon momentum carried by the quark, has a very simple interpretation in terms of the parton model. g_1 is the difference between a quark distribution with the helicity parallel to that of the proton and a quark distribution with the helicity antiparallel to the proton one, summed over the flavors of the quarks and antiquarks, and weighted by the square of the quark electrical charge:

$$g_1(x) = \frac{1}{2}\sum_f e_f^2 \Delta_L q_f(x) = \frac{1}{2}\sum_f e_f^2 [q_{f+}(x) - q_{f-}(x)] \qquad (1)$$

where $q_{f+(-)}(x)$ are the densities of quarks with flavor "f", charge e_f and helicity $+(-)$ in the proton of helicity $+$. These distributions can be measured with a target and a beam longitudinally polarized.

Similarly one can define distributions for the transverse spin of the quark in a proton polarized in the $\hat{x}$ direction.

The transverse spin distribution is then:

$$\Delta_T q_f(x) = q_f \uparrow (x) - q_f \downarrow (x) \qquad (2)$$

where $q_f \uparrow (x)$ and $q_f \downarrow (x)$ are distributions of quarks with transverse spins parallel or antiparallel to the transverse spin of the nucleon. Δ_Tq(x) is sometimes called $h_1(x)$ by analogy to $g_1(x)$. This transverse spin distribution can be studied in the Drell-Yan processes with both protons polarized transversely. In that case, to be sensitive to $\Delta_T q(x)$, it is necessary that either the sea quarks are polarized or polarized antiprotons are used. However, the sea quarks are probably not highly polarized and, at present, it is only possible to obtain polarized antiproton beams from the decay products of $\bar{\Lambda}$. Therefore, semi-inclusive DIS is necessary and complementary to Drell-Yan experiments.

Naïvely one would expect that $\Delta_T q(x) = \Delta_L q(x)$. However, because of relativistic internal motion and orbital momentum, $\Delta_T q(x)$ is expected to be different from $\Delta_L q(x)$. In a naïve covariant parton model where the baryon is composed of a quark and a scalar diquark, it can be shown that $\Delta_T q(x) = q_+(x)$ (2). The transverse spin distributions, which arise at the leading twist level, are therefore not redundant informations on the spin stucture of the proton and may help in understanding this highly debated topic.

Quark Polarimetry

It is well known that the quark flavor and the electric charge of the leading hadron are correlated to that of the fragmenting quark. Therefore, one can expect that the same behavior holds for the spin. Then the hadron polarization can be used to probe the quark polarization.

In the case of longitudinal polarization, it has been suggested that three particle correlations within a jet could be used (4). Recently this idea was rediscovered and called jet "handedness" and it was shown how it can be measured in e⁺e⁻ annihilation (5). Independently, Artru proposed to measure the transverse polarization of the quark by measuring the polarization of self-analyzing baryons from fragmentation (2). Recently Collins suggested that the azimuthal dependence of the leading hadrons in semi-inclusive DIS is sensitive to the transverse polarization of the quark (3).

Baryonic Polarimeter

The measurement of the hadrons polarization in semi-inclusive DIS gives information on the transversity distribution of the nucleons (2). The final baryon polarization, $\vec{P}_T^B$, can be related to the transversity distribution by:

$$\vec{P}_T^B = R \cdot \vec{P}_T^N \cdot \frac{\sum_f e_f^2 \cdot \Delta_T q_f(x) \cdot \Delta_T f_{q_f \to B}(z)}{\sum_f e_f^2 \cdot q_f(x) \cdot f_{q_f \to B}(z)} \cdot D_{nn} \qquad (3)$$

where: $\Delta_T f_{q_f \to B}(z) = f_{q_{f\uparrow} \to B\uparrow}(z) - f_{q_{f\uparrow} \to B\downarrow}(z)$, q_f = u, d, s...; and B is a self-analyzing baryon = $\Lambda, \Sigma...$ and R = rotation matrix about the normal to the scattering plane in the e-q scattering plane which brings the momentum of the target nucleon N, $\vec{p}_N$, along the one of the baryon, $\vec{p}_B$. $\vec{P}_T^N$ is the transverse polarization of N. D_{nn} is the depolarization parameter of electron quark scattering and z is the energy of the hadron divided by the virtual photon energy.

In this type of measurements, e.g. in ep→e'Λ+X, the polarization of the Λ is given by the product of two functions, the transversity distribution, $\Delta_T q_f(x)$ and its equivalent for the fragmentation, $\Delta_T f_{q_f \to B}(z)$. In order to extract the transversity distribution it is necessary to know $\Delta_T f_{q_f \to B}(z)$ which could be "calibrated" in an independent measurement.

The longitudinal polarization of the Λ has been measured at LEP at the Z^0 peak. A preliminary OPAL note indicates P_Λ = -29.3 ± 9.9 % for $z \geq 0.3$ (6), in good agreement with the expectation of -30% (7). At the Z^0 peak the quarks are longitudinally polarized and the strange quark polarization is about -92%. This means that the longitudinal spin transfer from the quark to the Λ may be larger than 1/3 since other mechanisms can reduce the magnitude of the spin transfer. Naïvely one could expect a similar value for the transverse spin.

At the Z^0 peak the quarks are not transversely polarized and effects of the transverse polarization cannot be observed, except in spin correlation between two jets, e.g. between Λ and Λ̄ in opposite jets. However, the yields for (Λ Λ̄) pair production are very low.

Mesonic Polarimeter

Another possibility to investigate transversity distribution consists in the measurement of the spin dependent asymmetry of the hadrons with respect to the jet axis (3,8). In this case the spin asymmetry is related to the transversity distribution by:

$$A(x, \phi_s, p_T, z) = P_t^N \cdot \frac{\sum_f e_f^2 \cdot \Delta_T q_f(x) \cdot \Delta_T f_{q_f \to h}(p_T, z) \cdot sin(-\phi_s)}{\sum_f e_f^2 \cdot q_f(x) \cdot f_{q_f \to h}(p_T, z)} \cdot D_{nn} \quad (4)$$

where $\Delta_T f_{q_f \to h}(p_T, z)$ describes the spin dependent fragmentation, P_t^N is the transverse target polarization and ϕ_s is the azimuthal angle of the final leading hadron with respect to the target polarization direction as shown in fig. 1. In the factorized parton model, the above functions can be separated and determined up to a normalization factor from the measured asymmetries.

Similarly to the Λ case, this asymmetry is related to the sum of the products of unknown functions. This single-hadron asymmetry can be investigated in reactions where the jet axis can be determined, e.g. in DIS with a transversely polarized target and an unpolarized beam since the transverse polarizations of the quarks and of the electrons are not correlated.

The experiment will be sensitive to the product of two terms; namely a structure function and a fragmentation function. Clearly it will be necessary to perform experiments in polarized hadronic production using different reactions to get a deeper insight into the structure functions. Looking at various hadrons, π^+, π^-, π^0 and kaons produced from polarized proton and deuteron targets is required to distinguish between various quark species and their relative contribution to the nucleon spin.

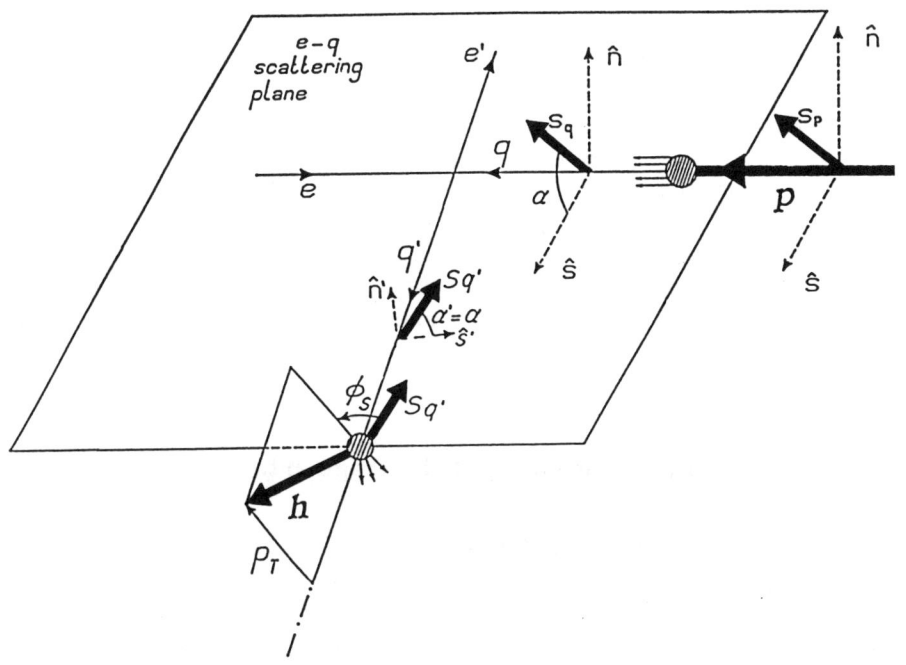

FIGURE 1. Quark polarization in semi-inclusive DIS within the framework of the parton model, in the hard process center of mass. The angle $\alpha(\alpha')$ between the spin vector $S_q(S_{q'})$ and the scattering plane is conserved since $D_{nn} = D_{ss}$, i.e. $\alpha = \alpha'$.

The HELP Proposal

The HELP (for Hadronic Electroproduction at LEP) collaboration has proposed (8) to install a polarized jet target in the LEP tunnel to perform a DIS experiment with a transversally polarized target with the unique LEP200 unpolarized beams in a parasitic manner.

Experimental Setup

A new setup has been designed for this semi-inclusive DIS experiment which consists of a two step spectrometer. The first spectrometer covers the small scattering angles up to 100 mrad for the high momentum scattered electrons and hadrons. It consists of a vertical dipole magnet (the former UA6 magnet), scintillating fiber trackers, T1...T8, an electromagnetic calorimeter and a Ring Imaging Cherenkov counter (RICH) with a gaseous radiator. The second spectrometer covers the larger angles up to 300 mrad with a solenoid, trackers, electromagnetic calorimeter and a RICH with liquid radiator. The whole setup is shown in fig. 2 with the target.

The spectrometer allows a very accurate measurement of the scattered electron momentum with low hadronic contamination for an accurate determination of the subprocess kinematics and the identification of the leading hadrons, π, K, Λ, p...

Analyzing Power of the Fragmentation

At present, we have no direct experimental information on the magnitude of the analyzing power of the fragmentation. However, if we assume that the single spin asymmetries observed in $p\uparrow p \to \pi + X$ reactions (9) are due to similar mechanisms, we can expect that the analyzing power is at least of the order of a few 10 %. Moreover, on the theoretical side, the string fragmentation model provides a mechanism for the Collins type asymmetry (10).

The LEP committee was concerned that the experiment relied on a sizeable analyzing power in the fragmentation of the struck quark jet and did not recommend the proposal. However this analyzing power should appear also in the form of a spin correlation in $e^+e^- \to 2$ jets at the Z pole. This should provide a more obvious signature of this spin dependent effect. At present, two-jet events are analyzed at LEP to determine the magnitude of asymmetries of spin dependent fragmentation functions. This analysis could be used to

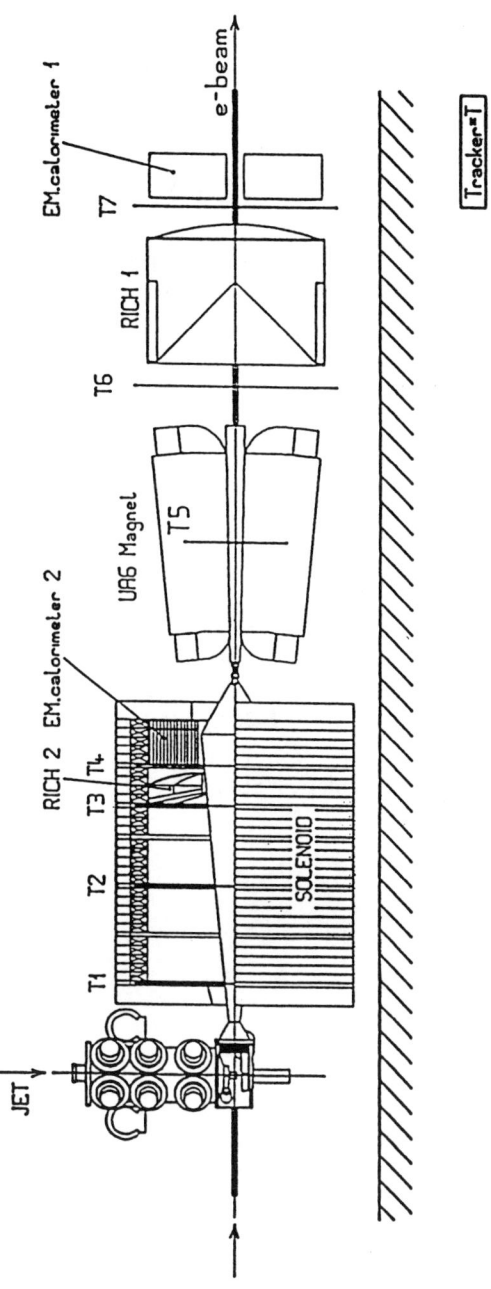

FIGURE 2. Side view of the HELP experimental setup.

determine the unknown scale factor and the size of the analyzing power of the fragmentation.

REFERENCES

1. J.P. Ralston and D.E. Soper, *Nucl.Phys.* **B152**, 109 (1979.
2. X. Artru in *Proceeding of the Polarized Collider Workshop*, University Park, Edts. J. Collins, S.F Heppelman, R.W. Robinett, AIP No.223, p. 176 (1990) and X. Artru and M. Mekhfi, *Z. Phys* **C45**, 669 (1990), *Nucl. Phys.* A532, **351c** (1991).
3. J. Collins, *Nucl. Phys.* **B396**, 161 (1993) and J. Collins et al., *Nucl. Phys.* **B420**, 583 (1994)
4. O. Nachtmann, *Nucl. Phys.* **B127**, 314 (1977).
5. A.V. Efremov et al., *Phys. Lett.* **B284**, 394 (1992).
6. OPAL Phys. Note *PN*112, Preliminary.
7. G. Gustafson and J. Häkkinen, *Phys. Lett.* **B303**, 350 (1993).
8. The HELP collaboration "A proposal for an Internal Jet-Target Experiment at LEP, CERN/LEPC 93-14, LEPC/P7, CERN/LEPC 94-1, LEPC/P7 Add. 1, and the ref. therein.
9. D.L. Adams et al., *Phys. Lett.* **B264**, 264(1991) and *Z. Phys.* **C56**, 181 (1992).
10. X. Artru, J. Czyzewski and H. Yabuki, LYCEN/9423, TPJU 12/94.

HADRON FORM FACTORS: SUMMARY

E. J. Beise, University of Maryland

INTRODUCTION

For many years electron scattering has been used as a probe of the structure of hadrons. Polarization observables add the possibility of determining small quantities which are difficult to isolate in other ways. The two parallel sessions summarized here focussed on the use of polarization observables to study electromagnetic properties of the neutron, neutral weak form factors, and transverse polarization in deep-inelastic scattering. Each topic is related to the contributions of sea quarks to the structure of hadrons.

ELECTROMAGNETIC PROPERTIES OF THE NEUTRON

The cross section for unpolarized elastic electron scattering from a nucleon target can be written in terms of the Sachs form factors, $G_E(Q^2)$ and $G_M(Q^2)$ which describe the internal charge and magnetization current distributions. Ultimately, $G_E(Q^2)$ and $G_M(Q^2)$ must be linked to the nucleon's internal constituents, valence quarks, sea quarks and gluons. Gorski, Grummer and Goeke [1] have calculated nucleon charge distributions using the Nambu-Jona-Lasinio model, and they indicate that although the proton charge distribution is mainly due to valence quarks, the neutron charge distribution is to a large extent determined by sea quarks, particularly at large radii ($r > 1$ fm). The neutron electric form factor, $G_E^n(Q^2)$, should therefore be sensitive to sea-quark contributions at low energy.

Due to the lack of free neutron targets, the study of neutron electromagnetic properties requires the use of *nuclear* targets. Electron scattering from an unpolarized deuterium target has been used, and the proton contribution is subtracted off with a theoretical model of the deuteron wave function. This method is problematic when one is trying to separate a very small G_E^n from a very large G_E^p. The most precise data of this type are in ref. [2], and the resulting uncertainty in G_E^n from variations in deuteron models is shown as the hatched area in figure 1(a).

The use of spin observables allows one to isolate the neutron contribution through interference terms. In the method used by Eden, *et al.*, [3], polarized electrons are scattered from an unpolarized deuterium target. The transverse polarization transfer coefficient of the scattered neutron, $D_{LS'}$, is measured. In the impulse approximation, this coefficient can be written in terms of the deuteron form factors

$A(Q^2)$ and $B(Q^2)$ as [4]

$$D_{LS'} = -\frac{(G_E^n/G_M^n)B(\theta_e, Q^2)}{A(\theta_e, Q^2) + (G_E^n/G_M^n)^2}, \qquad (1)$$

and is relatively insensitive to details of the deuteron wave function. The results of [3] are shown as the solid triangle in figure 1(a). This measurement was limited by statistics, due to the relatively low detection efficiency of the neutron polarimeter. In the parallel sessions, an upgraded polarimeter was discussed [5], which will be used in future measurements at Bates and CEBAF [6] to measure G_E^n in the range $0.5 < Q^2 < 1.5 \, (\text{GeV/c})^2$ with a precision of about 10%. $D_{LS'}$ has also recently been measured at Mainz ($Q^2 = 0.31 \, (\text{GeV/c})^2$) [7]; results are however not yet available.

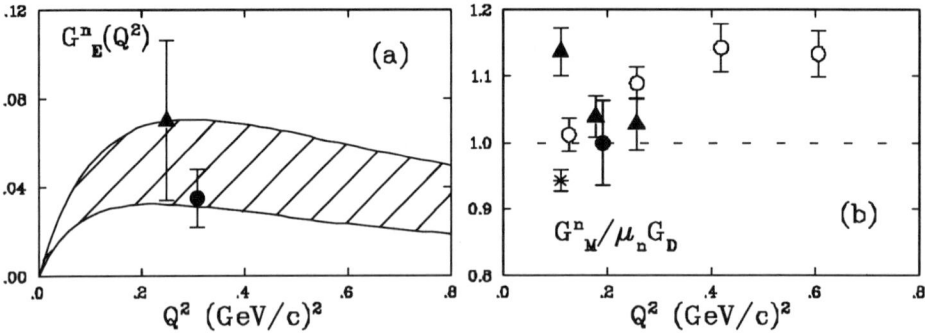

Figure 1: Recent measurements of (a): G_E^n (hatched area: variation in models as constrained by [2], circle: [7], triangle: [3]) and (b): G_M^n (solid triangles: [18], solid circle: [13], open circle: [19], star: [20]).

Another approach is to use quasielastic scattering of polarized electrons from *polarized* nuclear targets. The cross section for the scattering of polarized electrons from a polarized *nucleon*, can be written as $\sigma = \Sigma(Q^2) + h_e\Delta(Q^2, \theta^*, \phi^*)$ [8], where Σ is the unpolarized cross section and Δ is the spin-dependent part of the cross section, determined by the target polarization direction, (θ^*, ϕ^*), with respect to the direction of momentum transfer. By reversing the electron beam helicity h_e (or reversing the direction of target spin), the spin-dependent term will change sign and the asymmetry is

$$A = \frac{\sigma_R - \sigma_L}{\sigma_R + \sigma_L} = \frac{2\tau v_{T'} \cos\theta^* G_M^2(Q^2) + 2\sqrt{2\tau(1+\tau)}v_{TL'} \sin\theta^* \cos\phi^* G_M(Q^2)G_E(Q^2)}{(1+\tau)v_L G_E^2 + 2\tau v_T G_M^2} \qquad (2)$$

where v_K are kinematic factors and $\tau = Q^2/4M_N^2$. When the target is polarized parallel to the momentum transfer, $A \propto G_M^2$, and with the target polarized transversely, $A \propto G_M G_E$ ($\phi = 0°$ or $180°$).

Limitations arise from how well one can associate a polarized neutron with the polarization of the nuclear target. One method is to detect the outgoing neutron, but such experiments have not in the past been practical. Developments in polarized targets, polarized electron sources and the availability of 100% duty cycle beam make this technique more feasible. A proposed measurement at CEBAF was discussed [9], where a low current beam of polarized electrons will be scattered from a (radiation doped ND_3) polarized deuterium target. This target was recently used in the SLAC experiment E143, where it had a vector polarization of 30-35% with ∼50 nA incident beam [10]. SLAC also produced an 80% polarized electron beam at low average current using a strained GaAs crystal [11]. At CEBAF, these developments are expected to be implemented to measure G_E^n, to about 10%, in the same Q^2 range as the above polarimeter experiments.

In quasielastic electron scattering from polarized ^{3}He, to first order the scattering can be considered to be from a polarized neutron target because the two protons are paired off in a singlet state. The *quasielastic* asymmetry in scattering from a spin-1/2 nuclear target is (ω is the energy loss of the scattered electron)

$$A_{QE} = \frac{v_{T'} \cos\theta^* R_{T'}(Q^2,\omega) + v_{TL'} \sin\theta^* \cos\phi^* R_{TL'}(Q^2,\omega)}{(1+\tau)v_L R_L + 2\tau v_T R_T}. \quad (3)$$

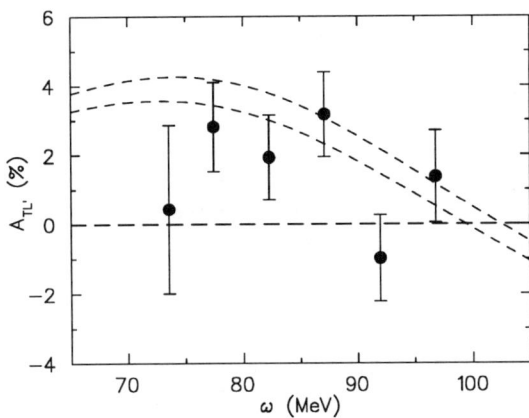

Figure 2: $A_{TL'}$ vs. ω in inclusive quasielastic electron scattering from polarized ^{3}He. (See ref. [14] for more details.)

In the approximation that $^3\vec{\text{He}}$ is a polarized neutron, $R_{T'} \to (G_M^n)^2$ and $R_{TL'} \to G_M^n G_E^n$. A few measurements of $^3\vec{\text{He}}(\vec{e},e')$ and $^3\vec{\text{He}}(\vec{e},e'n)$ have recently been published [12, 13, 14, 7], and this has stimulated theoretical effort to study the degree to which this simple approximation is true. Two measurements were presented at this conference. In the parallel session, Hansen [14] discussed a measurement of $R_{TL'}$ in $^3\vec{\text{He}}(\vec{e},e')$ at $Q^2 = 0.2$. In figure 2 is shown the measured asymmetry $A_{TL'}$ vs. ω, compared to a band representing the uncertainty in calculations using spin-dependent spectral functions. At the kinematics of the experiment, the uncertainty in the ^{3}He wave function introduces a larger variation in $A_{TL'}$ than the variation in models of G_E^n, making an extraction of G_E^n unreliable. The disagreement between the data and the calculations has been attributed to final state interactions, which Laget has shown can be sizable in $(\vec{e}, e'n)$ below $Q^2 \sim 0.3$ [15]. At higher Q^2 the variation in current models of G_E^n dominates the theoretical uncertainties, so $^3\vec{\text{He}}(\vec{e}, e')$ is more promising. In a plenary session, Reichert [7] presented a measurement of the ratio $A_{TL'}/A_{T'}$ in $^3\vec{\text{He}}(\vec{e}, e'n)$ at $Q^2 = 0.31$ (GeV/c)^2. The ratio is proportional to G_E^n/G_M^n, and the resulting G_E^n is shown as the solid circle in figure 1(a). Future experiments are planned at Mainz, Bates and CEBAF [7, 16]. By the next SPIN conference one can anticipate a substantially increased body of data on G_E^n, with higher accuracy than has previously been achievable.

Determination of G_E^n requires accurate knowledge of the magnetic form factor G_M^n, and in figure 1(b) is a compilation of recent data. Measurements of G_M^n using nuclear targets are less model dependent because $G_M^n \sim G_M^p$. Markowitz et al. [17] used *unpolarized* $d(e, e'n)$ to extract G_M^n (solid triangles). The most recent experiments (open circles, star), as presented by Mitchell [9], have determined the ratio $\sigma(e, e'n)/\sigma(e, e'p)$, which further cancels model and absolute luminosity uncertainties, but still requires accurate knowledge of the neutron detection efficiency. G_M^n was also measured using $^3\vec{\text{He}}(\vec{e}, e')$ by Gao et al. [13] (solid circle). Systematic uncertainties due to the nuclear corrections were small, and the error on the measurement is dominated by statistics.

As measurements with polarized nuclear targets become more precise, the nuclear structure of few body systems has to be understood at a detailed level if one is to use these targets to extract neutron electromagnetic properties. This is evident in comparisons between the deuteron and the 3-body system. Although several calculations can reproduce the unpolarized form factors for $A = 2$ and $A = 3$, no single calculation can simultaneously reproduce the form factors and the measured polarization observables in both systems.

FLAVOR DECOMPOSITION OF NUCLEON STRUCTURE

Another area of recent activity is in the study of the flavor structure of nucleons. In the standard model for electroweak interactions the electromagnetic and neutral weak vector interactions can be expressed as

$$V_\mu^\gamma = \sum \bar{q} \gamma_\mu q \qquad (4)$$

$$V_\mu^Z = \sum \bar{q} \left(\tfrac{1}{2} T_i - Q_i \sin^2 \theta_W \right) \gamma_\mu q . \qquad (5)$$

Using isospin symmetry and taking the interaction between proton states leads to a set of neutral weak form factors which depend on electromagnetic form factors and

$\sin^2 \theta_W$, with a remaining contribution due to $\bar{s}s$ pairs:

$$G^Z_{E,M} = \left(\frac{1}{4} - \sin^2 \theta_W\right) G^p_{E,M} - \frac{1}{4} G^n_{E,M} - \frac{1}{4} G^s_{E,M}. \tag{6}$$

The matrix element due to the axial vector current can similarly be expressed as $G^Z_A = \frac{-1}{2} g_A^{T=1} + \frac{1}{4} G^s_A$. Note that accurate knowledge of *neutron* electromagnetic form factors is required to determine $G^s_{E,M}$ in the proton.

The neutral weak form factors of the proton can be determined either by elastic ν-p scattering or from parity-violating electron scattering. Neutrino scattering, discussed in a plenary talk by Stancu [20], is predominantly sensitive to $G^s_A \propto \bar{s}\gamma_\mu\gamma_5 s$. PV electron scattering is more sensitive to the vector form factors $G^s_{E,M} \propto \bar{s}\gamma_\mu s$, and two parallel session speakers described ongoing and future measurements [21, 22].

In scattering a polarized electron beam from an unpolarized proton target, the parity-violating asymmetry is

$$A_{PV} = \frac{\sigma_R - \sigma_L}{\sigma_R + \sigma_L} = \frac{G_F Q^2}{\pi\alpha\sqrt{2}} \frac{\left[\varepsilon G^p_E G^Z_E + \tau G^p_M G^Z_M - \frac{1}{2}\left(1 - 4\sin^2\theta_W\right)\varepsilon' G^p_M G^Z_A\right]}{\varepsilon \left(G^p_E\right)^2 + \tau \left(G^p_M\right)^2} \tag{7}$$

where $\varepsilon = \left(1 + 2(1+\tau)\tan^2\frac{\theta}{2}\right)^{-1}$ and $\varepsilon' = \sqrt{(1-\varepsilon^2)\tau(1+\tau)}$. The first two terms dominate at forward angles and the latter two terms contribute at backward angles, although the term containing G^Z_A is suppressed by $(1 - 4\sin^2\theta_W)$.

The SAMPLE experiment at Bates [21, 23] seeks to determine G^s_M at $Q^2 = 0.1$ (GeV/c)^2. Longitudinally polarized electrons of 200 MeV are scattered in the backward direction, from a 40 cm hydrogen target, into an air Cerenkov detector with ten mirrors and phototubes placed symmetrically about the beam line.

At the SAMPLE kinematics the contribution from the axial vector term $G^p_M G^Z_A$ is about 20%, and it should be multiplied by an isovector weak radiative correction $(1-R^1_A)$ which is rather uncertain [24]. This results in a theoretical uncertainty in the extraction of G^s_M from a hydrogen measurement alone. In quasielastic scattering from deuterium at the same kinematics, the axial term contributes approximately the same fraction to the asymmetry, but the s-quark terms of the proton and neutron add incoherently and nearly cancel. The ratio of the asymmetry in hydrogen to that in deuterium is almost independent of R^1_A, and the resulting error in G^s_M is dominated by statistics. Further SAMPLE running with a deuterium target has recently been approved [25], and the expected 1σ limits on G^s_M and R^1_A are shown in figure 3.

Additional information about the flavor structure of nucleons can be found in the *strange electric* form factor. Since the proton has no net strangeness $G^s_E(0) = 0$, but the Q^2 dependence is unconstrained. Three experiments at CEBAF will investigate the Q^2 dependence of G^s_M and G^s_E using both proton and ^{4}He targets, as discussed by Beck [22]. One experiment will make a single measurement with a proton target

100   Hadron Form Factors

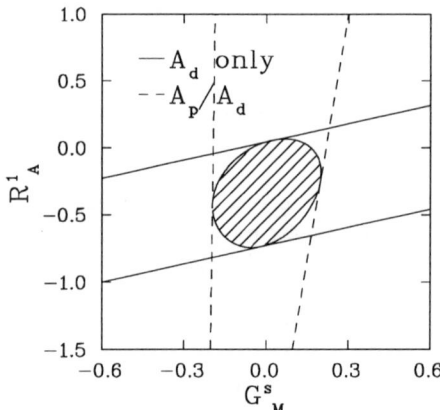

Figure 3: Expected 1σ limits on R_A^1 and G_M^s from a combined measurement on hydrogen and deuterium targets with SAMPLE.

in Hall A at $Q^2 \sim 0.5$ (GeV/c)^2 [26]. This would not allow separation of G_M^s and G_E^s, but it would determine if strange quarks effects are present to a significant amount (modulo present uncertainties in G_E^n). If large effects are seen, a more extended program of measurements on proton and ^{4}He targets is anticipated.

For experiment 91-017 [27], a dedicated superconducting toroidal spectrometer is planned in Hall C, with which one can make a series of $\vec{e}$-p measurements at forward and backward angles and do a Rosenbluth-type separation of G_M^s and G_E^s. Protons are detected at angles corresponding to forward scattered electrons in segmented scintillation counters, and the range $0.2 < Q^2 < 0.6$ (GeV/c)^2 is measured simultaneously. The apparatus will then be rotated by 180° so that backward-going electrons are detected. As in SAMPLE, an additional measurement of $d(\vec{e}, e')$ will constrain the term proportional to G_A^Z. The SU(3) singlet combination, $G^0 = G^p + G^n + G^s$, would be determined with good accuracy.

With a $(0^+, 0)$ target such as ^{4}He, elastic scattering occurs only through charge monopole terms, and only G_E^s can contribute to the PV asymmetry. Such an experiment is planned for CEBAF at $Q^2 = 0.6$ (GeV/c)^2 using the Hall A spectrometers [28]. Musolf and Donnelly [29] have argued that two measurements on ^{4}He at low and moderate Q^2 might ultimately constrain G_E^s to a higher level of precision than could be achieved with a proton target, but it is possible that medium effects could complicate the interpretation of G_E^s in ^{4}He compared to a proton target. Meson exchange corrections are expected to be small (few %) at low Q^2[30], and have recently been calculated [31] at $Q^2 \sim 0.6$ GeV^2 to contribute about 15% to the PV asymmetry.

Figure 4 summarizes the anticipated errors on G_E^s and G_M^s from SAMPLE and the upcoming CEBAF experiments. When combined with new data on neutron electromagnetic form factors, these results will substantially increase the body of experimental information available regarding the flavor structure of the nucleon at

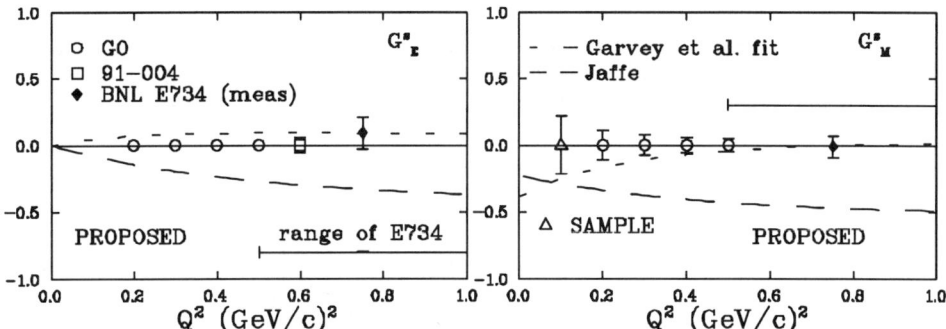

Figure 4: Expected results for $G_E^s(Q^2)$ and $G_M^s(Q^2)$ from parity violation experiments at CEBAF.

low energies.

POLARIZATION IN DEEP INELASTIC SCATTERING

Two additional topics in deep-inelastic scattering were covered in the parallel sessions on hadron structure. There has been considerable activity, both experimental and theoretical, regarding spin structure functions of the neutron, as demonstrated by several talks at this conference. As in elastic scattering, experimental information about the neutron requires the use of *nuclear* targets. This has stimulated both additional experiments using both $^3\vec{\text{He}}$ and $\vec{d}$, and theoretical work in understanding the nuclear corrections in such measurements. In the parallel session, a progress report on the HERMES experiment was presented [32]. HERMES will use polarized internal $\vec{p}$, $\vec{d}$ and $^3\vec{\text{He}}$ targets in the 30 GeV electron ring at HERA, and will therefore be able to directly compare measurements of g_1^n from deuterium and ^{3}He with the same apparatus.

The issue of nuclear corrections in deuterium was addressed by Umnikov [33]. Using a Bethe-Salpeter approach, he investigated relativistic effects in the extraction of g_1^n. His results indicated that at small Bjorken x, relativistic corrections are small and any differences between his calculation and nonrelativistic models can be attributed to differences the amount of D-state used in the deuteron wave function.

In semi-inclusive hadron production, it is possible to obtain information about the transverse polarization of quarks in hadrons. Vuaridel [34] presented a measurement proposed at CERN of the process $e\vec{p} \to e'hX$ using a transversely polarized target and an unpolarized 200 GeV beam at LEP. The asymmetry under reversal of the target spin has a dependence on the azimuthal angle ϕ_s of the produced hadron about the jet axis defined by parton kinematics.

$$A(x, \phi_s, p_T, z) \propto \left[\sum e_f^2 \Delta_T q \Delta_T f_q\right] \sin\phi_s. \tag{8}$$

The term Δ_q is the "transversity" distribution, the difference between quark polarization along or against the transversely polarized target. $\Delta_T f_q$ are spin-dependent fragmentation functions, also unknown. Analysis of data on $e^+e^- \to 2\text{-}jets$ is currently underway to estimate $\Delta_T f_q$ and thus the feasibility of an asymmetry measurement.

CONCLUSIONS

Polarized beams and polarized targets allow the use of spin observables to learn about nucleon structure. Several recently published experiments have added significantly to the body of available data on the structure of the *neutron*. Combined with upcoming neutral weak form factor measurements these date should help identify contributions of sea quarks to nucleon structure at low energies. All experiments to determine properties of the *neutron* require the use of nuclear targets, and as the experiments become more precise, they test the limits of theoretical models of few-nucleon systems.

This work was supported by NSF grant # PHY-9220690.

REFERENCES

1. A.Z. Gorski, F. Grummer, and K. Goeke, Phys. Lett. **B278**, 24 (1992).
2. S. Platchkov *et al.*, Nucl. Phys. **A510**, 740 (1990).
3. T. Eden, *et al.*, to be published in Phys. Rev. **C**. Oct. 1994.
4. R.G. Arnold, C.E. Carlson and F. Gross, Phys. Rev. **C23**, 363 (1981).
5. A. Lai, these proceedings.
6. Bates exp. 89-04, R. Madey, contact. CEBAF exp. 93-026, D. Day, contact. CEBAF exp. 93-038, R. Madey, contact.
7. E. Reichert, plenary talk in these proceedings. See also M. Meyerhoff *et al.*, Phys. Lett. **B327**, 201 (1994);
8. T.W. Donnelly and A.S. Raskin, Ann. Phys. **169**, 247 (Academic Press, NY, 1986).
9. J. Mitchell, these proceedings.
10. D. Day, plenary talk in these proceedings. See also references therein.
11. M. Woods, plenary talk in these proceedings. See also references therein.
12. C.E. Jones *et al.*, Phys. Rev. **C47**, 110 (1993); A.K. Thompson *et al.*, Phys. Rev. Lett. **68**, 2901 (1992);
13. H. Gao *et al.*, Phys. Rev. **C50**, R546 (1994).
14. J.O. Hansen, these proceedings. See also J.O. Hansen, submitted to Phys. Rev. Lett., Sept. 1994.
15. J.M. Laget, Phys. Lett. **B273**, 367 (1991).
16. CEBAF exp. 91-020, R. McKeown, contact. Bates BLAST proposal. Bates exp. 88-25, R. Milner and T. Chupp, contacts.
17. P. Markowitz, *et al.*, Phys. Rev. **C48**, R5 (1993).
18. E.E.W. Bruins, private communication.

19. J. Jourdan, proceedings of the Few Body Conference, Williamsburg, VA, June 1994.
20. I. Stancu, plenary talk in these proceedings.
21. M. Pitt, these proceedings.
22. D. Beck, these proceedings.
23. Bates experiment # 89-06, R. McKeown and D. Beck, contacts.
24. M.J. Musolf and B. Holstein, Phys. Lett. B242, 461 (1990).
25. Bates exp. 94-11, M. Pitt and E. Beise, contacts.
26. CEBAF exp. 91-010, P. Souder and J.M. Finn, contacts.
27. CEBAF exp. 91-017, D. Beck, contact.
28. CEBAF exp. 91-004, E. Beise, contact.
29. M.J. Musolf and T.W. Donnelly, Nucl. Phys. A546, 509 (1992).
30. M.J. Musolf and T.W. Donnelly, Phys. Lett. B318, 263 (1993).
31. M.J. Musolf, R. Schiavilla, and T.W. Donnelly, CEBAF preprint TH-94-10 (1994).
32. S. Pate, these proceedings. See also, H. Jackson, plenary talk in these proceedings.
33. A.Y. Umnikov, these proceedings. See also, A.Y. Umnikov and F.C. Khanna, preprint DFUPG 92/94.
34. B. Vuaridel, these proceedings. See also J. Collins, Nucl. Phys. B396, 161 (1993).

II. SYMMETRIES

Limit on the Electric Dipole Moment of ^{199}Hg

Blayne R. Heckel

Department of Physics
University of Washington, Seattle, WA. 98195

Abstract. The spin precession frequency of optically pumped ^{199}Hg atoms in an applied electric field has been measured to set a limit on the electric dipole moment of these atoms as a test of time reversal symmetry. Our result, $d(^{199}\text{Hg}) < 9.1 \times 10^{-28}$ e-cm, is the most precise experimental bound on an electric dipole moment and provides a new test for theories of time reversal symmetry violation in atomic systems.

Time reversal (T) invariance is a powerful concept that leads to a variety of constraints on physical processes, including the form of the S (scattering) matrix, the relative phases of certain competing transition matrix elements, and the absence of odd order electric and even order magnetic multipole moments of elementary particles. The apparent violation of T symmetry in the CP-violating (yet CPT-conserving) decays of the K_L meson and in the observed baryon excess in the universe has lead to a number of experimental searches for additional evidence of T-violation. One class of these experiments, attempts to detect an electric dipole moment of the neutron and of a variety of atoms, has proven to be especially fruitful. Recent electric dipole moment (EDM) limits on the neutron[1,2], cesium[3], thallium[4], thallium flouride[5], and our own mercury results[6] provide exacting tests of T-violation in hadronic, semi-leptonic, and leptonic interactions.

The mercury experiment was performed at the University of Washington by W.M. Klipstein, J.P Jacobs, S.K. Lamoreaux, E.N.Fortson, and the author over a period of 5 years. The results reported here have been submitted for publication recently and include additional data to that reported in ref.(6). We chose to study the mercury atom for several reasons. Sandars[7] first showed that the effects of T-violation can be enhanced in heavy atoms, scaling as roughly the square of the atomic number for nuclear spin dependent sources of T-violation. Because EDM measurements ultimately try to detect a change in the spin precession frequency due to the coupling of an EDM to an electric field, it is advantageous to select a species with a long spin relaxation lifetime to permit the highest possible frequency resolution to be achieved. The $^1S_0$

electronic structure of ground state mercury atoms provides a shield for the spin 1/2 nucleus of ^{199}Hg, allowing nuclear spin relaxation lifetimes in excess of 100 s in small vapor cells containing several hundred torr of buffer gas as well as the ^{199}Hg atoms. In addition, the mercury atoms can be polarized by optical pumping of the 254 nm intercombination line, and the precessing polarization can be detected by monitoring the transmission of the 254 nm light through the vapor cell. The mercury EDM measurement is most sensitive to nuclear and semi-leptonic sources of T-violation. In contrast, experiments that use atoms with unpaired electrons are primarily sensitive to leptonic T-violation, while the neutron EDM measurement is primarily sensitive to hadronic sources.

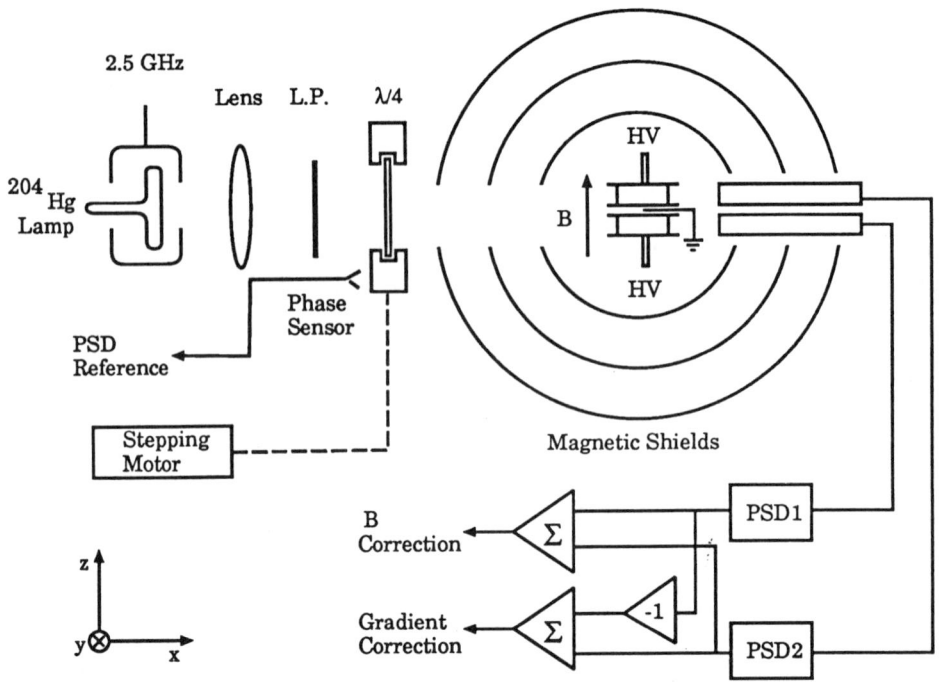

FIGURE 1. Schematic view of the mercury EDM apparatus

The apparatus for the mercury EDM experiment is shown in schematic form in Figure 1. Two fused silica vapor cells, approximately 2.5 cm in diameter and 1 cm high, that contain enriched ^{199}Hg along with 300 torr of nitrogen gas and 15 torr of carbon monoxide are placed in a uniform 10 mG magnetic field inside three layers of magnetic shielding. The endcaps of the cells are coated with a semiconducting layer underneath a thin layer of wax to permit the application of an electric field within the cell without spoiling the spin relaxation lifetime of the mercury atoms(8). Electric fields, aligned with the static magnetic field, are applied in opposite directions in the two cells. The difference between the spin precession frequencies in the two cells, $\delta\nu$, is then given by: $h\delta\nu = 4Ed$ where E is the electric field magnitude and d is the EDM. In addition, there will be a term from the magnetic field gradient, but by reversing both electric fields, only drifts in the magnetic field gradient can project onto the EDM signal.

As illustrated in Fig. 1, the resonant light to optically pump the ^{199}Hg atoms comes from a microwave driven discharge lamp that contains enriched ^{204}Hg. This light connects the F=1/2 ground state levels to an F=1/2 excited state multiplet. The light is first sent through a Brewster angle linear polarizer and then through a quarter wave retardation plate to produce circularly polarized radiation. Were a magnetic field applied along the light ($\hat{x}$) axis, then as the atoms absorbed angular momentum from the photons, the vapor would become transparent: selection rules prevent the polarized atoms from absorbing radiation. In this way, the transmitted light intensity becomes a measure of the component of atomic polarization along the light axis. For EDM measurements, the magnetic field is along the $\hat{z}$ axis, and the quarter wave plate is continuously rotated at a frequency ω_m. This produces a light beam whose circular polarization is given by $\hat{x}sin(2\omega_m t)$. The magnetic field is adjusted so that the Larmor frequency of the atoms, ω_L, is equal to $2\omega_m$. In this way a synchronously driven atomic polarization is created that precesses in the xy plane. In the presence of the light beam, the spin relaxation time of the mercury atoms is 20 s. The phase of the modulation on the transmitted light relative to the rotation angle of the quarter wave plate is then proportional to $2\omega_m - \omega_L$. We feed back the sum of the phase angles from the two cells to the magnetic field supply to lock the average ω_L to $2\omega_m$. We also use the phase difference between the cells to correct the magnetic field gradient. When electric fields are applied, an EDM would produce a phase difference between the cells that we detect by a change on the magnetic field gradient correction signal.

The electric field is changed every 1000 s in a $+ \; 0 \; - \; 0$ sequence. We have used electric field magnitudes of 5, 7, and 10 kV/cm. Each day, experimental parameters such as the electric field magnitude, magnetic field direction, linear polarizer axis, and rotation sense of the quarter wave plate would be changed. We report here the results of 299 days of data, split between sets taken with different pairs of vapor cells, and different orientations of the cells inside of the apparatus. The value we find for the EDM of ^{199}Hg is:

$$d(^{199}Hg) = (-1.1 \pm 2.4 \pm 3.6) \times 10^{-28} e - cm \qquad (1)$$

where the first error is statistical and the second is our systematic uncertainty. A detailed discussion of the data and error analysis can be found in a forthcoming article(9). This leads to a 95% confidence level upper limit on the mercury dipole moment of 9.1×10^{-28} e·cm.

The standard model explains the kaon CP-violation by a phase in the quark mixing matrix. In this model, EDMs arise in second order in the weak interaction, at the level of 10^{-32} e·cm, well below the present experimental bounds. Most extensions to the standard model allow EDMs that are first order in the weak coupling constant, well within the range of recent EDM experiments. In this sense, EDM measurements can be viewed as tests for physics beyond the standard model. Table 1 summarizes the constraints for several models of CP-violation that arise from our mercury result and the tightest constraints that are set by other EDM experiments. References for both the theories and experiments are cited within the table.

TABLE 1. Upper limits (95% confidence level) on T-violating interactions set by the ^{199}Hg result reported here ($d < 9.1 \times 10^{-28}$ e·cm), compared with the best current limits from other experiments.

T-Violating Parameters	Ref.	^{199}Hg Limit	Best Other Limit	
Hadronic:				
Schiff Moment: $Q_S(e \cdot \text{fm}^3)$	a	2.3×10^{-11}	1×10^{-9}	TlF[e]
$i\eta G_F(\bar{n}n)(\bar{n}\gamma_5 n)/\sqrt{2} : \eta$	a	1.7×10^{-3}	2×10^{-2}	TlF[e]
$i\eta_q G_F(\bar{q}q)(\bar{q}\gamma_5 q)/\sqrt{2} : \eta_q$	a	3.5×10^{-6}	4×10^{-5}	n[f]
Semileptonic:				
$iC_T \frac{G_F}{\sqrt{2}}(\bar{n}\gamma_5\sigma_{\mu\nu}n)(\bar{e}\sigma^{\mu\nu}e) : C_T$	ad	1.4×10^{-8}	5×10^{-7}	TlF[e]
$iC_S G_F(\bar{n}n)(\bar{e}\gamma_5 e)/\sqrt{2} : C_S$	a	7×10^{-7}	8×10^{-7}	Tl[g]
Leptonic:				
electron: $d_e(e \cdot \text{cm})$	a	6.3×10^{-26}	4×10^{-27}	Tl[g]
Gauge Model:				
QCD phase: $\bar{\theta}_{qcd}$	abc	1.4×10^{-9}	4×10^{-10}	n[f]
Supersymmetry: ϵ^{susy}	b	7×10^{-3}	1×10^{-2}	n[f]
	b		4×10^{-2}	Tl[g]
Multi-Higgs: ϵ^{higgs}	b	$0.7/\tan\beta$	$0.7/\tan\beta$	Tl[g]
Left-Right Sym.: x^{LR}	b	2.1×10^{-3}	1.3×10^{-2}	n[f]

[a] Ref. (10-14); [b] Ref. (15); [c] Ref. (16); [d] Ref. (17); [e] Ref. (5); [f] Ref. (1) taking $d_n < 1 \times 10^{-25}$ e·cm; [g] Ref. (4);

More than 50 years of EDM experiments have provided ever tighter bounds on this elusive quantity, and have greatly narrowed the range of possibilities for T-violating interactions. We believe that we can acheive another order of magnitude improvement in our mercury edm result, and hope to one day report something other than a new upper limit. We gratefully acknowledge the support of the National Science Foundation under Grant No. PHY-9206408.

REFERENCES

1. K. F. Smith, et. al., Phys. Lett. B **234**, 191 (1990).

2. I. S. Altarev, et. al., Phys. Lett. B **267**, 242 (1992).

3. S. A. Murthy, D. Krause, Jr., Z. L. Li, and L. R. Hunter, Phys. Rev. Lett. **63**, 965 (1989).

4. K. Abdullah, C. Carlberg, E. D. Commins, Harvey Gould, and Stephen B. Ross, Phys. Rev. Lett. **65**, 2347 (1990).

5. D. Cho, K. Sangster, and E. A. Hinds, Phys. Rev. Lett. **63**, 2559 (1989).

6. J. P. Jacobs, et. al., Phys. Rev. Lett. **71**, 3782 (1993).

7. G. E. Harrison, M. A. Player, and P. G. H. Sandars, Journal of Physics E, **4**, 750 (1971).

8. J. P. Jacobs, Ph.D. thesis, University of Washington, (1991) unpublished.

9. J. P. Jacobs, et. al., Phys. Rev. A, submitted for publication.

10. I. B. Khriplovich, Atomic Physics 11. *Proceedings of the Eleventh International Conference on Atomic Physics*, 113 (1988).

11. V. V. Flaumbaum, I. B. Khriplovich and O. P. Sushkov. Phys. Lett. **162B**, 213, (1985).

12. V. V. Flaumbaum, I. B. Khriplovich and O. P. Sushkov. Nuclear Physics **A449**, 750, (1986).

13. V. M. Khatsymovsky, I. B. Khriplovich, and A. S. Yelkhovsky. Ann. of Phy. **186**, 1 (1988). V. M. Khatsymovsky, I. B. Khriplovich, Phys. Lett. B, **296**, 219 (1992).

14. W. C. Haxton and E. M. Henley, Phys. Rev. Lett. **51**, 1937, (1983).

15. S. M. Barr, Int. J. Mod. Phys. A **8**, 209, (1993).

16. Kar Lee, Ph.D. thesis, University of Washington, (1994) unpublished.

17. A.-M. Mårtensson-Pendrill, Phys. Rev. Lett. **54**, 1153 (1985).

Search for Incomplete Parity-Violation in Leptonic and Semileptonic Weak Processes : Status and Perspectives

Jules Deutsch

Department of Physics, Université Catholique de Louvain
B-1348 Louvain-la-Neuve, Belgium

Abstract. We stressed the importance of searching for deviations from the full parity-violation postulated by the Standard Model in weak-interaction processes as a means of restoring parity at some new energy-domain and outlined a number of associated theoretical scenarios for new physics. Discussing various complementary experimental approaches to this problem, we described their status and perspectives. In this written version of the talk we restrict ourselves to some recent developments which were not discussed as yet in a recent review-paper (1).

INTRODUCTION

It was generally believed that the laws of physics are invariant under reflection and that the lack of mirror-symmetry of most processes is due to asymmetrical initial conditions (2). So it came as a surprise in 1957 and the following years that fundamental weak processes violate parity-symmetry (2). The experimentally observed violation is sizeable (3) and it was soon assumed that it is maximal indeed (3). This feature is then built in "by hand" in the $SU(2)_L \times U(1)$ structure of the Standard Model (3).

As the Standard Model has many ad-hoc features, a good fraction of the efforts in particle physics is devoted to test it (4) and to find eventually deviations pointing toward a more satisfactory fundamental theory. Because full parity-violation is not only an ad-hoc assumption but also surprising in view of our expectations, as outlined above, the efforts to invalidate it constitute a particularly important part of our searches for new physics beyond the Standard Model.

These efforts are guided by scenarios in which full mirror-symmetry is recovered at some higher unification-level, as will be outlined in the following section.

LEFT-RIGHT SYMMETRIC SCENARIOS

For the original references to these scenarios and their detailed discussion we refer the reader to refs. 5, 6 and 7.

Manifest left-right symmetric models constitute a minimal extension of the Standard Model based on a $SU(2)_L \times SU(2)_R \times U(1)$ gauge-group. A new charged gauge-boson W_2 of mass m_2 is introduced in addition to the known W_1 of mass $m_1 = 80.2$ GeV/c^2. The left (L) resp. right (R) coupling of the leptons to weak charged currents is described by a linear combination of the two mass-eigenstates as : $W_L = \cos(\zeta)W_1 + \sin(\zeta)W_2$ and $W_R = -\sin(\zeta)W_1 + \cos(\zeta)W_2$ where possible CP-violating imaginary phases are neglected (5). Left-right symmetry is then recovered in the high q^2 - limit and its presence is manifested even in low-q^2 processes as a deviation from full parity-violation. For $\zeta = 0$ this deviation scales as $(\delta)^2 = (m_1/m_2)^4$ and precision-experiments will explore increasing m_2 mass-domains and will eventually delineate a solution-region in the (δ,ζ)-plane which does not encompass the prediction of the Standard Model : $(\delta) = (\zeta) = 0$. Such constraints will be shown in the figure. It should be noted, finally, that the mixing-angle (ζ) is constrained by the relation : $\zeta < |\delta|$ (8).

These manifest left-right symmetric models introduce only two parameters to be constrained by experiment. Their extensions (5,6,7) enlarge the parameter-space and so will require - as we shall see - a greater wealth of complementary experiments to obtain useful constraints.

In this note we shall neglect CP-violation and the possibility of massive neutrinos which could also give rise to neutrino flavor-mixing (5,7). We shall also restrict our discussion to the strangeness (charm)-conserving processes of the (u,d) quark-sector alone and refer the reader to important results pertaining to the other sectors (7,9,10). As shown in ref. 7, the stringent constraints deduced in refs. 9 and 10 may not apply to the domains we consider here in extended left/right symmetric models.

With these restrictions, only two additional parameters are introduced : $r_g = (g_R/g_L)$, the ratio of the two gauge coupling-strengths g_L, g_R and $r_K =$

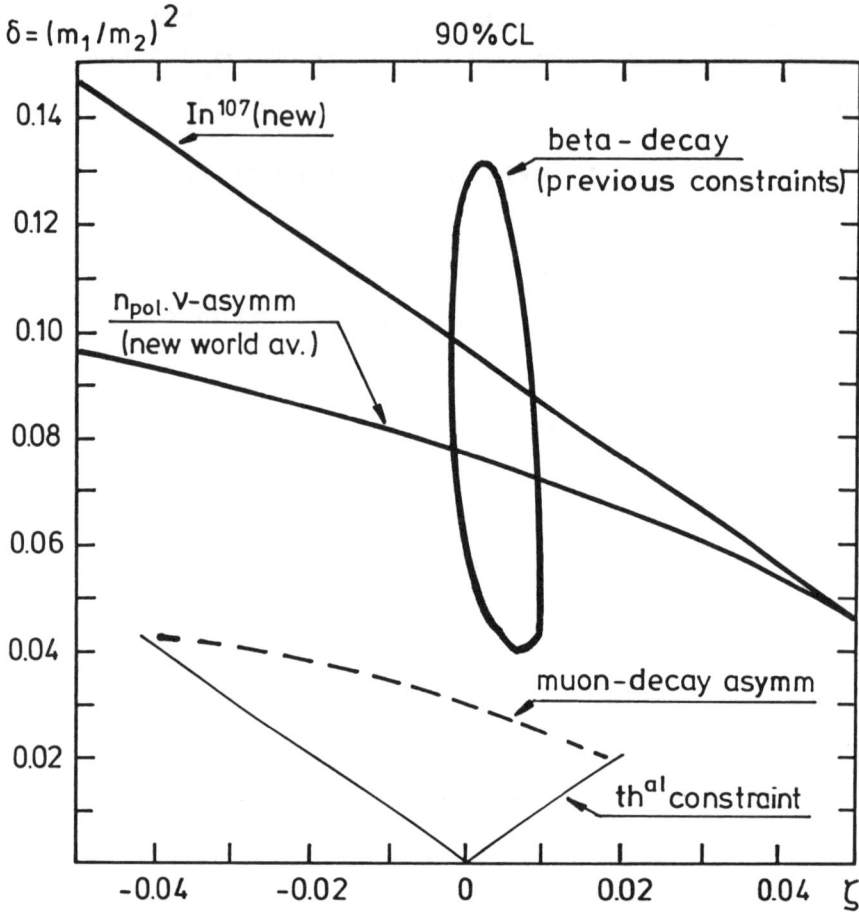

The figure illustrates the constraints on the two parameters of manifest left-right symmetric models from beta-decay (previous constraints) and muon-decay asymmetry at 90 % C.L. illustrating potential deviations from the Standard Model in semileptonic decays as explained in the text. It is also shown that new results, In[107] (new) and n_{pol} neutrino-asymm. (new world-average) exclude to some extent such a deviation. The line labeled "th[al] constraint" corresponds to the requirements discussed in ref. 8.

(U_{ud}^R/U_{ud}^L), where $U_{ud}^{R,L}$ are the (u,d)-matrix-elements of the Cabibbo-Kobayashi-Maskawa matrices pertaining to the R/L-sectors.

As will be stressed, the various experiments we shall discuss are complementary as they have different dependences in the (δ, ζ, r_g, r_K) parameter-space.

STATUS AND PROSPECTS OF EXPERIMENTS

Collider Experiments

Helicity-sensitive collider-experiments are rare as yet. We mentioned only the SLC-experiment performed with polarized electrons (11) to illustrate its sensitivity to helicity-anomalies such as the one which could be introduced by a second neutral gauge-boson $Z^{o'}$ required by the left-right symmetric models. Direct helicity-sensitive searches for the charged W_R will have to await higher luminosities for polarized electrons at HERA (12).

Indirect constraints on new charged gauge-bosons (of any helicity) W' can be deduced from $p\bar{p}$ collider experiments looking for anomalies in the energy-dependence of the (cross-section) x (branching-ratio) product to produce either jets (13) or a charged-lepton - neutrino-pair (14). The CDF-experiment, in particular, obtains a 95 % C.L. limit of 520 GeV/c^2 on the mass of a new charged gauge-boson (whether with (V+A) or (V-A) couplings) assuming it has gauge-coupling of standard strength and that its associated neutrinos do not decay inside the detector.

Recently P. Herczeg performed a critical comparison of the constraint provided by the collider-result with those of other origin within extended left-right symmetric models (16). Since in the (cross-section) x (branching ratio) product the decay branching-ratio of the produced W' is helicity-independent, the experiment constrains r_g^2 x r_K^2 x $f[(m_1/m_2)]$, where the function f, in addition to (m_1/m_2), also depends on kinematical variables ; for pedagogical reasons we assume here $\zeta = 0$.

Muon-Decay Experiments

Precision-experiments are available on the emission-asymmetry of positrons from polarized muons (17) and were extensively analyzed in various extensions of the left-right symmetric models (7). In our notations and assuming $\zeta = 0$ as above the deviation from full asymmetry at the positron energy end-point is sensitive to $2r_g^4 (m_1/m_2)^4 (1+r_K^2)$. A possible 2.4 standard deviation from zero of this quantity found by the authors of ref. 17 was attributed to uncontrolled systematic effects and illustrates the difficulty to perform high precision absolute measurements.

It was stressed recently (18) that a relative positron polarization measurement in the direction of the spin of the emitting system and opposite to it is very sensitive to deviations from the standard helicity structure. This method was readily applied to nuclear beta-decay (cfr. below) and can be applied also to muons (18,19) where the sensitivity-factor of 2 caracteristic of the asymmetry-measurement can be increased to about 100 (18,20) with the additional (experimental) advantage of the relative measurements over absolute ones. It should be noted also that this experiment is insensitif to small deviations from unity of the muon polarization and so constrains the combination $r_g^4(m_1/m_2)^4$ complementary to the above mentioned one.

Nuclear Beta-Decay

Helicity-anomalies in nuclear beta-decay are sensitive to $r_g^4 \, r_K^2 \, (m_1/m_2)^4$ assuming here $\zeta = 0$ for pedagogical purposes. As noticed by P. Herczeg, the combinations of parameters constrained by the other experiments discussed above are different from this combination and so the different experimental approaches are complementary in extended left-right symmetric models.

In the following we shall discuss briefly the constraints provided by nuclear beta-decay experiments using, for convenience, the two-parameter representation (δ,ζ) of the manifest left-right symmetric models.

We shall mention the new developments only and refer the reader for details and a discussion of the earlier status of the field to refs. 1, 21, 22.

Since absolute asymmetry - or polarization measurements were not performed as yet for pure Gamow-Teller transitions sufficiently fast to avoid problems with nuclear-structure dependent recoil-order terms (1), absolute beta-asymmetry measurements useful for our purposes were performed in the superallowed but mixed Fermi/Gamow-Teller beta-decay of n and Ne[19]. For these cases, the asymmetry depends also on the ratio of the two amplitudes which has to be extracted from a comparison of the corresponding ft-values to those of pure superallowed Fermi-transitions. Here, the analysis of ref. 1 in this respect is slightly updated, using for the neutron life-time the new world average of 887.0 ± 1.5 s and for the pure Fermi superallowed ft-value 3136.6 ± 4.6 s as suggested by the author of ref. 23. Including in the analysis the Ne^{19}-data (1,22), the "old" world-average of neutrino-asymmetry and the relative polarization results on Fermi and Gamow-Teller transitions, we obtain the near-elliptic region illustrated in the figure. At 90 % C.L., these data find a right-handed gauge-boson at $0.04 \leq \delta \leq 0.13$, i.e. having a mass of $220 \leq m_2$ (GeV/c^2) ≤ 400. In manifest left-right symmetric models (assumed for illustration) these limits are in contradiction with the ones extracted from muon-decay. It should be stressed, however, as stated above, that these two processes constrain different parameter-combinations of extended left-right symmetric models (16).

More importantly, the result is strongly influenced by a single electron-asymmetry measurement performed with polarized neutrons and so should be controlled by other independent experiments.

In the figure we illustrate the constraints provided by two such new experiments excluding the "upper" part of the parameter-plane.

The line labeled In^{107} (new) illustrates the first results of a relative beta-polarization measurement performed on polarized nuclei as discussed above (18,24).

This relative experiment excludes already part of the near elliptic parameter-region resulting from the earlier (mostly absolute) experiments and is actually continued with the hope to increase its precision so as to confirm the anomaly or to exclude it definitively.

A new measurement of the neutrino-asymmetry from polarized neutrons was performed at the Petersburg Nuclear Physics Institute by A. Serebrov and collaborators (25) with a particularly careful determination of the neutron-beam polarization using a novel polarimetric method. It should be noted in this respect that it is the determination of this polarization which was a delicate part of the earlier electron-asymmetry experiment. The new experiment obtains for the neutrino-asymmetry coefficient the value of $B = 0.9894 \pm 0.0083$ (25), much more precise than the old world-average of $B = 0.9957 \pm 0.0276$. In the figure we illustrate the strong constraint provided by the resulting new world-average of $B = 0.9899 \pm 0.0079$. As can be seen, most of the anomalous parameter-region is excluded by this new experiment.

PERSPECTIVES

Novel experiments are planned or in progress in the various complementary domains we discussed. At colliders DESY will undertake its direct search for W_R and CDF at Fermilab will improve its constraints on a possible W'. In the muon-decay sector a new asymmetry-measurement is planned at Triumf and a relative polarization measurement at PSI. In the nuclear beta-decay sector improvements on the absolute Ne^{19}-asymmetry are in progress at Princeton and for the neutron asymmetry are planned both at Gatchina and at Grenoble. As for relative measurements, an improvement of the In^{107}-experiment is in progress at Louvain-la-Neuve and a similar experiment on N^{12} is to be completed at PSI. Similar experiments are in preparation also using optical traps for Na^{21} and K^{37}.

I hope that I succeeded in conveying the interest and dynamism of this research-field encompassing various energy-domains.

ACKNOWLEDGMENTS

I wish to thank P. Herczeg and A. Serebrov for having kindly provided us with their results prior to their publication and P. Herczeg also for illuminating discussions. E. Thomas kindly helped in producing the figure and J. Govaerts with a critical reading of the manuscript.

REFERENCES

1. Deutsch J., Quin P., "Symmetry-tests in semileptonic weak interactions : a search for new physics", to appear in *Precision Tests of the Standard Electroweak Model*, World Science Advanced Series on Directions in High Energy Physics, Paul Langacker, Editor.
2. Genz H., Decker R., "Symmetrie une Symmetriebrechung in der Physik", Vieweg, 1981.
3. Cfr refs. e.g. in : Commins E.D., Bucksbaum P.H., *Weak interactions of leptons and quarks*, Cambridge Univ. Press, 1983.
4. Darriulat P., Conference Summary at the ICHEP 94, Glasgow, July 94, CERN-preprint PPE/94-159.
5. Herczeg P., *Phys. Rev.* **D34**, 3449 (1986).
6. Mohapatra R.N., "Unification and Supersymmetry", Springer Verlag, 1986, ch. 6.
7. Langacker P., Sankar S.U., *Phys. Rev.* **D40**, 1569 (1989).

8. Masso E., *Phys. Rev. Letters* **52**, 1956 (1984).
9. Beall G., Bander M., Soni A., *Phys. Rev. Letters* **48**, 848 (1982).
10. Imazato J. et al., *Phys. Rev. Letters* **69**, 877 (1992).
11. Abe A. et al., *Phys. Rev. Letters* **70**, 2515 (1993).
12. Schrempp F., DESY 93-096.
13. UA2 Collaboration, *Nuclear Physics* **B400**, 3 (1993).
14. CDF Collaboration, *Phys. Rev. Letters* **67**, 2609 (1991).
15. Langacker P. et al., *Phys. Rev. Letters* **68**, 2871 (1992).
16. Herczeg P., private communication, to be published.
17. Jodidio A. et al., *Phys. Rev.* D34, 1967 (1986) ; ibid D37, 237 (1988) and refs. therein.
18. Quin P.A., Girard T., *Phys. Letters* B229, 29 (1989).
19. Scheck F., *Physics Reports* **44**, 187 (1978).
20. Letter of Intent at PSI under preparation by a ETH-Louvain-PSI-Zürich collaboration, 1994.
21. Carnoy A.S. et al., *J. Phys. G. Nucl. Part. Phys.* **18**, 823 (1992).
22. Prieels R., International Symposium on Weak and Electromagnetic Interactions in Nuclei, Dubna, 16-21 June 1992, ed. TS.D. Vylov, World Scientific, Singapore, 1993, 528.
23. Wilkinson D.H., *Z. Phys.* A348, 129 (1994).
24. Severijns N. et al., *Phys. Rev. Letters* **70**, (1993) ; ibid 73, 611 (1994).
25. Serebrov A., private communication, to be published ; cfr also I.A. Kuznetsov et al., Gatchina preprint NP-63-1994-2005.

Study of Parity and Time-Reversal Violation in Neutron-Nucleus Interactions

Yi-Fen Yen,[1] J. D. Bowman,[1] B. E. Crawford,[2] P. P. J. Delheij,[3]
C. M. Frankle,[1] K. Fukuda,[4] C. R. Gould,[5] A. A. Green,[3]
D. G. Haase,[5] M. Iinuma,[6] J. N. Knudson,[1] L. Y. Lowie,[5] A. Masaike,[6] Y. Masuda,[7] Y. Matsuda,[6] G. E. Mitchell,[5] S. I. Penttilä,[1] H. Postma,[8] N. R. Roberson,[2] S. J. Seestrom,[1] E. I. Sharapov,[9] H. M. Shimizu,[7] S. L. Stephenson,[5] and V. W. Yuan[1]

(The TRIPLE Collaboration)

[1]Los Alamos National Laboratory, Los Alamos, New Mexico 87545, USA
[2]Duke University, Durham, North Carolina 27708-0305 and Triangle Universities Nuclear Laboratory, Durham, North Carolina 27708-0308, USA
[3]TRIUMF, Vancouver, British Columbia, Canada V6T 2A3
[4]Meiji College of Oriental Medicine, Japan
[5]North Carolina State University, Raleigh, North Carolina 27695-8202 and Triangle Universities Nuclear Laboratory, Durham, North Carolina 27708-0308, USA
[6]Kyoto University, Department of Physics, Kyoto 606-01, Japan
[7]National Laboratory for High Energy Physics, 1-1, Oho, Tsukuba 305, Japan
[8]University of Technology, P. O. Box 5046, 2600 GA, Delft, The Netherlands
[9]Joint Institute for Nuclear Research, 141980 Dubna, Russia

Abstract. The parity and time-reversal symmetries can be studied in neutron-nucleus interactions. Parity non-conserving asymmetries have been observed for many p-wave resonances in a compound nucleus and measurements were performed on several nuclei in the mass region of A~100 and A~230. The statistical model of the compound nucleus provides a theoretical basis for extracting mean-squared matrix elements from the experimental asymmetry data, and for interpreting the mean-squared matrix elements. The constraints on the weak meson-exchange couplings calculated from the compound-nucleus asymmetry data agree qualitatively with the results from few-body and light-nuclei experiments. The tests of time-reversal invariance in various experiments using thermal, epithermal and MeV neutrons are being developed.

INTRODUCTION

The weak parity non-conserving (PNC) nucleon-nucleon (N-N) interaction can be described in terms of meson-exchange potentials between strong and weak meson-nucleon couplings. The Desplanques, Donoghue and Holstein (DDH) theory (1) characterizes the PNC interaction with seven weak meson-exchange amplitudes and predicts a range of values of these coupling constants. It is a great experimental and theoretical challenge to determine these weak coupling constants.

Hadronic PNC phenomena have been observed (2) in few body ($A \leq 4$), light nucleus and compound nucleus (CN) systems. In the few-body system all the seven meson couplings are significant. The interpretation of results is direct, but asymmetries are of the order of 10^{-7} and experiments are difficult (2). The only experiments involving $A \leq 4$ that have provided results precise enough to constrain the PNC meson-exchange couplings are measurements of the longitudinal analyzing power (A_L) in $\vec{p}+p$ (3-5) and $\vec{p}+\alpha$ (6) scatterings. Experiments measuring A_L in $\vec{p}+d$, as well as the asymmetry (A_γ) and circular polarization (P_γ) of γ rays from neutron capture on protons or deuterons have been carried out. These measurements still lack sufficient precision (2). Experiments measuring PNC neutron spin-rotation in few-body systems are being pursued (7,8) in order to provide stronger constraints on the PNC meson-exchange couplings.

In light nuclei, the π and ρ (isospin zero) meson-exchange couplings, F_π and F_ρ, respectively, are dominant. The PNC experiments in light nuclei usually deal with special transitions involving closely-spaced parity doublets that have identical spins but opposite parities. Typical observables arise from the interference between a strong parity-conserving transition and a weak PNC transition. The large ratio of the transition strengths and small energy denominators of these nuclei lead to an amplification of 10^2-10^4 which brings the N-N PNC effects to a level of 10^{-5}-10^{-3}. A good example is the measurement of P_γ in ^{18}F (9-11) which is primarily sensitive to the F_π amplitude. The lifetimes and energy-splittings of the nuclear states in ^{18}F are well known. With some nuclear-structure parameters taken from nuclear model calculations, the ^{18}F results indicate a small F_π coupling. Other experiments measuring P_γ in ^{21}Ne and A_γ in ^{19}F have been carried out but the sensitivities are not yet sufficient to provide reliable constraints on F_π and F_ρ couplings. The sensitivity required for the PNC light-nuclei experiments is less stringent than that for the few-body experiments. However, the interpretation of the light-nuclei experimental results requires a detailed knowledge of the nuclear structure, which poses difficulties.

The chaotic nature of the compound nucleus results in a very large enhancement of the PNC asymmetries and allows the data to be interpreted using a statistical model of the compound nucleus. A compound nucleus (CN) is formed when a nucleus, A, captures a neutron to form an excited state of the A+1 nucleus. The cross-section for the scattering of low-energy neutrons from medium and heavy nuclei consists of closely spaced (~ 20 eV) extremely narrow (~ 0.1 eV) resonances. The narrowness of these resonances indicates that a long-lived intermediate state is formed, hence the term compound nucleus. One reason to study fundamental symmetries in the compound nuclear system is to understand how symmetry non-conservation in the effective N-N interaction manifests itself in the complicated many-body systems and to determine the strength of the nuclear weak force.

PNC phenomena of this type were first observed by Abov et al. (12). Recent advances in this regime have resulted from the experimental program of the TRIPLE collaboration (13-15). The statistical behavior of the CN observables results from the fact that the CN wave functions are superpositions of a large number ($N \sim 10^4$-10^6) of independent particle components. The weak interaction mixes the nearby s-wave (angular momentum $\ell = 0$, positive parity) and p-wave ($\ell = 1$, negative parity) CN states of the same total spin. The mixing matrix elements, $\langle \psi_s | H_w | \psi_p \rangle$, are represented by an ensemble of independent and Gaussian random variables. The useful information is contained in the average quantities, such as the mean-squared matrix element, defined as M^2:

$$\overline{\langle \psi_s | H_w | \psi_p \rangle^2} = M^2. \qquad (1)$$

The statistical model of the CN provides a theoretical basis for extracting mean-squared matrix elements from PNC experimental data and for interpreting the mean-squared matrix elements. In this picture only average properties of the nuclear wave functions need to be calculated.

In this work, we measured PNC asymmetries of numerous p-wave CN resonances and extracted M^2. This is an advance over the previous PNC studies (12) in the CN where only a single asymmetry per nucleus was measured. A theoretical model has been developed (15) to relate the M^2 to the weak meson-exchange coupling constants. With the M^2 determined from experiment and the theoretical work based on the statistical model, we obtain constraints on F_π and F_ρ. The constraints complement the results of the few-body and light-nuclei experiments. Furthermore, we measure M^2 in two mass regions (A~100 and A~230) to check the consistency of the constraints.

Parity Non-conserving Asymmetry in the Compound Nucleus

The observable in the TRIPLE experiments is the PNC longitudinal asymmetry. For the ith p-wave resonance, the PNC asymmetry (P_i) is

$$P_i = \frac{\sigma_+ - \sigma_-}{\sigma_+ + \sigma_-}, \tag{2}$$

where σ_+ and σ_- are the resonance cross sections for positive and negative helicity states of neutrons. The asymmetry is manifestly parity violating. The total cross section (σ_T) is proportional to the forward scattering amplitude $f(0°)$, which is expressed by

$$\sigma_T \propto f(0°) = f_0 + f_1 \vec{\sigma} \cdot \vec{I} + f_2 \vec{\sigma} \cdot (\vec{k}_n \times \vec{I}) + f_3 \vec{\sigma} \cdot \vec{k}_n. \tag{3}$$

Here, $\vec{k}_n$ is the neutron momentum, $\vec{\sigma}$ is the Pauli matrix for neutron spin, and $\vec{I}$ is the target spin. With an unpolarized target and polarized neutron beam, the cross-section contains only one spin-dependent term, $\vec{\sigma} \cdot \vec{k}_n$. The parity symmetry can be tested by measuring the difference in the total cross-section when the sign of the $\vec{\sigma} \cdot \vec{k}_n$ term is changed by flipping the neutron spin direction parallel or antiparallel to its momentum.

The PNC asymmetries in the CN are remarkably large because of a huge amplification arising from the high degree of level mixing in the CN. For two-state PNC mixing, the asymmetry (P_i) is a function of the mixing matrix element, the energy level splitting ($E_{s_j} - E_{p_i}$), and the ratio of the neutron width amplitudes of s-wave and p-wave resonances, $\sqrt{\Gamma_{s_j}^n}$ and $\sqrt{\Gamma_{p_i}^n}$, respectively,

$$P_i = 2 \frac{\langle \psi_{s_j} | H_w | \psi_{p_i} \rangle}{E_{s_j} - E_{p_i}} \sqrt{\frac{\Gamma_{s_j}^n(k_n)}{\Gamma_{p_i}^n(k_n)}}. \tag{4}$$

The high level density and a large ratio of the widths can give an overall enhancement of $\sim 10^6$. This makes the observable asymmetry as large as 10% in the CN system. Indeed, we have observed a PV asymmetry as large as 10% in ^{139}La and ^{232}Th.

TRIPLE EXPERIMENTS AND RECENT RESULTS

The TRIPLE collaborators have developed an apparatus to take advantage of the very high neutron flux available at the Los Alamos Neutron Scattering Center. The 500-μs-wide 800-MeV proton pulses from the Los Alamos Meson Physics Facility linac are injected into the Proton Storage Ring at the rate of 20 Hz and time compressed into triangular pulses 250-ns wide at the base. These pulses impinge on a tungsten target surrounded by a water moderator. The intensity of the epithermal neutron beam at neutron energy E is approximately given by

$$\frac{d^2n}{dE\,dt} = \frac{0.01}{E}\frac{I}{e}f\Omega \sim \frac{6.6\times 10^7}{E} \quad (1/\text{eV}/\text{s}), \tag{5}$$

where I is the average proton current (~70 μA), e is the quantum of charge, $f \sim 0.63$ is the fraction of the 13-cm by 13-cm moderator viewed by the collimator system, and $\Omega \sim 4\times 10^{-5}$ sr is the solid angle. It is the large neutron intensity that makes the measurements described here possible.

The apparatus (16) is designed to make simultaneous measurements of the longitudinal asymmetries of multiple p-wave resonances. The apparatus is shown schematically in Fig. 1. First, neutrons pass through a flux monitor (17) consisting of a pair of ionization chambers. The first chamber contains ^{3}He and the second ^{4}He at atmospheric pressure. The ^{3}He ion chamber responds to both neutrons and photons, while the ^{4}He ion chamber responds only to photons. The signals from each of these ion chambers are amplified by operational amplifiers and digitized by voltage-to-frequency converters. The neutron flux is measured for each burst (with 0.1% accuracy) by subtracting the digitized ^{4}He signal from the digitized ^{3}He signal.

The neutrons next pass through a polarized-proton polarizer (18). The protons in a sample of irradiated NH_3 are dynamically polarized by microwave pumping at a temperature of 1 K and in a longitudinal magnetic field of 5 T. The

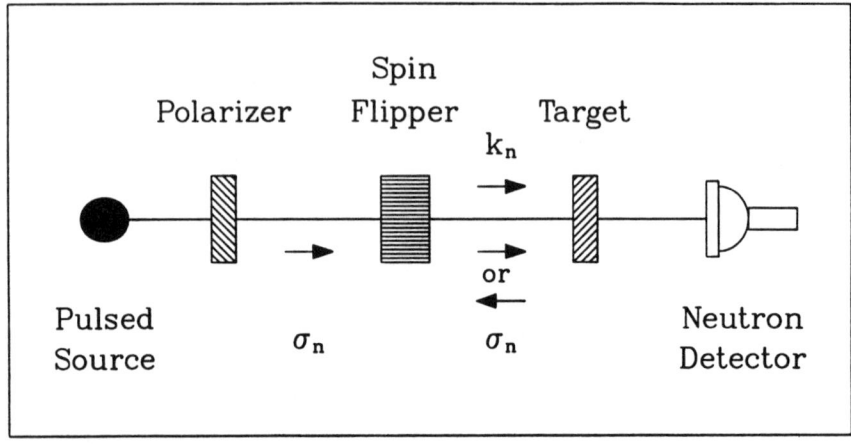

FIGURE 1. Schematic view of the apparatus in the PV experiment. Neutrons from the pulsed source pass through a polarizer, a cell of longitudinally polarized protons, where one helicity state is filtered out leaving a beam of longitudinally polarized neutrons. The neutron spin can be reversed by the spin flipper, a system of magnetic fields. After passing through the target, the neutrons are detected by a neutron detector. The neutron energy is measured by the time-of-flight between the pulsed source and the detector.

cross section for neutrons with spin parallel to the proton spin is much smaller than for neutrons with spin anti-parallel. The neutrons with anti-parallel spin are filtered out, producing a 70% polarized neutron beam for neutron energies up to ~50 keV.

It is necessary to reverse the neutron polarization over a wide range of neutron energies in order to study many resonances simultaneously. This is accomplished by a "spin flipper," (19) a system of magnetic fields located after the polarizer. The spin flipper has a solenoidal field along the beam direction that blends into the stray field of the polarizer. This 0.01-T longitudinal field smoothly decreases, changes direction at the spin-flipper midpoint, and increases to a value opposite in sign and equal in magnitude to the initial field. A longitudinally polarized neutron near the spin-flipper cylindrical axis experiences no torque and passes through the spin flipper with its spin direction unchanged. In order to reverse the neutron spin direction, a transverse field is applied. This field smoothly builds up and decreases over a one-meter transition length where the longitudinal field reverses direction. The moving neutron experiences a field that is initially parallel to its velocity and then rotates slowly to be anti-parallel over a distance of one meter. Since the Larmor frequency of the neutron spin is large compared to the rate at which the field direction rotates, the neutron spin follows the field direction adiabatically and is reversed. The neutron spin direction is controlled by turning the transverse field on or off. Calculations (19) show that the neutron spin is efficiently (better than 88%) reversed for neutron energies between 0.1 and 1000 eV. Thick target samples, $n\sigma_T \sim 3$ (n is the areal density) are placed in the downstream part of the spin-flipper.

The neutron detector (20), which is located 56 m from the spallation source, consists of an array of 55 ^{10}B-loaded liquid scintillator cells arranged in a hexagonal pattern. A neutron is captured by ^{10}B in the reaction n + ^{10}B → ^{7}Li + ^{4}He + 2.79 MeV which produces scintillation light detected by the photomultipliers. The neutron-boron capture cross-section is inversely proportional to the neutron velocity. The neutrons are first slowed down by elastic collisions with the hydrogen in the mineral oil base of the scintillator. When the neutron energy decreases, the chance for neutron-boron capture reaction increases. The hydrogen serves to confine the neutrons within the volume of the scintillator and increase the probability of the neutron-boron capture. An efficiency better than 90% was observed for neutron energies up to 1 keV. The measured efficiency is consistent with a Monte Carlo prediction.

The data acquisition system is operated in a hybrid digital-current mode. The 55 signals from the photomultiplier tubes viewing the scintillator cells are discriminated and the digital pulses sent to the data acquisition system located 150 m away. The digital signals are shaped and added together, and the analog sum is reconverted to digital signals by a transient digitizer. This unit periodically samples the combined analog signals and encodes the digitized signal into a 12-bit word. The resulting sequences of digitized signals are added into a 8192-channel summation memory located in CAMAC. The memory is read out by the data acquisition computer after every 200 beam pulses have been accumulated. The combined instantaneous rate of the 55 detectors is as large as 500 MHz.

Systematic Errors

Several techniques are used to reduce noise and systematic errors. One sixtieth of a second after the data from each beam pulse are encoded by the transient digitizer, the data from a beam-off time interval are encoded and then subtracted from the data in the summation memory. This procedure reduces the size of the 60-Hz noise by two orders of magnitude.

The neutron spin is reversed every ten seconds in the eight-step spin-flipper sequence NRRNRNNR, which eliminates drifts up to second order (16). For no reversal, N, the spin-flipper transverse field is off. For reversal, R, the transverse field direction is alternated so that, on the average, the stray transverse field is zero. This alternation of the transverse field direction eliminates the effects of the stray transverse field in first order. The spin direction is reversed independent of the spin flipper by reversal of the polarization direction of the protons in the polarizer. This is done by tuning the microwave pumping frequency to a different transition. The combination of spin-flipper and polarizer reversal eliminates any systematic errors from either reversal method on the detector in first order. The leading systematic effect from the combined reversal method is the product of the first-order effects in both. The first-order effect from the spin-flipper reversal is estimated to be less than 10^{-10} and the systematic effect from changing the microwave pumping frequency is thought to be zero.

Systematic errors were shown to be less than a few times 10^{-5} from *in situ* measurements on s-wave impurity resonances, while measured longitudinal asymmetries of p-wave resonances were typically 10^{-3} to 10^{-1}. Thus, we believe that the systematic errors are negligible compared to other errors or the size of non-zero asymmetries in the results reported below.

Recent TRIPLE Experimental Results

In 1993, the TRIPLE collaboration measured PNC longitudinal asymmetries in the following targets, ^{238}U, ^{232}Th, ^{127}I, ^{115}In, ^{113}Cd, and natural Ag (^{107}Ag and ^{109}Ag). The data are still under analysis, but many new PNC asymmetries have been observed. The time-of-flight spectrum of ^{238}U near the neutron energy of 65 eV is shown in Fig. 2. The CN resonances appear as dips in

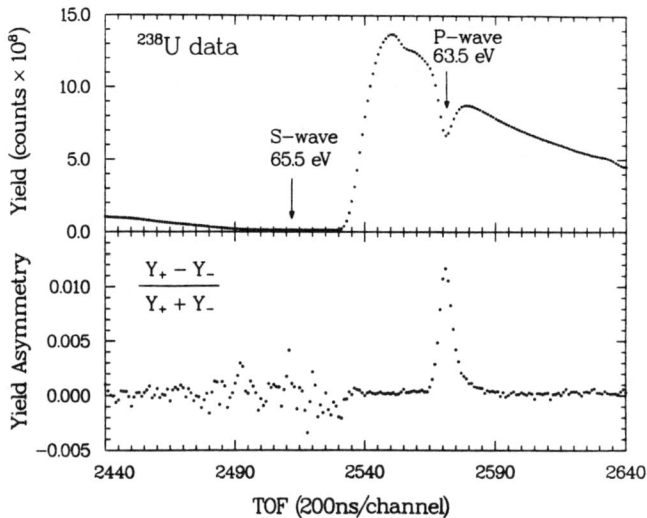

FIGURE 2. The yield and yield-asymmetry spectra of ^{238}U near the 65-eV region. A non-zero PNC asymmetry appears near the $p_{1/2}$ resonance at 63.5 eV while the asymmetry elsewhere is consistent with zero. The large fluctuation of the asymmetry near the $s_{1/2}$ resonance region is due to a lack of neutron counts caused by the large cross-section.

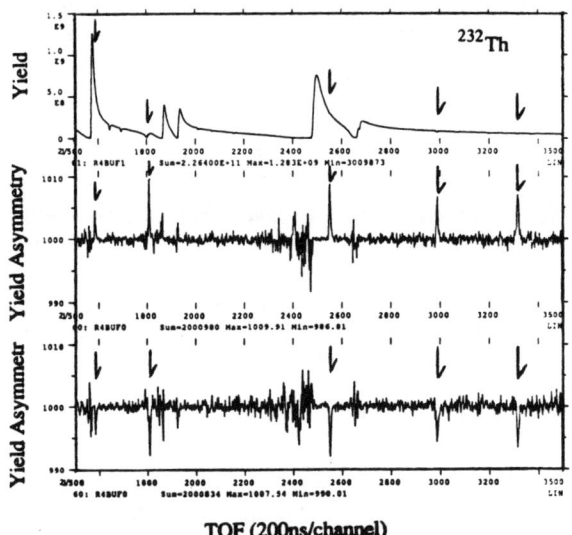

FIGURE 3. Multiple non-zero PNC asymmetries in ^{232}Th p-wave resonances (indicated by arrows). The yield spectrum is on the top. Yield asymmetries for opposite polarization directions of the polarizer are in the middle and bottom plots.

the spectrum because we measure neutron transmission. At 63.5 eV, there exists a small $p_{1/2}$ resonance which is nearly degenerate with a large $s_{1/2}$ resonance at 65.5 eV. A large PV asymmetry is observed. The yield asymmetry shown in Fig. 2 near the $p_{1/2}$ resonance at 63.5 eV has a statistical accuracy better than 1%. Near the $s_{1/2}$ resonance region, the asymmetry has a large fluctuation due to a lack of neutron counts caused by the large cross-section. The yield and asymmetry spectra of ^{232}Th are presented in Fig. 3. Multiple non-zero asymmetries were observed in an individual nucleus. The asymmetries change sign (between the middle and bottom plots of Fig. 3) when the polarization direction of the polarizer is reversed. The signal-to-noise ratios of the PNC asymmetries are large and the PNC asymmetries are apparent in the data.

The preliminary PNC asymmetries from the 1993 data are presented in Fig. 4. For all of the target nuclei, except for ^{113}Cd, asymmetries were obtained by measuring the neutron transmission. The ^{113}Cd data were obtained by measuring the capture-γ rays using BaF$_2$ scintillators (21). This detector system measures the total cross-section because the p-wave resonances decay almost completely through the (n, γ) channel. The data are plotted as the asymmetry multiplied by $\sqrt{E}$ in order to correct for the energy dependence of the angular momentum barrier. The error bars represent the quadratic sum of the statistical error and the fitting error. The crosses are the statistically significant ($P/\Delta P$(statistical) > 3) non-zero asymmetries. It is clear that significant non-zero asymmetries are observed in all of the nuclei. Parity non-conservation appears to be a universal phenomenon in the compound nucleus.

Interpretation of Data

The PNC asymmetry for the ith level, P_i, may be written as a perturbation series (22,23)

$$P_i = \sum_j \left[\frac{2}{E_{s_j} - E_{p_i}} \sqrt{\frac{\Gamma^n_{s_j}(E_{p_i})}{\Gamma^n_{p_i}(E_{p_i})}} \right] \langle \psi_{s_j} | H_w | \psi_{p_i} \rangle \qquad (6)$$

$$\equiv \sum_j A_{ij} \cdot V_{ij} ,$$

where E_i and E_j are the energies of the p-wave resonance and the admixed s-wave resonances, $V_{ij} \equiv \langle \psi_{s_j} | H_w | \psi_{p_i} \rangle$ is the matrix element of the PNC interaction between states i and j, and $\Gamma^n_{p_i}$ and $\Gamma^n_{s_j}$ are their neutron widths. The parameter $A_{ij} \equiv 2\sqrt{\Gamma^n_{s_j}/\Gamma^n_{p_i}}/(E_{s_j} - E_{p_i})$ is a function of resonance energies and neutron widths, known experimentally. According to the statistical model of the CN, the

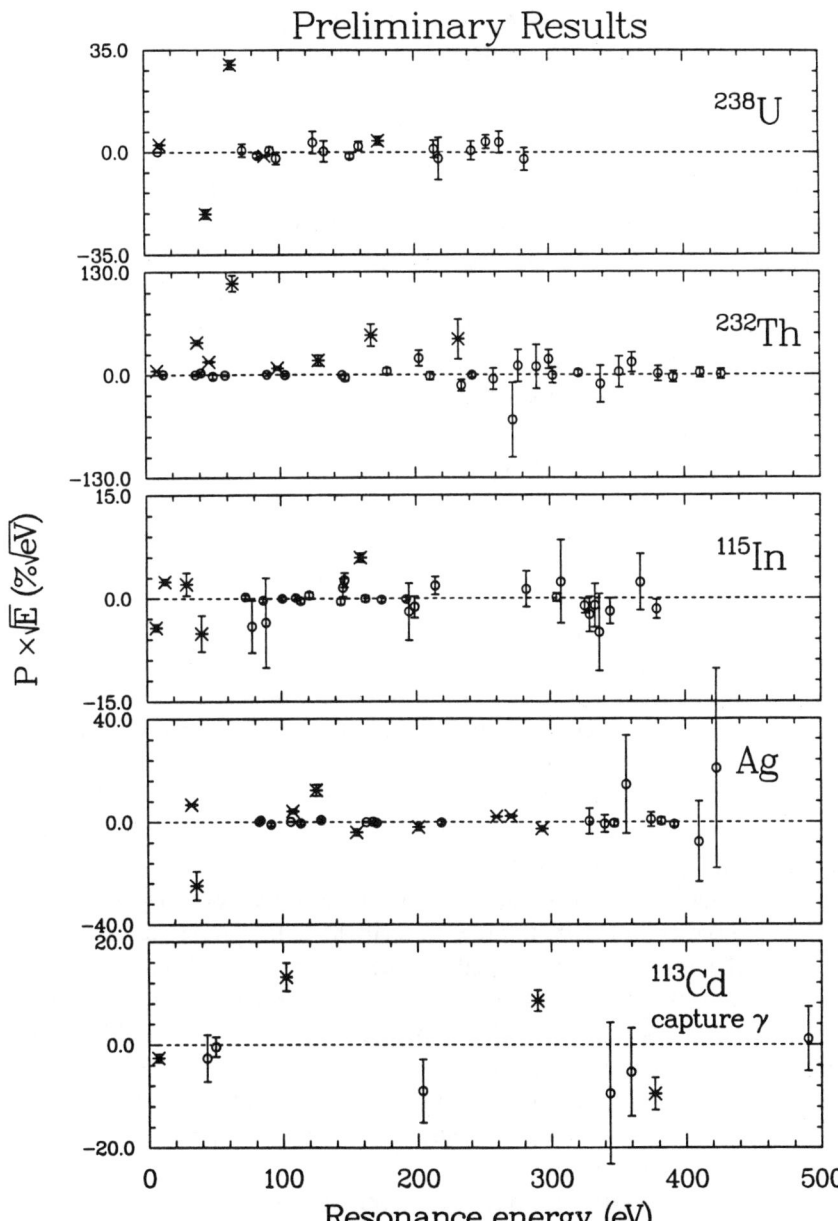

FIGURE 4. The preliminary result of PNC asymmetries multiplied by $\sqrt{E}$. The statistically significant ($P/dP(\text{statistical}) > 3$) non-zero asymmetries are labeled by crosses and the other data points by circles.

neutron-decay amplitudes and the mixing matrix elements are mean-zero Gaussian random variables. This implies that the measured asymmetries are also mean-zero with variance

$$\langle P_i^2 \rangle = A_i^2 M^2. \quad (7)$$

Here $M^2 = \langle V_{ij}^2 \rangle = \langle \psi_{s_j} | H_w | \psi_{p_i} \rangle^2$ and $A_i^2 = \sum A_{ij}^2$. We have done a likelihood analysis to determine the value of M^2 by including the experimental uncertainties and the unknown spin assignments.

As expected, the measured asymmetries (in Fig. 4) have large fluctuations and both positive and negative asymmetries are observed for all target nuclei, except for ^{232}Th. There is strong evidence that the asymmetries for ^{232}Th are not random and have a non-zero average. Theoretical models (24,25) are being developed to explain the non-zero average. Basically, the asymmetries in ^{232}Th can be expressed as two terms (24),

$$P_i = \sum_j \left[\frac{2\langle \psi_{s_j} | H_w | \psi_{p_i} \rangle}{E_{s_j} - E_{p_i}} \sqrt{\frac{\Gamma_{s_j}^n(E_{p_i})}{\Gamma_{p_i}^n(E_{p_i})}} \right] + B\sqrt{\frac{1\text{eV}}{E_i}}. \quad (8)$$

The first term fluctuates about zero with random signs. The second term is the non-zero average arising from correlations between the mixing matrix element, $E_{s_j} - E_{p_i}$, neutron width amplitudes of the p-wave resonance and far away s-wave resonances. A non-zero average of the PNC asymmetries is observed only in the ^{232}Th data (Fig. 4) presumably due to some particular nuclear structure in ^{232}Th (25). A two-dimensional likelihood analysis has been done for the ^{232}Th data in order to extract M and B values. The M value of the ^{232}Th data presented below is deduced from the two-dimensional likelihood analysis.

The M values determined by the likelihood analyses for ^{238}U, ^{232}Th, ^{115}In and ^{113}Cd are shown in Fig. 5. In order to relate M to the properties of the N-N interaction, the results are presented by the square-root of the PNC weak spreading width, $\sqrt{\Gamma^{PNC}} = \sqrt{2\pi M^2/d_J}$. The p-wave resonance level spacing, d_J (~ 10 eV), is roughly 1/N of the typical shell model level spacing (MeV). The mean-squared matrix element is proportional to 1/N of the matrix element squared of the effective N-N interaction. Therefore one expects that Γ^{PNC} is independent of the atomic mass of the compound nucleus. We can combine the data in the same mass region to obtain better statistics (Fig. 5 bottom).

A theoretical model (15) has been developed to relate the M^2 to the π-meson and ρ-meson exchange couplings, F_π and F_ρ, respectively,

$$M^2 = \alpha F_\pi^2 + \beta F_\pi F_\rho + \gamma F_\rho^2. \quad (9)$$

The coefficients α, β and γ are determined using the statistical model of the CN.

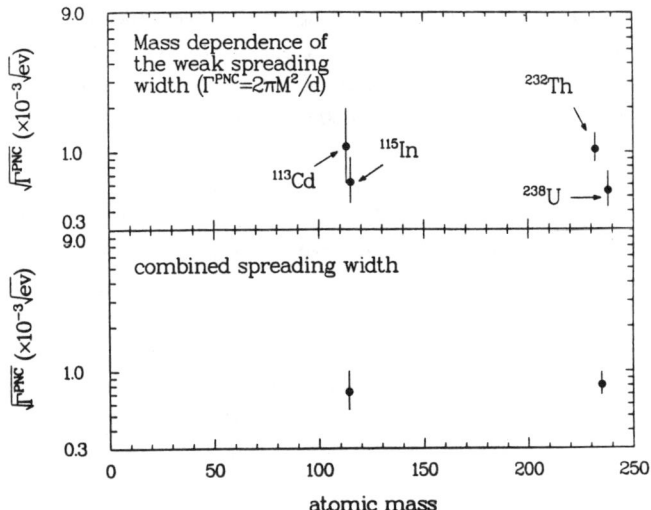

FIGURE 5. The preliminary result of the PNC weak spreading widths for mass A~100 and A~230 regions. The weak spreading width is expected to have a smooth mass dependence. The data near the same mass region are combined and shown at the bottom.

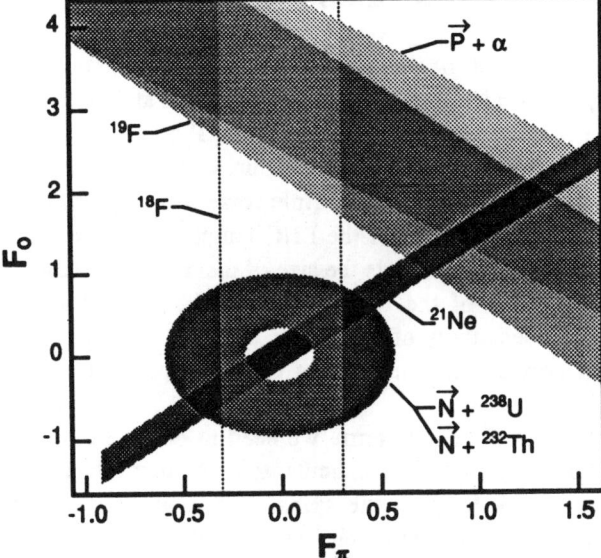

FIGURE 6. Constraints on the isovector π and isoscalar ρ weak meson-nucleon coupling constants, F_π and F_ρ, respectively, imposed by various experiments. (2,15)

Using the experimental value of M and the above relation, our results give constraints on F_π and F_ρ. From the M values at A~100 and A~230 regions, we can learn about the mass-dependence of the constraints and check the theory. Figure 6 shows an example of the constraints (the "donut" shape) calculated from the previous ^{238}U and ^{232}Th data, which have 20-fold worse statistical accuracy than the recent data. The range of values of the F_π and F_ρ predicted by the DDH theory (1) covers everything on the plot. The straight lines are the constraints given by PNC experiments in the few-body ($\bar{p} + \alpha$) and light-nuclei (^{18}F, ^{19}F and ^{21}Ne) systems. Among those experiments, the ^{18}F system has the most complete information. It indicates that F_π is small but the experiment is insensitive to F_ρ. The uncertainty of the constraints given by the ^{21}Ne experiment has not been included in this plot since it is believed to be larger than the uncertainties of the other light-nuclei experiments. With the new PNC data from this work and newly developed theory (26), we expect a larger donut that possibly covers the overlap region of the $\bar{p} + \alpha$ and ^{19}F predictions. The most exciting discovery of this work is that the very different theoretical frameworks for understanding the light-nuclei and compound nucleus PNC experiments agree qualitatively.

TEST OF TIME-REVERSAL INVARIANCE IN NEUTRON-NUCLEUS INTERACTION

A sensitive test of time-reversal invariance (TRI) can be carried out by observing the dependence of the total cross section at a p-wave resonance on the scalar triple product $\vec{\sigma} \cdot (\vec{k}_n \times \vec{I})$, which is the P-odd T-odd term in the neutron forward scattering amplitude (Eq. (3)). Bunakov and Gudkov (27) have argued that the enhancement factor for the triple correlation asymmetry should be of the same order of magnitude as for the PNC longitudinal asymmetry. Their studies indicated that for certain models the size of the triple correlation asymmetry may be as large as 10^{-3} of the PNC longitudinal asymmetry. Herczeg (28) has estimated that a sensitivity of 10^{-3} in the $\vec{\sigma} \cdot (\vec{k}_n \times \vec{I})$ correlation is comparable to the TRI test provided by the neutron electric dipole moment (EDM) of 10^{-25} $e \cdot$cm.

To test the $\vec{\sigma} \cdot (\vec{k}_n \times \vec{I})$ term, we need to orient the neutron spin, neutron momentum and target spin perpendicular to each other. The sign of the triple term can be changed by flipping the spins. The experimental apparatus (Fig. 7) includes a polarizer, upstream spin-rotator, polarized target, downstream spin-rotator, analyzer, and neutron detector. The addition of the analyzer eliminates a false asymmetry resulting from the final state interactions and therefore makes the experiment manifestly time-reversal symmetric. The target is polarized perpendicular to the reaction plane. The polarizer and analyzer have the same

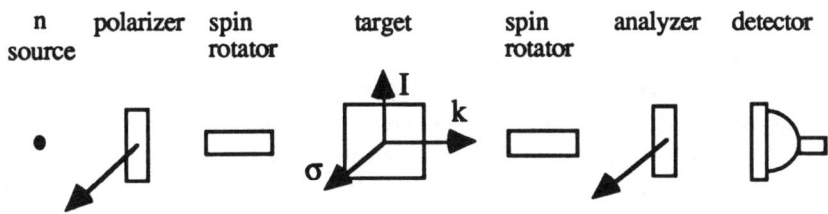

FIGURE 7. An experimental apparatus for the time-reversal invariance (TRI) test. The TRI asymmetry is measured by changing the sign of the triple correlation term, $\vec{\sigma} \cdot (\vec{k}_n \times \vec{I})$.

polarization directions which are perpendicular to both the neutron momentum and target polarization. The time-reversal condition can be accomplished by simultaneously flipping the spin-directions of the polarizer and analyzer. The TRI asymmetry will be measured at the 0.734-eV p-wave resonance in ^{139}La, where a large PNC asymmetry of 0.0955 ± 0.0035 has been observed (29).

Polarized ^{139}La targets have been developed by our collaborators at Kyoto University and KEK (30,31). A ^{139}La polarization of 20% has been achieved with a 5-cm^3 LaAlO$_3$ crystal polarized by the dynamic nuclear polarization technique in a magnetic field of 2.3 T and temperature of 1.5 K. The ^{139}La polarization as large as 70% may be obtained by using a dilution refrigerator for cooling. Two polarized ^{3}He systems will be built for the polarizer and analyzer. The ^{3}He polarization of 70% has been achieved recently at Los Alamos by optical pumping with high-power diode laser arrays (32). The time-reversal modulation can be done with the upstream and downstream spin-rotators (in seconds), with the polarizer and analyzer adiabatic-fast-passage (in hours) or with the ^{139}La polarization direction (in days). Based on the sensitivity of the PNC asymmetry measurements, it is estimated that in 2×10^6 s, a statistical sensitivity of about 10^{-3} in the ratio of $\langle \vec{\sigma} \cdot (\vec{k}_n \times \vec{I}) \rangle / \langle \vec{\sigma} \cdot \vec{k}_n \rangle$ can be achieved. This sensitivity level is comparable to the sensitivity of the neutron electric dipole moment measurements (28,33).

The triple correlation term can also be tested by measuring the asymmetry of neutron spin rotation through a polarized target. An experimental apparatus has been developed (34) to do such a measurement on a polarized LaAlO$_3$ target. The sensitivity in the ratio of $\langle \vec{\sigma} \cdot (\vec{k}_n \times \vec{I}) \rangle / \langle \vec{\sigma} \cdot \vec{k}_n \rangle$ from this experiment is expected (35) to be 3×10^{-4}, which is about one order of magnitude better than the neutron EDM measurement.

A test of the P-even and T-odd five-fold correlation term, $\vec{\sigma} \cdot (\vec{k}_n \times \vec{I})(\vec{k}_n \cdot \vec{I})$, has been carried out (36) using fast (6.7 MeV) neutrons. The TRI asymmetry is measured by polarized neutron transmission through a rotating, cryogenically aligned ^{165}Ho target. The P-even T-odd spin correlation coefficient, $A_5 = (1.1 \pm 2.3) \times 10^{-5}$, obtained in this measurement is consistent with time-reversal invariance. This sets a bound of 3.8×10^{-3} (99% confidence) on α_T, the ratio of T-odd to T-even couplings in the effective nucleon-nucleon interaction.

SUMMARY

Parity non-conservation (PNC) phenomena have been studied in compound nuclei in mass regions of A~100 and A~230. The chaotic nature of the compound nucleus results in a very large enhancement of the PNC asymmetries and allows the data to be interpreted using the statistical model of the compound nucleus. The mean-squared PNC mixing matrix element (M^2) was extracted from a set of PNC longitudinal asymmetry data of many p-wave resonances using the likelihood technique. Applying the statistical model of the compound nucleus, the M^2 can be related to the π and ρ meson-exchange couplings (F_π and F_ρ, respectively). With the PNC longitudinal asymmetry data in compound nuclei, constraints on F_π and F_ρ are calculated. The results agree qualitatively with the constraints given by PNC studies in the few-body and light-nuclei systems.

The time-reversal observables in the compound nucleus can have the same large enhancement as the PNC observables. Experiments to test the time-reversal invariance (TRI) in neutron-nucleus interactions are being developed. The triple correlation P-odd T-odd term can be tested by measuring the asymmetry of the compound nucleus cross-sections when the sign of the triple term is changed by flipping neutron or target spins. The triple term can also be tested by measuring the asymmetry of the neutron spin rotation through a polarized target. For these experiments, a polarized ^{139}La target was chosen because a large PNC asymmetry has been observed at the 0.734 eV p-wave resonance. With high energy (MeV) neutrons, the five-fold correlation P-even T-odd term has been tested by measuring the neutron transmission through a rotating aligned ^{165}Ho target. This experiment sets a bound of 3.8×10^{-3} on the ratio of T-odd to T-even couplings in the effective nucleon-nucleon interaction.

ACKNOWLEDGMENTS

This work was supported in part by the U.S. Department of Energy, Office of High Energy and Nuclear Physics, under grants No. DE-FG05-88-ER40441 and DE-FG05-91-ER40619, and by the U.S. Department of Energy, Office of Energy Research, under contract No. W-7405-ENG-36.

REFERENCES

1. B. Desplanques, J. F. Donoghue, and B. R. Holstein, *Ann. Phys.* **124**, 449 (1980).
2. E. G. Adelberger and W. C. Haxton, *Ann. Rev. Nucl. Part. Sci.* **35**, 501 (1985).
3. D. E. Nagle et al., *Proc. 3rd Int. Symp. on High Energy Physics with Polarized Beams and Polarized Targets*, Argonne, 1978, *AIP Conf. Proc.* **51**, 24 (1979).
4. R. Balzer et al., *Phys. Rev. C* **30**, 1409 (1984).
5. V. W. Yuan et al., *Phys. Rev. Lett.* **57**, 1680 (1986).
6. J. Lang et al., *Phys. Rev. Lett.* **54**, 170 (1985).
7. B. R. Heckel, University of Washington, private communication (1994).
8. J. D. Bowman, Los Alamos National Laboratory, private communication (1994).
9. C. A. Barnes et al., *Phys. Rev. Lett.* **40**, 840 (1978).
10. P. G. Bizetti et al., *Lett. Nuovo Cimento* **29**, 167 (1980).
11. G. Ahrens et al., *Nucl. Phys. A* **390**, 486 (1982).
12. Yu. G. Abov, P. A. Krupchitsky, and Yu. A. Oratovsky, *Phys. Lett.* **12**, 25 (1964).
13. J. D. Bowman, G. T. Garvey, M. B. Johnson, and G. E. Mitchell, *Ann. Rev. Nucl. Part. Sci.* **43**, 829 (1993).
14. J. D. Bowman et al., in *Proceedings of the Second International Workshop on Time Reversal Invariance and Parity Violation in Neutron Reactions*, ed. C. R. Gould, J. D. Bowman and Yu. P. Popov (World Scientific, Singapore, 1993), pp. 8-29.
15. M. B. Johnson, J. D. Bowman, and S. H. Yoo, *Phys. Rev. Lett.* **67**, 310 (1991).
16. N. R. Roberson et al., *Nucl. Inst. and Meth.* **A326**, 549 (1993).
17. J. J. Szymanski et al., to be published in *Nucl. Inst. and Meth.* (1994).
18. S. I. Penttilä et al., in these proceedings.
19. J. D. Bowman et al., to be submitted to *Nucl. Inst. and Meth.* (1994).
20. Yi-Fen Yen et al., in *Proceedings of the Second International Workshop on Time Reversal Invariance and Parity Violation in Neutron Reactions*, ed. C. R. Gould, J. D. Bowman and Yu. P. Popov (World Scientific, Singapore, 1993), pp. 210-219.
21. P. E. Koehler, Oak Ridge National Laboratory, private communication (1993).
22. V. P. Alfimenkov et al., *Nucl. Phys. A* **398**, 93 (1983).
23. J. R. Vanhoy et al., *Z. Phys. A* **333**, 229 (1988).
24. J. D. Bowman et al., *Phys. Rev. Lett.* **68**, 780 (1992).
25. N. Auerbach, J. D. Bowman, and V. Spevak, submitted to *Phys. Rev. Lett.* (1994).
26. M. B. Johnson, Los Alamos National Laboratory, private communication (1994).
27. V. E. Bunakov and V. P. Gudkov, *JETP Lett.* **36**, 329 (1982) and *Nucl. Phys.* **A401**, 93 (1983).
28. P. Herczeg, in *Proceedings of the Aqueduct Conference Center Workshop on Tests of Time Reversal Invariance in Neutron Physics*, ed. N. R. Roberson, C. R. Gould, and J. D. Bowman (World Scientific, Singapore, 1987), pp. 24-53.
29. V. W. Yuan et al., *Phys. Rev. C* **44**, 2187 (1991).
30. Y. Masuda et al., *Nucl. Phys.* **A504**, 269 (1989).
31. T. Adachi et al., *Nucl. Phys* **A577**, 433c (1994).
32. W. J. Cummings et al., in these proceedings.
33. V. P. Gudkov, *Phys. Rep.* **212**, 77 (1992).
34. Y. Masuda, in these proceedings.
35. Y. Masuda, in *Proceedings of the Second International Workshop on Time Reversal Invariance and Parity Violation in Neutron Reactions*, ed. C. R. Gould, J. D. Bowman and Yu. P. Popov (World Scientific, Singapore, 1993), pp. 126-134.
36. P. R. Huffman et al., in these proceedings.

Parity Violation in P-P Scattering at TRIUMF

J. Birchall[a], A.R. Berdoz[a], J.D. Bowman[b], J.R. Campbell[a], C.A. Davis[d], A.A. Green[a], P.W. Green[c], A.A. Hamian[a], D.C. Healey[d], R. Helmer[d], E. Korkmaz[c], L.R. Lee[a], C.D.P. Levy[d], R.E. Mischke[b], S.A. Page[a], W.D. Ramsay[a], S.D. Reitzner[a], G. Roy[c], P.W. Schmor[d], A.M. Sekulovich[a], J. Soukup[c], G.M. Stinson[c], T. Stocki[c], V. Sum[a], N.A. Titov[e], W.T.H. van Oers[a], A.N. Zelenskii[e]

[a] Dept. of Physics, University of Manitoba, Winnipeg, Manitoba, Canada R3T 2N2
[b] Los Alamos National Laboratory, Los Alamos, New Mexico, 87545, USA
[c] Dept. of Physics, University of Alberta, Edmonton, Alberta, Canada T6G 2N5
[d] TRIUMF, 4004 Wesbrook Mall, Vancouver, British Columbia, Canada V6T 2A3
[e] Institute for Nuclear Research, Russian Academy of Sciences, Moscow SU 117312

The weak nucleon-nucleon interaction has been investigated through a number of extremely precise measurements of the parity-violating longitudinal analyzing power $A_z = (\sigma^+ - \sigma^-)/(\sigma^+ + \sigma^-)$, where σ^+ and σ^- are the scattering cross-sections for + and − beam helicity. The quantity A_z can be decomposed in a model independent way into partial wave components, the angular distributions of which are determined by strong interaction phase shifts. Below 50-100 MeV, the dominant component is $A_z(^1S_0\text{-}^3P_0)$, which is due to mixing of S and P partial waves. Existing proton-proton measurements are shown in figure 1. The most precise measurements so far have been performed at 13.6 MeV[1] and 45 MeV[2]. A further experiment (TRIUMF experiment 497) is currently in preparation to measure A_z to similar precision ($\pm 2 \times 10^{-8}$) at 223 MeV where $A_z(^1S_0\text{-}^3P_0)$ averages to zero over the geometrical acceptance of the detector, thus isolating the $^3P_2\text{-}^1D_2$ component.

At low and intermediate energies, the weak nucleon-nucleon interaction is often described in terms of a meson exchange model (π, ρ, ω) involving one weak and one strong vertex. The strong meson-nucleon couplings are taken from a conventional meson exchange description of the strong nucleon-nucleon

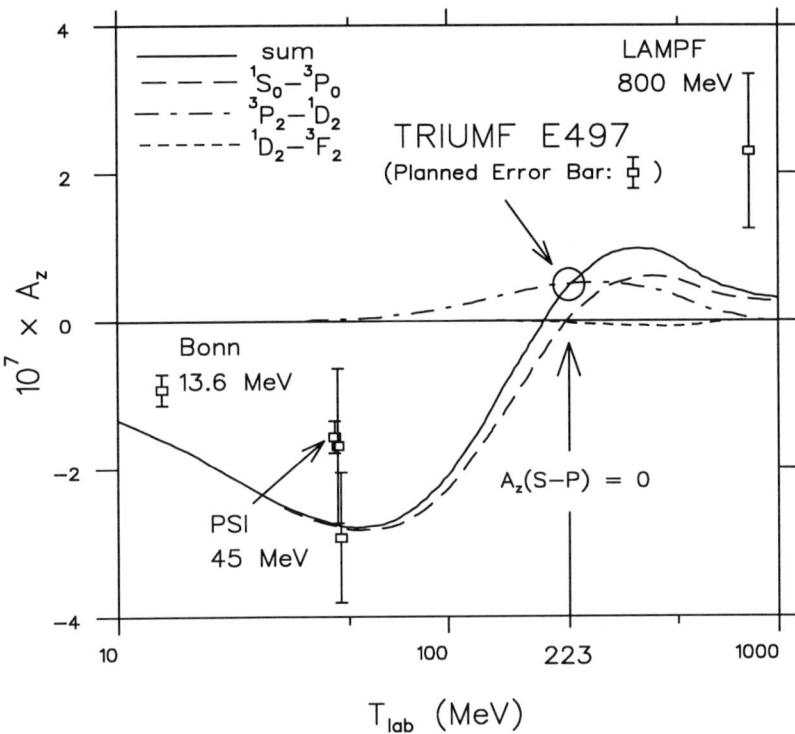

FIGURE 1: Parity-violating analyzing power as a function of energy. The calculations are by Driscoll and Miller (ref. 5) using a meson exchange model of the weak nucleon-nucleon interaction.

interaction (e.g. the Bonn potential). The weak couplings are predicted from the Weinberg-Salam model involving W and Z exchanges between quark constituents of the nucleon and meson. In the simplest version of this model where only single meson exchanges are allowed, there are two constraints that can be obtained from p-p scattering experiments. These determine effective ρ and ω weak couplings summed over isospin: $h_\rho^{pp} = (h_\rho^0 + h_\rho^1 + h_\rho^2/\sqrt{6})$ and $h_\omega^{pp} = (h_\omega^0 + h_\omega^1)$. The standard reference calculations of these constants by Desplanques, Donoghue and Holstein[3] (DDH) and by Feldman, Crawford, Dubach and Holstein[4] carry large uncertainties associated with the strong interaction. Extensions to the simple meson exchange model have been considered via the inclusion of multiple meson exchanges, Δ-isobar excitations, the exchange of parity-mixed meson states and the explicit inclusion of quark degrees of freedom.

In the context of the meson exchange model, $A_z(^1S_0$-$^3P_0)$ measured at low energy is sensitive to both h_ρ^{pp} and h_ω^{pp}, while $A_z(^3P_2$-$^1D_2)$ to be measured at

TRIUMF at 223 MeV is sensitive to h_p^{pp} alone. The measurement of A_z to the planned precision of $\pm 2 \times 10^{-8}$ will determine h_p^{pp} to $\pm 25\%$ if h_p^{pp} has the DDH predicted value. When the results of all possible calculations are taken into account, the theoretical value of A_z lies in the range $(0.5 - 2.4) \times 10^{-7}$ at 223 MeV.

In the TRIUMF 223 MeV measurement, A_z will be extracted from the helicity dependence of the proton scattering from a liquid hydrogen target by measuring the ratio of signals in transverse field ionization chambers (TRICs) positioned up- and downstream of the target. The TRICs, operated in current mode to cope with the large flux of protons, are designed for low noise, high linearity and minimum sensitivity to beam size and position. Low noise, by the use of a wide aperture and by snouts to range out spallation fragments from entrance and exit windows; high linearity and insensitivity to beam size and position through careful control of space charge effects and use of field shaping electrodes.

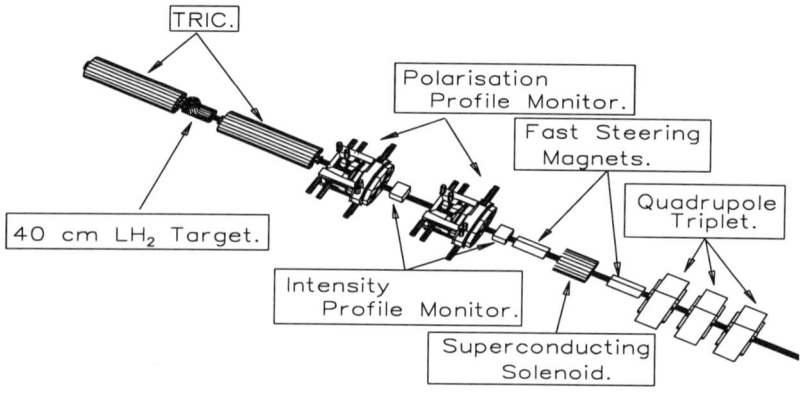

FIGURE 2: The parity beamline at TRIUMF

The parity beamline, which was commissioned at TRIUMF in August 1994, is shown in figure 2. Extensive diagnostic equipment has been developed to measure beam properties and control systematic errors: intensity and polarization profile monitors to measure the distribution of current and polarization within the beamspot; a fast-acting servo system to control beam direction and beam position on the scale of a few microns. The superconducting solenoid shown in the diagram is to induce a small transverse polarization in the beam for test measurements. Upstream of the apparatus shown in figure 1 is a combination S-D-S-D of longitudinal field solenoids (S) and magnetic dipole magnets (D) which together allow the production of longitudinally polarized beam on the parity beamline throughout the energy range of the TRIUMF cyclotron. Note that S-D alone would produce longitudinally polarized beam at only one energy for a given beamline.

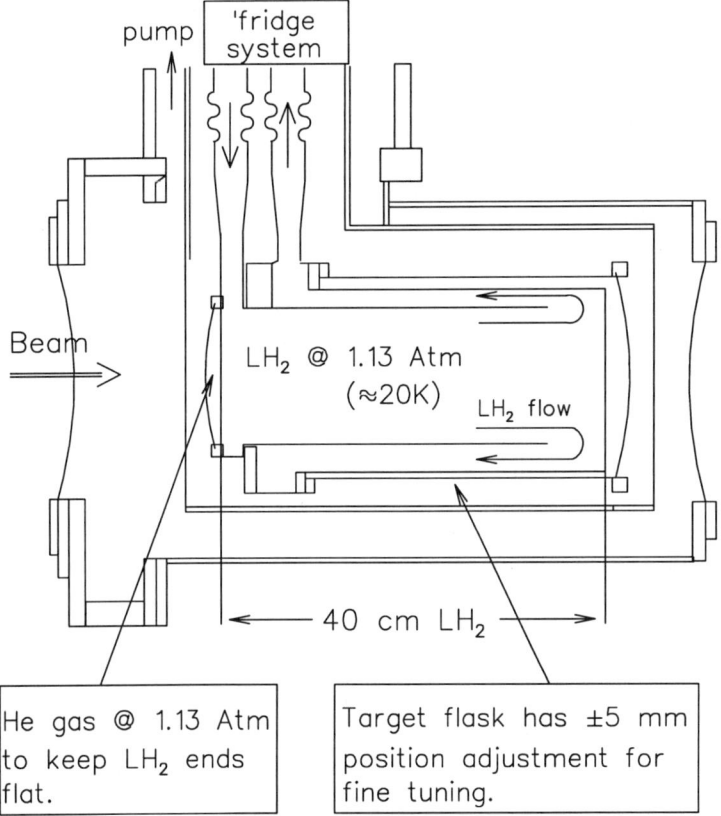

FIGURE 3: The liquid hydrogen target

The LH$_2$ target, shown in figure 3, was designed for rapid axial flow of the liquid hydrogen to avoid bubbling and maximize cylindrical symmetry. An innovative feature is the helium gas end caps which keep flat the inner windows which contain the liquid hydrogen. Flat windows are very important for reducing false parity signals due to beam movement as the effective thickness of the liquid hydrogen becomes independent of beam position on target. The target was installed on the beamline in August 1994.

The intensity profile monitors (IPMs) are dual function secondary electron emission devices. One component consists of split foils to measure beam current at either side of the split. Two of these split foil devices provide information on beam position (current median) in x and y at two points along the beamline. Beam position signals derived from the split foils are fed to a servo system based on ferrite core steering magnets that clamp beam position at

TABLE 1. Beam Property Requirements

Beam Property	Nominal DC Value	Helicity Correlated
I	500 nA	$\frac{\delta I}{I} \leq 10^{-5}$
E (LH$_2$ entrance)	222.4 MeV	≤ 0.5 eV
x, y	0.	$\leq 5\mu$m
σ_x, σ_y	5 mm	$\leq 0.05\mu$m
P_x, P_y	0.	≤ 0.002
$<xP_y>, <yP_x>$	0.	$\leq 0.2\mu$m

two points to within $\pm 5\mu$m. The servo system is effective at frequencies up to 1000 Hz. The second part of each intensity profile monitor consists of foil strips to measure in detail the distribution of current within the beamspot.

Because large false parity signals can be produced if the beam has a net transverse polarization or a non-zero polarization moment, polarization profile monitors (PPMs) are used. They contain thin CH$_2$ rods that are scanned at constant speed through the beam left-right and up-down. At each point, the asymmetry in the elastic scattering of protons from the rods is measured by sets of detectors, left-right and up-down. The PPMs determine the net transverse polarization and polarization moments of the beam in both x and y. A special feature is built into the polarimeter detection system that was also used for the $n-p$ charge symmetry experiments at TRIUMF – the solid-angle defining collimators are rotated to minimize false asymmetries all of the way across the beamspot. The idea is to compensate the angular variation of the $p-p$ scattering cross-section at points away from the axis of the polarimeters with an equal and opposite change in detector solid angle.

Monte Carlo simulations have been made of the parity detection system. Table 1 shows the resulting requirements for helicity-correlated beam properties. Control measurements and data-taking are expected to continue for two years.

Notes on Beam Property Requirements

1. The requirement on beam current comes from predictions of TRIC non-linearity. Much work has been done on the optically pumped polarized ion source (OPPIS). The helicity-correlated change in beam current is $\leq (0.9 \pm 0.8) \times 10^{-5}$ if the rubidium polarizations in the two spin states are matched to $\pm 0.5\%$. The next step will be to measure the false parity signal as seen by the equipment as a function of changes of beam current.
2. The requirement on beam energy modulation comes from coupling of energy modulation to emittance modulation from the cyclotron. The energy of the ion beam from the ECR source is stable to (0.08 ± 0.02) eV. The emittance coupling will be checked with new beam tunes.

3. Measurements show that there is an optimum setting of extraction voltage from the source that minimizes coherent beam position shifts. Additionally, the servo system locks the position of the beam current median to $\leq \pm 3\mu$m. An important part of the program of test measurements will be to find the "position neutral axis" of the equipment. That is, to find the alignment of TRICs, target and beam that minimizes the effects of beam position modulation.
4. The restriction on beam size modulation comes from beam loss in the downstream TRIC from the tails of the beam multiply-scattered in the LH$_2$ target. It has been found that modulation of beam size can be minimized by tuning the energy of the ion beam from the ECR source – in a way consistent with minimizing beam position effects. Changes of rms beam size coherent with spin flip of $\leq 0.01\mu$m have been achieved.

Parity Measurement at 450 MeV

A second measurement of A_z at TRIUMF is being planned at 450 MeV. As seen from figure 1, A_z reaches a peak near 450 MeV where $A_z(^1S_0\text{-}^3P_0)$ and $A_z(^3P_2\text{-}^1D_2)$ are of comparable size. Theoretical values of A_z lie in the range $(0.9 - 3.0) \times 10^{-7}$ at this energy. A 450 MeV experiment is sensitive to a very different combination of weak coupling constants than previous measurements. The two independent high precision measurements of A_z at TRIUMF will do much to constrain theoretical models of the weak nucleon-nucleon interaction in the intermediate energy range.

An analysis of the 450 MeV experiment suggests that it should be possible to achieve the required statistical error in A_z of $\pm 2 \times 10^{-8}$ in comparable running time to the 223 MeV measurement and to control all systematic errors at least as well with only minor changes to the experimental equipment. Systematic errors produced by transverse polarization components of the beam are expected to be somewhat larger than at 223 MeV, as the parity-allowed analyzing power is larger. The effect is offset, however, by the transverse polarization components being easier to measure, by a larger $p-p$ scattering probability at 450 MeV (5% as opposed to 4%) and decreased multiple scattering in the LH$_2$ target.

References

1. P.D. Eversheim et al., Phys. Lett. **256B**, 11 (1991) and private communication.
2. S. Kistryn et al., Phys. Rev. Lett. **58**, 1616 (1987).
3. B. Desplanques, J.F. Donoghue, and B.R. Holstein, Ann. Phys. (N.Y.) **124**, 449 (1980).
4. G.B. Feldman, G.A. Crawford, J. Dubach and B.R. Holstein, Phys. Rev. **C43**, 863 (1991).
5. D.E. Driscoll and G.A. Miller, Phys. Rev. **C39**, 1951 (1989).

Theoretical Overview of Parity Violation in pp Scattering

M. Shmatikov

Russian Research Center "Kurchatov Institute", 123182 Moscow, Russia

Abstract. Weak interaction of nucleons and, in particular, parity-violating effects in pp-scattering occur in the region where nonperturbative QCD mechanisms are operative. This makes them a unique tool for investigating the nature of strong interaction in the low-energy region. Various approaches to the decription of P-odd effects are reviewed. Their predictions for the forthcoming TRIUMF and Jülich experiments on measuring longitudinal analyzing power in pp-scattering are discussed.

Strong interaction at low energies and effective degrees of freedom

Parity-violating effects in the NN system result from the combined action of weak and strong coupling of nucleons. With two types of forces in action using the same degrees of freedom in both interaction sectors is mandatory. Weak interaction being a perturbation, the choice of effective degrees of freedom is dictated by strong coupling of hadrons.

"Basic" degrees of freedom in the QCD lagrangian are *current* (i.e. almost massless) quarks and gluons. They are operative, however, at small distances (or large momentum transfers) only. At some distance $r_{\chi SB}$ a phase transition occurs resulting in the spontaneous chiral symmetry breaking (χSB). Restructuring of the QCD vacuum causes remodelling of the strong interaction lagrangian. Now its degrees of freedom are *constituent* quarks, gluons and pions. At even larger distances r_{QCD} one more phase transition occurs: quarks and gluons merge into colorless baryons and mesons.

The nature and mechanisms of these two celebrated nonperturbative phenomena inherent to the QCD have not yet found a detailed theoretical description. At the same time, phenomenological analysis [1] shows that scales of phase transitions differ significantly from each other:

$$0.2 \; fm \approx r_{\chi SB} \ll r_{QCD} \approx 1 \; fm \qquad (1)$$

This inequality implies the existence of a rather broad range of distances where effective degrees of freedom are constituent quarks, pions and gluons.

The amplitude of weak NN scattering can be written generically in the form

$$M_W = \int \psi_f^\dagger \mathcal{L}^{PV} \psi_i \, d\tau, \qquad (2)$$

where $\psi_{i,f}$ are the wave functions of the NN system in the initial and final states, $\mathcal{L}^{PV}$ is the parity-violating weak-interaction lagrangian and $d\tau$ is the element of the phase-space volume. All the components of the amplitude M_W are to be expressed in terms of the effective degrees of freedom pertinent to the low-energy region under consideration. The longest-range part of weak NN coupling stems from the one-pion exchange. Due to isotopic selection rules P-odd exchange by a (neutral) pion is forbidden. Weak forces next in the range hierarchy are generated by a two-pion exchange. Their contribution, calculable without going into details of coupling at small internucleon distances, proves to make a minor fraction of the observed effect. Thus the magnitude of parity-violating effects is saturated by short-range NN coupling and the key question arises: what mechanisms govern strong and weak interactions of nucleons?

Weak interaction of hadrons: nucleons and mesons

Development of ideas concerning weak NN forces follows the evolution of understanding strong-interaction dynamics. Historically, the first approach to the problem of parity violation in the NN system based on the assumption that degrees of freedom which are operative in the low-energy region are nucleons and (π, ρ, ω) mesons. In this vein weak-interaction lagrangian in (2) is presented as a sum of (local) meson-exchange potential. Strong meson-nucleon coupling constants being extracted from OBEP-type NN potentials, the problem reduces to calculation of their weak counterparts. Various models applied to this end are catalogued and unified in the review [2]. In spite of many theoretical efforts, values of weak NNM coupling constants are ill-known. It is especially true for the coupling constants of vector mesons to nucleons which scale the value of the longitudinal analyzing power $\mathcal{A}_L$ in the pp-scattering and, as such, they should be considered rather as overall fitting parameters.

Resulting value of $\mathcal{A}_L$ depends also on the NN wave functions in (2) and, hence, on the choice of the strong nucleon potential. The most sophisticated realization of the outlined approach is materialized in [3], where Bonn potential was employed. Obtained results reproduce with reasonable accuracy experimental data on the $\mathcal{A}_L$ value at $T_{lab} = 14$ and 45 MeV. Predicted value for $T_{lab} \approx 220$ MeV, which is of interest in view of the forthcoming TRIUMF [4] and Jülich [5] experiments, is rather small making $\mathcal{A}_L \simeq (2 \div 3) \cdot 10^{-8}$.

The value of $\mathcal{A}_L$ may be contributed as well by intermediate-state inelastic processes of which weak $pp \to N\Delta$ is, apparently, the most important one. Parity-violating $NN\pi$ coupling constant is known to be small and even compatible with zero. There are theoretical grounds to expect that the weak $N\Delta\pi$ constant is also small. Then the parity-violating OPE amplitude of the (intermediate-state) Δ-isobar excitation is insignificant and some short-range mechanisms are to be invoked. A mechanism of this kind, suggested in [6], consists in parity-violating mixing of the ρ and A_1 mesons. Required strength constants of the $\Delta N A_1$ and $\Delta N \rho$ strong coupling are calculated by relating them, in the framework of the constituent quark model, to corresponding constants of coupling to the nucleon.

Excitation of the Δ-isobar by this mechanism produces noticeable effect ($\approx 30\%$) even at very low energies where experimental data are available. The most dramatic changes occur at the energy of the forthcoming TRIUMF experiment. Here the $\mathcal{A}_L$ value is completely dominated by the inelastic intermediate state and exceeds the contribution of elastic pp-amplitude by an order of magnitude amounting to $\mathcal{A}_L \approx (1 \div 2) \cdot 10^{-7}$.

Weak coupling of nucleons: quark degrees of freedom

Weak interaction of protons occurs in the region of small internucleon distances which is just the domain where effective degrees of freedom in the low-energy strong interaction lagrangian are (constituent) quarks and π-mesons (see (1)). Thus we are prompted to describe weak NN coupling in terms of these fundamental fields. The wave function of the NN pair treated as a $6q$-system reads then

$$\psi = \mathcal{A}\left\{\Phi_N \, \Phi_N \, \eta(\vec{R})\right\}, \tag{3}$$

where $\mathcal{A}$ is the antisymmetrization operator, Φ_N is the wave function of a $3q$-cluster with the quantum numbers of the nucleon, and $\eta(\vec{R})$ is the wave function describing relative motion of two clusters, $\vec{R}$ being the distance between their centers of mass. Each of the $3q$-cluster wave functions is in turn the product of three individual quark wave functions. The wave function $\eta(\vec{R})$ of the cluster relative motion is calculable by means of the resonating group method (RGM).

Self-consistent approach requires the weak interaction lagrangian in (2) to be expressed as well in terms of quark degrees of freedom. Weak coupling of current quarks is described by the celebrated Weinberg-Salam lagrangian:

$$\mathcal{L}^{PV} = \sum_i C_i \mathcal{O}_i. \tag{4}$$

Here C_i are some numerical constants and $\mathcal{O}_i$ are (4-fermion) operators having the current × current structure. Quark evolution from distances $\approx 1/m_W$ up to $r_{\chi SB}$ can be accounted for by the perturbative QCD. It results in some renormalization of C_i factors and emergence of new $\mathcal{O}_i$ operators (in particular, celebrated "penguin" terms). Still $\mathcal{O}_i$ operators preserve their structure as a product of two currents. Thus all the components of the weak-interaction amplitude are calculable in terms of quark degrees of freedom.

Making integration first over internal coordinates of quarks $\vec{r}_1 \ldots \vec{r}_6$ for a given distance $\vec{R}$ between clusters and then over separation between clusters $\vec{R}$ we get the value of $\mathcal{A}_L$ in the quark-model based approach. It would be relevant to emphasize that the obtained value contains no fitting parameters. It reproduces with reasonable accuracy experimental data at $T_{lab} = 14$ and 45 MeV. Predicted value of $\mathcal{A}_L$ at $T_{lab} \approx 220$ MeV makes about $2 \cdot 10^{-7}$ [7], exceeding its counterpart in the paradigm of P-odd meson-exchange potential [3] by about an order of magnitude. Note that the large $\mathcal{A}_L$ value is generated by the elastic pp-scattering only.

Inelastic channels can be readily incorporated in the quark picture of weak interaction substituting one of Φ_N wave functions in (3) by the $3q$-cluster wave function with the quantum numbers of the Δ-isobar. It is reasonable to expect that the contribution of the Δ-isobar excitation in the intermediate state will be comparable in magnitude to that of elastic channel.

The reason for the dramatic difference between results in the quark and meson-exchange pictures of weak NN interaction is as follows. The longitudinal analyzing power in the considered energy range is contributed by two partial waves with $J = 0$ and $J = 2$:

$$\mathcal{A}_L = \mathcal{A}_L^{(0)} + \mathcal{A}_L^{(2)}. \tag{5}$$

At very low energies the $\mathcal{A}_L^{(0)}$ dominates. It reads:

$$\mathcal{A}_L^{(0)} = -2k \frac{F_{J=0}^W \sin(\delta_S + \delta_P)}{\sin^2 \delta_S + \sin^2 \delta_P}, \tag{6}$$

where k is the c.m.s. momentum of colliding protons, $F_{J=0}^{W}$ is the amplitude of weak interaction in the $J=0$ partial wave and $\delta_S(\delta_P)$ is the scattering phase shift in the $^1S_0(^3P_0)$ state. The energy dependence of $\mathcal{A}_L^{(0)}$ is governed mainly by the behavior of the scattering phase shifts. In particular, $\mathcal{A}_L^{(0)}$ vanishes at $T_{lab} \approx 220$ MeV by virtue of vanishing of $\delta_S + \delta_P$ sum. Behavior of scattering phase shifts is predetermined by the strong interaction of nucleons, implying that the behavior of $\mathcal{A}_L^{(0)}$ is predetermined as well. However, this is not the case of the $\mathcal{A}_L^{(2)}$ component, which comes into play at energies $T_{lab} \geq 50$ MeV. This is the place where quark and meson-exchange models differ in the most noticeable way. Configuration which furnishes the dominant contribution to the parity-violating effect in the quark picture corresponds to "touching" or slightly overlapping of nucleons. Indeed, contact quark-quark coupling fades down when the nucleons are separated at distances about twice the nucleon quark core ($b \approx 0.5$ fm). At smaller distances the repulsive core is operative entailing smallness of NN wave functions and the weak amplitude as a whole. Hence, the magnitude of longitudinal analyzing power is saturated by a narrow region of internucleon distances $r_q \approx 1$ fm.

In the model of weak one-boson exchanges the integral (2) is saturated at distances which are controlled by inverse mass of the vector meson, i.e. at $r_m \approx 0.5$ fm. The difference between r_q and r_m proves to be crucial. In the meson-exchange model integrand (2) involves NN wave functions in the region where both repulsive core and centifugal barrier are operative in contrast to the quark picture wherein they are suppressed by the centrifugal-barrier effects only. In the latter case NN wave function grows with the increase of energy much more rapidly resulting in more steep increase of the $\mathcal{A}_L^{(2)}$ component.

Concluding, two alternative approach based on the meson-exchange potential and (mesonless) contact interaction of quarks yield similar results for the longitudinal analyzing power $\mathcal{A}_L$. Both of them predict the value of $\mathcal{A}_L$ at the energy of the forthcoming TRIUMF ($T_{lab} \approx 220$ MeV) and Jülich experiments about $2 \cdot 10^{-7}$. However, the mechanisms underlying large magnitude of the P-odd effect prove to be quite different. In the model of weak one-boson exchanges elastic pp-channel is of minor importance, while the main contribution comes from the intermediate-state excitation of the Δ-isobar. Quark-model based approach yields large $\mathcal{A}_L$ value already for the elastic channel. With the allowance of intermediate-state inelastic channels resulting value of $\mathcal{A}_L$ may be essentially larger. Expected values of longitudinal analyzing power in the forthcoming TRIUMF and Jülich experiments prove to be of the same order of magnitude as at the low energies. Large $\mathcal{A}_L$ being observed, it is not related to the weak coupling of the vector meson to the nucleon. Disentagling of strong-interaction dynamics might require measurements of $\mathcal{A}_L$ at even higher energies above the π-meson production threshold. There the $\mathcal{A}_L$ value will be saturated, in the meson-exchange model, by inelastic final-state events, in contrast to the quark model predicting comparable relative weight of the elastic and inelastic channels. Experiments in this energy region are also planned on both accelerators [4,5].

REFERENCES

1. Manohar, A., and Georgi, H., *Nucl. Phys.* **B234**, 189–212 (1984) .
2. Desplanques, B., Donoghue,J.F., and Holstein, B., *Ann. of Phys.* **124**, 449–495 (1980).
3. Driscoll D.E., and Miller G.A., *Phys.Rev.* **C40**, 2159–2167 (1989).
4. Birchall, J. *et al.*, Contributions to the 8th Int.Symp. on Polarization Phenomena in Nuclear Physics, 1994, pp. 6–7.
5. Eversheim, P.D. *et al.*, ibid. pp. 4–5.
6. Iqbal M.J., and Niskanen J.A., *Phys.Rev.* **C49**, 355–359 (1994).
7. Grach I., and Shmatikov M., *Phys.Lett.* **B316**, 467–471 (1993).

T-violation in Neutron Spin Rotation

Yasuhiro Masuda

National Laboratory for High Energy Physics
1-1 Oho, Tsukuba-shi, Ibaraki-ken, 305 Japan

ABSTRACT: A new T-violation experiment concerning the neutron-nucleus interaction is proposed. In this experiment, the neutron spin rotates around the nuclear spin by 180° from the forward to backward direction during propagation through a polarized nuclear target. The rotation direction and incident neutron spin are reversed at a suitable period. A T-violation effect is found in a change in the transition probability between neutron-spin states upon the reversal of the rotation direction as well as the one of incident neutron spin. The systematic error can be small compared with the T-violation limit bound by the neutron EDM measurement.

I. INTRODUCTION

T-violation in the neutron-induced reaction in the eV region is one of the recent topics concerning fundamental symmetry violations, since a large enhancement of the P-odd $\sigma \cdot k$ correlation term has been found in the p-wave resonance of the neutron-radiative-capture reaction.[1, 2, 3, 4] σ is the neutron spin and k the neutron momentum. Several kinds of T-violation experiments have been proposed. Among them, a measurement of the T-odd and P-odd $\sigma \cdot (k \times I)$ correlation term, where I is the nuclear spin, is most interesting, since it has the following three advantages. Firstly, a large enhancement is expected. Secondly, the effect is free from final state interactions, since the initial and final states are the same plane wave. Thirdly, the ratio of the T-odd effect to the P-odd effect is insensitive to the nuclear wave-function uncertainty. As a result, the measurement can be compared with other elementary-particle experiments, for example neutron electric-dipole-moment (EDM) measurements. Therefore, many experimental methods have been proposed to measure the T-odd and P-odd term.

At first, the measurement of neutron-spin rotation around $k \times I$ was proposed.[5] However, fake effects due to the coupling of two different kinds of spin-correlation terms, $\sigma \cdot k$ and $\sigma \cdot I$, disable one from carrying out an accurate measurement. Measuring the transmission asymmetry and polarization for the neutron spin in the $k \times I$ axis was also proposed.[6] According to the polarization-asymmetry theorem, fake effects are canceled out in the sum of the polarization and asymmetry.[7] A detailed balance in spin-flip processes induced by the $\sigma \cdot (k \times I)$ correlation term is a good candidate to obtain the T-violation effect without fake effects, since the two processes which are time-reversed processes for each other, are compared in its measurement.[6] In these measurements, however, it is still very difficult to obtain T-violation information beyond the present limit bound by the neutron EDM measurements.

Here, we discuss a new method, which breaks through the problem. The present method involves a double T-violating asymmetry. During the propagation through a polarized nuclear target, the neutron spin rotates by 180° around the nuclear spin from the forward to bacward direction and vice versa. A T-violation effect is found in a change in the transition probability between neutron-spin states upon the reversal of the rotation as well as the one of the incident neutron spin. The present method is a combination of the T-odd transmission asymmetry and the spin detaild balance. The size of the present T-violation effect is as large as the T-odd transmission asymmetry and polarization. There is no fake effect in the present method like that in the spin detaild balance. Furthermore, another systematic error which may be induced by a mis-adjustment between the neutron spin and the nuclear spin is also canceled in the present method. As a result, a considerable improvement is expected in the T-violation limit.

II. ROTATION ASYMMETRY AS A MEASURE OF T-VIOLATION

The propagation of a polarized neutron through a polarized nuclear target is discussed in terms of the forward scattering amplitude $f(0)$ and the density matrix.[8] The general form of the scattering amplitude for the polarized nucleus can be described as

$$f(0) = F_0 + F_1 \sigma_x + F_2 \sigma_y + F_3 \sigma_z$$
$$= F_0 + F_1 \sigma \cdot I + F_2 \sigma \cdot (k \times I) + F_3 \sigma \cdot k. \quad (1)$$

Here, σ is the Pauli spin operator.[6] F_i (i = 0, 1, 2, 3) are scattering amplitudes. After propagation through the target, the effect of the nuclear interaction is found in a phase shift of the neutron wave,

$$\Delta = \lambda l N f(0). \quad (2)$$

Here, λ is the neutron wavelength, l the neutron propagation-length and N the target-nucleus-number density. The density matrices of the incident neutron-spin states, which are parallel and anti-parallel to the neutron momentum which is along the z axis, are

$$\rho_\pm = 1/2 \cdot (1 \pm \sigma_z), \quad (3)$$

respectively. After propagation, the density matrices are modified in the presence of the phase shift as

$$\rho_f = F \rho_\pm F^\dagger, \quad (4)$$

where, F is

$$F = \exp(i\Delta). \tag{5}$$

The 180° neutron-spin rotation is described in terms of the density matrices and projection operators ($P_\pm$) for parallel and anti-parallel spin state to the z axis,

$$P_\pm = (1 \pm \sigma_z)/2. \tag{6}$$

The transitions from the parallel to anti-parallel state and from the anti-parallel to parallel state are descibed as follows:

$$\begin{aligned} R_{+-} &= \mathrm{Tr}(F\rho_+ F^\dagger P_-) \\ &= \exp(-2\mathrm{Im}(\phi_0))(\sin b/b)^2 \\ &\quad \{|\phi_1|^2 + |\phi_2|^2 + 2(\mathrm{Im}(\phi_1)\cdot \mathrm{Re}(\phi_2) - \mathrm{Re}(\phi_1)\cdot \mathrm{Im}(\phi_2))\}. \end{aligned} \tag{7}$$

and

$$\begin{aligned} R_{-+} &= \mathrm{Tr}(F\rho_- F^\dagger P_+) \\ &= \exp(-2\mathrm{Im}(f_0))(\sin b/b)^2 \\ &\quad \{|\phi_1|^2 + |\phi_2|^2 - 2(\mathrm{Im}(\phi_1)\cdot \mathrm{Re}(\phi_2) - \mathrm{Re}(\phi_1)\cdot \mathrm{Im}(\phi_2))\}, \end{aligned} \tag{8}$$

respectively. Here,

$$b = \sqrt{|\phi_1|^2 + |\phi_2|^2 + |\phi_3|^2}, \tag{9}$$

and ϕ_i are spin-dependent phase shifts, which can be represented by

$$\phi_i = \lambda l N F_i. \tag{10}$$

In the eq. (9), $|\phi_1|^2$ is dominant,

$$\begin{aligned} b^2 &= |\phi_1|^2 (1 + \delta_\phi). \\ &= (\mathrm{Re}(\phi_1))^2 (1 + r_1)(1 + \delta_\phi) \end{aligned} \tag{11}$$

Here,

$$r_1 = (\mathrm{Im}(\phi_1)/\mathrm{Re}(\phi_1))^2 \tag{12}$$

In the p-wave resonance of ^{139}La at $E_n = 0.734$ eV,

$$\text{Re}(\phi_1) \gg \text{Im}(\phi_1). \tag{13}$$

According to the theoretical prediction,[9] and a recent experiment concerning parity-violating neutron-spin rotation,[10]

$$\text{Im}(\phi_3) \gg \text{Re}(\phi_3). \tag{14}$$

As a result, the $|\phi_2|^2$ and $\text{Im}(\phi_1) \cdot \text{Re}(\phi_2)$ terms in R+- and R-+ can be neglected.

$$\begin{aligned} R_{+-} &= \exp(-2\text{Im}(\phi_0))\,(\sin b\,)^2\,(1-\delta_\phi)\,\{1 - 2\text{Im}(\phi_2)/[\text{Re}(\phi_1)\,(1+r_1) \\ &\qquad (1+\delta_\phi)]\} \\ &= \exp(-2\text{Im}(\phi_0))\,(\sin b\,)^2\,(1-\delta_\phi)\,\{1 - 2\text{Im}(\phi_2)/\text{Re}(\phi_1)\} \end{aligned} \tag{15}$$

and

$$R_{-+} = \exp(-2\text{Im}(\phi_0))\,(\sin b\,)^2\,(1-\delta_\phi)\,\{1 + 2\text{Im}(\phi_2)/\text{Re}(\phi_1)\}. \tag{16}$$

For simplicity, we neglect the correction $(1+r_1)(1+\delta_\phi)$ in the T-odd term.

Since the neutron spin rotates by 180° from the forward to backward direction around the nuclear polarization which is along the x axis and vice versa, the neutron spin is in the +y ($k \times I$) region and -y region during the propagation through the target, respectively. The imaginary part of the T-odd phase shift is found in neutron wave attenuation, and its sign is changed upon the reversal of the incident neutron spin. From this point of view, the present method is similar to the T-odd transmission asymmetry. The asymmetry of the transition probabilities, R_{+-} and R_{-+} is

$$\begin{aligned} A_S &= (R_{+-} - R_{-+})/(R_{+-} + R_{-+}) \\ &= -2\,\text{Im}(\phi_2)/\text{Re}(\phi_1). \end{aligned} \tag{17}$$

Eq. (16) is essentially the same as the spin-detailed balance. If we use a LaF$_3$ or LaAlO$_3$ single crystal of a 4-cm length as a polarized lanthanum target,[11,12] the value of $\lambda l N$ is

$$\lambda l N = 2.4 \times 10^{14} \tag{18}$$

and the value of A_S is

$$A_S = 1.4 \times 10^{-2}\,\eta P_I. \tag{19}$$

Here, η is

$$\eta = \text{Im}(\phi_2)/\text{Im}(\phi_3). \tag{20}$$

P_I is the nuclear polarization. The sign of the T-odd effect is also changed upon a rotation reversal. The asymmetry of the transition probability with respect to the rotation reversal, which is referred to as rotation asymmetry, is

$$A_r(+) = (R_{+-}(180°) - R_{+-}(-180°))/(R_{+-}(180°) + R_{+-}(-180°))$$
$$= -2\, \text{Im}(\phi_2)/|\text{Re}\phi_1|, \qquad (21)$$

when the incident spin is in the $+z$ direction. When the incident spin is in the $-z$ direction,

$$A_r(-) = (R_{-+}(180°) - R_{-+}(-180°))/(R_{-+}(180°) + R_{-+}(-180°))$$
$$= +2\, \text{Im}(\phi_2)/|\text{Re}\phi_1|. \qquad (22)$$

III The experimental scheme and systematic error

The experimental scheme is shown in Fig. 1. The neutron beam is polarized by either a polarized ^{3}He filter or a polarized proton filter. A neutron enters into a

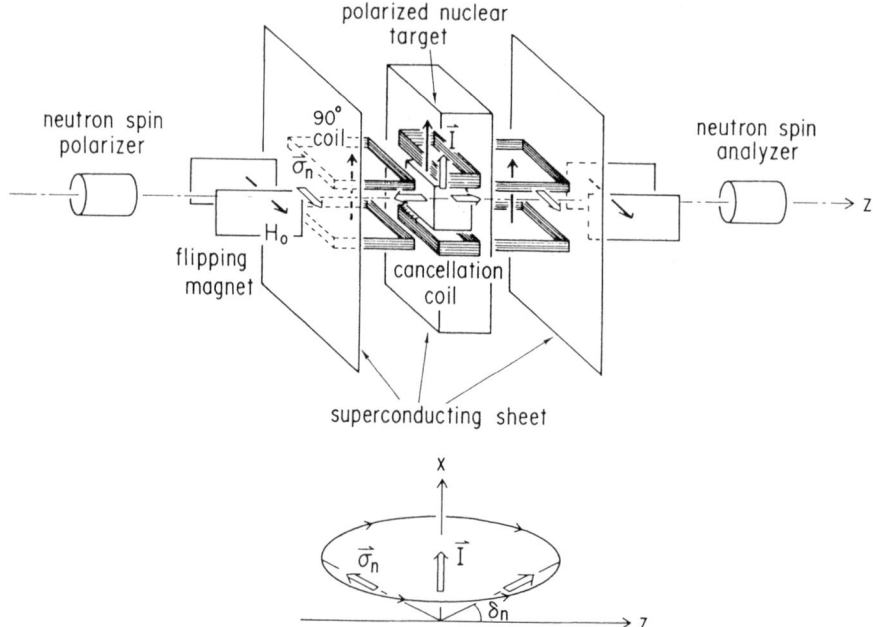

Fig. 1 Experimental scheme of the T-violating neutron spin rotation

target section through a flipper section. After propagation through the target, the neutron spin is analyzed by a ^{3}He neutron polarimater.[13]

The neutron spin is manipulated as follow. In the neutron path from the neutron polarizer to the flipper, the neutron spin is guided following an adiabatic passage. The magnetic field direction is along the y axis at the flipper and is reversed at a suitable period. The neutron spin is also reversed upon the reversal of the field direction. After the flipper, the neutron spin enters into a 90° coil through a superconducting sheet. In a superconductor, the magnetic field is zero due to the Meissner effect. The neutron sees a sudden decrease of the magnetic field to zero upon passing through the surface of the superconductor. The magnetic field penetration-depth of niobium is $\lambda(0)$ = 440 Å at T = 0 K. The neutron time of flight for the penetration depth is 3.7 x 10^{-12} sec at a neutron energy of E_n = 0.734 eV. The neutron spin does not follow this sudden change in the magnetic field, because the rate of the magnetic field change is very fast compared with the Larmor frequency which is 2.9 x 10^3 Hz/G. As a result the neutron spin enters into the 90° coil without any change in its direction. In the the 90° coil, the magnetic field direction is along the x axis. The neutron spin rotates from the y to z direction before entering into the polarized nuclear target. In the target section, a magnetic field (H_0) is applied in order to hold the nuclear polarization. In addition to the magnetic field, the neutron sees an field called "pseudomagnetic field (H^*)" due to the F_1 term during transmission through the target. Therefore, the neutron spin rotates in the sum of these fields, ($H_0 + H^*$). The neutron-spin rotation is controlled by adjusting the magnetic field so that the angle of rotation is either 180° and -180° during transmission. After transmission, the neutron spin rotates from the z to y direction in the second 90° coil and is then analyzed by a ^{3}He neutron polarimeter.

In the neutron-spin manipulation, detection of the rotation angle is very important. We have developed a rotation phase detector by using the Meissner effect.[14] The rotation phase of the neutron wave is converted to a projection angle of the neutron spin on the magnetic field. The projection component of the neutron spin is measured by the ^{3}He neutron polarimeter. In this method, an accuracy of $\delta\theta$ < 1° has been found to be possible. For a LaF$_3$ or LaAlO$_3$ single crystal and anuclear polarization of 50%, the pseudomagnetic field is ~1 kG. Neutrons of E_n = 0.734 eV rotate ~10 turns upon passage through a 1-kG magnetic field for a 4-cm length. A 1°-rotation adjustment means a field adjustment of $\delta H_0/H_0$ = 3 x 10^{-4}.

If we assume that the uncertainty of the rotation angle is $\delta\theta$, the uncertainty in the phase of the neutron wave, δ is

$$\delta = \delta\theta/2. \tag{23}$$

Then, the real part of ϕ_1 is

$$\mathrm{Re}(\phi_1) = \pi/2 + \delta. \tag{24}$$

In the measurement of A_s, the effect of the rotation uncertainty is cancelled, if the common coefficient, $\exp(-2\text{Im}(\phi_0))(\sin b)^2 (1 - \delta_\phi)$, is not changed upon incident neutron-spin flipping. The value of $(\sin b)^2$ may change during neutron-spin flipping, since it has a small polarization dependence,

$$b = \pi/2 + \delta + r_1/2 + \delta_\phi/2, \tag{25}$$
$$(\sin b)^2 = 1 - (\delta + r_1/2 + \delta_\phi/2)^2. \tag{26}$$

In Eq. (26), δ is dominant. If we assume that the value of $\delta\theta$ is $1°$ and the change in δ upon neutron-spin flipping is $0.01°$, then the error of the ratio η is

$$\delta\eta < 3 \times 10^{-4} \tag{27}$$

for a nuclear polarization of 50%.

In addition to the above mentioned errors, a systematic error may be induced due to any mis-alignment between the neutron polarization and the nuclear polarization. If the neutron-polarization direction slightly deviates by δ_n from $90°$ with respect to the nuclear polarization, a $\sin \delta_n$ component of the neutron polarization is held in the direction of the nuclear polarization during the transmission through the target. This mis-alignment induces spin-dependent absorption which is proportional to $\sin\delta_n \text{Im}(\phi_1)$. Since the sign of δ_n is reversed upon neutron-spin flipping, the effect of the mis-alignment remains in the value of A_s. However, the absorption effect is small and independent of the neutron-spin-rotation direction. Therefore, the mis-alignment effect is canceled in the rotation asymmetry(A_r). As a result, the systematic errors are greatly reduced, if we take a double asymmetry,

$$A_d = (A_r(+) - A_r(-))/(A_r(+) + A_r(-)). \tag{28}$$

We can expect a large T-violation effect in the T-odd transmission asymmetry and polarization. However, it is very difficult to reduce the fake effect in these measurements. Although the fake effect is canceled in the sum of the T-odd transmission asymmetry and polarization,[7] the two measurements are independent of each other; therefore, the fake effect does not necessarily have the same value in the two measurements. In the spin detailed balance, there is no fake effect, since any difference of spin-flip processes which are time reversed to each other is measured.[6] Cancellation of the Larmor and pseudomagnetic precessions makes the effect of the F_1 term on the spin flip small, so that the effect of the F_2 term increases in the spin detailed balance. However, the accuracy is limited, because we should measure any small difference between small spin-flip probabilities.[15] Also, any mis-alignment between the neutron spin and the nuclear polarization induces a sprrious effect. Based on this point of view, the present method has a great advantage. The value of A_s or A_r is as large as the T-odd transmission asymmetry and polarization. There is no fake effect as in the spin detailed balance. The mis-alignment effect is also canceled in the rotation asymmetry. As a result, a comparison with the neutron EDM measurement is possible.

The present experimental accuracy of the neutron EDM corresponds to the value of η as

$$\eta < 4 \times 10^{-3} \quad \text{(n EDM).[8,16]} \tag{29}$$

Therefore, the present method has possibility to obtain more accurate information than the neutron EDM.

The auther would like to thank Dr. V.P. Gudkov, Prof. K. Asahi and Dr. J.D. Bowman for their valuable discussion. He also expresses his thanks to his colleague concerning P-violation and T-violation experiments carried out at KEK for their indispensable collaboration. He also thanks Prof. H. Sugawara, Prof. N. Watanabe and Prof. K. Nakai for their warm encouragement.

[1] V.P. Alfimenkov et al., Nucl. Phys. A398(1983)93.
[2] Y. Masuda et al., Hyp. Int. 34(1987) 143; Nucl. Phys. A478(1988)737c; Nucl. Phys. A504(1989)269.
[3] S.A. Biryukov et al., Sov. J. Nucl. Phys. 45(1987)937.
[4] C.D. Bowman et al., Phys. Rev. C39(1989)1721.
[5] V.E. Bunakov and V.P. Gudkov, Z. Phys. A308(1982)363.
[6] L. Stodolsky, Physics Lett. B172(1986)5.
[7] P.K. Kabir, Phys. Rev. Lett. 60(1988)686.
[8] I.I. Grevich and L.V. Tarasov, "Low-Energy Neutron Physics", North Holland (1968).
[9] V.P. Gudkov, Phys. Rep. 212(1992)77.
[10] K. Sakai et al., Hyp. Int. 84(1984)199.
[11] Y. Masuda et al., KEK report 90-25 (1991) 1002.
[12] Y. Masuda et al., Proc. 18th INS Int. Symp., World Scientific(1991)293.
[13] H. Sato et al., Hyp. Int. 84(1994)205.
[14] Y. Masuda, Neutron Research 1(1993)53.
[15] Y. Masuda et al., Hyp. Int. 74(1992)149.
[16] P. Herczeg, Hyp. Int. 75(1992)127.

Parity Nonconservation in ^{207}Pb

M. Leuschner, B. Cain, J. E. Knott, A. Komives
W. M. Snow, J. J. Szymanski

Indiana University Cyclotron Facility, Bloomington, IN 47405

A. Andalkar, I. C. Girit, M. Petrov

Princeton University, Princeton, NJ 08540

J. D. Bowman

Los Alamos National Laboratory, Los Alamos, NM 87545

Abstract. Two experiments are currently underway to measure the single-particle weak mixing matrix element for the 1064 KeV transition in ^{207}Pb. One experiment measures the circular polarization of the 1064 KeV gamma ray emitted from an unpolarized source, while the other experiment measures the forward-backward asymmetry of gamma rays emitted from a polarized source. Analysis of the first set of polarized source data yields an upper limit of 23 eV for the single-particle weak mixing matrix element.

INTRODUCTION

The recent interest in parity violation in heavy nuclei has been inspired by the results of the Los Alamos TRIPLE collaboration. The TRIPLE collaboration has measured the asymmetry of longitudinally polarized epithermal neutrons (1-1000 eV) transmitted through unpolarized targets. From their first set of measurements on ^{232}Th and ^{238}U asymmetries of up to 10% were reported[1,2]. Even more surprising than the large absolute values was the observation that all seven asymmetries in ^{232}Th of at least $2-\sigma$ significance were positive in sign. In order to reconcile these experimental results with theory several calculations[3-6] have required parity violating matrix elements in the range $10 < V_{pnc} < 500$ eV, whereas previous experimental evidence suggests that these matrix elements are on the order of 1 eV or less.

In order to improve the understanding of this phenomenon experiments have been sought which satisfy three criteria. First, a heavy nucleus ($A > 200$) was required so that a meaningful comparison may be made to the ^{232}Th and ^{238}U results. It is assumed that the mechanism which is responsible for the sign effect suggested by the TRIPLE results is a global feature of parity violation in heavy nuclei. Second, a system with a relatively simple nuclear structure is desired to facilitate the theoretical interpretation of the results. The search for a candidate was therefore restricted among those nuclei that are one or fewer nucleons away from major shell closures. Finally, a system in which the parity violating effect can be observed from a radioactive decay rather than a scattering experiment was required. In such an experiment the interpretation of the data will not be hindered by the ambiguities inherent in nucleon-nuclear scattering phenomenology.

The most promising candidate satisfying these three conditions was found to be the ^{207}Pb nucleus. The $E_x = 1.63$ MeV $13/2^+$ state of ^{207}Pb is populated via electron capture by the ^{207}Bi nucleus ($t_{1/2} = 32.2$ years). The $13/2^+$ state decays with a M4 transition to a $5/2^-$ state at $E_x = 0.57$ MeV. The weak interaction produces an opposite parity admixture into the $5/2^-$ state, resulting in E4 multipole mixing into the 1064 KeV line. The interference of the M4 and E4 strengths produces a circular polarization of the 1064 KeV gamma ray from an unpolarized ^{207}Pb source, as well as a forward-backward asymmetry of the gamma ray angular distribution from a polarized source.

EXPERIMENT

The measurement of the circular polarization of the 1064 KeV gamma ray from an unpolarized source is underway at the Indiana University Cyclotron

158 Parity Nonconservation in ^{207}Pb

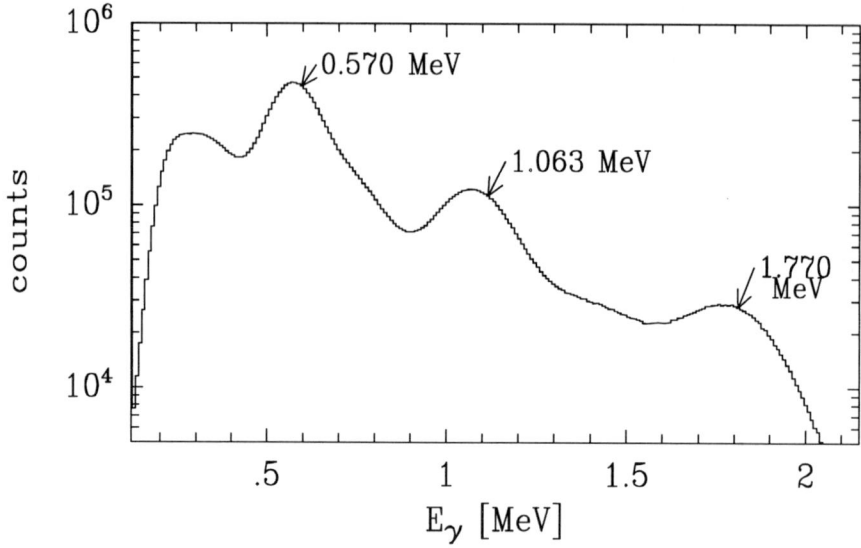

Figure 1 Gamma ray spectrum of ^{207}Pb arising from the decay of ^{207}Bi. The gamma rays are detected with intrinsic CsI crystals.

Facility (IUCF). The circular polarization is given by

$$P_\gamma = 2\frac{V_{pnc}}{\Delta E}\frac{<E4>}{<M4>}$$

where the energy spacing ΔE between the two admixed $f_{5/2}$ single particle states is expected to be about one major shell spacing ($\sim$ 6 MeV). The ratio $<E4>/<M4>$ has been calculated[7] to be 8.5. Assuming a PNC matrix element of 50 eV, the resulting circular polarization would be 1.44×10^{-4}.

The helicity of the 1064 KeV gamma rays are analyzed by a Compton polarimeter with a thickness of 3 interaction lengths. The transmitted gamma rays are detected in a pair of large intrinsic CsI crystals. A sample spectrum is shown in figure 1. Although the resolution of CsI is poor compared to other scintillators, it is still sufficient to clearly resolve the three gamma rays resulting from transitions between ^{207}Pb levels. The field direction in the polarimeter is reversed every 3 seconds to minimize systematic effects caused by time dependent drifts in the detectors and/or electronics.

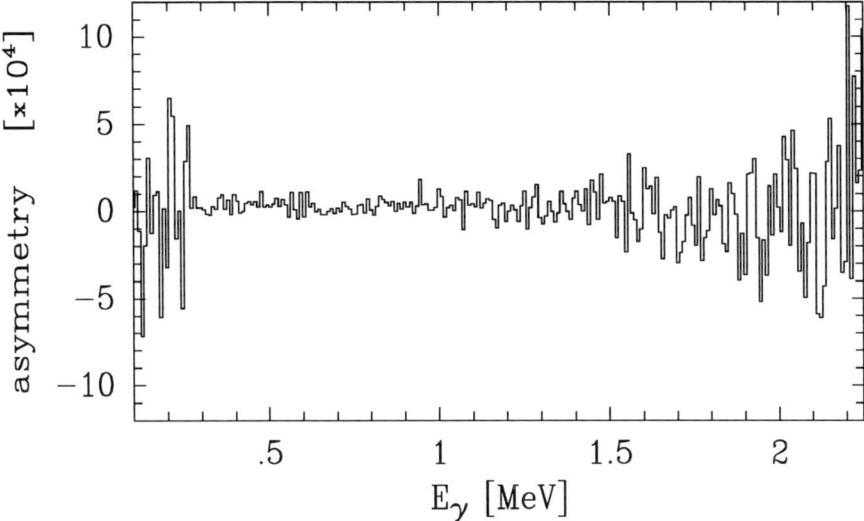

Figure 2. The gamma ray asymmetry for transitions in ^{207}Pb. The asymmetry is calculated separately for each of the 256 ADC channels.

Since the 1064 KeV gamma ray falls in the energy regime beyond which the photoelectric and compton scattering cross sections are large, and below which pair-production processes are important, the analyzing power of the polarimeter is rather low, somewhere in the range 1-1.5%. The product of the circular polarization and the analyzing power yields an asymmetry of 1.44×10^{-6}. In order to measure such a small asymmetry a high counting rate is required. To achieve the necessary counting rate a 4.2 mCu source has been ordered commercially. Folding in the 83% branching ratio to the $13/2^+$ state, 5% transmission through the polarimeter, and a 2.5% detection solid angle yields a counting rate of 166 KhZ/detector. Thus the statistical sensitivity of the experiment allows for a $1 - \sigma$ determination of a 50 eV matrix element after 17 days of data acquisition.

Monte Carlo simulations predict a signal/noise ratio of 1/1, leading to a total counting rate of 332 KhZ/detector. To handle such high rates a fast multi-channel analyzer was designed and built at IUCF. The count rate capacity of the acquisition system has been measured to be in excess of 2 Mhz. Systematic effects have been investigated using a weak ^{207}Bi source produced with the proton beam from the IUCF cyclotron. Figure 2 shows the results from a trial run using the weak ^{207}Bi source and a prototype polarimeter. The asymmetry is calculated separately for each ADC channel. The current overall sensitivity, including statistical and systematic effects, has been pushed below the 10^{-4} level. The experiment now awaits the delivery of the "production" 4.2 mCu ^{207}Bi source, expected by the end of this year.

The first set of measurements of the forward-backward asymmetry from a polarized source has been recently completed at Princeton University. The ^{207}Bi is polarized by the hyperfine field within a BiNi alloy. The Bismuth atoms are thermally diffused into the Nickel lattice by RF induction melting at a temperature of 1500 degrees Centigrade. The alloy is cooled to 7-10 mK using a ^{3}He-^{4}He dilution refrigerator, and the polarization direction is defined by the field applied by a superconducting magnet mounted within the dilution refrigerator. The emitted 1064 KeV gamma rays are detected by two germanium detectors placed at 0 and 180 degrees with respect to the applied magnetic field direction.

To date only modest success has been achieved in polarizing the ^{207}Pb. During the first two production runs, completed in January and May of 1994, a 13% polarization was measured from a 8 μCu source. Despite the low polarization, the polarized source experiment provides a higher sensitivity to the PNC matrix element than the circular polarization experiment. Assuming $V_{pnc} = 50$ eV, a backward-forward gamma ray asymmetry of $A = 1.15 \times 10^{-4}$ is expected. Analysis of the January and May data yields an asymmetry of $A < 5.3 \times 10^{-5}$, which corresponds to an upper limit $V_{pnc} < 23$ eV.

SUMMARY

Two experiments are currently underway to measure a weak mixing matrix element in the ^{207}Pb nucleus. Both experiments show exceptional promise and will likely continue to run and undergo development for several years before their potential is exhausted. It is expected that the experiments will eventually realize a sensitivity of 5 eV or less, thereby providing additional constraints on the weak coupling constants f_π and h_ρ.

REFERENCES

1. J. D. Bowman *et al.*, Phys. ReV. Lett. **65** 1192 (1990).
2. C. M. Frankle *et al.*, Phys. ReV. Lett. **67** 564 (1991).
3. J. D. Bowman *et al.*, Phys. ReV. Lett. **68** 780 (1992).
4. S. E. Koonin *et al.*, Phys. ReV. Lett. **69** 1163 (1992).
5. V. V. Flambaum, Phys. ReV. **C45** 437 (1992).
6. N. Auerbach, Phys. ReV. **C45** 514 (1992).
7. J. J. Szymanski *et al.*, Phys. ReV. **C49** 3297 (1994).

Measurement of charge symmetry breaking in np elastic scattering at 350 MeV[1]

R. Abegg*, A.R. Berdoz†, J. Birchall†, J.R. Campbell†, C.A. Davis*†,
P.P.J. Delheij*, L. Gan†, P.W. Green*‡, L.G. Greeniaus‡, D.C. Healey*,
R. Helmer*, N. Kolb‡, E. Korkmaz‡, L. Lee†, C.D.P. Levy*, J. Li‡,
C.A. Miller*, A.K. Opper‡, S.A. Page†, H. Postma§, W.D. Ramsay†,
J. Soukup‡, G.M. Stinson*‡, W.T.H. van Oers†, A.N. Zelenski*, J. Zhao†

*TRIUMF, 4004 Wesbrook Mall, Vancouver, B.C. Canada V6T 2A3
†University of Manitoba, Winnipeg, Manitoba, Canada R3T 2N2
‡University of Alberta, Edmonton, Alberta, Canada, T6G 2N5
§Technische Hogeschool, Delft, the Netherlands, 2600 GA

Abstract. TRIUMF Experiment 369, a measurement of charge symmetry breaking in np elastic scattering at 350 MeV, has completed data taking. Scattering asymmetries were measured with a polarized (unpolarized) neutron beam incident on an unpolarized (polarized) frozen spin target. Coincident scattered neutrons and recoil protons were detected by a mirror symmetric detection system in the center-of-mass angle range from 50° – 90°. A preliminary result for the difference of the zero-crossing angles, where analyzing powers cross zero, is $\Delta\theta_{cm} = 0.445° \pm 0.054°$(stat.)$\pm 0.051°$(syst.) based on fits over the angle range $53.4° \leq \theta_{cm} \leq 86.9°$. The difference of the analyzing powers $\Delta A \equiv A_n - A_p$, where the subscripts denote polarized nucleons, was deduced with $dA/d\theta_{cm} = (-1.35 \pm 0.05) \times 10^{-2}\mathrm{deg}^{-1}$ to be $[60 \pm 7\mathrm{(stat.)} \pm 7\mathrm{(syst.)} \pm 2\mathrm{(syst.)}] \times 10^{-4}$.

INTRODUCTION

The study of isospin symmetry breaking is of fundamental interest since it relates to the mass difference of the quarks of the first generation, the up and down quarks with $m_d - m_u > 0$, in addition to the well-understood electromagnetic interaction(1). Early evidence for charge independence breaking and charge symmetry breaking (CSB) came from the differences in the nucleon-nucleon $^1S_0$ scattering lengths(2) and from the binding energy differences of mirror nuclei(3). The latter is the well-known Okamoto-Nolen-

---
[1] Work supported in part by the Natural Sciences and Engineering Research Council of Canada.

Schiffer anomaly. Theoretical interpretations of these observations carry uncertainties due to the subtraction of electromagnetic interaction effects. CSB in the np system belongs to a different class of charge symmetry breaking. It has the advantage of the absence of the Coulomb interaction.

Charge symmetry in the np system leads to the complete separation of the isoscalar and isovector components of the np interaction. This in turn leads to the equality of the differential cross sections for polarized neutrons scattering from unpolarized protons and vice versa. As a result, $A_n(\theta) \equiv A_p(\theta)$ where A denotes the analyzing power and the subscript represents the polarized nucleon. A nonvanishing asymmetry difference is directly proportional to the isospin triplet-singlet, spin singlet-triplet mixing amplitudes and therefore direct evidence of a charge-asymmetric interaction, anti-symmetric under the interchange of the two nucleons in isotopic spin space.

The scattering matrix for np elastic scattering can be expressed in terms of the formalism of LaFrance and Winternitz(4) as

$$M(\vec{k}_f, \vec{k}_i) = \frac{1}{2}\Big\{(a+b) + (a-b)(\vec{\sigma}_1 \cdot \hat{n})(\vec{\sigma}_2 \cdot \hat{n}) + (c+d)(\vec{\sigma}_1 \cdot \hat{m})(\vec{\sigma}_2 \cdot \hat{m})$$
$$+ (c-d)(\vec{\sigma}_1 \cdot \hat{\ell})(\vec{\sigma}_2 \cdot \hat{\ell}) + e(\vec{\sigma}_1 + \vec{\sigma}_2) \cdot \hat{n} + f(\vec{\sigma}_1 - \vec{\sigma}_2) \cdot \hat{n}\Big\}. \quad (1)$$

Here, $\hat{\ell}, \hat{m}$ and $\hat{n}$ are unit vectors given as

$$\hat{\ell} = \frac{\vec{k}_i + \vec{k}_f}{|\vec{k}_i + \vec{k}_f|}; \quad \hat{m} = \frac{\vec{k}_f - \vec{k}_i}{|\vec{k}_f - \vec{k}_i|}; \quad \hat{n} = \frac{\vec{k}_i \times \vec{k}_f}{|\vec{k}_i \times \vec{k}_f|}; \quad (2)$$

with $\vec{k}_i$ and $\vec{k}_f$ the initial and final state center-of-mass nucleon momenta. The amplitudes a, b, c, d, e, and f are functions of center-of-mass energy E and scattering angle θ, with f the isotopic spin mixing amplitude. Written explicitly, the difference in the analyzing powers

$$\Delta A(\theta) \equiv A_n(\theta) - A_p(\theta) = \frac{2}{\sigma_0} \mathcal{R}e(b^* f), \quad (3)$$

is proportional to f. The quantity σ_0 is the differential cross section for the scattering of unpolarized neutrons from unpolarized protons.

The first measurement of charge symmetry breaking in np elastic scattering was performed at TRIUMF(5). The measurement of $\Delta A \equiv A_n - A_p$, at the zero-crossing angle of the average analyzing power, at an incident neutron energy of 477 MeV, yielded $\Delta A = (47 \pm 22 \pm 8) \times 10^{-4}$, an effect just over two standard deviations. More recently the results of a similar experiment at a neutron energy of 183 MeV performed at IUCF have been reported(6). The measured value of $\Delta A \equiv A_n - A_p$, averaged over the angular range $82.2° \leq \theta_{cm} \leq 116.1°$ over which $<A(\theta)>$ averages to zero, is $(34.8 \pm 6.2 \pm 4.1) \times 10^{-4}$, where as above the first error represents mainly the statistical uncertainty and the second error the systematic uncertainty. The IUCF result differs from zero by 4.5 standard deviations. It differs from the value expected from the electromagnetic spin-orbit interaction by 3.4 standard deviations. This difference represented the most unambiguous experimental evidence of charge symmetry breaking in the nuclear interaction.

Extracting an angular distribution of $\Delta A(\theta)$ is more difficult. This follows directly from the expression for the difference in the asymmetries for beam and target polarized, respectively, or

$$\epsilon_b(\theta) - \epsilon_t(\theta) = \Delta A(\theta)(P_b + P_t)/2 + <A(\theta)>(P_b - P_t), \qquad (4)$$

pointing to the need for calibration of the beam and target polarizations (P_b and P_t) with an accuracy unattainable at present. In the analysis of the IUCF experiment this difficulty was overcome by adjusting the ratio of (P_b/P_t) applying a minimal variance procedure to $\Delta A(\theta)$ over the angular range of the experiment(6). Following this procedure a twelve point angular distribution was obtained. The procedure does not work at 477 MeV where $\Delta A(\theta)$ and $<A(\theta)>$ have zero-crossing angles in close proximity and consequently the angular dependencies are no longer orthogonal. If the theoretical calculations were precise in their predictions of the zero-crossing angle of $\Delta A(\theta)$, one could in principle also determine $\Delta P/<P>$ with $\Delta P = P_b - P_t$ and $<P> = (P_b + P_t)/2$, and consequently the angular distribution of $\Delta A(\theta)$ would follow.

The measured analyzing power differences of the IUCF and TRIUMF experiments are well-reproduced by theoretical predictions based on meson exchange potential models, which indirectly incorporate quark level effects(7). The calculations include contributions from one photon exchange (the magnetic moment of the neutron interacting with the current of the proton), from the neutron-proton mass difference affecting charged one π and ρ exchanges, and from the more interesting isospin mixing $\rho^0 - \omega$ meson exchange. Some other smaller effects (like 2π-exchange not included in ρ-exchange) have also been evaluated(8). The effects of $\pi\gamma$ exchanges have not yet been calculated. The effects of inelasticity amount to about 10% at 800 MeV but are vanishingly small at lower energies(9).

EXPERIMENT

A new experiment at 350 MeV has been performed at TRIUMF in order to delineate the various contributions to CSB. The experiment is similar in most aspects to the earlier TRIUMF measurement at 477 MeV. Designed as a null measurement, the difference of the zero-crossing angles where the analyzing powers A_n and A_p cross zero, respectively, is determined. With the slope, $dA/d\theta$, from phase shift analyses, or determined experimentally, the difference of the analyzing powers is obtained. Figure 1 illustrates the method to determine ΔA from the measured difference of the zero-crossing angles.

A 350 MeV neutron beam was produced using the (p,n) reaction on deuterium. The proton beam, polarized up to 80% and normal to the scattering plane, was obtained from an optically-pumped polarized ion source and accelerated by the cyclotron to 369 MeV, with polarizations up to 78%. The proton beam had an intensity of about 2 μA and was incident on a 0.217 $\pm$ 0.004 m long liquid deuterium (LD$_2$) target. Figure 2 shows a schematic layout of the beam line and experimental set-up. The polarization, the energy, the position and direction of the proton beam were monitored throughout the experiment and controlled (in the case of position and direction) using a feedback system coupling two sets of split-plate secondary electron emission monitors (which determined the median of the intensity distribution) with steering magnets upstream in the beam transport line. At the two sets of split-plate secondary emission monitors (SEMs), the beam position was kept fixed with a standard deviation of $\pm$0.05 mm in both x and y intensity profiles. The proton beam polarization was measured by two polarimeters. The beam energy monitor, based on range determinations, allowed the beam energy to be kept constant with a standard deviation of less than 36 keV (through minute changes in rf of the cyclotron and stripper foil position).

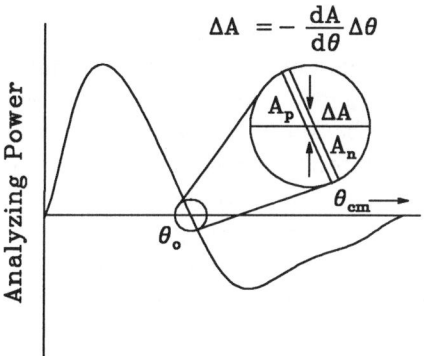

FIGURE 1. An illustration of the method employed to extract ΔA from the difference of the A_n and A_p zero-crossing angles for np elastic scattering. $dA/d\theta$ and $\Delta\theta_0$ are determined in the experiment.

The polarization was transferred from the proton to the neutron by making use of the large sideways to sideways polarization transfer coefficient $r_t(-0.88$ for $d(\vec{p},\vec{n})2p$ at 364 MeV). The proton beam polarization was rotated into the horizontal plane by a superconducting solenoid magnet. The neutrons passed a 3.3 m long, tapered steel collimator placed at 9° before impinging on a frozen spin type polarized proton target (containing butanol beads) positioned at 12.85 m from the center of the LD$_2$ target. The neutron beam polarization was rotated to the longitudinal direction and then to the normal direction by

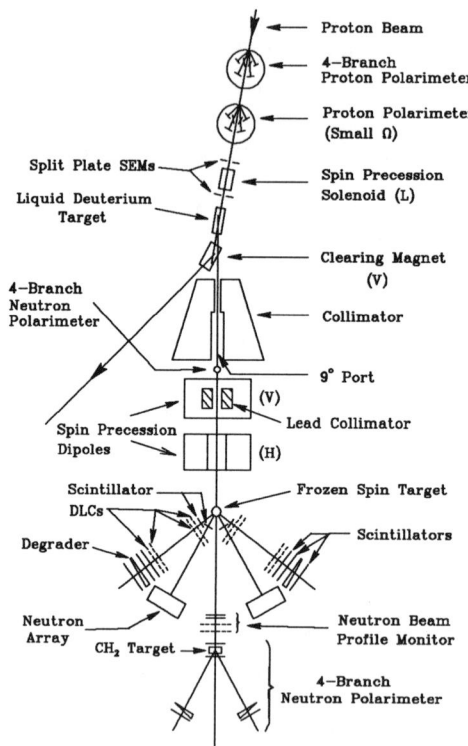

FIGURE 2. Schematic layout of the beam line and experimental setup.

a combination of two dipole magnets. The neutron beam polarization was monitored by two polarimeters, one at the exit of the collimator and the other 3.6 m downstream of the frozen spin target. The frozen spin target (FST) had typical polarizations of 80 to 90%. After being polarized, the spin of the "free" protons in the butanol beads was held "frozen" with a reduced, normal direction holding field of 0.22 T at a temperature of about 55 mK. For further details see Ref. (10).

The scattered neutrons were detected by scintillation detector arrays, each consisting of two banks of scintillator bars, one behind the other, with seven horizontally-stacked bars in each bank. Two additional bars with the same dimensions were placed vertically on each side beside the main arrays to enlarge the detection angle coverage. The recoil protons were detected by time-of-flight (TOF)/range telescopes, each consisting of a TOF start counter, four delay line wire chambers (DLCs), two scintillation counters, a wedge-shaped brass absorber and a veto scintillator. Only high energy protons from background processes can penetrate the absorber and trigger the veto counter. These events were rejected off-line by software. Figure 3 shows a detailed view of the detection system. To study the background (mainly due to quasi-elastic (n, np) scattering off the carbon of the butanol beads and other surrounding materials), the butanol beads were replaced by carbon beads having an equivalent target thickness with the rest of the target unchanged.

DATA ANALYSIS

The procedure of the data analysis was as follows: i) in order to identify possible sources of systematic errors, a study was made on a run-by-run basis of all system parameters recorded during data taking: beam energy, proton

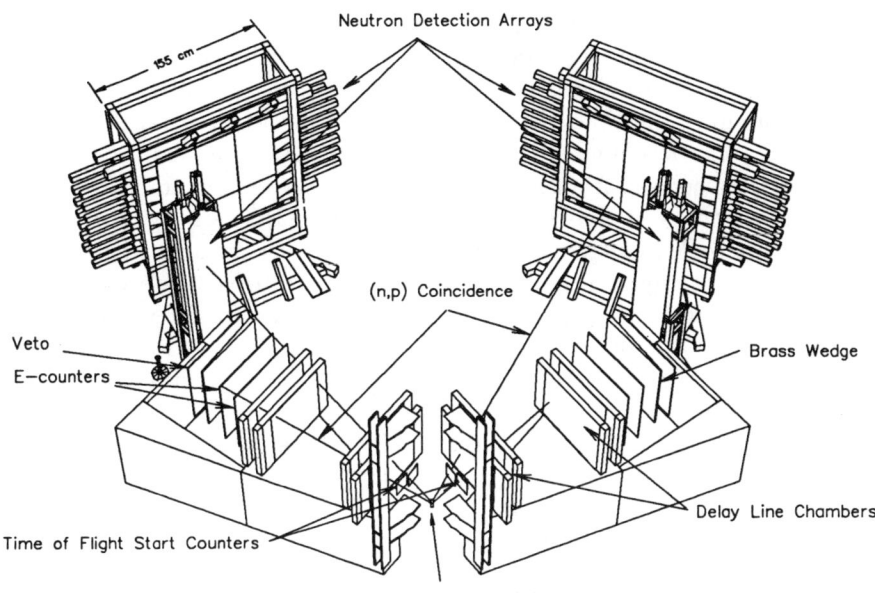

FIGURE 3. Detailed view of the detector system.

and neutron polarizations, SEM asymmetries, neutron profile, holding field strength and FST parameters; ii) elastic scattering np events were selected; iii) the asymmetry angular distributions and the zero-crossing angles were determined; and iv) corrections were made for contributions from quasi-elastic (n, np) background and for the effective average neutron beam energy difference of the polarized and unpolarized beam reflecting the energy-polarization correlation.

To select elastic scattering np events, proton and neutron tracks were reconstructed and their kinetic energies were calculated from their TOFs. Proton tracks were reconstructed from the information in the DLCs. At least one pair of coordinates (x and y) from each pair of DLCs (front and rear) was required. A loose cut of 40 mm on the total residuals of the proton tracks was applied to remove apparent multiple hits (about 3% of the total data). The proton energy was calculated from the TOFs between the TOF start counter and both scintillation counters. Event-by-event corrections were applied to the reconstructed proton angle and the proton energy to account for the proton deflection in the FST holding field, multiple scattering, and energy loss in the FST and detectors. The corrections were obtained from a Monte Carlo simulation of these processes. Neutron tracks were determined by their impact points in the neutron detector scintillator and the presumed scattering points in the FST. The hit position of a neutron at the scintillator and its arrival time were determined by the difference and the average, respectively, of the timing signals recorded on both ends of each bar. The neutron origin in the FST was assumed to be along the central normal axis (y), and y_{no} was determined by the proton track reconstruction. The neutron energy was calculated from its TOF. The software energy threshold of the bars was varied by applying various cuts on the ADCs after gain matching and pedestal subtraction. Four kinematic variables: opening angle, coplanarity angle, energy sum and horizontal momentum balance were used to test for np events. The momentum-dependent sigmas of these variables were obtained from their distributions and the chi-squares (χ_i^2) were calculated for each event. Various chi-square tests including the individual $\chi_i^2 \leq 6$, 7.5, 9 or the four variable combined $\chi_{sum}^2 \leq 10$, 15, 20 were applied to the data (Figs. 4a – 4e). Roughly 25% of the total data passed these cuts. The selected events were dominated by elastic scattering np events with a small contribution from the background.

Asymmetries were calculated based on the proton angle (θ_p) distributions (Fig. 4f). An "overlap" method(5) was used to obtain the asymmetries $\epsilon = \frac{r-1}{r+1}$, where $r = \sqrt{\frac{L+R-}{L-R+}}$ and zero-crossing angles (θ_0) were deduced from fits of the asymmetry angular distributions. To calculate ΔA from the measured $\Delta\theta_0$, $dA/d\theta$ is required. Because the $dA/d\theta$ from phase shift analyses show large discrepancies among different phase shift analyses and the FST polarization is known to about 2.5%(11), the experimentally determined $dA/d\theta$ is considered superior to the one from phase shift analyses.

The incident neutron beam energy was measured in two ways: the first was from the TOF between the cyclotron rf phase stabilized timing signal at the LD$_2$ target and the timing signal from the proton TOF start counter, the second was from the sum of the neutron and proton kinetic energies. To correct for the different effective average energy of the polarized and unpolarized neutron beam due to the energy dependence of the neutron polarization, Monte Carlo simulations were made (Figs. 5a – 5b). Comparisons with the experimental distributions were made after folding in the experimental resolution (taken to be a Gaussian). Good agreement was achieved as shown in Fig. 5c. The difference between the average effective neutron beam energy for polarized and unpolarized beams is calculated, based on these distributions, to be 0.4 MeV. A $d\theta_0/dE = -0.055°$/MeV deduced from phase shift analyses was used resulting in a correction of +0.022° (cm) for the zero-crossing angle of A_n.

Background data obtained with carbon beads replacing the butanol beads were analyzed in an identical fashion. To remove some hydrogen contamina-

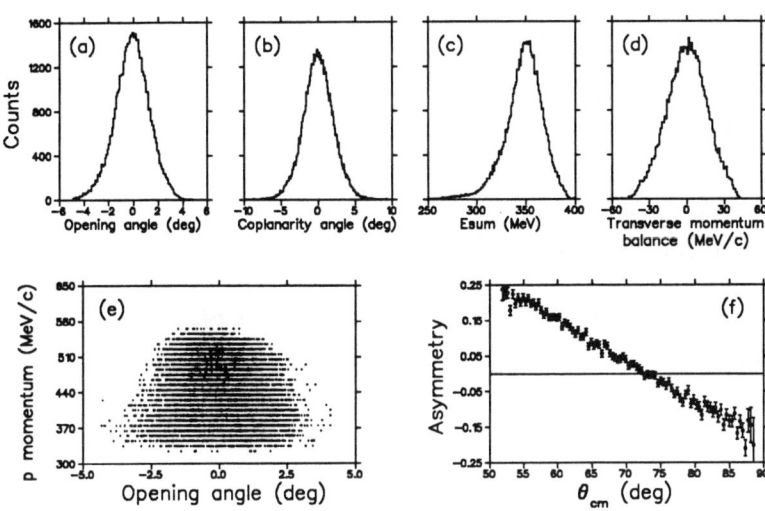

FIGURE 4. (a – d) kinematic variable distributions after $\chi_i^2 \leq 6$ cuts; (e) proton momentum versus opening angle distribution to determine the momentum dependence of the sigmas; (f) asymmetry distribution as function of the center-of-mass scattering angle.

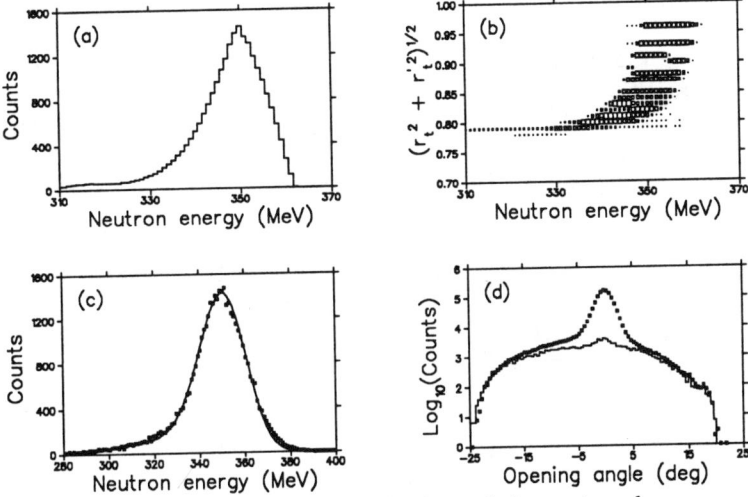

FIGURE 5. (a) and (b) simulated distributions of the neutron beam energy and neutron energy versus effective r_t for the reaction $D_2(\vec{p}, \vec{n})2p$; (c) comparison of the simulated and experimental measured neutron beam energy after applying a Gaussian distribution to the curve in (a) reflecting the experimental resolution; (d) opening angle distribution for the butanol target data, shown as the upper curve; the lower curve is obtained from the carbon target data. The carbon target data have been normalized to the butanol data by matching the tails of the distributions.

tion due to a hydrogen-containing resistor of the FST and super-insulation material around the target cell, various methods were used including applying additional cuts of $\chi_i^2 \geq 1$ or subtracting the hydrogen peak with normalized butanol target data. It was determined that 4% of the events which passed the $\chi_i^2 \leq 7.5$ cuts were from the background (Fig. 5d) and that the analyzing power of the background is $A_b = -0.004 \pm 0.007$. A correction of $\Delta A = (+1.6 \pm 2.8) \times 10^{-4}$ for background contribution was applied to the result based on this evaluation.

PRELIMINARY RESULT AND DISCUSSION

A preliminary result of the data analysis shows the difference of the zero-crossing angles to be $\Delta \theta_{cm} = 0.445° \pm 0.054°$(stat.) $\pm 0.051°$(syst.) based on fits over the angle range $53.4° \leq \theta_{cm} \leq 86.9°$. With $dA/d\theta_{cm} = (-1.35 \pm 0.05) \times 10^{-2}$ deg^{-1}, as determined from the measured asymmetries with the target polarized, the value of $\Delta A \equiv A_n - A_p$, is $[60 \pm 7(\text{stat.}) \pm 7(\text{syst.}) \pm 2(\text{syst.})] \times 10^{-4}$. The second systematic error reflects the uncertainty in $dA/d\theta_{cm}$.

The triptych in Fig. 6 shows the experimental results for $\Delta A \equiv A_n - A_p$ for np elastic scattering, at 183 MeV(6), at 350 MeV, and at 477 MeV(5). The horizontal lines present the theoretical predictions of Iqbal and Niskanen (IN)

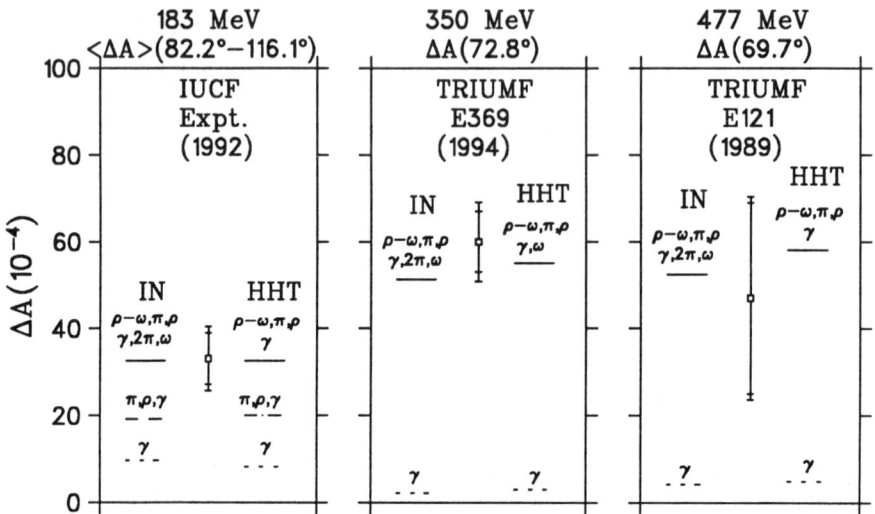

FIGURE 6. Experimental results for $\Delta A \equiv A_n - A_p$ for np elastic scattering at 183 MeV, 350 MeV and 477 MeV. The horizontal lines represent theoretical predictions of Iqbal and Niskanen (IN) and Holzenkamp, Holinde and Thomas (HHT).

and Holzenkamp, Holinde and Thomas (HHT). Note that the contributions corresponding to one photon exchange and to the np mass difference affecting charged π and ρ exchanges together suffice to give a theoretical prediction in agreement with the 350 MeV and 477 MeV TRIUMF results. This is because at these energies the angular distribution of the contributions due to $\rho^0 - \omega$ meson mixing crosses zero close to the zero-crossing angle of $<A(\theta)>$. The contribution due to 2π exchanges is small at all three energies. Reproducing the 183 MeV result from IUCF with the present calculations requires inclusion of the $\rho^0 - \omega$ meson mixing contribution, an approximately two standard deviation effect. Excluding the $\rho^0 - \omega$ meson mixing contribution will change the theoretical angular distributions of $\Delta A(\theta) \equiv A_n(\theta) - A_p(\theta)$ at 350 MeV and 477 MeV. A great deal of controversy has arisen regarding the role of $\rho^0 - \omega$ meson mixing in CSB(1). Consequently it is most important to extract any information regarding the angular distribution that can be obtained from the experimental data. More extensive data analysis is in progress and a more definitive systematic error for the 350 MeV data will be deduced.

REFERENCES

1. Miller, G.A., Nefkens, B.M.K. and Slaus, I., Phys. Rep. **194**, 1 (1990); Miller, G.A. and van Oers, W.T.H., in "Symmetries and Fundamental Interactions in Nuclei," ed. Haxton, W.C. and Henley, E.M. (World Scientific Publishing Co., Singapore), in press; and references therein.
2. Dumbrajs, O., et al., Nucl. Phys. B **216**, 277 (1983).
3. Nolen, Jr., J.A. and Schiffer, J.P., Ann. Rev. Nucl. Sci. **19**, 471 (1969).
4. LaFrance, P. and Winternitz, P., J. Physique **41**, 1391 (1980).
5. Abegg, R., et al., Phys. Rev. Lett. **56**, 2571 (1986); Phys. Rev. D **39**, 2464 (1989).
6. Knutson, L.D., et al., Phys. Rev. Lett. **66**, 1410 (1991); Vigdor, S.E., et al., Phys. Rev. C **46**, 410 (1992); Vigdor, S.E., private communication.
7. Williams, A.G., Thomas, A.W. and Miller, G.A., Phys. Rev C **34**, 756 (1987); Ge, L. and Svenne, J.P., Phys. Rev. C **33**, 417 (1986); **34**, 756(E) (1986); Holzenkamp, B.H., Holinde, K. and Thomas, A.W., Phys. Lett B **195**, 121 (1987); Iqbal, M.J. and Niskanen, J.A., Phys. Rev. C **38**, 838 (1988) and private communication.
8. Niskanen, J.A., Phys. Rev. C **45**, 2648 (1992).
9. Niskanen, J.A. and Vigdor, S.E., Phys. Rev. C **45**, 3021 (1992).
10. Abegg, R., et al., Nucl. Instrum. Methods A **234**, 11 (1985); **234**, 20 (1985); **254**, 469 (1987).
11. Abegg, R., et al., Nucl. Instrum. Methods A **306**, 432 (1991).

Overview of Charge Symmetry

Gerald A. Miller

Department of Physics, FM-15, University of Washington
Seattle, Washington 98195

Abstract. Charge independence and symmetry are approximate symmetries of nature. The observations of the small charge symmetry breaking effects and the consequences of those effects are reviewed. The effects of the mass difference between up and down quarks and the off shell dependence q^2 of ρ^0-ω mixing are stressed. We find that models which predict a strong q^2 dependence of ρ^0-ω mixing seem also to predict a strong q^2 variation for the ρ^0-γ^* matrix element, in contradiction with experiment.

INTRODUCTION

I will begin by stating the outline. First I define the terms charge independence and charge symmetry. Charge independence breaking of the $^1S_0$ nucleon-nucleon scattering lengths is discussed briefly. The bulk of the remainder is concerned with the breaking of charge symmetry (CSB). I shall review the evidence that the positive value of the light quark mass difference m_d-m_u plus electromagnetic effects accounts for CSB in systems of baryon number ranging from 0 to 208. A more detailed recent review is given in Ref. 1. See also Ref. 4.

DEFINITIONS

In the limit that m_d and m_u vanish and, ignoring electromagnetic effects the u and d quarks are equivalent. They form an isodoublet $\binom{u}{d}$. One may introduce the isospin operators $\vec{\tau}$ with $[\tau_i, \tau_j] = i\ \epsilon_{ijk}\tau_k, \tau_3|u>= |u>$ and $\tau_3|d>= -|d>$. The total isospin for a system of quarks is then $\vec{T} = \Sigma \vec{\tau}(i)/2$. In the limit in which each of m_d, m_u, α vanishes $[H, \vec{T}] = 0$. This vanishing, equivalent to the invariance under any rotation in isospin space is called charge independence. Charge symmetry requires only an invariance about rotations by π about the α axis in isospin space: $[H, P_{cs}] = 0$, with $P_{cs} = e^{i\pi T_2}$. P_{cs} converts u quarks into d quarks and vice versa: $P_{cs}|u>= -|d>, P_{cs}|d>= |u>$.

Henley's 1969 (2) review explained why it is important to distinguish between charge independence and charge symmetry.

There is a legitimate concern about the application of these concepts to reality. While each of m_d and m_u is less than 10 MeV, it is well known that $\frac{m_d}{m_u} \approx 2$; see the review (3). Thus one may wonder why any trace of charge independence would remain in nature. However the strong interaction effects of confinement cause the ratios governing charge independence breaking to be $\sim \frac{m_d-m_u}{300\text{ MeV}}$ or $\frac{m_d-m_u}{\Lambda_{QCD}}$ or $\frac{m_d-m_u}{4\pi f_\pi}$. The ~ 300 MeV can be thought of as arising from a constituent quark mass, bag model energy or quark condensate. Thus the effects of $m_d - m_u > 0$ are small, as are the electromagnetic effects. Thus charge independence holds approximately. This is well known, as hadronic and nuclear states are organized as isomultiplets.

The symmetry is not perfect and gives a unique opportunity to search for clues about the underlying dynamics. A prominent example is that the positive value of $m_d - m_u$ causes the neutron to be heavier than the proton.

NUCLEON-NUCLEON SCATTERING - $^1S_0$ CHANNEL

Charge independence $[H, \vec{T}] = 0$ imposes the equalities of the nucleon-nucleon scattering lengths $a_{pp} = a_{nn} = a_{np}$. But electromagnetic effects are large and it is necessary to make corrections. The results are analyzed, tabulated and discussed in Ref. 4. These are

$$\begin{aligned} a_{pp} &= -17.7 \pm 0.4 \text{ fm} \\ a_{nn} &= -18.8 \pm 0.3 \text{ fm} \\ a_{np} &= -23.75 \pm 0.09 \text{ fm} \end{aligned} \quad (1)$$

The differences between these scattering lengths represent CIB and CSB effects. There are very large percentage differences between these numbers which may seem surprising. But one must recall that that is the inverse of the scattering lengths that are related to the potentials. For two different potentials, V_1, V_2 the scattering lengths a_1, a_2 are related by

$$\frac{1}{a_1} - \frac{1}{a_2} = M \int dr\, u_1(V_1 - V_2)u_2 \quad (2)$$

where u_1 and u_2 are the wave functions. The differences between the inverse of the scattering lengths are small and furthermore (2)

$$\frac{\Delta a}{a} = (10 - 15)\frac{\Delta V}{V}. \quad (3)$$

One defines Δa_{CD} to measure the CIB, with

$$\Delta a_{CD} = \frac{1}{2}(a_{pp} + a_{nn}) - a_{np} = 5.7 \pm 0.3 \text{ fm}. \quad (4)$$

This corresponds to a charge dependence breaking of about 2.5% (2). The violation of charge symmetry is represented by the quantity

$$\Delta a_{CSB} = a_{pp} - a_{nn} = 1.5 \pm 0.5 \text{ fm.} \tag{5}$$

It is natural to use meson exchange models to analyze these low energy data. The longest range force arises from the one pion exchange potential OPEP, which also supplies significant breaking of charge independence. This is due to the relatively large mass difference. $\frac{m_{\pi^\pm} - m_{\pi^0}}{m_{\pi^0}} \approx 0.04$. One might worry about including the charge dependence of the coupling constants for neutral (g_0) and charged (g_c) pions. However $g_0^2 = g_c^2$ to better than about 1%, according to recent phase shift analyses of Bugg and Machleidt (6) and the Nijmegen (7) groups. One must also include the effects of the π mass difference in the two pion exchange potential TPEP. Henley & Morrison were the first to do that.

Some computations (2,7,8) of Δa_{CD} are displayed in Table 1. One can see that the agreement with the experimental value of $\Delta a_{CD} = 5.7 \pm 0.5$ fm is very good. There is room for a small contribution from quark effects. The net result is that the understanding of charge dependence has been rather good. However, more work is needed on channels other than $^1S_0$. See Ref. 1 for a discussion.

TABLE 1. Calculations of Δa_{CD}

	Henley, Morrison (2]) 1966	Ericson, Miller (7) 1983	Cheung, Machleidt (9) 1986
OPEP	3.5	3.5 ± 0.2	3.8 ± 0.2^a
TPEP (all)	0.90	0.88 ± 0.1	0.8 ± 0.1
Coupling Constants	b	0^c	
$\gamma\pi$		1.1 ± 0.4^d	1.1 ± 0.4^d
Total		5.5 ± 0.3	5.7 ± 0.5

All values of Δa are in fm.
a. This also includes the effects of $\pi\rho, \pi\omega$ and $\pi\sigma$ exchanges
b. HM showed that one could choose charge dependent coupling constants to describe the remainder of Δa_{CD}, but these were unknown
c. The effect of using charge dependent coupling constants tends to cancel if these are used consistently in OPEP and TPEP
d. This is an average (7) of the results of Refs. 9 and 10

CHARGE SYMMETRY BREAKING - $\rho^0\omega$ MIXING

The strongest and most prominent observation of charge symmetry breaking occurs in $\rho^0\omega$ mixing. The wave functions are given schematically as

$$|\rho^0> = \frac{1}{\sqrt{2}}\left(|u\bar{u}> -|d\bar{d}>\right)$$
$$|\omega> = \left(|u\bar{u}> +|d\bar{d}>\right), \qquad (6)$$

so that

$$<\rho^0|H|\omega> = \frac{1}{2} <u\bar{u}|H|u\bar{u}> -\frac{1}{2} <d\bar{d}|H|d\bar{d}>. \qquad (7)$$

This vanishes unless the Hamiltonian includes effects that distinguish between the u and d quarks. One simple example is the mass terms which contribute $m_u - m_d$. Thus the mixing matrix element is strongly influenced by the quark mass difference. Electromagnetic effects also enter, as we shall discuss.

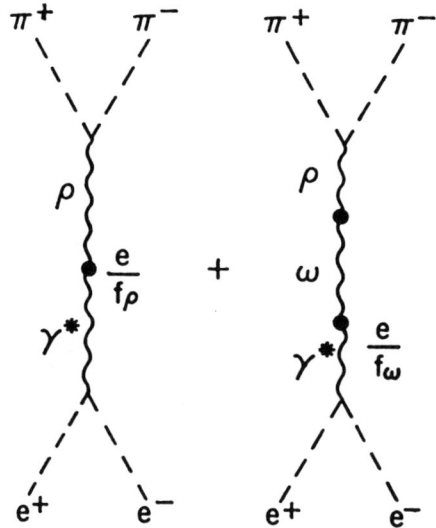

FIGURE 1. Amplitudes for $e^+e^- \to \pi^+\pi^-$

The effects of this matrix element are observed (11,12) in the annihilation process $e^+e^- \to \pi^+\pi^-$. The relevant diagrams are shown in Fig. 1 and the huge signal arising from the small widths of the ω-meson is displayed in Fig. 2. The mixing matrix element has been extracted (13) to be

$$<\rho^0|H|\omega> \approx -4500 \text{ MeV}^2. \qquad (8)$$

This matrix element includes the effect of the electromagnetic process depicted in Fig. 3. The quantities f_ρ and f_ω have been determined from the processes $e^+e^- \to \rho, \omega \to e^+e^-$. The most recent analysis (14) gives $<\rho^0|H_{em}|\omega> = 640 \pm 140$ MeV^2 so that the strong contribution ($H = H_{str} + H_{em}$) is given by $<\rho^0|H_{str}|\omega> \approx -5100$ MeV^2. Another notable feature is that the electromagnetic contribution to the $\rho\omega$-mixing self-energy is of the form

$$\Pi_{\rho\omega}^{em}(q^2) \sim \frac{e^2}{f_\rho f_\omega} \frac{1}{q^2} \tag{9}$$

where q^2 is the square of the vector meson four-momentum.

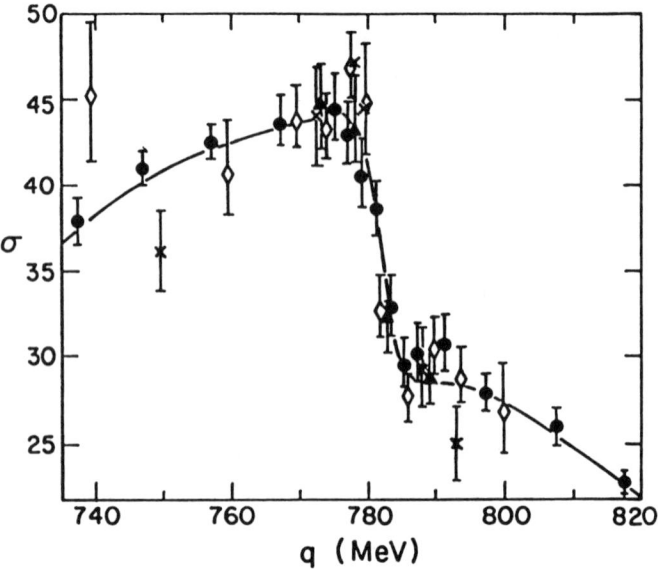

FIGURE 2. $\sigma(e^+e^- \to \pi^+\pi^-)$. These are the data introduced and summarized in Ref. 12.

It is natural to use the exchange of a mixed $\rho^0\omega$ meson as a mechanism for charge symmetry breaking nucleon-nucleon forces. This is shown in Figs. 4a and 4b. The electromagnetic contribution Fig. 4b is part of the long range, mainly Coulomb, electromagnetic interaction. The strong interaction term gives a nucleon-nucleon force of a medium range. This leads to a contribution to Δa_{CSB} of 1.4 fm, obtained by rescaling the Coon-Barrett (13) result by the ratio $1.11 = \left(\frac{5100}{4600}\right)$. This accounts for the observed effect $\Delta a_{CSB} = 1.5$ fm± 0.5 fm, while other effects seem small (13).

But this agreement with the experiment may not be satisfactory. A significant extrapolation is involved since $<\rho^0|H_{str}|\omega>$ is determined at $q^2 = m_\rho^2$, while in the NN force the relevant q^2 are spacelike, less than or equal to zero. Goldman, Henderson and Thomas (15) investigated the possible q^2 dependence of $<\rho^0|H_{str}|\omega>$ by evaluating the diagram of Fig. 6 using free quark propagators. They obtained a substantial q^2 dependence. The use of such a $<\rho^0|H_{str}|\omega>$ kills the resulting charge symmetry breaking potential. Very similar results were also obtained in the work of Refs. 16-19.

FIGURE 3. Electromagnetic contribution to ρ^0-ω mixing.

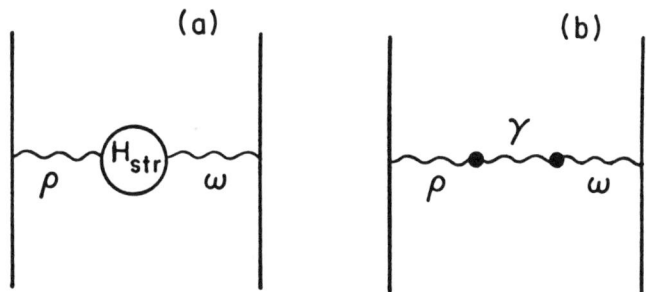

FIGURE 4. ρ^0-ω exchange contributions (a) Short range, strong interaction effect; (b) Long range, electromagnetic effect.

My opinion is that the charge symmetry breaking effects of the d-u mass difference in vector exchanges must persist, with little variation in q^2. However, I shall examine the consequences of the idea that $<\rho^0|H_{str}|\omega>$ does have a strong variation with q^2.

FIGURE 5. Quark model of ρ^0-ω mixing.

Consider the results of the "minimal" model of Krein, Thomas and Williams (17) which are displayed in Fig. 6. This work models confinement in terms of pole-less quark propagators. The rapid decrease of $<\rho^0|H_{str}|\omega>$ as q^2 is changed from time-like to space-like leads to a nearly vanishing CSB nucleon-nucleon interaction. But I stress that models which obtain the q^2 dependence of $<\rho|H_{str}|\omega>$ from the diagram Fig. 5 have an implicit prediction for the q^2 variation of the ρ-γ^* transition matrix element $e/f_\rho(q^2)$, see Fig. 7. My evaluation of this using the minimal model of Ref. 17 is shown in Fig. 8. I obtain similar results for the model of Ref. 19. A significant variation is seen, with a gain of a factor of 4 in the magnitude of $e/f_\rho(q^2)$. This is a noteworthy observation because $f_\rho(q^2)$ can be extracted from $e^+e^- \to \rho \to e^+e^-$ data at $q^2 = M_\rho^2$ and from the high energy $\gamma + P \to \rho^0 + \rho$ reaction at $q^2 = 0$. The results of many experiments are discussed in the beautiful review of Bauer, Spital, Yennie and Pipkin (20). They summarize $f_\rho^2(q^2 = M_\rho^2)/4\pi = 2.11 \pm 0.06$ and $f_\rho^2(q^2 = 0)/4\pi = 2.18 \pm 0.22$, as obtained from experiments at the CEA, DESY, SLAC and Cornell. Real photon data at γ energies from 3 to 10 GeV are used in the analysis.

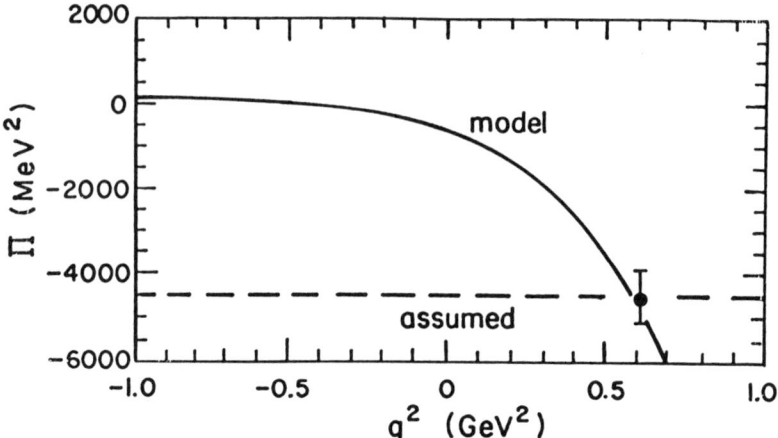

FIGURE 6. Model of Krein & Thomas - q^2 variation of $<\rho^0|H_{str}|\omega>$.

No variation of $f_\rho(q^2)$ with q^2 is found! This seems to be in strong disagreement with the consequences of the models of Refs. 15-19. The survival of such models seems to depend on finding a new way to account for the $\gamma + P \to \rho^0 + P$ data as well as for data on many γ-nucleon and nuclear reactions. A discussion of more recent work by O'Connell, Pierce, Thomas and Williams is given in Ref. 1.

For this article I shall assume that $< \rho^0|H_{str}|\omega >$ has little dependence on q^2. Then charge symmetry breaking in the $^1S_0$ channel is accounted for.

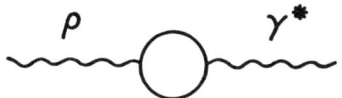

FIGURE 7. Quark model of the ρ^0-γ^* transition.

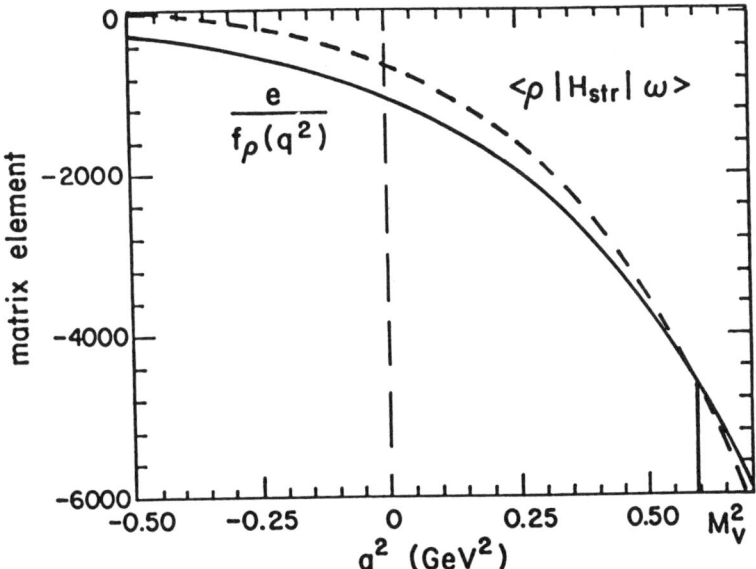

FIGURE 8. q^2 variation of $\frac{1}{f_\rho}$. The magnitude of f_ρ has been scaled to allow a comparison with the q^2 dependence of $< \rho^0|H_{str}|\omega >$.

CHARGE SYMMETRY BREAKING IN THE np SYSTEM

Searches for charge symmetry breaking in neutron-proton scattering offer an opportunity to find CSB effects not present in the nn or pp system. These class IV forces of Henley and Miller [22] have the form

$$V^{IV} = (\vec{\sigma}_1 - \vec{\sigma}_2) \cdot \vec{L}(\tau_1 - \tau_2)_3 A + (\vec{\sigma}_1 \times \vec{\sigma}_2) \cdot \vec{L}(\vec{\tau}_1 \times \vec{\tau}_2)_3 B \qquad (10)$$

where A and B are reasonable operators. The A term receives contributions from γ, and ρ^0-ω exchanges. B is dominated by π exchange effects. These operators cause the analyzing powers of polarized neutrons $A_n(\theta_n)$ and polarized

180 Overview of Charge Symmetry

protons to differ $A_p(\theta_p)$. Measurements (23,24) compare scattering with polarized neutron beam to neutron scattering on a polarized proton target. Time reversal invariance relates the latter measurement to A_p. These analyzing powers pass through zero at one angle θ_0 for the energy of TRIUMF (23) and IUCF (24) beams. If θ_0 for polarized neutrons differs from θ_0 obtained from polarized protons, then $\Delta\theta = \theta_0(n) - \theta_0(p) \neq 0$ and charge symmetry has been violated. Such observations were made in two beautiful experiments (23,24). The results presented in terms of $\Delta A = \frac{dA}{d\theta}\Delta\theta$, are shown in Fig. 9. The calculations use the Bonn meson-exchange potential so that all of the parameters governing the strong interaction are pre-determined. (Other calculations are discussed in Ref. 4.) The agreement between theory and experiment is very good. A pion exchange effect arising from the presence of the n-p mass difference in the evaluation of the vertex function dominates the 477 MeV measurement. The ρ^0-ω mixing term has a significant but non-dominating influence at 183 MeV. A separate contribution to this proceeding concerns the recent beautiful TRIUMF work at 350 MeV.

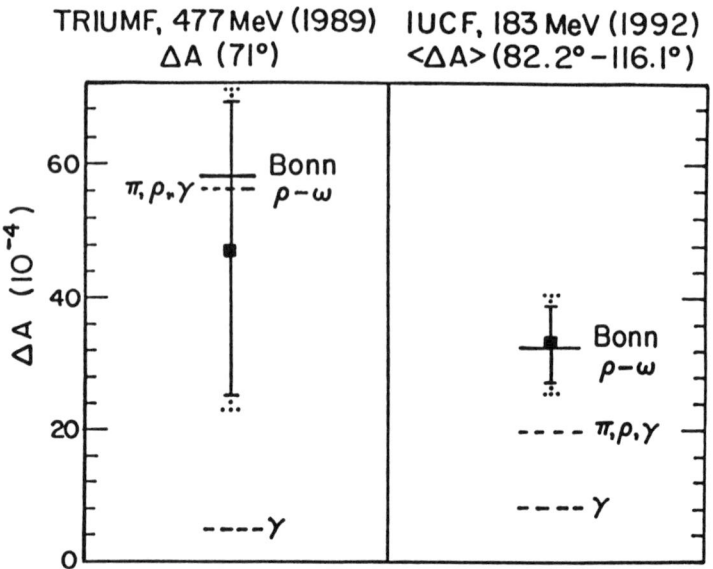

FIGURE 9. CSB in np scattering. This is after Fig. 3 of Ref. 24.

THE ^{3}He-^{3}H BINDING ENERGY DIFFERENCE

The ground states of ^{3}H and ^{3}He would have the same binding energy if charge symmetry holds. Instead $B(^3\text{H}) - B(^3\text{He}) = 764$ keV. The neutron rich system is more deeply bound. The bulk of the difference is due to the Coulomb interaction and other electromagnetic effects. The determination of the strong charge symmetry breaking relies on the ability to make a precise evaluation of such effects. The three body system is the best for such evaluations because the electromagnetic terms can be evaluated in a model independent way using measured electromagnetic form factors (27). Coon & Barrett used recent Saclay data to obtain

$$\Delta B(em) = 693 \pm 19 \pm 5 \text{ keV}, \tag{11}$$

where the first uncertainty is due to the determination of the form factors, and the second to the small model dependence of some relativistic effects. Similar values of $\Delta B(em)$ were obtained in Ref. (28). The difference between 764 and 693 is 71 keV, to be accounted for by charge symmetry breaking of the strong interaction. The use of $\rho^0 \omega$ exchange potential which reproduces Δa_{CSB} yields about 90 ± 14 keV in good agreement. The errors allow some room for other small effects such as $\pi\eta$ or $\pi\gamma$ exchanges.

NOLEN SCHIFFER ANOMALY

The mirror nuclei (N, Z) and $(N-1, Z+1)$ have the same binding energy, if charge symmetry holds. Nolen and Schiffer made an extensive analysis of the electromagnetic effects which dominate the observed binding energy difference. After removing the electromagnetic effects the neutron rich nuclei were more deeply bound (by about 7%) than the proton rich nuclei. Including additional detailed nuclear structure effects reduced the number to about 5%, see the review (4). A related problem occurs in understanding the energy difference between nuclei with $T > 1/2$ (^{48}Ca, ^{90}Zr, ^{208}Pb) and their isobaric analog states.

Blunden and Iqbal took up the challenge of seeing if a charge symmetry violating nucleon-nucleon potential, consistent with Δa_{CSB} could account for the missing 5% attraction. As shown in Table 2, it did. Actually I have rescaled the contributions due to $\rho^0 \omega$ mixing to reflect my present value of $< \rho^0 |H_{str}| \omega > = -5100$ MeV^2. The agreement is good but not perfect. Similar results have been obtained in Refs. (32) and (33).

The main point is that the anomaly is gone. CSB effects consistent with those observed in the NN system account for the bulk of the missing binding energy. There is some room for other effects such as nuclear-medium enhancements of the role of the d-u quark mass difference due to scalar effects (33-36). In any case the ultimate source of nuclear CSB is the light quark mass difference.

Note also that the use of CSB and CIB forces consistent with the NN data allows an explanation of the A dependence of non-Coulomb effects in the parent-analog mass differences (37). The use of such forces is now a standard part of shell model calculations (38).

TABLE 2. Blunden Iqbal calculation (see text)

A	orbit	Required CSB (keV)		Calc. CSB (keV)	
		DME	SkII	total	$\rho^0\omega$
15	$p_{3/2}^{-1}$	250	190	210	182
	$p_{1/2}^{-1}$	380	290	283	227
17	$d_{5/2}$	300	190	144	131
	$1s_{1/2}$	320	210	254	218
	$d_{3/2}$	370	270	246	192
39	$1s_{1/2}^{-1}$	370	270	337	290
	$d_{3/2}^{-1}$	540	430	352	281
41	$f_{7/2}$	440	350	193	175
	$1p_{3/2}$	380	340	295	258
	$1p_{1/2}$	410	330	336	282

The DME and SkII calculations are from Ref. 31.

SUMMARY

1. Charge independence breaking in the $^1S_0$ system is well explained (1,7,8).
2. Charge symmetry breaking is caused by the d-u quark mass difference $m_d - m_u > 0$, along with electromagnetic effects.
3. Measuring the e^+e^--$\pi^+\pi^-$ cross section at $q^2 \approx M_\omega^2$ allows an extraction of the strong contribution to the ρ-ω mixing matrix element $< \rho^0|H_{str}|\omega > \approx$ - 5100 MeV^2.
4. The TRIUMF (477, 350 MeV) and IUCF (183 MeV) experiments compare analyzing powers of $\vec{n}p$ and $\vec{p}n$ scattering and observe CSB at the level expected from π, γ and ρ^0-ω exchange effects.
5. The ρ^0-ω exchange potential accounts for $\Delta a_{CSB} = a_{pp} - a_{nn} = 1.5 \pm 0.5$ fm.
6. The use of such a potential accounts for the strong CSB contribution to the ^{3}He-^{3}H mass difference.
7. The use of potentials consistent with Δa_{CSB} and Δa_{CIB} accounts for formerly anomalous binding energy differences in mirror nuclei and in analog states.

The quark mass difference seems to be related to a large variety of phenomena in particle and nuclear physics. Most of the effects are well understood. Perhaps the next relevant question is why are there two light quarks with a slightly different mass?

REFERENCES

1. Miller, G.A., and Van Oers, W.T.H., "Charge Independence and Charge Symmetry," to be published in *Symmetries and Fundamental Interactions in Nuclei*, eds. E.M. Henley and W. Haxton, NUCL-TH 9409013.
2. Henley, E.M., in *Isospin in Nuclear Physics*, ed. Wilkinson, D.H. (North-Holland, Amsterdam,1969) 17; Henley, E.M. and Morrison, L.K., Phys. Rev. **141**, 148 (1966).
3. Gasser, J. and Leutwyler, H., Phys. Rpts. **87** 77 (1982).
4. Miller, G.A., Nefkens, B.M.K., and Slaus, I., Phys. Rept. **194** 1 (1990).
5. Bugg, D.V. and Machleidt, R., "πNN Coupling Constants from NN Elastic Data between 210 and 800 MeV", 1994 preprint NUCL-TH 9404017.
6. Bergervoet, R.R., van Campen, P.C., Rijken, T.A., and de Swart, J.J., Phys. Rev. Lett. **59** 2255 (1987); Bergervoet, J.R. *et al.*, Phys. Rev. **C41** 1435 (1991); Klomp, R.A.M., Stoks, V.G.J., and de Swart, J.J., Phys. Rev. **C44**, 1258 (1991).
7. Ericson, T.E.O. and Miller, G.A., Phys. Lett. **B132** 32 (1983); Phys Rev **C36** 2707 (1987).
8. Cheung, C.Y., and Machleidt, R., Phys. Rev. **C34**, 1181 (1986).
9. Chemtob, M., in *Interaction Studies in Nuclei*, eds. Jochim, H. and Ziegler, B. (North-Holland, Amsterdam, 1975) 487.
10. Banerjee, M., Univ. Maryland Technical Report No. 75-050 (1975).
11. Quenzer, A. *et al.*, Phys. Lett. **76B**, 512 (1978).
12. Barkov, L.M. *et al.*, Nucl. Phys. **B256** 365 (1985).
13. Coon, S.A., and Barrett, R.C., Phys. Rev. **C36**, 2189 (1987).
14. Langacker, P., Phys. Rev **D20**, 2983 (1979).
15. Goldman, T., Henderson, J.A., and Thomas, A.W., Few Body Systems **12**, 193 (1992).
16. Piekarewicz, J., and Williams, A.G., Phys. Rev. **C47**, R2462 (1993).
17. Krein, G., Thomas, A.W., and Williams, A.G., Phys. Lett. **B317**, 293 (1993).
18. Hatsuda, T., Henley, E.M., Meissner, T.H., and Krein, G., Phys. Rev. **C49**, 452 (1994).
19. Mitchell, K.L., Tandy, P.C., Roberts, C.D., and Cahill, R.T., Phys.

Lett. **B335**, 282 (1994).
20. Bauer, T.H., Spital, R.D., Yennie, D.R., and Pipkin, F.M., Rev. Mod. Phys. **50**, 261 (1978).
21. Henley, E.M., and Miller, G.A., in *Mesons in Nuclei*, eds. Rho, M. and Wilkinson, D.H. (North-Holland, 1979) 405.
22. Cheung, C.Y., Henley, E.M., and Miller, G.A., Nucl. Phys. A**305**, 342 (1978); A**348**, 365 (1980).
23. Abegg, R., *et al.*, Phys. Rev. Lett. **56**, 2571 (1986); Phys. Rev. D **39**, 2464 (1989).
24. Knutson, L.D., *et al.*, Nucl. Phys. A**508**, 185 (1990); Vigdor, S.E., Phys. Rev. C**46**, 410 (1992).
25. Holzenkamp, B., Thomas, A.W., and Holinde, K., Phys. Lett. **B195**, 121 (1987).
26. Miller, G.A., Thomas, A.W., and Williams, A.G., Phys. Rev. Lett. **56**, 2567 (1986); Williams, A.G., Miller, G.A., and Thomas, A.W., Phys. Rev. C**36**,1956 (1987).
27. Friar, J.L., Nucl. Phys. A**156**, 43 (1970); Faibre de la Ripelle, M., Fizika **4**, 1 (1972).
28. Brandenburg, R.A., Coon, S.A., and Sauer, P.U., Nucl. Phys. **294**, 305 (1978); Wu, Y., Ishikawa, S., and Sasakawa, T., Phys. Rev. Lett. **64**, 1875 (1990).
29. Nolen, J.A., and Schiffer, J.P., Ann. Rev. Nucl. Sci. **19**, 471 (1969).
30. Blunden, P., and Iqbal, J., Phys. Lett. **B198**, 14 (1987).
31. Sato, H., Nucl. Phys. A**269**, 378 (1976).
32. Suzuki, T., Sagawa, H, and Arima, A., Nucl. Phys. A**536**, 141 (1992).
33. Shahnas, M.H., 1994 preprint to be published Phys. Rev. C.
34. Krein, G., and Henley, E.M., Phys. Rev. Lett. **62**, 2586 (1989).
35. Hatsuda, T., Hogassen, H., and Prakash, M., Phys. Rev. Lett. **66**, 3851 (1991).
36. Saito, K., and Thomas, A.W., 1994 preprint NUCL-TH 9405009.
37. Schäfer, T., Koch, V., and Brown, G.E., Nucl. Phys. A**562**, 644 (1993).
38. Suzuki, T., Sagawa, K, and Van Gai, P., Phys. Rev. C**47**, 1360 (1993).
39. Ormand, W.E., and Brown, B.A., Phys. Lett. **B174**, 128 (1986).

An Experimental Test of Parity-Even Time Reversal Invariance with MeV Neutrons

P. R. Huffman[a,c], C. R. Gould[b,c], D. G. Haase[b,c], C. D. Keith[b,c],
N. R. Roberson[a,c], M. L. Seely[b,c] and W. S. Wilburn[a,c]

[a] *Physics Department, Duke University, Durham, NC 27708-0308*
[b] *Physics Department, North Carolina State University, Raleigh, NC 27695-8202*
[c] *Triangle Universities Nuclear Laboratory, Durham, NC 27708-0308*

Abstract. A new measurement of the parity-conserving, time reversal noninvariant (PC TRNI) fivefold correlation has been performed at Triangle Universities Nuclear Laboratory, using 6.57 MeV polarized neutrons and a cryogenically aligned holmium target. The PC TRNI spin correlation coefficient A_5 is measured to be $(1.1 \pm 2.3) \times 10^{-5}$, consistent with time reversal invariance. A bound of 3.8×10^{-3} (99% confidence) is extracted for the ratio of PC TRNI to TRI nuclear matrix elements.

INTRODUCTION

In this paper we present results from an improved search for parity-conserving, time reversal noninvariance (PC TRNI) recently performed at Triangle Universities Nuclear Laboratory. The measurement consists of a search for the fivefold correlation term $\vec{s} \cdot (\vec{I} \times \vec{k}) \vec{I} \cdot \vec{k}$ in the neutron-nucleus forward scattering amplitude (1). (Here, $\vec{s}$ is the spin of the neutron, $\vec{k}$ is the momentum of the neutron, and $\vec{I}$ is the spin of the target.) The first measurement of the fivefold correlation (FC) was reported by Koster et al (2). In the present work we use the $D(\vec{d}, \vec{n})$ source reaction and a higher target alignment to gain a factor of fifteen improvement in sensitivity to the FC term.

PC TRNI arises only as a second-order weak effect within the standard model, and as such gives rise to unobservably small effects ($\sim 10^{-15}$ according to Herczeg (3)). Nevertheless, experimental bounds on PC TRNI observables are much less stringent. The most precise direct experimental bound comes from detailed balance studies of the reaction $^{24}Mg(\alpha, p)^{27}Al$ and its inverse (4). In a series of measurements, relative differential cross sections were found to be equal to within 0.51%, and a limit of 3.5×10^{-3} (99% confidence) was extracted for the ratio of TRNI to TRI amplitudes (5). Indirect bounds of TRNI can be inferred from measurements of electric dipole moments (edm)

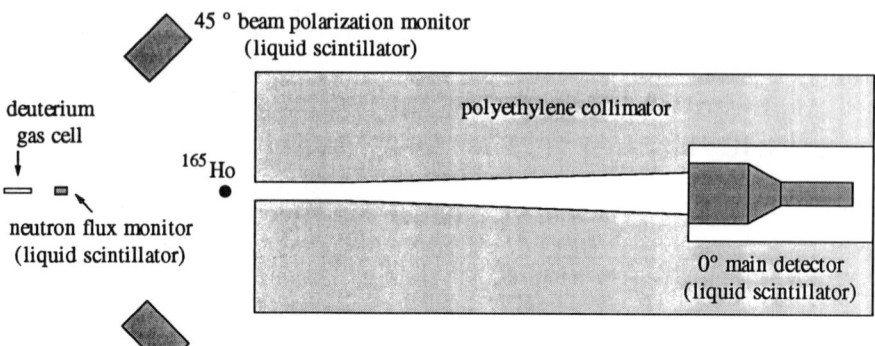

Figure 1: Experimental setup

of atoms and nucleons. Under the assumption that PC TRNI arises from $\rho^\pm$ exchange (6), Haxton et al (7) use bounds on the ^{199}Hg edm to give limits on α_T of 10^{-4}, where α_T is the ratio of T-violating to T-conserving nuclear matrix elements. A stronger bound (10^{-5}) is obtained from the neutron edm, but depends on the as yet unknown magnitude of the parity violating pion coupling constant.

EXPERIMENT

The FC term is measured via polarized neutron transmission through a rotating, cryogenically aligned ^{165}Ho target. The cross section for neutrons polarized parallel/anti-parallel (+/−) to the direction $\vec{I} \times \vec{k}$ is given by $\sigma^\pm = \sigma_0(1 \pm \sqrt{\frac{15}{8}} P_n \hat{t}_{20} A_5 \sin 2\theta)$ (8) where P_n is the polarization of the neutron beam, $\hat{t}_{20}$ is the tensor alignment of the holmium target with respect to the crystal symmetry axis, A_5 is the PC TRNI spin correlation coefficient, and θ is the angle between the holmium crystal alignment axis and the beam direction. The $\sin 2\theta$ angular signature of the FC term is uniquely isolated by flipping the neutron spin while simultaneously rotating the holmium target. A sequence of measurements of the transmission asymmetry $\epsilon_5(\theta) = [N^+(\theta) - N^-(\theta)]/[N^+(\theta) + N^-(\theta)]$ as a function of angle θ, can be fit to the form $C + D\sin 2\theta$ (C, D constants) to extract $A_5 = D/(\sqrt{\frac{15}{8}} P_n \hat{t}_{20} n \sigma_{tot})$ where n = 0.058 atoms/b is the target thickness and σ_{tot} is the total cross section (5.0 b at E = 6.57 MeV).

A schematic of the experimental arrangement is shown in Figure 1. The $D(\vec{d}, \vec{n})$ source reaction provides both higher neutron polarization ($P_n = 0.68$) and a more intense neutron flux ($N \sim 150\,\text{kHz}$) than the $T(\vec{p}, \vec{n})$ reaction used in Reference (2). An in-line monitor ($N \sim 300\,\text{kHz}$) measures neutron yield changes due to fluctuations in the beam intensity and tensor polarization. Finally, an improved thermal connection between the holmium target and the dilution refrigerator coldfinger provides increased tensor alignment within

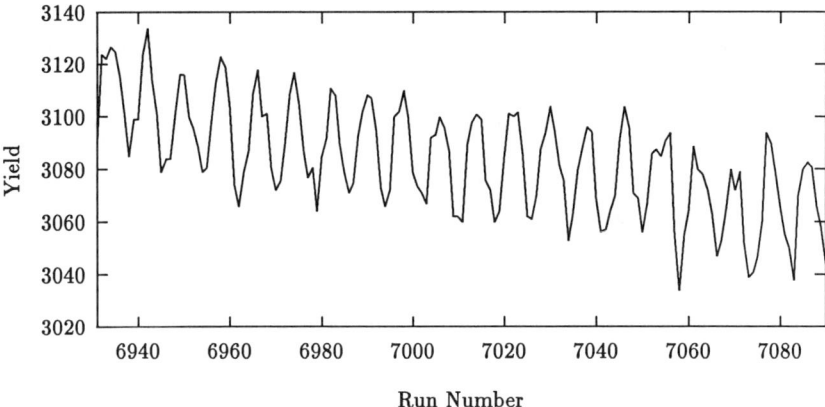

Figure 2: Deformation effect in the 9.5 MeV unpolarized neutron transmission yields

the holmium single crystal ($T \sim 120\,\text{mK}$, $\hat{t}_{20} = 1.44$, 95% of the maximum compared to 60% previously).

The cylindrical holmium target is rotated about its cylinder axis from $-180°$ to $+180°$ and back to $-180°$ in increments of $22.5°$. The c-axis of the crystal is perpendicular to the axis of the cylinder. The incident neutron spin is flipped every 100 ms. A run consists of 256 eight-step neutron spin sequences at a particular angle theta. Each spin sequence consists of neutron spin up (+) or down (−) in the time sequence $+--+-++-$. This eight step sequence removes any false asymmetries from linear or quadratic time dependant changes in detector efficiency. Each eight step sequence takes 800 ms, and a run lasts about four minutes. Polarization of the neutron beam is monitored by a pair of symmetrically located detectors placed at $45°$ to the left and right of the beam direction measuring the left-right asymmetry from the $D(\vec{d},\vec{n})$ reaction. Target alignment is measured by thermometry and confirmed by the deformation effect measurements discussed below.

DEFORMATION EFFECT RESULTS

To confirm the target alignment and crystal axis position, we first accumulate deformation effect data using an unpolarized neutron beam. The deformation effect in the total cross section is given by $\sigma_T = \sigma_0 + P_2(\cos\theta)\hat{t}_{20}\sigma_2 + \ldots$, where σ_2 is defined to be the deformation effect cross section. Transmission yields are shown in Figure 2 as a function of angle for 9.5 MeV neutrons. The oscillation is consistent with the expected $P_2(\cos\theta)$ variation in the yield. Deformation effect cross sections are extracted from fits to the data and are shown in Table 1 and Figure 3.

The measured deformation effect cross section values are consistent with previous measurements (9 – 12). These measurements confirm the target align-

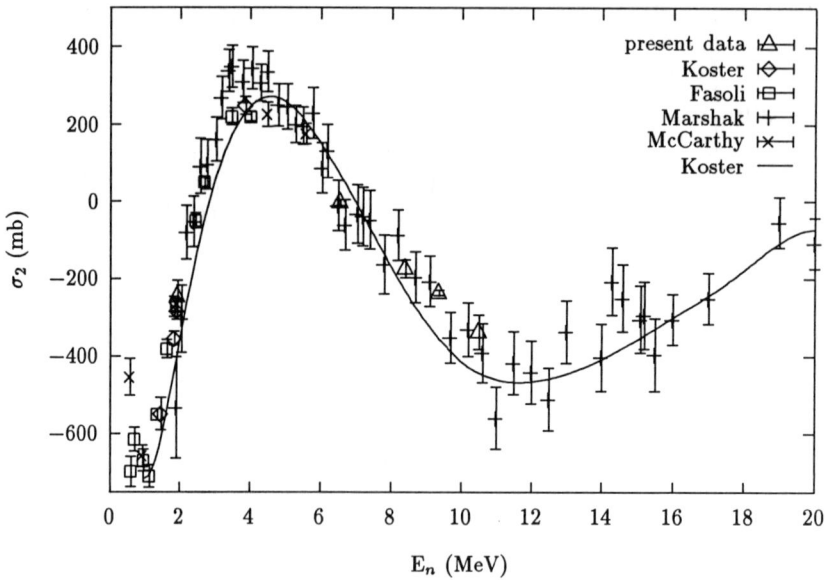

Figure 3: Deformation effect cross section as a function of energy. Data are taken from References (9 – 12). The solid line is a coupled channels optical model calculation of Koster (9).

ment obtained from thermometry and also confirm the crystal axis orientation.

Table 1. Deformation effect cross sections

Energy (MeV)	σ_2 (mb)
1.93	-246 ± 41
6.57	-3 ± 8
8.41	-174 ± 23
9.37	-237 ± 7
10.5	-338 ± 44

RESULTS

The FC experiment is carried out at 6.57 MeV, an energy where the deformation effect cross section is approximately zero. This minimizes angle-dependent inscattering and multiple scattering effects. The monitor-normalized asymmetry $\epsilon_5(\theta)$ for a sequence of 688 runs is shown in Figure 3. Each run corresponds to 256 spin flip sequences at a particular angle. Time reversal violation will appear as a $\sin 2\theta$ variation in the yield. As can be seen, there is no evidence for a $\sin 2\theta$ variation in the asymmetry, and a fit to these data

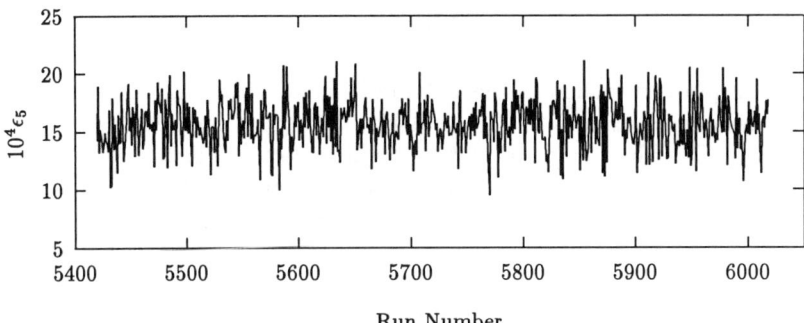

Figure 4: Monitor-normalized detector asymmetry $\epsilon_5(\theta)$ as a function of holmium alignment axis angle θ. Each run corresponds to four minutes of data at a given angle in the sequence $-180° \rightarrow +180° \rightarrow -180°$.

yields a coefficient $D = (2.9 \pm 5.9) \times 10^{-6}$. The PC TRNI spin correlation coefficient is $A_5 = (1.1 \pm 2.3) \times 10^{-5}$, consistent with time reversal invariance.

An upper limit on the ratio of T-violating to T-conserving ρ-nucleon coupling constants, $\bar{g}_\rho$ can be obtained from a coupled-channels, optical model calculation including a TRNI optical potential. Using the Simonius potential (6), a term having the form of the fivefold correlation has recently been derived from a microscopic folding model calculation (13). Including this potential alongside the strong optical potential in a coupled-channels code (14) allows a spin-correlation coefficient A_5 to be determined for a particular value of $\bar{g}_\rho$. From the measured asymmetry ($A_5 = (1.1 \pm 2.3) \times 10^{-5}$), one obtains a limit on $\bar{g}_\rho = 0.045 \pm 0.093$.

To compare to the values of α_T quoted by Haxton et al (7), requires scaling $\bar{g}_\rho$ by the ratio of the r.m.s. value of the non-zero NN matrix elements from the Simonius potential to that of the strong interaction ($\sim$ 3.6 keV / 300 keV). Using this prescription, one obtains a value of $\alpha_T \leq 3.8 \times 10^{-3}$ (99% confidence).

An improvement of 10-20 appears possible in this measurement. A new liquid nitrogen cooled, deuterium gas cell will provide six times the neutron flux than previously available and a four-segmented detector array will subtend about 15 times greater solid angle. Since count rates of about 2 MHz are anticipated in each detector, plastic scintillators must be used which will preclude pulse shape discrimination of neutrons from gamma rays. The gamma rays thus counted will dilute the final asymmetry by a small amount, but will not change the $\sin(2\theta)$ angular dependence of the neutron yield. A four-segmented, donut-shaped detector array located directly in front of the holmium target will provide an accurate neutron flux monitor as well as a monitor of the beam polarization. With about two weeks of data, a one-sigma bound of about 5×10^{-5} on α_T can be obtained.

SUMMARY

An improved version of the PC TRNI fivefold correlation experiment has been performed using 6.57 MeV polarized neutrons and an aligned holmium target. A PC TRNI spin correlation coefficient of $(1.1 \pm 2.3) \times 10^{-5}$ has been determined. An improvement of a factor of ten to twenty appears possible using an improved neutron source, a larger solid angle and segmented plastic detector.

ACKNOWLEDGMENTS

This work was supported in part by the U.S. DOE, Office of High Energy and Nuclear Physics, under contracts Nos. DEFG05-88-ER40441 and DEFG05-91-ER40619.

REFERENCES

1. For a recent overview of time reversal tests in neutron transmission see "Time Reversal Invariance and Parity Violation in Neutron Reactions," C. R. Gould, J. D. Bowman, and Yu P. Popov, eds. (World Scientific Publishing Co., 1994).
2. J. E. Koster, E. D. Davis, C. R. Gould, D. G. Haase, N. R. Roberson, L. W. Seagondollar, W. S. Wilburn, and X. Zhu, Phys. Lett. **B267**, 23 (1991).
3. P. Herczeg, Hyp. Int. **43**, 77 (1988).
4. E. Blanke *et al*, Phys. Rev. Lett. **51**, 355 (1983).
5. J. B. French, A. Pandey, and J. Smith in "Tests of Time Reversal Invariance in Neutron Physics", N. R. Roberson, C. R. Gould, and J. D. Bowman, eds. (World Scientific, 1987) p. 80.
6. M. Simonius, Phys. Lett. **B58**, 147 (1975).
7. W. C. Haxton, A. Höring and M. J. Musolf (to appear in Phys. Rev. D).
8. The expression for A_5 in Reference (2) contains an extra factor of two.
9. J. E. Koster, C. R. Gould, D. G. Haase, and N. R. Roberson, Phys. Rev. **C49**, 710 (1994).
10. U. Fasoli *et al*, Lett. Nuovo Cimento, **6**, 485 (1973).
11. H. Marshak *et al*, Phys. Rev. **C2**, 1862 (1970).
12. J. S. McCarthy *et al*, Phys. Rev. Lett. **20**, 502 (1968).
13. J. Engel, C. R. Gould, and V. Hnizdo (to be published).
14. V. Hnizdo and C. R. Gould, Phys. Rev. **C49**, R612 (1994).

TEST OF TIME-REVERSAL INVARIANCE IN PROTON-DEUTERON SCATTERING

P.D. Eversheim [a], F. Hinterberger [a], J. Bisplinghoff [a], R. Jahn [a],
J. Ernst [a], W. Kretschmer [b], H. Paetz gen. Schieck [c],
and H.E. Conzett [d]

[a] *Institut für Strahlen- und Kernphysik, Universität Bonn, D-53115 Bonn, Germany*
[b] *Physikalisches Institut, Universität Erlangen, D-91058 Erlangen, Germany*
[c] *Institut für Kernphysik, Universität Köln, D-50937 Köln, Germany*
[d] *Lawrence Berkeley Laboratory, Univ. of California, Berkeley, CA 94720, USA*

Abstract. A novel (P-even, T-odd) null test of time-reversal invariance is discussed that allows for a accuracy of 10^{-6}. The parity conserving time-reversal violating observable is the total cross-section asymmetry $A_{y,xz}$ of proton-deuteron scattering. The measurement is planned as an internal target transmission experiment at the cooler synchrotron COSY-Jülich.

INTRODUCTION

So far, CP-violation has only been observed in the neutral kaon system. In turn, evidence for the violation of time-reversal symmetry can be expected due to the CPT-theorem. Although the CP-violation could be accomodated by a complex phase of the Kobayashi-Maskawa-matrix (1) or the Θ-term (2) allowed by QCD, other explanations go beyond the standard model, like for instance the left-right symmetric models (3), the superweak interaction (4), or the extension of the Higgs sector (5). These extensions of the standard model may lead to interactions that are not related to the observed CP- or T-violation. Since the origin of the CP- or T-violation is not clear, further experimental tests of CP- or T-invariance outside the kaon system are necessary to probe the manifestation of the interaction responsible for the observed or possible new CP-violating effects.

Usually the parity conserving time-reversal invariance violating observable *P-even, T-odd* is tested by comparing the analyzing power A_y with the polarization P_y in the reversed reaction or by detailed balance. Since two observables have to be compared, the experimental accuracy (6) was limited to 10^{-3} - 10^{-2}. The accuracy can be increased by orders of magnitude if a null experiment is performed i.e. a non-vanishing value of one single observable proves that the symmetry involved is violated. An example of this kind of experiment is the measurement of the parity violating *P-odd, T-even* quantity A_z in proton-proton scattering (7), which has been measured to some 10^{-8} (Table 1).

Table 1. Comparison of accuracies of TRI and Parity-violation tests

Measurement	Remarks	Type	Ref.
Electric dipole moment of the neutron	$g_{PT} \leq 10^{-11}$	PT-odd	8)
$\gamma-\gamma$ correlation in ^{57}Fe	$\alpha_T \leq 5\cdot 10^{-6}$	T-odd	9)
P-A in p-p scattering	$g_T \leq 3\cdot 10^{-2}$	T-odd	10)
Detailed balance in p+^{27}AL↔^{4}He+^{24}Mg	$\alpha_T \sim g_T \leq 10^{-3}$	T-odd	11)
$\vec{n}$-transmission through ^{139}La	Hope for enhancement	PT-odd	12)
$\vec{n}$-rotation in ^{139}La	Enhancement $\sim 10^5$	P-odd	13)
A_z in $\vec{p}$-p scattering	Error: $\delta(A_z) \sim 2\cdot 10^{-8}$	P-odd	7)
$A_{y,xz}$ in $\vec{p}$-$\vec{d}$ scattering	This experiment	T-odd	

g: strength of T_{odd} NN potential
α: strength of an effective T_{odd} N-core potential

It has been proved (14) that there exists no null test of time-reversal invariance (TRI) in a nuclear reaction with two particles in and two particles out, except for forward scattering. On the basis of this latter statement Conzett (15) could show that a transmission experiment can be devised, which constitutes a TRI null test. He suggested to measure the total cross-section asymmetry $A_{y,xz}$ of vector polarized spin 1/2 particles interacting with tensor polarized spin 1 particles.

The observable $A_{y,xz}$ can be studied in the proton-deuteron system with the proton polarization along the y direction and the deuteron tensor polarization aligned along the x=z direction. The proton-deuteron system has the advantage of being a particularly simple system allowing still a direct analysis in terms of time-reversal violating (TRV) nucleon-nucleon potentials based e.g. on one meson exchange. In addition, the proton-deuteron system offers the opportunity to test simultaneously the p-p and the p-n interaction. According to an argument of Simonius (16), the n-p system is favoured over the p-p system as a TRI testing ground in view of the symmetry restrictions on possible TRV meson exchange processes. In principle, both systems can be tested with the intended experiment.

The TRI test can be performed at any beam energy. But since the TRV processes are of short-range nature - the long range contributions for these processes may be parameterized by ρ vector meson or f_1 axial-vector meson exchange (17) - the experiment should be performed at about 1 GeV.

THE QUANTITY OF INTEREST

A transmission experiment involving polarized particles is described by the generalized optical theorem (18):

$$\sigma_T = 4\pi/k \; \text{Im}(\text{Tr } \rho \cdot F(0)) \tag{1}$$

with: σ_T is the total cross section, k the wave number, ρ the density matrix, and $F(\vartheta)$ the scattering amplitude matrix for scattering at angle ϑ. The density matrix ρ reflects the experimental set-up, whereas the scattering matrix $F(0)$ of the forward scattering amplitude contains the physics, which is to be probed. In the following it is shown which observable conserves parity but violates time-reversal and that the time reversed situation is tested by flipping spins.

The discussion of the parity conserving TRV (*P-even, T-odd*) observable follows the arguments of Ohlsen (19). It is discussed in the projectile helicity frame. According to Ohlsen, the symmetry character of a polarization observable with $n_r = n_x + n_y + n_z$ indices can be determined by "counting rules". n_x, n_y, and n_z are the numbers of x, y, and z indices of the observable in question.

In detail the following counting rules apply for a P-even, T-odd, and R_z-even (invariant under rotation around the z-axis) quantity:

Time-reversal   : n_x has to be odd
Parity conservation : $n_x + n_z$ has to be even   (2)
R_z invariance   : $n_x + n_y$ has to be even

The minimal configuration fulfilling these conditions gives:

$$n_x = n_y = n_z = 1 \tag{3}$$

For a two particles in and two particles out reaction with a proton being the projectile, the target has to be at least a spin 1 particle, in order to offer a tensor polarization P_{xz} aligned in the x=z direction. Deuterons fulfill this requirement. Thus, the quantity of interest is the total cross-section asymmetry $A_{y,xz}$ for proton-deuteron scattering. $A_{y,xz}$ is measured by flipping the spins of the interacting particles. By flipping the spins in this particular system the time reversed situation can be prepared in two ways. This is shown in Fig. 1a-c by reversing all momenta and spins and exchanging the in- and out-going particles.

$A_{y,xz}$ is measured in a transmission experiment. This process is described by the transmission factor T(N):

$$T(N) = I(N)/I(0) = \exp\,-(\sigma_T\, N\, \rho d) \tag{4}$$

with I(0) is the intensity of the primary beam, I(N) is the intensity of the beam having passed N times the internal target with density ρ and thickness d, and σ_T is the total cross-section.

For the case of polarized particles σ_T has to be replaced by:

$$\sigma_T = \sigma_{y,xz} + \sigma_{Loss} = \sigma_o(1 + P_y P_{xz} A_{y,xz}) + \sigma_{Loss} \tag{5}$$

σ_o is the unpolarized total cross-section, σ_{Loss} is the loss cross-section, taking account of beam losses outside of the target, P_y and P_{xz} are the proton y-vector polarization and deuteron xz-tensor polarization, respectively. $A_{y,xz}$ is the quantity of interest, the total cross-section asymmetry.

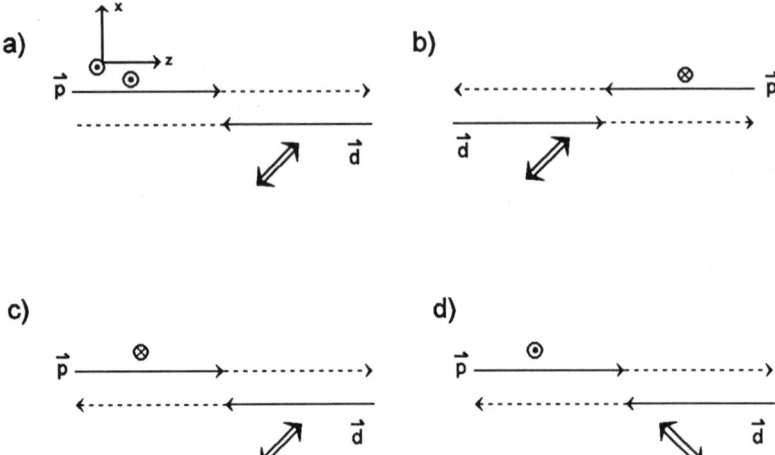

Fig. 1 Pictorial demonstration that a time-reversed situation is prepared by either a proton or a deuteron spin-flip. a) The basic system is shown. b) The time reversal operation is applied (momenta and spins are reversed and the particles are exchanged). In order to have a direct comparison between situation a) and b), two rotations $R_y(\pi)$ or $R_x(\pi)$ by 180° around the y- or x- axis are applied, leading to the situations c) and d), respectively. This is allowed, since the scattering process is invariant under rotations.
⊙ means proton spin up (y-direction), ⊗ proton spin down, and <=> stands for the deuteron tensor polarization

In order to become independent of σ_{Loss} the transmission asymmetry $\Delta T_{y,xz}$ can be introduced:

$$\Delta T_{y,xz} =: T^+ - T^- / [T^+ + T^-] =$$
$$= \exp{-(\chi^+)} - \exp{-(\chi^-)} / [\exp{-(\chi^+)} + \exp{-(\chi^-)}] \quad (6)$$

with T^+ is the transmission factor for the proton-deuteron spin-configuration $P_y > 0$ and $P_{xz} > 0$, T^- is the transmission factor for the time reversed situation, i.e. either $P_y < 0$ or $P_{xz} < 0$, and $\chi^{+/-}$ is the product of the factors (σ_T N ρd) with respect to the proton-deuteron spin-alignment. This gives:

$$\Delta T_{y,xz} = -\tanh{(\sigma_o \, N \, \rho d \, P_y \, P_{xz} \, A_{y,xz})} =: \tanh{(S \cdot A_{y,xz})} \quad (7)$$

Here S gives the sensitivity of the experiment with respect to the quantity of interst $A_{y,xz}$. Is the argument of the tanh in equation (7) small:

$$\Delta T_{y,xz} = -\sigma_o \, N \, \rho d \, P_y \, P_{xz} \, A_{y,xz} = S \cdot A_{y,xz} \quad (8)$$

With the help of equation (8) the total cross-section asymmetry $A_{y,xz}$ can be calculated.

THE EXPERIMENTAL SET-UP

Basically, the experimental set-up does only need equipment that is provided for other experiments, i.e. a polarized proton beam in COSY, an atomic beam source producing polarized protons or deuterons for internal target experiments, and an on-line current monitor (20) of high precision that is a standard diagnosis device for the operation of COSY. Therefore, in this experiment the COSY-ring is not only used as an accelerator, but also as a forward spectrometer and a detector.

For a conventional atomic beam target and a two weeks run a maximum sensitivity $S = -0.1$ can be expected. With the on-line current monitor, used for the usual COSY beam diagnostic, the beam current should be measurable to $5 \cdot 10^{-4}$ within 40 ms. Within 10 days there is time for about $2 \cdot 10^7$ subsequent measuments, increasing the precision of a $\Delta T_{y,xz}$ measurement to $1 \cdot 10^{-7}$. With the sensitivity $S = -0.1$, the TRV quantity $A_{y,xz}$ can then be measured to a precision of 10^{-6}.

DISCUSSION

It should be noted that this result seems to be independent of the proton beam intensity. However, the current measuring device in the COSY ring provides the accuracy of $5 \cdot 10^{-4}$ within 40 ms only for more than $1.3 \cdot 10^9$ protons circulating at 1.5 MHz in the ring. On the other hand, since the ring has the potential to be filled with up to $2 \cdot 10^{11}$ protons and eventually a target-cell may be added to the atomic beam target, the basic statistics can be improved to allow for a 10^{-6} accuracy of $A_{y,xz}$. Therefore, the handling of the systematic errors becomes increasingly important.

Since the time reversed state can be generated by even two spin configurations of beam and target, more artificial parameter correlations can be introduced than for instance in measurements of parity violation (PV) in p̄-p scattering (7). In PV measurements these artificial correlations are used to reduce systematic error contributions due to parameter fluctuations. Another advantage of the observable of interest $A_{y,xz}$ is, that all observables are suppressed that are not R_z- or P-even. Furthermore, the proton polarization p_z is very well prepared, since it has to be aligned to the magnetic field of the dipols in the COSY-ring. In the end, the most probable observable, which could fake a *P-even, T-odd* effect, is the asymmetry $A_{y,y}$, which is sensitive to proton and deuteron y-vector polarisation. Thus, special attention has to be payed to the alignment of the atomic beam target and to the production of vector polarized deuterons. It can be shown that none of these error sources discussed so far dominates the precision of the experiment at the 10^{-6} level.

Finally, a decisive advantage of this internal experiment shows up in equation

(8). In contrast to external experiments (21) the sensitivity S increases the more frequent the beam passes the target. Therefore, by combining the data in the analysis appropriately, the precision of the experimental result is improved more, than can be expected from the increased statistics alone.

REFERENCES

1) M. Kobayashi, T. Maskawa, Progr. Theor. Phys. **49** (1973) 652
2) R.J. Crewther et al., Phys. Lett. **88B** (1979) 123
3) J.C. Pati, A. Salam, Phys. Rev. **D10** (1974) 275
4) L. Wolfenstein, Ann. Rev. Nucl. Part. Sci. **36** (1986) 137
5) S.L. Glashow, Nucl. Phys. **22** (1961) 579
6) H.L. Harney et al., Nucl. Phys. **A518** (1990) 35
7) COSY Proposal #3
   P.D. Eversheim et al., Phys. Lett. **B256** (1991) 11
   S. Kistryn et al., Phys. Rev. Lett. **58** (1987) 1616
8) N.F. Ramsey, Ann. Rev. Nucl. Sci. **32** (1982) 211
9) M. Beyer, Nucl. Phys. **A493** (1989) 335 and
   N.K. Cheung, H.E. Henrikson, F. Boehm, Phys. Rev. **C16** (1977) 2381
10) C.F. Hwang et al., Phys. Rev. **119** (1960) 352
11) E.Blake et al., Phys. Rev. Lett. **51** (1983) 355
12) V.E. Bunakov, Phys. Rev. Lett. **60** (1988) 2250
13) Y. Yamaguchi, J. Phys. Soc. Jpn. **57** (1988) 1518
14) F. Arash, M.J. Moravcsik and G.R. Goldstein, Phys. Rev. Lett. **54** (1985) 2649
15) H.E. Conzett, 7th Intl. Conf. on "Pol. Phen. Nucl. Phys.", Paris (1990) 2D
16) M. Simonius, Phys. Lett. **B58** (1975) 147
17) M. Beyer, Nucl. Phys. **A560** (1993) 895
18) C. Bourrely, E. Leader, J. Soffer, Phys. Rep. **59** (1980) 95; note that their equation (3.41) is not the most general one, since ρ is in general not a real matrix
19) G.G. Ohlsen, Rep. Progr. Phys. **35** (1972) 717
20) K.B. Unser, CERN/SL/90 – 27 (BI)
21) Yuan et al.,Phys. Rev. Lett. **57** (1986) 1680

Tests of Discrete Symmetries Using Polarization

B.M.K. Nefkens

University of California, Los Angeles
Department of Physics, Los Angeles, CA 90024-1547

Abstract Some tests based on polarization observables of the discrete strong symmetries C, P, CP and T are discussed. Such tests can provide a practical way to search for physics beyond the Minimal Standard Model.
Other tests probe the nuclear discrete symmetries of isospin, charge symmetry and G-conjugation. Such symmetries are the consequence of the light-flavor symmetry of QCD and allow an assessment of the symmetry breaking due to $\pi^\circ - \eta$, $\rho - \omega$ and even $a^\circ - f$ meson mixing and ultimately lead to the determination of the up-down quark mass difference.

Introduction

The magnitude of a spin vector is not affected by the application of the discrete strong symmetries, P, C, CP and T or by the discrete, broken symmetries of isospin, charge symmetry and G-conjugation. This implies important constraints on the polarization in many reactions. One can search for physics beyond the Minimal Standard Model by testing the validity of the discrete strong symmetries.

Measurements of the intrinsic, that is the non-Coulombic, breaking of charge symmetry and G-conjugation is the only way to determine the up-down quark mass difference and provide valuable information on the validity of the modern nuclear potentials models with meson mixing and on low energy QCD theories and quark models.

CP and C invariance have not been tested well. The charge-conjugation operator $(\hat{C})$ changes colored quarks into anticolored antiquarks and charged leptons into their anti-leptons. Therefore, only neutral flavorless bosons, meson-antimeson pairs and lepton-antilepton pairs *in a well defined state of angular momentum* (ℓ) can be eigenstates of $\hat{C}$. The effects of the operators $\hat{P}, \hat{C}$ and $\widehat{CP}$ on various particles, fermion-antifermion pairs, e.g., $e^+e^-, p\bar{p}$ and $q\bar{q}$ and boson-antiboson pairs such as $\pi^+\pi^-$ and K^+K^- are summarized in Table 1.

Table 1: The eigenvalues of the operators $\hat{P}, \hat{C}$ and $\widehat{CP}$. $(f\bar{f})$ is a fermion-antifermion pair and $(b\bar{b})$ is a boson-antiboson pair in a well defined state of angular momentum (ℓ).

	$\hat{P}$	$\hat{C}$	$\widehat{CP}$
$(f\bar{f})$, e.g., $e^+e^-, p\bar{p}, q\bar{q}$	$(-1)^{\ell+1}$	$(-1)^{\ell+s}$	$(-1)^{s+1}$
$(b\bar{b})$, e.g., $\pi^+\pi^-, K^+K^-$	$(-1)^\ell$	$(-1)^\ell$	+
$\gamma, \rho^\circ, \omega, \psi, Z, K_1^\circ$	−	−	+
$\pi^\circ, \eta, \eta', K_2^\circ$	−	+	−

Discrete Strong Symmetries

Consider the decay $\eta \to \pi^\circ \mu^+ \mu^-$ or $K_L \to \pi^\circ \mu^+ \mu^-$ in the dilepton center of mass frame. The application of the charge conjugation operation changes the $\mu^\pm$ into $\mu^\mp$ but leaves the momentum and spin vectors unchanged. Thus, C invariance requires that the μ^+ polarization in $\eta \to \pi^\circ \mu^+ \mu^-$ decay is symmetric about 90°.

There are no direct tests of time reversal invariance because of the nature of the operator $\hat{T}$. However, there are many consequences of T invariance and this implies testing CP invariance by the CPT theorem. Again, consider the decay $\eta \to \pi^\circ \mu^+ \mu^-$. The amplitude constructed as $A_{1T} \sim \vec{\sigma}_{\mu^+} \cdot (\vec{p}_\mu^+ \times \vec{p}_\mu^-)$ is odd under T. The absence of such an amplitude requires that the μ^+ polarization vector must lie in the decay plane. There is a related test of CP and T invariance namely the absence of the term $A_{2T} \sim \vec{p}_\pi \cdot (\vec{\sigma}_\mu^+ \times \vec{\sigma}_\mu^-)$, requiring the simultaneous measurement of the μ^+ and μ^- polarizations. There is as yet no practical way for determining the μ^- polarization.

Consider the decay of a pseudoscaler meson into a lepton pair such as $K_L \to \mu^+ \mu^-$ or $\eta \to \mu^+ \mu^-$. CP and P invariance require that the longitudinal polarization (P_L) of the muon is zero. Explicitly, $P_L \equiv (N_R - N_L)/(N_R + N_L)$, where $N_{R(L)}$ is the number of right-(left)-handed positive muons. It has been calculated[1] that in the Minimal Standard Model (MSM) $P_L(K_L \to \mu^+ \mu^-) \simeq 7 \times 10^{-4}$, while $P_L(\eta \to \mu^+ \mu^-) \sim 0$. One can extend the very successful MSM without compromising its many agreements with various experimental tests, by hypothesizing the existence of two or more extra Higgs particles[2], leading to the prediction $P_L(K_L \to \mu^+ \mu^-) \simeq 1 \times 10^{-3}$. In a similar way a hypothetical leptoquark scaler that transforms under $SU(3)_C \times SU(2)_L \times U(1)_Y$ as $(3,1,-\frac{2}{3})$ leads to the prediction[3] $P_L(\eta \to \mu^+ \mu^-) \simeq 1 \times 10^{-3}$.

Broken Discrete Symmetries

An important contribution of nuclear physics to basic Modern Physics is the development of the concept of broken symmetry, specifically, charge symmetry (CS), isospin symmetry (IS) and G-conjugation symmetry (GS). These symmetries have the same origin in unbroken QCD: the interactions of the up and down quarks and antiquarks have the same strength. In the idiom of QCD this carries the fancy name "light flavor symmetry"[4]. Nuclear charge symmetry is the invariance of the Lagrangian to the interchange of the up and down quarks. G-conjugation symmetry is the invariance under interchange of the up with the anti-down and the down with the anti-up quark.

A CS, IS and GS transformation has no effect on spin. This implies a 90°-symmetry or antisymmetry relation among the analyzing powers and polarizations of isospin symmetries reactions, derived originally by Bilenkii *et al.*[6]. It is the logical extension of the Barshay-Temmer theorem[7], which governs the symmetry about 90° of the differential cross sections of isospin symmetric reactions. Consider the reaction $a + b \to c + d$ where particles a and b are members of an isospin doublet and the total isospin of particles c and d can have only one value of the total isospin. The classical nuclear physics example is $^3He(^3H, d)\alpha$, see Ref. [8]. The effect of a CS transformation is the interchange of ^3He and ^3H, leaving the deuteron and α particle unaffected. CS invariance implies symmetry about 90° of the tensor analyzing powers A_{yy} and A_{xx} and antisymmetry of the tensor A_{xz} and vector analyzing power A_y.

The up and down quarks have slightly different masses, $m_d - m_u \simeq 3$ MeV, they differ also in electric charge and magnetic dipole moment. Therefore, CS, IS and GS can only be approximate symmetries, called broken symmetries. The intrinsic CS breaking (CSB) is due to the mass difference between the up and down quarks. The long term objective of investigating CS, IS and GS breaking is the determination of the up and down quark mass difference at different matter densities and energies. Elementary particle physics provides several cases where there is a simple relation between CSB and $(m_d - m_u)$, for instance in the decay $\psi' \to \psi + \pi°$, see Refs. [4] and [5]. Nuclear physics is not in the same position because the intrinsic CSB is quite small. The standard technique is to describe CSB by a CSB potential that is driven by CSB meson mixing such as $\rho° - \omega$ and $\pi° - \eta$ mixing which have a one-to-one correspondence to $(m_d - m_u)$, see Ref. [5].

G-conjugation symmetry, (GS), has not been investigated much for lack of suitable systems containing antimatter, except at LEAR. To test GS consider the annihilation reaction $\bar{n}p \to \pi^+\pi°$, Refs. [6],[9],[10]. The isospin of initial and final state is $I = 1$. Conservation of angular momentum and parity require that $\ell_i = \ell_f \pm 1$ and $S_i = 1$, where S_i is the spin of the $\bar{n}p$ system. Since $G = (-1)^{\ell+S+I}$ and ℓ_i is even, we must have odd ℓ_f and $G = +1$. $\bar{n}$ and p form an G-spin doublet and π^+ and $\pi°$ are self conjugate under a G

transformation. Thus, by application of the Bilenkii et al., relations we find that the vector analyzing power in $\bar{n}p \to \pi^+\pi^0$ must be antisymmetric about 90°. An analogous study concerns the complimentary reaction $\bar{n}p \to \pi^+\eta$, in this case the final state must have even angular momentum. In both cases, $\bar{n}p \to \pi^+\pi^0$ and $\bar{n}p \to \pi^+\eta$, we expect GS violation as a consequence of $\pi^0 - \eta$ mixing. Studies of such violations can be of help to investigate the reaction mechanism of the $\bar{n}p$ interactions, in analogy to Conzett's analysis[11] of CS breaking in $^3He(^3H,d)\alpha$.

Next we discuss some polarization/asymmetry tests of CS which may be of interest to LISS if it can provide a tagged neutron beam. We have selected cases in which there are no Coulomb corrections to be made. There are further examples of the use of the Bilenkii et al., relations[6].

Consider the reactions $n + \vec{p} \to \eta + d$ and $n + \vec{p} \to \pi^0 + d$, the vector analyzing power must be antisymmetric about 90°. We expect deviations to occur at the opening of the η channel due to $\pi^0 - \eta$ mixing. Similarly for the reactions $n + \vec{p} \to \omega + d$ and $n + \vec{p} \to \rho + d$ in which a violation of the antisymmetry of the vector analyzing power is due to $\rho^0 - \omega$ mixing. One might even try to investigate the reactions $n + \vec{p} \to f_o + d$ and $n + \vec{p} \to a_o + d$ selecting the a_o and f_o by their decay into $K^\circ - \bar{K}^\circ$ systems with the objective of studying $a_o - f_o$ mixing.

We conclude by mentioning some tests of isospin invariance based on polarization/asymmetry measurements. In all cases the incident beam energy should be chosen such as to cover the opening of either the η, the ω or the f_o channel.

Consider the reactions $n + d \to \pi^0 + T$ and $n + d \to \pi^- +^3 He$, CS implies that the deuteron analyzing power should be the same. A drawback of this test is evaluating the correction for the mass differences in the final state. There exist already preliminary results from Saturne on a related pair of reactions $\vec{d} + p \to \pi^0 +^3 He$ and $\vec{d} + p \to \pi^+ +^3 H$.

Another nice test of IS is the validity of the triangle inequalities measured with a transversely polarized proton target:

$$\left[\sqrt{\sigma+\uparrow} - \sqrt{\sigma-\uparrow}\right]^2 \leq 2\sigma^\circ \uparrow \leq \left[\sqrt{\sigma+\uparrow} + \sqrt{\sigma-\uparrow}\right]^2$$

where $\sigma+\uparrow$ stands for $d\sigma/d\Omega[\pi^+p\uparrow \to \pi^+p]$, $\sigma^-\uparrow$ for $d\sigma/d\Omega[\pi^-p\uparrow \to \pi^-p]$ and $\sigma^\circ \uparrow$ for $d\sigma/d\Omega[\pi^-p\uparrow \to \pi^\circ n]$, where $\uparrow$ indicates a spin-up polarized proton target. Analogous expressions hold for the spin down cross sections. Some data already exists on this test[12].

One can also test the triangle inequalities for hyperon production, $\pi^-p \to K^\circ\Sigma$, $\pi^-p \to K^+\Sigma^-$ and $\pi^+p \to K^+\Sigma^+$. CSB is expected here as a consequence of $\Lambda - \Sigma^\circ$ mixing.

Work supported in part by the U.S. Department of Energy.

References

1. Herczeg, P., Phys. Rev. **D27**, 1512 (1983).

2. Geng, C.Q. and Ng, J.N., TRIUMF Report TRI-PP-90-64.

3. He, X-G. and McKellar, Phys. Rev. **D42**, 248 (1990).

4. Nefkens, B.M.K., Miller, G.A. and Slaus, I., in *Comments in Nuclear and Particle Physics*, **20**, 221 (1992).

5. Miller, G.A., Nefkens, B.M.K. and Slaus, I., Phys. Rep. **194**, 1 (1990).

6. Bilenkii, S.M. *et al.*, Sov. J. Nucl. Phys. **4**, 763 (1967).

7. Barshay, S. and Temmer, G., Phys. Rev. Lett. **12**, 728 (1964).

8. Haglund, R.F. *et al.*, Phys. Rev. **C16**, 2151 (1977).

9. Bertin, A. *et al.*, Phys. Lett. **B244**, 519 (1990).

10. Green, A.M. *et al.*, Helsinki preprint HU-TFT-91-24.

11. Conzett, H., in Proc. 4th Int. Symp. on Polarization Phenomena in Nuclear Reactions, ed. Grüebler, W. and König, V. (Birkhauser, Basel, 1976) p. 105.

12. Wightman, J. *et al.*, Phys. Rev. **D38**, 3365 (1988).

Reflections on Symmetries at SPIN '94

S.A. Page

Department of Physics, University of Manitoba, Winnipeg, Canada R3T 2N2

INTRODUCTION

In my view, the parallel sessions on 'Symmetries' were amongst the most stimulating sessions of this conference. Speakers reported on experimental tests of Charge Symmetry, Parity, and Time Reversal violation and their theoretical interpretation, spanning a wide range of energy scales and experimental techniques. I hope that this brief summary will whet the reader's appetite to explore the many contributed papers which follow.

CHARGE SYMMETRY VIOLATION

Charge Symmetry is an approximate symmetry of the strong interaction, and the small extent to which this symmetry is broken has been long established, as illustrated by familiar examples such as low energy scattering length differences in nucleon-nucleon scattering and binding energy differences of mirror nuclei. However, until recently, precision tests in very simple systems offering unambiguous theoretical interpretation have been elusive. For high precision tests, the np system is particularly attractive, since the Coulomb interaction does not contribute to the individual measurements which are compared to evaluate the degree of charge symmetry violation.

W.T.H. van Oers[1] reported a new result from TRIUMF which is the third in a series of high-precision measurements of charge symmetry breaking (CSB) in np scattering at intermediate energy based on a common experimental technique. The CSB test is performed by comparing the transverse spin dependent analyzing power $A(\theta)$ for $(\vec{n} + p)$ and $(n + \vec{p})$ elastic scattering. The scattering cross section for transversely polarized beam or target is given by: $\frac{d\sigma(\theta)}{d\Omega} = \frac{d\sigma^\circ(\theta)}{d\Omega}(1 + PA(\theta))$, where P is the beam polarization. The very small ($\simeq 10^{-3}$) difference between the analyzing powers, $\Delta A(\theta) = (A_n(\theta) - A_p(\theta))$ is a measure of charge symmetry breaking in the np system. The experiment can only be done to sufficient precision at one angle $\theta_\circ$ where $A(\theta) = 0$, and the result becomes independent of beam and target polarizations. ΔA is determined by comparing restricted angular distributions of $A_n(\theta)$ and $A_p(\theta)$ near the zero-crossing angles $\theta_{on} \simeq \theta_{op}$. The TRIUMF result at 350 MeV is in good agreement with theoretical predictions accounting for direct electromagnetic effects and the up-down quark mass difference. The three new np measurements are shown in figure 1; all are in good agreement with theory.

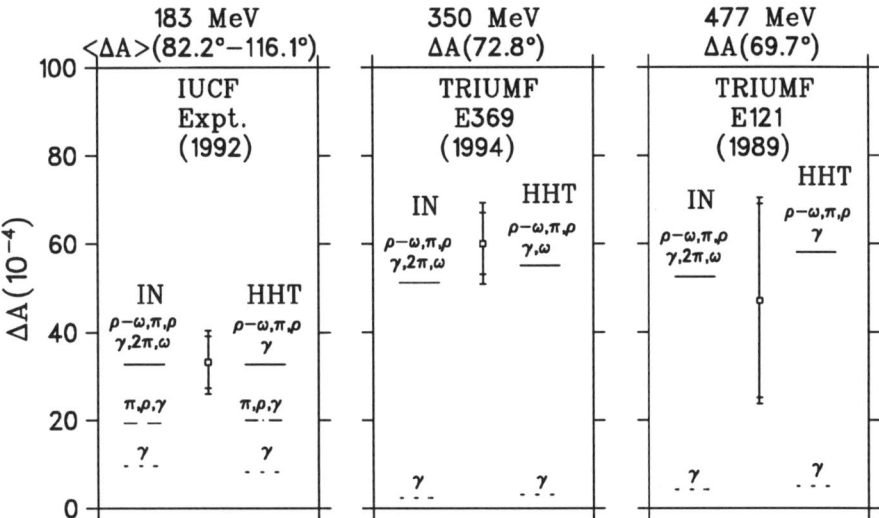

FIGURE 1: Comparison of high precision CSB tests in np scattering(1). The new TRIUMF measurement at 350 MeV is shown in the center panel. Theoretical predictions are indicated by the solid lines – see reference (1) for discussion.

G.A. Miller(2) outlined a consistent theoretical approach for calculating CSB effects in a variety of strongly interacting systems. The major contributions to CSB in the np experiments include electromagnetic effects, isospin mixed $\rho^0 - \omega$ exchanges, and π exchange, calculated using the Bonn meson-exchange potential. At present, a spirited debate is ongoing regarding the magnitude of the isospin mixing $\rho^0 - \omega$ contribution to a variety of CSB experimental results, via the q^2 dependence of the $\rho^0 - \omega$ mixing matrix element. Unfortunately, this contribution is just one of many to the value of ΔA in np scattering; at 350 MeV, the $\rho^0 - \omega$ contribution is negligible at θ_o, but it contributes significantly to the slope of the angular distribution of ΔA in that region. The TRIUMF group is currently exploring ways to extract this angular distribution, which suffers from the imprecision of beam and target polarization measurements, from their data, in hopes of shedding light on this important question.

PARITY VIOLATION

The study of the weak interaction in strongly interacting systems is perhaps equally challenging both to experiment and to theory. At low and intermediate energy, a meson exchange model of the weak nucleon-nucleon interaction has been the most successful at describing experimental data, with a minimum of 6 independent weak meson-nucleon coupling constants for π, ρ, and ω exchanges of different isospin character predicted with relatively large uncertainties using the Standard Model and QCD(3). Existing data can place constraints on the values of 4 weak meson-nucleon couplings, but not all data are consistent. The most stringent constraints are from a new generation of high precision measurements in $\vec{p}p$ scattering, which have determined the helicity dependence of the total scattering cross-section: $A_z = \frac{(\sigma^+ - \sigma^-)}{(\sigma^+ + \sigma^-)}$ to $\pm 2 \times 10^{-8}$ at low energy.

J. Birchall(4) presented a status report on an experiment underway at TRIUMF to measure A_z in $\vec{p}p$ scattering at 223 MeV with the goal of achieving a precision of $\pm 2 \times 10^{-8}$. The experiment will be performed in transmission geometry using parallel plate ionization chambers operated in current mode. In a partial wave decomposition, the dominant contribution to A_z below 100 MeV is due to the interference between S and P waves in the scattering amplitude. The TRIUMF experiment (E497) is designed to be carried out at 223 MeV, where the $^1S_0 - {}^3P_0$ amplitude averages to zero over the acceptance of the detectors, leaving the $^3P_2 - {}^1D_2$ amplitude as the sole contribution to A_z, as illustrated in figure 2. The $S - P$ term depends on weak ρ and ω exchanges with roughly equal weight, whereas the $P - D$ term depends on ρ exchange alone, enabling an independent constraint to be placed on the elusive weak meson-nucleon coupling constants from the 223 MeV experiment.

A new beamline and target have just been commissioned at TRIUMF for this challenging experiment, and many years of development work and optimization of the TRIUMF Optically Pumped Polarized Ion Source will be relied on for the very stringent requirements of systematic error control. The TRIUMF group plans to begin initial data taking in late 1994, and to follow this experiment with a second measurement at 450 MeV which would require only minor changes to the 223 MeV apparatus. (Plans are also underway at COSY to perform a parity violation measurement in $\vec{p}p$ scattering at 230 MeV, using a similar apparatus, as proposed by P.D. Eversheim et al.)

M. Shmatikov(6) presented an overview of the theoretical approaches that have been used to calculate the parity-violating asymmetry in $\vec{p}p$ scattering. A thorough, one-meson exchange model calculation performed by Driscoll and Miller(5) is illustrated in figure 2. Straightforward extensions to this model include the consideration of two-pion exchanges and Δ-isobar excitations, but both are expected to be small at low and intermediate energy. An additional mechanism involving parity-mixed $\rho - a1$ meson exchanges has been considered by Iqbal and Niskanen(7), but caution must be applied to avoid double-counting, since the form of this term is similar to a contribution to the conventional weak ρ-nucleon couplings. Shmatikov also presented the results of his own nonrelativistic quark model calculations up to 250 MeV. It is important to realize that there are no meson exchanges in this model, which contains no adjustable parameters in the prediction of A_z. Calculations based on these two contrasting approaches differ dramatically in the strength of the J=2 contribution $(^3P_2 - ^1D_2)$ which will be measured in the TRIUMF experiment, both in the scale and in the shape of the energy dependence, as illustrated in figure 2. A lively discussion ensued, focussing on the surprising result that this treatment of direct quark-vector boson exchanges leads to an interaction of longer range than ρ and ω exchange.

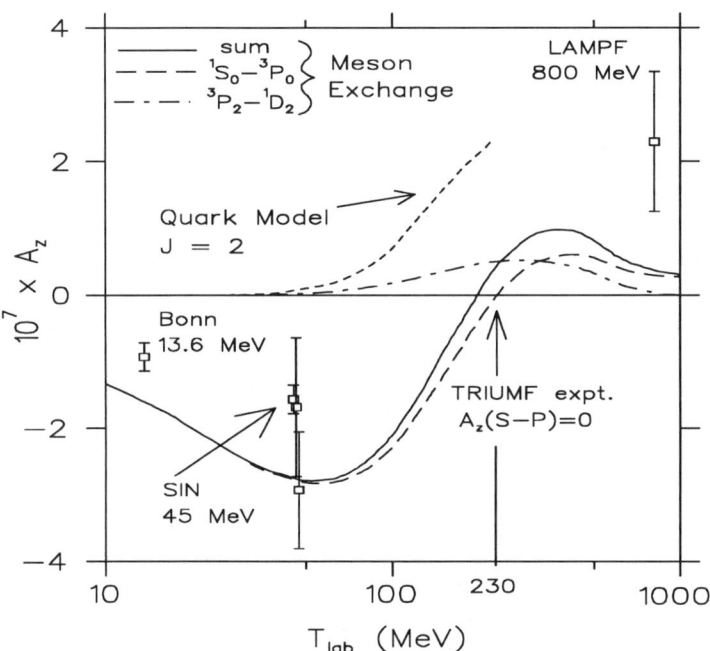

FIGURE 2: Energy dependence of A_z in $\vec{p}p$ scattering. A partial wave decomposition of A_z calculated by Driscoll and Miller(5) using the weak meson exchange model is shown. The $(^3P_2 - ^1D_2)$ contribution calculated in a nonrelativistic quark model by Shmatikov(6) is indicated for comparison.

Several years ago, new data from measurements of helicity dependence of low energy (≤ 500 eV) neutron transmission on heavy nuclear targets indicated a number of unusually large ($\simeq 10^{-1}$) parity-violating asymmetries(8). Perhaps the most tantalizing feature of the initial data was the tendency for most of the parity-violating asymmetries observed in ^{232}Th+$\vec{n}$ and ^{238}U+$\vec{n}$ resonances to have the same (positive) sign. Calculations which were able to account for the observed sign effect via compound nuclear statistical models in most cases invoked the existence of unexpectedly large single-particle parity mixing matrix elements, of order $\simeq 100$ eV, as compared with the matrix elements of less than 1 eV which have been measured in light nuclei.

M. Leuschner(9) reported on the status of two gamma ray experiments which were designed to measure a single-particle parity mixing matrix element in the nucleus ^{207}Pb with a sensitivity at roughly the 10 eV level. The advantage of ^{207}Pb is that its level structure is relatively simple, so that interpretation of the result can be formulated as a two-level mixing problem. The gamma ray of interest is emitted in a 1063 keV transition between the third ($J^{\pi} = \frac{13}{2}^{+}$) and second ($J^{\pi} = \frac{5}{2}^{-}$) excited states in ^{207}Pb. Parity mixing of the $\frac{5}{2}^{-}$ ($2f_{\frac{5}{2}}$ neutron hole) state with a $\frac{5}{2}^{+}$ ($3d_{\frac{5}{2}}$ neutron hole) state at 6 MeV higher excitation leads to circular polarization of the 1063 keV γ-rays (P_{γ}) or equivalently to a forward-backward asymmetry (A_{γ}) of the 1063 keV transition when the latter is fed by the β decay of polarized ^{207}Bi. Both measurements are technically challenging: for P_{γ}, the circular polarimeter sensitivity is $\simeq 1\%$, while for A_{γ} the nuclear polarization of the source which has been achieved to date at low temperature is limited to about 10%. Results thus far from the A_{γ} measurement are more precise and have been used to place an upper limit of $\simeq 21$ eV on the single-particle matrix element. With the new neutron resonance data presented by Y.-F. Yen(10) in a plenary talk, which suggest that the positive asymmetry trend is unique to ^{232}Th, emphasis has shifted to attempting to achieve the best possible precision in the ^{207}Pb experiments.

C. Horowitz(11) reported on calculations of the matrix element which account for relativistic effects in nuclear matter using a Hartree Fock approach. Small renormalizations ($\leq 10\%$) of the effective weak meson-nucleon couplings were found. The value of the matrix element is reduced by cancellations between π, ρ and ω contributions which are particularly sensitive to the treatment of short-range correlations. Including correlations in an approximate manner, the matrix element was estimated to be of order 0.05 eV; nonrelativistic calculations range from 0.4 to 1.2 eV. If the ^{207}Pb measurements can reach the level of ± 1 eV accuracy or better, they might be used to help constrain predictions of the weak meson-nucleon coupling constants. Unfortunately, the large energy gap in ^{207}Pb between the $\frac{5}{2}^+$ and $\frac{5}{2}^-$ one-hole states means that such a stringent constraint on the matrix element must come from a measurement which is many orders of magnitude more difficult than the neutron resonance experiments which motivated the original study.

TIME REVERSAL VIOLATION

Perhaps the most well-known experimental test of time reversal symmetry violation is the search for an electric dipole moment of the neutron, which is currently constrained to be less than 10^{-25} e cm. In a plenary session at this conference, B. Heckel(12) reported results of an experiment on ^{199}Hg atoms, placing an even smaller upper limit of less than 10^{-27} e cm on this elusive quantity. The electric dipole moment is odd under both parity (P) and time reversal (T). In contrast, the upper limits on direct P-even T violation are not nearly as well constrained – the most familiar and stringent limit in this category being the detailed balance test: ^{24}Mg$(\alpha,p)^{27}$Al. To date, no direct experimental evidence of time reversal violation has been found. Comparisons of the significance of different experiments rely on theoretical interpretations based on meson exchange models of P-odd, T-odd and P-even, T-odd interactions. The usual approach is to determine an upper limit for the ratio of (T-odd/T-even) nuclear matrix elements and attempt to relate this to an upper limit for the ratio of (T-odd/T-even) couplings in the underlying interaction.

P. Huffman(13) reported new results of a P-even time reversal test in polarized neutron transmission through an aligned nuclear target of ^{165}Ho. The experiment searches for a five-fold correlation that is P-even and T-odd, of the form: $(\vec{\sigma}.(\vec{I}\times\vec{k}))(\vec{I}.\vec{k})$ where $\vec{\sigma}$ is the neutron spin, $\vec{k}$ is the neutron momentum, and $\vec{I}$ is the direction of the nuclear spin alignment ($I = \frac{7}{2}$), achieved by cooling a crystalline target to low temperature. *(Recall that for an aligned target, $m_I = m_{-I}$)*. The neutron beam energy of 6.57 MeV does not correspond to any resonance in the compound nucleus. The neutron beam is polarized vertically, while the axis of alignment of the Ho crystal is in the horizontal plane, which contains the beam axis. The Ho target is physically rotated so that the axis makes an angle $\theta(t)$ to the direction of $\vec{k}$. A double asymmetry formed by simultaneously flipping the spin of the neutron beam while rotating the target gives a unique $sin(2\theta)$ signature for T-violation. The result is consistent with zero, placing an upper limit of 3.8×10^{-3} on the ratio of T-odd/T-even nuclear matrix elements, which is comparable to the limit of 3.5×10^{-3} set by ^{24}Mg$(\alpha,p)^{27}$Al. False effects arising from spin misalignments require two sequential parity-violating interactions, resulting in negligible corrections at this level of accuracy. The collaboration plans to continue the experiment with the aim of reducing this upper limit to $\leq 1 \times 10^{-4}$.

P.D. Eversheim(14) discussed another P-even test of T violation planned for COSY. The experiment is a null test which depends on a fivefold correlation using an aligned nuclear target – in this case, a polarized proton beam at 2 GeV will interact with a tensor polarized deuteron target in the COSY ring. The ring will act as a forward spectrometer with essentially zero acceptance for scattered beam, which is an important feature rendering the experiment insensitive to false effects associated with double scattering. The observable is denoted $A_{y,xz}$, where the cross-section for polarized beam and target is given by: $\sigma = \sigma_o(1 + P_y^p P_{xz}^d A_{y,xz})$. A time reversed state can be reached either by flipping the proton beam polarization or by rotating the deuteron tensor alignment axis by 90°, offering the possibility to invoke a double asymmetry test for systematic error suppression. The goal is to reach an accuracy of $\Delta A \leq 10^{-6}$; in contrast, P-odd, T-even asymmetries are expected to be an order of magnitude smaller. Many of the possible false T-violating systematic errors are ruled out in forward scattering at the 10^{-6} level, either by rotational or parity invariance.

Y. Masuda(15) reported on progress towards a P-odd time reversal test in the transmission of low energy polarized neutrons through a polarized ^{139}La target at KEK. (This experiment was also suggested by Y.F. Yen(10) in her plenary session talk.) The test is proposed at 0.734 eV neutron energy where a p-wave resonance shows a huge parity-violating asymmetry, $A_z = (9.55 \pm 0.35)\%$, enhanced by compound nuclear structure; this same mechanism is expected to enhance the basic T-odd interaction. The P-odd, T-odd transmission asymmetry for polarized neutrons interacting with a polarized nuclear target arises from a $\vec{\sigma}.(\vec{k} \times \vec{I})$ term in the forward scattering amplitude. The incident neutron spin is longitudinal at the entrance to a transverse, vertically polarized La nuclear target. In the target region, a combination of external and internal fields precess the neutron spin by exactly 180° so that it exits from the target with the opposite helicity. With this arrangement, both $<\vec{\sigma}.\vec{k}>$ and $<\vec{\sigma}.\vec{I}>$ are zero inside the target region. A double transmission asymmetry can be formed by combining measurements when the neutron spin is flipped upstream of the target for opposite target spin precessions, cancelling many systematic errors. The success of the experiment hinges in part on achieving high polarization of the Lanthanum nuclei over a reasonably large volume target; thus far, polarization of roughly 20% has been achieved in a moderate-sized crystal. The goal of this experiment is to achieve an upper limit of T-odd/P-odd matrix elements at the 10^{-3} level or better. To set a scale for comparison, Herczeg(16) has estimated an upper limit on the ratio of the dominant T-odd/P-odd pion coupling constants to be 4×10^{-3}.

SUMMARY AND FUTURE OUTLOOK

Results, progress and plans for many new experimental tests of Charge Symmetry, Parity and Time Reversal violation in strongly interacting systems were presented. While much has been accomplished in this active field, many interesting avenues remain to be explored. The concluding speaker, B.M.K. Nefkens(17), advocated the need for more tests of C and G-parity invariance using polarization as a probe to better determine the u-d quark mass difference, and gave examples of many possible reactions which could be used. The ambitious possibility of performing CP tests in η-meson decays, such as a measurement of the longitudinal polarization of muons in the rare decay branch $\eta \to \mu^+\mu^-$ (B.R. 6×10^{-6}) was also raised, although reaching the required level of accuracy might pose an incredible experimental challenge.

REFERENCES

1. Abegg, R. et al., "Measurement of Charge Symmetry Breaking in np Elastic Scattering at 350 MeV", *Proc. SPIN '94*
2. Miller, G.A., "Theoretical Overview of Charge Symmetry Violation", *Proc. SPIN '94*
3. Desplanques, B., et al., *Ann. Phys* **124**, 449 (1980)
4. Birchall, J. et al., "Parity Violation in P-P Scattering at TRIUMF", *Proc. SPIN '94*
5. Shmatikov, M., "Theoretical Overview of Parity Violation in p-p Scattering", *Proc. SPIN '94*
6. Driscoll, D.E. & Miller, G.A., *Phys. Rev.* **C39**, 1951 (1989)
7. Iqbal, M.J. & Niskanen, J., *Phys. Rev.* **C49**, 355 (1994)
8. Bowman, J.D., et al., *Ann. Rev. Nucl. Part. Sci* **43**, 829 (1993)
9. Leuschner, M. et al., "Parity Nonconservation in ^{207}Pb", *Proc. SPIN '94*
10. Yen, Y.-F., "Study of Parity and Time-Reversal Violation in Neutron-Nucleus Interactions", *Proc. SPIN '94*
11. Horowitz, C. & Yilmaz, O., "Relativistic Effects on Parity Violation in Nuclei", *Proc. SPIN '94*
12. Heckel, B., "Limit on the Electric Dipole Moment of the ^{199}Hg Atom", *Proc. SPIN '94*
13. Huffman, P. et al., "An Experimental Test of Parity-Even Time Reversal Invariance with MeV Neutrons", *Proc. SPIN '94*
14. Eversheim, P.D. et al., "Test of Time Reversal Invariance in Proton-Deuteron Scattering", *Proc. SPIN '94*
15. Masuda, Y., "Neutron Spin Rotation and P and T Violations", *SPIN '94*
16. Herczeg, P., "T-Violating Effects in Neutron Physics and CP Violation in Gauge Models", in *Proceedings of the Workshop on Tests of Time Reversal Invariance in Neutron Physics*, 1987, World Scientific, pp. 24-53
17. Nefkens, B.M., "Tests of Discrete Symmetries Using Polarization", *SPIN '94*

III. FEW BODY SYSTEMS

Proton-Proton Interactions with Polarized Internal Targets in Storage Rings

W. Haeberli

Department of Physics, University of Wisconsin
Madison, Wisconsin 53706, USA

Abstract. Results of the first two experiments that use polarized hydrogen targets in proton storage rings are discussed. Both experiments used storage cell targets, i.e. long, narrow, windowless target cells into which polarized hydrogen atoms from an atomic-beam source are injected and through which the circulating beam passes. In the first experiment, carried out at the test storage ring of the Max Planck Institut in Heidelberg, a 27 MeV beam of alpha particles was passed though the target cell to measure the target polarization and the target thickness, and to investigate possible problems arising from background and from target depolarization by the beam. The target was used for a test of principle of the spin filter (spin-selective attenuation) as a means to polarize the circulating beam. The second experiment, currently in progress at the IUCF proton storage ring, has demonstrated the feasibility of spin correlation experiments at 200 MeV by detecting recoil protons in coincidence with scattered protons at forward angles. Both experiments used an improved atomic-beam source for polarized atoms and achieved target thicknesses of about 3×10^{13} polarized H/cm^2 in a single hyperfine state of hydrogen.

INTRODUCTION

In nuclear and high-energy spin physics the possible advantages of polarized H (or D) gas targets in storage rings have been under discussion for about two decades. At the 1975 Symposium on Polarization Phenomena in Nuclear Reactions, Hanna (1) discussed earlier proposals to use the polarized beam produced by a conventional atomic-beam source as a target. The discussion included the suggestion to overcome the very low target thickness, which such a jet target presents to the beam, by injecting the polarized atoms into a bottle (storage cell), an idea proposed already at a Polarization Symposium a decade earlier by Haeberli (2) in connection with the enhanced production of polarized ions. At the SPIN80 Symposium, Dick et al. (3), described work on a polarized H jet that was planned to be used as an internal target in the CERN SPS, and Schüler (4) suggested the use of the storage cell target, which had been developed the year before at Wisconsin (5), as in internal polarized H target in electron rings. At a conference on "Nuclear Physics with Stored Cooled Beams" (Indiana University, 1984), Haeberli discussed the advantages of physics with internal polarized gas targets, particularly if storage cells are used to enhance the target thickness (6). However, the actual use of polarized internal targets in storage rings developed only slowly during the last few years, and the number of experiments can still be counted on the fingers of one hand. A first experiment in an electron ring used a jet of polarized deuterium atoms as an internal target (7), but in spite of the large circulating beam (up to 0.5 A) the small target thickness (about 10^{11} D/cm^2) severly limited the statistical accuracy of the experiment. More recently (8), a storage cell was used to increase the effective target thickness to 6×10^{11} D/cm^2. The increase in gain over the jet target was modest because the storage cell had a large aperture and because only a small part of the cell was viewed by the detector. In the work to be described below, the aim was to develop targets of much larger target thickness for use in proton storage rings.

ADVANTAGES AND PROBLEMS

The potential advantages of polarized internal gas targets in storage rings, compared to conventional polarized targets placed in the external beam of accelerators, have been enumerated several times. The most important advantages are:

a) gas targets are chemically and isotopically pure and thus avoid the background from unpolarized material in solid targets such as NH$_3$;

b) the target polarization can be reversed rapidly (msec);

c) for deuterium, large vector or large tensor polarization can be produced as the experiment requires;

d) a weak magnetic guide field (e.g. 0.5 mT) can be used if atoms are prepared in a pure spin state. Consequently, there is almost no deflection or spin precession of charged particles, and little obstruction of detectors by magnets. Helmholtz coils can be used to provide free choice of the polarization direction;

e) energy loss in the target is small, which is necessary for detection of low energy recoils;

f) small scattering angles (Coulomb-nuclear interference region) are accessible.

A major obstacle to the use of polarized H or D gas targets is that the available flux of polarized atoms provides only very marginal target thickness, unless a storage cell target is used. Since the density of atoms in the target cell decreases rapidly with increasing diameter of the cell, one is interested to use cells of small aperture. On the other hand, use of a small aperture target cell (say 1 cm diameter or less) raises a number of problems:

a) the restricted aperture causes problems with injection of particles into the ring and affect the lifetime of the stored beam;
b) background may result from beam halo striking the cell walls, since even for a thin walled cell the mass of the wall is about 10^9-times the mass of the target gas;
c) the gas load from directed gas flow out of the cell may affect ring operation;
d) the atoms may depolarize by wall collisions inside the cell, unless suitable (radiation resistant!) cell coatings can be found to prevent wall depolarization;
e) recoil detectors placed near the beam may suffer high rates or radiation damage by particles outside the ring acceptance, particularly during injection.

IMPROVED ATOMIC BEAM SOURCES

The target thickness that is achieved by injecting polarized atoms into a storage cell (Fig. 1) depends not only on the total intensity of the atomic beam but on the phase space density (brightness) of the beam, since the diameter of the entrance tube, and thus the leak rate of atoms out of the cell, can be made smaller if the atomic beam is well focussed. Work on high-brightness atomic beams was carried out in collaboration between the University of Wisconsin and the Max Planck Institut (Heidelberg). Initially, the plan was to use the Heidelberg target in an antiproton spin filter experiment (known as FILTEX) at LEAR, but more recently it was decided to use the Heidelberg atomic beam source for the HERMES experiment on electron deep inelastic scattering at DESY.

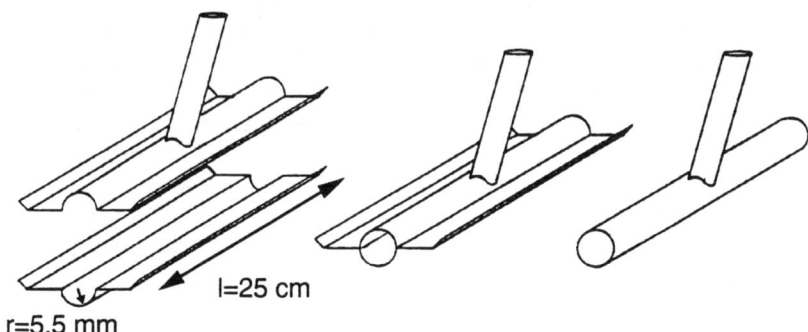

Figure 1. Storage cell used at MPI-Heidelberg.

The atomic beam apparatus developed at Wisconsin (9) and at Heidelberg (10) is based on careful studies of the intensity limitations in conventional atomic beam sources, in particular investigations of beam attentuation by gas scattering (intra-beam scattering and residual gas scattering), degree of dissociation as a function of flow rate and nozzle temperature, and proper matching of the magnet system to the velocity spectrum of the beam. An important new feature was the replacement of the conventional electromagnet sixpoles by permanent magnets, which permits not only higher pole tip fields (1.5 T, i.e. a gradient of up to 3.75 T/cm for the smallest pole tip radius used), but permits segmentation into short sections. The gaps between these sections provide for differential pumping and thus reduce beam attenuation. Beam transport calculations showed that a significant gain results from the use of a tapered aperture in the first magnet section to reduce chromatic abberations. With a distance of 26 cm between the end of the last magnet and the entrance to the feed tube of the target cell, a flux of 6.7×10^{16} polarized H/s was found to pass through the 1 cm diameter, 13 cm long feed tube. For applications in which the target has to operate in a weak magnetic field, it is advantageous to reject one hyperfine state by inducing an RF transition in the atomic beam before it enters the last set of sixpole magnets. The remaining beam is in a pure spin state (state 1, see ref. (11)), yielding a large polarization in a weak guide field (0.5 mT), but at the expense of a factor two loss in intensity.

TARGET TESTS AT TSR-HEIDELBERG

A target cell constructed of 0.2 mm thick aluminum was inserted into a scattering chamber installed at the Test Storage Ring in Heidelberg. The purpose was to study the behaviour of the target cell in the high vacuum of a storage ring, and to study possible deterioration of the target polarization as a result of prolonged exposure of the cell to beam currents of the order 1 mA. The target chamber (12) was a bakeable stainless steel vessel pumped with cryopanels (10,000 ℓ/s) and separated from the ring vacuum (10^{-10} mbar) by a three-stage differential pumping system. The target polarization was deduced from the asymmetry in the number of recoil protons at 23° that resulted from bombardment of the target with 27 MeV stored α-particles. The p-α analyzing power at this angle and energy is large (0.82) and known to better than 2% from previous experiments (13). The cell walls were thin enough for recoil protons (but not α-particles) to reach the detector.

The target cells were coated with a high-temperature version of Teflon (14), since tests at Wisconsin (15) had shown good polarization retention with this coating for cell wall temperatures down to about 100K. The cell originally installed was made to open by remote control ("clam shell cell", see left hand side of Fig. •1) to permit a large ring acceptance during beam injection. Fortunately, it was found possible to inject with the 11 mm aperture cell in place so that this feature was not necessary. The target polarization observed with this cell was about half the expected value. The suspected reason was diffusion of atoms into and out of the narrow gap between the two halves of the cell, where the surfaces were not well coated. The cell was replaced by the simple T-shaped cell shown on the right of Fig. 1. The two tubes were coated first and were then assembled by joining the two pieces with a high-vacuum epoxy applied to the outside of the joint. In order to enhance the target thickness, the cell was cooled to temperatures as low as 70K. Cooling was accomplished by attaching the cell at the upstream end to a cooled copper block and wrapping the cell with superinsulation to reduce heating by radiation.

The H target thickness of the 25 cm long cell was calculated from the p-α count rate, using the known cross section, beam current and detector geometry. As a second, independent measurement, the target thickness was deduced from the rate of energy loss of the circulating beam, based on the observation of the shift in revolution frequency (Schottky signal) when the electron cooling in the ring was turned off. The two measurements agreed to better than 10%. When two hyperfine states (states 1 and 2, maximum nuclear polarization in a weak magnetic field P=1/2; see ref. (11)) were injected into the target cell (no RF transitions on the atomic-beam source), the target thickness at 115K wall temperature was $(8.2 \pm 0.3) \times 10^{13} H/cm^2$, while for a single hyperfine state (state 1, P=1 independent of field) the target thickness was $(5.4 \pm 0.2) \times 10^{13} H/cm^2$. The target thickness showed the expected $T^{-1/2}$-dependence on cell temperature.

The target polarization was reversed about once a second by reversing the direction of a 0.5 mT vertical guide over the target. The cell wall temperature was varied between room temperature and 70K. The target polarization is constant between room temperature and about 110K, and drops rapidly for lower temperatures, in agreement with previously reported depolarization tests (15). The target polarization when a single hyperfine state was selected (state 1) was measured to be P = 0.82 ± 0.02, while for two hyperfine states (states 1 and 2) P = 0.46 ± 0.01. An important finding was that the polarization did not diminish with exposure of the target to beam currents up to 1 mA for several days. Also within the errors there was no dependence on beam current or on magnetic guide field in the range 0.2 to 1 mT. For details see ref. (16).

One of the most striking results of the target tests in the Heidelberg ring was the almost complete absence of background in the recoil-proton spectrum. In preparation of the tests in the TSR, the target cell had been exposed to the external α-beam of the Heidelberg tandem accelerator, using the same atomic-beam source and scattering chamber that was later installed in the ring. Every effort was made to reduce empty-cell background by most stringent collimation of the incident beam, but background remained severe (Fig. 2). In view of the very dilute target and the fact that the walls of the target cell have a mass about 10^9-times larger than the H target itself, even a very small beam halo is sufficient to explain the background. Since α-particles scattered by the cell wall have insufficient energy to reach the detectors, the background is caused primarily by ^{27}Al(p, α)-reactions. When the same target was installed in the ring, and the α-beam from the same accelerator was stored and cooled in the ring, a dramatic improvement in the background resulted, as shown on the right hand side of Fig. 2. The measurements shown in Fig. 2 required 200 hrs running time in the external beam compared to 9 min for the internal beam. With the internal target, an upper limit to the background was found to be <1%. The upper limit on the background was determined by replacing the hydrogen gas in the target with an amount of N_2 which causes the same amount of beam loss (beam heating) as the H target.

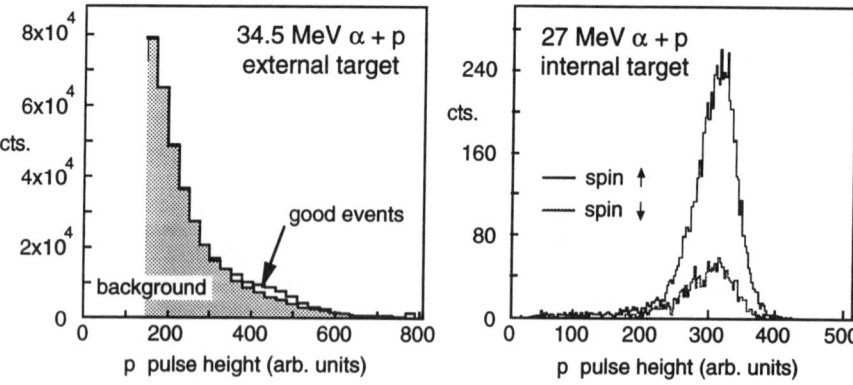

Figure 2. Proton energy spectrum from bombardment of a polarized hydrogen storage cell by α-particles at MPI-Heidelberg. Left: external beam and target; Right: internal beam and target.

SPIN FILTER

The idea to polarize a circulating beam of protons or antiprotons in a storage ring by spin-selective attenuation of the beam in a polarized internal target was first proposed by Conka (17) already in 1967. The idea was taken up later by the FILTEX collaboration (18) with the proposal to polarize the antiprotons in LEAR by passing the beam through a storage cell containing polarized hydrogen. The target described in the preceding section was used to test the principle of the spin filter with protons in the Heidelberg TSR. The interaction cross section between the circulating beam and the polarized target atoms for beam spin and target spin parallel and antiparallel, can be written, respectively as:

$$\sigma^+ = \sigma_0 + \sigma_1 \text{ for } \Uparrow\uparrow$$

$$\sigma^- = \sigma_0 - \sigma_1 \text{ for } \Uparrow\downarrow$$

where ⇑ indicates the direction of the target polarization, ↑, ↓ the direction of the beam spin. The cross section σ_1 (polarizing cross section) determines the rate of polarization buildup of the circulating beam, while σ_0 contributes to the loss of beam intensity with time (beam lifetime). Beam particles are lost if they scatter by an angle larger than the ring acceptance angle (4.4±0.5 mrad for the low-β straight section used for this experiment).

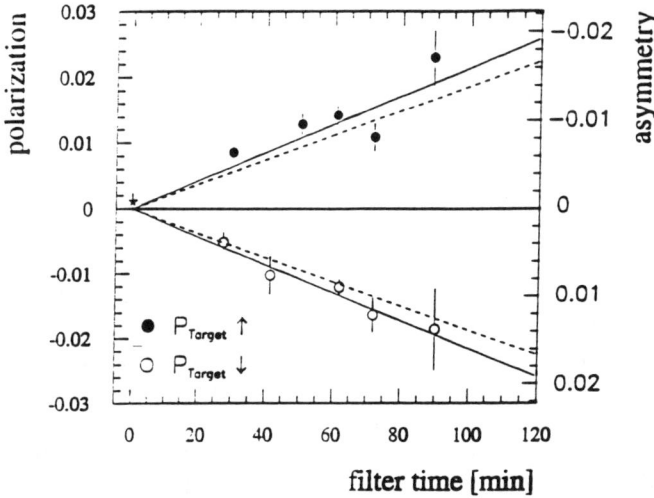

Figure 3. Spin filter: polarization vs. filter time.

In the experiment (19) as much as 1 mA of unpolarized 23 MeV protons were accumulated by multi-turn injection, after which the beam was left to circulate for periods of 30 min to 90 min without changing the direction of the target polarization. Subsequently, the count rate of scattered protons was measured in four detectors (left, right, up, down) while reversing the target polarization about once a second. This provided a sensitive measure of the extent to which the beam had become polarized, because pp scattering at this energy is strongly dependent on the relative spin orientation of beam and target ($A_{xx} = -0.93$). Figure 3 shows the measured beam polarization after filtering with target polarization up (solid dots) and with target polarization down (open circles). As expected, the beam polarization reverses sign with reversal of the target polarization. The solid line shows a fit to the data based on the known H target thickness $(5.3 \pm 0.3)10^{13} \text{H/cm}^2$ and known target polarization (0.80 ± 0.02), with the polarizing cross section as an adjustable parameter. The result, $\sigma_1 = (72.5 \pm 5.8)$ mb, can be compared to a recent calculation by H.O. Meyer (20), who pointed out important shortcomings in the previous analysis (19). The new analysis identifies three contributions to the polarizing cross section: spin-dependent removal of protons by scattering through angles exceeding the acceptance of the ring at the target (83 mb); polarization transfer to protons scattered by small enough angles to be retained in the ring (52 mb); and effect on the stored beam from interaction with the polarized electrons in the polarized H gas target, which is of opposite sign to the first two contributions (−70 mb). The sum of the three contributions (65 mb) is in good agreement with the experiment. For discussions of improvements in the method and the important role of the interaction of the beam with polarized electrons, see refs. (21,22).

INTERNAL TARGET AT THE IUCF COOLER

The work described here, which was carried out by the CE-35 collaboration (23) in the IUCF "Cooler", represents the first attempt to perform a pp spin-correlation experiment in a storage ring. The "Cooler", is a proton storage ring that permits the energy of the stored proton beam to be ramped up to 450 MeV. The beam is injected into the ring from a cyclotron, and the stored beam is cooled by interaction with electrons. Prior to the work described here, an experiment on pion production using an internal (unpolarized) jet target had been carried out, but no previous experience with stored polarized beam was available. Thus, while the polarized target was still under construction, experiments were carried out to establish the feasibility of the planned spin-correlation experiments. The individual steps of this development were the following:

a) Storage of a polarized beam in the Cooler in the presence of an internal target was demonstrated, and the polarization lifetime in the presence of electron-cooling was measured. It was found possible to avoid depolarization of the stored beam by careful attention to the ring tune. Under proper conditions no measurable depolarization was observed either with or without the (unpolarized) H jet target in the ring, i.e. the polarization lifetime is at least an order of magnitude longer than the beam lifetime.

b) It was found possible to detect recoil protons in coincidence with forward protons by placing silicon strip detectors some 5 cm from the H target. As a first application of the stored polarized beam, the pp analyzing power in the Coulomb-nuclear interference region was measured using the jet target (24).

c) To test the feasibility of a narrow-aperture polarized H storage cell, beam lifetime and background were measured to determine the optimum cell geometry. To permit use of the most restrictive aperture,

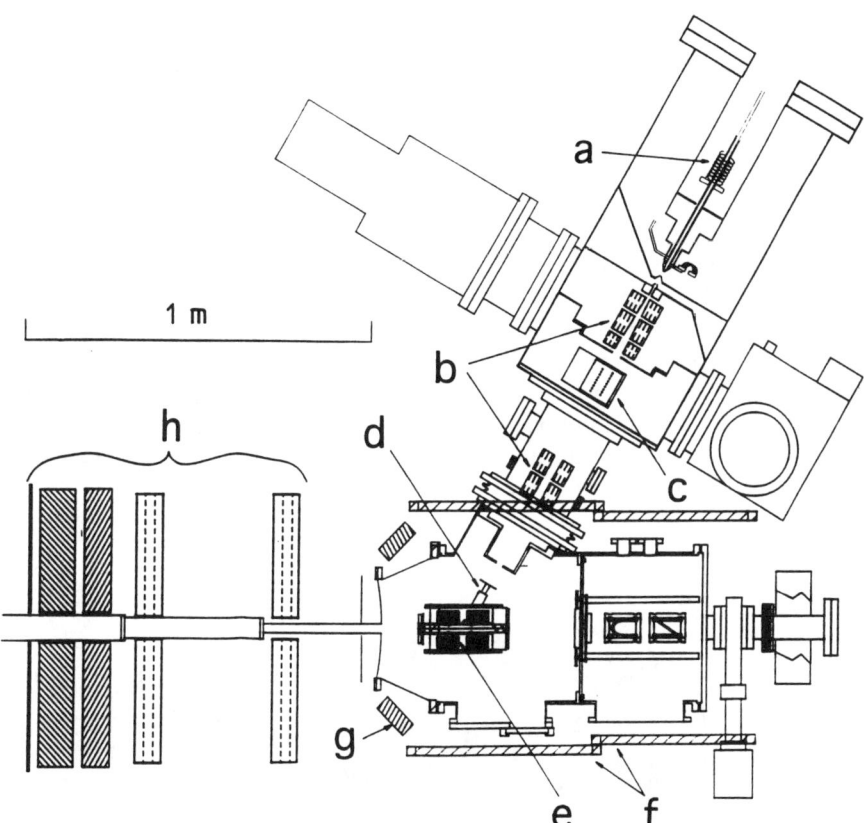

Figure 4. Wisconsin polarized H target installed on the IUCF Cooler. a) dissociator; b) spin separation magnets; c) RF spin flip transitions; d) entrance tube to target cell; e) Si-strip recoil detectors; f) guidefield and compensating coils; g) 45° scintillators; h) forward stack (wire chambers, scintillators).

the straight section with the lowest β-function was chosen (A-region). The results, which are described in ref. (25), led to a cell aperture of 8 mm × 8 mm.

d) The results of the above development were applied to the study of p–^{3}He spin correlation parameters in the IUCF Cooler by Lee et al. (26).

e) Prior to installation in the Cooler, the polarized target was tested at Wisconsin by observation of low energy pp spin correlation. A target polarization of P = 0.72 was achieved (27) after problems with magnetic field shaping between the atomic beam source and the target were solved.

f) the target was installed in June 1993. Experience with the target and first pp spin correlation experiments are described below.

A simplified scale drawing of the target and detector arrangement is shown in Fig. 4. The axis of the atomic-beam source is at an angle of 60° to the proton beam in order to provide space in the forward hemisphere for detectors. The scattering chamber has a thin-walled forward cone through which protons pass to the forward detector stack consisting of scintillation counters and wire chambers. The polarization direction of the H target atoms is determined by a guide field (0.5 mT) that is provided by coils external to the chamber. Three sets of coils provide polarization along ±x, ±y or ±z. The six target polarization states are turned on one after the other for 2 s each, so that the cycle repeats every 12 s. Ambient DC fields are reduced by superposing a DC component on the guide field current. The effect of the guide field on the circulating beam is reduced by x- and y-compensation coils placed just before and downstream of the target.

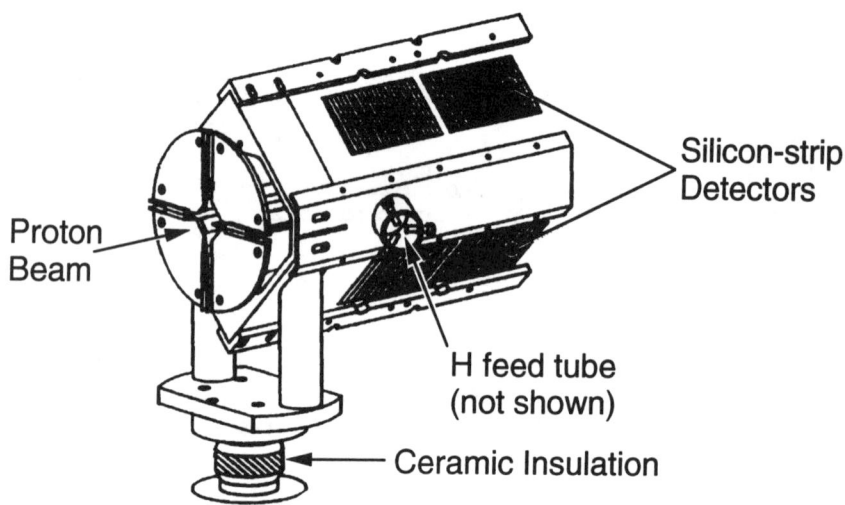

Figure 5. Target assembly for the polarized H target at IUCF

The target assembly (27) is shown in more detail in Fig. 5 (see also ref. (28)). A 25 cm long channel of cross section 8 mm × 8 mm is formed by four teflon foils (1 mg/cm^2, and more recently 0.5 mg/cm^2). Polarized H atoms enter at the center of the channel though a 1 cm diameter, 13 cm long feed tube. To permit tests with unpolarized target, a small Teflon tube connected to the feed pipe permits a flow of metered amounts of unpolarized H$_2$ into the cell. The silicon strip detectors (Fig. 5) are 4 cm × 6 cm each. The detectors have a nominal depletion depth of 1 mm. Each detector has 28 strips, connected to a position readout.

RESULTS: TARGET POLARIZATION, TARGET THICKNESS, BACKGROUND

As a first application of the internal polarized target, the CE-35 collaboration proposed to measure pp spin correlation parameters as a function of angle at 200 MeV. During the last few months, extensive experience has been gained with the target and detection equipment. It has been found possible to inject polarized beam into the ring in the presence of the target without noticeable radiation damage to the target foils or the silicon detectors. The detection system allows clean identification of pp coincidences. Figure 6a shows the energy deposited in one of the silicon detectors as as function of laboratory angle of the coincident forward proton. The knee represents the angle for which the recoil energy exceeds the thickness of the recoil detector. The hit pattern of the wire chambers (in coincidence with the recoil detectors) is shown in Fig. 6b. Both Figures show very few counts outside the expected locus for pp coincidence events. Additional studies with empty target cell, with nitrogen in the target cell and with beam heating, produced by nitrogen in another part of the ring, indicates that background is below 1% of the good events.

Beam and target polarization were determined from the count rate asymmetries in pp scattering for the angular range 6° to 18°, for which the average pp analyzing power $\langle A_y \rangle = 0.26$. It was found that a guide field of 0.5 mT over the target is sufficient to obtain the full target polarization (28). Horizontal and vertical target polarizations are the same within errors ($Q_x = Q_y = 0.75 \pm 0.01$). No evidence has been found for deterioration of the target polarization with exposure of the cell to the beam during one weak running periods. The Teflon films are replaced after each run since they become brittle and sometimes break when the target is vented to air. The polarization rise time of the target after sign reversal (< 40 ms) is limited by the rise time of the current in the guide field coils.

The target thickness was determined by comparing the count rate with polarized H in the target to the count rate observed when a metered amount of H_2 was admitted to the cell. Under good conditions, the target thickness is $(3.5 \pm 0.3) \times 10^{13} H/cm^2$, but decreases by some 30% after several days, after which the cold nozzle of the dissociator in the atomic beam source needs to be warmed up to evaporate deposits that build up on the nozzle. While experience (15,16) shows that the target thickness can be increased without loss in target polarization by cooling the cell walls to about 100K, for the present experiments a room temperature target is used for simplicity.

For suppression of systematic errors it is desirable that the target thickness and the target polarization be the same for target polarization up and down. The relative target thickness can be obtained to high accuracy from comparison of count rates for target spin up and down. Significant modulations of the target thickness (or target luminosity) with polarization reversal were observed initially. The reason for the modulation was that the fringe field of the guide field coils affected the transition probability of the RF transition in the atomic beam source, so that the rate of rejection of the (unwanted) state 2 atoms depended on the guide field currents. After compensation of the fringe field effect, the target luminosity is the same to an accuracy of 10^{-3} for target spin up and down, and the same for x, y, and z target spin direction. This assures that the RF transition probability remains unchanged when polarization directions are changed.

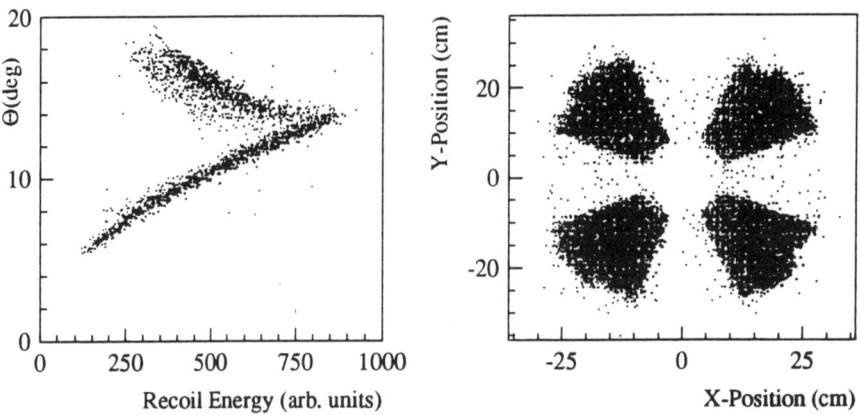

Figure 6. Pulse height in the silicon detector vs. angle of the forward proton as determined by the wire chambers (left). Hit pattern of the second wire chamber in coincidence with the recoil detectors (right). For both graphs, a small dot represents one pulse.

PP SPIN CORRELATION MEASUREMENTS

Measurements have been made of pp spin correlation as a function of angle between 6° and 18° at 200 MeV. Until recently, a typical measuring cycle consisted of beam injection (480 s), data taking without further injection (500 to 700 s), after which the beam was dumped, followed by an identical cycle with proton spin of the opposite sign. The beam current after 480 s injection has at times been as high as 200 µA, but more often is of the order 100 µA or less. Considerably higher beam currents are anticipated when beam from the new polarized-ion source (29) becomes available. With 100 µA beam, the coincidence rate in all detectors combined is about 10 Hz.

Data are analyzed in angular bins of 1°. An example of results obtained so far is shown in Fig. 7, which shows the measured pp analyzing power as a function of angle. The absolute normalization of the data points is adjusted to give a best fit to the solid line, which is the phase shift prediction of the C200 SAID phase shift analysis. The spin correlation coefficients have been deduced as well. The results are still preliminary and need to be checked for possible systematic errors. The present indications suggest good agreement with the predictions of the Nijmegen phase shift analysis. The final results of the experiment will yield A_{xx}, A_{yy} and A_{xz} to an accuracy of about 0.01.

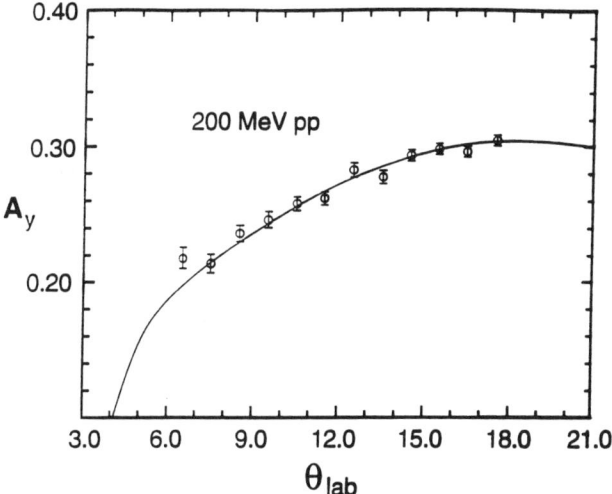

Figure 7. Analyzing power for pp scattering vs angle compared to a phase shift prediction (solid curve). The overall normalization factor of the measurements is chosen to give a best fit to the solid curve.

In spin correlation experiments with conventional polarized targets in an external beam of an accelerator, systematic errors are suppressed by frequent reversal of the beam polarization, while the target polarization is reversed at most a few times a day. In contrast, in a storage ring, the target polarization can be rapidly reversed, but the polarization of the beam stored in the ring maintains the same sign, so that one is forced to dump the beam in order to inject beam of opposite spin from the cyclotron. Dumping the beam is undesirable since it reduces the average luminosity of the experiment. Ideally, rather than dumping the beam, one would flip the spin of the stored protons, and would periodically make up for beam loss by injecting additional beam from the cyclotron. Tests of a spin flipper for a stored proton beam have been reported in ref. (30). The practical application of a spin flipper requires, however, that the flipper has a large flipping efficiency and a very high reliability, because the experiment has no immediate knowledge if the spin actually flipped. If not, the measurements are out of phase with the assumed sign of beam polarization and at best only tedious offline analysis can recover the data. It has been pointed out (see ref. 21) that the adjustment of the flipper is much less critical if the time structure of the beam is removed by turning off the buncher cavity in the ring, since this removes the side bands so that the flipper frequency can safely be swept over a wide frequency range. Measurements of the proton beam polarization remaining as a function of the number of flips (Fig. 8) indicate a flipping probability of 98.4%. During the most recent run the flipper was used successfully for routine data taking. Beam was injected for 300 s and measurements were taken for 760 s, during which time the beam spin was flipped twice such as to give equal luminosity for proton spin up and down. After each of theses cycles the beam current was built up again by a further 300 s injection. Beam was dumped only every two hours, However, often the beam disappeared for unknown reasons (instabilities in the ring). On the other hand, no instance of malfunctioning of the flipper was observed, so that this will most likely become the preferred mode of data acquisition.

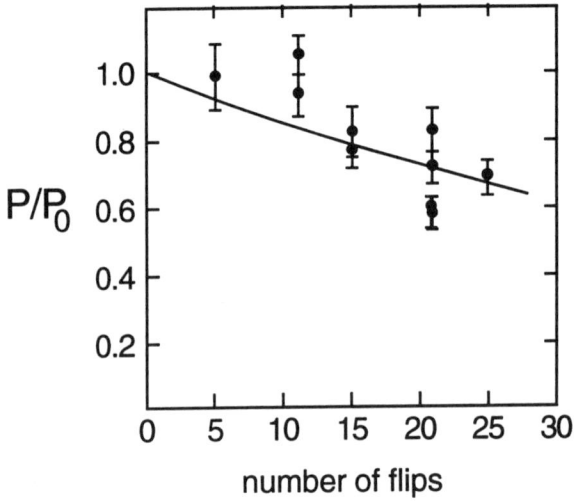

Figure 8. Beam polarization vs. number of spin flips.

PLANS

The work described above is the first step of a program at IUCF to study pp spin dependences in the energy range 100–450 MeV in the Cooler. The program is intended to lead to a study of axial currents in the reaction $p+p \to p+p+\pi°$ with polarized beam and target in the energy region from threshold to 400 MeV. This will require the development of longitudinal beam polarization (21) and the capability to ramp the ring in the presence of the target. Ramping needs to be done for increasing as well as decreasing energy so that after up-ramping the beam can be returned to the original energy to remeasure the beam polarization. In this process, a number of intermediate results will be obtained. Observation of the longitudinal spin correlation (A_{zz}) will allow a measurement of the longitudinal target polarization which otherwise has to be inferred from the transverse polarization. The development of up and down ramping will lead to spin correlation measurements as a function of energy and accurate beam polarization analyzers at higher energies. For these measurements the detection system will be expanded to cover a wider range of angles up to 45°. Finally, a new study is planned to determine whether the presence of a polarized target in the ring has a significant effect on the beam polarization lifetime.

OUTLOOK

Experience obtained so far shows that experiments with internal polarized targets in proton storage rings are feasible and have important advantages over experiments with conventional polarized targets. It has been shown that storage cell targets overcome the severe limitation in target thickness from polarized jet targets in that they increase the luminosity by about a factor 100. Thanks to the great improvement in beam quality that results from phase space cooling of the stored beam, the use of small aperture target cells in the MPI-Heidelberg and IUCF storage rings has been possible without introducing background problems and without serious interference with the operation of the ring. In addition, target polarization of 70–80% has been observed over prolonged periods without any deterioration from radiation damage to the cell walls.

The two targets described above have used improved atomic-beam sources to supply the target cells with polarized atoms. The target thicknesses of the two targets are similar (about $3 \times 10^{13} \text{H/cm}^2$ at room temperature), but can be increased to 10^{14}H/cm^2 by cooling the cell and by injecting two spin states instead of one. Injection of two spin states is favorable in situations where application of a strong guide field ($B \gg B_c$, where $B_c = 0.05$ T for H and 0.01 T for D) permits decoupling the hyperfine interaction. In low energy rings this is difficult (except for longitudinal polarization) but for high energy rings a large guide field may well be compatible or even advantageous to the experiment. In particular, a strong guide field will make the nuclear polarization of the target less susceptible to depolarization.

During the next few years, increased target thicknesses for polarized H or D gas targets can be expected. Development work at CERN aims to increase the flux from atomic beam sources by use of large-aperture superconducting sixpole magnets (31). The method of spin-exchange optical pumping has been shown to produce 2×10^{17} deuterium atoms/s with electron polarization of 80%, and up to 1×10^{18} atoms/s with lower polarization (32), but a measurement of the nuclear polarization still needs to be carried out. Whether the larger fluxes of atoms should be exploited to increase the target thickness in a storage cell or to increase the cell diameter (increased ring acceptance) depends on the application. Even for spin-exchange optical pumping the production rate of polarized atoms is still marginal if one wants to use the beam as a jet target. As an alternative, the group at Michigan (33) is proposing production of a very cold H atomic beam as a jet target of increased density. The importance of the recent progress lies in the promise of new physics in the future. The feasibility of experiments with internal polarized targets of useful luminosity is no longer in question. Experience gained at the CERN proton-antiproton collider shows that detectors can be placed within 2 mm of the beam, which indicates that the use of narrow target cells in high energy rings is not only feasible but quite free of background (34). A number of interesting problems that can be addressed by the use of internal polarized hydrogen or deuterium targets in storage rings at higher energies were discussed by Vigdor (35) at this conference. We conclude that polarized internal targets offer exciting opportunities for new experiments not accessible by conventional methods.

ACKNOWLEDGEMENTS

This work was supported in part by the U.S. Deparment of Energy and by the National Science Foundation.

REFERENCES

1. Hanna, S.S., "Some Recent and Possible New Applications of Polarized Beams", in *Proceedings of the Fourth International Symposium on Polarization Phenomena in Nuclear Reactions*, W. Grüebler and V. König, eds., Birkäuser (Basel, Switzerland, 1976) p.407.
2. Haeberli, W., "Sources of Polarized Negative Ions", in *Proceedings of the Second International Symposium on Polarization Phenomena of Nucleons*, P. Huber and H. Schopper, eds., Experientia Supplemement **12**, (Birkhäuser, Basel, Switzerland, 1966), p.64.
3. Dick, L., Jeanneret, J.B., Kubischta, W., and Antille, J., "The CERN Polarized Atomic Hydrogen Beam Target", in *Proceedings High-Energy Physics with Polarized Beams and Targets*, C. Joseph and J. Soffer, eds., Experientia Supplement Vol. **38**, (Birkhäuser, Basel, Switzerland, 1981), p.212.
4. Schüler, K.P. "A Proposed New Technique for Polarized Electron-Polarized Nucleon Scattering", in *Proceedings High-Energy Physics with Polarized Beams and Targets*, C. Joseph and J. Soffer, eds., Experientia Supplement Vol **38**, (Birkhäuser, Basel, Switzerland, 1981), p.460.
5. Barker, M.D. et al., "A Target of Polarized Hydrogen by Storage of Atoms in a Coated Pyrex Vessel", in *Proceedings of the Fifth International Symposium on Polarization Phenomena in Nuclear Physics* (Santa Fe, NM, 1980).
6. Haeberli, W. "Free and Stored Atomic Beams as Internal Polarized Targets", in *Proceedings of the conference on Nuclear Physics with Stored, Cooled Beams*, P. Schwandt and H. O. Meyer, eds, AIP Conf. Proc. **128** (American Institute of Physics, New York, 1985) p. 251.

7. Dmitriev, V.F. et al., Phys. Lett. **157B**, 143 (1985); M.V. Mostovoy et al., Phys. Lett. **188B**, 181 (1987).
8. Gilman, R. et al., Nucl. Instrum. Methods in Phys. Res. **A327**, 277 (1993);
9. Wise, T., Roberts, A.D., and Haeberli, W., Nucl. Instrum. Meth in Phys. Res. **A336**, 410 (1993).
10. Stock, F. et al., Nucl. Instrum. Meth in Phys. Res. **A343** (1994) 334; Stock, F., "The HERMES-FILTEX Target Source for Polarized Hydrogen and Deuterium", contribution to this conference.
11. Haeberli, W., Ann. Rev. Nucl. Sci. **17**, 373 (1967).
12. M. Düren et al., Nucl. Instrum. Meth. in Phys. Res. **A322**, 13 (1992).
13. Schwandt, P., Clegg, T.B., and Haeberli, W., Nucl. Phys. **A163**, 432 (1971).
14. Type 3170PTFE Teflon. Teflon is a trade name of the DuPont Corp.
15. Price, J.S., and Haeberli, W., Nucl. Instrum. Meth. in Phys. Res. **349**, 321 (1994).
16. Zapfe, K. et al., contribution to this conference.
17. Conka, P.L., Nucl. Instrum. Methods **63**, 247 (1968).
18. Steffens, E. et al., "Proposal for Measurement of Spin Dependence of $p\bar{p}$ Interaction at Low Momenta", *Conference on Physics with Antiprotons at LEAR in the ACOL Era*", p.245 (1985).
19. Rathmann, F. et al., Phys. Rev. Lett. **71**, 1379 (1993); see also Rathmann, F., "Polarizing a Stored, Cooled Proton Beam by Spin-Dependent Interaction with a Polarized Hydrogen Gas Target", contribution to this conference.
20. Meyer, H.O., Phys. Rev. **E50**, 1485 (1994);
21. Meyer, H.O., "On the Polarization of a Stored Beam", contribution to this conference.
22. Horowitz, C.J., and Meyer, H.O., "Polarizing stored beams by interaction with polarized electrons", contribution to this conference.
23. CE-35 Collaboration: Haeberli, W., Lorentz, B., Rathmann, F., Ross, M.A., Wise, T. (University of Wisconsin); Dezarn, W.A., Doskow, J., Hardie, J.G., Meyer, H.O., Pollock, R.E., von Przewoski, B., Rinckel, T., Sperisen, F. (Indiana University and IUCF), Pancell, P.V. (Western Michigan University).
24. Pitts, W.K. et al., Phys. Rev. **C45**, R1 (1992).
25. Ross, M.A. et al., Nucl. Instrum. **A326**, 424 (1993).
26. Lee, K. et al., Phys. Rev. Lett. **70**, 783 (1993).
27. Ross, M.A. et al., Nucl. Instrum. **A344**, 307 (1994).
28. Wise, T. et al, "Polarized Internal Gas Target for Hydrogen and Deuterium at the IUCF Cooler Ring", contribution to this conference.
29. Derenchuk, V. et al., "IUCF high intensity polarized ion source operation", contribution to this conference.
30. Phelps, R., contribution to this conference.
31. Dick, L., and Kubischta, W., "The CERN polarized atomic beam target project" in *High-Energy Spin Physics*, AIP Conf. Proc. **187**, p.1518 (1988).
32. Toporkov, D.K. et al., "Laser-driven internal polarized deuterium target for the VEPP-3 Electron Storage Ring", contribution to this conference; Coulter, K.P. et al., Phys Rev. Lett. **68**, 174 (1992).
33. Lupov, V.G., "Status of the Michigan-MIT ultra cold polarized hydrogen jet target", contribution to this conference.
34. Augier, C. et al., Phys. Lett. **B316**, 448 (1993).
35. Vigdor, S.E., "LISS: Planning for Spin Physics with Multi-GeV Nucleon Beams at IUCF", invited paper at this conference.

Few-Body Physics with Polarized Photons

A. M. Sandorfi
Physics Department, Brookhaven National Laboratory
Upton, New York, 11973, U.S.A.

I've been asked to give an overview of recent experiments with polarized photons. Polarization observables enhance interference effects that often remain hidden in spin-averaged, unpolarized measurements. Access to such interference effects allows us to tackle new problems and gain new insights. But, because the effects are often subtle, progress is almost always linked to calculations with reasonably sophisticated models. That automatically drives the attention to few-body systems where such calculations can be carried out with reasonable accuracy. There is quite a collection of new experiments with polarized γ-ray beams. I'm going to discuss the highlights from a few, focusing on those areas where polarization tells us something really new. Each of these experiments are the work of moderately large collaborations and their composition is given at the beginning of each section. We begin with a short review of polarized photon sources.

A New Generation of Sources

Polarized γ rays can be produced in a variety of ways. The backscattering of laser light from relativistic electrons results in the highest polarizations, either linear or circular. This is just Compton scattering in the rest frame of a free electron. But, when the electron energy is several GeV the Lorentz boost throws all of the photons back through almost 180° and collapses all the cross section into what in the laboratory amounts to a beam of γ rays. Angular momentum conservation guarantees that the backscattered γ ray carry the same polarization as the instant laser light and, since lasers are 100% polarized, so are the backscattered photons. Limitations in this method are associated with the difficulties in achieving the high luminosity in the laser-electron collision that is needed to produce a high γ-ray flux.

High photon fluxes are easy to produce by the bremsstrahlung of electrons in a high-Z radiator. If the initial electron beam is longitudinally polarized, then the resulting 0° bremsstrahlung is also circularly polarized. In the absence of electron polarization, although the bremsstrahlung radiation at 0° to the beam axis is unpolarized, away from the beam axis it carries a degree of linear polarization. Lastly, the coherent bremsstrahlung of electrons in single crystals such as diamond and silicon has also been used to produce linear polarization. Both this technique, and off-axis bremsstrahlung, require careful definition of the angles involved, and the resulting polarizations are limited by the electron-beam divergence.

These techniques have been exploited in experiments for almost 30 years. Work with polarized photons can be roughly divided into two epochs: experiments in the early era, and work with the new generation of sources. The early era I define as roughly being anything before 1990. Up until that point, one can find experiments with high flux, or with good energy resolution, or with high polarization, but never with all three of these characteristics together. The turning point has been the new generation of electron accelerators that are now coming on line, and with them come a new generation of photon sources, all of high flux

($\sim 10^7$ s^{-1}), high energy resolution from tagging, and promises of quite high polarizations.

Our Laser Electron Gamma Source (LEGS) Facility (1) at Brookhaven uses laser backscattering against a 2.5-2.8 GeV electron beam circulating in a storage ring of the National Synchrotron Light Source. We have been operating since 1990. The SAL Facility in Saskatoon has just started using off-axis bremsstrahlung, and the MAMI Facility in Mainz began work with coherent bremsstrahlung in diamond at the beginning of this year. Beams from coherent bremsstrahlung are expected from Bonn next year, and with laser backscattering from the GRAAL Facility in Grenoble, also sometime next year. CEBAF is just turning on, and there are plans to make polarized photons using every conceivable technique. With our head start of about 4 years, we have completed several experiments at Brookhaven, and all of the data I will be discussing were taken at the LEGS Facility. You should view this as a precursor to a whole new class of intense activity that can be expected to blossom in the next few years.

For the sake of the discussion of the experiments given below, it is helpful to understand the characteristics of the source. These are summarized in Fig. 1. The top panel shows a number of different curves, each one corresponding to the backscattered spectrum one would observe with the indicated choice of laser line and electron ring energy. In contrast to bremsstrahlung, laser backscattering produces a nearly-flat energy-independent spectrum. The resolution of our beam is

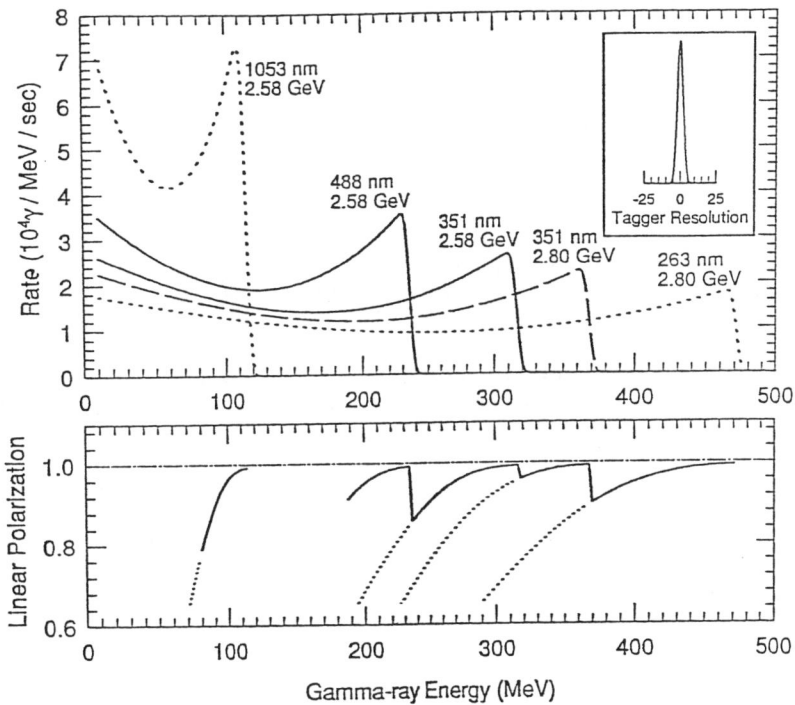

FIGURE 1. Spectra of γ-ray energies produced at LEGS for different combinations of laser and storage-ring energies (top). The degree of linear polarization is shown in the bottom panel.

about 5.5 MeV (inset to figure 1) as defined by tagging the scattered electrons. As shown in the lower panel, the polarization starts off at 100% at the maximum γ-ray energy. As the energy drops, so does the polarization. However, our practice is to change laser lines before the polarization drops too far so that, in effect, we follow the solid curve in the lower panel, thus keeping the polarization greater than about 85% throughout the entire tagging range.

The E2 N→Δ Transition
(LEGS Collaboration[1,2] – Exps. L2, L5, and L7)

I would like to start the physics discussion with the first many-body system, the proton. The issue here will be the breaking of spherical symmetry by spin dependent forces. This can be approached from a number of different paths. For simplicity, let me just consider the constituent quark model. There one starts with an infinite confining potential like an harmonic oscillator, to which is usually added another potential that breaks the degeneracy as much as one can while still preserving symmetry. At this point, one has a spectrum of states with even-orbital angular momentum. To go from here to the observed half-integral baryon spectrum, it is necessary to couple in the intrinsic spins of the quarks, which is accomplished through a tensor interaction. However, there is a consequence to this. Just as the N-N tensor interaction breaks the spherical symmetry of the deuteron, so too this tensor interaction breaks the symmetry of the nucleon. There are a very large number of papers investigating the nucleon's intrinsic deformation, and a wide range of predictions, which means that if we can measure a quantity related to this deformation we have a new way of discriminating between models.

The approach we take is from classic nuclear physics. For nuclei, we measure deformation with a B(E2) value, the E2 transition strength between the ground and first-excited states. We do the same thing for the nucleon. The first-excited state of the nucleon is the Δ(1232) resonance. This is excited primarily by a magnetic dipole transition. But if it is indeed deformed, there must be an E2 component buried amid the M1 strength. The Δ decays with the 99.4% branched to the π-nucleon channel. So, the most straightforward approach is to look at (γ,π) reactions. There are two complications here. One is that the predicted E2 strength is only a few % of the M1 transition. The other is that Born amplitudes can also produce a pion as the result of E2 absorption, without ever producing a Δ in the intermediate state, and model dependence necessarily enters into the background-resonance decomposition.

Since the deformation of interest is caused by a spin-dependent interaction, it should not be surprising to learn that signatures of this effect are enhanced in polarization observables. In pion photoproduction there are 16 possible observables one can consider for each charge channel. I've gone through the

---

[1] *LEGS Collaboration – Exps. L2,5*: Blanpied, G., Blecher, M., Caracappa, A., Djalali, C., Duval, M.-A., Giordano, G., Hoblit, S., Khandaker, M., Kistner, O.C., Matone, G., Miceli, L., Mize, W.K., Preedom, B.M., Sandorfi, A.M., Schaerf, C., Sealock, R.M., Thorn, C.E., Thornton, S.T., Vaziri, K., Whisnant, C.S., Zhao, X. and Moinster, M.A., (ref. 2).

[2] *LEGS Collaboration – Exp. L7:* Blanpied, G., Blecher, M., Caracappa, A., Djalali, C., Giordano, G., Hicks, K., Hoblit, S., Khandaker, M., Kistner, O.C., Matone, G., Miceli, L., Preedom, B.M., Rebreyend, D., Sandorfi, A.M., Schaerf, C., Sealock, R.M., Ströher, H., Tam, A., Thorn, C.E., Thornton, S.T., Whisnant, C.S. and Zhao, X, (ref. 3).

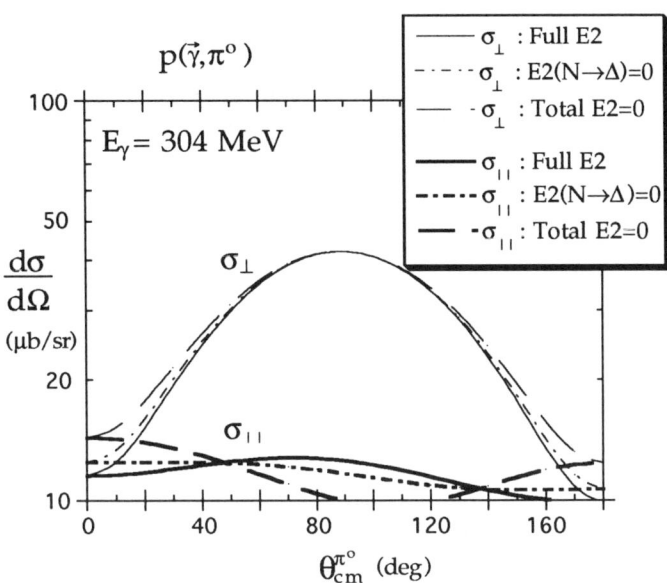

FIGURE 2. Cross sections for the $p(\vec{\gamma}, \pi^o)$ reaction, for parallel (||) and perpendicular (⊥) orientations of linear polarization, calculated (DMW – ref.4) with different E2 contributions.

exercise of testing the sensitivity of each one by artificially changing the E2 multipole and looking at the resulting predictions. With that, one can produce one's own *Michelin Guide* for the E2/M1 effect, and the most sensitive observable turns out to be π^0 production with linearly polarized photons. The cross sections for this process are shown in Fig. 2. Here there are two bands of curves. The upper band of 3 are calculated for the perpendicular geometry, where the electric vector of the linearly polarized photon is 90° to the reaction plane. The bottom 3 curves correspond to the parallel geometry. The long-dashed curves for both cases result from setting the total E2 to 0, Born as well as resonant components. The dot-dashed curves result from keeping the E2 part of the Born amplitudes while turning off the E2 part of the N→Δ transition, and the solid curves are the full calculations (4) of Davidson, Mukhopadhyay and Wittman (DMW).

The first thing to notice here is that the calculations for the perpendicular geometry are almost indistinguishable from one another over most of the angular range. All of the sensitivity to E2 comes in the parallel geometry. Notice also that these are plotted with a log scale. Since the perpendicular cross sections are much bigger than the parallel, they dominates the unpolarized reaction and render it quite insensitive to this problem. If we focus on the parallel set of curves, the second thing to note is that the contribution of the Born amplitudes is far from trivial.

There have been three measurements of this reaction at our facility over the last few years (2,3). The most recent (Exp. L7) has covered a large range of angles and energies. In this measurement, one of the decay photons from the π^0 was detected

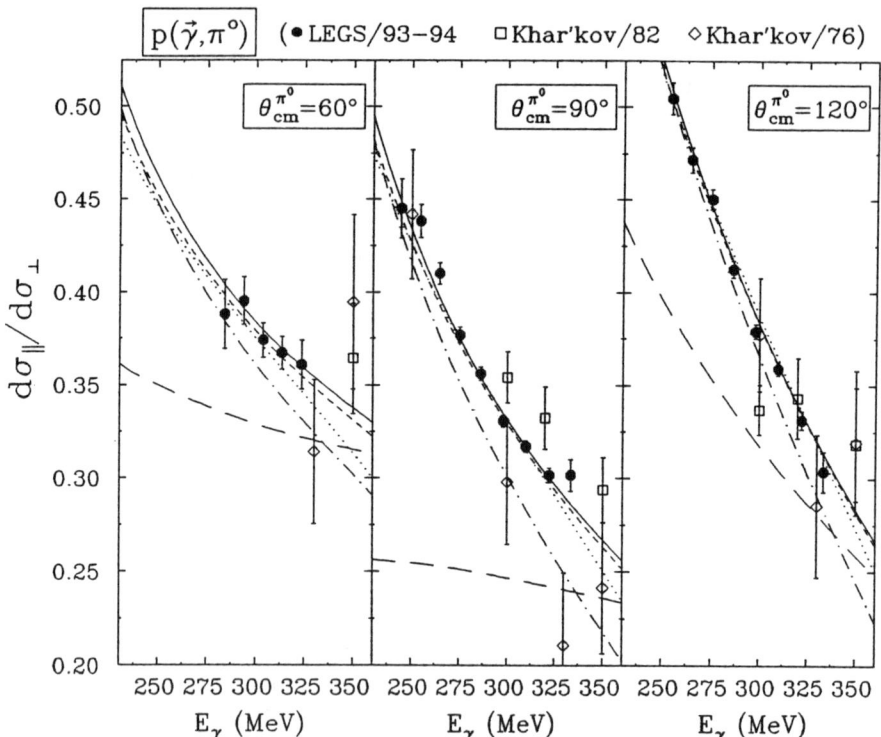

FIGURE 3. The E2-sensitive observable $\sigma_\parallel/\sigma_\perp$ for the $p(\vec{\gamma},\pi^0 p)$ reaction. New results from Exp. L7 are plotted as solid circles (3) and compared with previous measurements from Khar'kov (5,6). The curves are calculations using the model of DMW (4): the long-dashed curves are produced by setting the total E2 to 0; the dashed-dot curves correspond to E2($N \rightarrow \Delta$) = 0; the solid, short-dashed and dotted curves are the full calculations using the *Olsson*, *K-matrix* and *Noelle* decompositions of the amplitudes into resonant+background components.

in a high-resolution sodium-iodide spectrometer, together with the recoil protons whose trajectories were tracked through wire chambers and whose energies were measured, both by energy deposition and by time of flight, in an array of plastic scintillators. This resulted in a very large over determination of kinematics and an excellent separation of π^0 production from Compton scattering.

Since the perpendicular cross section is essentially a constant, at least insofar as the E2 transition strength is concerned, the ratio of parallel –to– perpendicular cross sections, $\sigma_\parallel/\sigma_\perp$, provides an E2 sensitive observable that is free from systematic effects. A sample of the new data is shown in Fig. 3 (solid points) for 3 different π^0 center-of-mass angles. Also shown are previous data taken at Khar'kov using bremsstrahlung in diamond crystals (open symbols).

The long-dashed lines in the figure result from turning off all the E2, total E_{1+} ($\tau=3/2)\rightarrow 0$ in the DMW calculation. Although these are quite far from the data, most of this "E2 signal" results from interferences with E2 components of the Born amplitudes and is quite uninteresting. The dashed-dot curves are obtained by including the full Born contribution while setting the resonant part of the E2 strength to zero. It is the differences between these dot-dashed curves and the data that represent the E2 *signal* of interest, and the sensitivity is maximal at 90°. Modeling the N→Δ transition requires a decomposition of each of the amplitudes into resonant and background terms, and this decomposition is not unique. Three choices for this decomposition have been discussed by DMW: *Olsson's method*, in which the background part of the T-matrix is made separately unitary; a *K-matrix method*, in which the decomposition is made in terms of K-matrix elements; and *Noelle's method* in which the decomposition is made in the pion phase shifts and carried through to photoproduction via unitary. We have refitted the parameters of the DMW model to our new polarization data. Since all of the analysis is not yet complete, we have included previously-measured unpolarized data from other laboratories. The results are shown in Fig. 3 as the solid (*Olsson*), short-dashed (*K-matrix*), and dotted (*Noelle*) curves, respectively. All three are equally good representations of the $\sigma_\parallel/\sigma_\perp \pi^0$ data. Compared to other observables, the *Olsson* and *K-matrix* results are essentially equivalent while the *Noelle* prediction is somewhat worse for the π^+ asymmetry. The resonant E2/M1 extracted from these fits range between -2.6% and -2.8%. This procedure will be repeated when the analysis of the full data set has been completed. The use of a single data set in the determination of model parameters will minimize effects of systematic uncertainties.

The unitarization procedures of the DMW model inherently include the effects of pion rescattering. As a result, it's most appropriate to compare these E2/M1 values with those models of hadron structure that contain pion fields. The negative phase of the mixing ratio implies an oblate intrinsic shape, and the value of about –2.7% is close to the low-end of predictions coming from Skyrme models of the nucleon, -3 to -5%, but somewhat larger than chiral-bag predictions, -1 to -2% (7). It is also a factor of 2 larger than the *accepted* value found in the literature of a couple of years ago, and this reflects the impact of the new polarization data (4,8).

Comparing the E2/M1 = –2.7 % value to the constituent quark model provides a certain added dimension. This extracted value is about 5 times larger than the model prediction (about -0.5 %). While both the nucleon ground state and the Δ are constructed predominantly from orbital angular momentum $l=0$ components, the tensor interaction introduces a small $l=2$ admixture, ~0.4% for the N and ~1.9% for the Δ (9). The new E2/M1 results shows that the $l=2$ components are *much larger* than expected. This is corroborated by the EMC-SMC/E142-E143 deep inelastic scattering measurements (10) which have deduced that only about 30 % of the proton's spin is carried by $l=0$ quarks. The rest has to reside in orbital angular momentum, either of quarks or gluons, and in a net polarization of the gluon sea. (Some of the effects of the glue appear *implicitly* through the tensor interaction which is assumed to be mediated by one-gluon exchange). A significant orbital contribution to the nucleon spin is reflected in an E2 component to the N→Δ transition that is much larger-than-expected in the non-relativistic quark model. Although it may at first seem surprising to think of ~300 MeV photons probing the internal nucleon structure, it must be remembered that orbital angular momentum is the exclusive property of an extended object, and as such must be studied at comparable wavelengths. (As already noted, chiral-bag and Skyrme models predict E2/M1 values that are much closer to our new results. Pion fields play a major role in these models and, since pions are just $q\bar{q}$ pairs coupled to intrinsic spin 0, it is tempting to associate the larger E2/M1 values with increased orbital contributions. Nonetheless, it must be remembered that, since relativity is an essential ingredient in these models, there are only two gauge-invarient components to the nucleon spin – the spin of $l=0$ quarks, and the *sum* of orbital and $l=0$ glue contributions.)

Deuteron Photodisintegration
(LEGS Collaboration[3] – Exps. L1 and L3)

Going up in mass and down in the number of interacting particles, we now come to the photodisintegration of the deuteron, which for this discussion we view simply as an n–p pair. The first issue here is the sensitivity of the photon polarization to the N-N tensor force. The nuclear tensor force comes about largely from one-pion exchange. Most of the observables sensitive to the tensor interaction (eg. the quadrupole moment, the asymptotic-D/S ratio) are essentially controlled by the πNN coupling constant which is now reasonably stable. However, the same cannot be claimed for the form factor at the πNN vertex. The traditional parameter used to constrain the short-range part of the N-N tensor force is the phase shift (ε_1) for $\Delta l = 2$ transitions in n-p scattering. However, there is still quite a spread in the values of this parameter extracted from multipole analyses.

The beam-polarization asymmetry in the deuterium photodisintegration reaction provides an important constraint on the short-range part of the nuclear tensor interaction. Calculations by Arenhövel and collaborators from Mainz were quite successful in describing the results around 200 MeV (11,12). For example, figure 4

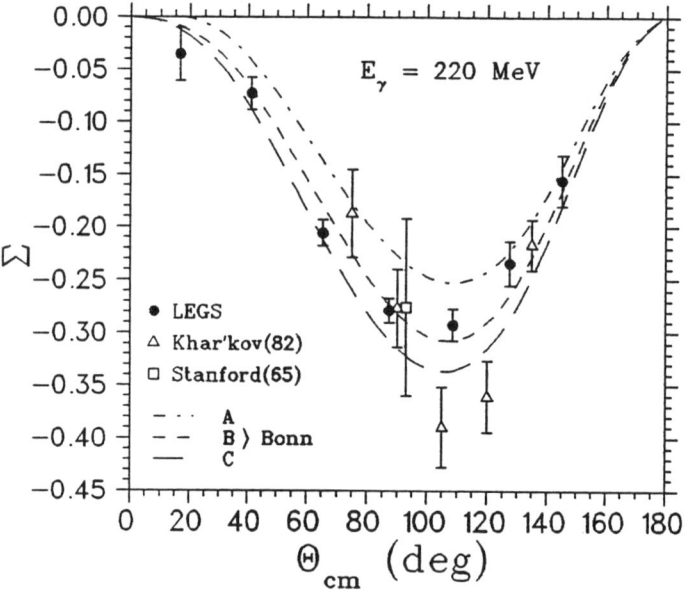

Figure 4. Angular distribution of the beam-polarization asymmetry in $D(\vec{\gamma},p)n$ at 220 MeV. The solid points are the combined results of Exps. L1 and L3; open-square and diamonds are from refs. 14 and 15, respectively. The curves are calculated (11,12) for three different versions of the Bonn OBEPQ potential which differ in the short range tensor force (13).

---

[3] *LEGS Collaboration – Exps. L1,3*: Blanpied, G., Blecher, M., Caracappa, A., Djalali, C., Duval, M.-A., Giordano, G., Hoblit, M., Kistner, O.C., Matone, G., Miceli, L., Mize, W.K., Preedom, B.M., Sandorfi, A.M., Schaerf, C., Sealock, R.M., Thorn, C.E., Thornton, S.T., Vaziri, K. and Whisnant, C.S., (ref. 11).

shows the new data compared to calculations using 3 different versions of the Bonn OBEPQ potential (13) which differed only in the form-factor at the πNN vertex. This is a convenient set of potentials to consider because it's essentially only the short range part of the tensor interaction that's changing between them. Using these NN potentials in the D(γ,p) calculations, one finds that the unpolarized cross section is totally insensitive to the tensor interaction, while the new polarization asymmetry data (solid points) shows quite a useful sensitivity. Previously published data are also shown (open symbols), but their scatter has been rather large. The new data favor the OBEPQ-B potential.

In the calculations of figure 4 the N-N and N-Δ interactions were treated as indistinguishable. At higher energies, near the peak of the delta (265 MeV in this channel), a better approach is to treat the interactions separately, solving coupled Schrödinger equations. In this way, it is possible to study the NΔ interaction about which very little is known. Recent work along these lines has suggested rather significant differences between the γNN and γNΔ couplings (16), although not all the D(γ,p) results are equally well reproduced. High precision, in both the asymmetry as well as the cross section, provides an important constraint since the NΔ-NN differences manifest themselves quite differently in the two observables.

Three Nucleon Currents in ^{3}He
(L4 Collaboration[4])

Photon absorption on light nuclei can provide a microscopic look at mesonic and nucleonic currents. In particular, mass-3 is the simplest system in which 3-body correlation effects might be present and for which exact calculations can be performed. Unpolarized measurements (17,19) have suggested that, when the two protons carry a significant fraction of the total energy, 3N absorption mechanisms involving all 3 nucleons dominate.

We have studied the $γ+^3$He→ppn reaction in a large detector array. Protons and neutrons were detected above the target, with an array of scintillators having both large polar and azimuthal coverage, in coincidence with another proton detected below the target in a second array having large polar but narrow azimuthal coverage. (For 3-body final states, the narrow azimuthal range in one of the arms is needed if you are going to ask questions about spin. You've got to define the parallel and perpendicular polarization directions relative to something, and, in this case, we use the plane containing the beam axis and this bottom set of detectors.) The momentum distributions of protons detected below the target at 100° (in the array with a narrow azimuthal range), integrated over the full acceptance of the upper array for the accompanying proton and neutron, are shown in Fig. 5 for an average beam energy of 270 MeV (18). The (γ,pn) cross sections (open circles) are dominated by a large quasi-deuteron component and peak at 470 MeV/c, the expected momentum from deuteron dissociation with the extra proton acting as the spectator. The dotted and solid curves are recent calculations, performed by J.-M. Laget of Saclay, and include contributions from 1N, 2N and 3N photoabsorption

---

[4] *L4 Collaboration*: Adams, G.S., Audit, G., Baghaei, H., Caracappa, A., Clayton, W.B., D'Angelo, A., Duval, M.-A., Giordano, G., Hoblit, S., Kistner, O.C., Laget, J.-M., Lindgren, R., Matone, G., Miceli, L., Mize, W.K., Moinester, M.A., Ruth, C., Sandorfi, A.M., Schaerf, C., Sealock, R.M., Smith, L.C., Stoler, P., Tedeschi, D.J., Teng, P.K., Thorn, C.E., Thornton, S.T., Vaziri, K., Whisnant, C.S. and Winhold, E.J., (refs. 17,18).

amplitudes, as shown schematically in the legend. The 2N dominates in (γ,pn) and addition of 3N components adds very little. The undetected p is truely a spectator.

The analogous 2N mechanism in pp absorption is depressed since the di-proton has no dipole moment and charge meson exchange currents cannot contribute. The (γ,pp) cross sections (solid squares in the upper panel) are a factor of 6 lower than those of the (γ,pn) channel, and theoretical predictions that exclude 3N amplitudes (dashed-dot curve) are a factor of 4 lower still, although they exhibit a similar momentum dependence. The addition of 3N absorption to the theory (dashed curve) produces results that are in very good agreement with the data.

Qualitatively similar results have been observed in unpolarized reactions with ^{3}He at other laboratories (19). However, a recurring objection can be made that, since the shapes of the predictions with (dashed), and without (dashed-dot), the 3N absorption mechanism are so similar, uncertainties in the N-N potential might allow

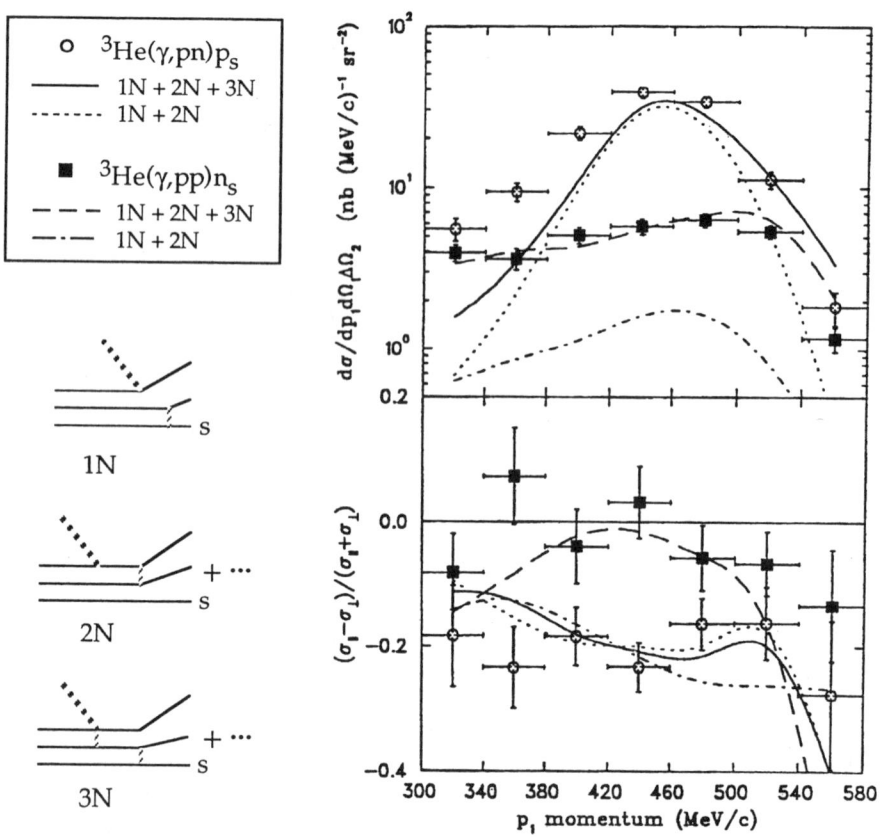

Figure 5. The momentum distributions of protons detected at $\theta_p = 100°$, integrated over the full acceptance (24° to 144°) for the accompanying n or p, at a γ-beam energy of 270 MeV. Cross sections are shown in the top-right panel (and asymmetries in the bottom) for $^3He(\vec{\gamma},pn)$, open circles, and $^3He(\vec{\gamma},pp)$, solid squares. The legend for the calculations is shown to the left (ref. 18).

for some adjustment in the overall normalization of the 2N contribution. Experiment L4 has provided the first polarization information on these reaction channels, and the beam polarization asymmetry (lower panel) completely rules out this scenario. The predicted (γ,pp) asymmetry from 2N amplitudes (dashed-dot curve) is large and negative. It is, in fact, quite similar to that observed in deuteron photodisintegration and to that of the ^{3}He (γ,pn) measurements. Increasing the 2N component would only make the asymmetry more negative. In contrast, the asymmetry data (solid squares), as well as the predictions including 3N absorption (dashed curve), are nearly zero. There can be no preferred direction when all 3 nucleons are involved.

Some of the highlights of this experiment have appeared in a recent issue of Physical Review Letters (18). This same issue also contained a report by the TAGX collaboration from Japan (20). They have studied the same reactions over similar angular and energy ranges, with a similar detector, using the same momentum thresholds, and have come to nearly *opposite* conclusions. In their work they have decomposed neutron momentum distributions into two components, the lower of which they have argued is due to E2 photo-absorption on a pp pair. They have also argued that final state interactions are small, implying that the di-proton absorption mechanism is quite significant.

The polarization asymmetry sheds light on this matter. Because of the symmetry

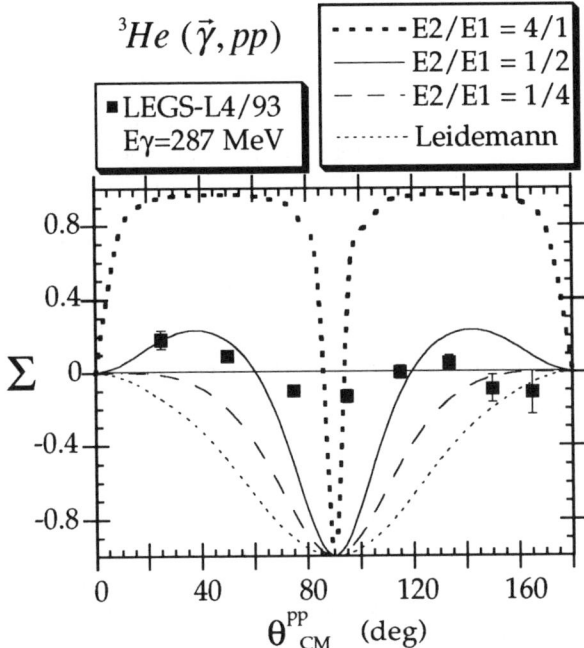

Figure 6. The asymmetry for di-proton kinematics in $^3He(\vec{\gamma}, pp)$ from Exp. L4 (18), compared with calculations assuming varying amounts of E2 excitation.

between the identical particles of the di-proton, M(J_{odd}) transitions are not allowed, E(J_{even}) transitions can only occur with channels spin zero, and E(J_{odd}) or M(J_{even}) transitions can only only occur for channel spins one. The result of this is that interfering terms of opposite parity all vanish, and any significant contribution from E2 absorption results in an asymmetry that is large and positive near 45º and 135º. This is illustrated in Fig. 6 where I've varied the amount of E2 relative to the next most probable allowed multipole, E1. When E2 dominates, the asymmetry is merely +1 for much of the angular range and only becomes negative when the E2 component becomes quite small. The dotted curve is a calculation by Leidemann (21) in which he has imposed the symmetry of the di-proton on his model for the D(γ,p) reaction. The largest multipole that survives the symmetrization is electric dipole, the E2 component being quite small, and the resulting asymmetry is negative. From Exp. L4 (solid points), there is clearly no evidence for a large positive asymmetry that would be consistent with a large E2 contribution to the pp channel. Another consequence of the symmetry of the di-proton is that the parallel cross section is always proportional to powers of cos θ and vanishes at 90º. Thus, if you really have the di-proton, the asymmetry must go to -1 at 90º. The deviation from that is the measure of the final state interactions and, as is quite evident in Fig. 6, these are quite substantial.

The Spin Dependent Compton Amplitude

There are several other new experiments which I haven't touched upon (22), but I'd like to close this overview with a look to the future. Here, the obvious choice is an experiment that is being given a high priority at every one of the new polarized photon facilities. The experiment consists of evaluating sum rules for the spin-dependent nucleon-Compton amplitude by measuring reaction cross sections with circularly polarized γ-rays incident on polarized protons and polarized neutrons.

The forward Compton cross section for initial and final photon polarizations $\vec{\varepsilon}$ and $\vec{\varepsilon}'$ has been given by Gell-Mann, Goldberger and Thirring (GGT-ref. 23) as,

$$A(\omega) = f(\omega^2)\vec{\varepsilon}\cdot\vec{\varepsilon}' + i\omega\, g(\omega^2)\vec{\sigma}\cdot(\vec{\varepsilon}\times\vec{\varepsilon}') \quad , \tag{1}$$

where $\vec{\sigma}$ is the target spinor. The spin independent, $f(\omega^2)$, and spin dependent, $g(\omega^2)$, amplitudes can be expanded in a Taylor series,

$$\begin{aligned} f(\omega^2) &= f(0) + f'(0)\,\omega^2 + O(\omega^4) \\ g(\omega^2) &= g(0) + g'(0)\,\omega^2 + O(\omega^4) \end{aligned} \quad , \tag{2}$$

where the prime-symbol indicates differentiation with respect to ω^2. Here, $f(0)$ is the Thomson scattering limit that corresponds to wiggling the nucleon up and down with the electromagnetic field of the photon. $f'(0)$ is identified with the sum of the electric and magnetic polarizabilities of the target and measures how the constituents of the target rearrange themselves in response to the applied electromagnetic field. In analogy, $g'(0)$ is referred to as the spin-dependent polarizability of the nucleon. The GGT dispersion relations provide a sum rule for this spin polarizability (γ),

$$\gamma \equiv g'(0) = \frac{1}{4\pi^2} \int_{\omega_o}^{\infty} \frac{\sigma_{\frac{1}{2}} - \sigma_{\frac{3}{2}}}{\omega^3} d\omega \quad . \tag{3}$$

Here $\sigma_{1/2}$ and $\sigma_{3/2}$ are the total reaction cross sections measured with the photon helicity oriented opposite to the nucleon polarization or parallel to it, respectively. Another GGT dispersion relation can be written down for $g(0)$,

$$g(0) - g(\infty) = \frac{1}{4\pi^2} \int_{\omega_o}^{\infty} \frac{\sigma_{\frac{1}{2}} - \sigma_{\frac{3}{2}}}{\omega} d\omega \quad . \tag{4a}$$

Low, Gell-Mann and Goldberger (LGG-ref. 24) have evaluated $g(0)$ in terms of κ, the anomalous magnetic moment of the target,

$$g(0) = -\alpha \kappa^2 / 2m^2 \quad . \tag{4b}$$

Drell and Hearn, and independently Gerasimov (DHG-ref. 25), have noted that under the assumption $g(\infty) = 0$, a sum rule is obtained that relates these polarized reaction cross sections to the anomalous magnetic moment of the target,

$$DHG \equiv \int_{\omega_o}^{\infty} \frac{\sigma_{\frac{1}{2}} - \sigma_{\frac{3}{2}}}{\omega} d\omega = -\frac{2\pi^2 \alpha}{m^2} \kappa^2 \quad . \tag{4c}$$

In writing eqns. (3) and (4c), it is implicitly assumed that the integrals converge.

Whisnant, Khandaker and I have recently evaluated these integrals (26), using the FA93 photo-pion multipole analysis of the VPI group (27) to predict the spin dependent cross sections. This multipole set extends out to 1.7 GeV, and the results of this exercise are shown in Fig. 7 as a fraction of this upper limit. The $N\pi$

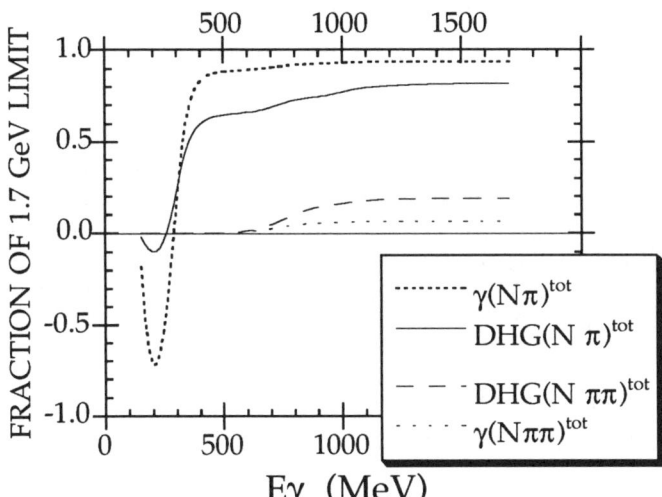

Figure 7. The fraction of the spin-polarizability and DHG integrals (26), as a function of the upper limit of integration, compared with their values computed up to 1.7 GeV from the VPI-FA93 multipoles (27).

Table 1. Multipole predictions from ref. 26 for the γ and DHG sum rules on the proton + neutron, compared with χPT predictions (28) for $\gamma \equiv g'(0)$ and with the magnetic moment predictions for DHG.

	χPT (rel-1-loop + Δ)	FA93 (VPI-SAID)	$-\frac{1}{2}(\kappa_p^2 + \kappa_n^2)\frac{2\pi^2\alpha}{m^2}$
$\gamma_{V+S} = \frac{1}{2}(\gamma_p + \gamma_n)$	-98×10^{-6} fm^4	-86×10^{-6} fm^4	
$DHG_{V+S} = \frac{1}{2}(DHG_p + DHG_n)$		-225×10^{-4} fm^2	-219×10^{-4} fm^2

contributions come directly from the multipole analysis of single pion-production data, while the N$\pi\pi$ components are produced by scaling the Nπ predictions with the known 2π branching ratios of baryon resonances. All of the curves appear to have converged above about 1.2 GeV.

The result of these calculations for the average of the proton and neutron sum rules are summarized in Table 1. (In the literature the DHG sum rule is usually broken up into vector, V, scalar, S, and mixed vector-scalar, VS, components. The proton-neutron average corresponds to the VS sum rule.) The FA93 prediction of -225×10^{-4} fm^2 (or µb) for DHG_{V+S} shows impressive agreement (2.7%) with the value expected from the anomalous magnetic moments, -219×10^{-4} fm^2. For the spin polarizability, comparisons can be made with Chiral perturbation theory (χPT). Bernard et al. have reported relativistic-χPT calculations for γ that include all diagrams to the one-loop order, excluding those involving a delta (28). The delta effects are large, and they have estimated the contributions from the dominant diagrams. The result, -98×10^{-6} fm^4, is within 14% of the FA93 evaluation. This is very encouraging agreement, although the final word must await a calculation which includes all delta contributions to the same order.

Since in χPT the delta contributions for the neutron and the proton are the same, the uncertainties in the one-loop calculation disappear when we consider the difference, $\frac{1}{2}(proton - neutron)$. These results are summarized in Table 2. Now, the relativistic one-loop value for γ is in impressive agreement with the FA93 evaluation. However, in sharp contrast, the predictions for $DHG_{VS} = \frac{1}{2}(DHG_p - DHG_n)$ are very different from the magnetic moment values – both in sign and a factor of 4 in magnitude.

The results in Table 2 are quite puzzling. In principle, the different energy weightings of the γ_{VS} and DHG_{VS} integrals admit the possibility that contributions above the 1.7 GeV limit of the VPI data base could bring the DHG_{VS} value up to that

Table 2. Multipole predictions from ref. 26 for the difference sum rules, 1/2(proton − neutron), compared with χPT predictions (28) for $\gamma \equiv g'(0)$ and with the magnetic moment predictions for DHG.

	χPT (rel-1-loop)	FA93 (VPI-SAID)	$-\frac{1}{2}(\kappa_p^2 - \kappa_n^2)\frac{2\pi^2\alpha}{m^2}$
$\gamma_{VS} = \frac{1}{2}(\gamma_p - \gamma_n)$	-52×10^{-6} fm^4	-48×10^{-6} fm^4	
$DHG_{VS} = \frac{1}{2}(DHG_p - DHG_n)$		-65×10^{-4} fm^2	$+15 \times 10^{-4}$ fm^2

expected by the sum-rule, without appreciably affecting γ_{VS}. However, 1.7 GeV is already so large that such $\Delta\sigma_{VS} = \frac{1}{2}\{[\sigma_{1/2} - \sigma_{3/2}]_p - [\sigma_{1/2} - \sigma_{3/2}]_n\}$ differences would have to be huge in order to overcome the $1/\omega$ energy weighting. For example, $\Delta\sigma_{VS} = 200$ µb between 2 and 3 GeV, which would require a prominent but as yet unidentified resonance; or a constant level of $\Delta\sigma_{VS} = -20$ µb extending out to 100 GeV, which would be much larger than the contributions of the resonance region in which the isospin structure of the Δ and N* states can be expected to enhance the proton-neutron difference. Apart from such scenarios, which seem highly unlikely, there are only two other possibilities. **Either a)** *both* **the two-loop corrections to the spin polarizability are large** *and* **the existing multiples are wrong, or b) modifications to the DHG sum rule are needed to fully describe the isospin structure of the nucleon.**

It is possible that the two-loop corrections to the χ^{PT} calculations for γ are large. Although this is not usually the case for χ^{PT} expansions, it would not be without precedence. This would imply that the multipoles would have to be modified. Changing the multipoles would potentially affect the predictions for a variety of observables. Although it is possible to imagine complicated alterations that would leave cross sections and single-polarization observables largely unaffected, several double-polarization observables would have to change significantly. As yet, there are no measurements of these quantities.

Alternatively, if the multipoles are basically correct, then the DHG sum rule requires at least a modification of the form,

$$DHG_{VS} \rightarrow DHG_{VS} - C \tag{5}$$

and the simplest choice for the correction factor would be $C = 2\pi^2[g_p(\infty) - g_n(\infty)]$, with $g_p(\infty) \approx -g_n(\infty) \approx 2\mu b$. In other words, contrary to the original DHG assumptions, g_p and g_n would tend to nearly equal but opposite constants at high energy. The physical origins of such constants would be quite interesting, although other choices for C have also been proposed (29,30). The DHG_{VS} integral is just the $Q^2 = 0$ limit of the Bjorken sum rule integral (31), and $C = 0$ has been assumed in modeling its Q^2 evolution (32-34). The Q^2 dependence of a possible non-zero $g(\infty)$ remains to be considered.

This is a challenging question that will be pursued in planned experiments at many different laboratories. Since the key physics issues with the least model dependence are in the proton-neutron difference, each of which involves cross section differences themselves, considerable care must be taken by the experimenters to minimize systematic uncertainties. Our approach at LEGS will be to carry out the proton and neutron measurements *simultaneously* with a new $\vec{H} \cdot \vec{D}$ polarized target (35). Some of the technical details of this experiment, and in particular the new target, have been discussed at this conference by Honig (36) and by Whisnant (37).

This work has been performed under Contract No. AC02-76CH00016 with the United States Department of Energy. Support for experimental programs at the LEGS facility by the US National Science Foundation, the Istituto Nazionale di Fisica Nucleare, the Deutsche Forschungsgemeinschaft, and the US-Israel Binational Science Foundation, is gratefully acknowledged.

REFERENCES

1. Sandorfi, A.M., *et al.*, IEEE Trans. Nucl. Sci. **30**, 3083 (1983); Thorn, C.E., *et al.*, Nucl. Instrum Methods Phys. Res., **A285**, 447 (1989).
2. *LEGS Collaboration*, Blanpied, G., *et al.*, Phys. Rev. Lett. **69**, 1880 (1992).
3. *LEGS Collaboration*, Sandorfi, A.M., *et al.*, Few-Body Systems, Suppl. **7**, 317 (1994).
4. Calculations with the model of Davidson, R., Mukhopadhyay, N., and Wittman, R., Phys. Rev. **D43**, 71 (1991).
5. Ganenko, V.B., *et al.*, Yad. Fiz. **23**, 310 (1976); Sov. J. Nucl. Phys., **23**, 162 (1976).
6. Belyaev, A.A., *et al.*, Yad. Fiz. **35**, 693 (1982); Sov. J. Nucl. Phys., **35**, 401 (1982).
7. Mukhopadhyay, N., private communication; and in *Topical Workshop on Excited Baryons –1988, Troy, NY*, eds. Adams, G., Mukhopadhyay, N., Stoler, P., 205 (World Scientific, Singapore, 1989).
8. Davidson, R. and Mukhopadhyay, N., Phys. Rev. Lett., **70**, 3834 (1993); Sandorfi, A.M. and Khandaker, M., Phys. Rev. Lett., **70**, 3835 (1993).
9. Giannini, M., Rep. Prog. Phys. **54**, 453 (1991), and references contained therein.
10. See the papers by Windmolders, R. and Day, D. in these proceedings.
11. *LEGS Collaboration*, Blanpied, *et al.*, Phys. Rev. Lett. **67**, 1206 (1991).
12. Schmitt, K.M. and Arenhövel, H., Few Body Systems. **7**, 95 (1989); These calculations used the NN potentials from ref. (13) which assumed a larger value for the πNN coupling constant than is currently accepted. Although this can change the predicted D(γ,p)n cross sections, it will have little effect on the beam-polarization asymmetry.
13. Machleidt, R., Adv. Nucl. Phys. **19**, 189 (1989).
14. Liu, F.F., Phys. Rev. **138**, B1443 (1965).
15. Gorbenko, V.G., et al., Nucl. Phys. **A381**, 330 (1982).
16. Wilhelm, P. and Arenhövel, H., Phys. Lett. **B318**, 410 (1993).
17. *L4 Collaboration*: Ruth, C., *et al.*, Phys. Rev. Lett. 72, 617 (1994).
18. *L4 Collaboration*: Tedeschi, D.J., *et al.*, Phys. Rev. Lett. **73**, 408 (1994); Tedeschi, D.J., "Exclusive Photodisintegration of ^{3}He by Polarized Photons", Report LEGS-**93T4.**
19. Audit, G., *et al.*, Phys. Lett. **B227**, 331 (1989); *ibid*, Phys. Rev. **C44**, 575 (1991); *ibid*, Phys. Lett. **B312**, 57 (1993); Sarty, A.J., *et al.*, Phys. Rev. **C47**, 459 (1993).
20. *TAGX Collaboration*: Emura, T., *et al.*, Phys. Rev. Lett., **73**, 404 (1994).
21. Leidemann, W., private communication.
22. For example, see the paper by Hicks, K., in these proceedings.
23. Gell-Mann, M., Goldberger, M. and Thirring, W., Phys. Rev. **95**, 1612 (1954).
24. Low, F., Phys. Rev. **96**, 1428 (1954); Gell-Mann, M. and Goldberger, M., Phys. Rev. **96**, 1433 (1954).
25. Drell, S.D. and Hearn, A.C., Phys. Rev. Lett. **16**, 908 (1966); Gerasimov, S.B., Sov. J. Nucl. Phys. **2**, 430 (1966).
26. Sandorfi, A.M., Whisnant, C.S. and Khandaker, M., Phys. Rev. D (in press).
27. Workman, R.L. and Arndt, D., private communication; the Scattering Analysis Interactive Dialin (**SAID**) program, available by TELNET to VTINTE (1993).
28. Bernard, V., Kaiser, N., Kambor, J. and Meißner, Ulf-G., Nucl. Phys. **B388**, 315 (1992).
29. Kawarabayashi, K. and Suzuki, M., Phys. Rev. **152**, 1383 (1966).
30. Chang, L.N., Liang, Y. and Workman, R.L., Phys. Lett. **B329**, 514 (1994).
31. Bjorken, J.D., Phys. Rev. **148**, 1467 (1966); *ibid*, **D1**, 1376 (1970).
32. Anselmino, M., *et al.*, Sov. J. Nucl. Phys. **49**, 136 (1989).
33. Burkert, V. and Li, Z., Phys. Rev. **D47**, 46 (1993).
34. Ji, X., Phys. Lett. **B309**, 187 (1993); Ji, X. and Unrau, P., Phys. Lett. **B333**, 228 (1994).
35. *LEGS Spin Collaboration*: Babusci, D., et al., BNL internal report (1994).
36. Honig, A, *et al.*, paper in these proceedings.
37. Whisnant, C.S., *et al.*, paper in these proceedings.

THE SPIN STRUCTURE OF ^{3}HE

Richard G. Milner
Department of Physics and Laboratory for Nuclear Science,
Massachussets Institute of Technology, Cambridge, MA 02139

Abstract. Faddeev calculations predict that the ground state spin of the ^{3}He nucleus is dominated by the neutron. Recently, experiments at IUCF, Bates, Mainz and SLAC have obtained for the first time experimental information on the ^{3}He ground state spin-dependent momentum distribution and on the electromagnetic form factors of the neutron in ^{3}He. The experiments are described and the data presented.

INTRODUCTION

The ^{3}He nucleus has several properties which make the study of its spin particularly interesting. The three body system is unique, in that although it is relatively tightly bound, essentially exact Faddeev solutions in non-relativistic approximation of the ground state have been obtained using a variety of two-nucleon potentials. In addition, unlike a heavy nucleus where the total spin is usually determined by only a few valence nucleons, the spin of ^{3}He involves all the nucleons in the nucleus. Further, Faddeev calculations predict that the ground state spin of the ^{3}He nucleus is dominated by the neutron. This property has motivated great interest in the use of polarized ^{3}He as an effective neutron target. Finally, because the ^{3}He atom has a closed electron structure, the polarized atom does not depolarize with high probability when colliding with container walls. Thus, targets suitable for scattering experiments have been constructed (1). Recently, several significant measurements on the polarized ^{3}He nucleus have been carried out. In addition, the spin-dependent spectral function, calculated from a Faddeev solution to the ^{3}He ground state, has become available.

THE ^{3}HE SPIN-DEPENDENT SPECTRAL FUNCTION

Non-relativistic Faddeev calculations (2,3) of the three body bound state predict the following components to dominate the ^{3}He ground state wave function: a) a spatially symmetric S-state, accounting for $\sim$90% of the spin-averaged wave function, has the ^{3}He spin entirely due to the neutron with the two protons in a spin singlet state; b) a D-state due to the tensor force accounts for $\sim$ 8% of the spin-averaged wave function and has the three nucleon spins dominantly oriented opposite to the ^{3}He nuclear spin; c) a mixed-symmetry configuration of the nucleons, the S'-state, arises from spin-momentum correlations (4) and accounts for $\sim$1.5% of the spin-averaged wave function. All other components are predicted to be negligibly small.

In the non-relativistic approximation using Faddeev techniques, the S and S' state contributions to the spectral function are found to be maximum for zero nucleon momentum while the D-state contribution is greatest for larger momenta (2,3).

The nuclear structure information is contained in the spin-dependent spectral function (3) $S_{\sigma_A}^N(E, t, \mathbf{p})$ defined as the probability density of finding a nucleon N of isospin t with separation energy E, momentum $\mathbf{p}$ and spin σ_N parallel (antiparallel) to the $\overrightarrow{^3\text{He}}$ spin indicated by $\sigma_A = +(-)$. The spectral function has the general form (3)

$$S_{\sigma_A}(E, \mathbf{p}, t) = \frac{1}{2}\{f_0(E, p, t) + f_1(E, p, t)\sigma_N \cdot \sigma_A +$$

$$f_2(E, p, t)[(\sigma_N \cdot \hat{\mathbf{p}})(\sigma_A \cdot \hat{\mathbf{p}}) - \frac{1}{3}\sigma_N \cdot \sigma_A]\} . \quad (1)$$

The spin-averaged contribution f_0 and the two spin-dependent contributions f_1 and f_2 are scalar functions which depend only on the magnitude of $\mathbf{p}$. The effects of the Coulomb interaction have been neglected. The momentum distribution for a nucleon of isospin t with spin parallel (anti-parallel) to the nuclear spin is defined as

$$\rho_{\uparrow(\downarrow)}(p, t) \equiv \int dE\, S_{\sigma_A = +(-)}(E, p, t) . \quad (2)$$

The spin dependent momentum distribution is then defined as

$$N(p, t) \equiv \frac{\rho_\uparrow(p, t) - \rho_\downarrow(p, t)}{\rho_\uparrow(p, t) + \rho_\downarrow(p, t)} \quad (3)$$

and is plotted for each nucleon isospin state in figure 1.

The neutron in ^{3}He can exist only in a state where the two protons are unbound (the three-body configuration.) However, the proton in ^{3}He can be found in a state where the other nucleons form a deuteron (the two-body configuration) or as individual nucleons (the three-body configuration.) In fig. 1(b) both two- and three-body pieces and the sum are shown for the proton. Physically, the distributions in figure 1 can be interpreted as the probability that a neutron or proton in ^{3}He has its spin directed parallel to the nuclear spin as a function of the nucleon momentum. Thus, the neutron in figure 1(a) at low p has its spin completely parallel to the nuclear spin but at high p, where the D-state is sizable, the neutron spin can be found with large probability directed opposite to the nuclear spin. In the case of the proton, from figure 1(b) it can be seen that the individual two- ($N^2(p)$) and three-body ($N^3(p)$) contributions to the total proton spin dependent

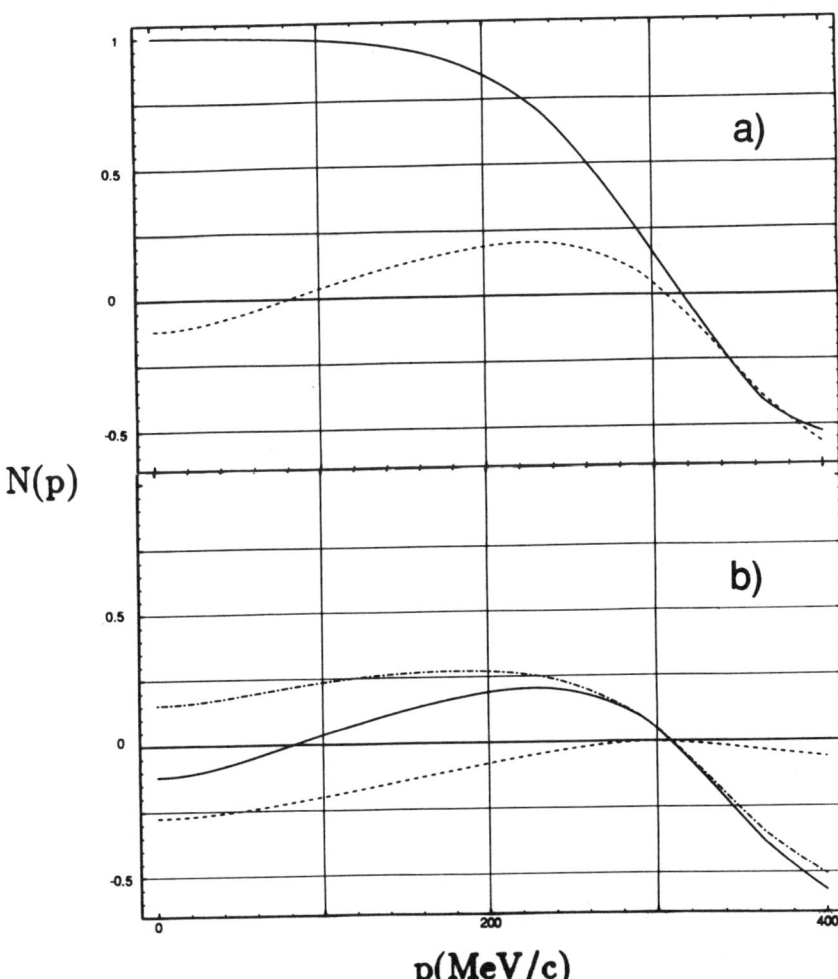

Figure 1. The spin-dependent momentum distribution of (a) the neutron (solid line) and the proton (dashed line) plotted vs. nucleon momentum p. In (b) the two-body contribution (dashed), the three-body contribution (dot-dashed) and the total (solid) proton distributions are plotted vs. p. The spectral function is from reference (3) and uses the Paris NN-potential.

momentum distribution are sizable and of opposite sign. (Note that even in the pure S-state $N^3(p=0) = -N^2(p=0) \approx 0.25$.) The addition of the S'-state yields for the complete wave-function $N^2(p=0) + N^3(p=0) = -0.12$. As with the neutron, at high p the presence of the D-state causes the proton spin to be predominantly directed opposite to the nuclear spin.

EXPERIMENTAL MEASUREMENT OF THE SPECTRAL FUNCTION

Quasielastic spin-dependent knockout of the constituent nucleons of ^{3}He with good resolution in the energy and momentum of the initial state nucleon offers the most direct experimental approach to constrain the spectral function. This has been successfully carried out for the first time by the CE-25 experiment using a 197 MeV polarized proton beam at the Indiana University Cyclotron Facility (IUCF) Cooler Ring (5,6).

Consider an incident polarized proton of 4-momentum ($T_{inc} + M$; $\mathbf{p}_{inc}$) scatters from a nucleon in ^{3}He resulting in a proton of 4-momentum ($T_1 + M$; $\mathbf{p}_1$) and a second nucleon with 4-momentum ($T_2 + M$; $\mathbf{p}_2$). In the plane wave impulse approximation (PWIA) (7) the two nucleons are ejected without secondary scattering from the residual nucleus implying that the missing momentum $\mathbf{p}_m \equiv \mathbf{p}_{inc} - \mathbf{p}_1 - \mathbf{p}_2$ can be identified with the initial momentum of the struck nucleon, $\mathbf{p}$. The recoil system is either a deuteron (two-body breakup) or two unbound nucleons (three-body breakup) and in PWIA the missing energy, $E_m \equiv T_{inc} - T_1 - T_2 - T_{recoil}$, is identified with the separation energy. If PWIA is a good approximation to the scattering process, then information on the spectral function can be directly extracted.

The spin-dependent differential cross section with both beam and target spins oriented normal to the scattering plane can be written as (8)

$$\sigma = \sigma_0(1 \pm P_b A_{00n0} \pm P_t A_{000n} + P_b P_t A_{00nn}), \qquad (4)$$

where P_b and P_t are the polarizations of the beam and target, the + and − distinguish between beam left and right regions of the scattering plane, σ_0 is the unpolarized cross-section, A_{00nn} is the spin-correlation parameter, and A_{00n0} and A_{000n} are the beam and target analyzing powers, respectively. In PWIA the target spin observables $A_{00in}^{3,N}$ (i=0 or n) for ^{3}He(p,pN) scattering (N is a proton p or neutron n) can be related to the N(p,p) elastic scattering observables, A_{00in}^N, extracted from phase shift analyses (9), by

$$A_{00in}^{3,N} = A_{00in}^N \cdot P^N \quad \text{and} \quad P^N = \frac{S_+^N - S_-^N}{S_+^N + S_-^N}, \qquad (5)$$

where P^N is a function of p_m and E_m and is interpreted as the polarization of a nucleon in ^{3}He. Comparisons (10) of unpolarized (p,2p) and (e,e'p) data indicate that rescattering effects of initial and final state nucleons may be large at high p_m. In addition, the Faddeev calculations predict little influence of the D-state at low p_m so that $P^n = 1$ and thus spin observables consistent with free scattering should be observed if PWIA holds.

Previous measurements of spin dependent $^3\overrightarrow{\text{He}}(\vec{p},2p)$ and $^3\overrightarrow{\text{He}}(\vec{p},pn)$ quasielastic scattering at 220 (11) and 290 MeV (12,13) incident proton energies disagreed significantly with PWIA calculations. In particular, $A_{000n}^{3,n}$ was observed to be strongly suppressed. These experiments were carried out for small proton scattering angles (27.5° at 290 MeV, 34° at 220 MeV), i.e., low momentum transfer with relatively small acceptance in angle and p_m. In contrast, the experiment at IUCF measured asymmetries over a wide range of scattering angles (21°-67°), or equivalently from low to high momentum transfer, and over an extended range in p_m by making use of large acceptance detector arrays.

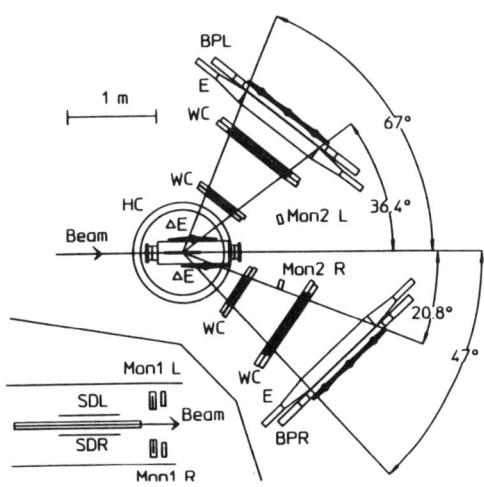

Figure 2. The schematic layout of the experimental apparatus.

The CE-25 experiment at the Indiana University Cyclotron Facility (IUCF) Cooler Ring (14) used a polarized ^{3}He internal gas target (15). The target and beam were polarized normal to the scattering plane and data were acquired at a beam energy of 197.5±0.1 MeV. Figure 2 shows a schematic layout of the experimental apparatus. A large acceptance detector arm was located on each side of the target to detect two nucleons in coincidence, one on each side of the beam. Each arm consisted of a set of 300 μm silicon microstrip detectors (SDL,R), a 3 mm thick ΔE plastic scintillator, two pairs of wire chambers (WC), six 100×10×15 cm³ scintillator E bars with a neutron efficiency of ~15% and an additional 9.5 cm thick plane of plastic scintillators (BPL,R). The time of flight resolution (FWHM) was ~2 ns, the p_m resolution (FWHM) was 30 MeV/c, the E_m resolution (FWHM) was 20 MeV and the background rate of scattered events from material other than polarized ^{3}He was below 1%.

The stored beam intensity varied from 100 to 50 μA with a lifetime of ~1000 s in the presence of target gas. The target (16) had an average polarization of 0.46 with normalization uncertainty of ±0.02 and a thickness of 1.5×10^{14} atoms/cm^2 resulting in a typical luminosity of 6×10^{28}cm^{-2}s^{-1}. The average beam polarization was 0.72 with a normalization uncertainty of 0.01.

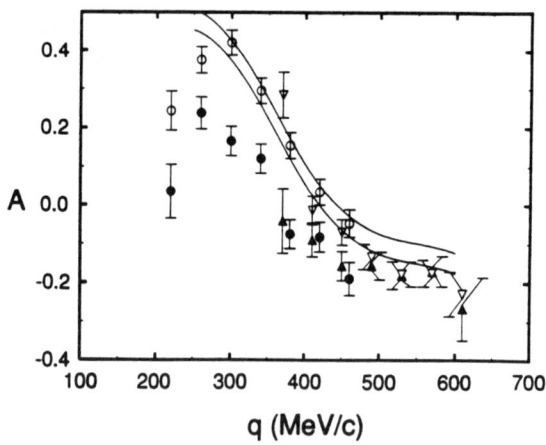

Figure 3. The target (filled symbols) and beam (open symbols) analyzing powers, $A^{3,n}_{000n}$ and $A^{3,n}_{00n0}$, respectively, for ^{3}He(p,pn) at $|p_m| < 100$ MeV/c as a function of the 3-momentum transfer to the struck neutron $|q|$. Data where the proton scattered to the right (left) detector and neutrons to the left (right) detector are indicated by circular (triangular) symbols. The error bars include statistical and systematic errors added in quadrature. The curves correspond to the upper and lower bounds of the PWIA prediction allowed by various phase shift solutions for the free observables.

The spin observables were extracted as a function of various kinematic variables for both ^{3}He(p,2p) and ^{3}He(p,pn) reactions. The beam and target analyzing powers in ^{3}He(p,pn) scattering should be equal in the limit $p_m \to 0$, if corrections to the PWIA are small. $A^{3,n}_{00n0}$ and $A^{3,n}_{000n}$ for $p_m < 100$ MeV/c are shown as a function of the magnitude of the 3-momentum transfer $|q|$ of the struck neutron in Figure 3. For data with $|q| > 500$ MeV/c the neutron polarization, extracted by calculating the ratio of target to beam analyzing power, is $0.94\pm.08\pm0.12$, where the errors are statistical and systematic (polarization plus luminosity uncertainties), respectively. This result is consistent with PWIA and the full polarization predicted by the Faddeev calculations. In contrast, at low momentum transfer there is a sizable deviation between the two observables, in agreement with the TRIUMF data (11,12,13) taken at $|q|\sim 370$ MeV/c. This $|q|$ dependence may be evidence of a significant spin-dependent final-state interaction of the recoiling neutron for low momentum transfer. Such an effect has been predicted in calculations

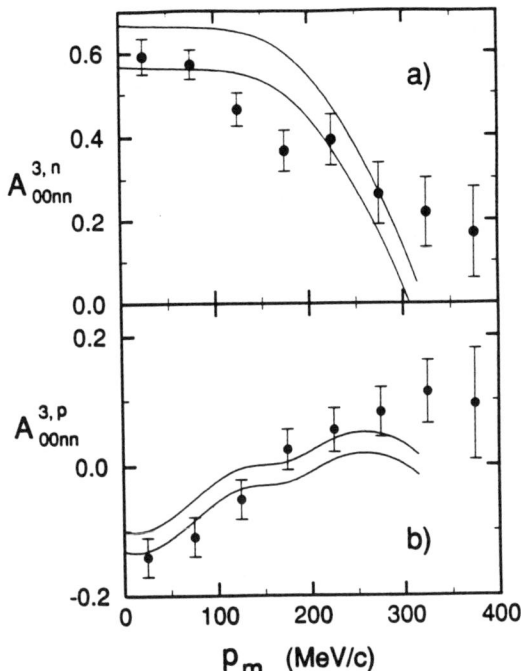

Figure 4. The missing momentum distribution of a) $A^{3,n}_{00nn}$ in ^{3}He(p,pn) for $|q|$ >500 MeV/c, b) $A^{3,p}_{00nn}$ in ^{3}He(p,2p). The error bars include both statistical and systematic errors added in quadrature. The curves correspond to the upper and lower bounds of the PWIA prediction allowed by various phase shift solutions for the free observables.

of spin-dependent quasielastic (e,e'n) scattering (17). As in the TRIUMF measurements, the present (p,2p) observables were found to be consistent with expectations from the Faddeev calculations and PWIA.

The good agreement between $A^{3,n}_{000n}$ and $A^{3,n}_{00n0}$ at high momentum transfer indicates that further investigation of the nucleon polarization is possible again at low missing momentum, by applying the PWIA model of eqn. (5) to the spin correlation data. Figure 4 shows a) $A^{3,n}_{00nn}$ for protons scattered to the left detector, and b) $A^{3,p}_{00nn}$ as a function of p_m for $|q| > 500$ MeV/c. The curves represent the upper and lower bounds of the PWIA predictions as generated from a number of phase shift solutions extracted from SAID (9) averaged over the detector acceptance for $|q| > 500$ MeV/c and are in good agreement with the data. Using eqn. (4) and the free scattering observables the polarization of the nucleons at low p_m can be extracted to yield $P^n(p_m = 0)$ = 0.98±0.06 ±0.05 ±0.08, in good agreement with the polarization extracted from the analyzing powers, and $P^p(p_m = 0)$= −0.16±0.01±0.03±0.004. The Faddeev calculations (2,3) both predict $P^n(p_m = 0) = 1.00$ and $P^p(p_m = 0)$

= -0.12, in good agreement with the data. In these calculations the full polarization of the neutrons reflects the dominance of the S state and the small negative polarization of the protons reflects the effects of the S' state, both at low p_m.

SPIN-DEPENDENT ELECTRON SCATTERING FROM POLARIZED ^{3}HE

Introduction

The Faddeev calculations show that polarized ^{3}He can act as an effective polarized neutron. Thus, a number of experiments have been carried out or proposed to measure the neutron electromagnetic form factors in quasielastic electron scattering (18,19,20,21)) and the neutron spin structure functions in deep-inelastic electron scattering from polarized ^{3}He (22). A basic premise in the extraction of neutron form-factors from spin-dependent electron scattering from polarized ^{3}He is that both the nuclear ground state spin structure and the scattering mechanism are well understood.

Inclusive Quasielastic Scattering

Elastic electron scattering from a spin-$\frac{1}{2}$ target is particularly straightforward (23). If both beam and target are polarized, two additional response functions beyond the familiar R_L and R_T appear in the cross section: a transverse response, $R_{T'}$, and an interference between longitudinal and transverse multipoles, $R_{TL'}$. In terms of the response functions, $R_K(Q^2,\omega)$, the asymmetry can be written as (23)

$$A = -\frac{\cos\theta^* v_{T'} R_{T'} + 2\sin\theta^* \cos\phi^* v_{TL'} R_{TL'}}{v_T R_T + v_L R_L} \quad (6).$$

As discussed in Ref. (23) θ^* and ϕ^* are the polar and azimuthal angles defining the direction of the target spin with respect to the momentum transfer $\vec{q}$, the v_K are kinematic factors, ω is the electron energy loss, and $Q^2 \equiv |\vec{q}|^2 - \omega^2$. By orienting the target spin at $\theta^* = 90°$ ($\theta^* = 0°$), one can measure an asymmetry $A_{TL'}$ ($A_{T'}$) proportional to $R_{TL'}$ ($R_{T'}$).

In general, inclusive quasielastic electron scattering involves a sum over the complete spectral function. In PWIA the inclusive quasielastic $R_{TL'}$ ($R_{T'}$) response for ^{3}He contains a contribution proportional to $G_M^n G_E^n$ ($G_M^{n\,2}$). Blankleider and Woloshyn, using closure to sum over final states (2), originally predicted that $A_{TL'}$ would be strongly sensitive to G_E^n in inclusive spin-dependent electron scattering from polarized ^{3}He. This was confirmed by Ciofi degli Atti et al. (24) who studied the effect of nuclear binding using a full spin-dependent spectral function to describe the ^{3}He ground state.

However, a more sophisticated analysis in the framework of PWIA has been recently performed (3), which agrees with the previous $A_{T'}$ predictions but points out an inconsistency in the previous calculations of $A_{LT'}$. In particular, the recent work (3) predicts, at low $Q^2 \sim 0.2$ $(\text{GeV}/c)^2$ a dominant proton contribution to $A_{LT'}$ which greatly reduces the sensitivity to G_E^n.

Simultaneous measurements of $A_{T'}$ and $A_{TL'}$ were carried out at the MIT-Bates Linear Accelerator Center using a 370 MeV longitudinally polarized, pulsed electron beam and a polarized ^{3}He gas target. The average beam and target polarizations were $(36.5 \pm 1.1)\%$ and $(38.0 \pm 1.5)\%$, respectively. The target spin was oriented at 42.5° with respect to the beam direction. A total beam charge of 5114 μA-hours was accumulated. The scattered electrons were detected in the MEPS ($A_{T'}$) and OHIPS ($A_{TL'}$) spectrometers using detector packages consisting of drift chambers, planes of plastic scintillators, and a gas Čerenkov detector in each spectrometer. The trigger was formed by a scintillator coincidence, the Čerenkov information being used in software to reject pions. The MEPS spectrometer was positioned at a scattering angle of 91.4° to the left of the beam and had a central momentum of 250 MeV/c corresponding to $Q^2 = 0.19$ $(\text{GeV}/c)^2$ and $\theta^* = 8.9°$. The OHIPS spectrometer was positioned at a scattering angle of 70.1° to the right of the beam and had a central momentum of 286 MeV/c, corresponding to $Q^2 = 0.14$ $(\text{GeV}/c)^2$ and $\theta^* = 87°$. The MEPS spectrometer had a momentum acceptance of $\pm 10\%$ and an extended target acceptance of 2 cm. The OHIPS spectrometer had a momentum acceptance of 10% and the target length viewed by the spectrometer was collimated with slits to 10 cm along the beam. Background scattering of beam halo from the copper target walls was measured at regular intervals with the target cell empty and was approximately 5% of the full target yield. As a check on the experimental procedure, samples of elastic data were taken at regular intervals with OHIPS. The resulting elastic asymmetry of 29.9±3.9% agrees well with the prediction of 32.1% calculated from elastic form factor data (25).

$A_{T'}$ and $A_{TL'}$ have been extracted from the detected spin-dependent yield as a function of the electron energy loss ω. To account for the dilution from unpolarized scattering, the result was divided by the product of the beam and target polarizations. Corrections were applied for the empty target background, the elastic radiative tail, and quasielastic radiative effects. The contribution of helicity-correlated background and efficiency variations to the asymmetry was found to be negligible. The measured asymmetries are shown in Figs. 5 and 6.

In figure 5 $A_{T'}(\omega)$ is shown along with calculations at the kinematics of the experiment by Salmè et al. (26) and Schulze et al. (27). The difference between the two calculations arises from the different wave functions and form factor parametrizations used in the calculations. The data are in good

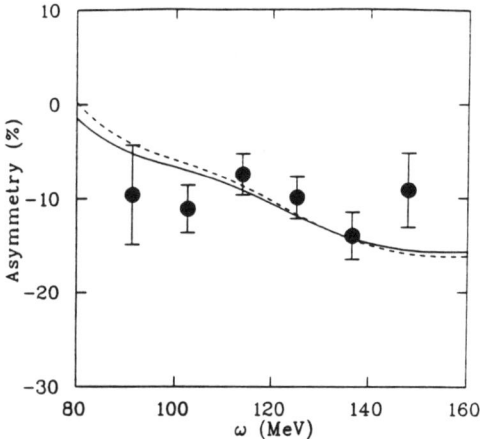

Figure 5. The transverse asymmetry $A_{T'}$ as a function of electron energy transfer ω. The solid circles are the data points from Bates (21) with statistical uncertainties only. The dashed line is the calculation by Salmè et al. (26), and the solid line is the calculation by Schulze et al. (27).

agreement with both calculations. Combining all the data yields a value of $-10.23 \pm 1.11 \pm 0.56$ %. The theoretical calculations give -9.85% (26) and -10.09% (27). Since the data are in good agreement with PWIA a value for $(G_M^n)^2$ can be determined. The Bates data yield $(\frac{G_M^n}{\mu_n G_D})^2 = 0.998 \pm 0.17 \pm 0.059 \pm 0.030$, with the uncertainties corresponding to statistics, systematics, and model dependence, respectively.

In figure 6 $A_{TL'}(\omega)$ is shown along with curves which represent the upper and lower bounds of the PWIA model, as determined in a detailed study (28). The curves result from consideration of three NN interaction potentials, three nucleon form factor parametrizations, and a series of nucleon off-shell prescriptions. Combining all the data over the experimental energy acceptance ($72 < \omega < 99$ MeV) yields a value of $A_{TL'} = 1.52 \pm 0.55 \pm 0.15$ %. The predictions of the PWIA model range from 2.1% to 2.9%. Thus, the data are low with respect to PWIA at the level of 1-2.5 σ. The suppression may be evidence of the presence of final-state interactions (FSI), as suggested by recent calculations of the unpolarized response R_L at similar Q^2 (29,30,31).

Exclusive Quasielastic (e,e'n) Scattering

Measurement of exclusive spin-dependent (e,e'n) scattering from polarized ^{3}He can greatly enhance the sensitivity to the neutron form factors while adding the complication of detecting a recoiling neutron. If the PWIA

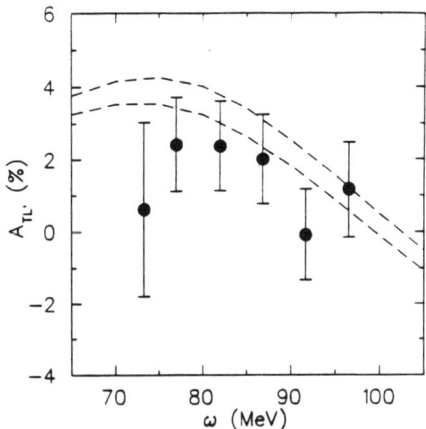

Figure 6. The transverse-longitudinal asymmetry $A_{TL'}$ as a function of electron energy transfer ω. The solid circles are the data points from Bates (28) with statistical uncertainties only. The curves represent the upper and lower bounds of the PWIA calculation (28).

is valid for (e,e'n) quasielastic scattering from polarized ^{3}He then

$$\frac{G_E^n}{G_M^n} = \sqrt{\tau + \tau(1+\tau)\tan^2(\frac{\theta}{2})} \cdot \frac{A_{TL'}}{A_{T'}} \qquad (7) .$$

Thus, in PWIA a measurement of the ratio $\frac{A_{TL'}}{A_{T'}}$ together with a knowledge of G_M^n can yield new information on the appicability of PWIA and on G_E^n. This has been the approach of the Mainz group and they have extracted information on G_E^n from spin-dependent electron scattering from polarized ^{3}He (20). However, they do not report any information on the energy and momentum of the initial neutron.

The experiment was carried out using 855 MeV polarized electrons from the MAMI accelerator. The electrons were incident on a high density optically pumped external polarized ^{3}He target (32). Non-magnetic detectors were used to detect both scattered electrons and neutrons. The electrons were detected in a segmented lead-glass calorimeter with an angular resolution of 0.3° and an energy resolution $\frac{16\%}{\sqrt{E}}$. In addition, a focussing Cerenkov detector allowed rejection of events originating from the end windows of the target. Neutrons were detected in several layers of plastic scintillators which formed a time-of-flight spectrometer. Veto counters in front of the scintillator walls discriminated against charged particles. The neutron detectors were carefully shielded. Kinematic cuts were applied which resulted in an average momentum transfer $Q^2 = 0.31$ $(GeV/c)^2$. Data were taken with the target spin oriented for maximum sensitivity to $A_{T'}$ and $A_{TL'}$, respectively. Combining all the data yielded $A_{T'} = -7.40 \pm 0.73$ % and $A_{TL'} = 0.89 \pm 0.30$ %. Using the dipole form for G_M^n, G_E^n was determined to be $0.035 \pm 0.012 \pm 0.005$. The datum is compared with data from elastic electron-deuteron scattering (33) in figure 7.

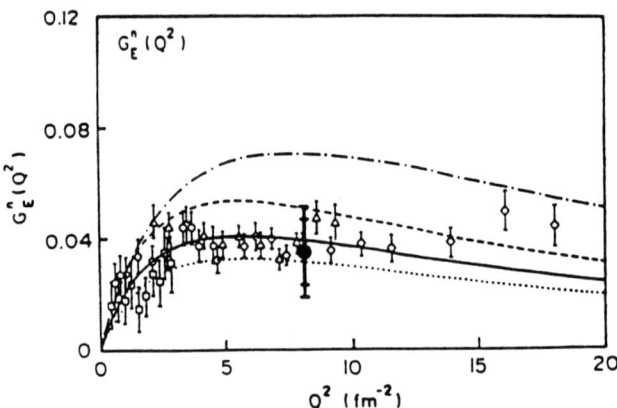

Figure 7. Comparison of G_E^n extracted from the ratio $\frac{A_{TL'}}{A_{T'}}$ measured at Mainz (20) with existing data (33). The solid curve gives the two parameter best fit to the plotted data points for G_E^n as deduced using the Paris potential. In addition, the corresponding two parameter fits using the RSC (dotted), Argonne V14 (dashed) or Nijmegen (dash-dotted) potentials are depicted.

Deep Inelastic Spin Dependent Scattering

In the parton model of deep inelastic electron scattering the virtual photon scatters from the pointlike quark constituents of the nucleon. Experimentally this occurs when $Q^2 > 1\ (\text{GeV/c})^2$ and the hadronic final state mass is greater than 2 GeV. $x \equiv \frac{Q^2}{2M\nu}$ is the fraction of the nucleon spin carried by the quark. Recent spin dependent deep inelastic scattering data on the proton from CERN and SLAC (34,35,36) indicate that the fraction of the nucleon spin carried by the quarks is only $\sim 30\%$ and the strange quarks have an appreciable polarization. This result represents a violation of the Ellis-Jaffe Sum Rule (37). Further, the fundamental Bjorken Sum Rule (38) relates measurements on the proton to those on the neutron in a model independent way. Thus, it is important to make measurements on the neutron.

At SLAC the spin dependent structure function of the neutron in ^{3}He was determined by measurement of the asymmetry in scattering of longitudinally polarized electrons from an external high density target of polarized ^{3}He. Approximately 4×10^8 inclusive events were obtained at beam energies of 19 - 26 GeV in the x range from 0.03 to 0.6 with $Q^2 > 1$ $(GeV/c)^2$. The scattered electrons were detected in two fixed angle spectrometers at 4.5° and 7°. The beam and target polarizations were 39% and 35%, respectively.

Figure 8 shows the measured $g_n^1(x)$ data on polarized ^{3}He from SLAC. The integral of the neutron spin structure function was found to be -0.022 ± 0.011 and was used to conclude that the total quark contribution to the nucleon spin is 0.57 ± 0.11. This is in good agreement with the Ellis-Jaffe Sum Rule prediction.

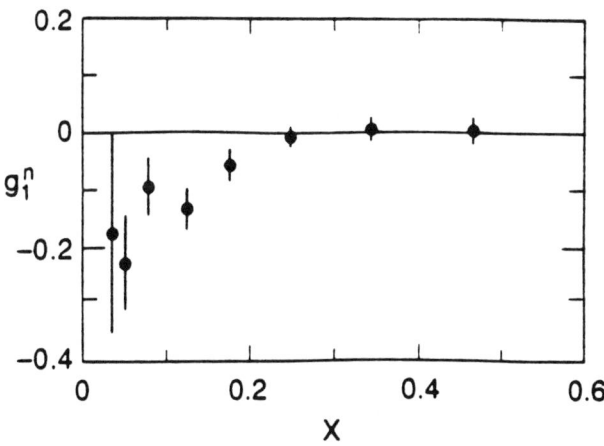

Figure 8. The neutron spin structure function $g_1^n(x)$ as measured at SLAC in spin dependent deep inelastic scattering from polarized ^{3}He (22).

SUMMARY

The availability of the full spin-dependent spectral function of Schultze and Sauer has provided a solid basis to compare the PWIA model with spin-dependent scattering data from polarized ^{3}He at intermediate energies. The CE-25 data have illustrated that the earlier TRIUMF measurements were taken in a kinematic regime where PWIA was not valid for (p,pn) scattering. By extending the angular range at 197 MeV to a region where good agreement is obtained with PWIA (i.e. $Q^2 > 0.25$ $(GeV/c)^2$), the first direct information on the spectral function has been obtained. The neutron and proton polarizations at $p_m = 0$ are in good agreement with Faddeev predictions. In addition, the shapes are qualitatively consistent with those expected from PWIA.

The inclusive scattering data from Bates have also found reasonable agreement with PWIA. The $A_{T'}$ asymmetry has yielded a value for $G_M^n(Q^2 = 0.19$ $(GeV/c)^2)$ in good agreement with values obtained on the deuteron.

The $A_{TL'}$ asymmetry is low with respect to PWIA at the level of 1-2 σ, which may not be significant. The PWIA analysis indicates that inclusive spin-dependent scattering from polarized ^{3}He should be able to constrain G_E^n at higher $Q^2 \sim 0.6$ (GeV/c)^2 to $\pm 20\%$. The exclusive scattering data from Mainz show good agreement with PWIA and the extracted value of G_E^n is in good agreement with that obtained from unpolarized elastic electron-deuteron scattering. The deep inelastic data from SLAC give a value for the quark contribution to the neutron spin which is significantly higher than that obtained for the proton.

Over the next several years, the quasielastic measurements will be extended to higher Q^2 at Bates and Mainz. These experiments should allow more precise determinations of G_E^n than are currently available from unpolarized deuterium data. In addition, experiments are planned which should provide much more information on the initial state nucleon. The use of a large acceptance detector in conjunction with polarized internal targets should be a major advantage in the future investigation of the spin structure of the ^{3}He ground state (39). In the deep inelastic regime further measurements at SLAC (40) and at HERA (41) should shed light on the spin structure of the nucleon.

ACKNOWLEDGEMENTS

I wish to thank R.-W. Schultze and P.U. Sauer for making their spin-dependent spectral function available and I acknowledge discussions with T.W. Donnelly, J.-O. Hansen, K. Lee, S. Nagorni, M.A. Titko, and W. Turchinetz. The author's research is supported in part by the U.S. Department of Energy, Nuclear Physics Division, under cooperative agreement No. DE-FC02-94ER40818, by a Presidential Young Investigator Award from the National Science Foundation and by the MIT Sloan Fund.

REFERENCES

1. T.E. Chupp, R.J. Holt, and R.G. Milner, Ann. Rev. Nucl. Part. Science **44**, 1994, to be published.
2. B. Blankleider and R.M. Woloshyn, Phys. Rev. C**29**, 538 (1984).
3. R.W. Schulze and P.U. Sauer, Phys. Rev. C **48**, 38 (1993).
4. J. Friar, in *Electronuclear Physics with Internal Targets and the BLAST Detector*, eds. R. Alarcon and M. Butler, (World Scientific, Singapore, 1993), p. 210.
5. M.A.Miller et al., submitted to Phys. Rev. Lett.
6. C. Bloch et al., to be published in Nucl. Instr. and Meth.
7. S. Frullani and J. Mougey, Adv. Nucl. Phys., **14** (1984).
8. J. Bystricky et al., J. Phys. (Paris) **39**, 1 (1978).

9. R.A. Arndt et al., Scattering analysis dial-in program (SAID), Phys. Rev. D **45**, 3995, (1992).
10. M.B. Epstein *et al.*, Phys. Rev. C **32**, 967 (1985).
11. E.J. Brash *et al.*, Phys. Rev. C **47**, 2064 (1993).
12. A. Rahav *et al.*, Phys. Lett. **275**, 259 (1992).
13. A. Rahav *et al.*, Phys. Rev. C **46**, 1167 (1992).
14. R.E. Pollock, Ann. Rev. Nucl. Part. Sci. **41**, 357 (1991).
15. K.Lee *et al.*, Phys. Rev. Lett. **70**, 738 (1993).
16. K.Lee *et al.*, Nucl. Instr. and Meth. A **333**, 294 (1993).
17. J.-M. Laget, Phys. Lett. B **276**, 398 (1992).
18. C. E. Woodward *et al.*, Phys. Rev. Lett. **65**, 698 (1990); C. E. Jones *et al.*, Phys. Rev. C **47**, 110 (1993).
19. A. K. Thompson *et al.*, Phys. Rev. Lett. **68**, 2901 (1992).
20. M. Meyerhoff *et al.*, Phys. Lett. **B327**, 201 (1994).
21. H. Gao *et al.*, Phys. Rev. C **50**, R546, (1994).
22. P.L. Anthony *et al.*, Phys. Rev. Lett. **71**, 959 (1993).
23. T. W. Donnelly and A. S. Raskin, Ann. Phys. (N.Y.) **169**, 247 (1986).
24. C. Ciofi degli Atti, E. Pace, G. Salmè, Phys. Rev. C **46**, R1591 (1992).
25. J. S. McCarthy *et al.*, Phys. Rev. C **15**, 1396 (1977).
26. G. Salmè, C. Ciofi degli Atti, and E. Pace, in *Proceedings of the 6th Workshop in Nuclear Physics at Intermediate Energies*, ICTP Trieste, 1993 (World Scientific, Singapore 1993); G. Salmè, private communication.
27. R.-W. Schultze, private communication.
28. J.-O. Hansen *et al.*, LNS MIT preprint 94-80, to be published.
29. E. van Meijgaard and J.A. Tjon, Phys. Rev. C **45**, 1463 (1992).
30. S. Ishikawa, H. Kamada, W. Glockle, J. Golak, and H. Witala, preprint, July 1994.
31. J. Carlson and R. Schiavilla, private communication.
32. G. Eckert *et al.*, Nucl. Instr. and Meth. **A320**, 53 (1992).
33. S. Platchkov et al., Nucl. Phys. **B510**, 740 (1990).
34. J. Ashman *et al.*, Nucl. Phys. **B328**, 1 (1989).
35. D. Adams *et al.*, Phys. Lett. **B329**, 399 (1994).
36. D. Crabb, Plenary Talk, Conference on Intersections of Nuclear and Particle Physics, St. Petersburg, Florida, 1994.
37. J. Ellis and R.L. Jaffe, Phys. Rev. **D9**, 1444 (1974).
38. J.D. Bjorken, Phys. Rev. **148**, 1457 (1966); **D1**, 1367 (1970).
39. Bates Large Acceptance Spectrometer Toroid proposal, 1991.
40. SLAC experiment E154, E. Hughes spokesman.
41. HERMES proposal, DESY No. PRC-90-01.

New Results in Nucleon-Nucleon Scattering at Low Energies

Werner Tornow

Department of Physics, Duke University and Triangle Universities Nuclear Laboratory, Durham, North Carolina 27708

Abstract. Recent progress in both NN phase-shift analyses and NN potential model studies below $T_{lab} = 100$ MeV will be discussed. New information on the influence of non-locality and three-nucleon forces on the binding energy of ^{3}H will be reviewed. Progress in experimental NN tensor force studies will be presented. Consequences of the recent determinations of the πNN coupling constant on our understanding of meson-exchange based NN potential models will be pointed out.

NN PHASE-SHIFT ANALYSES

During the last few years considerable progress has been achieved in parametrizing NN scattering data in terms of phase shifts. In 1990 the Nijmegen group reported an accurate phase-shift analysis of pp scattering data below T_{lab} =350 MeV (1). This analysis was subsequently extended to include also np data, and results of combined pp+np analyses were published in 1991 and 1992 (2,3). Finally, a long standing goal in NN interaction studies has been achieved: in 1993 the Nijmegen group reported the first np phase shift analysis that did not rely on any input from pp scattering (4). Again, the analysis covered the energy range from 0 to 350 MeV. Like the Nijmegen analyses mentioned above, an excellent $\chi^2/N_{data} \approx 1$ was achieved. Here, N_{data} denotes the number of scattering data.

A phase-shift analysis based on np data only is not straight forward. In the case of pp scattering only the isovector partial waves contribute, whereas for np scattering also the isoscalar partial waves have to be parametrized. In addition, np data are in general not as accurate as pp data. Therefore, in the past, except for the $^1S_0$ phase shift, it was common practice in np phase-shift analyses to parametrize only the isoscalar partial waves and to use for the isovector np phase shifts the results obtained from the analysis of pp data. In some cases the so obtained np isovector phase shifts were slightly modified to account for charge-independence breaking. Of course, this procedure makes these np phase shifts more model dependent than the associated pp phase shifts.

As is well known, phase-shift analyses do not provide a unique solution. Nevertheless, all existing phase-shift analyses more or less come up with about the same phase shifts. This apparent uniqueness is a consequence of the imposed condition that the long-range part of the NN interaction is described by the one-pion-exchange (OPE) tail, i.e., all partial waves with high total angular momentum J are assumed to be well known, and only a relatively small number of lower partial waves needs to be parametrized.

The Nijmegen Analysis

In the Nijmegen work the short-range ($r \leq 1.4$ fm) part of the interaction in the lower partial waves ($J \leq 4$) is represented by energy dependent square wells. For all partial waves the long-range part is made up of a nuclear and an electromagnetic potential. The nuclear part consists of the OPE tail and the tail of the heavy-boson-exchange contribution of the Nijmegen potential (5):

$$V_N = \frac{M}{E} V_{OPE}(f_\pi, m_\pi) + f(S) V_{HBE} \quad (1)$$

with $f(S=0) = 1.8$ and $f(S=1) = 1.0$. Here S denotes the total spin of the two-nucleon system, M is the nucleon mass and $E = (M^2 + q_0^2)^{1/2}$ with $q_0^2 = MT_{lab}/2$. The local OPE potential is given by

$$V_{OPE}(f_\pi, m_\pi) = \frac{f_\pi^2}{3}\left(\frac{m_\pi}{m_s}\right)^2 \frac{e^{-m_\pi r}}{r}\left[\sigma_1 \cdot \sigma_2 + S_{12}\left(1 + \frac{3}{m_\pi r} + \frac{3}{(m_\pi r)^2}\right)\right] \quad (2)$$

where f_π is the πNN coupling constant and m_π the pion mass Furthermore, $m_s = m_{\pi\pm} = 139.568$ MeV, and σ_1, σ_2 and S_{12} denote the spin and tensor operators, respectively. For pp scattering one uses

$$V_{OPE}^{pp} = V_{OPE}(f_{\pi°}, m_{\pi°}) \quad (3)$$

with $m_{\pi°} = 134.974$ MeV. For np scattering one must distinguish between isospin T=0 and T=1:

$$V_{OPE}^{np}(T) = -V_{OPE}(f_{\pi°}, m_{\pi°}) + 2(-1)^{T+1} V_{OPE}(f_{\pi\pm}, m_{\pi\pm}) \quad (4)$$

Considering the pp+np analysis the Nijmegen group assumes that charge independence is broken by the electromagnetic interaction, by the neutral-pion and charged-pion mass difference, and by the neutron-proton mass difference. As mentioned already, an important exception is the $^1S_0$ partial wave, where the pp and np $^1S_0$ partial waves are parametrized independently. The actual phase shifts are obtained by using the potentials discussed above in a Schrödinger equation. Predictions are calculated for observables of interest, and from the comparison with experimental data χ^2 is determined and minimized by adjusting the parameters of the square well potentials and the πNN coupling constants. One may criticize the Nijmegen approach as being too model dependent. Solutions which deviate from meson theory for r>1.4 fm are ruled out from the beginning. On the other hand one may argue that the χ^2 values achieved by the Nijmegen group speak for themselves: all three analyses (pp, pp+np, and np) have $\chi^2/N_{data} \approx 1$. For future reference, the Nijmegen pp+np analysis will be denoted by NI93.

The VPI Analysis

Historically, the VPI group focuses in their phase-shift analyses on a larger energy range (6). The recent SM94 analysis extends to 1.6 GeV for pp scattering and to 1.3 MeV for np scattering (7). Since the procedure used by the VPI group has not changed significantly during the last years only a very brief summary will be given here. In the VPI analysis partial waves with orbital angular momentum $L\geq 6$ are described by OPE in Born approximation and the associated phase shifts are not varied in the analysis. The phase shifts with $L\leq 5$ are determined from the NN data. Below pion-production threshold the phase shifts are parametrized by an energy-dependent K-matrix. For uncoupled partial waves L the K-matrix is given by

$$K_L(T_{lab}) = B_L^{OPE}(T_{lab}) + \sum_{i=1}^{N} \alpha_{L_i} F_{L_i}(T_{lab}) , \quad (5)$$

where B_L^{OPE} is the partial-wave projected OPE amplitude in Born approximation and $F_{L_i}(T_{lab})$ are, except for constants, Legendre functions of the second kind. Since the phase shifts δ_L are related to the K-matrix by

$$\tan\delta_L(T_{lab}) = K_L(T_{lab}) , \quad (6)$$

effectively each partial wave is parametrized individually in terms of Yukawa functions $(\alpha_L e^{-mr})/r$ of different ranges, defined by multiples of the pion mass.

In order to compare with the recent NI93 analysis, the VPI group performed an analysis covering the energy range 0-400 MeV (7). This energy restricted version of the SM94 analysis is denoted by VZ40. The combined pp+np solution has $\chi^2/N_{data} \approx 1.4$. Inspecting the phase shifts, one finds that, in general, the NI93 and VZ40 phase shifts agree fairly well. The most striking disagreement is noticed for ε_1 and to a smaller degree also for $^1P_1$. We will come back to this observation later.

The VPI group performed an interesting exercise. The data set was pruned by discarding all data with χ^2 contributions greater than 7, i.e., 2.6% of all data used in the original analysis were removed. The χ^2/N_{data} decreased to 1.1 with virtually no detectable change in the resulting phase shifts. It would be interesting to compare the χ^2 values of the Nijmegen and VPI pp+np analyses using exactly the same data base. This exercise would shed some light on how much phase-shift results are biased by different theoretical model assumptions.

The VPI group confirmed the recent finding of the Nijmegen group that the np data can now be fitted separately without any input from pp scattering. All pp data were removed from the VZ40 data base and the solution was adjusted to produce an optimum fit to the np data. A stable solution was found, reducing χ^2 for the np data by 4% compared to the combined pp+np analysis.

Summarizing the discussion on recent progress in phase-shift analyses, it is fair to say that we now have reached the stage where the overwhelming majority of all available NN data can be fitted with $\chi^2/N_{data} \approx 1$. This is good news, but it does not help, for example, a few-body theoretician who calculates 3N scattering observables using a realistic NN interaction and who wants to learn something from the comparison of the calculated results with experimental data.

NN POTENTIAL MODELS

Ideally, NN potential models should be fitted to pp data as well as to np data and again, ideally, they should describe all NN scattering data and the deuteron properties with $\chi^2/N_{data} \approx 1$. In reality, the best potentials of the 1980's have $\chi^2/N_{data} \approx 2$ and they were constructed by fitting np data for T=0 states and either np or pp data for T=1 states. Therefore, it is not surprising that potential models which were fitted only to np data (like Urbana v_{14} (8), Argonne v_{14} (9), full Bonn (10) and Bonn B (11)) do in general not describe the pp data as well, even after corrections for the Coulomb interaction and the $^1S_0$ phase shift have been applied. By the same token, potentials fitted to pp data in T=1 states (like Reid soft core (12), Nijmegen soft core (5) and the parametrized Paris potential (13)) do not fit the np data very well, especially since the quality of the np data has increased considerably over the last decade. Due to strong interest from the few-body community, some of the older NN potential parameters have recently been updated. Stoks et al. reported an updated version of the Nijmegen soft core potential (NY93) which gives a much better description of the np data than the older version (14). It now fits all NN data with $\chi^2/N_{data}=1.87$. This result still deviates considerably from the ideal value $\chi^2/N_{data} \approx 1$ and those achieved in the phase-shift analyses discussed above. However, considering that 39 free parameters were used in the Nijmegen phase-shift analysis NI93 compared to 15 in NY93, this result is not surprising. Since the χ^2/N_{data} value of NY93 agrees closely with those of the Paris pp (2-350 MeV) and Bonn np and pp (15) potentials, it appears safe to state that present day meson-exchange based NN potential models lack some important physics.

Recent theoretical progress has concentrated on the cutoff parameter in the meson-NN form factors. In order to describe the deuteron, meson-exchange NN potential models require a rather large value for the cutoff parameter $\Lambda_{\pi NN}$ (≥ 1.3 GeV in a monopole parametrization), in disagreement with nucleon structure calculations where values around 0.8 GeV or even lower have been obtained (soft form factor). It has recently been shown by Haidenbauer et al. that it is possible to construct a one-boson-exchange potential in which the meson-NN form factors are quite soft (16). However, it was necessary to introduce an effective π' meson mass to increase the short-range tensor force. The recent Bochum NN potential (Ruhrpot) is based on an interesting approach (17). The potential is based on meson exchange at long distances and on direct NN interactions, coming from the intrinsic nucleon structure, at short distances. This procedure not only avoids the problem of large cutoff parameters in the form factors, but it provides a good description of the NN data as well ($\chi^2/N_{data}=1.68$ for np scattering up to 300 MeV). Some progress has also been achieved in constructing NN potentials in the frame-work of the non-relativistic constituent quark model (18). Reasonable fits have been obtained to data below $T_{lab}=250$ MeV. The same statement holds for a NN potential ($T_{lab} \leq 100$ MeV) based on an effective chiral Lagrangian (19).

Turning to phenomenological NN potential models it is interesting to point out that Stoks et al., constructed three different potentials with $\chi^2/N_{data} \approx 1$: a non-local Reid-like Nijmegen potential, a local version, and an updated regularized version of the Reid soft core potential (14). These potentials are fitted to the same database with about the same number of free parameters as the Nijmegen pp+np phase-shift analysis (NI93).

Very recently, Wiringa et al. produced an updated Argonne v_{14} potential (20). In addition to the 14 charge-independent operator components of the original model, four charge-dependent operators and the complete electromagnetic interaction have been included. With 40 adjustable parameters this new version (Argonne v_{18}) has χ^2/N_{data}=1.1 for the Nijmegen pp and np database in the range 0-350 MeV. The Argonne v_{18} potential is a compromise between meson-exchange based NN potential models and the recent purely phenomenological Reid like potentials of Stoks *et al.*

Quantum inversion of the relativistic Schrödinger equation (a differential form of the Blankenbecler-Sugar equation) offers an appealing method for determining NN potentials from experimental data. Recently, the Hamburg group solved the Gelfand-Levitan and Marchenko fundamental inversion equations and obtained energy independent and local r-space potentials for various NN phase-shift solutions, including NI93 and VPI SM94 (21, 22).

NEW RESULTS

Now let me compare results of the two most recent NN phase-shift analyses: the VPI VZ40 and the Nijmegen NI93 pp+np analyses. Inspecting the isovector pp phase shifts one recognizes that below T_{lab}=100 MeV all relevant phase shifts agree within less than 2.5%, except for the $^3P_0$ phase shift above 50 MeV. Figure 1 shows the percentage difference between VZ40 and NI93 for the three worst cases: $^1S_0$, $^3P_0$ and $^3P_1$. Although the 9% discrepancy of $^3P_0$ at 100 MeV is surprising we will immediately move on to the np phase shifts where much larger differences exist. According to Fig. 2 the percentage difference between VZ40

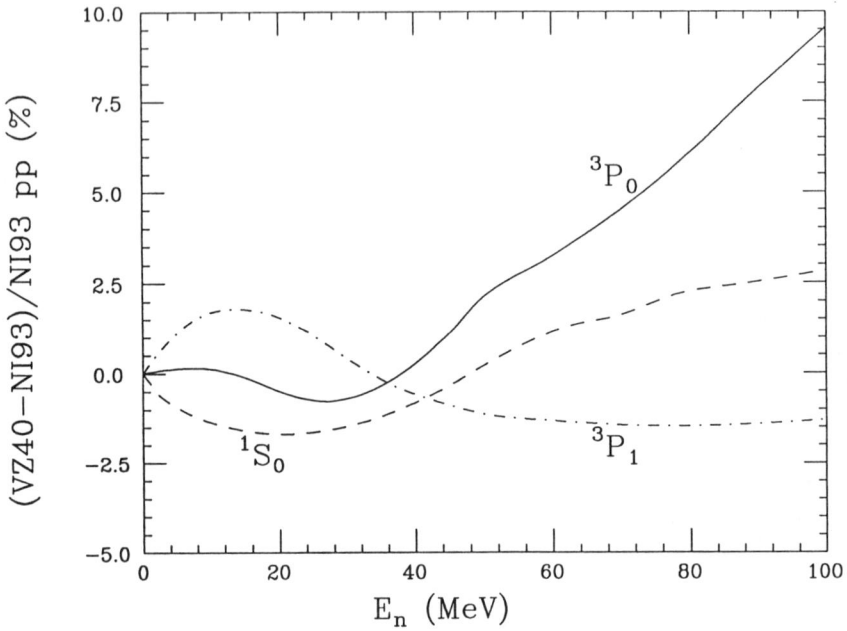

Figure 1. Percentage difference between the VPI VZ40 (0-400 MeV) and the Nijmegen NI93 (0-350 MeV) pp+np phase-shift analysis for the T=1 pp phase shifts $^1S_0$, $^3P_0$, and $^3P_1$.

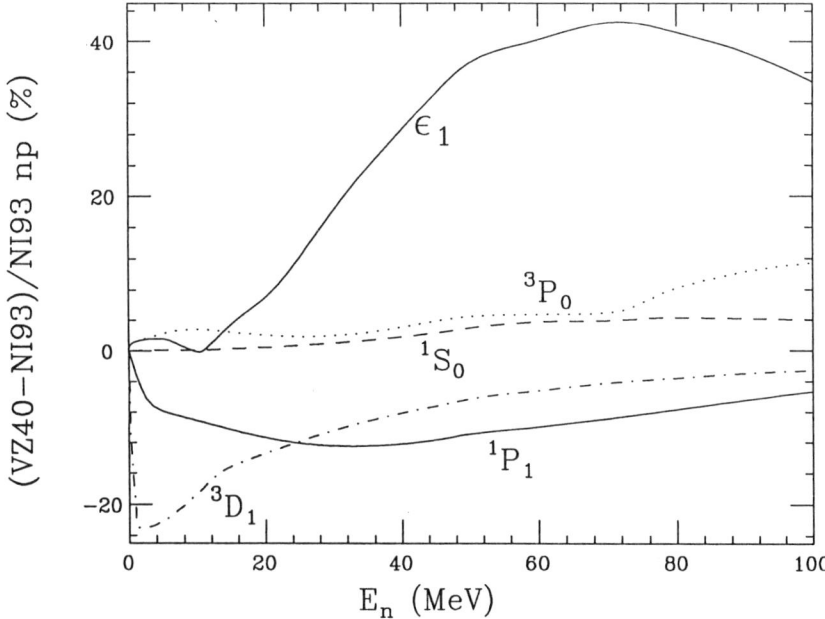

Figure 2. Same as Fig. 2 for T=0 (ϵ_1, $^1P_1$, $^3D_1$) and T = 1 ($^1S_0$, $^3P_0$) np phase shifts.

and NI93 is as large as 40% for the isoscalar $^3S_1$-$^3D_1$ mixing parameter ϵ_1. This mixing parameter is a measure of the NN tensor force in phase-shift analyses (i.e. on-shell). In NN potential models the D-state probability P_D of the deuteron is a measure of the tensor force (on- and off-shell). P_D and ϵ_1 are proportional to each other.

Non-locality and ^{3}H binding energy

It has been established quite some time ago that there exists a strong correlation between the ^{3}H binding energy and P_D. Fig. 3 clearly shows that potential models with low P_D, like Bonn A (11), reproduce the experimental ^{3}H binding energy much better than potentials, with a larger P_D, like Paris. About 0.3 MeV of the difference between the Bonn A, B and C (11) and the Paris (13) potential binding energy predictions is due to specific non-locality properties of the full Bonn potential (23). (Bonn A, B, and C are one-boson exchange approximation of the full Bonn potential (10)). Using quantum inversion of the Bonn B potential, it has recently been shown that the local inversion potential of Bonn B gives a binding energy of -7.84 MeV for ^{3}H compared to -8.14 MeV for the original, non-local Bonn B potential (24). According to the relation displayed in Fig. 3 the inversion potential of Bonn B must have a larger P_D. In fact, it is 5.81%, compared to 5.0% of the original Bonn B potential. It is now well established that all local NN potentials that fit NN data accurately, underbind ^{3}H by 0.8 MeV if the charge dependence of $^1S_0$ is properly taken into account.

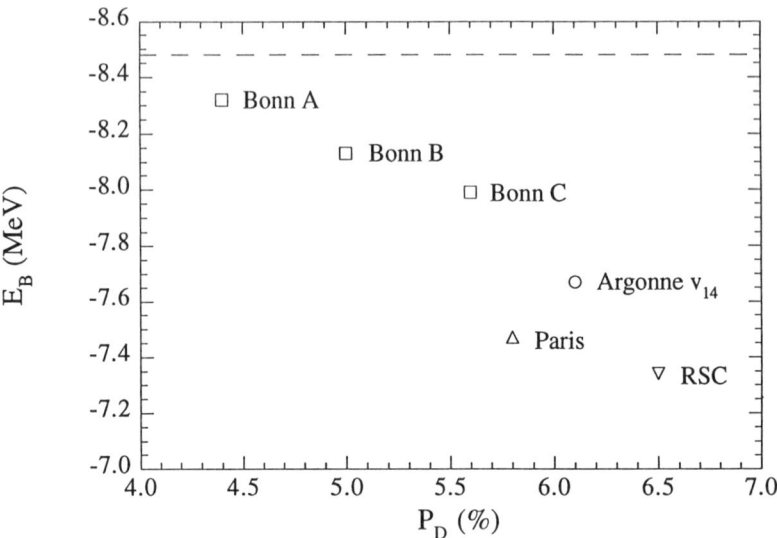

Figure 3. Relation between the D-state probability P_D of the deuteron and the calculated ^{3}H binding energy for various potential models. The experimental ^{3}H binding energy is -8.48 MeV.

^{3}H binding energy and three-nucleon forces

The issue of three-nucleon force (3NF) effects in ^{3}H binding energy calculations is still quite controversial. Taking Δ degrees of freedom explicitly into account, Pickelsimer et al. demonstrated in a comprehensive and consistent ^{3}H binding energy calculation that the repulsive dispersive effects of the 3NF on the NN force practically cancel the attractive 3NF in ^{3}H (25). In the more traditional calculations dispersive effects are not taken into account. An example is the recent calculation of Stadler et al. (26). They combined the full Tucson-Melbourne (TM) 3NF ($\pi\pi$, $\rho\pi$ and $\rho\rho$ exchange) with the Paris and Nijmegen potentials and found that the calculated binding energies were close to the experimental value of -8.48 MeV (27). A second criticism deals with inconsistencies associated with the choice of the meson-nucleon form factor cutoff parameters. Stadler et al. used soft cutoff parameters in the TM 3NF and hard cutoff parameters in the NN potentials. Nevertheless, it cannot be ruled out completely at the present time that small 3NF effects play a role in ^{3}H. As has been argued by van Kolck, QCD implies that 3NF effects contribute at the 5% level of the NN force contribution (28). I think that the 3NF issue will be settled first in the 3N breakup system. There the experimenter has the freedom to position detectors in the laboratory at any given location to study any given 3N configuration of interest. Such studies are already underway at TUNL, where the nn and np scattering lengths a_{nn} and a_{np} are being determined simultaneously from the nd breakup reaction in a kinematically complete geometry for different production angles of the nn and np pairs. Of course, a_{np} is known accurately from free np scattering and new data for a_{nn} from the π^-d capture reaction have recently been taken at LAMPF and TRIUMF. Therefore, accurate information is available to which the TUNL results can be compared. This seems to be the only practical way to address the question of 3NF effects in the 3N systems.

NN tensor force

As shown above, non-locality is one way of producing a smaller P_D and therefore, a larger ^{3}H binding energy. 3NF effects may also help to increase the ^{3}H binding energy slightly. But the old paradigm still holds: a smaller ε_1 is accompanied by a smaller P_D. The questions arises: Is there any room for a smaller ε_1 in the energy range of interest for the ^{3}H binding energy, say 0-60 MeV? Recall the large discrepancy for ε_1 between the VPI VZ40 and NI93 analyses shown in Fig. 2. For clarity, Fig. 4 presents data for ε_1 up to 60 MeV in comparison to VZ40 and NI93. Why are both analyses so different? The main reason is the Basel/PSI datum of $\varepsilon_1 = 2.9° \pm 0.3°$ at 50 MeV reported in 1991 (29). (Actually, the data were taken at 67.5 MeV, but the associated analysis was performed at 50 MeV). The experiment involves the scattering of polarized neutrons from polarized protons with neutron and proton spins either parallel or antiparallel to the momentum direction of the incoming neutron beam. The observable is the spin-correlation parameter $A_{zz}(\theta)$. In the PSI/Basel experiment, the zero-crossing angle of A_{zz} was determined. This powerful technique does not require the accurate knowledge of the neutron beam and proton target polarization. According to Machleidt and Slaus, there is no way that realistic meson-exchange based NN potential models can yield an ε_1 value as large as claimed by the Basel/PSI group (30). Potential models predict values around 2.0°

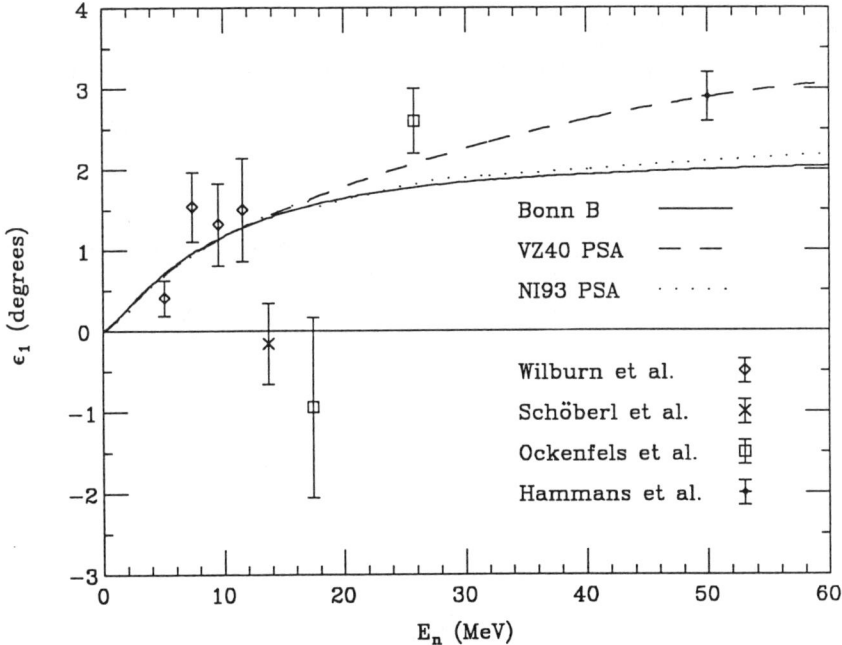

Figure 4. Data for the isoscalar $^3S_1$-$^3D_1$ mixing parameter ε_1 in comparison to the Bonn B NN potential model and the VPI VZ40 and Nijmegen NI93 phase-shift analyses.

at 50 MeV. It is fair to state that the Basel/PSI result for ε_1 is so large because the $^1P_1$ phase shift found in the analysis is so small. There exists a strong correlation between $^1P_1$ and ε_1 for this particular observable in the angular range studied by the Basel/PSI group: a smaller, i.e., less negative value for $^1P_1$ increases ε_1. Unfortunately, $^1P_1$ is after ε_1 the least well determined T = 0 NN phase shift. In the Basel/PSI analysis the value of -9.4° was found for $^1P_1$, while NN potential models prefer values between -10.5 and -11.0° at 50 MeV. Using the latter value for $^1P_1$ a much smaller result could be obtained for ε_1 ($\approx 2°$). Therefore, potential models have no problem with the Basel/PSI result for A_{zz}. In order to remove the ambiguity between ε_1 and $^1P_1$ the Basel /PSI group very recently took data for $A_{zz}(\theta)$ at forward angles; their results are not available yet. As can be seen from Fig. 5, the solid curve based on $\varepsilon_1=2°$ and $^1P_1=-10.5°$ gives about the same zero-crossing angle as the dashed-dotted curve based on $\varepsilon_1=3°$ and $^1P_1=-9.2°$. However, at angles forward of 70° both curves are quite different. In an attempt to further constrain $^1P_1$ the Basel/PSI group recently measured also the relative np differential cross section at 67 MeV (31). Summarizing the ε_1 situation at energies around 50 MeV one can state that there are no indications of an unusually small value for ε_1. In the contrary, the Basel/PSI and the VZ40 result as well as the single energy phase-shift analysis of Henneck et al. predict even larger values than obtained from any realistic NN potential model (32).

Since there are no data available between 50 and 100 MeV and since 50 MeV is about the upper limit for energies of interest for the ^3H binding energy problem, we will now concentrate on energies below 50 MeV. The datum at 25.8 MeV (see Fig. 4) obtained by Ockenfels et al. in 1991 at Bonn from a np polarization transfer $K^y{}_y(\theta_{cm}=130°)$ experiment is $\varepsilon_1=2.6°\pm 0.4°$ (33). This result is also quite large in comparison to potential model predictions. It should be noted that the np $K^y{}_y(\theta)$ is insensitive to $^1P_1$. In a subsequent experiment, Ockenfels et al. obtained a value for $K^y{}_y(\theta_{cm}=133°)$ at 17.4 MeV (34). Probably due to the somewhat unexpected experimental result, Ockenfels et al. did not report a value for ε_1. Our analysis of their result is shown in Fig. 4 with a fairly large error bar.

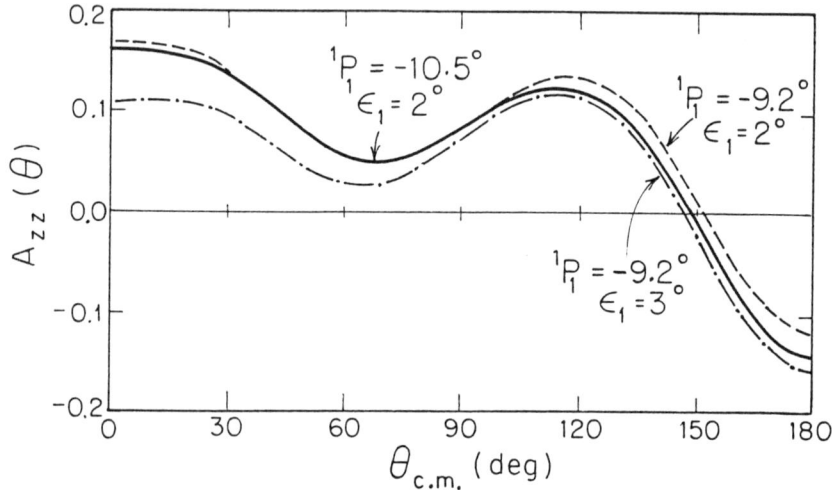

Figure 5. Sensitivity calculations for $A_{zz}(\theta)$ in np scattering at 50 MeV using different values for $^1P_1$ and ε_1.

Nevertheless, this value seems to support the low result for ε_1 obtained at 13.7 MeV by the Erlangen/Tübingen group in 1988 (35). Here, the observable was $A_{yy}(90°)$ and both the scattered neutrons and the associated recoil protons were detected in coincidence. In order to check on the conjecture of a low-energy ε_1 problem around 15 MeV, data were taken at TUNL at even lower energies. The TUNL group concentrated on transverse total cross-section difference $\Delta\sigma_T$ measurements. The transverse total cross-section difference is defined by

$$\Delta\sigma_T = \sigma(\uparrow\downarrow) - \sigma(\uparrow\uparrow), \tag{7}$$

where the first arrow refers to the target spin and the second to the projectile spin. Transverse (T) means that the spins are transverse with respect to the momentum direction of the incident neutron beam. The observable $\Delta\sigma_T$ is determined by measuring the asymmetry ε in neutron transmission through a polarized proton target as the neutron spin is reversed:

$$\varepsilon = \frac{N(\uparrow\uparrow) - N(\uparrow\downarrow)}{N(\uparrow\uparrow) + N(\uparrow\downarrow)} = \tanh\left(\frac{1}{2}P_n P_T \omega \Delta\sigma_T\right) = \frac{1}{2}P_n P_T \omega \Delta\sigma_T \tag{8}$$

Here, P_n is the neutron beam polarization, P_T the proton target polarization and finally, ω is the target thickness. In a first series of experiments the zero-crossing energy of $\Delta\sigma_T$ was determined. Again, in such a type of measurement the influence of uncertainties associated with the absolute magnitude of the neutron beam and proton target polarization are greatly reduced. Fig. 6 shows data for

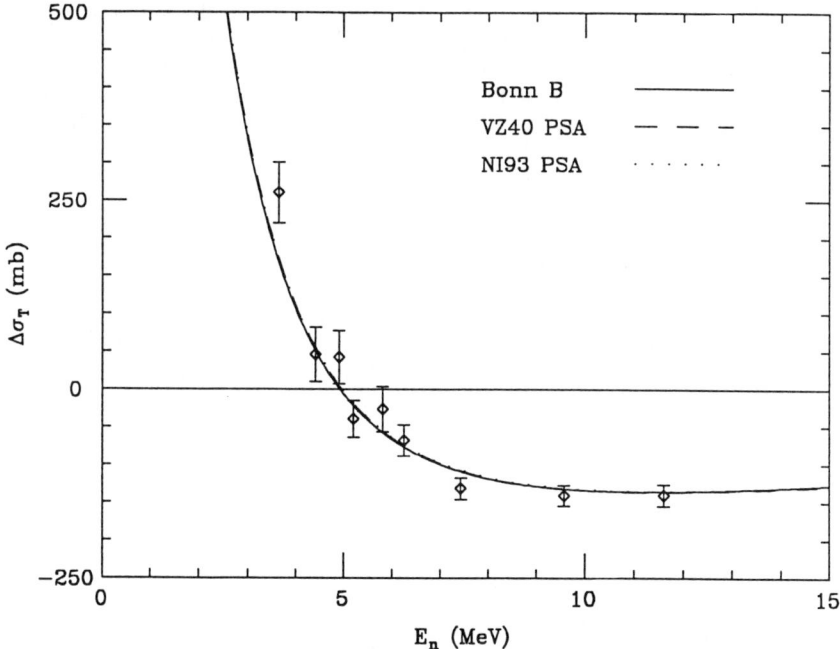

Figure 6. Data and predictions for the np total cross-section difference $\Delta\sigma_T$.

$\Delta\sigma_T$ near the zero-crossing energy of 5 MeV (36). In order to obtain a result for ε_1 the Bonn B potentials was used and its value for ε_1 was modified to reproduce the experimentally determined zero-crossing energy. In Fig. 4 the TUNL result is shown near 5 MeV (36). The error bar includes all known uncertainties. The result is slightly lower than the Bonn B potential model and the NI93 phase-shift analysis prediction. It should be pointed out that $\Delta\sigma_T$ is also practically insensitive to $^1P_1$. Subsequently the TUNL group took data at slightly higher energies.* The associated results for ε_1 are shown in Fig. 4 between 7.5 and 11.5 MeV. Again, the error bars include statistical and all known systematic uncertainties. The TUNL data seem to rule out the negative ε_1 values obtained from the experiments of Refs. (34) and (35). Therefore, it seems now safe to state that a low-energy ε_1 problem does not exist.

Of course, there is always a caveat. Like A_{yy}, A_{zz} and Ky'_y the observable $\Delta\sigma_T$ depends also on the $^1S_0$ phase shift. As stated already earlier, in phase-shift analyses of pp and np data, the np $^1S_0$ phase shift is obtained from the pp $^1S_0$ phase shift in a purely phenomenological way. Of course, charge independence is broken in $^1S_0$, as documented by the values for the pp and np scattering length a_{pp}=-17.3±0.3 fm and a_{np}=-23.748±0.010 fm. Therefore, a well defined anchor point exists for both the pp and np $^1S_0$ phase shift at zero energy. The energy dependence of $^1S_0$ is then parametrized by an effective-range expression. However, contrary to pp scattering where accurate data exist at low energies, it is not clear at all for np scattering how high one can go in energy with the effective range parametrization. The only np observable available at low energies is the np total cross section σ_T, but this observable is used and needed to determine the isoscalar $^3S_1$ phase shift. Therefore, data for the longitudinal total cross-section difference $\Delta\sigma_L$ are required to remove the sensitivity of $\Delta\sigma_T$ to the $^1S_0$ phase shift. Only the difference $\Delta\sigma_T - \Delta\sigma_L$ is a true measure of the NN tensor force. Such measurements will be conducted at TUNL in 1995 where a new polarized proton target based on dynamically polarized butanol has been installed. The TUNL group plans to improve the accuracy of their present $\Delta\sigma_T$ measurements by a factor of 2 to 3 and then to extend the measurements up to 20 MeV. Subsequently data for $\Delta\sigma_L$ will be taken in the 5-20 MeV energy range. In principle, energies up to about 40 MeV are accessible at TUNL.

πNN Coupling Constant

According to Eqs. 2 and 3 the neutral-pion NN coupling constant f_{π^0} and the charged-pion NN coupling constant $f_{\pi^\pm}$ can be treated as free parameters in pp and np phase-shift analyses. In the literature, either the pseudovector coupling constant f_π or the pseudoscalar coupling constant $g_\pi^2/4\pi$ is used. The two notations are related by

$$\frac{g_\pi^2}{4\pi} = \left(\frac{2M}{m_\pi}\right)^2 f_\pi^2 \approx 181 f_\pi^2 \ ,$$

---

* Because of an error in calculating the target polarization from the measured data, the values of $\Delta\sigma_T$ reported in (36) at 7.43, 9.57, and 11.60 MeV are incorrect. They have to be multiplied by $(\cos 54°)^{-1}$. This yields to an increase of the values for ε_1 reported in (36) at the energies given above.

where M denotes the average nucleon mass. Before 1987 it was commonly accepted that the πNN coupling constant was well known. From $\pi^{\pm}$p data, Koch and Pietarinen determined $g_{\pi}^2\pm/4\pi$ =14.28 ± 0.18 (37). Kroll analyzed pp data by means of forward dispersion relations and obtained $g_{\pi^0}^2/4\pi$ =14.52 ± 0.40 (38). In 1987 the Nijmegen group extracted a first value for $g_{\pi^0}^2/4\pi$ from a pp phase-shift analysis that deviated considerably from the previous result (39). Their present value is $g_{\pi^0}^2/4\pi$ = 13.47 ± 0.11 (40). Subsequently, Arndt et al. reanalyzed $\pi^{\pm}$p data and obtained $g_{\pi}^2\pm/4\pi$ = 13.31 ± 0.13, also in clear disagreement with the previously accepted value (41). Finally, the Nijmegen group determined $g_{\pi}^2\pm/4\pi$ from an analysis of np and $\bar{p}$p data (42,43). Their recent result is $g_{\pi}^2\pm/4\pi$ = 13.54 ± 0.05 (40). In agreement with the previous findings, there is no indication of any charge-independence breaking but the new results are about 6% smaller than the values used prior to 1987.

In their recent NN phase-shift analyses, the VPI group does not treat the coupling constants as free parameters anymore. As indicated above, the VPI group determines $g_{\pi}^2\pm/4\pi$ from $\pi^{\pm}$p scattering and assumes charge independence. Although the χ^2 minimum obtained from the combined VPI SM94 pp+np analysis is quite shallow compared to the one found for the $\pi^{\pm}$p data, the results for $g_{\pi}^2/4\pi$ coincide very well. However, in contradiction to the Nijmegen results, the individual pp and np phase-shift analyses seem to indicate a substantial amount of charge-independence breaking in the πNN coupling constants (7).

The Nijmegen group determined the πNN coupling constant from NN scattering data only. The NN bound state, the deuteron, was not included in the analysis. However, the OPE mechanism is dominant in the deuteron. For example, the quadrupole moment Q_d of the deuteron and the aymptotic D-state to S-state ratio A_d/A_s are caused by the NN tensor force which is generated mostly by OPE. The deuteron properties are known to better than 2%. Modern NN potential models which are all based on the old and large value $g_{\pi}^2/4\pi \approx 14.4$ for the πNN coupling constant, in general do reproduce the deuteron properties within their experimental uncertainties. Therefore, a 6% change in the πNN coupling constant is of great importance for NN potential models.

As has been shown by Machleidt and Li, the reduced neutral πNN coupling constant does not cause a problem for describing the deuteron properties accurately (44). It is the small charged πNN coupling constant that causes a severe problem. The Q_d of the deuteron cannot be reproduced by present realistic πNN potential models with $g_{\pi}^2\pm/4\pi \leq 13.9$, unless new tensor force generating mechanisms are introduced in meson-exchange based NN potential models (45). This conclusion makes an accurate experimental determination of the NN tensor force, i.e., ε_1, even more important.

Although the Nijmegen group repeatedly pointed out that there exists no single observable that can determine the πNN coupling constant (it is the bulk of the NN data that determines πNN), let us consider a single observable: the analyzing power $A_y(\theta)$ in np scattering at low energies. We focus on np and not on pp scattering, since we are concerned with $g_{\pi}^2\pm/4\pi$. At sufficiently low energies only S- and P-waves contribute to the observable. Since only the P-waves are peripheral waves, they are expected to mainly probe the long-range part of the interaction, i.e., the interaction region governed by OPE. As is well known, $A_y(\theta)$ is determined by ^{3}P-waves and therefore $A_y(\theta)$ must be sensitive to $g_{\pi}^2\pm/4\pi$. Recent sensitivity studies performed for $A_y(\theta)$ in np scattering by Machleidt and Li support this intuitive picture (44). In Fig. 7 Model I (solid curve) is based on Bonn B, i. e., it uses $g_{\pi}^2/4\pi$ =14.4 and for the ρ meson the large vector-to-tensor

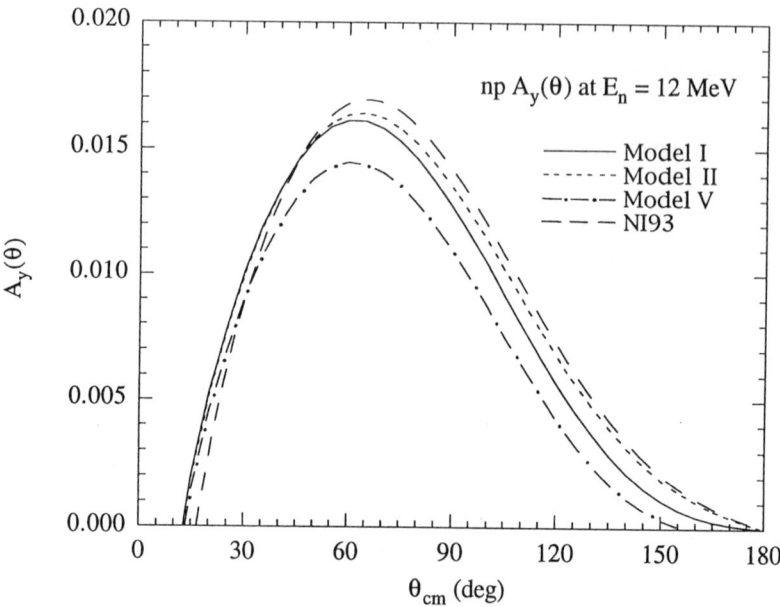

Figure 7. Model predictions of Machleidt and Li for the np analyzing power $A_y(\theta)$ at $E_n = $ 12 MeV in comparison to the Nijmegen NI93 phase-shift analysis result. Model I: $g_{\pi 0}^2/4\pi$ = $g_{\pi\pm}^2/4\pi$ =14.4; Model II: $g_{\pi 0}^2/4\pi$= $g_{\pi\pm}^2/4\pi$=13.5 and Model V: $g_{\pi 0}^2/4\pi$=13.5 and $g_{\pi\pm}^2/4\pi$ =14.4.

ratio κ_ρ=6.1 (actually, it slightly differs from Bonn B where the average of the neutral-pion and charged-pion mass is used while Model I uses the charged-pion mass). This model reproduces Q_d (0.278 fm^2 compared to the "experimental" value of 0.276 fm^2, where meson-exchange and relativistic contributions have been subtracted from the measured result of 0.2859 fm^2). Model II (short-dashed curve) gives a somewhat larger $A_y(\theta)$ and it is based on $g_{\pi 0}^2/4\pi$=13.5 = $g_{\pi\pm}^2/4\pi$ and κ_ρ=6.1 and fails to describe Q_d: it gives Q_d=0.266 fm^2. One could lower κ_ρ to 3.7, the so-called vector-meson dominance value; this increases Q_d to 0.274 fm^2, but now the pp $^3P_0$ and $^3P_2$ phase shifts are not reproduced anymore. Therefore, this trick is not a solution. Finally, Model V (dashed-dotted curve) gives a much reduced $A_y(\theta)$ compared to the two other models. It assumes $g_{\pi 0}^2/4\pi$ =13.5 and $g_{\pi\pm}^2/4\pi$ = 14.4, i.e., it is based on charge-independence breaking. This model (with κ_ρ=6.1) fits pp scattering data very well and, due to its larger $g_{\pi\pm}^2/4\pi$, also Q_d: it gives 0.275 fm^2. One may argue that the study presented in Fig. 7 is model dependent. To obtain some idea of a possible model dependence, the long-dashed curve in Fig. 7 gives the NI93 prediction for $A_y(\theta)$ (i.e., $g_{\pi 0}^2/4\pi$ =13.5 = $g_{\pi\pm}^2/4\pi$). Comparing the long-dashed and short-dashed curves we notice, that the model dependence is not too large. The very recent, preliminary TUNL data (46) definitely rule out Model V, the only of the 3 models that describes the deuteron properties correctly and simultaneously uses the now widely accepted smaller value for $g_{\pi 0}^2/4\pi$. Therefore, large charge-independence breaking in the coupling constants is not an issue. Since the TUNL data are slightly lower in magnitude than the NI93 prediction, a somewhat larger value for $g_\pi^2\pm/4\pi$ than obtained by the Nijmegen group cannot be ruled out.

Summarizing, I think we have to get used to the fact that the present meson-exchange picture of the NN interaction is approaching its limits and that new ideas are necessary to describe the wealth of accurate new experimental information.

ACKNOWLEDGMENTS

This work was supported in part by the U.S. Department of Energy, Office of High Energy and Nuclear Physics, under Grant No. DEFG05-91-ER40619.

REFERENCES

1. Bergervoet, J.R. et al., Phys. Rev. C **41**, 1435-1452 (1990).
2. Klomp, R.A.M., Stoks, V.G.J., and de Swart, J.J., Phys. Rev. C **44**, R1258-R1261 (1991).
3. Klomp, R.A.M., Stoks, V.G.J., and de Swart, J.J., Phys. Rev. C **45**, 2023-2026 (1992).
4. Stoks, V.G.J., Klomp, R.A.M., Rentmeester, M.C.M., and de Swart, J.J., Phys. Rev. C **48**, 792-815 (1993).
5. Nagels, M.M., Rijken, T.A., and de Swart, J.J., Phys. Rev. D **17**, 768-776 (1978).
6. Arndt, R.A., Roper, L.D., Workman, R.L., and McNaughton, M.W., Phys. Rev. D **45**, 3995-4001 (1992) and references therein.
7. Arndt, R.A., Strakovsky, I.I., and Workman, R.L., "An updated analysis of NN elastic scattering data to 1.6 GeV", preprint (1994).
8. Lagaris, I.E. and Pandharipande, V.R., Nucl. Phys. **A359**, 331-348 (1981).
9. Wiringa, R.B., Smith, R.A., and Ainsworth, T.L., Phys. Rev. C **29**, 1207-1221 (1994).
10. Machleidt, R., Holinde, K., and Elster, Ch., Phys. Rep. **149**, 1-89 (1987).
11. Machleidt, R., Adv. Nucl. Phys. **19**, 189-376 (1989).
12. Reid, R.V. Jr., Ann Phys. (N.Y.) **50**, 411-448 (1968).
13. Lacombe, M. et al., Phys. Rev. C **21**, 861-873 (1980).
14. Stoks, V.G., Klomp, R.A.M., Terheggen, C.P.F., and de Swart, J.J., Phys. Rev. C **49**, 2950-2962 (1994).
15. Haidenbauer, J., and Holinde, K., Phys. Rev. C **40**, 2465-2472 (1989).
16. Haidenbauer, J., Holinde, K., and Thomas, A.W., Phys. Rev. C **49**, 2331-2336 (1994).
17. Plümper, D., Flender, J., and Gari, M.F., Phys. Rev. C **49**, 2370-2378 (1994).
18. Machavariani, A.I., Straub, U., and Faessler, A., Nucl. Phys. **A548**, 592-612 (1992).
19. Ordonez, C., Ray, L., and van Kolck, U., Phys. Rev. Lett. **72**, 1982-1985 (1994).
20. Wiringa, R.B., Stoks, V.G.J., and Schiavilla, R., "An accurate nucleon-nucleon potential with charge-independence breaking", preprint (1994).
21. Sander M., and von Geramb, H.V., "Nucleon-nucleon potentials for VPI-SM94-(version 2) and Nijmegen-NY93 phase shifts", presented at the 8th International Symposium on Polarization Phenomena in Nuclear Physics, Bloomington, IN, September 15-22, 1994.
22. Schröder, B.C., Sander, M., and von Geramb, H.V., "Comments about the Nijmegen-NY93 and VPI-SM94-(V1,V2) phase shifts", presented at the 8th International Symposium on Polarization Phenomena in Nuclear Physics, Bloomington, IN, September 15-22, 1994.
23. Machleidt, R., private communication (1994).
24. Gibson, B.F., Kohlhoff, H., von Geramb, H.V., and Payne, G.L., "Inversion Potential Analysis of the Nuclear Dynamics in the Triton", preprint (1994).
25. Picklesimer, A., Rice, R.A., and Brandenburg, R., Phys. Rev. Lett. **68**, 1484-1487 (1992).

26. Stadler, A., Adams, J., Flemming, H., and Sauer, P.U., "Triton calculations with π-and δ-exchange three-nucleon forces", preprint (1994).
27. Coon, S.A., and Glöckle, W., Phys. Rev. **C23**, 1790-1802 (1981).
28. van Kolk, U., Phys. Rev. C **49**, 2932-2941 (1994).
29. Hammans, M. et al., Phys. Rev. Lett. **66**, 2293-2296 (1991).
30. Machleidt, R., and Slaus, I., Phys. Rev. Lett. **72**, 2664-2665 (1994).
31. Götz, J. et al., Nucl. Phys. **A541**, 467-473 (1994).
32. Henneck, R., Phys. Rev. C **47**, 1859-1875 (1993).
33. Ockenfels, M. et al., Nucl. Phys. **A526**, 109-130 (1991).
34. Ockenfels, M. et al., Nucl. Phys. **A534**, 248-254 (1991).
35. Schöberl, M. et al., Nucl. Phys. **A489**, 284-302 (1988).
36. Wilburn, W.S. et al., Phys. Rev. Lett. **71**, 1982-1985 (1993).
37. Koch, R., and Pietarinen, E., Nucl. Phys. **A336**, 331-346 (1980).
38. Kroll, P., in "Phenomenological Analysis of Nucleon-Nucleon Scattering", Physics Data Vol. 22-1, H. Behrens and G. Ebel, eds. Fachinformation-Zentrum, Karlsruhe (1981).
39. Bergervoet, J.R., van Campen, P.C., Rijken, T.A., and de Swart, J.J., Phys. Rev. Lett. **59**, 2255-2258 (1987).
40. Stoks, V., Timmermans, R., and de Swart, J.J., Phys. Rev. C **47**, 512-520 (1993).
41. Arndt, R.A., Li, Z.J., Roper, L.D., and Workman, R.L., Phys. Rev. Lett. **65**, 157-158 (1990).
42. Klomp, R.A.M., Stoks, V.G.J., and de Swart, J.J., Phys. Rev. C **44**, R1258-R1261 (1991).
43. Timmermans, R.G.E., Rijken, T.A., and de Swart, J.J., Phys. Rev. Lett. **67**, 1074-1077 (1991).
44. Machleidt, R., and Li, G.Q., "Constraints on the NN Coupling Constant from the NN System", presented at the 5th International Symposium on Meson-Nucleon Physics and the Structure of the Nucleon, Boulder, Colorado, September 6-10, 1993.
45. Machleidt, R., and Sammarruca, F., Phys. Rev. Lett. 66, 564-567 (1990).
46. Braun, R.T. et al., "Neutron-Proton Analyzing Power at 12 MeV and Charged πNN Coupling Constant", contributed paper, presented at the 8th International Symposium on Polarization Phenomena in Nuclear Physics, Bloomington, IN, September 15-22, 1994.

New Results in Nucleon-Nucleon Scattering at Intermediate Energies

H. Spinka

High Energy Physics Division
Argonne National Laboratory, Argonne, Illinois 60439, USA

Abstract. Many np elastic scattering spin observables have recently been measured between kinetic energies of about 500 and 1100 MeV at Saclay and LAMPF. These data are summarized and some new results are presented. Evidence for structure in pp observables near 2100 MeV is reviewed, and new data in this energy region are shown from SATURNE.

A) INTRODUCTION

Nucleon-nucleon (N-N) interactions have had a significant impact on a wide variety of topics in nuclear and particle physics. These include the understanding of the nuclear force, nucleon scattering from nuclei, and the allowable quark configurations, such as perhaps 6-quark states.

Nucleon-nucleon elastic scattering is one of the most basic reactions involving the strong interaction. The N-N elastic amplitudes for many years have served as stringent constraints on theoretical models describing the nuclear force at intermediate energies. Such models have included meson exchange, Skyrme, and QCD-inspired parton models. The differences in inelastic channels for the isospin-0 ($I = 0$) and isospin-1 ($I = 1$) N-N interactions have been significant in the development of some of these models. The nucleon-nucleon reactions are becoming important for the study of spin effects of the nucleon constitutents, since the nucleon contains relatively few valence quarks, and since such spin effects are partially masked in meson-nucleon amplitudes by the zero spin of both pions and kaons.

Measurements of N-N interactions have been useful for calculations of the scattering of nucleons from nuclei [1]. At present, these calculations either directly use the free N-N amplitudes from phase shift analyses of the experimental data [2-5], or use the results of theoretical models, fit to the N-N amplitudes, to predict the effective nucleon-nucleon interaction inside nuclei. In the future, one goal will be to understand nucleon scattering from nuclei in terms of quark and gluon interactions. Free N-N scattering will provide a stringent test case in the work to achieve this goal.

The third motivation for N-N spin measurements concerns the observation of energy dependent structures seen in various reactions and spin parameters. For example, there has been considerable controversy [6-13] about the interpretation of the resonance-like behavior of the $^1D_2$ and $^3F_3$ partial waves seen in phase shift analyses. Evidence from a number of experiments has also suggested structure near a kinetic energy of 2100 MeV in pp reactions. This energy is well above the thresholds for production of low-lying N*, Δ, and Λ states. It will be important to try to confirm these structures, and to understand their origin, since 6-quark states are predicted near this energy; see Sec. C.

Assuming the usual symmetries for the strong interaction, there are five $I = 0$ and five $I = 1$ nucleon-nucleon elastic scattering amplitudes at each angle and energy. Therefore, at least ten spin observables must be measured for pp scattering and ten additional observables for np scattering to obtain unique values for both the real and imaginary parts of these amplitudes. Note that pp and nn scattering are pure $I = 1$ channels, whereas the np scattering amplitudes are mixtures of $I = 0$ and $I = 1$ channels,

$$Ampl(np) = \frac{1}{2}[Ampl(I = 0) + Ampl(I = 1)],$$

at each angle and energy. The factor of 1/2 in the equation above indicates that the $I = 0$ amplitudes will generally be more poorly determined than the $I = 1$ amplitudes, assuming that the pp and np spin observables are measured to the same accuracy. On the other hand, np measurements are more difficult because polarized neutron beams are usually of lower intensity and have a broader energy spread than proton beams, neutron counters normally have significantly less than 100% efficiency and poorer spatial resolution than charged particle detectors, etc.

At intermediate energies, a large amount of new data in the 1980's on pp spin observables has permitted a reasonable determination of the $I = 1$ amplitudes in terms of phase shifts and inelasticities up to $\sim$ 1700 MeV. In addition, the pp elastic scattering amplitudes have been found in a model-independent fashion over part of the scattering angle range at 19 energies up to $T_{lab} = 5.14$ GeV or $P_{lab} = 6.00$ GeV/c. This includes: a) model independent amplitude analyses [14-16] up to 800 MeV that have been performed and published at 7 energies, b) a determination [17] of the pp amplitudes at 11 somewhat coarsely-spaced energies between 830 and 2700 MeV by the Saclay nucleon-nucleon collaboration, and c) a model independent amplitude determination [18] and phase shift analysis [19] that were performed at the single energy of 5.14 GeV. By contrast, many fewer $I = 0$ model independent amplitude determinations have been performed, and the number of angles is small compared to the $I = 1$ case.

The spin observables in this paper will be described in terms of the spin directions shown in Fig. 1. The directions are $\vec{N}$, normal to the scattering plane, $\vec{L}$, longitudinal along the beam or outgoing particle directions, and $\vec{S} = \vec{N} \times \vec{L}$ in the scattering plane. The observables will be denoted $(B,T;S,R)$ with B the beam, T the target, S the forward scattered, and R the recoil spin direction, and "0" denotes that the spin direction is not measured or is unpolarized. Other notations [20] will also be used to assist the reader to translate between the various expressions used by different groups for the same spin observables.

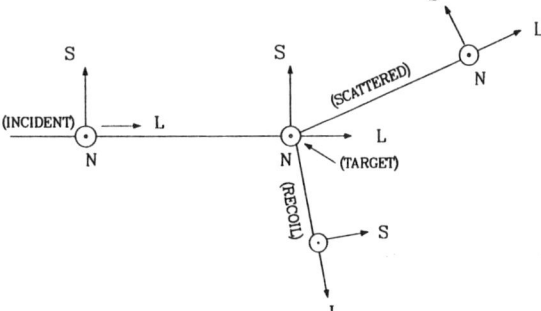

FIGURE 1. Definition of the spin directions for the beam, target, forward-scattered, and recoil particles for nucleon-nucleon elastic scattering.

N: NORMAL TO THE SCATTERING PLANE
L: LONGITUDINAL DIRECTION
S= N x L IN THE SCATTERING PLANE

B) np SCATTERING AT LAMPF AND SATURNE

Several years ago, a number of authors [21,22] noted the poor state of knowledge of I = 0 nucleon-nucleon elastic amplitudes or phase shifts above a kinetic energy of $\sim$ 500 MeV. Problems were also indicated [23,24] in inelastic I = 0 cross sections.

Within the past few years, a huge amount of new np elastic and total cross section spin data have been published from extensive programs at Los Alamos and Saclay. The collaborating institutions at SATURNE included the DAPNIA and LNS departments, CE-Saclay, France; DPNC, University of Geneva, Switzerland; Freiburg University, Germany; LNP-JINR, Dubna, Russia; University of California at Los Angeles; C.E.N.B., Gradignan, France; and the HEP division, Argonne National Laboratory. Collaborators on the LAMPF experiments included physicists from LAMPF; University of Central Arkansas; University of Colorado; Earlham College; University of Montana; New Mexico State University; Rice University; Rudjer Boskovic Institute in Zagreb, Croatia; Rutgers University; Texas A&M University; University of Texas at Austin; Washington State University; University of Wuppertal, Germany; and the HEP division, Argonne National Laboratory.

The new LAMPF data were collected in three different experimental arrangements at the energies 484, 567, 634, 720, and 788 MeV. Polarized neutrons were produced by forward charge exchange of polarized protons on a liquid deuterium target. One set of experiments used the new high intensity optically pumped polarized ion source, and consisted of a liquid hydrogen target, a magnetic spectrometer and carbon rescattering polarimeter for the outgoing proton [25], and a neutron counter array [26]. These neutron counters were only used to provide a check on the results, and especially the background subtraction. The measured np spin observables included $P = A_{N0} = A_{oono} = (N, 0; 0, 0)$, $K_{NN}, K_{SS}, K_{LS}, K_{SL}$, and $K_{LL} = K_{ok''ko} = (L, 0; 0, L)$ [27-29]; only three of the last four observables are independent. These experiments also permitted an improved measurement of the neutron beam polarization [30], which deviated by $\sim 12\%$ from previous estimates. Some data recently collected on P and K_{NN} over a broader angular range than in [29] are presently being analyzed.

A second setup used beam from the old, lower intensity Lamb-shift polarized ion source, and consisted of a polarized proton target and a large aperture magnetic spectrometer. The polarized target magnet prevented the simultaneous detection of neutrons. The measured spin observables included $C_{SS}, C_{LS} = C_{SL}$, and $C_{LL} = A_{ookk} = (L, L; 0, 0)$. All the elastic scattering data from these experiments are analyzed, including corrections for the new beam polarization measurements, and are shown in Fig. 2. Some of the data have been published [31,32], and a model independent I=0 amplitude determination is planned using the LAMPF data at $\sim$ 484, 634, and 788 MeV.

The third setup was nearly the same as the second, except that the magnetic spectrometer was replaced with a set of neutron counters [26] in the beam at 0°. The total cross section difference between antiparallel and parallel longitudinally polarized neutron beam and proton target, $\Delta\sigma_L(np) = \sigma_{tot}(\overset{\leftarrow}{\rightarrow}) - \sigma_{tot}(\overset{\rightarrow}{\rightarrow})$, was measured at five energies. The results were recently published [33] and are shown in Fig. 3.

The majority of the new SATURNE data were collected using a free polarized neutron beam produced by stripping polarized deuterons. The neutron beam struck a frozen spin polarized proton target. Outgoing protons were measured with scintillation counters and wire chambers, and outgoing neutrons were detected either in a neutron counter array or in a pair of wire chambers after conversion in a carbon block [34]. Either the polarization of the outgoing protons was determined by rescattering in a carbon polarimeter, or the proton momentum was measured in a magnetic spectrometer for these results.

A large number of spin parameters were measured at energies of 800, 840, 860, 880, 910, 940, 1000, 1080, and 1100 MeV [35-39]. These included

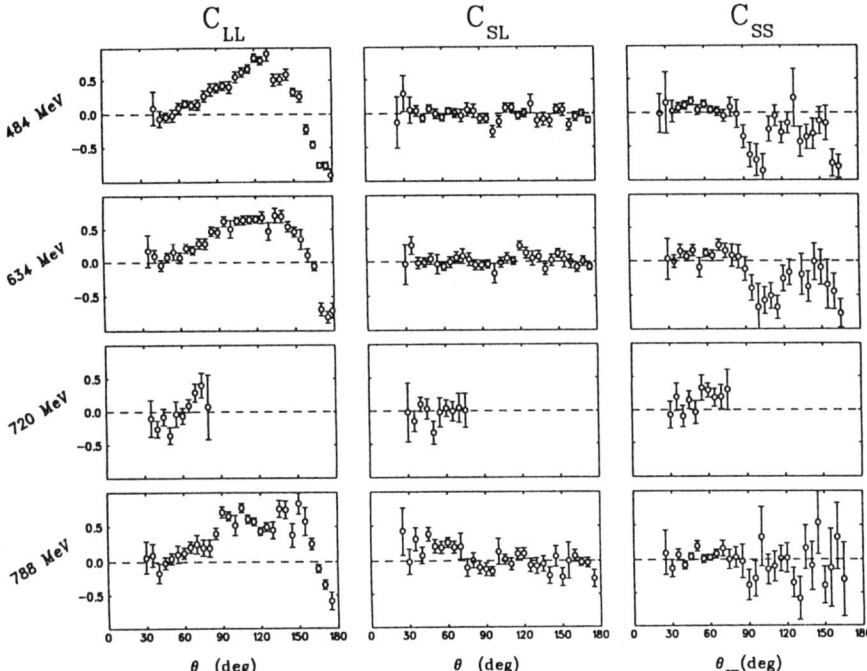

FIGURE 2. Final values for the spin parameters $C_{SS} = (S,S;0,0)$, C_{LS}, and C_{LL} with a polarized neutron beam on a polarized proton target at LAMPF. Some of these results were extracted from measurements of mixtures of the observables.

P, C_{NN}, C_{SL}, C_{LL}, D_{NN}, $D_{LS} = D_{os"ok} = (0, L; 0, S)$, K_{NN}, K_{LS}, K_{SS}, $H_{SLN} = N_{onsk} = (S, L; 0, N)$, and H_{LLN}. A model independent amplitude determination [40] is in progress using these data.

The total cross section differences $\Delta\sigma_L(np)$ and $\Delta\sigma_T(np)$ (which is defined analogously for transverse spins) were measured [41,42] using nearly identical sets of counters before and after the polarized target instead of the charged particle and neutron detectors used for the elastic scattering measurements. The $\Delta\sigma_L(np)$ results from Saclay and from PSI [43] are shown in Fig. 3. Good agreement among the three data sets can be seen. The I=0 cross sections can be obtained from $\Delta\sigma_L(I = 0) = 2\Delta\sigma_L(np) - \Delta\sigma_L(pp)$ and similarly for $\Delta\sigma_T$; see Fig. 4. These expressions follow from the generalized optical theorem, where $\Delta\sigma$ is proportional to the imaginary part of a forward elastic amplitude. It is surprising that the size of the energy dependent structure seen in $\Delta\sigma_L(I = 0)$ is comparable to that for $\Delta\sigma_L(I = 1)$, even though the strong $NN \to N\Delta$ channel cannot contribute to I=0, and the inelastic I=0 cross section is much smaller than the I=1 inelastic cross section at these energies. A possible interpretation in terms of a dibaryon state is described in [33].

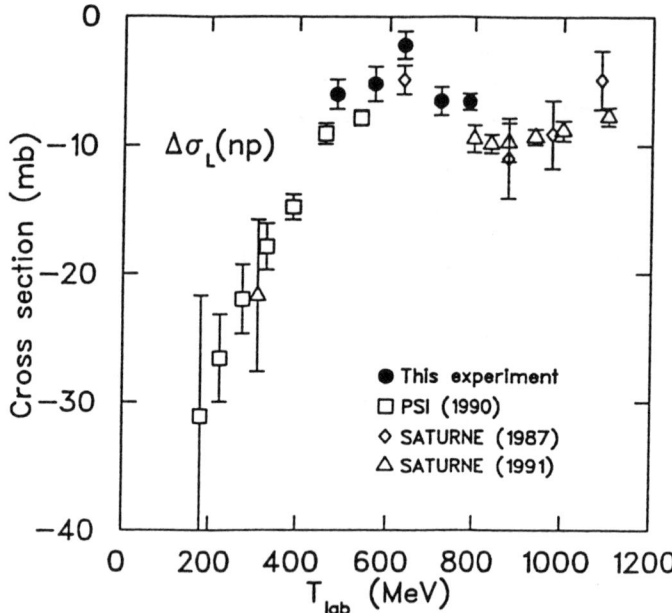

FIGURE 3. Total cross section difference in longitudinal spin states for np interactions from [33,41,43]. In all cases, a free polarized neutron beam and a polarized proton target were used.

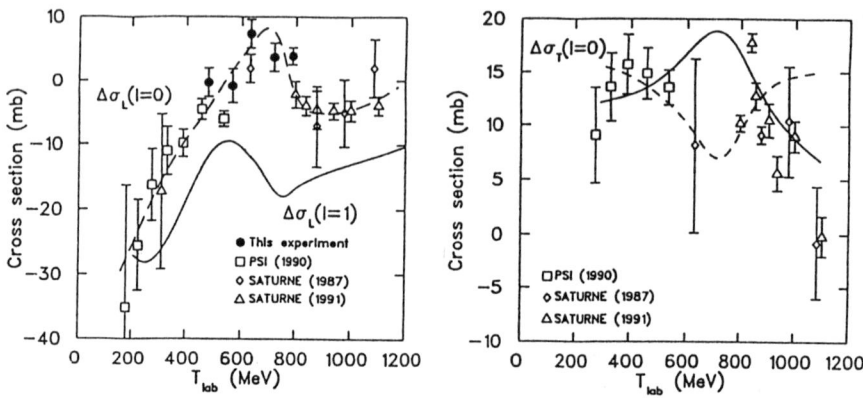

FIGURE 4. Total I=0 cross section differences in longitudinal and transverse spin states obtained from pp and np measurements in Los Alamos, Saclay, and other laboratories. The left figure shows $\Delta\sigma_L(I=0)$ data. The dashed curve is drawn to guide the eye, whereas the solid curve is drawn through the I=1 data. The right figure presents the $\Delta\sigma_T(I=0)$ results, and the two curves are from models in [33].

The improvement in the np data base can be illustrated by plotting the total number of data points with statistical errors up to some size as a function of that size. The very numerous, and sometimes conflicting, differential cross section and polarization parameter measurements are excluded. The results for kinetic energies between 400 and 900 MeV for pp and np spin observables up to 1984, for np data up to 1994, and for np data between 400 and 1100 MeV in 1994 are shown in Fig. 5. The impressive increase in new data is immediately apparent for the whole range of statistical errors.

Additional np measurements are in progress at TRIUMF [44], and a major new program is under way at PSI [45]. In addition, pp elastic scattering experiments are being performed at the Indiana cooler [46], and are being prepared for COSY.

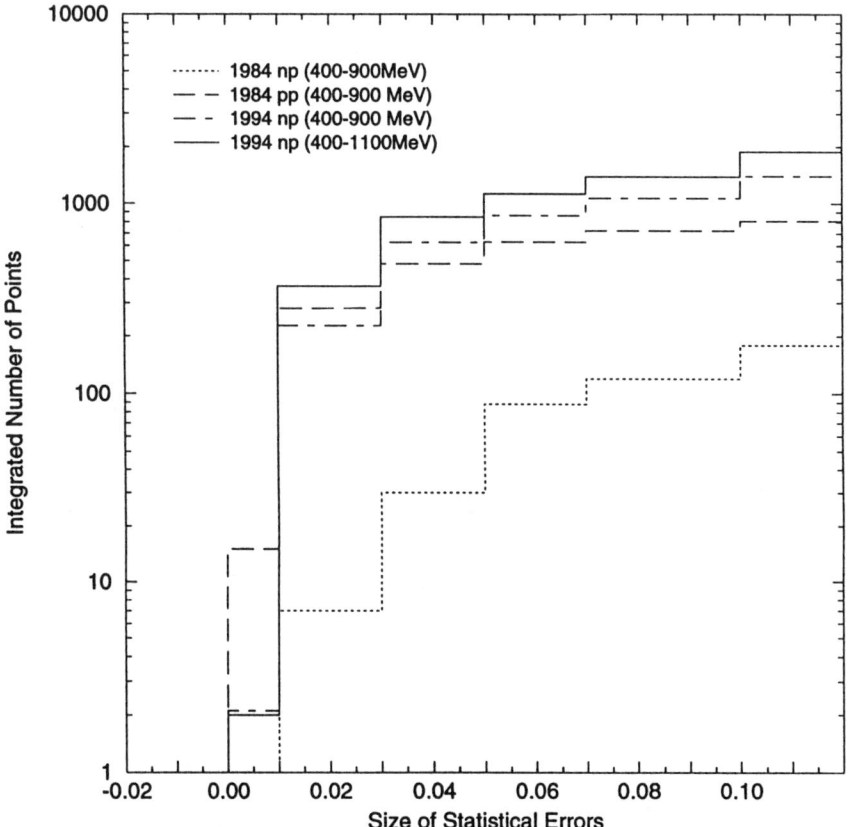

FIGURE 5. Integrated number of data points as a function of the statistical error for nucleon-nucleon elastic scattering spin observables except the polarization parameter and the differential cross section.

Knowledge of the I=0 nucleon-nucleon phase shifts has increased substantially over the past few years. The agreement between various phase shift analyses is expected to improve as the new np data are incorporated into the data bases for these analyses.

C) pp SCATTERING AT SATURNE

A number of early theoretical approaches suggested that dibaryon states should exist. These included models based on SU(3) and SU(6) symmetries [47,48] and on Regge ideas [49,50]. A variety of experimental evidence appeared to support these theories, such as a "bump" in the pp total cross section [51] and breaks in the slope of the 90° pp differential cross section [52,53]. Later, theoretical arguments were presented, based on Regge ideas and duality, that the N-N channel is "exotic" and without resonances [54]. As a consequence, the physics community strongly believed there were no dibaryons except deuteron-like objects. This belief was strengthened by the fact that the interpretation of many experimental data in terms of dibaryon states was often ambiguous. For example, energy dependent structure was often near strong inelastic thresholds, or of insufficient statistics, or nonreproducible, etc.

More recently, QCD-inspired bag and cloudy-bag models [55-58], constituent quark models [59,60], and diquark cluster models [61] have appeared which also predict dibaryon states. The masses for the expected lowest lying

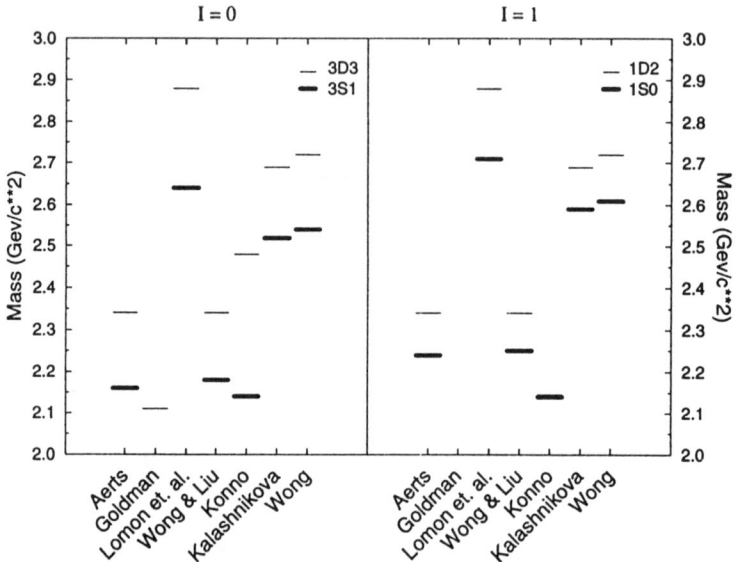

FIGURE 6. Predicted masses for the $^3S_1$, $^3D_3$, $^1S_0$, and $^1D_2$ dibaryon states in various quark models [55,57-61]. Generally, these are the lowest lying states in the models.

states differ for these models; see Fig. 6. Several of the recent models predict the lowest I=1 dibaryon states to have masses above 2.4 GeV.

It seems that a whole range of QCD-inspired models are able to predict 6-quark states, while there appear to be none that forbid dibaryons. If 6-quark states do not exist, then there is probably something wrong or missing from the present models. The missing ingredient could have extremely important consequences for the understanding of nuclear matter. Therefore, further experiments are needed to clarify whether dibaryons exist.

For dibaryon masses above about 2.4 GeV, there are many inelastic thresholds present, and the widths of N* and Δ states cause many of these thresholds to effectively overlap. As noted before, phase shift analyses are not reliable in this energy range at this time. Yet, many energy dependent effects are observed in a variety of reactions; see Fig. 7. For example, the pp total cross section data indicate a bump [51], polarization measurements at the 4-momentum transfer squared u=0 in the pp → πd reaction exhibit large changes [62,63], and the Saclay pp amplitude determinations [17] show a sizeable variation at T_{lab} = 2100 MeV, compared to 1800 and 2400 MeV. For this last case, $C_{SL} = -A_{xz} = A_{oosk}$ = (S,L;0,0) data near $\theta_{cm} = 45°$ from Saclay [64,65] and the ZGS [66] are shown in Fig. 7. There is also a suggestion of structure in $\Delta\sigma_L(pp)$ from ZGS data [67,68].

Recently, measurements were performed of the spin observable C_{NN} in pp elastic scattering at a set of 14 finely spaced energies at Saclay [69]. These results show a sharp drop in $C_{NN}(90°)$ with energy. Predictions of this spin observable by González, LaFrance, and Lomon in their cloudy bag model [58] appear reasonably consistent with the observations. This model has been quite successful in reproducing the pp phase shifts at lower energies.

The drop in $C_{NN} = A_{oonn}$ at $\theta_{cm} = 90°$ may be related to changes in spin singlet wave(s), since at that angle,

$$(1 - A_{oonn})\frac{d\sigma}{d\Omega} \sim |Singlet\ Ampl.|^2.$$

Thus, if a singlet wave becomes quite strong or resonates, A_{oonn} would decrease. At this time it is unclear whether the changes seen in the direct amplitude reconstruction [17] or in the polarization for pp → πd [62,63] are related to the behavior of spin singlet wave(s). However, an increase in the strength of such waves at $\theta_{cm} = 0°$ would appear as a peak in $\Delta\sigma_L(pp)$, as is suggested by the data in [67,68].

Two recent runs at Saclay by the N-N collaboration were devoted to rechecking the earlier results in [69] and to extending them in energy. The runs in November and December, 1993 measured a variety of two and three spin observables at T_{lab} = 1800 and 2100 MeV to clarify the behavior of the pp amplitudes reported earlier in [17]. Included were new data for P and C_{NN} at these energies and at 1850 and 2040 MeV. The runs in May and

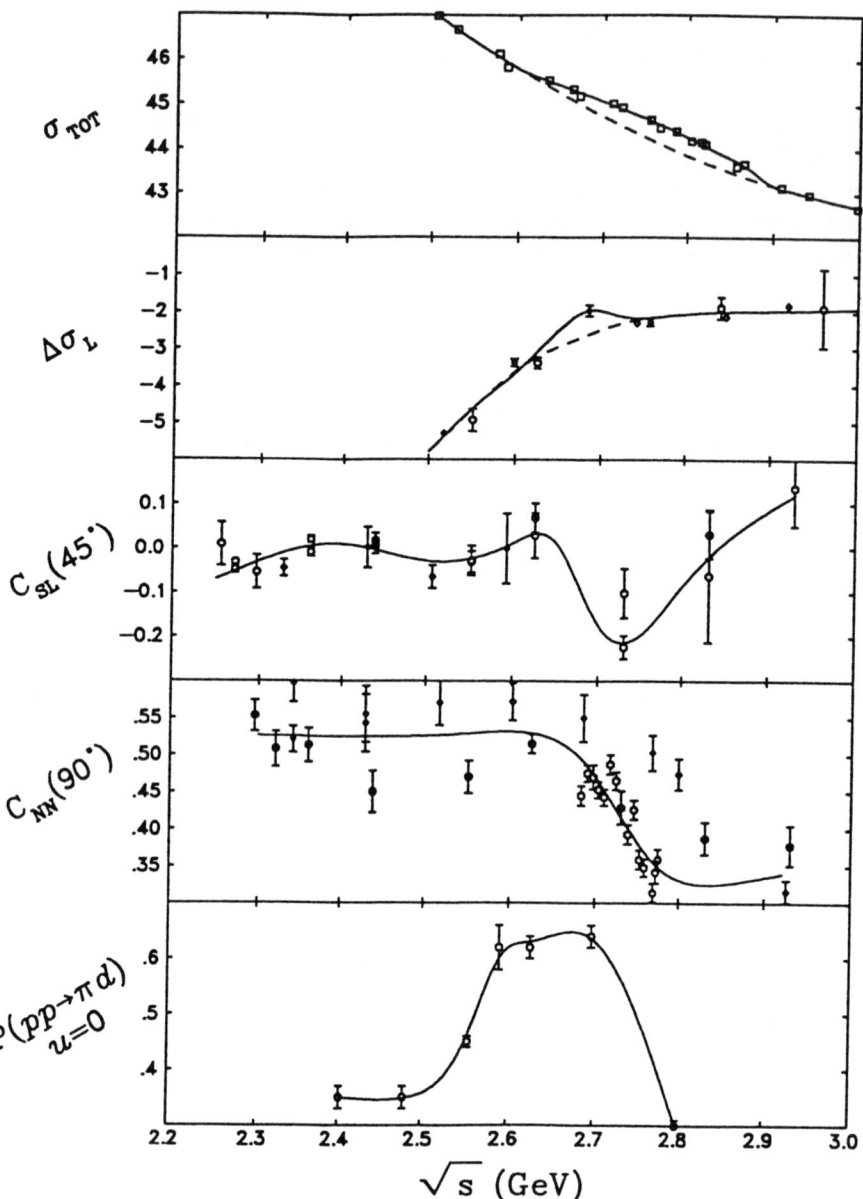

FIGURE 7. Collection of various data showing energy dependent behavior near $\sqrt{s} = 2.7$ GeV. The pp total cross section data are from [51], $\Delta\sigma_L(pp)$ from [67,68], the pp elastic scattering spin observables C_{SL} and C_{NN} from [64-66] and [69-72] respectively, and polarization at u=0 for the $pp \to \pi d$ reaction from [62,63]. All curves are drawn to guide the eye.

June, 1994 measured P and C_{NN} as well, rechecking several energies reported in [69] and extending the data in coarse steps up to 2650 MeV. If the structure seen near T_{lab} = 2100 MeV is caused by a $^1S_0$ dibaryon (see Fig. 6), then the $^1D_2$ state is expected near 2500-2600 MeV, and another drop in C_{NN} should be seen.

A number of experimental improvements were made in these two runs. Additional counters and electronics for monitoring beam polarization and intensity were employed. A new material was used for the polarized target which gave a faster polarizing time. A new data acquisition system was added; this system has improved capabilities to monitor the hardware performance. Checks of the incident beam angle were performed at a number of energies with the polarized target magnetic field turned off. The target polarization direction was reversed between each energy change, and data were recorded with "zero" beam polarization state at each energy. These changes should allow better control of systematic errors compared to the earlier data.

A preliminary analysis of data at a few energies has been performed, but not all systematics checks have been included. The data are normalized to the same beam polarimeter analyzing power as the published data [69]. The November, 1993 and May, 1994 preliminary results agree well with the earlier measurements at 2040 MeV as seen in Fig. 8; data at 2120 MeV agree similarly. Somewhat poorer agreement is observed at 2220 MeV (Fig. 9) and 2160 MeV. The origin of the differences is not understood at this time.

Preliminary results for $C_{NN}(90°)$ are given in Fig. 10 as well as published results from [69,70]. Although the agreement in Figs. 8-10 appears quite good, it is premature to conclude that the energy dependent structure is confirmed until all the new checks of systematic effects have been carefully studied. A reanalysis of the data in [69] may also be required. In any case, it is anticipated that the new measurements will clarify the past Saclay observations on structure in this energy region.

D) CONCLUSIONS

A large amount of new np elastic scattering and total cross section spin data have recently been analyzed. This includes about 850 new points between 480 and 790 MeV from LAMPF and 1600 new points between 470 and 1100 MeV from SATURNE in the past four years (or about 760 and 950 points excluding P). These new results will permit model independent I=0 amplitude determinations for the first time at 8 energies and a number of scattering angles. Phase shift analyses for I=0 partial waves will also be greatly improved.

Several different spin observables and reactions suggest energy dependent structure near $\sqrt{s}$ = 2.7 GeV or T_{lab} ~ 2100 MeV in pp elastic scattering; see Fig. 7. For example, a sizeable change in the pp elastic amplitudes

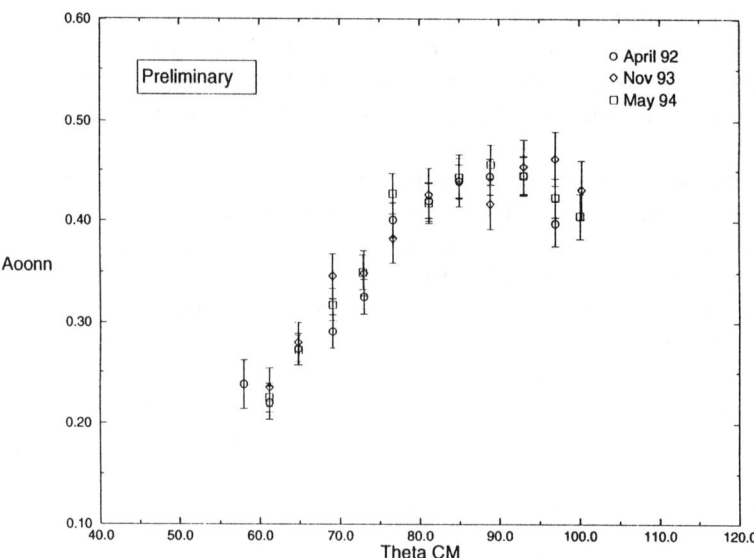

FIGURE 8. Comparison of C_{NN} data for pp elastic scattering at 2040 MeV collected in different runs as a function of angle. The Nov., 1993 and May, 1994 results are preliminary.

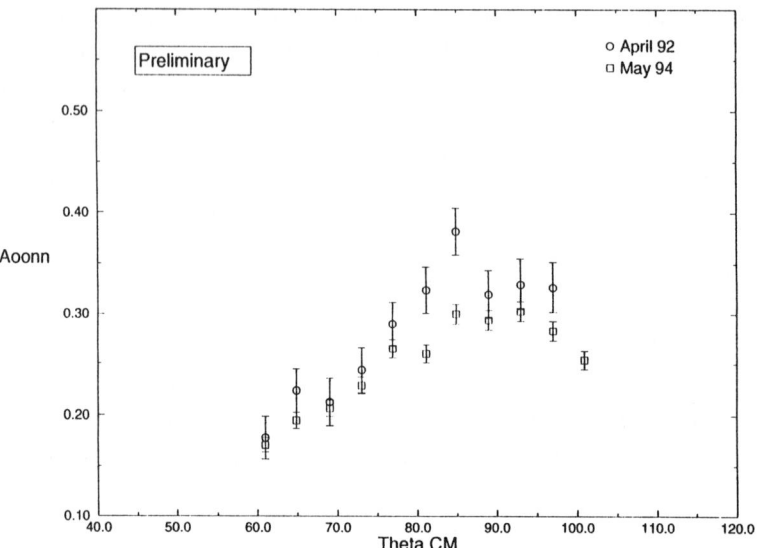

FIGURE 9. Comparison of C_{NN} data for pp elastic scattering at 2220 MeV collected in different runs as a function of angle. The May, 1994 results are preliminary.

between 1800, 2100, and 2400 MeV was observed in a model independent amplitude reconstruction [17]. Additional data were collected at the first two beam energies to clarify these results. Another example is the drop in the pp spin observable $C_{NN}(90°)$ seen near 2100 MeV at SATURNE [69]. Several energies have been repeated and the energy range extended in recent experiments. Using certain assumptions, the new and published C_{NN} results seem to agree reasonably well, but many checks of systematic effects are still required of the new data. Once these are completed, the status of the drop in $C_{NN}(90°)$ should also be considerably clarified.

An increase in the strength of a spin singlet wave is compatible with the decrease in C_{NN} shown in Fig. 10. This behavior is expected in some quark models, where the lowest lying I=1 dibaryon state is expected to be in the $^1S_0$ partial wave. In particular, the cloudy bag model of Lomon et al. [58] suggests that a $^1S_0$ dibaryon state should occur near $\sqrt{s} = 2.7$ GeV. On the other hand, if no dibaryons are present in the Saclay energy range, this would call into question the assumptions that went into the quark models in [55-61]. In particular, there may be serious implications for the understanding of nuclear matter.

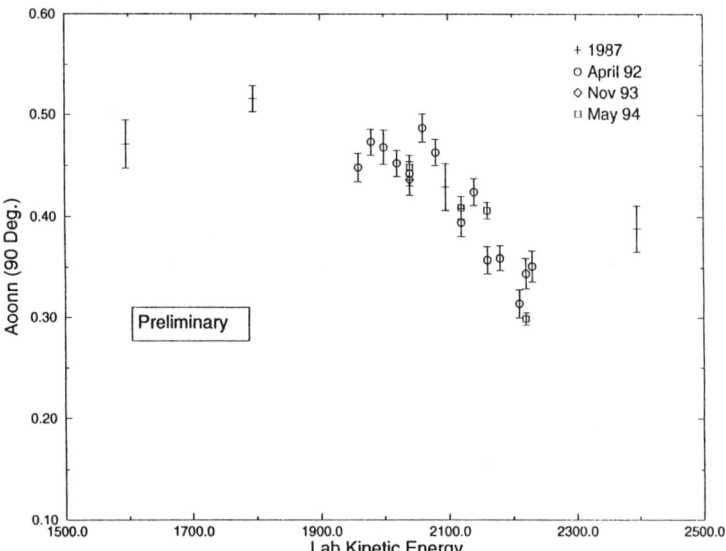

FIGURE 10. Measured values of $C_{NN}(90°)$ data for pp elastic scattering as a function of lab energy. The Nov., 1993 and May, 1994 results are preliminary.

Acknowledgements

I wish to express my gratitude to my colleagues in the LAMPF and Saclay N-N collaborations and at Argonne National Laboratory for helpful suggestions during the course of this work. I am also thankful for help with some of the figures by C. Allgower, M. Beddo and D. Lopiano. This work was supported in part by the U.S. Department of Energy, Division of High Energy Physics, Contract W-31-109-ENG-38.

References

1. For example, see S.J. Wallace, Ann. Rev. Nucl. Sci. **37**, 267 (1987) and the talk of T. Taddeucci at this conference.
2. J. Bystricky et al., J. Phys. (Paris) **48**, 199 (1987); **51**, 2747 (1990); and F. Lehar et al., *ibid.* **48**, 1273 (1987);.
3. D.V. Bugg, Phys. Rev. **C41**, 2708 (1990).
4. Y. Higuchi et al., Prog. Theor. Phys. **86**, 17 (1991); N. Hoshizaki and T. Watanabe, *ibid.* **86**, 321, 327 (1991); Y. Higuchi et al., *ibid.* **88**, 1019 (1992).
5. R.A. Arndt et al., Phys. Rev. **D45**, 3995 (1992).
6. C.E. Waltham et al., Nucl. Phys. **A433**, 649 (1985).
7. A.B. Wicklund et al., Phys. Rev. **D35**, 2670 (1987).
8. R.L. Shypit et al., Phys. Rev. Lett. **60**, 901 (1988); **61**, 2385 (1988); Phys. Rev. **C40**, 2203 (1989).
9. I.I. Strakovskii et al., Yad. Fiz. **40**, 429 (1984) [trans. Sov. J. Nucl. Phys. **40**, 273 (1984)].
10. M.G. Ryskin and I.I. Strakovsky, Phys. Rev. Lett. **61**, 2384 (1988).
11. T.-S.H. Lee, Phys. Rev. **C40**, 2911 (1989).
12. N. Hiroshige et al., Mod. Phys. Lett. **A5**, 207 (1990).
13. N. Hoshizaki, Phys. Rev. **C45**, 1424 (1992); Prog. Theor. Phys. **89**, 245, 251, 563, 569 (1993).
14. E. Aprile et al., Phys. Rev. Lett. **46**, 1047 (1981); R. Hausammann et al., Phys. Rev. **D40**, 22 (1989).
15. M.J. Moravcsik et al., Phys. Rev. **D31**, 1577 (1985).
16. M.W. McNaughton et al., Phys. Rev. **C41**, 2809 (1990).
17. C.D. Lac et al., J. Phys. (Paris) **51**, 2689 (1990).
18. I.P. Auer et al., Phys. Rev. **D32**, 1609 (1985).
19. M. Matsuda et al., Phys. Rev. **D33**, 2563 (1986).
20. J. Bystricky et al., J. Phys. (Paris) **39**, 1 (1978).
21. C. Lechanoine-Leluc et al., J. Phys. (Paris) **48**, 985 (1987).
22. R.A. Arndt, Phys. Rev. **D37**, 2665 (1988); R.A. Arndt et al., *ibid.* **D35**, 128 (1987).
23. B.J. VerWest and R.A. Arndt, Phys. Rev. **C25**, 1979 (1982).
24. J. Bystricky et al., J. Phys. (Paris) **48**, 1901 (1987).
25. R.D. Ransome et al., Nucl. Instrum. Meth. **201**, 309, 315 (1982); M.W. McNaughton et al., *ibid.* **A241**, 435 (1985).
26. R. Garnett et al., Nucl. Instrum. Meth. **A309**, 508 (1991).
27. M.W. McNaughton et al., Phys. Rev. **C44**, 2267 (1991).
28. K.H. McNaughton et al., Phys. Rev. **C46**, 47 (1992).
29. M.W. McNaughton et al., Phys. Rev. **C48**, 256 (1993).
30. M.W. McNaughton et al., Phys. Rev. **C45**, 2564 (1992).
31. W.R. Ditzler et al., Phys. Rev. **D46**, 2792 (1992).
32. T. Shima et al., Phys. Rev. **D47**, 29 (1993).
33. M. Beddo et al., Phys. Lett. **B258**, 24 (1991); Phys. Rev. **D50**, 104 (1994).

34. J. Ball et al., Nucl. Instrum. Meth. **A327**, 308 (1993).
35. J. Ball et al., Nucl. Phys. **A559**, 477 (1993).
36. J. Ball et al., Nucl. Phys. **A559**, 489 (1993).
37. J. Ball et al., Nucl. Phys. **A559**, 511 (1993).
38. J. Ball et al., Z. Phys. **C61**, 579 (1994).
39. J. Ball et al., Nucl. Phys. **A574**, 697 (1994).
40. C. Lechanoine-LeLuc and F. Lehar, Rev. Mod. Phys. **65**, 47 (1993).
41. J.M. Fontaine et al., Nucl. Phys. **B358**, 297 (1991).
42. J. Ball et al., Z. Phys. **C61**, 53 (1994).
43. R. Binz et al., Nucl. Phys. **A533**, 601 (1991).
44. Talk by C.A. Davis at this conference.
45. Talk by B. Vuaridel at this conference.
46. Talk by S. Bowyer at this conference.
47. R.J. Oakes, Phys. Rev. **131**, 2239 (1963).
48. F.J. Dyson and N.-H. Xuong, Phys. Rev. Lett. **13**, 815 (1964).
49. L.M. Libby, Phys. Lett. **29B**, 345 (1969).
50. S. Graffi et al., Lett. Nuo. Cim. **2**, 311 (1969).
51. R.F. George et al., Phys. Rev. Lett. **15**, 214 (1965); D.V. Bugg et al., Phys. Rev. **146**, 980 (1966).
52. R.C. Kammerud et al., Phys. Rev. **D4**, 1309 (1971).
53. L.M. Libby and E. Predazzi, Lett. Nuo. Cim. **2**, 881 (1969).
54. For example, see C. Schmid, Phys. Rev. Lett. **20**, 689 (1968).
55. A.Th.M. Aerts et al., Phys. Rev. **D17**, 260 (1978).
56. C.W. Wong and K.F. Liu, Phys. Rev. Lett. **41**, 82 (1978).
57. T. Goldman et al., Phys. Rev. **C39**, 1889 (1989).
58. P. LaFrance and E.L. Lomon, Phys. Rev. **D34**, 1341 (1986); P. González and E.L. Lomon, *ibid.* **D34**, 1351 (1986); P. González, P. LaFrance, and E.L. Lomon, *ibid.* **D35**, 2142 (1987).
59. C.W. Wong, Prog. Part. Nucl. Phys. **8**, 223 (1982).
60. Yu.S. Kalashnikova et al., Yad. Fiz. **46**, 1181 (1987) [trans. Sov. J. Nucl. Phys. **46**, 689 (1987)].
61. N. Konno et al., Phys. Rev. **D35**, 239 (1987).
62. R. Bertini et al., Phys. Lett. **162B**, 77 (1985).
63. R. Bertini et al., Phys. Lett. **203B**, 18 (1988).
64. F. Perrot et al., Nucl. Phys. **B296**, 527 (1988).
65. J.M. Fontaine et al., Nucl. Phys. **B321**, 299 (1989).
66. I.P. Auer et al., Phys. Rev. Lett. **51**, 1411, 1814(E) (1983).
67. I.P. Auer et al., Phys. Rev. **D34**, 2581 (1986).
68. I.P. Auer et al., Phys. Rev. Lett. **62**, 2649 (1989).
69. J. Ball et al., Phys. Lett. **B320**, 206 (1994).
70. F. Lehar et al., Nucl. Phys. **B294**, 1013 (1987).
71. D. Miller et al., Phys. Rev. **D16**, 2016 (1977).
72. A. Lin et al., Phys. Lett. **74B**, 273 (1978).

NEUTRON-PROTON ANALYZING POWER AT 12 MEV AND CHARGED πNN COUPLING CONSTANT

R. T. Braun[a], W. Tornow[a], D. E. González Trotter[a], C. R. Howell[a],
R. Machleidt[b], C. D. Roper[a], F. Salinas[a], H. R. Setze[a], R .L. Walter[a]
[a] *Department of Physics, Duke University and Triangle Universities Nuclear Laboratory, Durham, North Carolina 27708, USA*
[b] *Department of Physics, University of Idaho, Moscow, ID 83843, USA*

Abstract. Recent reanalysis of scattering data by the Nijmegen group has led to new values for the πNN coupling constants, $g_{\pi^\circ}^2/4\pi$ and $g_{\pi\pm}^2/4\pi$, about 6% smaller than the previously accepted values. The impact of this finding is far reaching. Since the neutron-proton $A_y(\theta)$ is dominated at low energies by the one-pion-exchange mechanism, accurate np data should provide unique information as to the magnitude of $g_{\pi\pm}^2/4\pi$. Using a new experimental setup consisting of a shielded neutron source, a five-pair neutron detector array, a n-^{4}He polarimeter, and an intense polarized source with fast spin-flipping capability, we have measured a 15 point angular distribution of the neutron-proton $A_y(\theta)$ at and incident neutron energy of 12 MeV to a statistical accuracy of 5×10^{-4}. We will discuss the data taking procedures, the analysis, and the corrections applied to the data. Preliminary results will be presented.

INTRODUCTION

Since 1987 the value for the πNN coupling constants has been a matter of controversy. At that time the Nijmegen group performed a partial-wave analysis of pp scattering data and reported a value for the neutral πNN coupling constant significantly lower than the previously accepted value[1]. The most recent result for the neutral πNN coupling constant is $g_{\pi^\circ}^2/4\pi = 13.47 \pm 0.11$[2]. Subsequent reanalysis of the np and $\bar{p}p$ data by the Nijmegen group has yielded a value for the charged pion-coupling constant that is in agreement with the lower neutral value, $g_{\pi\pm}^2/4\pi = 13.54 \pm 0.05$[3,4]. Additionally this lower value is in agreement with an independent analysis by Arndt of the $\pi^\pm p$ data[5]. The previously accepted values for the pion-coupling constants are those obtained by Kroll[6] for the neutral-pion coupling constant, $g_{\pi^\circ}^2/4\pi = 14.47 \pm 0.11$ and by Koch and Pietarinen[7] for the charged-pion, $g_{\pi\pm}^2/4\pi = 14.54 \pm 0.05$. The newer values are about 6% smaller, and this difference has serious consequences for the bound state properties of the deuteron. According to Machleidt and Li[8], NN potentials using the smaller charged-pion coupling constant are unable to reproduce the quadrupole moment of the deuteron; a value ≥ 14 is required, unless new tensor-force generating physics is introduced.

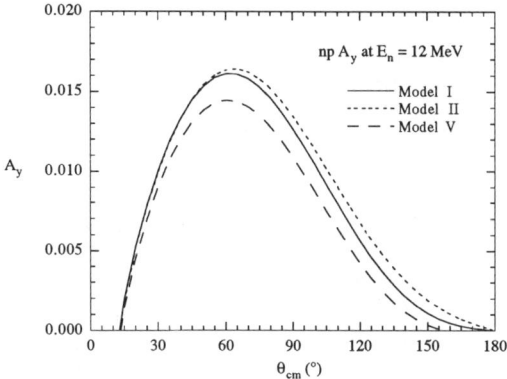

Figure 1: Different coupling constant model predictions for np $A_y(\theta)$.

To better fix the value of the charged-pion coupling constant, better np scattering data is necessary. One observable of particular interest is the np $A_y(\theta)$. Small changes in the pion coupling constant have considerable effect on the ^3P_J NN phase shifts, and at low energies the magnitude of $A_y(\theta)$ is dominated by the ^3P_J phase shifts. Previous low energy np $A_y(\theta)$ by Weisel et al.[9] at TUNL and Holslin et al.[10] at Wisconsin played an important role in the Nijmegen phase shift analysis. With that in mind we performed a new series of experiments at TUNL to measure the np $A_y(\theta)$ to an accuracy of 5×10^{-4}, a factor of two more accurate than the best previous measurements.

The sensitivity of the np $A_y(\theta)$ to different choices for the $\pi^\pm$NN can be seen in Figure 1. Here we show calculations using a modified Bonn B NN potential, prepared by Machleidt and Li[8], in which they have varied both πNN coupling constants between the old and new values. Following the convention in Ref.8 they differ as follows: Model I uses the 'old' large values for the pion coupling constant, $g^2_{\pi\pm}/4\pi = g^2_{\pi^\circ}/4\pi = 14.4$. Model II uses the 'new' small values, $g^2_{\pi\pm}/4\pi = g^2_{\pi^\circ}/4\pi = 13.5$. The third curve, Model V, is a 'compromise' model, assuming a charge-dependent coupling constant with a large $g^2_{\pi\pm}/4\pi = 14.4$ and a small $g^2_{\pi^\circ}/4\pi = 13.5$. The sensitivity of $A_y(\theta)$ to these changes is clear. Distinguishing between these predictions however requires data for which both the statistical and systematic uncertainties are small and for which the multiple scattering and finite geometry corrections are well known.

EXPERIMENTAL SETUP

Polarized deuterons were produced by the Atomic Beam Polarized Ion Source (ABPIS) at TUNL and accelerated through the FN Tandem Van de Graaff. Typical deuteron currents for this experiment were 1.5 μA. In order to

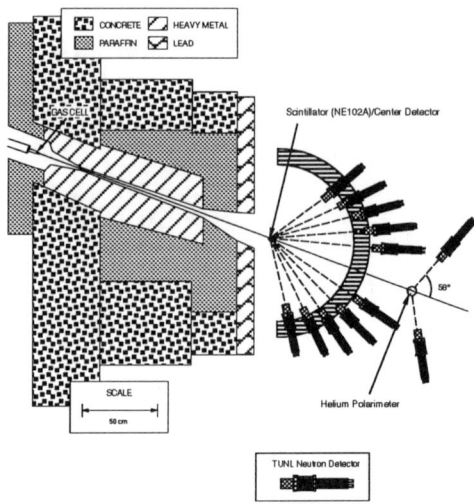

Figure 2: Experimental setup for TUNL neutron-proton analyzing power measurements.

reduce instrumental asymmetries, the deuteron polarization, and consequently the neutron polarization, was reversed at a rate of 10 Hz by means of RF transition units on the polarized ion source. The spin was reversed in an eight-step sequence (↑↓↓↑↓↑↑↓) designed to cancel systematic drifts to second order in time.

At the target area, polarized neutrons of mean energy 12 MeV and energy spread of ± 0.2 MeV were produced at 0° via the polarization transfer reaction $^2H(\vec{d},\vec{n})^3He$. The setup for the experiment is shown in Figure 2. Five pairs of liquid scintillator (NE213) detectors were arranged around the hydrogen scatterer, a plastic scintillator (NE102A) attached via a small light guide to a photomultiplier tube (PMT), which served as both a target and a center detector. A shielding wall between the neutron source and the scatterer provided tight collimation of the beam, preventing it from striking the center detector PMT as well as reducing the background seen by all detectors. Data was taken at 15 angles over a range of $\theta_{lab} = 16 - 72°$ in 4° steps, with a separation of 12° between each detector for each set of angles. The side detectors were positioned betweeen 50 and 70 cm from the scatterer, with the flight path chosen to equalize the count rate between the forward and backward angles as well as to minimize the cross-talk between the detectors.

The neutron polarization was continuously monitored by a sixth pair of detectors, via the $^4He(\vec{n},n)$ reaction, for which the analyzing power is well known. The neutron beam was incident on a high-pressure, helium gas cell which had a PMT attached to both its top and bottom. This polarimeter

Experimental geometry	(cm)
Gas cell-scatterer distance	180
Scatterer-side detector distance	50-70
Scatterer-helium polarimeter distance	114
Helium polarimeter-side detector distance	40
Scatterer dimensions (diameter × height)	1.9 × 3.8
Neutron detector dimensions (width × depth × height)	4.3 × 7.5 × 11.9

Table 1: Experimental geometry

typically provided us with a 1% determination of the beam polarization over the course of a 45 minute run. For each set of np scattering angles measured 250-300 runs were required. The neutron polarization averaged $\overline{p_y} = 0.57$ over the course of the experiment.

Data was taken in event mode to allow for future analysis, and was sorted into both one and two-dimensional spectra. For each event the proton recoil energy in the center detector, the neutron time-of-flight from the center detector to a side detector, the neutron detector pulse height, the neutron/gamma pulse shape (PSD), and the spin state were recorded, allowing us to perform complicated sorting and to eliminate much of the background.

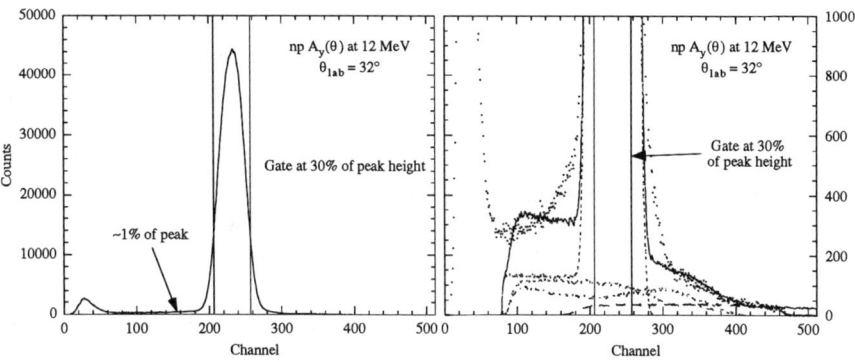

Figure 3: Data at 32°. The left side shows the center detector pulse height spectrum with gates set at 30% of the peak height. The right side is a magnified view of the same data with the Monte Carlo simulations superimposed to show the multiple scattering contributions. The solid curve is the sum of all calculated contributions while the short-dashed curve is the single-scattering contribution.

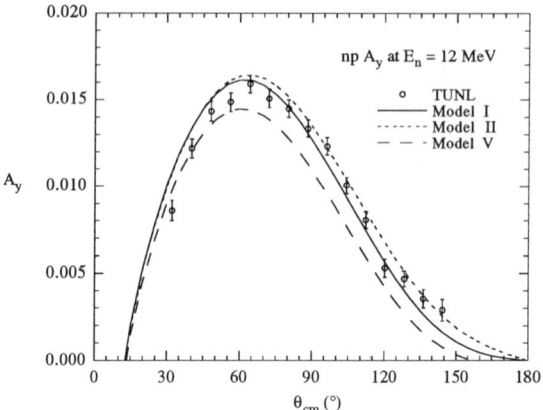

Figure 4: Preliminary results for np $A_y(\theta)$. The curves are as in Figure 2.

DATA ANALYSIS

We have measured $A_y(\theta)$ for np scattering at 12 MeV for 15 angles from $\theta_{cm} = 32 - 144°$ to a statistical accuracy of $\pm 5 \times 10^{-4}$. In order to correct for finite geometry and multiple scattering effects, sophisticated Monte Carlo simulations of the experiment have been performed. The left side of Figure 3 shows a typical proton-recoil spectrum, here at $\theta_{lab} = 32°$ with the gates set at 30% of the peak height. The right side of Figure 3 shows a magnified view of the same spectrum (dots) with the Monte Carlo calculations superimposed over the data, showing the contributions of the various multiple scattering processes. The solid curve shows the sum of all the multiple-scattering calculations together with the single-scattering calculation while the short-dashed curve shows the calculated single-scattering contribution alone. From these calculations the ratio of the single-scattering contribution of the data was determined.

An additional, important correction modeled was the Polarization Dependent Detector Efficiency (PDDE). This proccess is a double-scattering process in the side detectors where the neutron scatters first from ^{12}C and then from a proton. Due to resonances in the n-^{12}C $A_y(\theta)$, neutrons in certain energy ranges can be preferentially scattered in such a way that for two detectors, one will have an increased efficiency and the other will be decreased; flipping the spin state does not cancel this effect. To model this scattering process, knowledge of the n-^{12}C $A_y(\theta)$ below 8 MeV is crucial. We have performed calculations based on existing n-^{12}C phase shifts and are preparing detailed measurements of n-^{12}C $A_y(\theta)$ in the region of interest to further improve the accuracy of our np $A_y(\theta)$ analysis. The present status of the analysis is shown in Figure 4.

θ_{lab}	θ_{cm}	Raw A_y	Background	PDDE	Final A_y	Final dA_y
16.0	32.1	0.0073	0.0014	-0.0001	0.0086	0.0006
20.0	40.1	0.0110	0.0013	-0.0001	0.0122	0.0006
24.0	48.2	0.0135	0.0008	0.0000	0.0143	0.0006
28.0	56.2	0.0135	0.0013	0.0001	0.0149	0.0005
32.0	64.2	0.0152	0.0008	-0.0001	0.0159	0.0005
36.0	72.2	0.0135	0.0016	0.0000	0.0151	0.0005
40.0	80.2	0.0165	-0.0001	-0.0019	0.0145	0.0005
44.0	88.2	0.0136	0.0002	-0.0005	0.0133	0.0005
48.0	96.2	0.0114	0.0001	0.0008	0.0123	0.0005
52.0	104.2	0.0126	-0.0021	-0.0005	0.0100	0.0005
56.0	112.2	0.0076	0.0015	-0.0010	0.0081	0.0005
60.0	120.2	0.0044	0.0003	0.0006	0.0053	0.0005
64.0	128.2	0.0041	-0.0002	0.0008	0.0047	0.0005
68.0	136.2	0.0039	-0.0003	0.0000	0.0036	0.0005
72.0	144.3	0.0032	-0.0002	-0.0001	0.0029	0.0006

Table 2: Experimental results showing contribution of the corrections.

ACKNOWLEDGMENTS

This work was supported in part by the U. S. Department of Energy, Office of High Energy and Nuclear Physics, under Grant No. DEFG05-91-ER40619.

REFERENCES

1. Bergervoet, J. R. et al., *Phys. Rev. Lett.* **59**, 2255-2287 (1987).
2. Stoks, V., Timmermans, R., de Swart, J. J. *Phys. Rev.* **C47**, 512-520 (1993).
3. Klomp, R. A. M., Stoks, V. G. J., de Swart, J. J., *Phys. Rev.* **C44**, R1258-R1261 (1991).
4. Timmermans, R. G. E., Rijken, T. A., de Swart, J. J., *Phys. Rev. Lett.* **67**, 1074-1077 (1991).
5. Arndt, R. A. et al., *Phys. Rev. Lett.* **65**, 157-158 (1990).
6. Kroll, P., in Phenomenological Analysis of Nucleon-Nucleon Scattering, Physics Data, Vol. **22**-1, eds. H. Behrens and G. Ebel (Fachinformationszentrum Karlsruhe, 1981).
7. Koch, R. and Pietarinen, E. *Nucl. Phys.* **A336**, 331-346 (1980).
8. Machleidt, R. and Li, G. Q., π-N Newsletter No. **9**, 37 (1993).
9. Weisel, G. J. et al., *Phys. Rev.* **C46**, 1599 (1992).
10. Holslin, D. et al., *Phys. Rev. Lett.* **61**, 1561 (1988).

A Measurement of the Spin Transfer Observable $D_{NN'}$ for p+p Elastic Scattering at T_p=200 MeV

S.M. Bowyer, S.W. Wissink, A.D. Bacher, T.W. Bowyer,
S. Chang, W. Franklin, J. Liu, J. Sowinski,
E.J. Stephenson, and S.P. Wells

Indiana University Cyclotron Facility, Bloomington, Indiana 47408

W.K. Pitts

University of Louisville, Louisville, Kentucky, 40292

D.V. Bugg

Queen Mary College, London, UK

Abstract Recent analyses of NN and πN scattering data have resulted in values for the πNN coupling constant which are significantly smaller than those obtained prior to 1987. These controversial results prompted us to investigate the usefulness of high-quality spin measurements towards resolving this issue. We found that the normal component spin transfer observable $D_{NN'}$ for p+p elastic scattering is very sensitive to g_o^2, particularly at small angles. We have therefore determined precise values of $D_{NN'}$ for this reaction for θ_{lab}= 5.0°, 7.2°, 8.4°, 9.7°, 11.8°, 14.6°, 18.8°, 24°, 30°, and 38° at an incident beam energy of 200 MeV. The forward-going proton was detected in the IUCF K600 spectrometer and the coincident recoil proton was detected in a Si/CsI detector telescope. Our preliminary $D_{NN'}$ values are reproduced reasonably well by Arndt's C200 solution, the Nijmegen PWA93, and the Nijmegen I potential, but differ severely from both Arndt's SM94 global solution and the predictions of the Bonn potential.

INTRODUCTION

In most models of the nucleon-nucleon interaction, an important parameter, and one which appears explicitly in meson-exchange formulations, is the πNN coupling constant, g_π^2. The strength of this coupling dictates much of the long-range behavior of the hadronic component of the N-N force, especially the tensor terms. Recent determinations of the values of both the neutral, g_o^2 [1], and the charged, g_c^2 [2], coupling constants, based primarily on the analysis of NN and πN scattering data, respectively, have been significantly smaller than values determined prior to 1987. These smaller coupling constants do not appear to agree with the value of g_π^2 required in meson-exchange models in order to yield the correct deuteron quadrupole moment and asymptotic D/S-state ratio [3]. These inconsistencies motivated our collaboration to investigate the sensitivity of g_π^2 to high precision spin observable data that could be obtained at IUCF. We found that the normal-component spin transfer observable $D_{NN'}$ for p+p elastic scattering is very sensitive to g_o^2, particularly over the angular range in which the $(\sigma_1 \cdot q)(\sigma_2 \cdot q)$ term in the NN scattering amplitude crosses zero. (This is the spin-dependence exhibited by pion exchange contributions in a simple one-boson exchange model.) Over this angular range, the momentum transfer q ranges from $\sim$ 0.3–0.8 fm^{-1}, which is sufficiently large that Coulomb effects are small, but sensitivity to multi-pion or heavy meson contributions are minimal. Of the existing $D_{NN'}$ data in the 150 MeV to 300 MeV energy range, none bracket this crossover (which occurs near $\theta_{lab} \approx 9°$), and all have relatively large statistical errors and normalization uncertainties.

EXPERIMENTAL DETAILS

All data acquisition for this work was completed in August 1993. $D_{NN'}$ measurements were made at 10 angles (θ_{lab} = 5.0°, 7.2°, 8.4°, 9.7°, 11.8°, 14.6°, 18.8°, 24°, 30°, and 38°). Sufficient statistics were acquired so that total errors (statistical plus systematic) of $\delta D_{NN'} \leq 0.014$ were achieved at all but the largest angle.

In our experimental setup, both outgoing protons from p+p elastic scattering were detected simultaneously, and their energies were accurately measured. The forward-going proton was momentum analyzed in the K600 spectrometer focal plane, FP [4]. A septum magnet was used for all measurements forward of 15° [5]. After passing through the FP, the polarization of the forward-going proton was measured in the focal plane polarimeter, FPP [4]. The recoil proton was detected in a Si/CsI detector telescope placed inside our scattering chamber. The Si detector measured 4.0 cm wide × 6.0 cm high, and consisted of 7 horizontal strips which were ganged together to form three sections consisting of the top three strips, the middle strip, and the bottom three strips. Relative coincidence rates in the three sections provided a very accurate gauge of the vertical position of the beam spot on target. Careful positioning of the beam spot on target was especially important for the small angle measurements because of the severe kinematic magnification of the ϕ acceptance of the spectrometer for the coincident flux. The 500 μm thickness of the Si was sufficient to provide total energy information for protons from $\approx$ 1 to 8 MeV. For protons of higher energy, a ΔE

signal resulted from the Si. Behind the Si μ-strip detector was a tapered CsI crystal, measuring 3.8 cm deep and 1.5 cm wide × 4.0 cm high on the front face. This crystal was optically glued to a 2" PMT attached to a compact, water-cooled base. The CsI crystal was sufficiently thick to stop protons up to 100 MeV. The combination of these two detectors in series gave the recoil telescope a dynamic energy range of ∼1–100 MeV.

We chose to use a solid CH_2 target in this experiment because the advantages over using a liquid hydrogen target far outweighed the disadvantages. Using a solid target (as opposed to an extended liquid target) allowed us to know our event origin very well. The excitation spectrum of ^{12}C, simultaneously obtained in the focal plane, resulted in enough kinematic information to provide an absolute determination of the scattering angle to within a few ×0.01°. By comparing characteristics of ^{12}C peaks and the H peak, we could constantly monitor beam and target properties with good sensitivity. The challenge of using a CH_2 target was in distinguishing p+p elastic events from quasifree (p,2p) knockout. Since the minimum proton separation energy in ^{12}C is 16 MeV, the combined energy resolution of our FP and our recoil telescope of < 5 MeV was more than sufficient to distinguish the two types of events.

ANALYSIS

During data collection, no hardware coincidence requirement was made between the FP and the recoil telescope, but the recoil telescope energy and timing information, if any, was read out for every good event in the FP. Figure 1 shows two FP spectra. The one on the top is a CH_2 singles spectrum in which various ^{12}C peaks are identified, as well as the p+p peak. The bottom spectrum is identical, except that it was incremented in the replay software only if a second proton fired the recoil detector(s) in tight timing coincidence and deposited the correct amount of energy. The counts to the right of the p+p peak in the bottom spectrum originate from p+p events in which the forward-going proton underwent slit-edge scattering in the entrance collimator of the K600 spectrometer. These spectra show that we are able to cleanly identify p+p elastic events.

For recoil protons below 8 MeV, timing and energy requirements had to be satisfied for only the Si detector. For recoil protons above 8 MeV, however, timing and energy requirements had to be satisfied for both the Si and the CsI detectors. "Accidentals" were tagged as those events which consisted of a forward-going proton within the hydrogen peak in the FP spectrum, and a second proton which deposited the proper amount of energy in the recoil detector(s), but for which the relative timing of the two protons differed from "true" coincidences by exactly one beam burst.

The FPP consists of a carbon analyzer followed by two sets of paired x-y

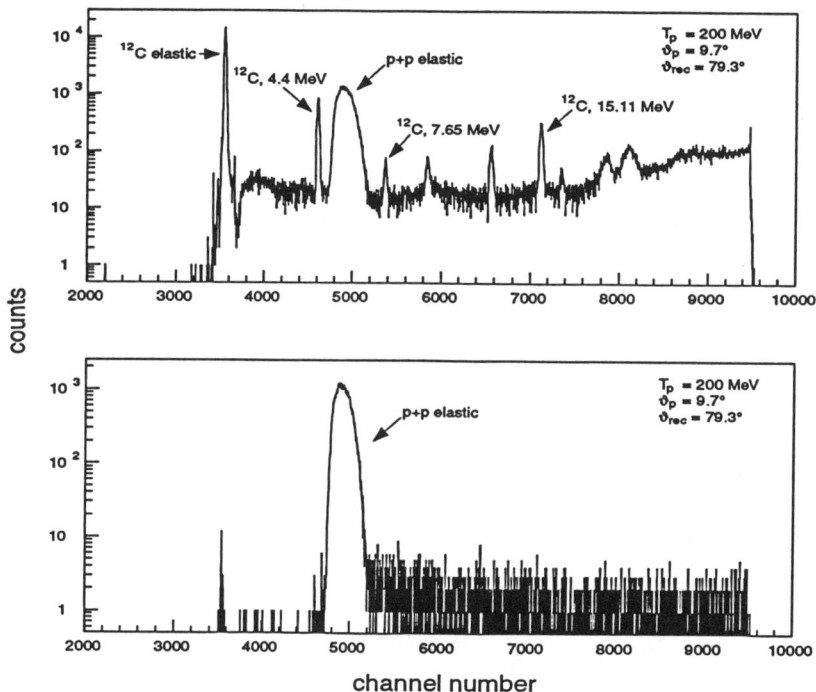

FIGURE 1. (top) Excitation spectra for 200 MeV protons incident on a CH$_2$ target at $\theta_p=9.7°$ ($\theta_{rec}=79.3°$). (bottom) Same as top spectrum, but with the software requirement that a proton fired the recoil telescope with the proper kinematic energy and timing for p+p elastic scattering.

multiwire proportional chambers backed by two scintillator planes, the ΔE-detector and the E-detector. In our analysis, we include primarily those events which scattered elastically in the FPP carbon analyzer. The vertex and angle of scattering inside the analyzer were determined for every event, and the energy losses, assuming p+^{12}C elastic scattering, through all the FPP materials were calculated. The differences between the calculated energy loss and the measured energy loss inside the ΔE-scintillator, DEDIFF, and the E-scintillator, EDIFF, were plotted against each other. The elastic scattering loci in these spectra were clearly distinguishable. Gates were drawn around the elastic scattering loci using an algorithm that could be applied consistently to all the data, and the number of counts within these gates constituted the FPP final yields. The EDIFF versus DEDIFF spectra were incremented for "true" p+p events and decremented for "accidental" p+p events.

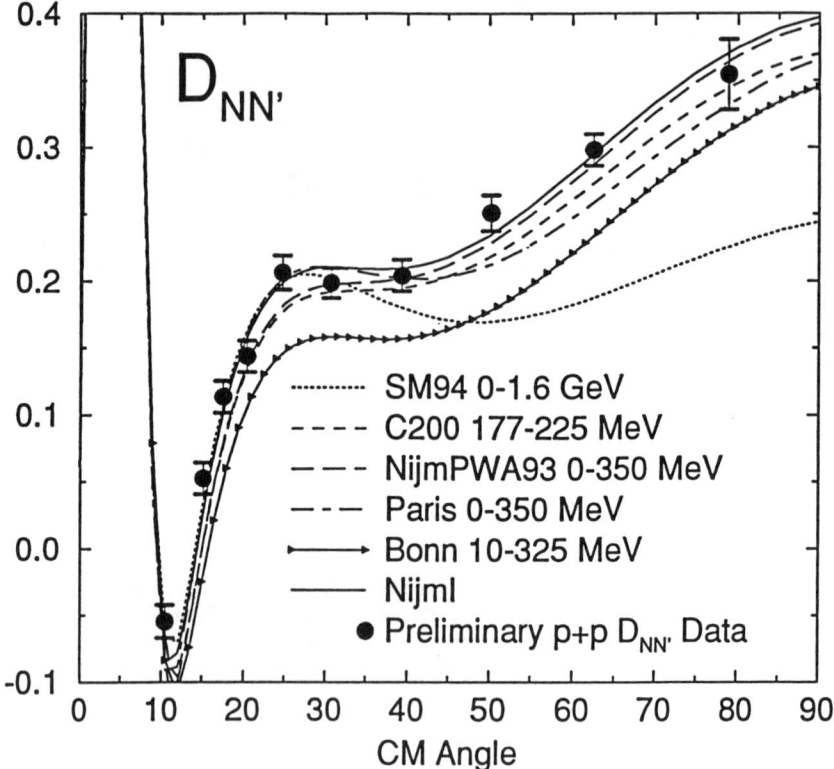

FIGURE 2. Preliminary $D_{NN'}$ data plotted with curves of three phase-shift solutions (SM94, C200, and NijmPWA93) and three potentials (Paris, Bonn, and NijmI). See the text for more details on these curves.

The minimization of effects due to potential sources of systematic errors was of paramount concern in both the design of the experiment and the analysis of the data. One of the most obvious sources of possible systematic error was the value used for the effective analyzing power of the FPP, A_{FPP}. To reduce this uncertainty, we carried out a calibration of A_{FPP} over an energy range of ~120 MeV to 200 MeV [6]. This energy range covers all energies incident on the FPP due to protons from p+p elastic scattering for θ_{lab}=5.0° to 38°. All software cuts and analysis procedures were performed identically for both the calibration data and the $D_{NN'}$ data.

There were several systematic error checks available to us in the data. The ^{12}C elastic peak was present on the focal plane from 5° to 14.6°, and for both the ^{12}C elastic peak and the p+p peak the induced polarization, P, and the analyzing power, A_y, were measured. These two observables should be equal; and since they are sensitive to different possible sources of systematic error, their comparison is an important means of systematic error evaluation. We also measured $D_{NN'}$ for ^{12}C elastic scattering, which must be exactly 1. Our preliminary results indicate that $P-A_y$ is zero and $D_{NN'}$ is 1 within statistics at all angles.

PRELIMINARY RESULTS

Our preliminary $D_{NN'}$ data is shown in Fig. 2. We do not consider this data to be "final" since all systematic error estimates have not been completed. We do not expect the final values of $D_{NN'}$ to differ from those shown here, although the errors may increase slightly. Also shown in Fig. 2 are the following: Arndt's global phase shift solution for 0-1.6 GeV (SM94) [7]; Arndt's local phase shift solution for 177 MeV to 225 MeV (C200) [7]; the Nijmegen p+p partial-wave analysis (NijmPWA93) [8]; the Paris potential (Paris) [9]; the Bonn potential (Bonn) [10]; and the Nijmegen I potential (NijmI) [11]. The C200, NijmPWA93, and NijmI curves agree reasonably well with the data, while the SM94 and Bonn curves do not. It is not yet clear how these various model predictions and phase-shift solutions will be affected once our results have been incorporated into the p-p database.

REFERENCES

[1] J.R. Bergervoet, P.C. van Campen, R.A. Klomp, J.L. de Kok, T.A. Rijken, V.G.J. Stoks, and J.J. de Swart, Phys. Rev. C41, 1435 (1990).
[2] R.A. Arndt, Z. Li, L.D. Roper and R.L. Workman, Phys. Rev. Lett. 65, 157 (1990).
[3] R. Machleidt and F. Sammarruca, Phys. Rev. Lett. 66, 564 (1991).
[4] A.K. Opper, Ph.D. dissertation, Indiana University, 1991.
[5] S.P. Wells, Ph.D. dissertation, Indiana University, 1994.
[6] S.M. Bowyer et al., contribution to this Symposium.
[7] Generated from R.A. Arndt's SAID program at VPI 10/30/94.
[8] R.A.M. Klomp (private communication, 1994). See also V.G.J. Stoks, R.A.M. Klomp M.C.M Rentmeester, and J.J. de Swart, Phys. Rev. C48, 792 (1993).
[9] Generated from R.A. Arndt's SAID program at VPI 10/30/94.
[10] Generated from R.A. Arndt's SAID program at VPI 10/30/94. See also R. Machleidt, K. Holinde, and C. Elster, Phys. Rep. 149,1 (1987).
[11] R.A.M. Klomp (private communication, 1994). See also V.G.J. Stoks, R.A.M. Klomp C.P.F. Terheggen, and J.J. de Swart, Phys. Rev. C49, 2950 (1994).

Zero-crossing Angle of the np Analyzing Power Below 300 MeV

C.A. Davis[a,b], R. Abegg[a,c], A.R. Berdoz[b], J. Birchall[b],
J.R. Campbell[b], L. Gan[b], P.W. Green[c], L.G. Greeniaus[c],
R. Helmer[a], E. Korkmaz[c], J. Li[c], C.A. Miller[a], A.K. Opper[c],
S.A. Page[b], W.D. Ramsay[b], A.M. Sekulovich[b], V. Sum[b],
W.T.H. van Oers[b], J. Zhao[b]

[a] TRIUMF, 4004 Wesbrook Mall, Vancouver, B.C., Canada V6T 2A3
[b] University of Manitoba, Dept. of Physics, Winnipeg, MB, Canada R3T 2N2
[c] University of Alberta, Dept. of Physics, Edmonton, AB, Canada T6G 2N5

To improve significantly upon the current knowledge of the nucleon-nucleon interaction requires experimental results of higher accuracy as input to phase shift analyses of scattering data. A particularly sensitive experimental test at intermediate energies is obtained by determining the scattering angle, θ_{zx}, at which the $n-p$ analyzing power crosses zero. Present phase shift solutions[1] disagree by as much as ±1° for the value of θ_{zx}. We have recently completed an experiment at TRIUMF (E498) to measure θ_{zx} to an accuracy of ±0.25° at four neutron beam energies. In this energy range, current phase shift predictions for θ_{zx} range from 99° to 80°. In contrast, above 300 MeV, θ_{zx} decreases much more slowly, reaching 69.74° at 477 MeV as measured by Abegg et al.[2].

The TRIUMF measurements made use of the neutron beam facility and specialized detectors that were assembled for the high precision measurements of charge symmetry breaking in the n-p system[3,4,5], see Fig. 1. Measurements of θ_{zx} were carried out at neutron beam energies of 174.3, 202.4, 216.5 and 260.3 MeV (all ∼ ±300 keV). Cuts on the opening angle, coplanarity, and x momentum balance of the outgoing n-p pair, as well as a cut on the difference between the energies of the incident neutron and the outgoing particles, determined from time-of-flight, serve to identify n-p elastic scattering events, see Fig. 2. These cuts are optimized to reduce the background contamination from $C+n \to p+n+X$ reactions to $\sim$0.2-0.3%, as

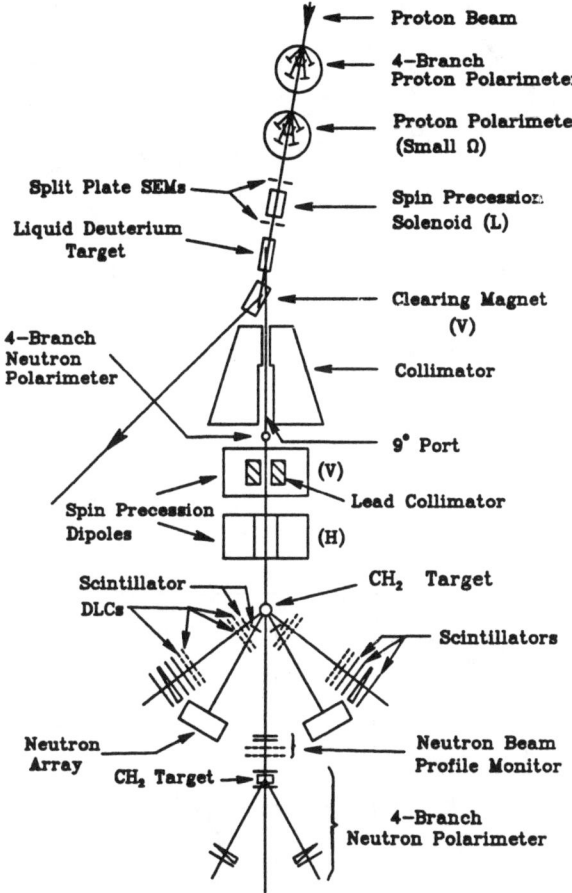

FIGURE 1. Layout of the beamline and experimental apparatus.

determined from background measurements performed with carbon targets. The background-corrected n-p elastic data are then used to determine the scattering asymmetry (analyzing power × beam polarization) as a function of scattering angle, see Fig. 3. The shape of the scattering asymmetry is first fit to the phase shift predictions[1] using a simple cubic expression:

$$\epsilon(\theta) = s[(\theta - \theta_{zx}) + a(\theta - \theta_{zx})^2 + b(\theta - \theta_{zx})^3]$$

with a and b fixed to the values determined from SAID, the curve is fit to the data allowing s and θ_{zx} to vary. Such a fit is shown in Fig. 3.

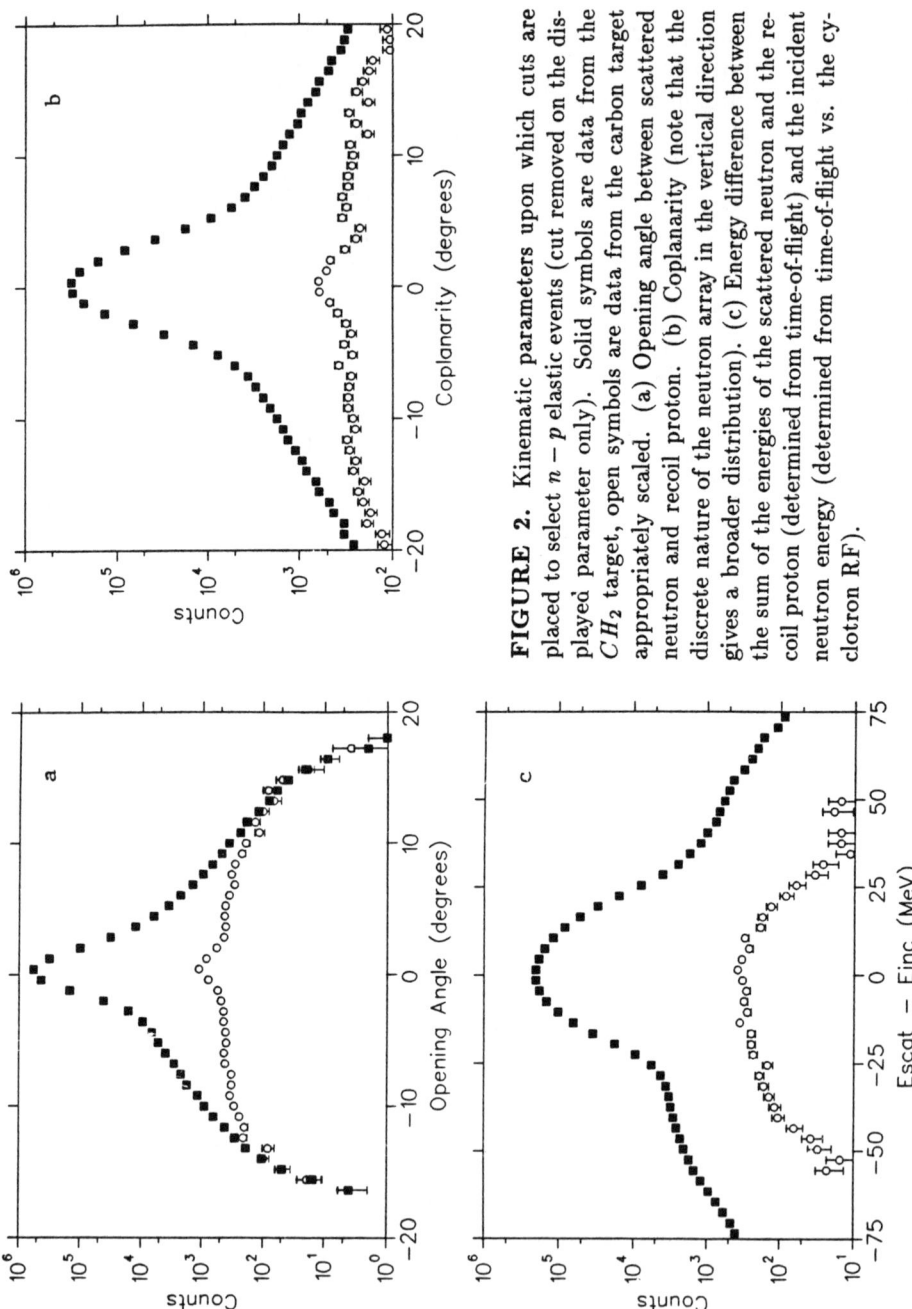

FIGURE 2. Kinematic parameters upon which cuts are placed to select $n-p$ elastic events (cut removed on the displayed parameter only). Solid symbols are data from the CH_2 target, open symbols are data from the carbon target appropriately scaled. (a) Opening angle between scattered neutron and recoil proton. (b) Coplanarity (note that the discrete nature of the neutron array in the vertical direction gives a broader distribution). (c) Energy difference between the sum of the energies of the scattered neutron and the recoil proton (determined from time-of-flight) and the incident neutron energy (determined from time-of-flight vs. the cyclotron RF).

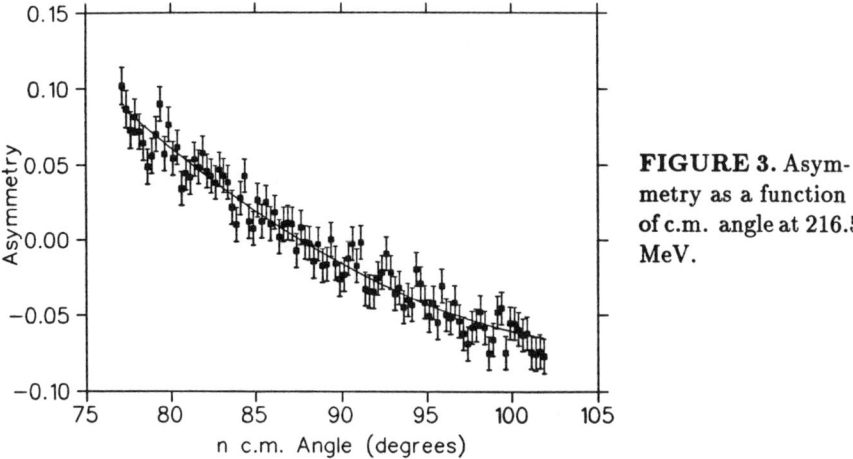

FIGURE 3. Asymmetry as a function of c.m. angle at 216.5 MeV.

The beam energy calibrations are being carefully studied, as θ_{zx} varies strongly with neutron beam energy in this energy range, with a slope varying between $-0.35°MeV^{-1}$ (174.3 MeV) to $-0.13°MeV^{-1}$ (260.3 MeV). The proton energy was monitored with a range counter telescope (BEM) [3] which was calibrated, at the highest energy by comparison to $np \to d\pi^0$ near threshold [6] kinematics, and at the other energies by comparing proton elastic

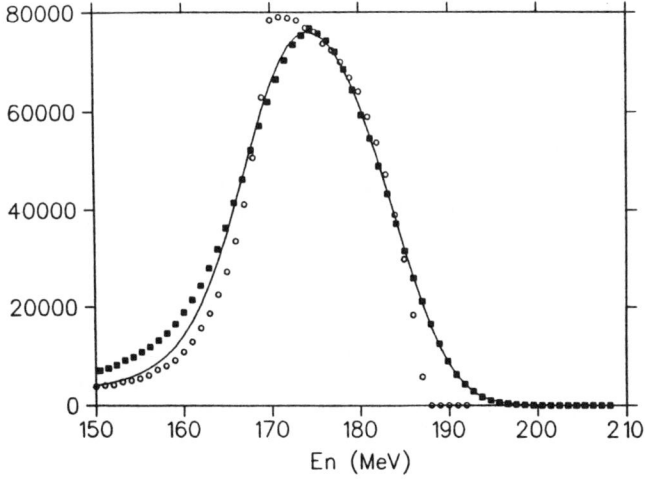

FIGURE 4. Neutron beam energy profile for the lowest energy. The open symbols are the Monte Carlo prediction. The solid line is this prediction corrected for cross section, detector acceptance and timing resolution. The solid symbols are the data.

scattering from carbon peak positions using the TRIUMF Medium Resolution Spectrometer at the various energies to the calibrated (highest) energy. A Monte Carlo prediction for the neutron beam energy, taking into account the kinematics of the $d(\vec{p},\vec{n})pp$ production reaction [7], energy losses in the deuterium production target, as well as the geometry of the 3.4 m long collimator immediately downstream of the liquid deuterium target, is shown in Fig. 4 for 174.3 MeV. This predicted beam energy distribution, corrected for detector acceptance and efficiency and convoluted with a Gaussian detector response function, is shown compared to the data in Fig 4. The Monte Carlo can also be used to predict and account for variations of average energy and polarization within the neutron beam profile.

Additional Monte Carlo studies are currently in progress to evaluate the systematic error contribution due to multiple scattering of the scattered proton. These will be used to correct the final results, accounting for the dependence of θ_{zx} on cuts in opening angle between the recoil proton and scattered neutron used to define elastic scattering events.

The results shown in Fig. 5 are preliminary, as effects of systematic errors, including multiple scattering of the recoil proton, corrections to the energy calibration, and even Charge Symmetry Breaking, continue to be studied. As θ_{zx} changes rapidly over the energy range, Fig. 5 (b) shows the *difference* between the data and the SAID [1] SP94 fit. The results are particularly sensitive to the $^3S_1$, ϵ_1, $^3D_1$, and $^3D_3$ phase shifts, and seem to indicate that the curvature of θ_{zx} is not as pronounced as the present phase shift fit of Arndt et al. would indicate.

[1] R.A. Arndt et al., Interactive Dial-In Program SAID. C. Leluc-Lechanoine, private communication.
[2] R. Abegg et al., *Phys. Rev. C* **40**, 2406-2409 (1989).
[3] R. Abegg et al., *Nucl. Instr. & Meth.* **A234**, 11-19 (1985); ibid, 20-29.
[4] R. Abegg et al., *Phys. Rev. D* **39**, 2464-2483 (1989).
[5] R. Abegg et al., *Nucl. Instr. & Meth.* **B79**, 318-321 (1993).
[6] D.A. Hutcheon et al., *Nucl. Phys.* **A535**, 618-636 (1991).
[7] D.V. Bugg and C. Wilkin, *Nucl. Phys.* **A467**, 575-620 (1987).
[8] J. Sowinski et al., "The Normalization of N-N Polarization Observables Near 180 MeV", in *IUCF Sci. and Tech. Report 1989-90*, 9-11 (1990).

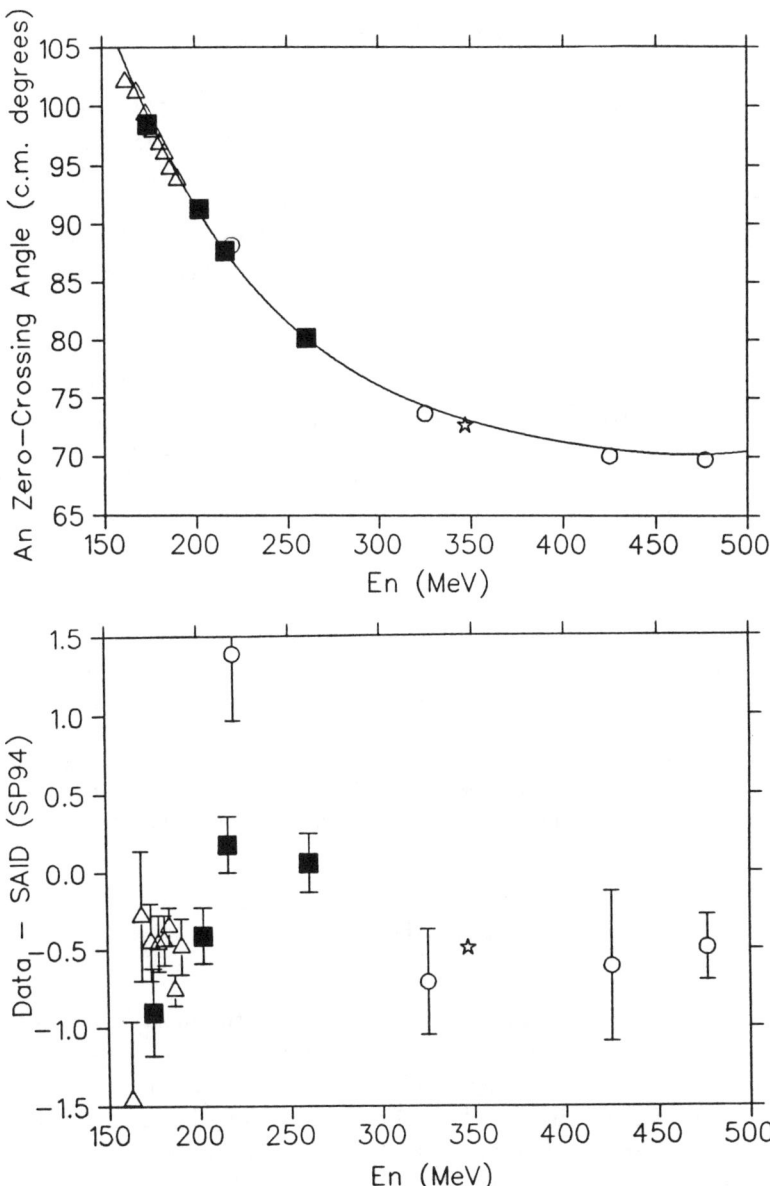

FIGURE 5. (Top) The present data (squares), data from IUCF [8] (triangles), data from Ref. [2] (circles), and preliminary data from the most recent TRIUMF CSB experiment [5] (also see contribution to the Symmetry session of this Conference) are compared to the SAID prediction of Ref. [1]. Note that all but the present data have an energy uncertainty of typically 2 MeV (compared to the present ~300 keV). (Bottom) The same data with the SAID (SP94) predictions subtracted.

Spin Observables in Neutron-Proton Elastic Scattering

A. Ahmidouch[a], J. Arnold[b], B. van den Brandt[e], M. Daum[e],
Ph. Demierre[a], R. Drevenak[c], M. Finger[cd], M. Finger Jr.[c],
J. Franz[b], N. Goujon[a], P. Hautle[e], Z. Janout Jr.[a], W. Hajdas[e], E. Heer[a],
R. Hess[a], R. Koger[b], J.A. Konter[e], H. Lacker[b], C. Lechanoine-LeLuc[a],
F. Lehar[f], S. Mango[e], Ch. Mascarini[a], D. Rapin[a], E. Rössle[b],
P.A. Schmelzbach[e], H. Schmitt[b], P. Sereni[b], M. Slunecka[c], R. Stachetzki[b],
A. Teglia[ae] and B. Vuaridel[a]

[a] *DPNC Université de Genève, CH-1211 Genève 4, Switzerland*
[b] *Fakultät für Physik der Universität Freiburg, D-79104 Freiburg, Germany*
[c] *Joint Institute for Nuclear Research, Dubna, Russia*
[d] *Faculty of Math. and Phys., Charles University, Prague, Czech Republic*
[e] *PSI, Paul Scherrer Institut, CH-5234 Villingen, Switzerland*
[f] *DAPNIA, CEN-Saclay, F91191 Gif-sur-Yvette, France*

Abstract. We describe here two experiments presently running at PSI using the NA2 polarized neutron beam. They are devoted to the measurement of 2- and 3-spin observables in np elastic scattering for kinetic energies from 230 to 590 MeV with a center of mass angular range from 60 to 180 degrees. The goal is to determine the five NN scattering amplitudes for isospin 0 in a model independent way. Preliminary results for K_{OSKO} and K_{OSSO} spin-transfers are presented.

Introduction

The nucleon-nucleon elastic scattering is one of the most basic reactions involving the hadronic interaction. It can be described for each isospin state, $I=0$ and $I=1$, by five scattering amplitudes which are complex functions of the angle and of the energy. A lack of data exists for the np channel below 600 MeV. Therefore the $I=0$ phase shifts remain unprecise. The aim of two new experiments installed on the new polarized neutron beam line at PSI is to study the spin dependence in np elastic scattering between 260 and 560 MeV.

In particular, it is planned to measure at least eleven different spin observables in a large angular range ($60° < \theta_{cm} < 160°$) to complete the database and to perform a direct reconstruction of the $I=0$ amplitudes at different energies. This large number of observables is required to remove possible ambiguities. Furthermore, this overdetermination is of great importance to limit the systematic uncertainties.

NA2 Beam line

The intense polarized neutron beam of the PSI is created by the reaction $C(\vec{p},\vec{n})X$ at $0°$ using a high intensity polarized proton beam ($10\mu A$). It has unique features: continuous energy up to about 600 MeV, polarization between 15% at 260 MeV and 40% above 450 MeV, intensity $5 \times 10^6 \vec{n} s^{-1} cm^{-2}$ at 13.7m from the production target, neutron energies measured by time-of-flight using the rf-signal from the accelerator, neutron beam polarization reversed every second with possible orientation in any direction.

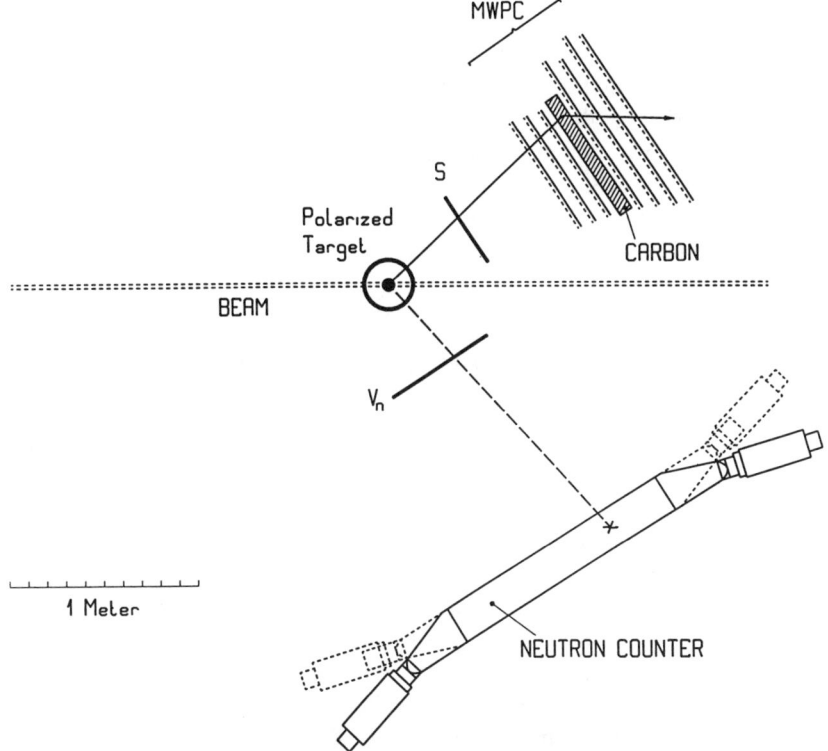

FIGURE 1. Schematic layout of Experiment I (top view).

Experiment I

The first experiment is devoted to the measurement of 2- and 3-spin observables of the $\vec{n}\vec{p}$ elastic scattering using a polarized proton target located at 13.7m from the neutron production target. The experimental layout is shown in fig.1. Both the scattered neutron and the recoil proton are detected. Moreover the recoil proton polarization is measured.

The 100 cm^3 polarized target works in frozen spin mode at a temperature of about 50mK. A relaxation time $\geq$ 1000 hrs was achieved with polarization values between 60 to 80%. A holding field of 0.8 T is generated by superconducting coils located inside the cryostat. It allows the polarization vector to be vertical or horizontal.

The scattered neutrons are detected by a hodoscope made of 11 plastic scintillator bars of $8 \times 20 \times 130$ cm^3. Each bar is viewed by two XP2040 photomultipliers. The horizontal coordinate of the neutron is given by the time difference between the two photomultipliers.

A polarimeter is used to detect the recoil protons and to measure their transverse polarization by rescattering on a carbon target. The proton's trajectories before and after the second scattering are detected by multi-wire proportional chambers (MWPC) with 2 mm wire spacing. A hardwired second level trigger, associated with the chambers, rejects, within 2 μs, the events with a rescattering angle $\theta_c \leq 3°$.

Both the neutron detector and polarimeter are installed on movable platforms allowing to cover a large angular range. Two angular set-ups are used for the measurements. They cover 60° to 120° (respectively 100° to 160°) in the center of mass.

Experiment II

The second experiment is shown in fig. 2. It is located at 24 m from the neutron production target and is devoted mainly to the measurement of 1- and 2-spin parameters A_{OONO}, K_{ONNO}, K_{OKSO} and K_{OKKO} of the $\vec{n}p$ elastic scattering in the angular range between 130° and 180°, in the center of mass, using a liquid hydrogen target. The magnetic spectrometer analyses the momentum of the recoiling protons and the polarimeter measures the transverse components of the spin. This experimental configuration allows to analyze longitudinal components of the spin as well, since the horizontal components $\hat{s}$ and $\hat{k}$ are rotated around the vertical field of the magnet. Both the spectrometer and the polarimeter are equipped with drift chambers. The scintillation counters *Start* and *Stop* are used for triggering and for time of flight (TOF) measurement. Thanks to the TOF and momentum analysis, a very good identification is obtained for the recoil particle. This allows to measure also the $\vec{n}p \rightarrow \pi^\circ d$ channel.

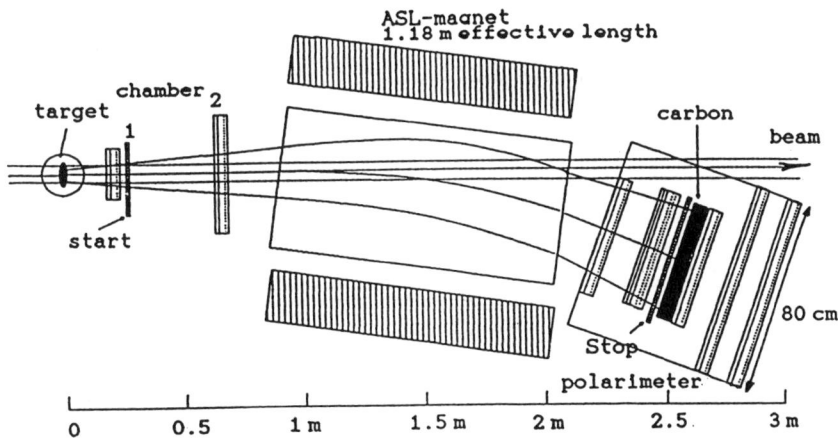

FIGURE 2. Layout of Experiment II.

Preliminary Results

Data taking with the full set-up for both experiments has started in 1993 and is still in progress. For experiment I, the energy-angle domain is divided into 5×15 bins and data are taken in order to achieve the following statistical accuracy:
±0.03 for the spin correlation A_{OONN}, A_{OOSK} and A_{OOKK}.
±0.04 for the depolarization D_{ONON} and D_{OSOK}.
±0.04 to 0.08 for the spin transfers K_{ONNO}, K_{OSSO} and K_{OSKO}.
±0.08 to 0.15 for the 3-spin parameters N_{OSSN} and N_{OSKN}.

Some preliminary results for K_{OSSO} and K_{OSKO} are shown in fig. 3 and 4. The other available data from TRIUMF (1) and from LAMPF (2) are also shown as well as phase shift predictions of Arndt (3) (dashed dotted line) and Saclay-Geneva (4) (dashed line).

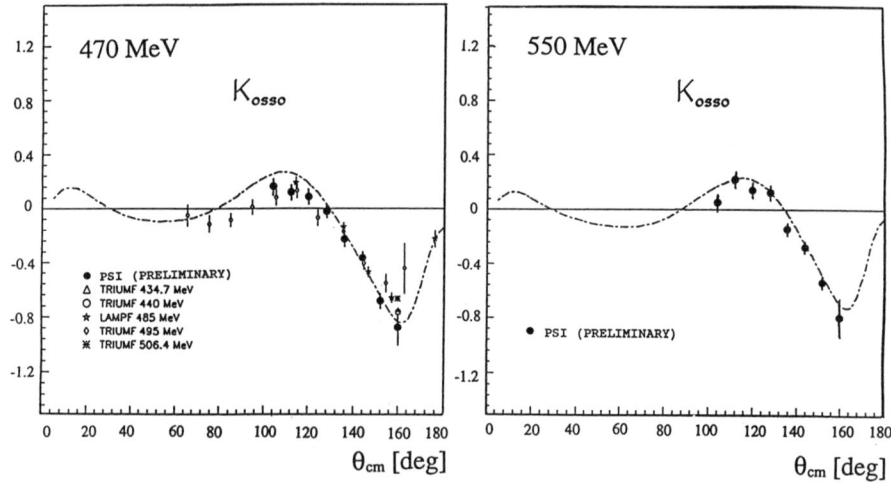

FIGURE 3. Preliminary results for K_{OSSO} spin transfer. plotted as a function of the cm angle for two energies. The curves are predictions of Arndt PSA.

This work is funded by the *Fonds National Suisse de la Recherche Scientifique* and by the German *Bundesminister für Forschung und Technologie* (BMFT) under the contract No 06 FR 652. Some of us also thank PSI and JINR Dubna for travel support.

REFERENCES

1. D.Axen et al., *Phys. Rev.* **C21**, 998 (1980).
2. M. MacNaughton et al., *Phys. Rev.* **C46**, 47 (1992).
3. SAID code, solution SM93.
4. J. Bystricky et al., *private communication*, 1993 solution.

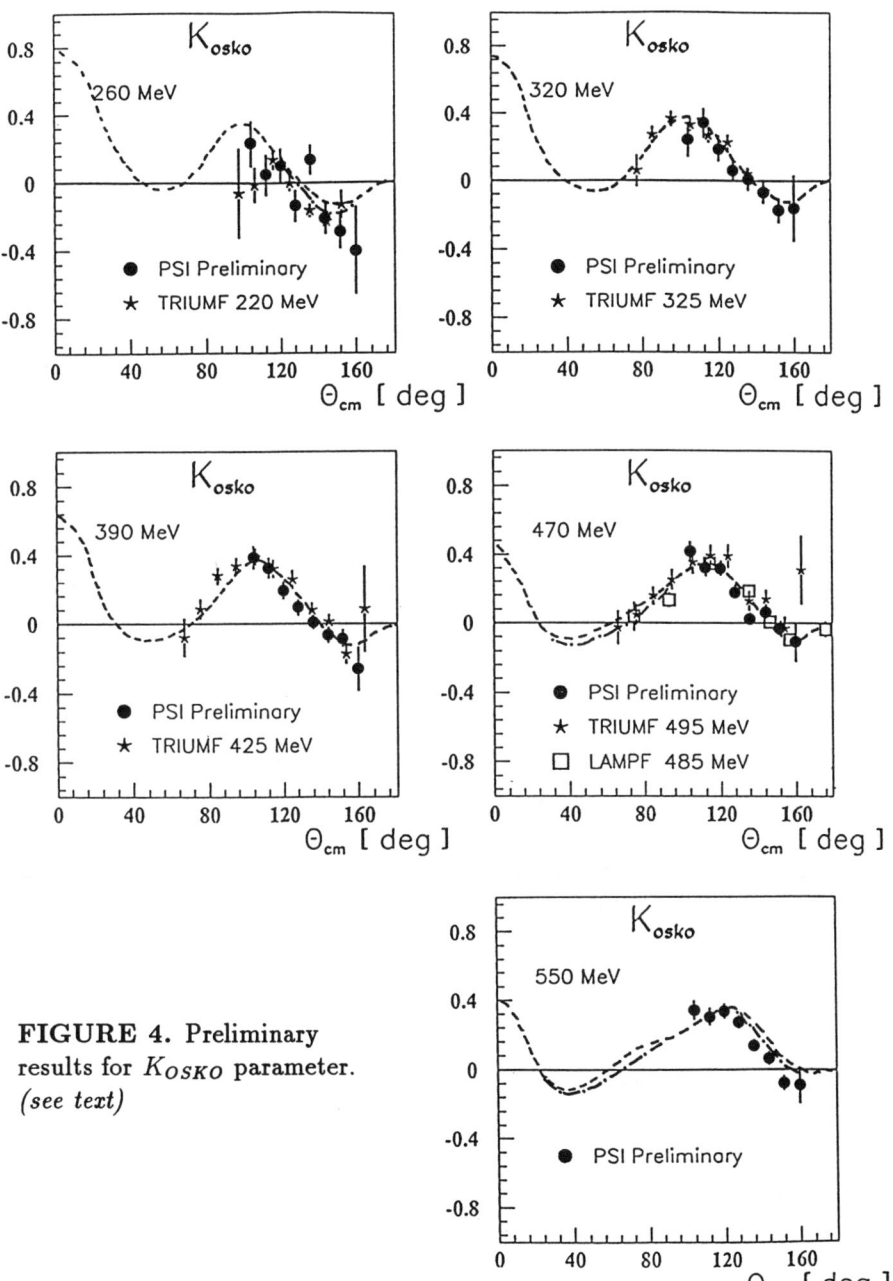

FIGURE 4. Preliminary results for K_{OSKO} parameter. *(see text)*

$\Sigma^+ + p$ scattering experiment at KEK

J.K.Ahn[a], B.Bassalleck[b], M.S.Chung[c], W.M.Chung[d], H.Enyo[c],
T.Fukuda[e], H.Funahashi[c], Y.Goto[c], A.Higashi[e], M.Ieiri[f], M.Iinuma[c],
K.Imai[c], Y.Itow[c], G.D.Kim[g], Y.D.Kim[f], J.M.Lee[f], A.Masaike[c],
Y.Matsuda[c], S.Mihara[c], K.Okada[h], I.S.Park[a], Y.M.Park[i], N.Saito[c],
Y.M.Shin[j], K.S.Sim[a], R.Susukita[c], R.Takashima[k], F.Takeuchi[h], P.
Tlustý[l], S.Yamashita[c], S.Yokkaichi[c], M.Yoshida[c]

[a] $Korea Univ., Seoul, Korea,$ [b] $Univ. of New Mexico, U.S.A.$
[c] $Kyoto Univ., Kyoto, Japan$ [d] $Yonsei Univ., Seoul, Korea$
[e] $INS, Univ. of Tokyo, Japan$ [f] $KEK, Tsukuba 305, Japan$
[g] $Pusan Univ., Pusan, Korea$ [h] $Kyoto Sangyo Univ., Kyoto, Japan$
[i] $Kyungsung Univ., Pusan, Korea$ [j] $Univ. of Saskatchewan, Canada$
[k] $Kyoto Univ. of Education, Kyoto, Japan$
[l] $Nuclear Physics Institute, Czech Republic$

Abstract The new low energy hyperon-nucleon scattering data have been long waited since the low statistics data were obtained in 1960s from the bubble chamber experiments. At KEK, an experiment to measure $\Sigma^+ + p$ elastic scattering has been performed utilizing the novel scintillating fiber technology. In this report we present the status of the analysis on this experiment and the preliminary total cross section data of $\Sigma^+ + p$ scattering.

1 Introduction

The baryon-baryon interaction at low and intermediate energy has been studied mainly for nucleon-nucleon interactions. On the contrary, our present understanding on the hyperon-nucleon strong interactions are relatively poor due to the insufficient experimental data as summarized in reference [1]. Theoretical study using one boson exchange(OBE) model with built-in SU(3) symmetry coupling constants[2] made comparisons with available data. Both Nijmegen and Jülich model agrees well with the total cross section data. QCD based calculations including medium and long range interaction with one boson exchange also agrees well with the total cross section data [3]. But they largely disagree in the prediction of spin observables like polarization and spin transfer parameters for which almost no data is available. It is clear that one should pursue to obtain the spin dependence of the interaction to test various models. This should shed some lights on the short range part of the baryon-baryon interaction and SU(3) symmetry in coupling constants.

Until present the shortage of the free space scattering data is mainly due to the difficulty to provide the low energy hyperon beam and detect the scattering events due to the short track length(a few cm) of the particles involved. Since the scattering events are quite rare (order of 0.1 %), obtaining higher statistics is also challenging.

2 $\Sigma^+ + $ p scattering experiment

The main purpose of the hyperon scattering experiment performed at KEK, was to obtain Σ^+, $\Lambda+$ p elastic scattering data in higher momentum region up to 800 MeV/c. The detector system consisted of the scintillating fiber live track detector, and chambers and magnets for beam particles(π^+) and outgoing particles(k^+). Scintillating fiber detector has good spatial resolution and wide coverage of the solid angle(intrinsically 4 π).

The experiment was performed with π^+ beam at 1.6 GeV/c momentum incident to the scintillating fiber target detector. The outgoing K^+s were tagged by the scintillator hodoscope coincidence making the rough cut in the mass of the outgoing particle. This second level trigger reduces the trigger rate down to 10-20 per spill appropriate to the speed of the CCD readout, and the fraction of K^+ was about 50 % of all trigger. The production vertex, scattering vertex, and decay points could be identified by the track images inside the scintillating fiber detector and used to fully reconstruct the events. For details of the geometry and readout scheme of the detector, see the reference [5, 6].

3 Analysis and result

The chambers before and after the target were analyzed to give the missing mass distribution which is shown in the Figure 1. The ground state of Σ and $\Sigma(1385)$ can be identified as broadened by the contribution from carbon nuclei.

The rms resolution in the missing mass for the production from the proton target is estimated to be about 12 MeV. The contamination in (π^+, K^+) events by other particles rather than kaons is less than 0.5 %.

The track residual from the fitted straight line for π^+ tracks inside the scintillating fiber block indicates that the overall position resolution of the detector is about 290 μm. Figure 2 shows the distribution of the track length of Σ^+ particles. The solid line is the distribution of the events identified as Σ^+ by eye-scanning, and the dashed line is that of Monte Carlo simulation.

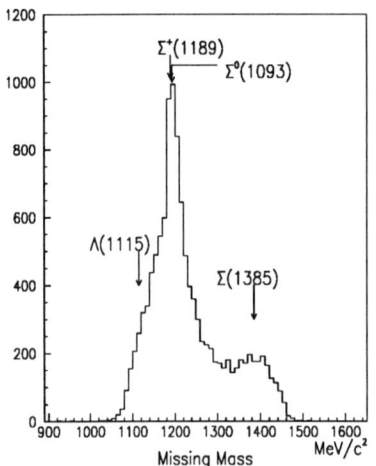

Figure 1: Missing mass distribution from (π^+, K^+) momentum analysis

Figure 2: Σ^+ flight length distribution.

The minimum track length identifiable (about 5mm) is mainly due to the fiber size, hit density, and the large size of the cluster from Σ^+ which is not the minimum ionizing particle. For more detailed analysis of the scintillating fiber detector tracking, see the reference [7].

3.1 Event selection

The total number of (π^+, K^+) events we collected was about 1.4×10^6. 54% of these, about 7.4×10^6 satisfies the missing mass cut (1150-1250 MeV/c^2 for Σ^+ from proton target). The global pattern recognition using the concept of the so called *Hough transformation* [4, 7] has been applied to further reduce the number of event to analyze. For Σ^+ production from proton target events, first the existence of 4 tracks(π^+, K^+, Σ^+, and decay track) is required. The efficiency of this automatic selection was about 80 % compared to the eye-scanning. Secondly, coplanarity of less than 0.25 radian is required for both in-plane and out-of-plane angles from the expected two-body kinematics to select Σ^+ production from proton target only.

To help more quantitative analysis, Monte Carlo simulation has been performed to generate the event image itself. The simulation follows the detailed detection system ; the geometry of our target, scintillating fiber light output, quantum efficiency of 1st stage IIT, 1st stage IIT position resolution, cluster formation and amplification in 2nd and 3rd stage of IIT, and CCD digitization. The measured values for cluster size and position resolution has been used for the input parameters in the simulation. Possible hadronic interactions are also simulated with the help of GEANT code. The eye-scanning of the simulated Σ^+ + p events shows that the eye-scanning efficiency is about 50 % overall momentum region in our experiment.

3.2 total cross section

The events thus selected were eye-scanned to identify Σ^+ + p scattering pattern requiring recoil proton track longer than 4mm. Same conditions are applied to the simulated events and eye-scanned at the same time to estimate the eye-scanning efficiency. The momentum of recoil proton and scattering angle were obtained, and kinematics were checked at the scattering vertex. 18 scattering events were identified from the eye-scanning of part of the data, and the preliminary total cross section is presented in Table 1 with the old data from Eisele et al. [8]

The preliminary data suffers from the statistics mainly due to the minimal track length identifiable which is about 5mm and the number of Σ^+ produced. To obtain the polarization, we need at least several hundred scattering events, and we are continuing the R&D to increase the statistics.

4 Summary

We have performed a new experiment aiming to measure differential cross section and polarization of Σ^+ + proton scattering. About 1.6×10^5 identifiable Σ^+ events are collected. The analysis using automatic pattern recognition was developed to select Σ^+ production events. From the preliminary analysis we

Table 1: The preliminary total cross section with part of the data analyzed

	Momentum (MeV/c)	No. of events	cross section
Eisele et.al	140-150	4	123 ± 62
ref. [8]	150-160	13	104 ± 30
	160-170	35	92 ± 18
	170-180	69	81 ± 12
This work	200-400	5	148 ± 66
	400-600	11	64 ± 20
	600-800	2	46 ± 33

obtained total cross section with 18 identified scattering events. The number of events are too small to obtain the polarization observables at present. Though further analysis will be done to increase the statistics, we are currently investigate the hardware improvement for the next beam time.

One of the authors(Y.D. Kim) acknowleges the support of Japan Society of Promotion of Sciences(JSPS).

References

[1] C.B. Dover and H. Feshbach, Ann. Phys, 198, 32 (1990)
[2] M.M. Nagels et al., Phys. Rev. D15 (1977) 2547; D20(1979) 1633; Maessen et al., Phys. Rev. C40 (1989) 2226; K. Holinde, Nucl. Phys. A547 (1992) 255c
[3] U. Straub et al., Nucl. Phys. A483 (1998) 686
[4] R.O. Duda et al., Comm. ACM 15, 1 (1972) 11
[5] J.K. Ahn et.al., "$\Sigma^+ + p$ scattering experiment", Conference proceedings of "Frontiers of High energy spin physics", 1993
[6] M. Ieiri in "Properties and interactions of hyperons" edited by Gibson et al., p17, World Scientific (1994)
[7] Y. Goto ibid p143
[8] F. Eisele et al., Phys. Lett. 37B, 204 (1971)

Cross Sections and Analyzing Powers of $p(\vec{p}, \pi^+)d$ Very Near Threshold

P. Heimberg, R.E. Segel, F-J. Chen
Northwestern University, Evanston, Illinois 60201
K. Ackerstaff, R.D. Bent, J. Blomgren, H.O. Meyer
H. Nann, B.v. Przewoski, T. Rinckel, A. Zhuralev
Indiana University Cyclotron Facility, Bloomington, Indiana 47408
M.A. Pickar
University of Kentucky, Lexington, Kentucky 40506
G. Hardie, P. Pancella
Western Michigan University, Kalamazoo, Michigan 49008
E. Jacobsen
Princeton University, Princeton, New Jersey 07544
J.D. Brown
Yale University, New Haven, Connecticut 06511

The s-wave strength of the pp→ dπ^+ reaction in the threshold region is a direct measurement of the axial exchange charge of the two nucleon system [1]. Far from the Δ resonance the reaction is dominated by non-resonant pion production mechanisms such as pion brehmsstrahlung, s-wave rescattering, and heavy meson exchange currents [2-5]. Additionally, the reactions pp→ dπ^+_{s-wave}, s-wave $\pi^\pm p \to \pi^\pm p$, and $\gamma p \to \pi^+ n$ are all related by symmetries of the strong interaction [6,7]. Because the Δ resonance is so strong, p-waves are prominent down to well within 1 MeV of threshold. The analyzing power, which is a result of interference between the s- and p-wave amplitudes, gives a good measure of the p-wave contribution to the cross section. From the measurements of σ_{tot}, $\frac{d\sigma}{d\Omega}$, and A_y sufficiently close to threshold and from Watson's theorem the s-wave amplitude can be determined.

The IUCF Cooler, with its extremely well-defined beam energy, very thin internal targets and high quality polarized beam, was an ideal facility for carrying out such measurements. The π^+ were detected and up to an energy well above the region of interest they are confined to a cone about the beam direction. In the very near threshold region the opening angle is approximately proportional to η ($= \frac{p^{cm}_\pi}{m_\pi c}$) and reaches a value of 11.5° at $\eta = 0.08$,

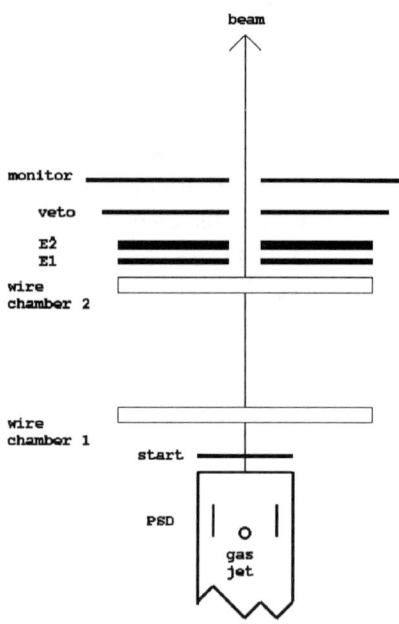

Figure 1: Experimental Layout - Top View

which is 1 MeV (lab), or 0.5 MeV (cm), above threshold. In fact, measuring the opening angle of the pions gives the best calibration (±70 keV) of the beam energy.

The measurement was carried out in the G-region of the IUCF Cooler ring using a 70% polarized proton beam and a hydrogen gas jet target. The detector stack (figure 1) contained left-right symmetric plastic scintillators which measured dE/dx and E of the forward going particles. Any particles that punched though the ΔE and E counters were vetoed by another set of plastic scintillators. Particle directions were determined by two double-planed wire chambers each of which gave both vertical and horizontal positions. The luminosity and beam polarization were measured by detecting the protons from pp elastic scattering. Elastic events required a coincidence between the forward "monitor" scintillator and a position sensitive silicon detector (PSD) near 90°.

The system covered an angular range of 4 to 19 degrees. Figure 2 shows the accepted portion of the kinematic loci for various values of η. The maximum angle corresponds approximately to 90° in the center of mass. At these energies the deuterons coming from pp$\to$ dπ^+ have a maximum lab angle $\simeq$ 1 degree and therefore did not enter the detector stack. Particles were

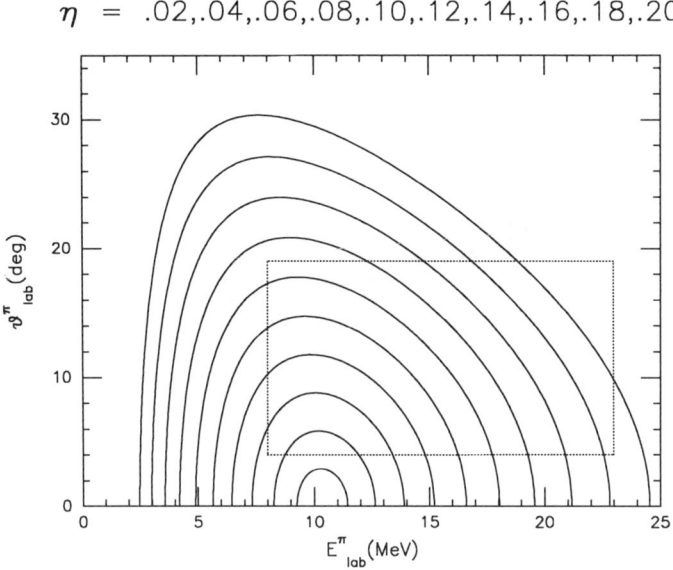

Figure 2: Kinematic Loci in θ vs E for η between 0.02 and 0.20. The dotted line marks the detector acceptance.

identified primarily by their ΔE vs E and time-of-flight vs E signatures, although $\pi \to \mu$ pulse shape discrimination was also used. Figure 3 shows pions (encircled) cleanly distinguished from low energy and reaction tail protons.

Cross sections and analyzing powers were measured at 9 energies between η =0.03 and 0.20. The lowest energy corresponds to $T_\pi^{CM} \sim 70$ keV; the lowest previous energy at which analyzing powers were measured is more than 1 MeV above threshold [8]. The energy range that could be covered was limited by the angular acceptance of the detector stack. Even at the lowest energy there were about 12 events/min and background varied between 3% and 15%.

Corrections for the $\pi-d$ Coulomb interaction were made using a deuteron wave function calculated from the Ried Soft Core potential. Because the pion wave function is nearly constant over the extent of the deuteron, this correction is, to a good approximation, model-independent. After these corrections are made, the cross section is extrapolated to be $(210 \pm 23_{sys.})\eta$ μb in the region immediately above threshold, (figure 5, figure 6) which is consistent with that reported for the np$\to$ dπ^0 reaction [9] (after accounting for isospin rotation).

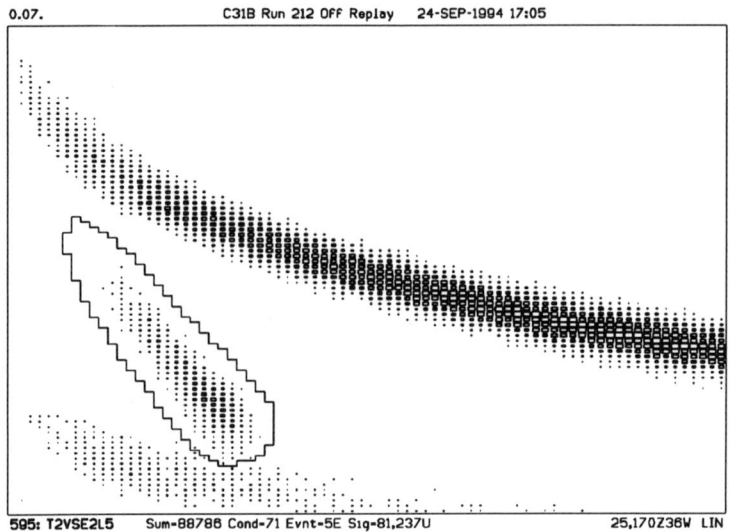

Figure 3: Time-of-flight between the "start" and "E2" scintillators. $\eta = 0.07$.

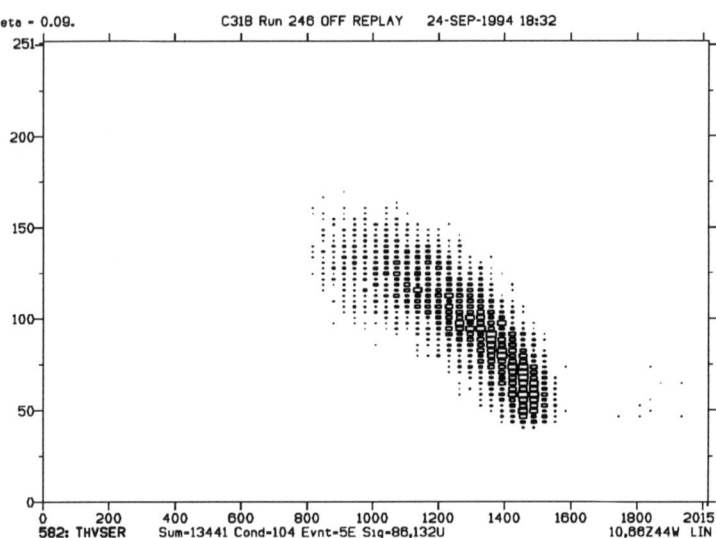

Figure 4: θ_{lab} vs E_{lab} for pions of $\eta = 0.094$.

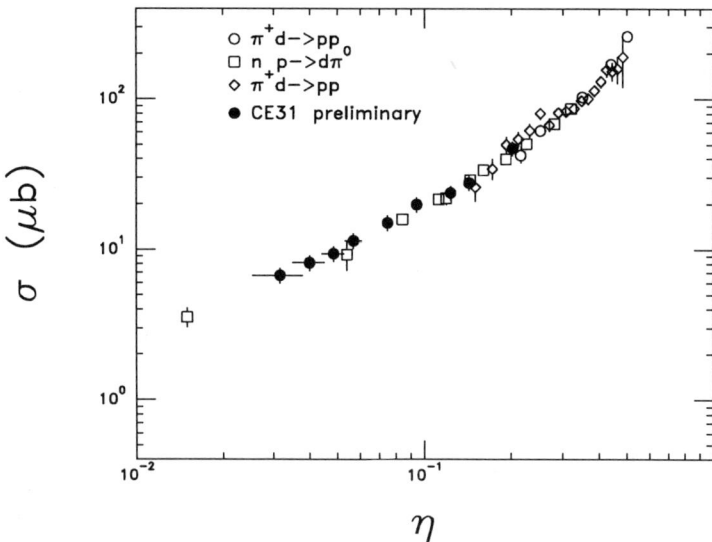

Figure 5: Total cross section for pp → dπ⁺ vs η. Solid circles are preliminary CE31 results.

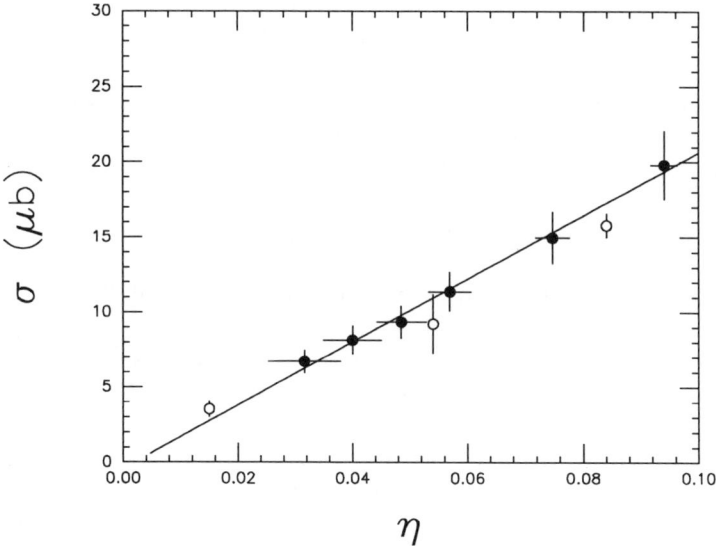

Figure 6: Total cross section in the very near threshold region. Solid circles are CE31. Open circles are for np → d π^0 [ref. 9].

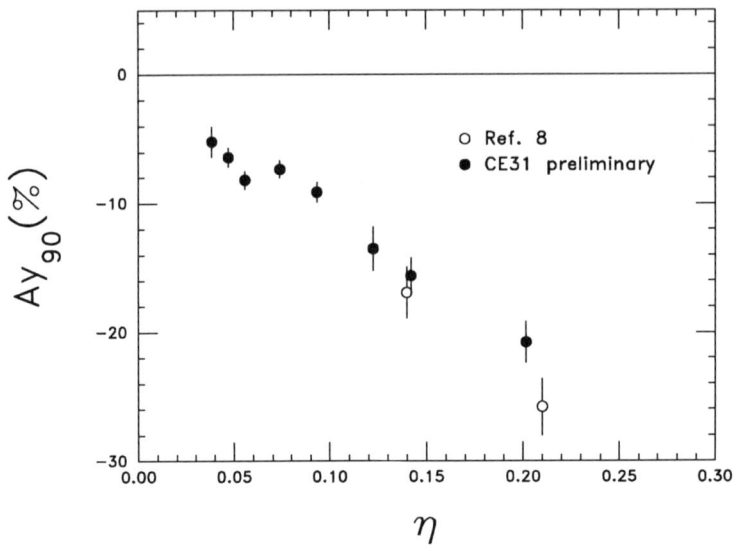

Figure 7: Analyzing Power at 90 deg in the center-of-mass vs η

The analyzing power follows a $\sin\theta_{cm}$ distribution, as expected, with a 90° value of approximately -1.2η. Even at the lowest point the analyzing power differs significantly from zero (figure 7).

These preliminary results are consistent with the calculated factor of two enhancement of the s-wave cross section from heavy meson exchange currents [2] and hopefully will prompt a calculation of the analyzing power within this formalism.

This work was supported by the National Science Foundation.

1. E. M. Nyman and D.O. Riska, Phys. Lett. B **215**, 1 (1988)29.
2. C.J. Horowitz, Phys. Rev. C, **48**, 6 (1993)2920.
3. T-S.H. Lee and D.O. Riska, Phys. Rev. Lett., **70**, 15 (1993)2237.
4. C.J. Horowitz, H.O. Meyer, David K. Griegel, Phys. Rev. C, **49**, 3 (1994)1337.
5. D.S. Koltun and A. Reitan, Phys. Rev., **141**, 4 (1966)1413.
6. C.M. Rose,Jr., Phys. Rev., **154**, 5 (1967)1305.
7. J.S. Spuller and D.F Measday, Phys. Rev. D, **12**, 11 (1975)3550.
8. E. Korkmaz, et. al., Nucl. Phys., **A535** (1993)637.
9. D.A. Hutcheon, et. al., Nucl. Phys., **A535** (1993)618.

Polarization Observables for the p-d Breakup Reaction and the Nuclear Three-Body Force

L. D. Knutson

University of Wisconsin, Madison, Wisconsin 53706

Abstract. It is shown that nuclear three-body potentials involve spin operators of a type not allowed for ordinary two-body NN interactions. The effect of these operators has been studied for proton induced deuteron breakup, and it is suggested that certain polarization observables for breakup reactions with non-co-planar geometry may have an enhanced sensitivity to the three-body forces. Preliminary results from an experiment to measure the longitudinal analyzing power in non-co-planar p-d breakup are described.

NUCLEAR THREE-BODY FORCES

It is generally believed that three-body forces play a small but possibly significant role in nuclei and complex nuclear systems. The evidence for existence of nuclear three-body forces is primarily circumstantial in nature. For example, it is well established (1) that most realistic NN potentials predict triton binding energies which are well below the measured value of 8.48 MeV, typically by around 1 MeV. Reasonable three-body potentials (derived from theoretical models) are well capable of producing the additional needed binding energy (2).

While there is some experimental evidence to support the existence of three-body forces, most of what we know about the details of these potentials comes from theoretical models. It is understood that many-body potentials must be present at some level in nuclear systems since nucleons are composite particles. In meson exchange models [see for example Refs. (3,4)] the internal structure can give rise to three-nucleon (3N) potentials through the coupling to excited states of the nucleon. A typical 3N force diagram might involve excitation of the delta resonance in a two-pion exchange process that involves all three nucleons. By employing models of this kind, it is possible to construct explicit 3N potentials, complete with spin and isospin dependence [see Refs. (3,4)].

Unfortunately, attempts to test these 3N potential models by comparing theoretical predictions against experimental observation have generally been inconclusive. Typically what one finds, for any given observable, is that the changes produced by the 3N potential are of roughly the same size as the differences between predictions obtained with different 2-body NN interactions. In other words, uncertainties in the NN interaction obscure the effects of the 3N force.

Our goal in the present work is to see whether it is possible to identify observables that have enhanced sensitivity to 3N forces. An important step in this process is to identify features of 3N potentials that are in some sense distinctive. For example, one might look for spin operators of a kind that are present in the 3N potential and not present in the ordinary NN potential.

To describe the three body system we adopt the traditional Jacobi coordinates

$$\mathbf{x} = \mathbf{r}_3 - \mathbf{r}_2 \tag{1}$$

and

$$\mathbf{y} = \frac{1}{2}(\mathbf{r}_2 + \mathbf{r}_3) - \mathbf{r}_1. \tag{2}$$

Since 3N forces depend on the coordinates of all three particles, the potential may contain operators that involve both $\mathbf{x}$ and $\mathbf{y}$ as well as the spins, σ_1, σ_2 and σ_3, of the three nucleons. For example, it is possible for the 3N potential to contain operators of the form

$$Q = (\sigma_1 \times \sigma_2) \cdot (\mathbf{x} \times \mathbf{y}). \tag{3}$$

This is an example of what we shall refer to as a "rank-1" interaction. To classify interactions according to this scheme, the operator in question is written as a contraction of spherical tensors,

$$Q = \sum_m (-)^m \Lambda_L^m R_L^{-m}, \tag{4}$$

where Λ_L involves the spin operators and R_L the spatial coordinates. An operator composed of spherical tensors of order L is then referred to as a rank-L interaction.

Since the components of a vector cross product transform under rotations as a spherical tensor of rank 1 we see that the operator of Eq. (3) is a rank-1 interaction. More explicitly, one can easily demonstrate that the operator R in this case is just proportional to the $B_{11,1}$ "bipolar harmonic",

$$B_{l_x l_y, L}^m(\hat{\mathbf{x}}, \hat{\mathbf{y}}) \equiv \sum_{\lambda_x \lambda_y} \langle l_x \lambda_x, l_y \lambda_y | Lm \rangle Y_{l_x}^{\lambda_x}(\hat{\mathbf{x}}) Y_{l_y}^{\lambda_y}(\hat{\mathbf{y}}). \tag{5}$$

In addition to the operator Q of Eq. (3), the two-pion exchange 3N potential involves a number of other complex space/spin operators which are at least partially of rank-1. The common feature of all rank-1 interactions is that they transform as vectors under rotations of the space coordinates. On the other hand, the 3N interactions conserve parity, and since the spin operators do not change sign under parity inversion, the space tensors must also be even under the parity operation. In other words, the spatial operators are all axial-vector in nature, which means that they can be expressed in terms of bipolar harmonics with $L = 1$ and with $l_x + l_y$ even.

It should be clear that operators of this class are not permitted for 2-body potentials, for the simple reason that one cannot construct a spatial axial vector from a single coordinate.

Let us now consider what effect rank-1 interactions might have on the wave function of a 3-nucleon system. For the simple case in which the particles occupy a state which is predominantly a positive-parity relative S-state (as for example in the triton), a rank-1 operator acting on the dominant L=0 component will generate a positive-parity P-state component. This suggests that experiments which probe the small P-wave components of the triton or ^{3}He wave functions might be particularly sensitive to three-body forces.

Turning now to the continuum states and in particular to the p-d breakup reaction, our goal is to identify observables that that will be sensitive to wave function components analogous to the positive-parity P-state components of the triton. For the breakup reaction the initial state of the system includes all possible values of L. However, the initial state is predominantly of natural parity, and it is expected that the rank-1 operators should mix in wave function components that have unnatural parity.

POLARIZATION OBSERVABLES FOR p-d BREAKUP

There are a great many polarization observables that could, in principle, be measured in p-d or n-d breakup. For the first order observables one has all of the usual proton and deuteron analyzing powers – A_y, iT_{11}, T_{20}, T_{21}, and T_{22}. In addition, if one considers the possibility of observing non-co-planar final states (configurations in which the final state momentum vectors do not all lie in a single plane) there are additional first order observables (5).

It is well known that parity conservation places restrictions on the number of non-zero analyzing powers for reactions with two-body final states. If one adopts the conventional coordinate frame in which $\hat{z}$ is along the incident beam direction ($\mathbf{k}_i$) and $\hat{y}$ is along $\mathbf{k}_i \times \mathbf{k}_f$ parity conservation requires that the reaction analyzing powers satisfy

$$T_{kq} = (-)^k T_{kq}^*. \qquad (6)$$

This result follows from a symmetry property of the helicity amplitudes that can be established [see for example Ref. (5)] by considering the effect of an operator, $Pe^{i\pi J_y}$, consisting of a rotation of 180° about the y-axis followed by a parity inversion. For a two-body final state, this operation returns all momentum vectors to their original values but reverses the helicities, leading to the desired symmetry.

For co-planar three-body final states the theorem of Eq. (6) still holds. However, for a non-co-planar final state it is easily seen that there is no choice of y-axis for which the operation $Pe^{i\pi J_y}$ returns all momenta to their original values. As a result, Eq. (6) is replaced by a relationship of the form (5)

$$T_{kq}(\mathbf{k}_i; \mathbf{k}_1, \mathbf{k}_2) = (-)^k T_{kq}^*(\mathbf{k}_i; \mathbf{k}_1', \mathbf{k}_2'). \qquad (7)$$

Here $\mathbf{k}_1$ and $\mathbf{k}_2$ represent the momenta of the two detected nucleons, while $\mathbf{k}_1'$ and $\mathbf{k}_2'$ are the reflections of $\mathbf{k}_1$ and $\mathbf{k}_2$ in the x-z plane.

We see that for non-co-planar reactions parity conservation gives rise to a relationship between the analyzing power for a particular geometry and the corresponding quantity for the reflected geometry. However, the new feature is that none of the analyzing powers are required to be zero. Thus for example, the analyzing power for longitudinally polarized protons (A_z or T_{10}) may in general be non-zero for non-co-planar breakup.

At this point one is led to speculate that there could be a connection between these new polarization observables and the rank-1 operators that are present in the nuclear 3-body potential. Recall that the rank-1 potentials are made possible by the existence of two independent coordinates in the 3N system, and that these potentials are characterized by having a distinctive (axial vector) behavior under rotations and reflections of the spatial coordinates. Similarly, we see that in a 3-body breakup state the presence of two independent momentum vectors makes possible the existence of new polarization observables which again have unusual behavior under combined rotations and reflections.

This connection has been investigated in detail for the longitudinal analyzing power. What we find is that in an angular momentum expansion, the leading contributions to A_z involve cross terms between the (presumably) dominant natural parity amplitudes and smaller unnatural parity amplitudes that are thought to be strongly influenced by the rank-1 interactions. Thus, although the connection is somewhat tenuous, there is some reason the expect that measurements of A_z may show an enhanced sensitivity to 3N forces.

EXPERIMENTAL RESULTS

With these ideas in mind we decided to undertake an experiment to measure A_z for proton induced deuteron breakup. Since this observable had never been measured previously, our goal was to carry out the experiment on a short time scale and to obtain results of moderate statistical precision. For the most part, the experiment was done using existing apparatus, detectors and electronics. The experiment was carried out at the University of Wisconsin Tandem Accelerator Laboratory using polarized protons at an energy of 9 MeV. For the target we used a chamber filled with deuterium gas at a pressure of 200 Torr. Particles were detected with silicon surface barrier detectors which were located inside the chamber in sealed cans to prevent damage from the deuterium gas. The two protons in the final state were detected in coincidence at angles $\theta_1 = 30°$ and $\theta_2 = 60°$ in the lab frame. The 60° detectors were located left and right of the beam, while the 30° detectors were above and below. There were two detectors at each location (left, right, up, down) giving a total of 16 30°–60° coincident pairs. The detectors of each coincidence pair were separated in azimuthal angle by 67.5°, 82.5°, 97.5°, or 112.5°.

Clean event identification was obtained by recording, for each event, the energy of the detected particles and the relative arrival time at the detectors. For each coincident pair, the relative timing spectrum showed a clear peak. By imposing a cut on the timing peak and subtracting random coincidences, the 2-dimensional energy spectra obtained showed the expected kinematic band with no discernable background other than residual statistical fluctuations from the accidental subtraction.

The experiment was completed in August, 1994, and the analysis of the data is still in progress. Although the final results are not yet available, some conclusions can already be made. First and foremost, we find that the measured values of A_z are very small. Plots of A_z as a function of distance along the kinematic curve show results consistent with zero, with statistical uncertainties on the order of ±0.01. If one averages over the entire kinematic curve, the values of A_z are still consistent with zero with uncertainties of typically ±0.003 for each $\Delta\phi$.

In parallel with the experiment, new Faddeev calculations have been carried out by W. Glockle and H. Witala (6) for n-d breakup at the kinematic conditions employed in our experiment. These calculations predict values of A_z of typically 0.002 in magnitude, just below the statistical limit of our experiment.

More disturbing is the observation that inclusion of 3N potentials does not have a large effect on A_z. It is not yet well understood why the conclusions drawn on the basis of the arguments presented above are not supported by the detailed calculations. One possibility is that the rank-1 parts of the 3N potential may be too weak to produce significant effects. Another point to note is that our arguments exploit only the spatial dependence of the rank-1 potentials. It could well happen that if one were to study the detailed spin dependence as well, it would be concluded that the potentials do not directly affect A_z. On the positive side, we should remember that breakup reactions and in particular non-co-planar breakup reactions offer a great wealth of possible observables. Thus, although our initial attempt to identify observables with enhanced sensitivity to 3N forces has not been successful, we believe that the basic approach is sound.

ACKNOWLEDGEMENTS

The experiment described in this paper was performed in collaboration with E.A. George, M.K. Smith, J. Frandy and Y. Zhou. These individuals are all richly deserving of credit for successful completion of a complex experiment on a very short time scale. I am also indebted to H. Witala and W. Glockle for performing the Faddeev calculations, once again on an extremely short time scale.

This work was supported in part by the National Science Foundation under grant number PHY-9316221.

REFERENCES

(1) C. R. Chen, G. L. Payne, J. L. Friar, and B. F. Gibson, *Phys. Rev. C* **31**, 2266 (1985).
(2) C. R. Chen, G. L. Payne, J. L. Friar, and B. F. Gibson, *Phys. Rev. C* **33**, 1740 (1986).
(3) S. A. Coon, M. D. Scandon, P. C. McNamee, B. R. Barrett, D. W. E. Blatt, and B. H. J. McKellar, *Nucl. Phys.* **A317**, 242 (1979).
(4) H. Coelho, T. K. Das, and M. R. Robilotta, *Phys. Rev. C* **28**, 1812 (1983).
(5) L. D. Knutson, *Nucl. Phys.* **A198**, 439 (1972).
(6) W. Glockle and H. Witala, private communication.

Status and Future of Polarization Phenomena Investigations in Backward Elastic Deuteron-Proton Scattering

I.M. Sitnik and V.P. Ladygin

Joint Institute for Nuclear Researches, 141980 Dubna, Russia

Abstract. Recent measurements of the polarization observables in the $dp \to pd$ reaction at Dubna and Saclay are reviewed. Effects due to the polarization of both colliding particles are considered. It is shown that new polarization observable is sensitive to possible P-wave component in the deuteron. Such measurements are planned now at Dubna.

INTRODUCTION

The dp backward elastic scattering is one of the classic reactions to check different models of the deuteron structure.

In general case the dp elastic scattering can be described by 12 comlex amlitudes. That means one need to measure at least 23 polarization observables to reconstruct all of them. Due to P-invariance and total helicity conservation the backward elastic scaterring can be described by only four independent complex amplitudes for the following transition $\lambda_d, \lambda_p \to \lambda'_d, \lambda'_p$,

$$\begin{aligned}
F_{0+\to 0+} &= g_2(s), \\
F_{++\to ++} &= g_1(s) + g_4(s), \\
F_{-+\to -+} &= g_1(s) - g_4(s), \\
F_{0+\to +-} &= -\sqrt{2} g_3(s),
\end{aligned}$$

where g_1–g_4 are so named scalar amplitudes. And only 7 polarization observables must be measured to reconstruct all of them.

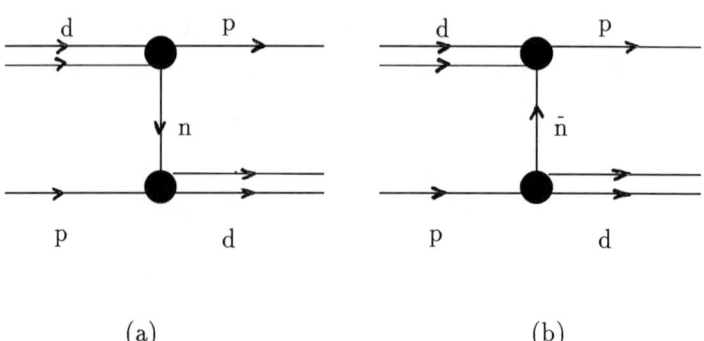

FIGURE 1. The IA diagrams for the $p(d,p)d$ reaction: **(a)** exchange by the neutron; **(b)** exchange by the antineutron.

Usually this reaction is considered as the one-neutron-exchange (ONE) process (Figure 1a). Often this approach is qualified as Impulse Approximation (IA). Only two real amplitudes are needed in this approach, when we are considering the deuteron as superposition of only S- and D-states of two nucleons. In this case the connection between such polarization observable as T_{20} and the deuteron wave function (DWF) components is given by

$$T_{20} = -\sqrt{2} \cdot \frac{x^2-1}{x^2+2} \qquad (1)$$

where $x = 1 + b(s)/a(s)$, $a(s) = u + w/\sqrt{2}$ and $b(s) = -3w/\sqrt{2}$.

The polarization transfer coefficient from the deuteron to the proton (κ_0) is connected with the DWF components by

$$\kappa_0 = \frac{3 \cdot x}{x^2+2} \qquad (2)$$

One can eliminate x and obtain connection between κ_0 and T_{20} which does not depend from the particular DWF[1].

$$\kappa_0^2 + (T_{20} + \frac{1}{2\sqrt{2}})^2 = \frac{9}{8} \qquad (3)$$

But as it will be shown below this approach is far from reality. So, the next step is an attempt to describe this process by four but still real amlitudes. This approach can be connected (in frameworks of the IA) with the deuteron as superposition with additional P-states (Figure 1b). Such components are inevtable when relativistic consideration of the deuteron[2] is used; they also emerge in models based on the quarks count[3]. Only 3 independent polarization observables must be measured to reconstruct all amplitudes in this case.

STATUS AND FUTURE OF INVESTIGATIONS

The momentum distributions of fragments extracted from the dp backward scattering[4] and electrodisintegration data[5] (Figure 2) demonstrate, on the one hand, a substantial discrepancy with the predictions using standard DWF in the IA and, on the other hand, a good agreement with one another, which gives serious motivations to search for the explanation of the observed effects not only in deviation from the IA but also in nonadequate standard DWF.

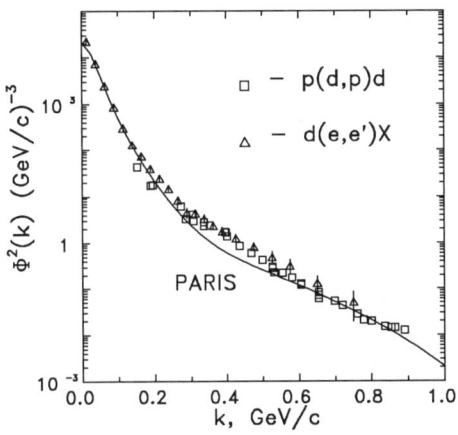

FIGURE 2. The momentum distributions of nucleons extracted from from the dp backward scattering[4] and electrodisintegration data[5].

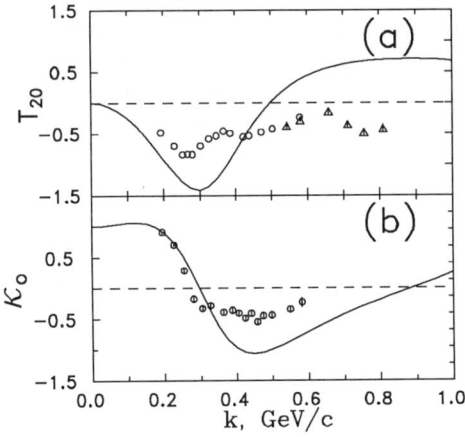

FIGURE 3. (a) T_{20} data[6,7] versus Infinite Momentum Frame variable k; (b) κ_0 data[7] versus k.

The Dubna-Virginia-Saclay collaboration was conceived to measure polarization observables in the $dp \to pd$ reaction.

The tensor analyzing power T_{20} was measured at Dubna[6] and (remeasured) at Saclay[7] (Figure 3a). The κ_0 data obtained at Saclay[7] are presented in Figure 3b.

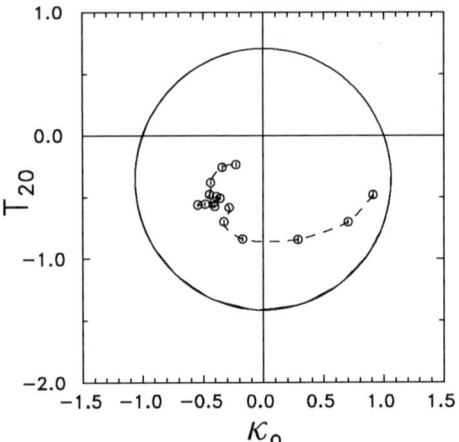

FIGURE 4. Correlation between T_{20} and κ_o in backward elastic scattering.

The principal contradiction of the data with the curve(3) on the correlation plot of T_{20} and κ_0 (Figure 4) demonstrate that at least one of two assumptions (ONE diagram dominates, the DWF contains S- and D- wave only) far from reality. One needs to measure new polarization observables to say more. They can be obtained using a polarized proton target.

In this case the differential cross section depends on mutual spin orientation of initial protons and deuterons. Here we'll mention the expression for transversal spin correlation effect for the DWF with only S- and D-waves:

$$C_{N,N,0,0} = \frac{2 \cdot x^3}{(x^2 + 2)^2} \qquad (4)$$

All mentioned above polarization observables are considered in general case and in more simple case of additional P-waves in ref.[8]

The measurements of this observable now are planned at Dubna. As it is seen from Figure 5, the new observable is very sensitive to the contribution of possible P-states in the deuteron.

The common analysis of new observable together with ones measured earlier, will allow to reconstruct all amplitudes of the $dp \to pd$ process if they are purely real. It will allow one to obtain general restriction for different deuteron models and for the reaction mechanism.

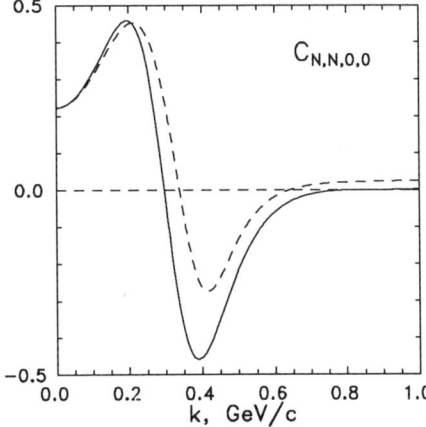

FIGURE 5. Spin correlation $C_{N,N,0,0}$ calculated for standard DWF[9] (solid line) and for DWF with additional P-wave components[2] (dashed line).

If the imaginary parts of the amplitudes are not negligible, in the complete experimental program it is necessary to obtain data for the polarization observables containing the $Im(g_i \cdot g_k^*)$ contributions. Needed combinations appear for triple correlations of vector polarizations:

$$\vec{S}_1 \times \vec{S}_2 \cdot \vec{S}_3, \quad \vec{n}\vec{S}_1 \cdot \vec{n}\vec{S}_2 \times \vec{S}_3 \qquad (5)$$

and so on, where $\vec{S}_i$ is the polarization vector of an i-particle. We would like to stress that polarization transfer measurements in case all three paricles have various combination of parallel or antiparallel spins, does not provide needed information.

More detailed consideration of all these questions is given in ref.[8]

ACKNOWLEDGMENTS

The authors would like to express their thankfulness to A.M.Baldin, J.Durand, B.A.Khachaturov, N.B.Ladygina, F.Lehar, A.de Lesquen, N.M. Piskunov, C.F.Perdrisat, V.Punjabi for useful discussions and stimulating interest in the present work. They are grateful to M.P.Rekalo for join investigation of spin effects in the $dp \rightarrow pd$ reaction. One of the authors (I.M.S.) thanks of the Advisory Committee of this Conference for very kind reception.

REFERENCES

1. Kuehn, B., Perdrisat, C.F., Strokovsky, E.A., "Correlation between polarization observables in inclusive deuteron breakup", presented at Int.Symp."Dubna Deuteron-93", Dubna, Russia, Sept.14-18, 1993.
2. Buck, W.W., and Gross, F., *Phys. Rev.* **D20**, 2361-2379 (1979).
3. Glozman, L.Ya.,Neudatchin, V.G., and Obukhovsky, I.T., *Phys. Rev.***C48**, 389-401 (1993).
4. Berthet, P. et al.,*J.Phys.G.: Nucl.Phys.* **8**, L111-L116 (1982).
5. Bosted, P., et al., *Phys. Rev. Lett.***49**, 1380-1383 (1982).
6. Azhgirey, L.S., et al., "Measurements of T_{20} in Backward Elastic dp Scattering at Deuteron Momenta 3.5-6 GeV/c", in *Abstracts of 14-th Int. IUPAP Conf. on Few Body Problems in Physics*, Willjamsburg, USA, May 26-31, 1994, pp. 22-25.
7. Punjabi, V., et al., "T_{20} and κ_0 in Deuteron Backward Elastic Scattering", in *Abstracts of 14-th Int. IUPAP Conf. on Few Body Problems in Physics*, Willjamsburg, USA, May 26-31, 1994, pp. 161-164.
8. Sitnik, I.M., Ladygin, V.P., and Rekalo, M.P., JINR preprint E1-94-23, Dubna, 1994, to be published in *Sov. Journ. Nucl. Phys. N^o12* (1994).
9. Lacombe, M., et al., *Phys.Lett.***B101** 139-140 (1981).

Determination of the asymptotic D- to S-state ratio for the triton and ^{3}He via ($\vec{d}$,t) and ($\vec{d}$,^{3}He) reactions

Z. Ayer, B. Kozlowska*, R.K. Das[†], H.J. Karwowski and E.J. Ludwig

Department of Physics and Astronomy, University of North Carolina at Chapel Hill, Chapel Hill, NC 27599-3255, USA and
Triangle Universities Nuclear Laboratory, Durham, NC 27708-0308, USA
*Present address: Physics Department, University of Silesia, Katowice, Poland
[†]*Present address: Radiation Oncology Center, Washington University School of Medicine, St. Louis, MO 63110, USA*

Abstract. Angular distributions of tensor analyzing powers (TAPs) have been measured in sub-Coulomb (d,t) reactions and in (d,^{3}He) reactions on several targets and at several energies. The data are compared with predictions of full finite-range distorted-wave Born approximation (DWBA) calculations from which the asymptotic D- to S-state ratios, η_t and $\eta(^3He)$ for the triton and ^{3}He, respectively, are extracted. A best fit to the (d,t) data yields η_t = -0.0411 ± 0.0013 ± 0.0012, while the best-fit value of $\eta(^3He)$ extracted from the (d,^{3}He) data is -0.0380 ± 0.0026 ± 0.0011.

INTRODUCTION

Measurements of tensor analyzing powers (TAPs) in deuteron-induced transfer reactions provide unique information about the three-nucleon system, especially the tensor-force component of the nucleon-nucleon (NN) interaction which provides a large portion of the binding (1). TAPs calculated for (d,t) and (d,^{3}He) reactions strongly depend on the D-state amplitude of the n + d and p + d components in the trinucleon wave functions, respectively. A measure of this D-state amplitude which can be reliably determined from experiment is the asymptotic D- to S-state ratio, η (2). Theoretical predictions of η for the trinucleon systems have improved dramatically over the last decade (3, 4). An accurate determination of η provides a test of calculations of few-nucleon systems using realistic NN forces. For ^{3}He, a practical application of this study stems from the considerable interest that exists in using a polarized ^{3}He target as a polarized neutron target. The presence of a D-state in ^{3}He dilutes the neutron polarization, since in this configuration the neutron spin is anti-parallel to the ^{3}He spin (5).

EXPERIMENTAL PROCEDURE

Angular distributions of the TAP A_{zz} were measured in $(\vec{d},t)$ reactions on ^{119}Sn, ^{149}Sm and ^{206}Pb targets at sub-Coulomb deuteron energies of 5.25, 6 and 10 MeV, respectively. $A_{zz}(\theta)$ was measured in $(\vec{d},^3\text{He})$ reactions on ^{63}Cu, ^{89}Y and ^{93}Nb targets at E_d = 8, 10.5 and 12 MeV, respectively. For ^{93}Nb the TAP A_{yy} was also measured. The experiments were performed at Triangle Universities Nuclear Laboratory (TUNL) using an atomic-beam polarized ion source and an FN tandem accelerator. The targets used in the experiments were isotopically enriched self-supporting foils, the thicknesses of which depended on the reaction being studied and the energies of the particles of interest. Outgoing tritons and ^{3}He particles were detected by three pairs of ΔE-E silicon surface-barrier detector telescopes arranged symmetrically on both sides of the incident beam. Measurements were made mainly at backward angles where the D-state effects on the TAPs are the greatest. A three-detector polarimeter located downstream from the target and utilizing the $^3\text{He}(\vec{d},p)^4\text{He}$ reaction was used to monitor continuously the beam polarization.

DATA ANALYSIS

Analysis of the $(\vec{d},t)$ Data

The A_{zz} data obtained were analyzed using full finite-range DWBA calculations. Since DWBA theory is used to determine η_t, it is important to maintain certain conditions to maximize the reliability of the theory. Reactions at sub-Coulomb energies with Q values close to zero tend to provide these conditions (6). In addition to the data presented in this paper, previously obtained $A_{zz}(\theta)$ data (6) at different sub-Coulomb energies have been re-analyzed. The targets, deuteron energies and their percentage below the Coulomb barrier that were used in the present determination of η_t are listed in Table 1.

The TAP A_{zz} was calculated using a modified version of the finite-range

TABLE 1. Sub-Coulomb $(\vec{d},t)$ reactions investigated.

Target	E_d (MeV)	% below V_C*
^{95}Mo	7†	13
^{119}Sn	5.25, 6† & 7†	41, 33 & 22
^{149}Sm	6 & 8†	43 & 24
^{206}Pb	10	21

*V_C represents the Coulomb barrier for deuterons calculated as $V_C = Z_t*1.44/(r_c*(A_t^{1/3}+2^{1/3}))$ MeV.
†Measurements of Das et al. (6)

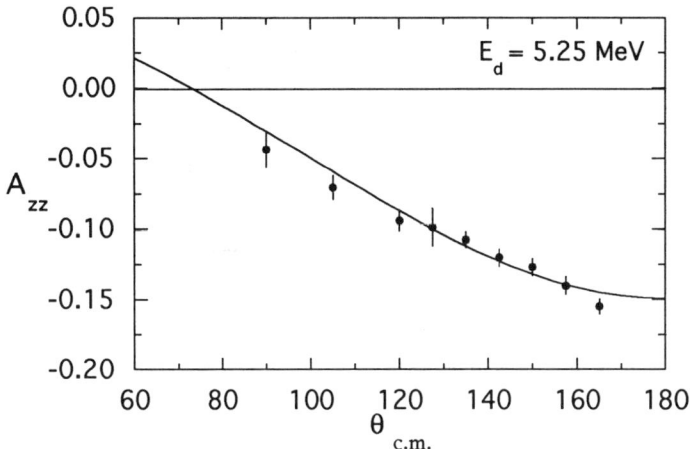

FIGURE 1. Angular distribution of A_{zz} for ^{119}Sn(d,t) at E_d = 5.25 MeV. The solid curve is the best fit to the data and was obtained for η_t = -0.0374.

DWBA code PTOLEMY (7) which allows the inclusion of deuteron-nucleus optical-model tensor potentials in the entrance channel. The real and imaginary central and real spin-orbit potentials for the deuteron were calculated from the global potential formulae of Daehnick *et al.* (8), and exit channel potentials were taken from the work of Becchetti and Greenlees (9).

In addition to the central and spin-orbit parts of the deuteron potential, the nuclear tensor potential (U_{TR}) was also considered. The proper determination of its parameters is particularly important since calculations of TAPs in $(\vec{d},t)$ reactions are quite strongly affected by its choice. The folding model (FM) as proposed by Keaton and Armstrong (10) has been commonly used to generate these parameters for sub-Coulomb energy deuterons. Since $(\vec{d},d)$ data on targets and at energies appropriate for the present work are very limited and the parametrization of Ref. (10) is found to provide inadequate fits to much of the low-energy elastic-scattering data (11, 12), a new approach was attempted. Instead of arbitrarily scaling the FM parameters, we have followed the prescription of Santos (13) to generate tensor-potential parameters. For details of this procedure the reader is referred to Ref. 14.

For each set of data a best-fit value of η_t was determined by using a χ^2-minimization procedure which is described in Ref. 14. An example of an angular distribution of A_{zz} and a best-fit calculation is shown in Fig. 1.

Various sources of uncertainty that contribute to the error in the value of η_t were carefully considered. In addition to statistical errors three main sources of error were taken into account. The first of these is due to the choice of central and spin-orbit parameters used in the DWBA calculations. Variations of these parameters that were consistent with ^{208}Pb$(\vec{d},d)^{208}$Pb data at E_d = 10 MeV (15) were taken as the limits of acceptable values that could be used in the DWBA calculations. The effect of choice of the tensor-potential parameters and the errors

TABLE 2. Values of $\eta \pm \Delta\eta$ extracted from the tensor-analyzing-power data measured in the ^{93}Nb$(\vec{d},^3\text{He})^{92}$Zr reaction for different sets of optical potentials.

Potential Set	^{93}Nb(A_{zz} data)	^{93}Nb(A_{yy} data)
Daehnick et al.	-0.0356 ± 0.0047	-0.0384 ± 0.0044
Lohr & Haeberli	-0.0349 ± 0.0054	-0.0370 ± 0.0045
Perrin et al.	-0.0354 ± 0.0047	-0.0384 ± 0.0046
TUNL	-0.0337 ± 0.0097	-0.0319 ± 0.0075
Final Value	-0.0356 ± 0.0047	-0.0384 ± 0.0047

introduced in the value of η extracted is discussed in Ref. 14. The final source of error considered is that due to effects that were not investigated such as virtual excitations in the triton wave function. Based on calculations for (d,p) reactions (16) an uncertainty of 3% in the value of η_t was assumed for this effect. These three sources of error were then combined in quadrature with the statistical error for each η_t determination to yield a final error for each individual determination.

Analysis of the $(\vec{d},^3\text{He})$ Data

As in the case for $(\vec{d},t)$ reactions described above, for $(\vec{d},^3\text{He})$ reactions it is also desirable that the incident deuteron energy be below the Coulomb barrier. However, the rapidly decreasing cross section for sub-Coulomb $(\vec{d},^3\text{He})$ reactions (<< 1 µb/sr) renders analyzing-power measurements unfeasible. Therefore in the present case, the deuteron energies were chosen as a compromise between performing the experiments where the cross sections were adequate and allowing some small sensitivity of the calculations to optical-model parameters (OMPs) in the entrance channel. The present targets were chosen since they have large spectroscopic factors for proton pickup and involve unique j transfers.

The TAP data were analyzed using the computer program PTOLEMY. Table 2 contains the results of η extracted from the best fits to the TAP data for ^{93}Nb. For the entrance channel, three different sets of global deuteron potentials were used: the sets of Daehnick et al. (8), Lohr and Haeberli (17), and Perrin et al. (18). For ^{93}Nb, the cross section, vector analyzing power A_y and tensor analyzing powers A_{yy} and A_{zz} were measured in the elastic scattering at E_d = 12 MeV. An optical-model analysis of these data which included a tensor potential of the folding-model type yielded a set of OMPs which were then used to analyze the ^{93}Nb$(d,^3\text{He})^{92}$Zr TAP data. This set of OMPs is denoted by TUNL in Table 2. For the exit channel, the global ^{3}He potentials of Ref. 9 were used.

Figure 2 contains the $A_{zz}(\theta)$ data obtained for ^{93}Nb and also shows the best-fit calculations for each potential set. The calculation which includes the tensor potential (TUNL) yields the worst fit to the data. In particular, there is a shift in the zero crossover in A_{zz} towards forward angles by about 6° due to the inclusion of the tensor potential. Attempts to improve the agreement by varying

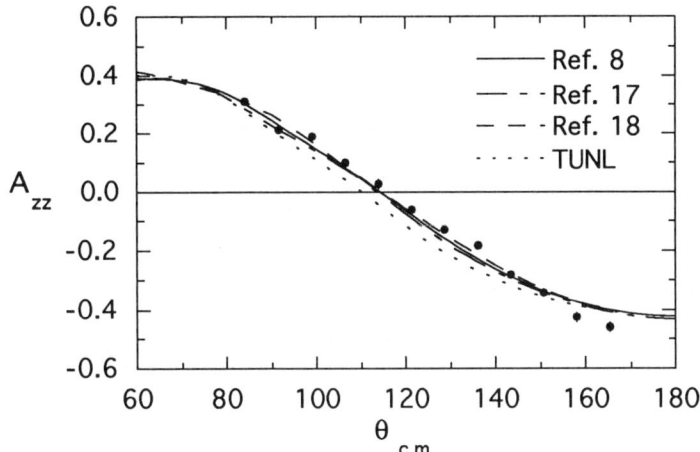

FIGURE 2. Azz as a function of c.m. angle for ^{93}Nb(d,^{3}He)^{92}Zr at E_d = 12 MeV. The curves are the best fits to the data using the potential sets described in the text.

the magnitudes of the real and imaginary parts of this potential failed. No attempts, however, have been made to improve the agreement by changing the shape of this potential from the folding-model prescription. Previously, Frick et al. (19) had shown that if in the case of ^{16}O($\vec{d}$,d) at E_d = 20 MeV, the imaginary part of the tensor potential is kept fixed and the shape of the real part is varied, then predictions for the TAP $T_{21}(\theta)$ are found to be nearly identical. Angular distributions of TAPs in deuteron elastic scattering do not show sensitivity to the shape of the deuteron tensor potential while reaction observables appear best fit when this potential is set to zero when using the standard folding model shape to describe data at energies above the Coulomb barrier. Other shapes will be attempted in future analyses.

For each set of TAP data the best-fit value of η was taken to be the value obtained using the potential set of Daehnick et al., hereafter referred to as η_τ. To calculate the error in η we first calculated the deviations of the η values obtained using the other potential sets from η_τ. We then weighted each deviation with the inverse square of the error in each η value and combined this in quadrature with the error in η_τ. For ^{93}Nb this is shown in the last row in Table 2.

For the ^{63}Cu and ^{89}Y A_{zz} data, no attempt was made to include tensor potentials in the analysis. Only the three sets of global potentials were used. For ^{89}Y, we also used the OMPs of Karp et al. (20) obtained from fits to cross-section and A_y data at E_d = 12 MeV. Using the procedure described above we obtain η values of -0.0401 ± 0.0070 and -0.0394 ± 0.0055 for the ^{63}Cu and ^{89}Y data, respectively.

RESULTS AND CONCLUSIONS

To obtain a final value of η_t we computed a weighted average of the seven individual η_t determinations, which yields a value of -0.0411 ± 0.0013. There is also a systematic uncertainty of 3% in the beam polarization which introduces an additional error in η_t. We thus obtain a final value of -0.0411 ± 0.0013 ± 0.0012, which agrees with the measurement of Ref. 21 (-0.0431 ± 0.0025). Our measurement agrees with the theoretical calculations of the Sendai group (4) (-0.0440 ± 0.0004), but is slightly lower than the predictions of the Los Alamos group (3) (-0.046 ± 0.001).

For ^{3}He, a weighted average of our four individual η determinations, yields a value of -0.0380 ± 0.0026. Taking into account the systematic uncertainty of 3% in the beam polarization, we obtain a final η value of -0.0380 ± 0.0026 ± 0.0011, which agrees with the measurement of Ref. 22 (-0.035 ± 0.006) obtained by an independent method. Recent theoretical predictions of η are -0.043 ± 0.001 by the Los Alamos group (3) and -0.0414 ± 0.0004 by the Sendai group (4). Some questions still exist, however, as to the role of the deuteron-nucleus tensor potential.

If a ratio of η_t/η_{3He} is calculated, we obtain η_t/η_{3He} = 1.08 ± 0.12 in excellent agreement with the results of the Los Alamos (1.06 ± 0.034) and Sendai (1.06 ± 0.014) groups.

This work was supported by the U.S. Department of Energy through Grant No. DE-FG05-88ER40442.

REFERENCES

1. Ericson, T.E.O. and Rosa-Clot, M., Ann. Rev. Nucl. Phys. 1985 **35**, 271-294 (1985).
2. Eiro, A.M. and Santos, F.D., J. Phys. G, Nucl. Part. Phys. **6**, 1139-1173 (1990).
3. Friar, J.L., Gibson, B.F., Payne, G.L. and Lehman, D.R., Phys. Rev. C **37**, 2859-2868 (1988).
4. Wu, Y., Ishikawa, S. and Sasakawa, T., Few-Body Systems **15**, 145-188 (1993).
5. Friar, J.L., et al., Phys. Rev. C **42**, 2310-2314 (1990).
6. Das, R.K., et al., Phys. Rev. Lett. **68**, 1112-1115 (1992).
7. Macfarlane, M.H. and Pieper, S.C., Argonne National Laboratory Report No. ANL-76-11, Rev. 1 (unpublished), modified by R.P. Goddard (1980).
8. Daehnick, W.W., Childs, J.D. and Vrcelj, Z., Phys. Rev. C **21**, 2253-2274 (1980).
9. Becchetti, F.D. and Greenlees, G.W., *Polarization Phenomena in Nuclear Reactions*, Madison, Wisconsin, University of Wisconsin Press, 1971, pp. 682-683.
10. Keaton, P.W. and Armstrong, D.D., Phys. Rev. C **8**, 1692-1701 (1973).
11. Kammeraad, J.E. and Knutson, L.D., Nucl. Phys. **A435**, 502-522 (1975).
12. Rodning, N.L. and Knutson, L.D., Phys. Rev. C **41**, 898-909 (1990).
13. Santos, F.D., Z. Phys. **A295**, 73-77 (1980).
14. Kozlowska, B., et al., submitted for publication to Phys. Rev. C.
15. Murayama, T., et al., Nucl. Phys. **A486**, 261-270 (1988).
16. Tostevin, J.A. and Johnson, R.C., Phys. Lett. **85B**, 14-16 (1979).
17. Lohr, J.M. and Haeberli, W., Nucl. Phys. **A232**, 381-397 (1974).
18. Perrin, G., et al., Nucl. Phys. **A282**, 221-242 (1977).
19. Frick, R., et al., Z. Phys. A **319**, 133-141 (1984).
20. Karp, B.C., et al., Nucl. Phys. **A457**, 15-44 (1986).
21. George, E.A. and Knutson. L.D., Phys. Rev. C **48**, 688-698 (1993).
22. Vuaridel, B., et al. , Nucl. Phys. **A499**, 429-452 (1989).

Measurement of Spin Observables in Quasielastic Scattering of Polarized Protons from Polarized ^{3}He

M. A. Miller[4], K. Lee[3], A. Smith[1], J.-O. Hansen[3*], C. Bloch[1],
J. F. J. van den Brand[4], H. J. Bulten[4], D. DeSchepper[3],
R. Ent[3†], C. D. Goodman[1], W. W. Jacobs[1], C. E. Jones[4*],
W. Korsch[3‡], L. H. Kramer[3], M. Leuschner[1], W. Lorenzon[7],
N. C. R. Makins[3*], D. Marchlenski[5], H. O. Meyer[1],
R. G. Milner[3], J. S. Neal[4], P. V. Pancella[6], S. F. Pate[3],
W. K. Pitts[2], B. von Przewoski[1], T. Rinckel[1], G. Savopulos[1],
J. Sowinski[1], F. Sperisen[1], E. R. Sugarbaker[5], C. Tschalär[3],
O. Unal[4], T. P. Welch[3§] and Z-L. Zhou[4]

[1] Indiana University Cyclotron Facility, Bloomington, Indiana 47405
[2] University of Louisville, Louisville, Kentucky 40292
[3] MIT-Bates Linear Accelerator Center and Laboratory for Nuclear Science, Massachusetts Institute of Technology, Cambridge, Massachusetts 02139
[4] University of Wisconsin-Madison, Madison, Wisconsin 53706
[5] The Ohio State University, Columbus, Ohio 43210
[6] Western Michigan University, Kalamazoo, Michigan 49007
[7] TRIUMF, British Columbia, Canada V6T 2A3

Abstract. Measurements of analyzing powers and spin correlation parameters for $^3\vec{\mathrm{He}}(\vec{p},2\mathrm{p})$ and $^3\vec{\mathrm{He}}(\vec{p},\mathrm{pn})$ quasielastic scattering have been carried out using a polarized internal ^{3}He target and a stored, polarized proton beam. They were made over a wide kinematic range with a proton energy of 197 MeV and a large acceptance detector system which allowed the simultaneous measurement of both reactions. At momentum transfers sufficiently high to minimize rescattering effects, the (p,pn) results are in agreement with PWIA calculations. The (p,2p) results were found to be less sensitive to kinematic conditions and are also in agreement with calculations.

---

*Present address: Argonne National Laboratory, Argonne, IL 60439
†Present address: CEBAF, Newport News, VA 23606
‡Present address: Caltech, Pasadena, CA 91125
§Present address: Oregon State University, Corvallis, OR 97331-6507

INTRODUCTION

The polarized ^{3}He nucleus is the subject of considerable interest to the nuclear and particle physics communities. As a three-body system, its wave function can be calculated exactly with Faddeev techniques. Such calculations indicate that the ^{3}He ground state is dominated by an S-state with anti-aligned proton spins and the spin the neutron aligned with the nuclear spin. Thus it is widely believed that polarized ^{3}He can be used to study the spin structure of the neutron. There has been much experimental verification of various aspects of calculated three-body wave functions, however, prior to the measurements reported here, it was found that the spin observables of the $^3\vec{\text{He}}(\vec{p},2p)$ and $^3\vec{\text{He}}(\vec{p},pn)$ reactions did not agree with those calculated from Faddeev generated wave functions [1–3]. Our measurements were performed to further investigate this disagreement and to expand on the range of kinematics covered by the existing data.

The measurements were carried out by the CE25 collaboration at the IUCF Cooler storage ring using the Cooler's polarized proton beam, an internal optically pumped, metastability-exchange polarized target [4] and a large acceptance detector array. The combination of the internal target, the stored beam and the large acceptance allowed for high luminosity and enabled us to simultaneously record data from the $^3\vec{\text{He}}(\vec{p},2p)$ and $^3\vec{\text{He}}(\vec{p},pn)$ reactions.

The detector system was designed to measure the momentum of two outgoing particles and thus reconstruct the missing momentum and energy of the reaction. This was done by using two ΔE-E detector arms to detect particles in coincidence, one on either side of the beam axis. Two pairs of crossed wire chamber planes were situated between the E and ΔE detectors to provide trajectory reconstruction for charged particles. In the case on neutrons, trajectories were calculated from position information derived from photomultiplier tubes placed at each end of the scintillator bars that formed the elements of the E-detector. Combining time-of-flight data with the trajectories provided a measure of the momentum of the outgoing particles in the reactions. A detailed description of the experimental techniques can be found in Ref. [5].

CALCULATIONS OF SPIN OBSERVABLES

In quasielastic scattering, it is usual to interpret the data in terms of the plane-wave impulse approximation (PWIA) or some other spectator model. This is an attractive approach because it leads to a simple factorization of cross sections and spin observables into parts that depend on nuclear structure and parts that depend on the free-scattering observables for the reaction. The cross section is factored as $\sigma = k\sigma_N^\star S^N(p,E)$, where $\sigma_N^\star$ is the free-scattering cross section for (p,p) or (p,n) scattering, depending on the quasielastic reaction, $S^N(p,E)$ is the spectral function for ^{3}He and k is a kinematic factor. The spectral function is the probability of finding a nucleon N with momentum p and separation energy E in the ^{3}He nucleus. This factorization is expanded to include spin by using a spin-dependent spectral function $S^N(\mathbf{s},\mathbf{p},E)$ and spin-dependent free-scattering cross sections, where $\mathbf{s}$ is the spin of the nucleon with respect to the nuclear spin and $\mathbf{p}$ is the momentum vector of the knocked out nucleon. The spin-dependent cross section is written in the form $\sigma = \sigma_0(1 + P_b A^N_{n0} + P_t A^N_{0n} + P_b P_t A^N_{nn})$ where σ_0 is the unpolarized cross section given above, P_b and P_t are the beam and target polarization, respectively, A^N_{n0} is the beam analyzing power, A^N_{0n} is the target analyzing power and A^N_{nn} is the spin correlation parameter. The n's in the subscripts indicate that both beam and target polarizations are normal to the reaction plane according to the notation convention presented in Ref. [6]. In terms of the ^{3}He spectral function, the spin observables can be written as

$$A^N_{ij} = A^{N,\star}_{ij} \times \left[\frac{S^N_+(p,E) - S^N_-(p,E)}{S^N_+(p,E) + S^N_-(p,E)}\right]$$

where $A^{N,\star}_{ij}$ is the free-scattering spin observable and i and j take on the values 0 and n as in the cross section expression. The spectral function S^N has been written as $S^N_\pm(p,E)$ with the $\pm$ indicating that it is evaluated for nucleons with spin parallel and anti-parallel to the nuclear polarization, respectively. The use of p instead of $\mathbf{p}$ indicates a summation over the acceptance of our detector. The superscript N represents the knocked out nucleon. Since we detect the outgoing nucleons in the plane normal to the polarization, the $\mathbf{p}$ acceptance is weighted heavily toward nucleons in that plane as well. The term in square brackets contains the nuclear structure information and is the effective polarization of the nucleon with the nucleus.

RESULTS AND CONCLUSIONS

In PWIA, the missing momentum, p_m, is identified as the initial momentum of the struck nucleon and the missing energy, E_m, as the separation energy of the struck nucleon. If PWIA holds, one expects the beam and target analyzing powers to be identical. These analyzing powers are plotted for the ^{3}He(p,pn) reaction as a function of momentum transfer in Fig. 1. The data have been summed over all E_m, while only those events with p_m less than 100 MeV/c have been included. This low-momentum sample, which includes more than 80% of the data, is dominated by the spatially symmetric S-state part of the wave function, which has the highest effective polarization of the neutron within the ^{3}He nucleus. Under these conditions, one finds that the measured analyzing powers are consistent only at momentum transfers above 500 MeV/c. The curves shown in Fig. 1 are PWIA calculations using the spectral function of Ref. [7] and free-scattering spin correlation parameters calculated with the

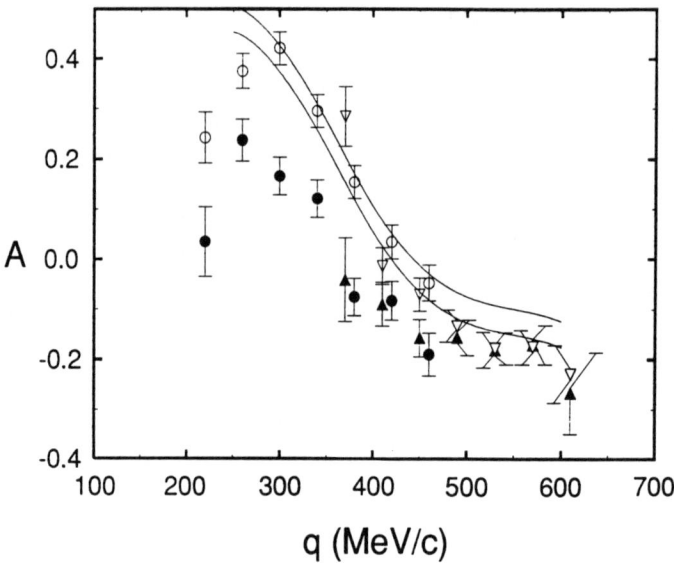

Figure 1: The three-momentum transfer (q) dependence of the beam (open symbols) and target (filled symbols) analyzing powers in $^3\vec{\text{He}}(\vec{p},\text{pn})$ scattering. The curves are PWIA calculations described in the text.

SAID program [8]. The width of the band of calculations is due to the spread in SAID observables calculated from different phase shift sets. We interpret this q dependence as evidence of spin-dependent final state interactions of the recoiling neutron at low momentum transfer.

Spin correlation parameters, A_{nn}^n and A_{nn}^p, are shown as a function of missing momentum in Fig. 2. The curves were calculated in the same manner as those in Fig. 1. The (p,2p) data shown on the bottom half of the figure have been summed over missing energy, while the $p_m < 100$ MeV/c condition has been applied to the (p,pn) data, along with a requirement that the momentum transfer be greater than 500 MeV/c. This last requirement is based on the q-dependence discussed above, and, as evidenced by the agreement between the data and calculations, results in a data set that does fit the quasielastic interpretation.

In summary, we have measured beam and target analyzing powers and spin correlation parameters for the (p,2p) and (p,pn) knock-out reactions on ^{3}He with polarized beam and target. The observables are in agreement with PWIA calculations in the (p,2p) case and in the (p,pn) case after applying kinematic conditions to the data that select the most quasielastic conditions. Further details of the experiment and analysis can be found in Ref.s [9], [10] and [4].

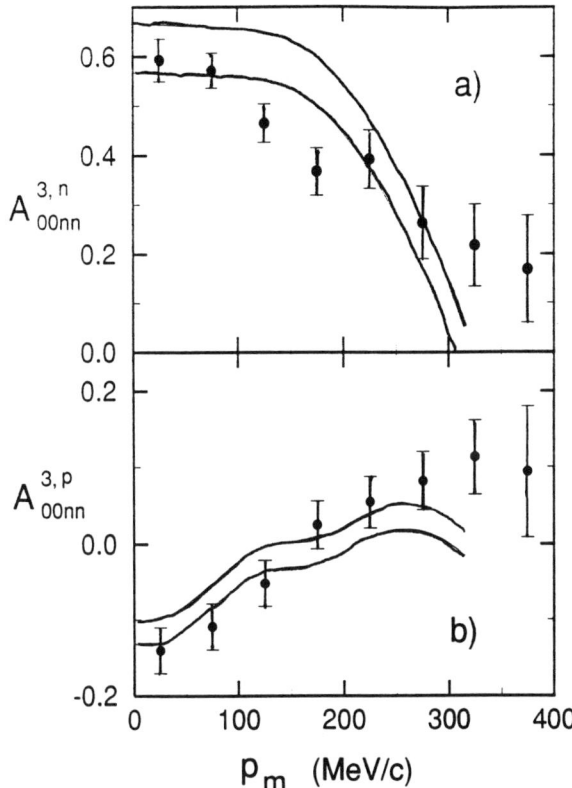

Figure 2: The spin correlation parameter A^N_{nn} versus missing momentum p_m. a) A^n_{nn} for the $^3\vec{\text{He}}(\vec{p},\text{pn})$ reaction. b) A^p_{nn} for the $^3\vec{\text{He}}(\vec{p},2\text{p})$ reaction. The curves are PWIA calculations described in the text.

ACKNOWLEDGMENTS

We acknowledge discussions with M.A. Titko and T.W. Donnelly at MIT and thank B. Blankleider, R. Woloshyn, R. Schulze, and P. Sauer for use of their three body spectral functions. We thank the staff of IUCF, in particular J. Doskow, for their efforts and J. Kelsey and B. Wadsworth and associates (MIT) for polarized target technical support. This work is supported in part by the National Science Foundation under Grants No. PHY-9015957 (IUCF), PHY-9316221 (Wisconsin), PHY-9108242 (Ohio State) and the NSF Research Opportunity Award Program (Louisville and W. Michigan) and by the Department of Energy under Contracts No. DE-AC02-76ER03069 (MIT) and W-31-109-ENG-38 (Argonne). RGM acknowledges a Presidential Young Investigator Award from NSF.

References

[1] E. J. Brash et al., *Phys. Rev.* **C47**, 2064 (1993).

[2] A. Rahav et al., *Phys. Lett.* **B275**, 259 (1992).

[3] A. Rahav et al., *Phys. Rev.* **C46**, 1167 (1992).

[4] K. Lee et al., *Nucl. Instr. and Meth.* **A333**, 194 (1993).

[5] C. Block et al., *Nucl. Instr. and Meth.*, accepted for publication.

[6] J. Bystricky, F. Lehar, and P. Winternitz, *J. Phys. (Paris)* **39**, 1 (1978).

[7] R.-W. Schulze and P. U. Sauer, *Phys. Rev.* **C48**, 38, (1993).

[8] R. Arndt, SAID - Scattering Analysis Interactive Dial-up. Computer Program.

[9] K. Lee et al., *Phys. Rev. Lett.* **70**, 783 (1993).

[10] M. A. Miller et al., submitted to *Phys. Rev. Lett.*, (1994).

Elastic π+ Scattering on Polarized ³He

M. A. Espy,[a] S. P. Blanchard,[b] J. E. Brash,[c] B. Brinkmöller,[d]
G. R. Burleson,[b] W. J. Cummings,[e] B. J. Davis,[a] D. Dehnhard,[a]
P. P. J. Delheij,[e] C. M. Edwards,[a] O. Häusser,[e] R. Henderson,[e]
B. K. Jennings,[e] M. K. Jones,[c] B. A. Lail,[b] J. L. Langenbrunner,[a]
B. Larson,[f] W. Lorenzen,[e] K. Maeda,[g] C. L. Morris,[h] B. Nelson,[b]
J. M. O'Donnell,[a] M. A. Palarczyk,[a] B. K. Park,[b] S. I. Penttilä,[h]
D. R. Swenson,[h] D. Thiessen,[e] D. Tupa,[h] and Q. Zhao[b]

[a]*University of Minnesota*, [b]*New Mexico State University*, [c]*Rutgers University*,
[d]*Universität Karlsruhe*, [e]*TRIUMF*, [f]*Ohio University*, [g]*Tohoku University*,
[h]*Los Alamos National Laboratory*

Abstract. Asymmetries, A_y, for π^+ elastic scattering from polarized ³He were measured in the P_{33} resonance region, where the π-nucleon force is strongly energy and isospin dependent. Preliminary results at energies of 142 and 180 MeV show large positive A_y near 90°, shifted to larger angles than predicted. At angles near 50° there is evidence for negative asymmetries of about -0.2 in contrast to several theoretical predictions that use up-to-date Faddeev wave functions and a first-order optical potential. Inclusion of a Δ-neutron spin-spin interaction gives an excellent fit to the data at 180 MeV.

We have measured analyzing powers or asymmetries, A_y, for elastic pion scattering on polarized ³He at incident energies near the P_{33} π-nucleon (Δ(1232)) resonance. Asymmetry data on a nucleus of well-known structure such as ³He are needed in order to test reaction models without the complications of nuclear structure. Recent work on pion elastic scattering from polarized 1p-shell nuclei of spin $J=1/2$ had revealed large discrepancies between theoretical predictions and experiment. For example, experiments on ¹⁵N(1) and ¹³C(2,3) measured generally small values of A_y whereas theory(4-6) predicts that A_y should be large.

Theoretical A_y for elastic scattering from 1p-shell nuclei show a strong dependence(6,7) on the details of the nuclear structure, implying that information on the nuclear spin density is contained in the data. However, as indicated by the failure of current reaction models to reproduce the measured A_y, either the π-nucleus interaction and specifically its spin-dependent part is not yet understood, or the nuclear wave functions of these nuclei are not sufficiently well known, or both. ^{3}He is ideal for a study of the π-nucleus reaction model because reliable wave functions have been obtained by rigorous Faddeev calculations.(8-10)

Asymmetry measurements on polarized ^{3}He became possible with the recent development of the high-density, optically pumped ^{3}He gas target at TRIUMF. This target was used at TRIUMF in a pion scattering experiment and very large values of A_y were found in π^+ scattering at 100 MeV.(11) At this energy the asymmetry is rather insensitive to the nuclear wave function and shows only a slight dependence on the reaction model. However, for energies at and above the P_{33} resonance, the asymmetries are predicted to become increasingly sensitive to the details of the reaction model.

Thus there is considerable interest in an extension of these measurements to energies at and above the P_{33} resonance. These measurements could not be done at TRIUMF because of the lack of a suitable spectrometer and the relatively low beam fluxes at these higher energies. Therefore, the TRIUMF target (Fig. 1) was set up in the P^{3}E area at LAMPF where high beam fluxes are available at P_{33} resonance energies and above. The scattered pions were detected with the Large Acceptance Spectrometer (LAS). Measurements were taken at incident energies T_π = 100, 142, 180, and 256 MeV and laboratory scattering angles ranging from 40° to 100°. The data are currently being analyzed.

The ^{3}He gas was contained in a cylindrical glass cell, about 4.8 cm in diameter and 6.5 cm in length. Target cells were made of quartz glass that was about 1.5 mm thick at the cylindrical cell walls and 0.4 mm thick at the hemispherical endcaps (where the pion beam entered and exited the cell). The cells were filled with 6-7 atm of ^{3}He gas, a trace of Rb, and a small amount of N$_2$

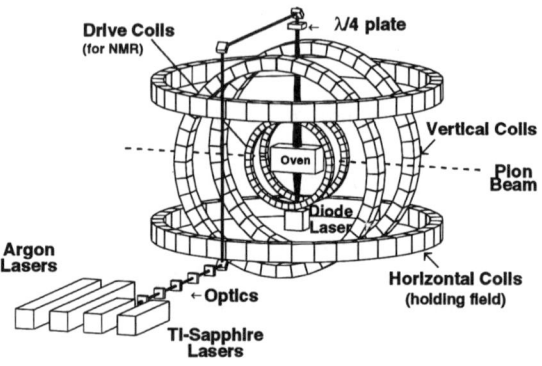

Figure 1. The TRIUMF target apparatus.

which served as a buffer gas. 8-10 W of polarized laser light at 795 nm (the D1 transition in Rb) was used to polarize the Rb atoms in the target cell. The electron spin polarization of the Rb was transferred through Rb-^{3}He collisions to the ^{3}He nucleus by the contact hyperfine interaction. The target cell was heated continuously in the target oven (Fig. 1) to a temperature of approximately 175 °C in order to achieve the required Rb vapor density for the optical pumping. When the glass cell was hot, small amounts of ^{3}He leaked from the cell. Therefore, the pressure in the cell and the cell temperatures were monitored periodically and the time periods when the cells were hot were recorded so that a correction for the pressure loss could be made. The target apparatus(12) was modified during the experiment by the addition of a diode laser(13) array. The diode laser added to the optical pumping power and significantly increased the polarization after one of the Argon lasers (see Fig. 1) failed. ^{3}He polarization was typically 35 to 45%, sometimes reaching 50%.

Since the helicity of the laser light determines the direction of the target polarization, the orientation of the ^{3}He spins (and thus the sign of A_y) was determined by use of a liquid crystal which transmits only left-hand circularly polarized light. The horizontal coils provide a holding field for the polarization and the vertical coils are used to change the direction of the polarization. The magnitude of the polarization was measured using the nuclear magnetic resonance (NMR) technique of adiabatic fast passage (AFP)(12). Absolute normalization factors for the NMR signals were obtained by comparing the NMR signals from the ^{3}He with the weak signals from the protons in a water-filled cell of the same dimension.

The LAS uses a magnetic quadrupole doublet, a magnetic dipole, and several sets of two-dimensional wire chambers in order to identify the scattered pions and to measure their momenta. The front wire chambers, located between the quadrupole doublet and the dipole, allow traceback of the scattered particle trajectories to a plane that intersects the center of the target. This plane is perpendicular to the central ray of the LAS. The projections of the reaction

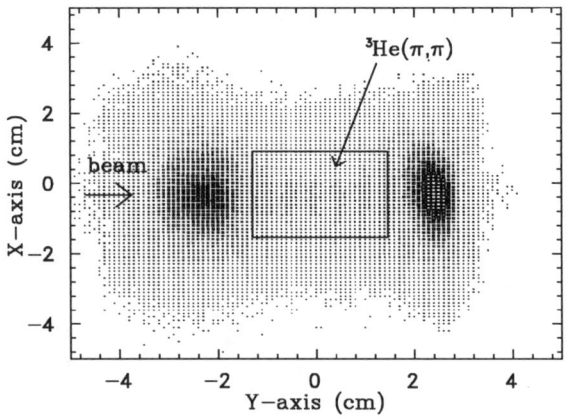

Figure 2. Projections of the reaction vertices onto the X-Y plane at the target. The high concentrations of events are from the endcaps of the cell. (See text).

vertices onto this x-y plane were used to discriminate between events from ^{3}He and the glass in the endcaps (Fig. 2) and the top and bottom of the cylindrical cell wall. Events from the left and right sides of the cylindrical wall could not be eliminated by software cuts. Thus the beam halo striking the sides of the cell was reduced by a conical lead collimator that was specially machined to match the beam divergence.

From a preliminary replay of the data, the top and center panels of Fig. 3 show missing mass spectra measured at 180 MeV and $\theta_{lab} = 50°$, normalized to integrated beam flux for the two target spin orientations. The target polarization parallel to the norm of the reaction plane is indicated by (↑), the one antiparallel by (↓). The ³He elastic peak has a width of about 4 MeV (FWHM). The difference spectrum is shown in the bottom panel of Fig. 3.

The asymmetries were obtained using

$$A_y = \frac{N_\uparrow - N_\downarrow}{N_\uparrow + N_\downarrow} \frac{1}{p}. \quad (1)$$

Here $N_\uparrow$ and $N_\downarrow$ are the normalized number of counts in the ³He elastic peak with the ³He spins oriented "up" and "down", respectively. p is the target polarization. A large negative asymmetry is apparent in the region of the elastic peak from ³He (Fig. 3, bottom panel, Missing Mass = 0).

Preliminary experimental and theoretical angular distributions of A_y for π^+ elastic scattering at T_π = 142 MeV and at T_π = 180 MeV are presented in Fig. 4. The theoretical curves were obtained using the distorted wave impulse approximation (DWIA) and state-of-the-art Faddeev wave functions(8-10) for ³He.

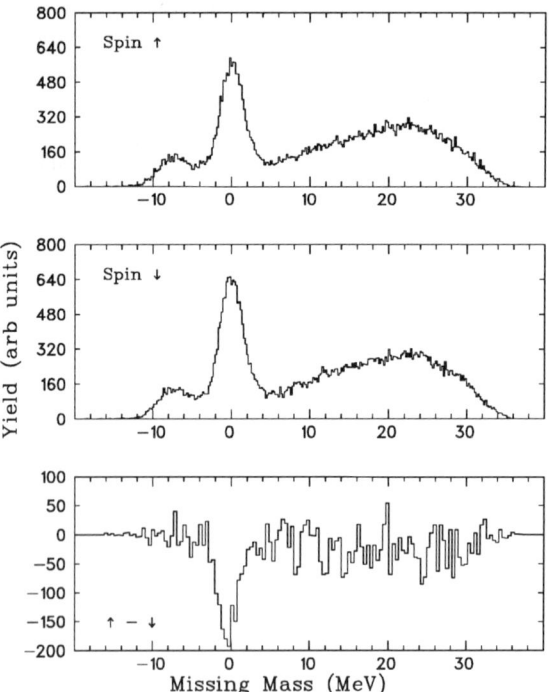

Figure 3. Typical normalized yield spectra for elastic scattering of π^+ from polarized ³He at T_π = 180 MeV and θ_{lab} = 50°. Top panel: target spin parallel (↑), center panel: target spin antiparallel (↓) to the normal to the reaction plane. The difference spectrum is shown in the bottom panel.

Neither the first-order calculations of Ref. 14 which include multiple scattering (Fig. 4, solid lines), nor those of Ref. 15 and 16 (not shown) give a satisfactory description of the data. We observe large positive asymmetries near 80° as predicted by the first-order calculation, but the maximum of A_y is shifted towards larger angles. At scattering angles near 60° the asymmetries are found to be negative at 142 and 180 MeV, in contradiction to any of the conventional model calculations which predict positive asymmetries between 100 and 180 MeV. At 256 MeV, the asymmetries (not shown) are negative at all measured angles, probably a result of multiple scattering which becomes more important at higher energies.

The theoretical asymmetry in pion scattering from a spin 1/2 nucleus can be written in terms of the complex spin-independent (F) and spin-dependent (G) scattering amplitudes as $A_y = 2\,\text{Im}(F \times G^*) / (|F|^2 + |G|^2)$. When the P_{33} resonance dominates the elementary π-nucleon interaction, the isospin coupling Clebsch-Gordan coefficients result in much larger scattering amplitudes, F and G, for π^+ elastic scattering on protons than on neutrons. In the impulse approximation, the π^+ interacting with one of the paired-off protons in a $(1s)^3$ ground state of ^{3}He cannot lead to a spin-dependent amplitude. G results only from scattering from the unpaired neutron. F has a large component from scattering from the protons and a small one from scattering from the neutron.

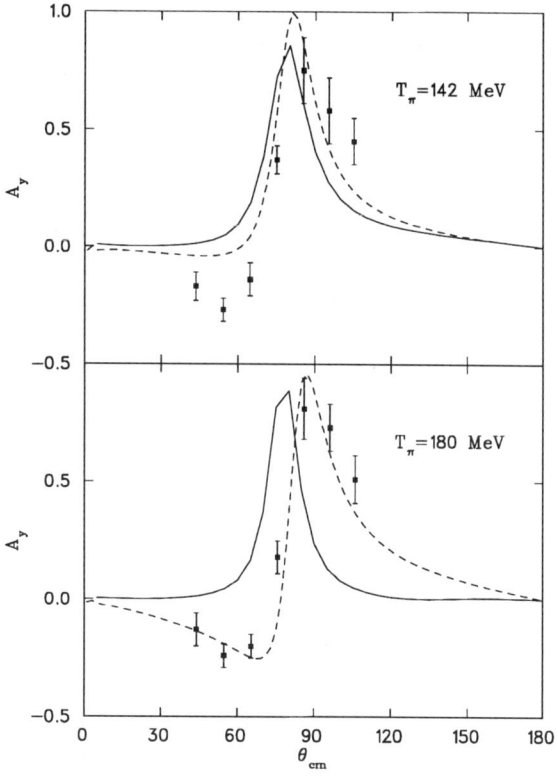

Figure 4. Preliminary asymmetry angular distribution for elastic π^+ scattering from polarized ^{3}He at $T_\pi = 142$ MeV and 180 MeV. Solid lines: First-order calculations of Ref. 14. Dashed lines: Calculations with inclusion of a Δ-N spin-spin interaction.

Nevertheless, the A_y can be large near the minimum of F where G can have a relative maximum. Qualitatively, this simple model gives asymmetries similar to those of the first-order optical model calculations (Fig. 4, solid lines).

If the intermediate Δ^{++}, generated with very high probability in π^+ scattering on one of the two protons, interacts with the neutron, a large second-order contribution to G arises. The magnitude of this second-order contribution to G has been investigated using a meson exchange model for the Δ-neutron interaction. The π, ρ, ω, σ and η mesons were included. For the π and ρ mesons the Pauli exchange diagrams are also important. The meson-Δ couplings were obtained from the meson-nucleon couplings by use of SU(3) symmetry. Gaussian single particle wave functions were used. Correlations were included by multiplying the single particle wave function by 1 minus a Gaussian that depends on the relative distance of the interacting particles. The rms radius of the density was kept fixed at the experimental ^{3}He value and the width of the Gaussian was varied. The best fit for 180 MeV was obtained with a range of the correlation Gaussian that was half that of the single particle Gaussian. The contributions from the direct ρ and ω exchange largely cancel. The remaining ρ and ω contribution is very sensitive to short range correlations. The η exchange is not very significant.

The fit with this model is excellent at 180 MeV (dashed lines in Fig. 4). The negative asymmetries near $60°$ and the shift of the positive maximum towards larger angles are reproduced. However, the energy dependence of the asymmetry is not predicted correctly; at 142 MeV we miss the experimental data near $60°$. Further theoretical and experimental work is needed. Of particular interest is a proposed measurement of the asymmetry for π^- scattering. There the first-order term is much larger than the second-order term and the effect from the Δ-nucleon interaction on the asymmetry is predicted to be negligible.

REFERENCES

1. R. Tacik et al., Phys. Rev. Lett. **63**, 1784 (1989).
2. Yi-Fen Yen et al., Phys. Rev. Lett. **66**, 1959 (1991).
3. J. T. Brack et al., Phys. Rev. C **45**, 698 (1992).
4. R. Mach and S. S. Kamalov, Nucl. Phys. **A511**, 601 (1990).
5. S. Chakravarti et al., University of Minnesota, Annual Report 1990, and D. Dehnhard et al., Few-Body Systems, Suppl. **5**, 274 (1992).
6. P. B. Siegel and W. R. Gibbs, Phys. Rev. C **48**, 1939 (1993), and private communications.
7. Yi-Fen Yen, PhD Thesis, U. Minnesota, 1991, and Yi-Fen Yen et al., Phys. Rev. C **50**, 897 (1994).
8. R. A. Brandenburg, Y. E. Kim, and A. Tubis, Phys. Rev. C **12**, 1368 (1975).
9. J. L. Friar et al., Phys. Rev. C **34**, 1463 (1986).
10. D. R. Tilley, H. R. Weller, and H. H. Hasan, Nucl. Phys. **A474**, 2 (1987).
11. B. Larson et al., Phys. Rev. Lett. **67**, 3356 (1991), and B. Larson et al, Phys. Rev. C **49**, 2045 (1994).
12. B. Larson et al., Phys. Rev. A **44**, 3108 (1991).
13. TRIUMF Annual Report, 1993, and W. Cummings et al., in these Proceedings.
14. S. S. Kamalov, L. Tiator, and C. Bennhold, Phys. Rev. C **47**, 941 (1993), and C. Bennhold, B. K. Jennings, L. Tiator, and S. S. Kamalov, Nucl. Phys. **A540**, 621 (1992).
15. P. B. Siegel and W. R. Gibbs, Phys. Rev. C **48**, 1939 (1993), and priv. comm.
16. S. Chakravarti, C. M. Edwards, D. Dehnhard, and M. A. Franey, Few-Body Systems, Suppl. **5**, 267 (1992).

Polarization Transfer in p-d Scattering at 22.7 MeV

W. Kretschmer[a]
POLAR Collaboration[b]

[a] *Universität Erlangen-Nürnberg, D-91058 Erlangen, Germany*

Abstract

For the elastic proton-deuteron scattering polarization transfer observables have been measured at E_p = 22.7MeV both for proton to proton and proton to deuteron polarization transfer. The new data are compared with rigorous three-body calculations using Paris, Nijmegen, Bonn-A and Bonn-B potentials.

Introduction

The Boson exchange model has been very successfull in the description of the nucleon-nucleon (NN) interaction since it reproduces a huge amount of NN data over a wide energy range. Despite the fact that the main features of the interaction appear to be well determined by NN scattering experiments several important questions still remain open and consequently leave room for the potentials to disagree in detail. Examples of such problems are the interaction in the ^3P_J states possibly displaying charge dependence and charge symmetry breaking and the $^3S_1 - ^3D_1$ mixing parameter ϵ_1 which is a measure of the NN tensor force. In the low energy regime the proton-proton ^3P_J phase shifts are well determined by high precision analyzing power and polarization transfer measurements [1] in p-p scattering; the situation is worse for the corresponding

---

[b]POLAR Collaboration: B. Aumüller, A. Glombik, G. Martin, K. Mümmler, G. Suft, R. Weidmann: *Universität Erlangen-Nürnberg*. M. Bruno: *INFN, University of Bologna, 40126 Bologna, Italy*. M. Clajus: *Physics Dept., UCLA, Los Angeles, CA 90024, USA*. P. M. Egun, W. Grüebler: *Institut für Mittelenergiephysik, ETH, Hönggerberg, CH-8093 Zürich, Switzerland*. W. Glöckle, H. Witala: *Institut für theoretische Physik, Ruhr-Universität Bochum, 44780 Bochum, Germany*. G. Mertens: *Physikalisches Institut, Universität Tübingen, 72076 Tübingen, Germany*. P. Hautle, P. A. Schmelzbach: *Paul Scherrer Insitute, F1, Accelerator Division, CH-5232 Villigen, Switzerland*. I. Šlaus: *Institute Rudjer Bošković, 41000 Zagreb Croatia*.

n-p phase shifts and for the mixing parameter ϵ_1 since the experiments are less accurate and since there are more free parameters. The accessible accuracy for NN data seems to be limited, therefore a possible alternative may be provided by the study of the 3N system [2] using rigorous 3N calculations [3] with realistic NN potentials as input. The spin structure of the 3N systems is 1/2 + 1 providing a large number of spin observables. Our attempt is to measure a nearly complete set of proton-deuteron (p-d) observables at 22.7 MeV to obtain additional information on the NN-interaction from a combined analysis of all observables. For this reaction we present here data for proton to proton polarization transfer [4] and for proton to deuteron vector and tensor polarization transfer observables compared to rigorous three-body calculations using Paris, Nijmegen, Bonn-A and Bonn-B potentials.

Experiment

The angular distributions of the polarization transfer observables in the reactions ^{2}H$(\vec{p},\vec{d})^1$H and ^{2}H$(\vec{p},\vec{p})^2$H were measured with the polarized proton beam of the PSI injector cyclotron at an energy of $E_{lab} = 22.7$ MeV. The typical beam intensity was 600nA – 800nA with a polarization of 76.0% – 80.0%. The proton polarization was monitored online using a ^{12}C polarimeter located upstream the deuterium gas target. The calibration of this polarimeter has been performed using the elastic p-^{4}He scattering, for which the absolute value of the analyzing power is known very well [5]. The direction of the proton spin can be rotated by a set of solenoids and bending magnets. In the primary scattering chamber the protons were scattered in a deuterium gas target pressurized to 7.5 bar absolute and cooled to 77 K by LN$_2$. The unscattered proton beam was stopped in a Faraday cup. The sign of the beam polarization was reversed every few seconds after accumulating a certain amount of protons in the Faraday cup. This method allows the determination of K_k^{ij} (k = x, y, z) from the ratio of the detector counting rates and avoids systematic errors arising from instrumental asymmetries. For a background reduction and for an increase of the solid angle the ejectiles (p or d) from the elastic p-d scattering were focussed into the polarimeter by a magnetic quadrupole triplet located between the primary scattering chamber and the polarimeter. The polarization of the scattered protons was measured via the elastic scattering on ^{4}He or ^{12}C depending on their energy [4], whereas the polarization of the scattered deuterons was deduced from the ^{3}He$(d,p)^4$He reaction. Protons from this latter reaction were easily distinguished from the background due to the high Q value of 18.4MeV. The deuteron polarimeter [6] consists of a ^{3}He gas cell pressurized to 12 bar absolute and cooled to 14 K, and of nine dE-E scintillator telescopes positioned at $\theta_{lab} = 0°$, 25° and 45° in the arrangement left, right, up and down. This polarimeter has the unique feature of measuring simultaneously many spin observables with high efficiency. An event is classified as a proton from the reaction ^{3}He(d, p)^{4}He if the correct energy signal from a dE-E telescope and a correct time-of-flight signal is determined. The time-of-flight is measured between the detector at the exit window of the primary scattering chamber and the passing detector in front of the entrance of the ^{3}He cell.

Results and Discussion

For the proton to proton polarization transfer the observables $K_y^{y'}, K_x^{x'}$ and $K_z^{x'}$ were measured in the angular range of $40° \leq \theta_{c.m.} \leq 130°$. The angular distributions of $K_y^{y'}$ and $K_z^{x'}$ are shown in figs. 1 and 2 compared with theoretical predictions based on the Argonne AV14, Bonn A and B, Paris and Nijmegen potentials. The $K_x^{x'}$ data, not shown here, are well described by all five potentials, whose predictions for that observables are very similar. For $K_y^{y'}$ the best description of the data is given by the Nijmegen potential, whereas the situation for $K_z^{x'}$ is not so clear, since the predicted position of the maximum at $\theta_{c.m.} \approx 125°$ differs from the experimental result. As discussed in more detail in [4], both observables are sensitive to an interplay of the $^3S_1 - ^3D_1$ tensor force and the $^1P_1$ force component, which may be used for a fine tuning of the potentials. Just modifying the Nijmegen potential by replacing its $^1P_1$ force component by the one of of Bonn B leads to a better agreement for $K_z^{x'}$ and keeps the overall good description of $K_y^{y'}$. The new measurements of proton to deuteron polarization transfer covers the angular range of scattered deuterons from $15° \leq \theta_{lab} \leq 40°$, corresponding to c.m. angles $100° \leq \theta_{c.m.} \leq 150°$. The counting statistics is of the order of $1\% - 2\%$. We present here the results for 120°-150°, the analysis of the 100° data requires a new calibration of the deuteron polarimeter, since in this case our measurement is out of range of the original calibration [6]. As an example of the new data base the polarization transfer observables $K_y^{y'}, K_y^{x'z'}$ and $K_z^{zz'}$ are shown in figs. 3 to 5 compared with Faddeev calculations using the Paris and Nijmegen potentials, the predictions of the Bonn potentials, which are between the Paris and Nijmegen predictions, are omitted for clarity of presentation. For all observables except for $K_z^{y'z'}$ the prominent structure around $\theta_{c.m.} = 125°$ is very well described by the calculations, supporting the validity of the theoretical approach despite the neglect of the Coulomb interaction. Additional information of the underlying NN potentials may be obtained from the description of the observables $K_y^{x'z'}$ and $K_z^{y'z'}$ [7], which are most sensitive to the NN force.

Supported by BMFT (FRG), PSI (CH) and by the European Community

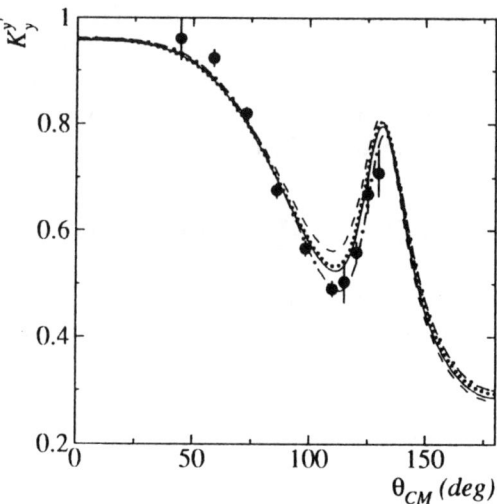

Fig. 1: $K_y^{y'}$ data compared with Faddeev calculations based on AV14(—), Paris(- - -), Bonn B($\cdots$) and Nijmegen ($-\cdot-$) potentials

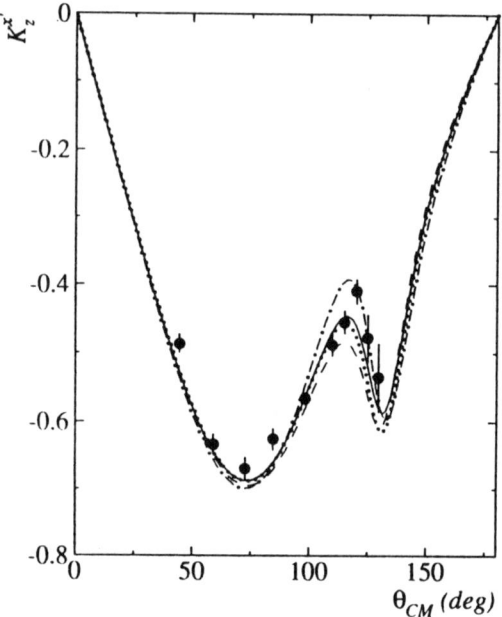

Fig. 2: $K_z^{x'}$ data compared with Faddeev calculations using Bonn A(—), Paris(- - -), AV14($\cdots$) and Nijmegen ($-\cdot-$) potentials

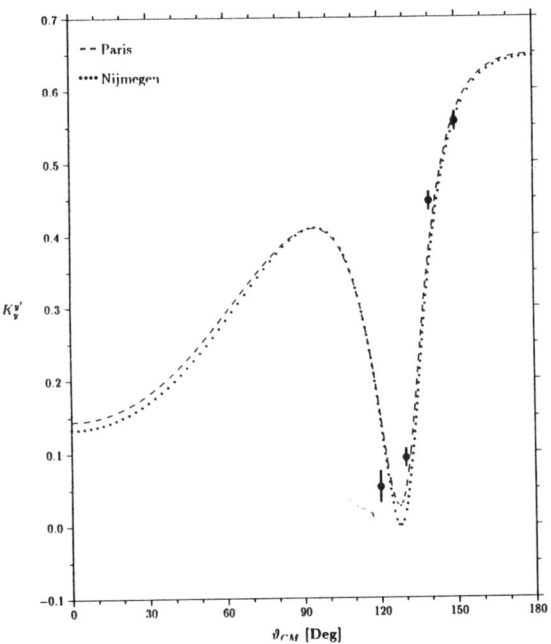

Fig. 3: Proton to deuteron polarizations transfer $K_y^{y'}$ compared with Faddeev calculations using Paris (- - -), Nijmegen ($\cdots$) potentials

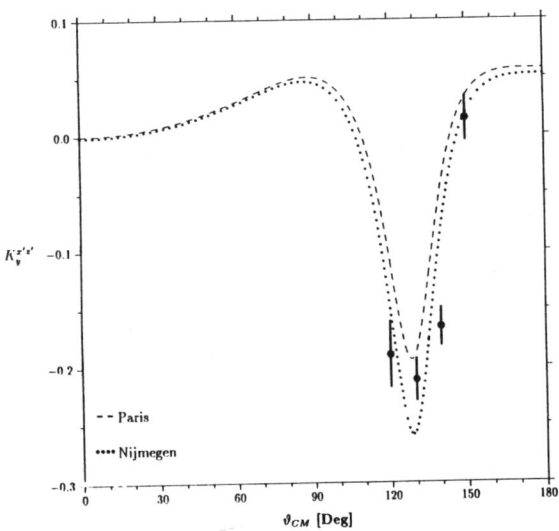

Fig. 4: The same as in fig. 3, now for $K_y^{z'z'}$

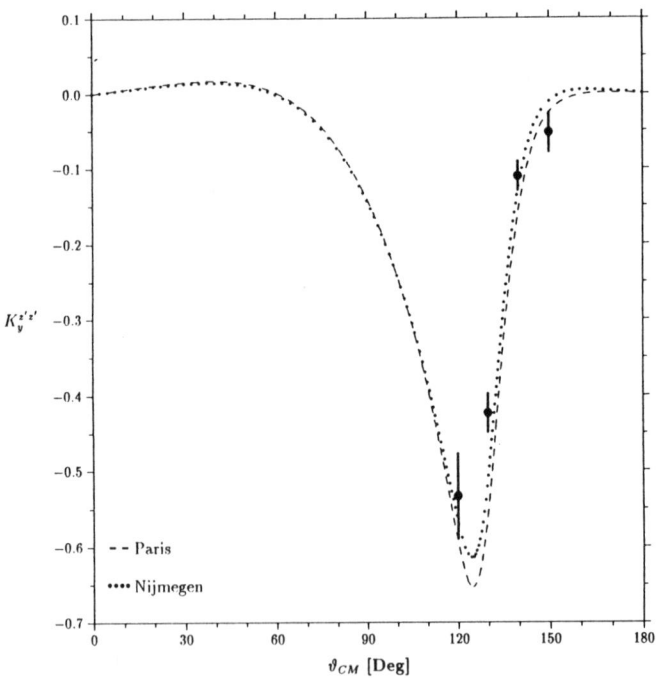

Fig. 5: The same as in fig. 3, now for $K_y^{x'z'}$

References

[1] W. Kretschmer et al., *Phys. Lett.* **B 328**(1994)5

[2] I. Šlaus, R. Machleidt, W. Tornow, W. Glöckle and H. Witala, *Comments Nucl. Part. Phys.* **20** (1991) 85

[3] H. Witala, W. Glöckle and T. Cornelius, *Nucl. Phys.* **A496** (1989)

[4] M. Clajus et al., submitted to *Phys. Rev.* **C**

[5] M. Clajus et al., *Nucl. Instr. Meth. in Phys. Res.* **A281**(1989)17

[6] W. Grübler et al., *Nucl. Instr. Meth. in Phys. Res.* **A262** (1987) 307

[7] A. Glombik et al., *Proc. 14th Int. IUPAP Conference on Few Body Problems in Physics*, Williamsburg 1994, to be published in *Nucl. Phys.*

Few Body Systems

Ana M. Eiró

Departamento de Física e Centro de Física Nuclear da Universidade de Lisboa, Campo Grande, 1700 Lisboa, Portugal

Abstract: We report twelve experiments involving few nucleons systems. The general purpose of the reactions with two or three nucleons is learning about barion-barion interaction, providing experimental constraints to model parameters in the scope of meson exchange theories, and making decisive contibutions to phase shift analysis. Information on reaction mechanisms when using the ^{3}He as a polarized target is, the goal of the reactions reported that involve four particles.

INTRODUCTION

In the few-body parallel session there were reported 12 experiments performed at very different energies, from 5 MeV to few GeV, having as a general purpose the better understanding of the barion-barion interaction. The on-shell behaviour of the interaction can be tested with experiments involving two nucleons, while the off-shell effects are prodminantly tested in the three body systems. Polarization experiments can in principle enhance certain small components of the interaction providing experimental constraints to solve the ambiguities that still exist between realistic potentials. In the last couple of years, the technique of producing polarized targets has greatly improved, and it was possible to produce high intensity beams of polarized neutrons, allowing the measurement of all sorts of spin-observables in the two-nucleon system. From the theoretical point of view, the level of sophistication that 3-body calculations have achieved, is acting as a tremendous encourage for experimentalist to get accurate data involving three nucleons that can really help to distinguish between potential models.

The papers presented were divided in two groups. The first group deals with NN reactions, making decisive contibutions to phase shift analysis. The second group deals with reactions involving three or four nucleons to obtain information on the 3-nucleon system, 3-body force and different reaction mechanisms.

TWO BARYONS

The recent determinations of the πNN coupling constant obtained by the Nijmegen group (1), $g_{\pi 0}^2/4\pi=13.47\pm0.11$ and $g_{\pi\pm}^2/4\pi=13.54\pm0.05$, are still a matter of controversy. The low value of the charged coupling constant, raises a problem at the level of some realistic potentials like Bonn, that in order to reproduce the bound state properties of the deuteron, namely those that depend directly from the tensor component of the interaction, imposes (2) a minimum value for the charged coupling constant to be 14. This feature is strongly connected with the strength of the ρ meson, in particular with the ratio $k\rho$ between the tensor and vector coupling constants, which in Bonn potential has to be kept of the order of 6.1 in order to reproduce the ^3P_J phase shifts.

In the attempt of contributing to the solving of this problem, R.T.Braun repported a measurement of A_y in n-p elastic scattering at 12 MeV performed at TUNL. High precision measurements of A_y for low energies n-p scattering are particularly valuable in placing constraints on the NN interactions associated with the $^3P_{0,1,2}$ phase shifts and they are expected to mainly probe the long-range part of the interaction, being extremely sensitive to the πN coupling constant, as seen in Fig.1 of ref.3. The results, shown in Fig.4 do not favour the model with different values for the neutral and charged coupling constant, indicating the need for some new ingridient in the potential.

One should not separate this question from the consideration of the type of form factor and the value of the cutt-off parameter used. There is now a considerable body of evidence, particularly related with the construction of three body force (4), sugesting that the $NN\pi$ vertex form factor is much softer than has been conventionally assumed in NN models. In the framework of OBE, this problem can be resolved (5) by the introduction of a heavier π' meson (with a mass of 1.2 GeV) that will enhance the short range isovector tensor force. Additional strength in the coupling constant of this π' meson can possibly account for the loss in tensor component implied by the lowering of the charged πNN coupling constant.

All this controverse about the πNN coupling constant motivated the investigation of the usefulness of high-quality spin measurements towards the resolving of this issue. A measurement of the normal component spin transfer observable $D_{NN'}(*)$ for the pp elastic scattering was recognize as particularly sensitive to $g_{\pi 0}^2$. An experiment performed at IUCF with polarized protons at Ep=200 MeV was reported by Sonya Bowyer (6).The results are shown in Fig.2 together with the predictions from three phase-shifts analysis and three realistic potentials. Interestingly the three phase-shifts curves differ significantly from one another, and the best agreements are obtained for the Nijmegen PWA93 and for Nijmegen I potential (7). This new potential, which contains momentum dependent terms, is consistent with the experimental low values of both the neutral and the charged πNN coupling constant and, using also a weak ρ reproduce well the bound state properties of the deuteron.

Another important issue in our current knowledge of the NN interaction is the complete understanding of charge symmetry breaking (CSB) that can give insights into the details of the interaction at short distances. This may be particularly important in forming the bridge between quantum chromodynamics and the effective NN interaction represented by one-boson exchange.

Charge symmetry is a weak symmetry, which requires only that the Hamiltonean H is invariant under reflection in the x-y plane of isopsin, i.e. [H, $e^{i\pi I_2}$] meaning that there is mixing of isopsin states, namely I=0 and I=1. Only the so-called class IV potentials in the classification of Henley and Miller (8) can contribute to the CSB in n-p elastic scattering. At different energies, the different mechanisms that contribute to these potentials have very different relative importance (9), motivating high precision data taking throughout the entire intermediate energy range. The observable sensitive to charge symmetry is the difference between the analyzing powers associated with the neutron spin, $A_n(\theta)$, and with the proton spin, $A_p(\theta)$. From the data previously obtained at TRIUMF (10) and at IUCF (11), it is clear that there are inconsistencies between the different PSA predictions and some ambiguities in the interpretation .

To contribute to the clarification of this question, measurements of the zero crossing angle of the n-p analyzing power below 300 MeV are underway at TRIUMF. The experiment (E498), reported by Davis (12), at bombarding energies of 174.3, 202.4, 216.5 and 260.3 MeV, seems to indicate some variation relative to the predictions of the PSA of Arndt (13). It would be very interesting to compare this results with predictions obtained with the new Argonne potential AV18 (14), which explicitly includes charge dependent and charge asymmetry terms.

Throughout the intermediate energy region there is still some inconsistencies in the description of the polarization observables using different phase shifts analysis, because there is not enough accurate spin dependent data to give stable solutions. The nucleon-nucleon elastic scattering can be described, for each isospin state I=0 and I=1, by five scattering complex amplitudes. From pp elastic scattering data, it is possible to establish phase shifts for the I=1 isospin state, but there is still lack of data at several energies for the np channel which originates that the I=0 phase shifts remain unprecise.

A major new program is underway at PSI, where the aim is to study the spin dependence in elastic np scattering between 260 and 560 MeV, and to determine the five NN amplitudes for I=0 in a model inependent way. The experimental set up will allow to obtain measurements of eleven spin observables, namely spin correlation (*) A_{00NN}, A_{00SK} and A_{00KK}, depolarization D_{0N0N} and D_{0S0K}, spin transfer K_{0NN0}, K_{0SS0} and K_{0SK0} and finally 3-spin parameters N_{0SSN} and N_{0SKN}. Preliminary results for K_{0SS0} and K_{0SK0} obtained at kinetic energies of 260, 320, 390, 470 and 550 MeV were reported by Vuaridel (15). At 550 MeV, where there is no other available data, there is significant differences in the predictions from PSA of Arndt (13) and Saclay-Geneva (16) suggesting the need of the inclusion of these data in the database.

To extend the scope of meson exchange models to the interpretation of baryon-baryon interaction with built-in SU(3) symmetry, request experimental data of hyperon nucleon elastic scattering. Both Nijmegen and Jülich model agrees well with the total cross section data, as well as QCD calculations including medium and long range interactions with one boson exchange. However these models largely desagree in the predictions of the few existing spin observables. The main purpose of the hyperon scattering experiment performed at KEK is to obtain Σ^+ and to measure differential cross section and polarization for the $\Sigma^+ + p$ elastic scattering. The results reported by Kim (17) were obtained for momentum of the recoil proton between 200 and 800 MeV/c and, from the 18 events identified, the total cross section was calculated. Is is expected that in the future the number of events will increase and allow to obtain polarization observables.

The pion production cross section near the threshold of the $NN \rightarrow NN\pi$ reactions, producing essentially S-wave pions, is a direct measurement of the axial exchange charge of the two nucleon system. Within the DWIA, the contributions to the cross section arise from one-body terms, pion rescattering terms and heavy meson exchange contributions (MEC). Recent measurements (18) of the $pp \rightarrow pp\pi^0$ were explained by the enhancement produced by the σ meson exchange current and furthermore it was prooved that the pion rescattering contribution is negligible (19). An important further test of this mechanism is provided by the bound state production where the reaction sample different spin and isospin matrix elements, and includes the important contribution from the D state of the deuteron (20) originating that the pion rescattering term becomes the major contribution.

An experiment of $pp \rightarrow d\pi^+$ reported by Heimberg (21) was carried out at IUCF, where they measure the cross section and analyzing power, for values of η ($=p_\pi^{cm}/m_\pi c$) = 0.03 to 0.2 corresponding to energies of the pion from 70keV to 2.8MeV, well below the previous measurements. The A_y, which is an interference between the S-wave with the P-wave amplitudes, was shown to be approximately -1.2η and the value obtained for the cross section was $(210\pm23)\eta$ μb, consistent with the one obtained by Hutcheon (18).

THREE AND FOUR NUCLEONS

The complete understanding of the interactions between nucleons includes the knowledge of the three nucleon forces (3NF). Although it is generally accepted the existence of 3NF, there is no direct experimental evidence and most of what we know from the details of these potentials comes from theoretical models. The goal of the paper presented by Knutson (22) was to identify observables in the three nucleon system with enhanced sensitivity to 3NF. It was shown that some components of the operator that describe 3NF are axial vectors, namely rank-1 operators that acting on the dominant L=0 component will generate positive parity P-state component. For a no-coplanar reaction, parity conservation relates analyzing powers for a particular geometry and the corresponding quantities for the reflected one. The logitudinal analyzing power A_z was identified as arising from interference between the dominat natural parity amplitudes and smaller uunatural parity ones, and therefore should show some sensitivity to 3NF. The experiment reported, pd→pnn in a no-coplanar geometry, was carried out at Wisconsin at proton energy of 9 MeV, showed very small values of A_z, consistent with zero. A Faddeev calculation carried out by Glockle and Witala predict for A_z, using the same kinematical conditions, a value of 0.002, but desapointingly the inclusion of 3NF does not have large effect on the observable.

The three nucleon system provides an independent tool to test several aspects of the NN interaction. Large discrepancies have been observed between the results of three nucleon Faddeev calculations and data (23) for $A_y(\theta)$ for elastic neutron-deuteron scattering, and to eliminate those discrepamcies a precise determination of the n-p $^3P_{0,1,2}$ phase shifts is needed. In an experiment, performed at Erlangen and reported by Kretschmer (24), a complete set of measurements of vector and tensor polarization transfer coefficients in a p-d elastic scattering were obtained at Ep=22.7 MeV. In particular it was mesured the proton polarization transfer in the $^2H(\vec{p},\vec{p})^2H$ and proton deuteron polarization transfer in the $^2H(\vec{p},\vec{d})^1H$. The measurements show (24) a resonable overall agreement with different calculations obtained with different realistic potential models providing an excellent test for these rigorous calculations in the break up channel. Although it is possible to establish that certain observables (*), like $K_y^{y'}$ for example, are clearly influentiated by the $^3S_1 - {}^3D_1$, it is not possible to distinguish between different interaction models used in the calculations.

Elastic scattering of protons and deuterons, if performed at higher energies, can provide other type of information, namely on the structure of the deuteron. If one considers backward elastic elastic scattering dp→pd, parity invariance and total helicity conservation implies that only only four independent complex amplitudes are needed to describe the process. Using the Impulse Approximation, and a non-relativistic deuteron wave function (DWF), it is possible to express the analyzing power T_{20} and the polarization transfer coefficient κ_0 as a function of the D/S state ratio of the deuteron. In an experiment reported by Sitnik (25) carried out at Dubna and Saclay, measurements of T_{20} and κ_0 were performed, up to an internal momentum of 0.570 GeV/c, showing desagreement with the theoretical predictions. To investigate if the reason of this discrepancy is the DWF used, it is proposed the measurement of the transversal final spin correlation (*) C_{NN00} which is predicted to have quite different behaviour depending on the use of a standard non-relativistic DWF or on a relativistic DWF that includes P states. These states are a pure relativistic and it will be therefore extremely interesting to obtain an experimental signature of such an effect.

There are several extremely accurate claculations (26) for the bound state of ^{3}H and ^{3}He which reproduce well three nucleon bound state properties. The experimental information on D-state parameters of three nucleon system is a crucial information for the utilization of ^{3}He target as a polarized neutron target. Using transfer reactions ($\vec{d}$,t) and ($\vec{d}$,^{3}He), Ayer et al (27) obtained a value for the D/S state ratio for the triton and ^{3}He. The experiment was run at TUNL at subcoulomb energies for the beam on ($\vec{d}$,t) reactions, using targets of ^{95}Mo, ^{119}Sn, ^{149}Sm and ^{206}Pb. The value of η_t was computed from an average of seven individual determinations leading to η_t=-0.0411±0.0013±0.0012 in very good agreement with the last experimental determination (28) and the theoretical predictions (26). For the case of the ^{3}He the situation is not so clear, due to the need of operating above the coulomb barrier and the sensitivity shown in the results to the description of tensor potentials in the incoming channel. The value obtained was η_{3He}=-0.0380± 0.0026±0.0011 providing a ratio of η_t/η_{3He}=1.08±0.12 in excellent agreement with the theory. It is nevertheless important to clarify the role of the tensor potentials in the analysis of these transfer reactions.

Getting rid of the uncertainties from nuclear stucture that occur in heavy targets motivates the study of elastic scattering of pions in ^{3}He, to learn about the mechanism of the reaction and the π nucleus interaction, particularly its spin dependent part. Using LAMPF beam of π^+ on a polarized ^{3}He gas target of TRIUMF, the cross sections and A_y were measured at T_π = 100, 142, 180 and 256 MeV in an experiment reported by Cummings (29). It is expected that for energies at and above the P_{33} resonance the asymmetries will become increasingly sensitive to the details of the reaction model. Theoretical curves obtained from DWIA do not agree with the data (Fig.4 of ref.29), even when rescattering terms are included. Bearing in mind that in the π^+ scattering from protons there is a very high probability for the formation of an intermediate Δ^{++}, this state was considered in the calculations. Using a meson exchange model for the Δ-neutron intearction including $\pi, \rho, \omega, \sigma$ and η mesons, a very good fit was obtained for the data at 180 MeV. However the energy dependence of the asymmetry was not predicted correctly suggesting the need for further experimental and theoretical work. It will be interesting to measure π^- scattering, as in the Δ resonance region π^-n interaction is about 10 times stronger that π^+n, originating a much larger first order effect and a consequent negligible effect of the Δ.

In some recent experiments at TRIUMF (30) of polarization observables for the (p,2p) and (p,pn) knock-out reactions on ^{3}He the results differ considerably form those for free pp scattering, and are predicted to be sensitive to the spin momentum distribution of protons in ^{3}He. Because it is widely beleived that polarized ^{3}He can be used to study the spin structure of the neutron, this matter deserve further investigation. With the purpose of expanding the kinematics covered by the existing data, measurements of analyzing powers and spin correlation parameters for $^3\vec{\text{He}}(\vec{p},2p)$ and $^3\vec{\text{He}}(\vec{p},pn)$ quasi elastic scattering at proton energy of 197 MeV, were obtained at IUCF and were reported by Miller (31). Within plane wave impulse approximation (PWIA) polarization observables can be simply related with the free-scattering spin observables, through spin-dependent spectral function, which are well known in the case of ^{3}He. If PWIA holds, one expects that the beam and the target analyzing powers should be identical, and from the free scattering spin correlation parameters calculated from phase shifts and the spectral functions, theoretical predictions can be compared with data. Unlike the previous results, this data were found to be in good agreement with PWIA, probably due to the particular kinematic chosen. It will be important to carry out a calculation where rescattering terms and final state interaction in the unobserved pair could be included.

(*) We made the option of keeping the notations of the authors of the papers presented. For clarification of notations see N. Hoshizaki, Proc. Sixth Int. Symp. Polar. Phen. in Nucl. Phys., Osaka, 1985, J. Phys. Soc. Jpn. 55 (1986) Suppl. p.549-552.

REFERENCES

1. Klomp, R.A.M. et al., Phys. Rev. **44**, R1258-R1261 (1991); Timmermans, R.G.E. et al., Phys. Rev. Lett. **67** 1074-1077 (1991).
2. Machleidt, R. and Sammarruca, F., Phys. Rev. Lett. **66**, 564-567 (1991).

3. Braun, R.T. et al., *Neutron-proton analyzing power at 12 MeV and charged πNN coupling constant*, Proceedings from SPIN94.
4. Sasakawa, T., et al., Phys. Rev. Lett. **68**, 3503- (1992).
5. Holinde, K. and Thomas, A.W., Phys. Rev. **C42**, R1195-1199 (1990).
6. Bowyer, S.M., et al. *A Measurement of the spin transfer observable D_{NN} for the Elastic Scattering at Tp=200MeV*, Proceedings from SPIN94
7. Stocks, V.G.J. et al., Phys. Rev. **C49**, 2950-2962 (1994).
8. Henley, E.M. and Miller, G.A., edited by M. Rho and D.H. Wilkinson, *Mesons in Nuclei*, Amsterdam, North-Holland, 1979, Vol.1, p.405.
9. Williams, A.G., Thomas, A.W., Miller, G.A., Phys.Rev. **C36**, 1956-1967 (1987).
10. Abegg, R. et al., Phys. Rev. **C40**, 2406-2409 (1989); Phys. Rev. **C39**, 2464-2483 (1989).
11. Vigdor, S.E. et al., Phys. Rev. **C46**, 410-448 (1992).
12. Davis, C.A. et al., *Zero crossing angle of the np analyzing power below 300MeV*, Proceedings from SPIN94.
13. Arndt, R.A., Interactive Dial-In Program SAID at VPI, SP94.
14. Wiringa, R.B., Stoks, V.G.J. and Schiavilla, R., *An accurate nucleon-nucleon potential with charge-independent breaking*, preprint (1994).
15. Ahmidouch, A. et al., *Spin observables in neutro-proton elastic scattering*, Proceedings from SPIN94.
16. Bystricky, J. et al., *private communication*, 1993 solution.
17. Ahn, J.K. et al., $\Sigma^+ + p$ scattering experiment at KEK, Proceedings from SPIN94.
18. Hutcheon, D.A. et al., Nucl. Phys. A535, 637- (1991);Meye, H.O. et al., Nucl. Phys. **A539** 633-661 (1992);
19. Lee, T.S.H. and Riska,D.O., Phys. Rev. Lett. **70** (1993) 2237-2240; Horowitz, C.J. et al., Phys. Rev. **C49** (1994) 1137-1346.
20. Horowitz, C.J., Phys. Rev. **C48** 2920-2925 (1993).
21. Heimberg, P. et al., *Cross section and analyzing powers of $p(\vec{p},\pi^+)d$ very near threshold*, Proceedings from SPIN94.
22. Knutson, L. D., *Polarization observables for the p-d breakup reaction and tha nuclear three-body force*, Proceedings from SPIN94.
23. Glöckle, W., Witala, H. and Cornelius, Th., Nucl. Phys. **A508**, 115c-130c, (1990).
24. Kretschmer, W., *Polarization transfer in p-d scattering at 22.7 MeV*, Proceedings from SPIN94.
25. Sitnik, I.M. and Ladygin, V.P., *Status and future of polarization phenomena investigations in backward elastic deuteron-proton scattering*, Proceedings from SPIN94.
26. Wu, Y., Ishikawa, S. and Sasakawa, T. Few-Body Systems, **15** 145-188 (1993); Friar, J.L. et al., Phys. Rev. **C37**,2859-2868 (1988).
27. Ayer, Z. et al., *Determination of the asymptotic D- to S-state ratio for the triton and ^{3}He via (d,t) and (d,^{3}He) reactions*, Proceedings from SPIN94.
28. George, E.A. and Knutson, L.D., Phys. Rev. **C48** 688-698 (1993).
29. Espy, M.A. et al., *Elastic π^+ scattering on polarized ^{3}He*, Proceedings from SPIN94.
30. Rahav,A.et al.,Phys.Rev.**C46** 1167-1177 (1992);Brash,E.J. et al.,Phys.Rev.**C47** 2064-2076 (1992);
31. Miller, M.A. et al., *Measurement of spin observables in quasi elastic scattering of polarized protons from polarized ^{3}He*, Proceedings from SPIN94.

IV. INTERMEDIATE-ENERGY HADRON-INDUCED REACTIONS

The Nuclear Spin-Isospin Response to Quasifree Nucleon Scattering

T.N. Taddeucci

Los Alamos National Laboratory, Los Alamos, NM 87545

Abstract. The Neutron-Time-of-Flight (NTOF) facility at LAMPF has been used to measure complete sets of polarization-transfer coefficients for quasifree $(\vec{p},\vec{n})$ scattering from ^{2}H, ^{12}C, and ^{40}Ca at 494 MeV and scattering angles of 12.5°, 18°, and 27° ($q = 1.2, 1.7, 2.5$ fm^{-1}). These measurements yield separated transverse ($\boldsymbol{\sigma} \times \mathbf{q}$) and longitudinal ($\boldsymbol{\sigma} \cdot \mathbf{q}$) isovector spin responses. Comparison of the separated responses to calculations and to electron-scattering responses reveals a strong enhancement in the spin transverse channel. This excess transverse strength masks the effect of pionic correlations in the response ratio.

INTRODUCTION

Mesonic fields in the nucleus may reveal their presence through collective effects on the quasifree nuclear response. In the $\pi + \rho + g'$ model of the residual particle-hole interaction, the pion field at moderate momentum transfers (1–2 fm^{-1}) produces a spin-longitudinal interaction ($\boldsymbol{\sigma} \cdot \mathbf{q}$) that is attractive, and the exchange of rho mesons produces a transverse interaction ($\boldsymbol{\sigma} \times \mathbf{q}$) that is repulsive. Much interest was generated by an early prediction that an interaction with these characteristics would lead to an enhancement and softening (shift toward lower energy transfer) of the quasifree isovector longitudinal spin response and a quenching and hardening (shift toward higher energy transfer) of the quasifree isovector transverse spin response [1].

The quasifree isovector transverse spin response can be measured with the (e, e') reaction, which is about 97% isovector because of the relative magnitudes of the isovector and isoscalar nucleon magnetic moments. Most analyses of quasifree (e, e') data lead to the conclusion that the isovector $1p$–$1h$ response is indeed quenched. Electromagnetic probes such as the electron are not sensitive to the longitudinal spin response, however. The first information about this response was obtained from $(\vec{p}, \vec{p}')$ measurements performed more than ten years ago with the High Resolution Spectrometer (HRS) at LAMPF

[2, 3]. After corrections for the isoscalar contribution to this reaction, these measurements yield the ratio of the longitudinal and transverse isovector spin responses. The surprising result obtained from analysis of these data was that there is no apparent enhancement of the spin longitudinal response relative to the spin transverse response.

The purely isovector $(\vec{p},\vec{n})$ reaction is better suited for study of the isovector nuclear response. The construction of the Neutron-Time-of-Flight (NTOF) facility at LAMPF was strongly motivated by the need to follow up the initial $(\vec{p},\vec{p}')$ measurements with similar and easier-to-interpret measurements using the $(\vec{p},\vec{n})$ reaction. Measurements of the quasifree response using polarized beam commenced at NTOF in 1990. The first results obtained with the $(\vec{p},\vec{n})$ reaction at 494 MeV and 18° (1.7 fm^{-1}) confirmed the $(\vec{p},\vec{p}')$ analysis by revealing no enhancement in the ratio of spin longitudinal to spin transverse responses, this time without the problematic corrections for isoscalar contributions [4]. In addition, the $(\vec{p},\vec{n})$ measurements provided the first look at the separated responses, rather than just the ratio [5].

Comparison of the transverse responses for ^{12}C and ^{40}Ca to (e,e') results at the same momentum transfer seemed to show good agreement between the two probes. This strengthened the conclusion that the longitudinal response, associated with the nuclear pion field, was the source of the problem. This perplexing result has fueled several diverse theoretical efforts that explore, separately, the effects of distortions and coupling to the Δ [6, 7, 8], in-medium changes in the nucleon-nucleon (NN) amplitudes and coupling constants [9, 10], and changes in the residual particle-hole interaction arising from density-dependence of hadron masses [11].

Unfortunately, the initial comparisons of the separated $(\vec{p},\vec{n})$ responses to electron-scattering and to theoretical responses were incorrect by a factor of two because of misunderstandings about the proper normalization of the electron data and theoretical response definitions. Revised comparisons now reveal that the transverse response as measured by the $(\vec{p},\vec{n})$ reaction appears to be much larger than that obtained in electron-scattering measurements. Analysis of the original $(\vec{p},\vec{n})$ data and new data obtained at lower and higher momentum transfer shows that expected enhancements in the longitudinal spin response, if present, are largely overshadowed by an excess of strength in the transverse channels.

EXPERIMENTAL METHOD

A detailed description of the NTOF facility at LAMPF and pertinent experimental techniques can be found in the report of the first quasifree polarization transfer measurement [5]. This initial measurement (18°) took place in 1990 when the new optically pumped polarized ion source (OPPIS) first came on-

TABLE 1. Quasifree $(\vec{p}, \vec{n})$ measurements at $E_p \simeq 494$ MeV.

θ_{lab}	ω_{free} (MeV)	ω_{QF} (MeV)	q_{lab} (fm^{-1})	Δq (fm^{-1})	targets
12.5°	28.9	53	1.21	1.19–1.50	CD$_2$, C
18.0°	58.1	82	1.72	1.70–1.87	CD$_2$, C, Ca
27.0°	121.7	138	2.52	2.52–2.63	CD$_2$, C, Ca

line for production use. The new data presented here were obtained during the 1992 (12.5°) and 1993 (27°) running periods at LAMPF. Regrettably, 1993 marks the end of polarized-beam experiments at LAMPF.

The NTOF detector/polarimeter consists of four position-sensitive scintillator planes. The collection area of each plane is approximately 102×107 cm^2. The planes are grouped into front and back pairs separated by approximately 1.4 m. The first three planes are stainless-steel tanks filled with liquid scintillator (BC-517s, H:C=1.7). The fourth plane is a set of ten plastic scintillator (BC-408) bars. Incident neutron energy is determined by time of flight to the front pair of detector planes, which also serve as polarization analyzers. Time, position, and pulse-height information from the front and back pairs of planes is used to kinematically select $n + p$ interactions. Neutron polarization is determined from the azimuthal intensity distribution of the $n + p$ events. Elastic (or $p + C$ quasielastic) $(\vec{n}, n)$ and $(\vec{n}, p)$ events are identified and sorted separately. The detection efficiency of the system in polarimetry mode is about 6×10^{-3} for the (n, p) channel and about 1.5×10^{-3} for the (n, n) channel. The effective analyzing power for each channel is about 0.15 and 0.23, respectively.

Complete sets of polarization-transfer coefficients were measured for $(\vec{p}, \vec{n})$ reactions on CD$_2$, natural C (98.9% ^{12}C), and natural Ca (96.9% ^{40}Ca) with an average beam energy of 494 MeV and a neutron flight path of 200 m. Overall energy resolution was about 2 MeV. Typical beam intensities were in the range from 50–100 nA, with beam polarization in the range from 0.50–0.65. Data for the ^{2}H(p, n) reaction were obtained from the cross-section-weighted difference of the CD$_2$ and C results. This subtraction is accurate to better than 3%. Cross sections were normalized relative to the ^{7}Li$(p, n)^7$Be(g.s.+0.43-MeV) transition at 0°, for which the cross section is $\sigma_{c.m.}(0°) = 27.0$ mb/sr [12].

A summary of the quasifree measurements is presented in Table 1. The momentum transfer q_{lab} corresponds to the peak of the quasifree distribution for ^{12}C$(\vec{p}, \vec{n})$ and ^{40}Ca$(\vec{p}, \vec{n})$, which is about $\omega_{QF} - \omega_{free} \approx 20$ MeV higher than the energy loss for free scattering. Because the measurements are made at a fixed angle, the momentum transfer is not constant, but varies slightly with energy loss across the spectrum. The range of values Δq corresponds to $\omega = 30$–150 MeV for $\theta = 12.5°$ and 18°, and $\omega = 30$–200 MeV for $\theta = 27°$.

DATA REDUCTION

The spin responses are obtained from $(\vec{p}, \vec{n})$ cross-section and polarization-transfer data by transforming the laboratory-frame polarization-transfer coefficients $\{D_{SS'}, D_{NN'}, D_{LL'}, D_{SL'}, D_{LS'}\}$ into a special set $\{D_0, D_n, D_q, D_p\}$ of c.m.-frame observables [13]. The c.m. coordinates are defined so that $\hat{n}$ is perpendicular to the reaction plane, $\hat{q}$ is along the direction of momentum transfer, and $\hat{p} = \hat{n} \times \hat{q}$. From these c.m. observables four responses (R_0, R_n, R_q, R_p) corresponding to the spin operators $\sigma_0, \sigma \cdot \hat{n}, \sigma \cdot \hat{q}$, and $\sigma \cdot \hat{p}$ can be obtained. The $(\vec{p}, \vec{n})$ responses are defined by

$$R_j = \frac{1}{N} \frac{1}{(2J_i+1)} \sum_{i,f} |\langle f | \sum_{k=1,A} \sigma_{jk} \tau_k^- e^{-i\mathbf{q}\cdot\mathbf{r}_k} | i \rangle|^2 \delta[\omega_{\mathrm{cm}} - (E_f - E_i)] . \tag{1}$$

and are normalized such that

$$\int R_i(q,\omega)\, d\omega = 1 \quad \text{as} \quad q \to \infty . \tag{2}$$

The responses to the two transverse operators $\sigma \cdot \hat{n}$ and $\sigma \cdot \hat{p}$ are identical [14] and can be equated to the response R_T to the transverse operator $(\sigma \times \hat{q})/\sqrt{2}$.

In a factorized impulse-approximation model, the relationship between the measured cross section and polarization transfer and the spin responses is given by

$$ID_0 = N_{\mathrm{eff}} C_K (|A|^2 R_0 + |C|^2 R_n) , \tag{3}$$
$$ID_n = N_{\mathrm{eff}} C_K (|C|^2 R_0 + |B|^2 R_n) , \tag{4}$$
$$ID_q = N_{\mathrm{eff}} C_K |E|^2 R_q , \tag{5}$$
$$ID_p = N_{\mathrm{eff}} C_K |F|^2 R_p , \tag{6}$$

where I is the double-differential cross section, C_K is a kinematic factor, and N_{eff} is a distortion factor represented as an effective number of neutrons. The distortion factor is assumed to be spin-independent and has values in the range $N_{\mathrm{eff}} \simeq 2.2$–$2.4$ for $^{12}\mathrm{C}(\vec{p},\vec{n})$ and $N_{\mathrm{eff}} \simeq 4.5$–$5.0$ for $^{40}\mathrm{Ca}(\vec{p},\vec{n})$. The NN amplitudes in these equations are from the standard KMT representation [15]:

$$M_{12} = A + C(\sigma_1 + \sigma_2)\cdot\hat{n} + B\sigma_1\cdot\hat{n}\sigma_2\cdot\hat{n} + E\sigma_1\cdot\hat{q}\sigma_2\cdot\hat{q} + F\sigma_1\cdot\hat{p}\sigma_2\cdot\hat{p} . \tag{7}$$

In the analysis of the data, the NN amplitudes are obtained from an optimal-frame transformation that provides the best factorization from the transition matrix [13]. The main effect of this transformation is to split the spin-orbit (C) amplitude into two unequal pieces, and is important only away from the peak of the quasifree distribution. Some other complications that are not explicitly accounted for in this simple factorized model are multistep contributions to the inclusive spectrum, spin-dependent distortions, and medium modification of the NN amplitudes.

The longitudinal response R_q and transverse response R_p are extracted from the data in a straightforward way by dividing the partial spin cross sections by the corresponding product of distortion and kinematic factors and NN amplitudes. The non-spin response R_0 and the transverse response R_n can in principle be obtained by matrix inversion of Eqs. (3)–(4). However, the amplitudes A and C are small and introduce large uncertainties into the formal solution for R_0 and R_n. For the case of R_n, a better approach is to take advantage of the relative sizes of the C and B amplitudes and rewrite Eq. (4) in the form

$$ID_n = N_{\text{eff}} C_K |B|^2 R_n \left(1 + \frac{|C|^2 R_0}{|B|^2 R_n}\right) . \tag{8}$$

For momentum transfer of $q \simeq 1.7$ fm^{-1}, the quantity in parentheses has a value of about 1.07 (assuming that the response ratio R_0/R_n is of order unity). Large uncertainties in the ratio in parentheses thus translate into relatively small uncertainties in the extracted response R_n. The ability to obtain this independent transverse response is very important. Spin-dependent distortions should have different effects upon the the in-plane and out-of-plane polarization observables and partial cross sections, while the underlying responses R_n and R_p should be identical. The consistency of the R_n and R_p responses extracted from the data thus puts some limits on the importance of spin-dependent distortions.

In the context of the factorized impulse approximation, the longitudinal-to-transverse response ratio R_q/R_p is a robust quantity in the sense that several common theoretical and experimental normalization factors cancel out. This ratio can be obtained from Eqs. (5)–(6) in the form

$$\frac{R_q}{R_p} = \frac{D_q/D_p}{|E/F|^2} . \tag{9}$$

The amplitude ratio $|E/F|^2$ can be obtained from NN phase-shift solutions. This is the method used in the previous analysis of the 18° data [5]. Alternately, for energy loss near ω_{free} the amplitude ratio can be replaced by the ratio $(D_q/D_p)_{^2\text{H}}$ for $^2\text{H}(\vec{p},\vec{n})$. This then gives the response ratio entirely in terms of measured quantities:

$$\frac{R_q}{R_p} = \frac{D_q/D_p}{(D_q/D_p)_{^2\text{H}}} . \tag{10}$$

This approach relies on the assumption that the response ratio R_q/R_p for $^2\text{H}(\vec{p},\vec{n})$ is near unity. Recent calculations indicate that this assumption will be closely met if the ^2H observables are obtained by integrating over the smallest

TABLE 2. Ratio of longitudinal to transverse polarization transfer for ^{2}H$(\vec{p},\vec{n})$ at 494 MeV.

θ_{lab}	ω (MeV)	D_q/D_p
12.5°	25–40	0.653 ± 0.055
18.0°	50–75	1.70 ± 0.11
27.0°	109–139	5.34 ± 0.65

possible region centered on the peak of the ^{2}H$(\vec{p},\vec{n})$ quasifree distribution [14, 16]. This procedure minimizes effects from the initial deuteron D-state and from tensor correlations in the 2p final state and gives the closest measure of the free $|E/F|^2$ ratio. Integration regions with widths of 15, 20, and 30 MeV have been used in the present analysis of the ^{2}H data at 12.5°, 18°, and 27°, respectively. The width of the region was increased in approximate proportion to the momentum transfer to account for the spreading of the quasifree distribution. The polarization-transfer ratios for ^{2}H$(\vec{p},\vec{n})$ are shown in Table 2.

RESULTS

The response ratios obtained from the data-to-data ratio [Eq. (10)] are displayed in Fig. 1. The results for ^{12}C$(\vec{p},\vec{n})$ and ^{40}Ca$(\vec{p},\vec{n})$ are essentially identical. Theoretical ratios have been calculated in a distorted-waves impulse-approximation (DWIA) model employing random-phase-approximation (RPA) responses generated with a $\pi + \rho + g'$ interaction ($g' = 0.6$) [6, 7, 8, 17]. Delta-hole (Δ–N^{-1}) contributions are included according to the standard universality ansatz, for which $g'_{NN} = g'_{N\Delta} = g'_{\Delta\Delta}$ and $f_{\pi NN} = 2.0 f_{\pi N\Delta}$. Two cases are shown: the solid lines correspond to calculations employing the full RPA response, and the dotted lines correspond to setting the residual interaction to zero (free response). The free-response calculations give a good description of the data at all three angles. While this result highlights the possible importance of distortion effects, the disagreement with the full RPA ratios also suggests that some important physics is being missed by describing the reaction entirely in terms of single-particle responses.

A clearer understanding of the response ratios is obtained by examining the separate responses and comparing them to appropriate benchmarks. Three comparisons are especially interesting: transverse spin responses obtained from electron scattering, the free Fermi-gas response, and RPA responses. The separate R_q and R_p responses for ^{12}C$(\vec{p},\vec{n})$ are shown in Fig. 2, and those for ^{40}Ca$(\vec{p},\vec{n})$ are show in Fig. 3. These responses are obtained assuming a

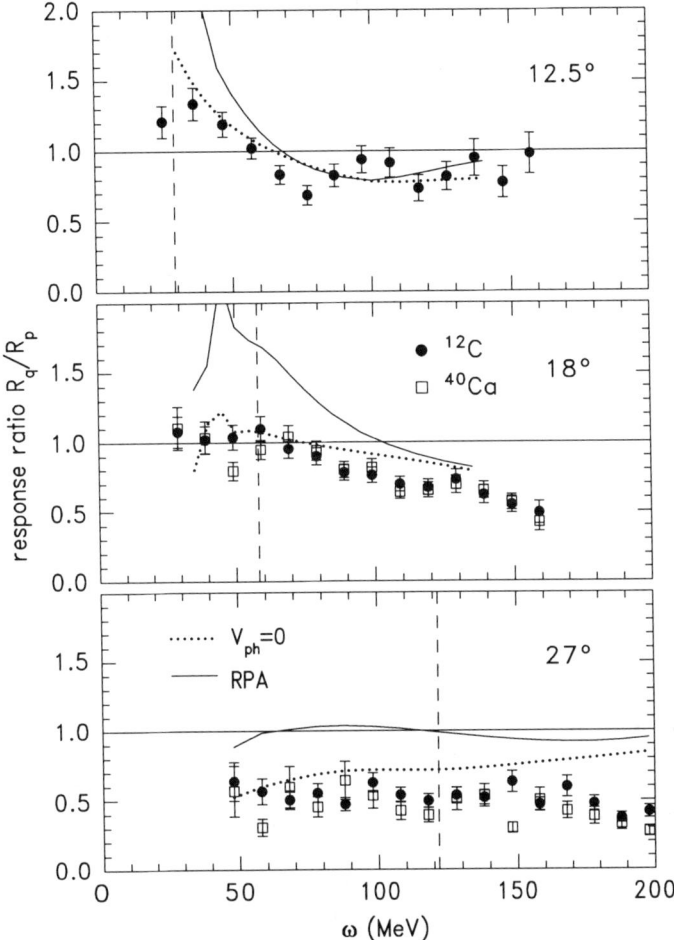

FIGURE 1. Longitudinal-to-transverse response ratios for $^{12}C(\vec{p},\vec{n})$ (solid circles) and $^{40}Ca(\vec{p},\vec{n})$ (open boxes). The ratios are calculated as the ratio of spin observables $(D_q/D_p)/(D_q/D_p)_{^2H}$ for ^{12}C or ^{40}Ca with respect to ^2H, with the ^2H values determined from a narrow region centered on the energy loss for free scattering (dashed vertical lines). The solid lines represent analogous ratios calculated in a RPA+DWIA model. The dotted lines represent DWIA calculations with the residual interaction set to zero (free responses).

common value for N_{eff} in both spin channels. The dotted curves represent the free Fermi-gas response R_{FG} calculated at a Fermi momentum of $k_F = 192$ MeV/c. This value corresponds to the average density $\langle \rho \rangle = 0.34\text{–}0.36\rho_0$ sampled by the $(\vec{p}, \vec{n})$ probe at this energy. The Fermi-gas response has been shifted by 18 MeV for ^{12}C and by 15 MeV for ^{40}Ca to account for the ground-state Q-value.

The solid curves in Figs. 2–3 represent ^{12}C(e, e') responses at momentum transfers of $q = 250$, 350, and 500 MeV/c [18] and ^{40}Ca(e, e') responses at 330 and 500 MeV/c [19, 20, 21]. These responses have been converted to per-nucleon responses according to

$$\frac{4\pi}{M_T} S_T \simeq \frac{A}{2} \left(\frac{q}{2m} \right)^2 (\mu_p - \mu_n)^2 G_M^2 R_T, \tag{11}$$

where $\mu_p = 2.79$, $\mu_n = -1.91$, G_M is the nucleon magnetic form factor, and A is the target nucleon number. The transverse isovector electron response is defined by

$$R_T = \frac{1}{A} \frac{1}{(2J_i + 1)} \sum_{if} |\langle f| \sum_{k=1,A} \frac{\boldsymbol{\sigma}_k \times \hat{\mathbf{q}}}{\sqrt{2}} \tau_k^z e^{-i\mathbf{q} \cdot \mathbf{r}_k} |i\rangle|^2 \delta[\omega - (E_f - E_i)], \tag{12}$$

and has the same integral normalization as the $(\vec{p}, \vec{n})$ response. Equation (11) ignores isospin-mixing effects, the small contribution from the isoscalar response, and the small convection current contribution [24]. With these approximations the response R_T corresponds to the spin operator $(\boldsymbol{\sigma} \times \hat{\mathbf{q}})/\sqrt{2}$. This is the proper normalization for comparison to the R_p and R_q $(\vec{p}, \vec{n})$ responses, and is a factor-of-two smaller than in our previous comparison to the 18° data [5].

Because it is simple to calculate and has a well-defined integral, the free Fermi-gas response serves as a useful baseline for comparison [22, 23]. The integral of this response is equal to $(y/2)(3 - y^2)$ for $y < 1$ and is unity for $y > 1$, where $y = q/(2k_F)$. The integrals of the experimental responses can be easily estimated by comparison. The Fermi-gas response is in good qualitative agreement with the main features of the data, particularly when compared to the longitudinal response. However, it is well known from analysis of electron-scattering that comparison to the free response can be very misleading. In the (e, e') reaction, a quenched single-particle transverse response and compensating higher-order contributions result in a total response very close in magnitude to the free response near the peak of the quasifree distribution [24, 25].

A more realistic theoretical response is given by the RPA. A comparison to the data is shown in Fig. 4 for ^{12}C$(\vec{p}, \vec{n})$ and ^{40}Ca$(\vec{p}, \vec{n})$ at 18°. The RPA responses shown here are the same as those used in the ratios of Fig. 1. These

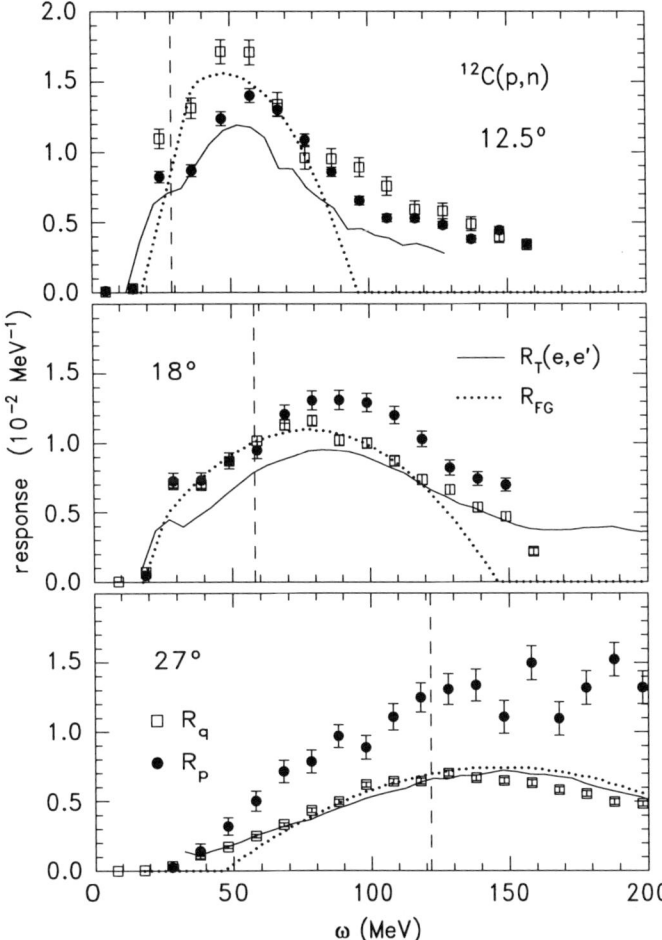

FIGURE 2. Longitudinal R_q (open boxes) and transverse R_p (solid circles) spin responses for $^{12}C(\vec{p},\vec{n})$ compared to $^{12}C(e,e')$ transverse spin responses R_T for $q = 250$, 350, and 500 MeV/c (solid lines) [18]. The dotted lines represent the free Fermi-gas response. The dashed vertical lines mark the energy loss for free scattering.

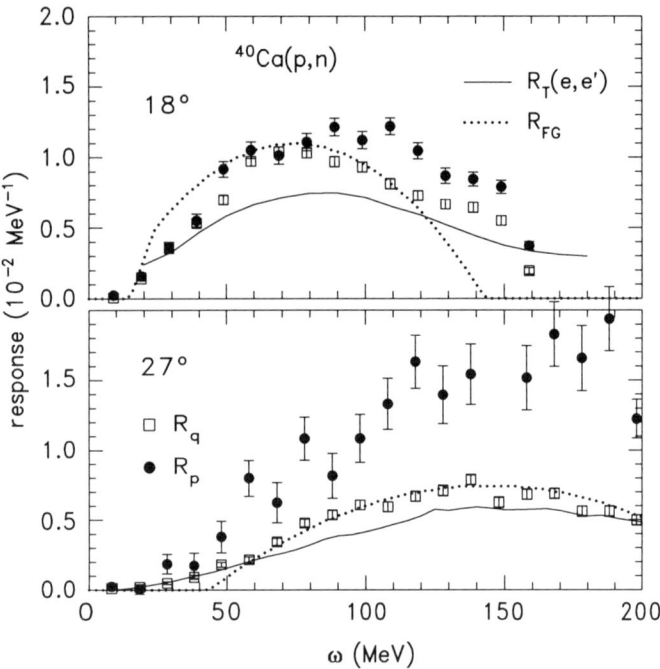

FIGURE 3. Longitudinal R_q (open boxes) and transverse R_p (solid circles) spin responses for $^{40}\text{Ca}(\vec{p}, \vec{n})$ compared to $^{40}\text{Ca}(e, e')$ transverse spin responses R_T for $q = 330$ and 500 MeV/c (solid lines) [19, 20, 21]. The dotted lines represent the free Fermi-gas response. The dashed vertical lines mark the energy loss for free scattering.

responses were shown in our previous analysis of the 18° data [5], but the normalization in that earlier comparison was too high by a factor of two because of a misunderstanding regarding an isospin operator ($\tau_{-1} = \sqrt{2}\tau^-$) omitted from the response definitions [13]. In this corrected comparison, the longitudinal RPA response is in reasonably good agreement with the data, but the transverse RPA response is about a factor-of-two too small.

As noted earlier, an important experimental question is the consistency of the results for the two independent transverse responses R_n and R_p that can be obtained from $(\vec{p}, \vec{n})$ quasifree scattering. A comparison of these two responses is made in Fig. 5 for $^{12}\text{C}(\vec{p}, \vec{n})$ and in Fig. 6 for $^{40}\text{Ca}(\vec{p}, \vec{n})$. The agreement between the two responses is extremely good. This agreement may serve as a useful constraint in testing more complicated reaction models, such as the DWIA, that in principle can take into account the different spin-dependent distortion effects in each channel. A theoretical comparison of this sort has not yet been done.

The transverse $(\vec{p}, \vec{n})$ responses R_p and R_n are larger than the corresponding

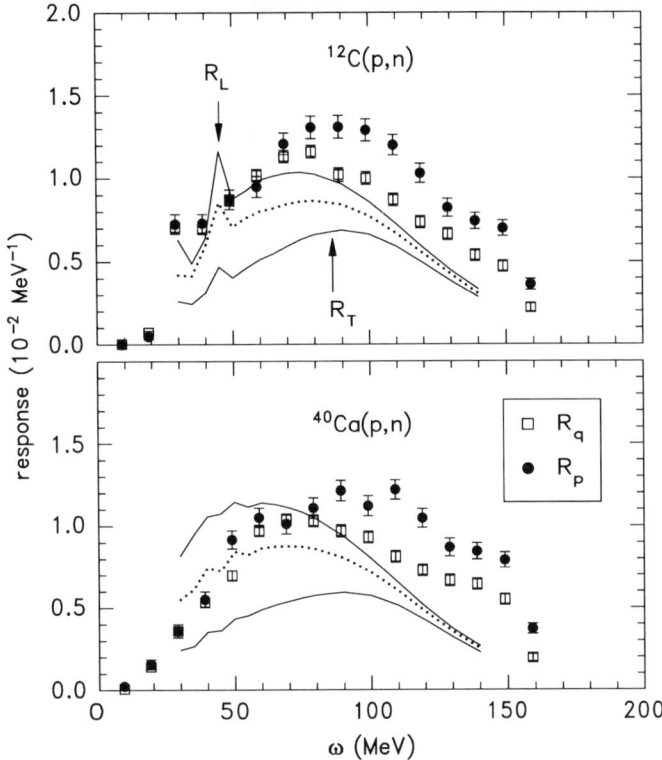

FIGURE 4. Longitudinal R_q (open squares) and transverse R_p (solid circles) responses for $^{12}\text{C}(\vec{p},\vec{n})$ at 494 MeV and 18° compared to longitudinal R_L and transverse R_T RPA responses. The dotted line represents the free response obtained by setting the residual interaction to zero.

transverse (e,e') response R_T at all momentum transfers. They are twice as large as the electron response at 27° ($q = 2.5$ fm^{-1}). At this angle the transverse $(\vec{p},\vec{n})$ strength is also twice the free $1p$–$1h$ (Fermi-gas) strength. This excess of strength may signal the presence of higher-order contributions such as multiple scattering or $2p$–$2h$ excitations. Contributions from $2p$–$2h$ configurations (including meson-exchange currents) are believed to contribute significantly to the electron transverse response [24, 25]. The present data suggest that such effects may be even more important for $(\vec{p},\vec{n})$ reactions, particularly at the highest momentum transfer.

The above comparisons show that expected enhancements in the longitudinal spin response are largely masked by an excess of strength or cross section in the transverse $(\vec{p},\vec{n})$ channel. Understanding the separated responses or spin-dependent partial cross sections is therefore at least as important as study of the ostensibly simpler response ratio in seeking evidence for collective effects

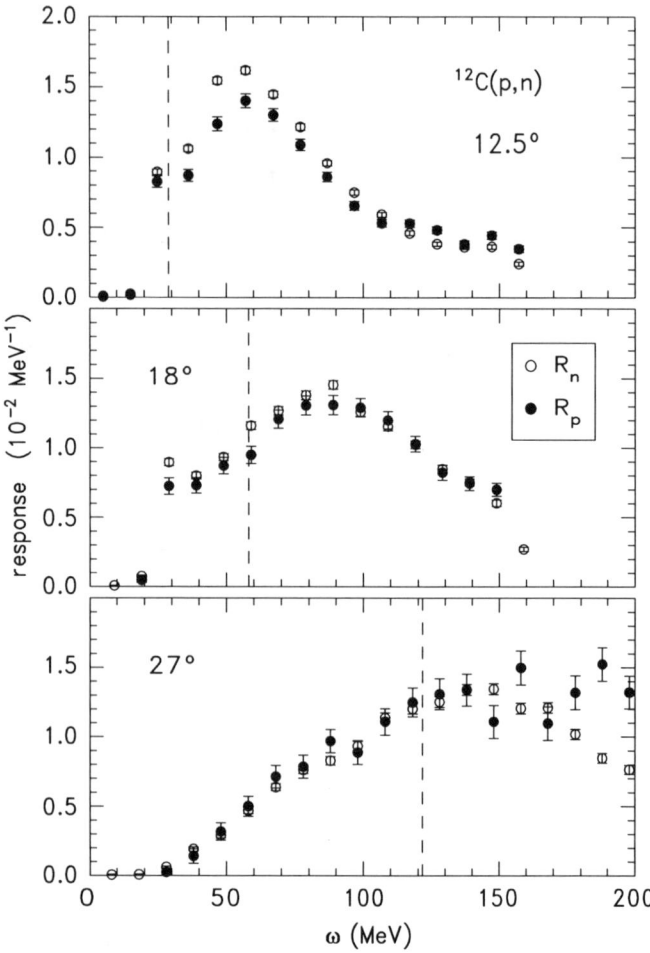

FIGURE 5. Transverse R_n (open circles) and transverse R_p (solid circles) responses for ^{12}C$(\vec{p},\vec{n})$ at 494 MeV.

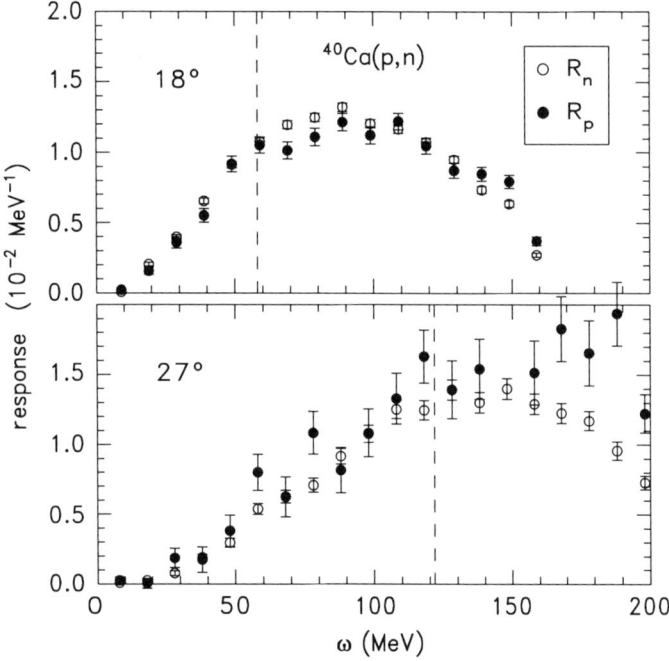

FIGURE 6. Transverse R_n (open circles) and transverse R_p (solid circles) responses for ^{40}Ca$(\vec{p},\vec{n})$ at 494 MeV.

from mesonic fields in the nucleus. Some important questions that remain to be addressed are the medium dependence of the NN amplitudes (the free values have been assumed in the present analysis), contributions from multiple scattering (and its spin dependence), contributions from higher-order processes such as $2p - 2h$ excitations, and the role of spin dependence in the distortion factor.

ACKNOWLEDGEMENTS

Credit for the data presented here is shared with the rest of the NTOF collaboration, most recently consisting of: B.A. Luther, L.J. Rybarcyk, R.C. Byrd, J.B. McClelland, D.L. Prout, S. DeLucia, D.A. Cooper, D.G. Marchlenski, E. Sugarbaker, B.K. Park, Thomas Sams, C.D. Goodman, and J. Rapaport. The RPA calculations shown here were supplied by M. Ichimura and K. Kawahigashi, and I would especially like to thank M. Ichimura for many long and useful E-mail conversations regarding the definition of the responses and their correct comparison to the RPA and (e, e') results.

This work was supported in part by the National Science Foundation and the U.S. Dept. of Energy.

REFERENCES

[1] W.M. Alberico, M. Ericson, and A. Molinari, Nucl. Phys. **A379**, 429 (1982).

[2] T.A. Carey, K.W. Jones, J.B. McClelland, J.M. Moss, L.B. Rees, N. Tanaka, and A.D. Bacher, Phys. Rev. Lett. **53**, 144 (1984).

[3] L.B. Rees, J.M. Moss, T.A. Carey, K.W. Jones, J.B. McClelland, N. Tanaka, A.D. Bacher, and H. Esbensen, Phys. Rev. C **34**, 627 (1986).

[4] J.B. McClelland et al., Phys. Rev. Lett. **69**, 582 (1992).

[5] X.Y. Chen et al., Phys. Rev. C **47**, 2159 (1993).

[6] M. Ichimura, in *Proceedings of the Fifth French-Japanese Symposium on Nuclear Physics*, 1989, edited by K. Shimizu and O. Hashimoto, p. 136 (unpublished).

[7] M. Ichimura, in *Proceedings of the Sixth Franco-Japanese Colloquium on Nuclear Structure and Interdisciplinary Topics*, Saint-Malo (France), Oct 6–10, 1992, edited by N. Alamanos, S. Fortier, and F. Dykstra, p. 49 (unpublished).

[8] Munetake Ichimura, Kimiaki Nishida, and Ken Kawahigashi, in *Proceedings of the International Symposium on Spin-Isospin Responses and Weak Processes in Hadrons and Nuclei*, Osaka, Japan, 8–10 March 1994, (unpublished).

[9] C.J. Horowitz and J. Piekarewicz, Phys. Lett. B **301**, 321 (1993).

[10] M. Ericson, Phys. Rev. C **49**, R2293 (1994).

[11] G.E. Brown and J. Wambach, Nucl. Phys. **A568**, 895 (1994).

[12] T.N. Taddeucci et al., Phys. Rev. C **41**, 2548 (1990).

[13] Munetake Ichimura and Ken Kawahigashi, Phys. Rev. C **45**, 1822 (1992); **46**, 2117(E) (1992).

[14] Atsushi Itabashi, Kazunori Aizawa, and Munetake Ichimura, Prog. Theor. Phys. **91**, 69 (1994).

[15] A.K. Kerman, H. McManus, and R.M. Thaler, Ann. Phys. (NY) **8**, 551 (1959).

[16] V.R. Pandharipande, J. Carlson, Steven C. Pieper, R.B. Wiringa, and R. Schiavilla, Phys. Rev. C **49**, 789 (1994).

[17] M. Ichimura, K. Kawahigashi, T.S. Jørgensen, and C. Gaarde, Phys. Rev. C **39**, 1446 (1989).

[18] P. Barreau et al., Nucl. Phys. **A402**, 515 (1983); P. Barreau et al., Note CEA-N-2334, CEN Saclay, 1983 (unpublished).

[19] M. Deady et al., Phys. Rev. C **28**, 631 (1983).

[20] M. Deady, C.F. Williamson, P.D. Zimmerman, R. Altemus, and R.R. Whitney, Phys. Rev. C **33**, 1897 (1986).

[21] Z.E. Meziani *et al.*, Phys. Rev. Lett. **54**, 1233 (1985).

[22] G.F. Bertsch and O. Scholten, Phys. Rev. C **25**, 804 (1982).

[23] H. Esbensen and G.F. Bertsch, Phys. Rev. C **34**, 1419 (1986).

[24] W.M. Alberico, M. Ericson, and A. Molinari, Ann. Phys. (NY) **154**, 356 (1984).

[25] W.M. Alberico, A. Molinari, A. De Pace, M. Ericson, and Mikkel B. Johnson, Phys. Rev. C **34**, 977 (1986).

Spin Polarization and Weak Decays of Hypernuclei

H. EJIRI

Dept. Phys., Osaka Univ., Toyonaka, Osaka 560, JAPAN
Research Center For Nucl. Phys., Osaka Univ., Ibaraki, Osaka 570, JAPAN

Abstract. Spin-polarized Λ-hypernuclei in wide excitation-energy (E_x) and angular-momentum ($J\hbar$) ranges are produced by direct and multistep-compound (π, K^+) and (K^-, π) reactions with GeV/c π and K beams. The intrinsic spin of Λ in the polarized Λ-hypernucleus is well polarized. The compound hypernuclei produced by (π^+, K^+) reactions decay by emittng particles (x) and γ-rays, finally populating bound Λ-hypernuclei. These (π, $K^+ x\gamma$) reactions are useful for porducing various kind of spin-polarized Λ-hypernuclei. Non-mesonic weak decays from polarized Λ-hypernuclei show a large asymmetry with respect to the polarization axis. It suggests contributions of heavy mesons (quarks) in the weak process associated with the strangeness change. Polarized hypernuclei are used for studying electroweak and strong interactions of hyperons in the nuclear medium.

1. SPIN-POLARIZATIONS OF HYPERNUCLEI

1.1 Elementary Process for Hyperon Production

Strange-quark (s-quark) exchange reactions of $K^- + N \to Y + \pi$ are used for producing Λ and Σ hyperons. Here the s-quark is transferred from the kaon to the hyperon. Defferential cross-sections are as large as a few mb/sr at forward angles. The momentum transfer is small, and becomes zero at around P_K=0.3 ∼ 0.6 GeV/c. The polarization becomes negligible at that momentum region. On the other hand the Λ polarizations at $P_K \sim$ 1.2 GeV/c and 1.6 GeV/c are $P_\Lambda \sim -1$ and $+1$.

Creation of a strange-quark pair of $s\bar{s}$ is used to produce hyperons. (π^+, K^+), (γ, K^+) and (p, K^+) are possible reactions. Here the s-quark transferred to a nucleon is used to produce the hyperon, while the $\bar{s}$-quark is used to produce K^+ at the final channel. The differential cross-section for the (π, K^+) reaction is of the order of 0.1 ∼ 0.4 mb/sr, which is about one tenth of the cross section for the (K^-, π^-) recation. The momentum transfers in (π, K^+) and (γ, K^+) are large. Then one gets large polarizations of hyperons.

The (γ, K^+) reaction cross-section is much smaller than those of the (π, K^+) and (K^-, π) reactions because the electromagnetic interaction is involved. The momentum transfer q is large, and accordingly the polarization of Λ is finite.

1.2 Direct (K^-, π^-), (π^+, K^+), (γ, K^+) and (p, K^+) Reactions for Hypernuclei[1,2,3]

The (K^-, π^-) reaction is used to populate low-spin hypernuclear states because of the low momentum transfer of $q < 0.1$ GeV/c. In particular, the reaction with $q \sim 0$ at P_K≈0.3−0.6 GeV/c excites preferentially substitutional states with $[j_n^{-1}, j_\Lambda]_{0^+}$, where a neutron in a j orbit is replaced by a Λ in the same orbit. Differential cross-sections and polarizatioins for $^{12}C(K^-, \pi^-)_\Lambda^{12}C$ are shown in Fig.1. Hypernuclear states excited by the (K^-, π) reaction are mostly natural parity states because of the small spin flip amplitude. Polarization of the Λ spin in the 0s orbit is negative (positive) at P_K=1.1(1.5)GeV/c, reflecting polarizations of the Λ spin in the elementary process[4]. The small intensity for K^- beam is well compensated by the large cross-section.

The momentum transfer in the (π^+, K^+) reaction is as large as $q\sim 0.4$ GeV/c. Thus the (π^+, K^+) reaction is used to populate various kind of hypernuclear states in wide angular and excitation regions. In particular, high-spin states of $[j_N^{-1}, j_\Lambda]_J$ with $J \sim l_n + l_\Lambda$, are preferentially excited[2]. Large spin-polarizations[5,6] are expected. Unnatural parity states are also excited although their cross-sections of the (π^+, K^+) reactions are smaller by $1 \sim 2$ orders of magnitude than those of the (K^-, π^-) reactions because of the $s\bar{s}$ production and of the large momentum transfer. However, the pion beam intensity is larger by typically 2-orders of magnitude than the kaon beam intensity. Thus the (π^+, K^+) reactions have extensively been used for hypernuclear spectroscopy.

The momentum transfer in the (γ, K^+) reaction is large as in the (π^+, K^+) reaction. Thus high-spin states with spin-polarization are expected to be excited. One unique feature of the (γ, K^+) reaction is a large spin-flip amplitude. Then unnatural parity states are excited as well as natural parity status[7]. Cross-sections of (γ, K^+) reactions are smaller by almost $3 \sim 4$ orders of magnitude than the (K^-, π^-) reaction because the interaction involved is the electromagnetic one. On the other hand high-intensity photons are obtained from high-intensity electron accelerators. Recently the (γ, K^+) reaction on ^{12}C has been studied in the quasi-free scattering region[8]. The momentum transfer in the (p, K^+) reaction is around GeV/c. Thus hypernuclear production is less likely by the direct process.

1.3 Multistep-compound Reactions

The (π^+, K^+) reaction proceeds predominantly by the quasi-free scattering (QFS) process. The excitation energy of the QFS region is about 30-120 MeV. Λ produced by the (π^+, K^+) reaction in the QFS region has a large momentum of $p \sim 0.4$ GeV/c. It escapes mainly from the nucleus by the escape process $\Gamma^\uparrow$, and decays by M-process of $\Lambda \to \pi^- + p$, $\pi^0 + n$.

The Λ produced in the QFS region is partly trapped in the nucleus, forming a Λ-hypernucleus by the spreading process $\Gamma^\downarrow$. The escape ($\Gamma^\uparrow$) and spreading ($\Gamma^\downarrow$) processes of Λ in the QFS region were studied[9] by measuring the weak decay vertext point (Z_d) with respect to the (π^+, K^+) reaction vertex point (Z_r). In case that Λ is trapped in the nucleus, it decays in the hypernucleus. Thus Z_d is nearly equarl to Z_r. On the otherhand, in case that Λ escapes from the nucleus, it flies outside the nucleus for it's mean-life. Then one gets $Z_d \approx Z_r + l$ with l being the mean flight length of Λ. The Z_d was measured by investigating energy deposits in the individual segments of the active plastic target, and Z_r by analyzing the π^+ and K^+ trajectries. The measured energy distribution in the segmented target is found to be shifted from the reaction point. It indicates about 15% of Λ in the QFS region is trapped into the nucleus. It gives the ratio $\Gamma^\downarrow / (\Gamma^\downarrow + \Gamma^\uparrow) \approx 0.15$. This ratio is also consistent with the ratio of the non-mesonic weak-decay protons and the mesonic decay pions by the weak-decay counter in the QFS region.

The hypernuclei produced by this process have large excitation energy of Ex ≈ 50 MeV. They deexcite by emmiting nucleons (x) and γ-rays, finally producing bound hypernuclei.

It is interesting to note that hypernuclei can be produced by the (π^+, K^+) reaction in the QFS region with the large cross-section of σ(HY.QFS) ≈ 0.15 σ(QFS), where σ(QFS) is the QFS cross-section.

The multistep compound process of the $(\pi^+, K^+ x\gamma)$ with x being nucleons/ nuclear-clusters are used for populating various kinds of hypernuclear states.

The multistep compound processes of $(p, K^+ x\gamma)$, $(d, K^+ x\gamma)$, $(\gamma, K^+ x\gamma)$, etc. are also promising since the direct processes bound are reduced because of the large momentum mismatching. The spreading and rescattering processes were also discussed in ref.9 and 10.

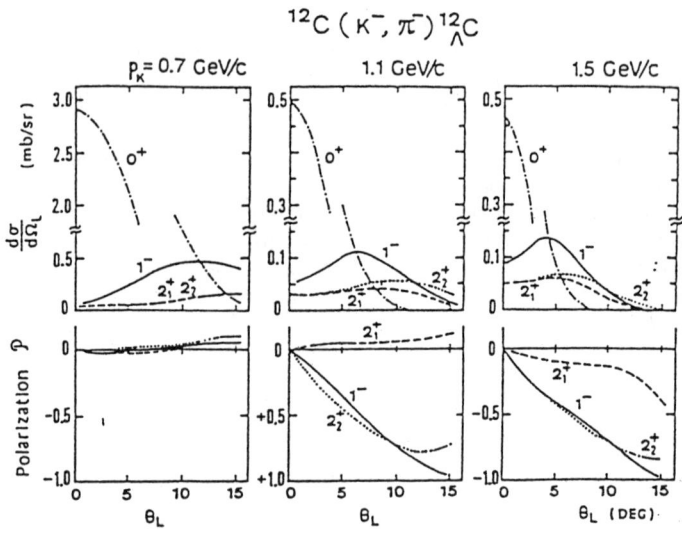

Fig.1. Calculated cross-sections and polarizations for the $^{12}C(K^-, \pi^-)^{12}_\Lambda C$ reactions at P_K = 0.7 GeV/c, 1.1 GeV/c and 1.5 GeV/c. (from Ref.2, 4)

$_\Lambda A(J^\pi)$	j_n^{-1}, j_Λ	$d\sigma/d\Omega$	$P(J_i)$	$_\Lambda D(J_f^\pi)$	$J_c \cdot s_\Lambda$	$P(J_f)$	P_Λ	$\bar{P}_\Lambda$
$^{56}_\Lambda Fe(6^+)$	$f_{7/2}^{-1}f_{5/2}$	6.5	−0.44	$^{54}_\Lambda Fe(3^-)$	$f_{7/2}s_{1/2}$	−0.40	+0.30	
	$f_{7/2}^{-1}f_{7/2}$	2.1	+0.03			+0.03	−0.02	+0.22
$^{56}_\Lambda Fe(5^-)$	$f_{7/2}^{-1}d_{3/2}$	4.9	−0.46	$^{55}_\Lambda Fe(1/2^+)$	$0 \cdot s_{1/2}$	−0.28	−0.28	
	$f_{7/2}^{-1}d_{5/2}$	1.9	+0.09			−0.05	−0.05	−0.22
$^{28}_\Lambda Si(4^+)$	$d_{5/2}^{-1}d_{3/2}$	7.2	−0.49	$^{26}_\Lambda Mg(2^+)$	$d_{5/2}s_{1/2}$	−0.44	+0.29	
	$d_{5/2}^{-1}d_{5/2}$	2.9	−0.02			−0.02	+0.01	+0.21
$^{28}_\Lambda Si(3^-)$	$d_{5/2}^{-1}p_{1/2}$	3.6	−0.54	$^{27}_\Lambda Al(*)$	$* \cdot s_{1/2}$	−0.48	+0.34	
	$d_{5/2}^{-1}p_{3/2}$	2.5	−0.17			−0.15	+0.11	+0.25
$^{12}_\Lambda C(2^+)$	$p_{3/2}^{-1}p_{1/2}$	3.2	−0.62	$^{11}_\Lambda B(5/2^+)$	$3 \cdot s_{1/2}$	−0.57	+0.43	
	$p_{3/2}^{-1}p_{3/2}$	3.2	+0.02			−0.02	+0.01	+0.22

Table.1. Polarization $P(J_f)$ and Λ-spin polarization P_Λ for Λ hypernuclear state with spin J_f produced by the $A(\pi^+, K^+ xnypr)_\Lambda D$ reaction through the compound (resonant) state of $[j_n^{-1}, j_\Lambda]$ with J^π. $\bar{P}_\Lambda$ is the average value for the two adjacent resonant states. $d\sigma/d\Omega$ is the (π^+, K^+) cross section in units of μb/ sr at $\theta_K = 15°$. $P(J_i)$ is the polarization of the compound state. $P_K = 1.05$ GeV/c. (Ref.5)

1.4 Polarization of Λ Hypernuclei Produced By (π, K^+) and (K^-, π) Reactions

Spin-polarization and spin-alignment in hypernuclei can be produced by two factors in (π, K^+) and (K^-, π) reactions at GeV/c region[5,6]. One is due to the distortion (absorption) of the initial (projectile) and the final (ejectile) meson waves, and other is due to the spin dependent elementary process in the strangeness exchange (π, K^+) and (K^-, π) reactions. The distortion (absorption) gives rise to the polarization of the transferred orbital angular momentum introduced by the (π^+, K^+) or (K^-, π^-) reaction. Combining the polarization of the transferred orbital angular momentum and the polarization of the Λ-spin, polarizations of Λ hypernuclei produced by (π^+, K^+), and $(\pi^+, K^+ x n y p \gamma)$ reactions on ^{56}Fe, ^{28}Si and ^{12}C are evaluated in terms of a statistical model[5]. The final hypernuclear states populated through resonant states with $[(j_n = l + 1/2)^{-1}(j_\Lambda = l - 1/2)]J = 2l$, which are strongly excited by (π^+, K^+) reactions on $(j_n = l + 1/2)$ shell closed nuclei, have large polarizations of the order of 30 ~ 50%. The calculated polarizations are shown in table 1. The corresponding Λ-spin polarizations are of the order of 20 ~ 40%. It is important to note that Λ hypernuclei are produced also by $(\pi^+, K^+ x n y p \gamma)$ reactions through highly excited compound Λ-hypernuclei where spin-polalizations are still kept well. Note that low energy nucleons and γ-rays emitted from compound nuclei are mostly s and p waves and thus do not disturbe the polarization[5].

The elementary process of the (π, K^+) reaction at 1 ~ 1.7 GeV/c produces the Σ hyperon. One can expect finite polarizations of the Σ spin and the orbital angular momentum, and accordingly polarized Σ hypernuclei. The conversion process of $\Sigma + N \to \Lambda + N$ and the successive internuclear cascade process may lead to a polarized Λ hypernucleus.

The (K^-, π^-) reaction in the higher momentum region of K^- above 1 GeV/c is useful[4]. The features of $P(K) = 1.1$ GeV/c and 1.5 GeV/c look very similar except for the sign of main polarization. Another merit of using higher momentum K^- is that the K^- beam intensity becomes higher due to large survival rate in the beam channel.

2. WEAK DECAYS OF Λ-HYPERNUCLEI

2.1 Weak Decay Machanisms

The ground state of the Λ-hypernucleus decays through the weak process with a strangeness-change of $\Delta S = 1$. This is essentially the weak decay of the Λ. There are the mesonic(M) decay and the non-mesonic(NM) decay, as shown in Fig. 2.

$$\text{M} - \text{decay} \quad \Lambda \to p + \pi^-, \ n + \pi^0, \tag{1}$$

$$\text{NM} - \text{decay} \quad \Lambda + n \to n + n, \ \Lambda + p \to p + n. \tag{2}$$

NM-decay is a unique and major decay process of the Λ in hypernuclei heavier than mass ~7, where M-decay is substantially reduced by Pauli-blocking because of the small momentum of the nucleon emitted in the M-decay.

Distortion of the pion wave gives rise to the high-momentum component of the pion, and accordingly to the high-momentum one of the proton. It is shown that the decay rate of $\Lambda \to p + \pi^-$ survives even in the heavy mass region because of the large Coulomb and nuclear distortions of the π^- wave[2]. The decay rate is of the order of % in the medium and heavy nuclei.

The NM-decay involves a large energy of the mass-difference, $(M_\Lambda - M_N)c^2$ ~0.18 GeV, and a large momentum transfer of $q \sim 0.4$ GeV/c. Consequently the short-range correlation of the hyperon-nucleon system is crucial. Thus meson-exchange processes up to heavy mesons and the possible 6-quark process can play important roles in the NM-decay[12-18]. The meson-exchange diagrams are shown in Fig. 3.

390 Spin Polarization and Weak Decays

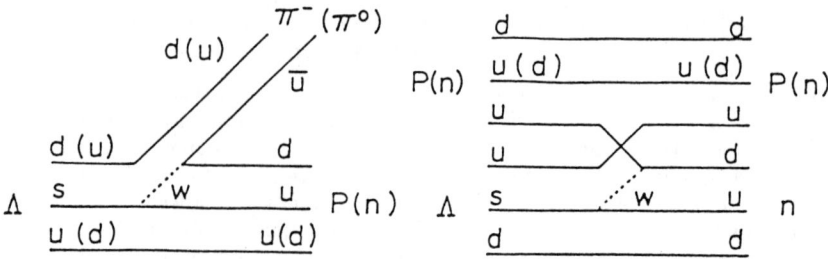

Fig.2. Weak decay diagrams for mesonic decays (left-hand side), and for non mesonic decays (right-hand side).

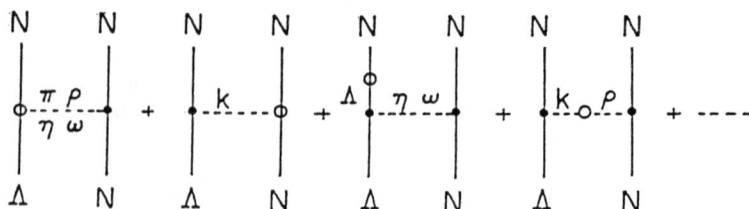

Fig.3. Digrams for non-mesonic weak decays of $\Lambda n \to NN$. Open circles are weak vertex points.

Let us take $^1S_0$ and $^3S_1$ for the initial $\Lambda - N$ pair of the NM decay. There are six amplitudes of a, b, c, d, e and f for the $^1S_0 \to {}^1S_0$, $^1S_0 \to {}^3P_0$, $^3S_1 \to {}^3S_1$, $^3S_1 \to {}^3D_1$, $^3S_1 \to {}^1P_1$, $^3S_1 \to {}^3P_1$, respectively. Then the angular distribution of the NM-decay proton from the state with Λ-spin polarization P_Λ is written as[2]

$$W(\theta) = \frac{1}{4}(a^2 + b^2) + \frac{3}{4}(c^2 + d^2 + e^2 + f^2) + \frac{\sqrt{3}}{2}(\sqrt{2}c + d) \cdot f \cdot P_\Lambda \cdot P_1(cos\theta)$$
$$= A_0[1 + A_1 P_1(cos\theta)], \qquad (3)$$

where θ is the angle of the decaying particle with respect to the polarization axis. The asymmetry A_1 is written as $A_1 = P_\Lambda a_1$. The observed asymmetry $A_1 = P_\Lambda a_1$, together with the calculated value for P_Λ, leads to a value for a_1, casting new light on the weak-decay mechanism. So far experimental studies on the weak decay have been made mainly on the decay rates and the branching ratios[13].

The amplitudes in eq.(3) depend on the NM-decay mechanism. The asymmetry is a consequence of interference of the isospin I=1 amplitude (f) and I=0 one (c and d). It is thus sensitive to the magnitudes and the relative phase of I=0 and 1 amplitudes.

2.2 Weak Decays of Polarized $^{12}_\Lambda$C and $^{11}_\Lambda$B

Weak decays from polarized hypernuclei produced by (π^+, K^+) reactions at $P_\pi =$ 1.05 GeV/c were studied for the first time[18]. The $^{12}C(\pi^+, K^+)$ reaction was used to populate low lying states with $p_n^{-1}s_\Lambda$ in the particle-bound region(B), medium excitation($E_X \sim$10MeV) states with $p_n^{-1}p_\Lambda$ in the Λ-bound region (ΛB), and high excitation ($E_X \sim$20MeV) states with $p_n^{-1}d_\Lambda$ and $s_n^{-1}s_\Lambda$ in the unbound region. The low lying excited states feed the $^{12}_\Lambda C(1_1^-)$ ground state by γ transitions. There are 0^+ and 2^+ states in the medium excitation region. They are proton-unbound and Λ-bound, and thus populate the $^{11}_\Lambda$B ($J = 5/2^+$) ground state by proton and γ emissions. The high excitation region with $E_X \sim$ 20MeV, while it is above the Λ-emission threshold, will be referred to as the quasi-bound(QB) region because the Λ escape is suppressed by the centrifugal potential barrier. Then, while the Λ may be emitted through the Λ-escape process, it is possible that the QB region deexcites by emitting particles such as protons, neutrons, ^{3}He and alpha, leaving a lighter hypernucleus(LH).

The (π^+, K^+) reactions were momentum-analyzed by the PIK spectrometer, and the weak decay π^- and p were measured by the decay counters made of plastic(ΔE) and NaI(E)detectors. Asymmetries A_1 were measured for the M-decay pions(π) and for the NM-decay protons(p) from the $^{12}_\Lambda$C and $^{11}_\Lambda$B ground states and from the Λ or LH associated with the QB region.

The experimental values for A_1^p/k are plotted versus the theoretical values for P_Λ in Fig. 2. The energy of the NM-decay p is so large that many final states in a large(angular) momentum space are available for the NM-decay. Thus the asymmetry A_1^p can be expected to be nearly independent of the nuclear core spins and of the nucleon configurations in the $^{12}_\Lambda$C, $^{11}_\Lambda$B and LH ground states. Thus one may obtain a common value for the asymmetry parameter of $a_1^p = -1.3 \pm 0.4$ for the p-shell nuclei, as shown in Fig.4 by the dotted region. The NM-decay asymmetry is sensitive to the decay mechanism, i.e. to the mesons involved in the weak decay process[12-17]. The pion-mediating process gives a small value of $|a_1^p| =$ 0.30, while inclusion of all relevant heavy mesons in the process produces a large value $|a_1^p| = 0.65$. The observed value of $a_1^p \sim -1.0$ suggests importance of heavier mesons. A recent relativistic finite nucleus calculation[16] for the pion-mediating NM-decay produces an asymmetry parameter, whose sign is consistent with the present experiment. Recently P_Λ has been calculated[19]. The obtained value for a_1^p depends on calculations of P_Λ.

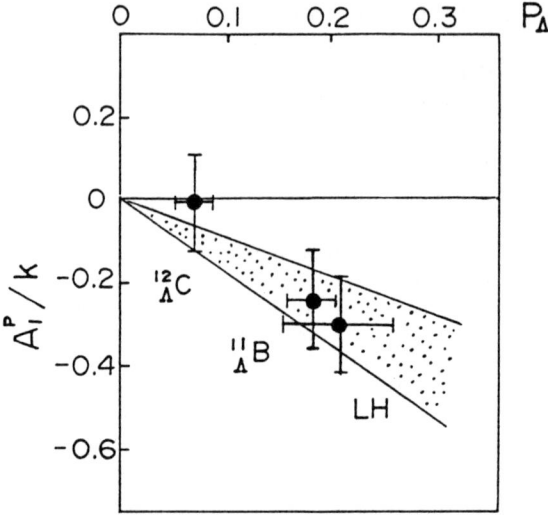

Fig.4. The observed asymmetries $A_1/k \equiv a_1^p P_\Lambda$, with $k = 0.8$ being the attenuation factor due to the internuclear cascade process, versus the evaluated polarizations P_Λ (see text). The dotted region of $a_1^p = -1.3 \pm 0.4$ is derived from the data. Note that the asymmetry for the M-decay π from the ΛB region is an effective value, including contribution from the QB region. (Fig.4 in Ref.18)

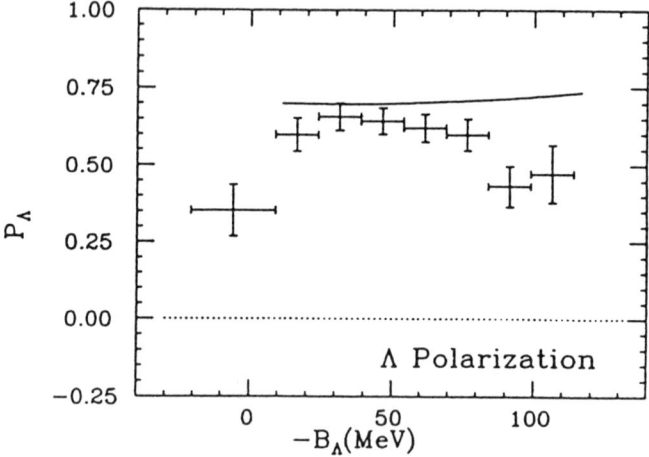

Fig.5. Polarization obtained from the measured asymmetry asuming free Λ production. The caluculated polarization is shown by a solid line. (Ref. 20)

2.3 Spin Polarization of Λ in the (π^+, K^+) Quasi-Free Scattering Region

The asymmetry of M-decay pions of quasi-free Λ procured by the $^{12}\text{C}(\pi^+,K^+)$ reaction was studied. The Λ spin-polarization was deduced from the observed asymmetry, as shown in Fig.5. The Λ polarization is found to be consistent with the elementary process of the quasi-free Λ production in the nucleus. A small reduction of the polarization may be due to the contribution from the M-decay of the Λ hypernucleus in the quasi-free region.

3. DISCUSSIONS

Spin-polarized Λ-hypernuclei are produced by (K^-,π) and (π,K^+) and $(\pi,K^+p\gamma)$ reactions in the GeV/c region. The Λ-spin S_Λ is polarized parallel or anti-paralleled to the hypernuclear spin, depending on $J_Y = J_C + S_\Lambda$ or $J_Y = J_C - S_\Lambda$, where J_Y and J_C are the hypernuclear spin and the core nuclear spin, respectively.

Finite spin-polarizations and finite asymmetries of weak decays open new fields of the hypernuclear spectrocopy for studying weak mechanisms associated with strangeness, electromagnetic moments and hypernuclear structures.

The (π^+,K^+) reactions feeding Λ in the bound, quasi-bound and QFS regions are all used for Λ-hypernuclear production with finite spin polarization. In particular, Λ-hypernuclear production of the QFS Λ with the large monumentum of $q \sim 0.5$ GeV/c suggests that the other (p,K^+), (γ,K^+), (K^-,π) reactions with large P can also be used to produce Λ-hypernuclei through Λ-spreading (re-scattering) process, i.e. the multistep-compound process.

The NM- and M-weak decays of $^5_\Lambda\text{He}$ porduced by the ^6Li (π^+,K^+,p) reaction were measured for further quantitative studies of weak decays of polarized hypernuclei. Studies of $^5_\Lambda\text{He}$ has following merits. The polarization of $^5_\Lambda\text{He}$ is obtained experimentaly by measuring the asymmetry of the M-decay π^-. The nuclear effect is very small because $^5_\Lambda\text{He}$ is very light. The ΛN in $^5_\Lambda\text{He}$ is in a pure relative-s orbit. The 1 GeV/c π^+ beam from the KEK 12 GeV PS and the super conducting kaon spectrometer(SKS) were used. The data anlysis is now under progress.

A part of the present work has been presented at a conference in Aug. 1994[21].

4. ACKNOWLEDGEMENT

The author thanks his colleagues of the ECHO group, Prof. T. Kishimoto, Dr. H. Noumi, Mr. S. Ajimura, and others for valuable discussions and collaborations.

5. REFERENCES

1. C. B. Dover and E. Walker, *Phys. Rep.* **89** (1982) No1; and refs. therein.

2. H. Bando, T. Motoba, and J. Žofka, *Int. J. Mod. Phys.* **A5** (1990) 4021; and refs. therein.

3. H. Ejiri, *Nucl. Phys.* **A574** (1994) 311c

4. T. Kishimoto, H. Ejiri, and H. Bando, *Phys, Lett.* **B232** (1989) 24.

5. H. Ejiri, T. Fukuda, T. Shibata, H. Bando, and K.-I. Kubo, *Phys. Rev.* **C36** (1987) 1435. H. Ejiri, T. Kishimoto, and H. Noumi, *Phys. Lett.* **B225** (1989) 35.

6. H. Bando et al, *Phys. Rev.* **C39** (1989) 587. M. Sotona and J. Žofka, *Prog. Theor. Phys.* **81** (1989) 160.

7. C. B. Dover, and D. J. Millener, *BNL*-44651 (1990) 1.

8. K. Maeda, *Nucl. Phys.* (1994).

9. S. Ajimura, H. Ejiri, T. Kishimoto, H. Noumi et al., Proc. Int. Symp. Spin Isospin Responces and Weak Processes in Hadrons and Nuclei, Osaka (1994), ed H. Ejiri, *Nucl. Phys.* **A577** (1994) 263c.

10. O. Shoult, *Private communication* (1992); Proc. Nuclear Reaction Mechanisms, Varenna, June 1994, ed, E. Gadiori.

11. W. Cassing, et al., Proc. Nuclear Reaction Mechanisms, Varenna, June 1994, ed, E. Gadiori.

12. H. Ejiri, *AIP Conf.* **No.224** (1990), ed. B. F. Gibson et al. p185, 260.

13. C. B. Dover and G. E. Walker, *Phys. Rep.* **89** (1982) 148. 1. P.D. Barnes et al., *Nucl. Phys.* **A450** (1986) 430C. P. D. Barnes, *AIP Conf. Proc.* **No.224** (1990), ed. B. F. Gibson et al., p86. J. Szymanski et al., *Phys. Rev.* **C43** (1991) 849.

14. E. Oset and L. L. Salcedo, *Nucl. Phys.* **A443** (1985) 704.

15. J.F. Dubach, *Nucl. Phys.* **A450** (1986) 71c.

16. A. Ramos et al., *Nucl. Phys.* **A554** (1992) 703.

17. M. Oka et al., Proc. Int. Symp. Spin Isospin Responces and Weak Processes in Hadrons and Nuclei, Osaka, Japan, ed. H. Ejiri et al., *Nucl. Phys.* to be Published (1994).

18. H. Ejiri et al., KEK experiment E160 *OULNS 89-1* (1989). S. Ajimura, H. Ejiri, A. Higashi, T. Kishimoto et al., *Phys. Lett.* **B282** (1992) 293.

19. K. Itonaga et al., Proc. 5th Int. Symp. Mesons & Light Nuclei, Prague (1991). K. Itonaga et al., *Phys. Rev.* **C49** (1994) 1045.

20. S. Ajimura et al., *Phys. Lett.* **63** (1992) 2137.

21. H. Ejiri, Proc. Int. Conf. Physics with GeV Particle Beams, Jülich, Aug. 1994.

MEASUREMENT OF SPIN ROTATION PARAMETERS IN PROTON ELASTIC SCATTERING FROM ^{58}Ni AT E_p=300 MeV

A. Tamii,[a] H. Akimune,[b] I. Daito,[b] M. Fujiwara,[b] K. Hatanaka,[b]
K. Hosono,[b] T. Inomata,[c] M. Nakamura,[a] T. Noro,[b] H. Sakaguchi,[a]
S. Toyama,[a] M. Yamagoshi,[a] M. Yoshimura,[a] and M. Yosoi[a]

[a] *Department of Physics, Kyoto University, Kyoto 606, Japan*
[b] *Research Center for Nuclear Physics, Osaka University, Ibaraki, Osaka 567, Japan*
[c] *Department of Physics, Osaka University, Toyonaka, Osaka 560, Japan*

Abstract. We have measured cross sections and analyzing powers of elastic proton scattering from ^{12}C, ^{58}Ni, and ^{120}Sn at energies from 200 MeV to 400 MeV. In addition, spin rotation parameters have been measured for ^{58}Ni at E_p=300 MeV using the newly constructed focal plane polarimeter for the spectrometer Grand Raiden at RCNP. Results of calculations based on the relativistic impulse approximation are in good agreement with the measured spin observables, although the differential cross sections are overestimated.

INTRODUCTION

Recently optical potential theories have been greatly developed in the intermediate energy region. Some theories now reproduce observables in elastic scattering of polarized protons. One of them is the relativistic impulse approximation (RIA). In the intermediate energy region, the spin rotation parameters have been measured at LAMPF, TRIUMF and IUCF, since they play an important role in developing recent theories. Unfortunately, the measurements of the spin rotation parameters are quite few at energies from 200 MeV to 400 MeV, where the nuclear potentials change from attractive to repulsive. In addition, the data are almost limited to the targets of ^{40}Ca and ^{208}Pb. In order to discuss the medium effects on meson-nucleon coupling constants, more systematic data are necessary. Therefore we start measurements of observables in elastic proton scattering including the spin rotation parameters.

EXPERIMENT AND RESULTS

We start a series of experiments to measure differential cross sections and analyzing powers of elastic proton scattering with the spectrometer Grand Raiden at the Research Center for Nuclear Physics (RCNP). Until now we have taken data for ^{12}C and ^{58}Ni at E_p=200, 300, and 400 MeV, and for ^{120}Sn at E_p=200 and 300 MeV. They are plotted in figures 1, 2, and 3, in comparison with the predictions of the RIA.

As the next step of our experiments we have measured Q-parameters of elastic proton scattering from ^{58}Ni at forward scattering angles. In order to obtain Q-parameters, we applied the double scattering technique employing the focal plane polarimeter (FPP), which has been newly constructed for the spectrometer Grand Raiden. Although the construction of the FPP has not been completed, the polarization transfers of a single peak can be measured. Therefore, we have performed our experiment with the FPP system for the test purpose.

The polarized proton beam was accelerated to 52.9 MeV by the AVF cyclotron and injected to the ring cyclotron. At the beam line between the two cyclotrons, the polarization axis of the proton beam was rotated from the vertical direction to the horizontal direction with a superconducting solenoid magnet. The proton beam was accelerated to 300 MeV by the ring cyclotron and extracted in the single turn extraction mode. The beam was transported to the experimental room and focused on the ^{58}Ni target. The beam intensity was about 1 nA and the beam polarization was about 70%. Momenta of scattered protons were analyzed with the spectrometer Grand Raiden. The positions and incidence angles of scattered protons were measured using two vertical drift chambers (VDC's) at the focal plane. Protons were subsequently scattered again by a carbon block with the thickness of 6 cm. The outgoing angles of protons were ray-traced by two MWPC's. In the off-line analysis, the horizontal polarizations of scattered protons were determined from the vertical asymmetries in this second scattering.

In order to deduce polarizations from the measured asymmetries we used the effective analyzing power of the FPP system, which was determined from the data of this experiment. By employing the special dipole magnet called "DSR" of the Grand Raiden, we can use two other focal planes (FP+,FP-) in addition to the standard focal plane (FP0). Polarization transfer can be measured with good accuracy and without discrete ambiguities by analyzing both the data taken at the FP+ and FP-. The effective analyzing power of the FPP system is simultaneously determined owing to the redundancy of data. The obtained effective analyzing power is consistent with the values estimated from the data measured at SIN by Besset et al. (1) and Aprile-Giboni et al. (2) within statistical errors. The detailed information on the FPP system is given elsewhere in this proceedings by Yosoi et al. (3).

The obtained Q-parameters are plotted in figure 4.

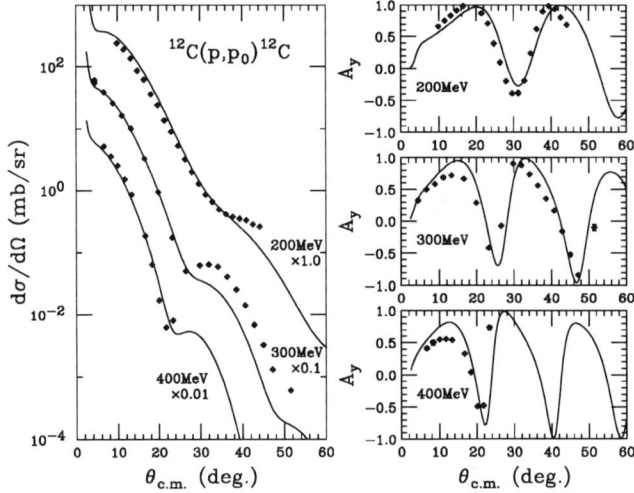

FIGURE 1. Angular distributions of the cross sections and analyzing powers of elastic proton scattering from ^{12}C at E_p = 200, 300, and 400 MeV. The solid curves are the results of the RIA calculations using the code by Horowitz et al.

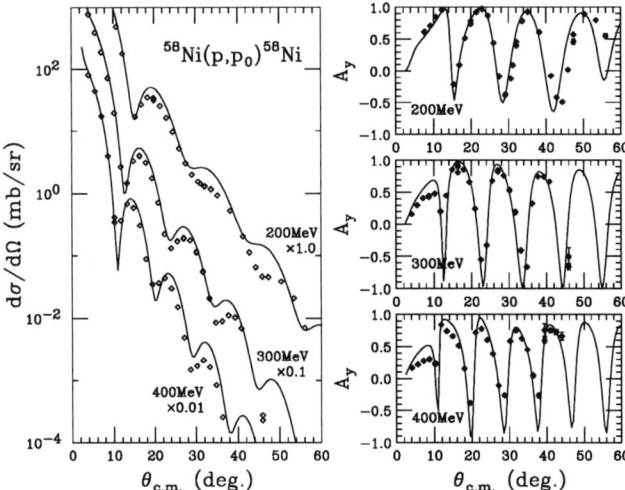

FIGURE 2. Angular distributions of the cross sections and analyzing powers of elastic proton scattering from ^{58}Ni at E_p = 200, 300, and 400 MeV. The solid curves are the results of the RIA calculations.

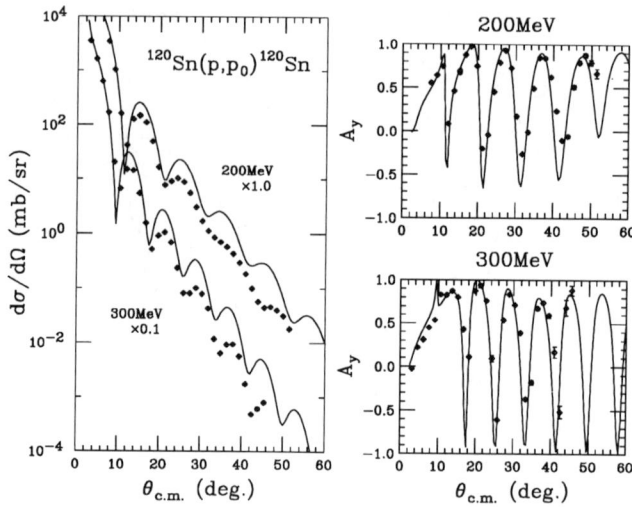

FIGURE 3. Angular distributions of the cross sections and analyzing powers of elastic proton scattering from ^{120}Sn at $E_p = 200$ and 300 MeV. The data are compared with the results of the RIA calculations.

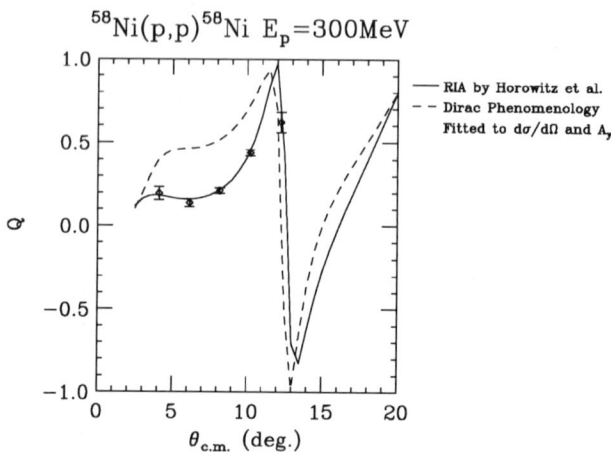

FIGURE 4. Obtained Q-parameters of elastic proton scattering from ^{58}Ni at $E_p = 300$ MeV at the angles of 4, 6, 8, 10, and 12 degrees, compared with the predictions of the RIA model (solid line) and the Dirac phenomenology (dashed line). The potentials of the Dirac phenomenology are determined to reproduce only the cross sections and analyzing powers.

DISCUSSION

As can be seen in figure 4, the experimental data remarkably differ from the results of calculations from the optical potentials, which are determined with Dirac phenomenology to reproduce only the cross sections and analyzing powers. This fact suggests that there are some difficulties in determining the optical potentials phenomenologically from only cross sections and analyzing powers. In other words, the third observables, Q-parameters, are essential to obtain the correct optical potentials. In figures 1, 2, and 3, the predictions of the RIA calculation using the code by Horowitz et al. (4) are compared with the experimental data. The scalar and baryon density distributions in target nuclei are calculated with the Dirac Hartree method in this code. Optical potentials in the Dirac equation are calculated using the RIA from these density distributions and the meson-nucleon coupling constants. It takes into account the effects of the particle exchange and the Pauli blocking approximately. As shown in the figures, the RIA calculations have succeeded to reproduce the analyzing powers for ^{58}Ni and ^{120}Sn at all measured energies. They overestimate the differential cross sections. For the case of the ^{12}C target the predictions of analyzing powers become worse. As can be seen in figure 4, the calculations of the RIA well reproduce the measured Q-parameters. It has been said that the RIA predicts the spin observables well but it overestimates the differential cross sections (5). A similar trend is observed in the case of ^{58}Ni. The reason of this is not well understood at present.

We have calculated the observables with the various artificial values of the meson-nucleon coupling constants using the RIA code. Figure 5 shows one of the results of the calculations. The spin observables are found to be quite robust against the changes of the coupling constants of σ and ω mesons when we apply the same multiplication factor to both the two coupling constants, although the differential cross sections somewhat change. On the other hand all the observables are modified enormously when we change the two coupling constants independently. Similar behaviors are seen in the case of various targets and incident energies. The other mesons cause relatively small change to observables. Therefore the relative ratio between the coupling constants of σ and ω mesons used in the RIA code seems to reflect the actual value in the nuclear medium.

We expect the RIA calculation can be used as a tool to investigate the medium modifications of the meson-nucleon coupling constants in nuclei by fitting their values to reproduce all the observables.

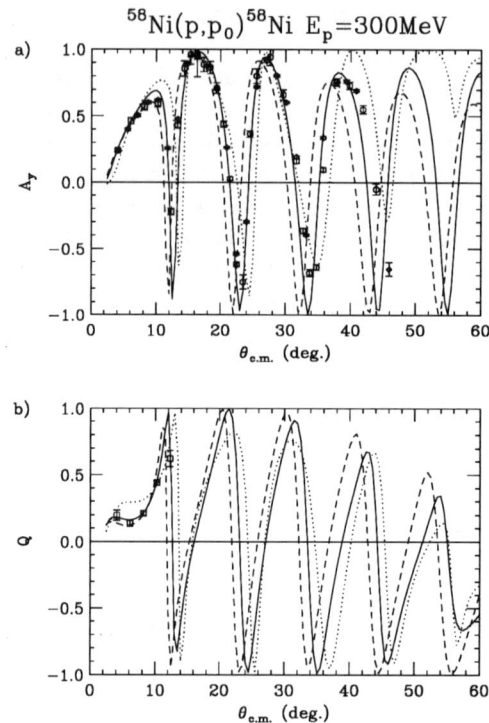

FIGURE 5. The change of analyzing powers (a) and Q-parameters (b) when the coupling constants of σ and ω mesons are multiplied by the same constant factor using the RIA calculation code by Horowitz et al. The solid curves are the results of the original RIA calculations. The dashed (dotted) curves are the results obtained by multiplying both the coupling constants of σ and ω mesons by the same factor of 1.5 (0.5).

REFERENCES

1. Besset, D., et al., Nucl. Inst. and Methods **166**, 379 (1979).
2. Aprile-Giboni, E., et al., Nucl. Inst. and Methods **215**, 147 (1983).
3. Yosoi, M., et al., the present proceedings.
4. Horowitz, C. J., et al., Computational Nuclear Physics 1 Nuclear Structure, edited by K. Langanke, J. A. Maruhn, and S. E. Koonin (Springer, Hamburg, 1991).
5. Murdock, D. P. and Horowitz, C. J., Phys. Rev. **C35**, 1442 (1987).

FRAGMENTATION OF "STRETCHED" 6⁻ STRENGTH IN ^{28}Si$(\vec{p},\vec{p}')^{28}$Si

J. Liu, E.J. Stephenson, A.D. Bacher, S.M. Bowyer,
S. Chang, C. Olmer, S.P. Wells, and S.W. Wissink

Indiana University Cyclotron Facility, Bloomington, Indiana 47408

J. Lisantti

Centenary College of Louisiana, Shreveport, Louisiana 71104

Abstract Using a 198.3 MeV polarized proton beam, measurements have been made of the $d\sigma/d\Omega$, A and $D_{NN'}$ for the ^{28}Si$(\vec{p},\vec{p}')^{28}$Si reaction to locate the 6⁻ strength for comparison with large basis shell model calculations. Based on both $d\sigma/d\Omega$ and A data, six states above 15 MeV were determined to carry 6⁻ strength. The measurements of $D_{NN'}$ were used as a meter of isospin mixing. The shell model calculations do not match the experimental results.

INTRODUCTION

A "stretched" particle-hole transition excites a particle from the orbital of the largest angular momentum in one major shell into the orbital of the largest angular momentum in another major shell, with the particle and hole angular momenta in the final state coupled the largest possible value. These excited states, for the even-even nuclei typically studied, have negative parity and total angular momentum $J = j_p + j_h$, with $j_p = l_p + 1/2$ and $j_h = l_h + 1/2$.

For ^{28}Si, the 6⁻, T=0 state at 11.58 MeV and the 6⁻, T=1 state at 14.36 MeV are such "stretched" states. With this spin and parity in a $1\hbar\omega$ basis, only the stretched configuration $1f_{7/2}1d_{5/2}^{-1}$ can contribute to a one-step M6 transition from the ground state. These are therefore believed to be transitions of relatively simple structure. However, the "stretched" 6⁻ one-particle one-hole strength mixes with other configurations and is fragmented among many smaller states at higher excitation. The one-step nature of the (p,p') reaction at intermediate energies will strongly favor excitation of these fragments only through the $1f_{7/2}1d_{5/2}^{-1}$ component, so comparisons of $d\sigma/d\Omega$ and A for higher energy transitions with data for the states at 11.58 MeV and 14.36 MeV will help us to identify other "stretched" 6^- states. Such comparisons fail for states of smaller spin where multiple configurations contribute because of the sensitivity of (p,p') reactions to the individual structure amplitudes. The values of $D_{NN'}$, which are substantially different between T=0 and T=1 states, may also provide isospin information.

A large basis shell model calculation for ^{28}Si has been published.[1] It allows for all configurations of the type $(d_{5/2}, s_{1/2})^{11-n} d_{3/2}^m f_{7/2}$ up to $n=4$. This calculation successfully explains the observed strength of the 11.58 MeV 6^- (T=0) and 14.36 MeV 6^- (T=1) states, and predicts fragmentation regions for both the 6^-, T=0 and 6^-, T=1 states. A state with an excitation energy of 17.3 MeV has been identified to be a 6^-, T=1 state in a ^{28}Si(e,e')^{28}Si experiment.[2] A ^{28}Si(p,n)^{28}P experiment[3] also reveals fragments of the 6^- strength. Seven states were reported to be 6^-, T=1 candidates.

EXPERIMENT

The data was acquired at the Indiana University Cyclotron Facility using the K600 magnetic spectrometer and its associated focal plane and focal plane polarimeter detectors. The data included measurements of $d\sigma/d\Omega$, A and $D_{NN'}$ for excited states in ^{28}Si at laboratory angles of 23°, 29°, 35°, and 41° covering the momentum transfer range from about 250 MeV/c to 440 MeV/c. A larger range of elastic scattering $d\sigma/d\Omega$ and A measurements were made to constrain the distorting potentials for DWBA calculations. The target used in the experiment was a natural ^{28}Si target with a thickness of 8.4±0.6 mg/cm^2.

The measured excitation energy ranged from a few MeV to about 18.5 MeV. After adjusting the beam focus and dispersion, an energy resolution of about 50 keV (FWHM) was maintained throughout the run.

A standard atomic beam polarized ion source offered normal beam polarization with a typical value of 0.75. The average beam line polarization was monitored during data acquisition using p+d elastic scattering in two polarimeters upstream of the K600 target.[4]

The ratio of polarization magnitudes of spin up and down was periodically measured using p+^{4}He elastic scattering[5] between the injector and the main stage cyclotron.

The analyzing power of the K600 focal plane polarimeter has been calibrated between 170 MeV and 200 MeV, and fitted to the form

$$A_{FPP} = a_0 + a_1(E' - E_0) + a_2(E' - E_0)^2, \quad (1)$$

where E_0=184.03, a_0=0.46066, a_1=0.0041489, and a_2=−0.0000207, and E'

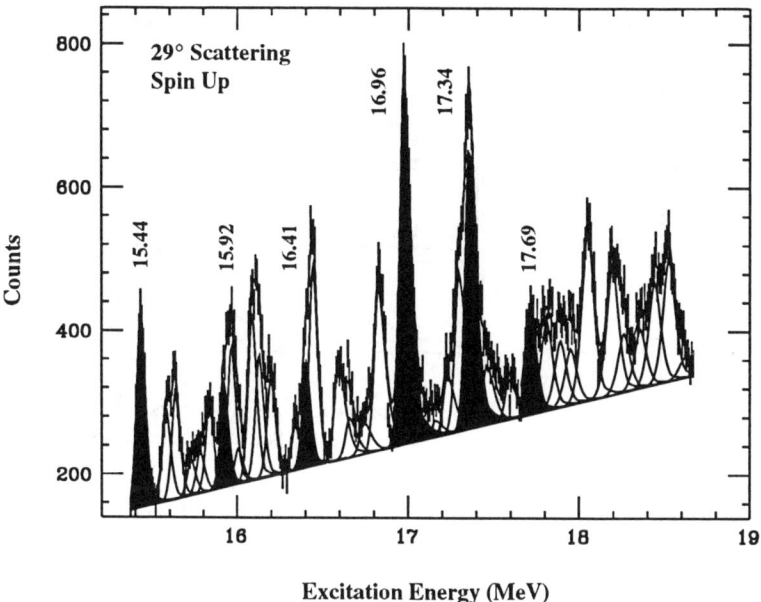

Figure 1. K600 focal plane energy spectrum for ^{28}Si$(\vec{p},\vec{p}')^{28}$Si with the stretched 6^- candidates marked.

is the scattered proton energy. A subsequent calibration over a larger energy range is given by another contribution to this conference.

RESULTS

Focal plane spectra were generated for both spin states based on a known fraction of all events, or on satisfaction of the energy and angle criteria associated with the focal plane polarimeter.

A peak fitting routine using Gaussian shapes with independent left and right widths and exponential tails was used to extract the peak sums from these spectra. The peak shape was determined by matching to an isolated state. The background of the energy spectra was from the particle knock-out continuum, and was represented by a spin-dependent piecewise linear function. States at higher excitation were not analyzed due to the spreading of the state width and the difficulties determining an accurate quasielastic background. Fig. 1 shows the fitted K600 focal plane spectrum at 29° with spin up.

The energy scale was determined by matching a quadratic function of momentum to the focal plane detector positions. The states used for matching were 5^- at 9.70 MeV, 3^- at 10.18 MeV, 6^- at 11.58 MeV and 6^- at 14.36 MeV for 23°, 4^+ at 4.62 MeV, 3^- at 6.88 MeV, 5^- at 9.70 MeV, 6^-

404 Fragmentation of "Stretched" 6⁻ Strength

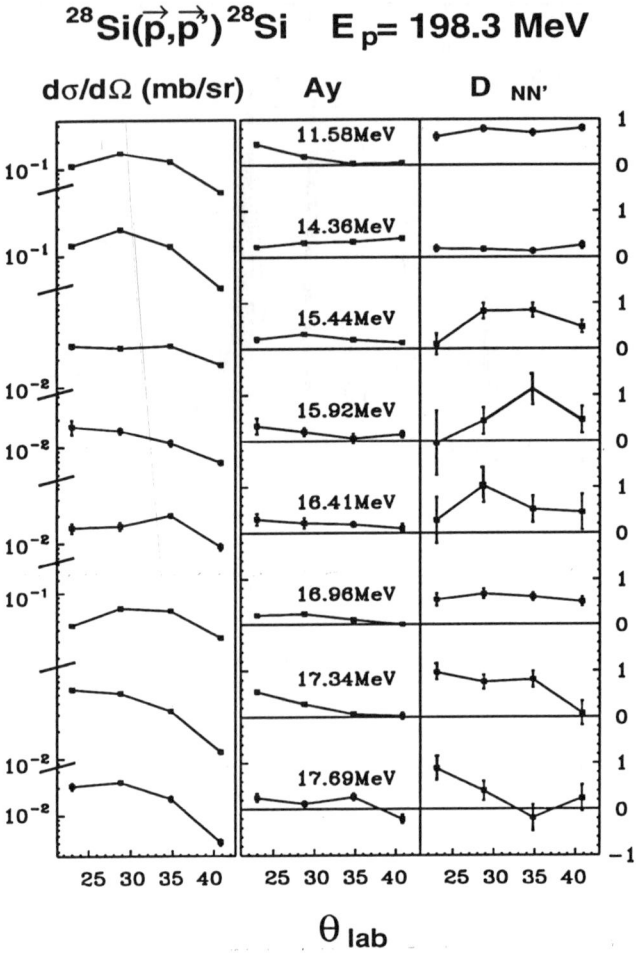

Figure 2. Measured $d\sigma/d\Omega$, A and $D_{NN'}$ values for the ^{28}Si$(\vec{p},\vec{p}')^{28}$Si 6⁻ transition candidates. Excitation energy is noted for each transition.

at 11.58 MeV and 6⁻ at 14.36 MeV for 29° and 35°, and 2⁺ at 1.78 MeV, 4⁺ at 4.62 MeV, 3⁻ at 6.88 MeV, 5⁻ at 9.70 MeV, 6⁻ at 11.58 MeV and 6⁻ at 14.36 MeV for 41°. The quadratic coefficient was relatively small. All identified state energies were well determined by the matching with errors less than 10 keV.

Since a natural silicon target was used in this experiment, the ^{29}Si and ^{30}Si isotopic abundances of 4.67% and 3.10% respectively made it possible to have some ^{29}Si and ^{30}Si states mixed with dominant ^{28}Si states. In order to judge if a state was associated with ^{28}Si, the peak positions of the state were used to calculate the excitation energy at all four scattering angles and the energies checked for consistency.

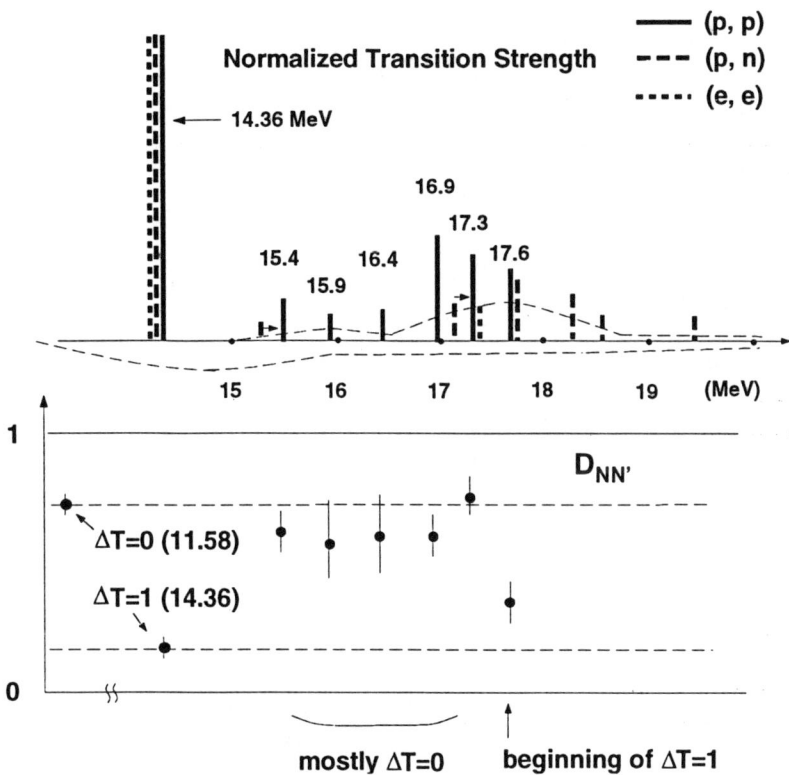

Figure 3. Relative transition strengths of (p,p'), (p,n') and (e,e') states normalized to the T=1 transition at 14.36 MeV. These may be compared to the predicted strenght distributions[1] for T=1 (above the axis) and T=0 (below). Associated $D_{NN'}$ values are given.

Over 100 peaks were fitted and identified to be ^{28}Si states. The cross sections and analyzing power of the 6^- transitions at 11.58 MeV and 14.36 MeV were consistent with former results.[6]

In addition to the two known 6^- states, we identified six other states on the basis of $d\sigma/d\Omega$ and A which appear to carry 6^- strength. The measured $d\sigma/d\Omega$, A and $D_{NN'}$ values for these states are shown in Fig. 2.

The two states at 15.92 and 16.41 MeV are small peaks resting under larger states at higher excitation energy (see Fig. 1), thus the errors in the extraction of $D_{NN'}$ are larger. Some variation was allowed in both cross section and analyzing power in making the comparison. All the six states appear to be consistent with a $\Delta L = 5$ transition. They are unlikely to be 4^- states because 4^- states are mainly dominated by $\Delta L = 3$ transitions. All the states have analyzing power values above zero except the 17.69 MeV state at 41°. A 5^-, T=0 transition typically shows analyzing power values close to a straight line which crosses zero at 370 MeV/c with a negative slope. An assignment of 5^-, T=1 cannot be excluded.

Fig. 3 shows the cross sections of the six candidate states normalized to the 6^-, T=1 at 14.36 MeV (solid), and the same thing for spectroscopic factors taken from both (p,n) (long dashed,[3]) and (e,e') (short dashed,[2]) reactions which are only sensitive to the isovector channel. The (p,n) energies may need to be shifted upward to compensate for systematic energy calibration errors. Experiments that include the 6^- states in ^{28}Si as well as the $\frac{1}{2}^- \to \frac{9}{2}^+$ transition in ^{13}C[7], the 4^- states in ^{16}O[8], and the $3^+ \to 0^+$ transition in ^{10}B[9] all indicate a large shift in $D_{NN'}$ values from near unity for isoscalar and neutron-like transitions to close to zero for isovector transitions. While DWBA calculations often give $D_{NN'}$ values more negative than the data, they clearly suggest that this range bounds the variations of $D_{NN'}$ with isospin. Shown in Fig. 3 are values of $D_{NN'}$ averaged over the four angles pictured in Fig. 2 to improve statistical precision. At left are the standard $D_{NN'}$ values for $\Delta T=0$ and $\Delta T=1$; and values between these extremes are probably isospin mixed. The averaged $D_{NN'}$ values of the eight 6^- transitions are shown below on the same energy scale. The first five candidate states have large values of $D_{NN'}$ that suggest they are mostly isoscalar, a conclusion consistent with the observation that the (p,p') strength exceeds the purely isovector (p,n) and (e,e') observations (including especially the state at 17.3 MeV). (Three of the first four states may also appear as L=3, T=0 states in the ^{27}Al$(\alpha,t)^{28}$Si work of Yasue.[10]) The 17.6 MeV state is mostly isovector, and this follows also from the near equality of the (p,p') and (p,n) strengths. The distribution of $\Delta T=0$ and $\Delta T=1$ strength from Carr[1] is shown by the curves below and above the baseline respectively. While the $\Delta T=1$ distribution agrees well with the (p,n) results, the (p,p') data shown here suggests that the $\Delta T=0$ distribution reaches its maximum at least 2 MeV above the prediction.

REFERENCES

[1] J.A. Carr, *et al.*, Phys. Rev. Lett. **62** 2249 (1989).
[2] S. Yen, *et al.*, Phys. Lett. B **289** 22 (1992).
[3] N. Tamimi, *et al.*, Phys. Rev. C **45** 1005 (1992).
[4] S.P. Wells, *et al.*, Nucl. Instum. and Meth. **A235**, 205 (1993).
[5] P. Schwandt, *et al.*, Nucl. Phys. **A163**, 432 (1971).
[6] C. Olmer, *et al.*, Phys. Rev. C **29** (1984) 361.
[7] W. Schmitt, private communication.
[8] A.K. Opper, Ph.D. thesis, Indiana University, 1991.
[9] H. Baghaei, *et al.*, Phys. Rev. Lett. **69** 2054 (1992).
[10] M. Yasue, *et al.*, Nucl. Phys. **A391**, 377 (1982).

Simultaneous Measurements of $(\vec{p}, \vec{p}')$ and $(\vec{p}, p'\gamma)$ Observables for the 15.11 MeV, 1^+, T=1 State in ^{12}C at 200 MeV

S.P. Wells[1,*], S.W. Wissink[1], A.D. Bacher[1], J. Beene[2], G.P.A. Berg[1], F. Bertrand[2], A. Betker[1], S.M. Bowyer[1], S. Chang[1], C. Foster[1], W. Franklin[1], M. Halbert[2], K. Hicks[3], D. Horen[2], J. Lisantti[4], J. Liu[1], P. Mueller[2], D. Olive[2], W. Schmitt[1], E.J. Stephenson[1], D. Stracener[2], and R. Varner[2]

[1] *Indiana University Cyclotron Facility, Bloomington, Indiana 47405*
[2] *Oak Ridge National Laboratory, Oak Ridge, Tennessee 37831*
[3] *Ohio University, Athens, Ohio 45701*
[4] *Centenary College of Louisiana, Shreveport, Louisiana 71134*

Abstract. The extension of singles $(\vec{p}, \vec{p}')$ studies to include coincident $(\vec{p}, p'\gamma)$ measurements can provide new information about the effective nucleon–nucleus interaction. We have therefore made simultaneous measurements of $(\vec{p}, \vec{p}')$ and $(\vec{p}, p'\gamma)$ observables for the 15.11 MeV, 1^+, T=1 state in ^{12}C at an incident beam energy of 200 MeV. Contained in these coincident observables are determinations of the sideways and longitudinal analyzing powers, D_{0S} and D_{0L}, which must vanish in singles measurements due to parity conservation. With this large data set on a transition with a relatively simple spin sequence, we have, in principle, enough information for a complete determination of the scattering amplitude.

Studies of $(\vec{p}, \vec{p}')$ spin observables have provided much valuable information about the NN interaction inside the nuclear medium (1). There are, however, at most eight independent quantities that can be determined in singles $(\vec{p}, \vec{p}')$ measurements. For transitions to discrete final states of non-zero spin J, additional information on the scattering amplitude can be obtained by studying the polarization state of the recoil nucleus. This can be done via measurements of the angular correlation between the scattered proton and the particle(s) emitted in the nuclear deexcitation, as in reactions of the type $(\vec{p}, p'\gamma)$, in which the polarizations of the outgoing proton and photon are not detected. Such measurements are now technically quite feasible. Of particular interest are studies of transitions with simple spin sequences, such as $0^+ \to 1^+$; in this case, it has been recently shown (2) that certain $(\vec{p}, p'\gamma)$ measurements, when combined with the complete set of $(\vec{p}, \vec{p}')$ observables discussed above, provide sufficient information to completely specify the scattering amplitude in a model-independent manner. We have therefore performed simultaneous measurements of the $(\vec{p}, \vec{p}')$ spin-transfer observables and the $(\vec{p}, p'\gamma)$ coincident observables for study of the 15.11 MeV, 1^+, T=1 state in ^{12}C at a proton bombarding energy of 200 MeV, using the septum magnet mode (3) of the IUCF K600 spectrometer and its associated focal plane polarimeter (FPP) to detect the scattered protons, and a system of BaF$_2$ detectors (4) surrounding the K600 target to detect the coincident γ–rays.

Because eleven quantities must be determined to fully describe the scattering amplitude for a $0^+ \to 1^+$ transition, one sees that the $(\vec{p}, p'\gamma)$ observables provide information which is not accessible in $(\vec{p}, \vec{p}')$ singles measurements. In general, the scattering amplitude for (p, p') transitions can be divided into two parts: terms which couple the proton spin to unit vectors that lie in the reaction plane; and terms in which the proton spin couples to the unit vector normal to the scattering plane. It is the relative phases between these two sets of terms that can not be determined via $(\vec{p}, \vec{p}')$ measurements alone. A bit of algebra reveals (2) that if the coincident γ-rays are detected at some angle out of the reaction plane, yet not normal to it, then the photon polarization tensor will have negative parity. In this case, the sideways and longitudinal analyzing powers D_{0S} and D_{0L} (or A_x and A_z, respectively), which must vanish identically in singles measurements if parity is conserved, may be non-zero in $(\vec{p}, p'\gamma)$ studies. Contained in these asymmetries are the relative phases just discussed. By placing three of the BaF$_2$ arrays at an angle of 45° out of the scattering plane, we were thus able to map out the angular distribution of these previously unmeasured asymmetries.

Simultaneous measurements of $(\vec{p}, \vec{p}')$ singles spin-observables and $(\vec{p}, p'\gamma)$ coincident observables were made at proton scattering angles of $\theta_p = 5°$, 8°, 11°, and 15°. At each of the angles studied, data were taken with three of the BaF$_2$ arrays arranged in two different geometric configurations; the fourth array was always positioned directly above the target. The two configurations allowed for a larger number of $(\vec{p}, p'\gamma)$ observables to be investigated, though each set of measurements placed different constraints on the orientation of the incident beam polarization vector.

For the first set of observables, which required that the incident proton beam be polarized vertically (normal to the scattering plane), the three BaF$_2$ arrays were positioned *in* the scattering plane at angles of $\theta_{lab} = 60°$, 100°, and 140° on beam right. In this configuration, we made measurements of the singles observables $d\sigma/d\Omega$, A_y, P, and $D_{NN'}$, and simultaneously measured the coincident double-differential cross section $d^2\sigma/d\Omega_\gamma\, d\Omega_p(\hat{k})$ and coincident asymmetry $A_y(\hat{k})$ as a function of photon angle in the scattering plane. (Here $\hat{k}$ is the momentum direction of the emitted photon.) We also measured a coincident cross section and asymmetry using the BaF$_2$ array placed directly above the target. In this case, both observables can be expressed, in a model-independent way, in terms of the normal-component $(\vec{p}, \vec{p}')$ spin-transfer coefficients (2). In particular, the coincident asymmetry, measured when the photon is emitted along the normal to the scattering plane, takes the form (2)

$$A_y(n) = -\frac{(P - A_y)}{(1 - D_{NN'})} \ . \tag{1}$$

Since we were simultaneously measuring these normal-component singles observables in the FPP, we thus obtained a valuable internal consistency check on the data. Shown in Fig. 1 are the measured values of the coincident yield asymmetry as described above, plotted as a function of the center-of-mass proton scattering angle, θ_{CM}. Also shown is the particular combination of $(\vec{p}, \vec{p}')$ observables presented in Eq. 1, deduced from the FPP yields. The excellent agreement seen between these two completely independent measurements of the same physical quantity gives us confidence that our measurements of both $(\vec{p}, \vec{p}')$ and $(\vec{p}, p'\gamma)$ observables are reasonable and consistent.

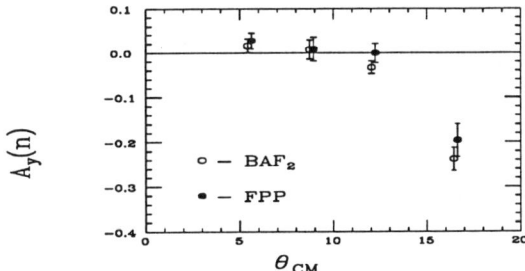

Figure 1. Comparison of the directly measured normal component coincident asymmetry when the photon is emitted normal to the scattering plane and the combination of $(\vec{p}, \vec{p}\,')$ observables given in Eq. 1 plotted versus θ_{CM}.

For the second set of observables, we used an incident proton beam which had its polarization vector rotated (via the high-energy beamline spin-precession solenoids) *into* the scattering plane. We chose to make measurements with the incident proton polarization in three different orientations, at in-plane angles of $\Phi = 53°$, $117°$, and $169°$ with respect to the incident beam direction. With these polarization directions, we were able to determine two linear combinations of the four in-plane spin-transfer coefficients $(D_\lambda \equiv D_{LL'}\sin\alpha + D_{LS'}\cos\alpha$ and $D_\sigma \equiv D_{SL'}\sin\alpha + D_{SS'}\cos\alpha$, where $\alpha = 264°$ is the angle of spin precession experienced by the scattered proton flux in the dipole field of the K600 spectrometer) using the FPP. Also for this second data set, with the incident polarization vector horizontal, we supported the three BaF$_2$ arrays at an angle of $45°$ out of the scattering plane. In this case, the BaF$_2$ arrays were positioned on beam right at angles (projected into the scattering plane) of $\theta_{lab} = 41°$, $88.3°$, and $140°$. In this configuration, the BaF$_2$ arrays were used to map out an angular distribution of the coincident yield asymmetries for the measurement of the sideways and longitudinal analyzing powers D_{0S} and D_{0L} discussed earlier. The use of three different incident beam polarization orientations allowed us to separate the sideways and longitudinal pieces, and also provided another check of internal consistency, in that only two orientations were actually needed. This is because the asymmetry can have only a sinusoidal dependence on both the incident proton spin orientation and the direction of the emitted photon, and therefore four independent coefficients are all that is necessary to completely describe the asymmetry mapped out in the cone $45°$ out of the scattering plane. Shown in Fig. 2 are these coefficients plotted as a function of θ_{CM}. These observables represent the asymmetry that would be measured if the beam were polarized purely along the longitudinal or sideways directions (upper or lower graphs, respectively) and if the photon was emitted in the average-momentum ($\hat{K}$) or momentum-transfer ($\hat{q}$) planes (left or right graphs, respectively). Also included on this plot are two distorted-wave impulse approximation (DWIA) calculations [5], performed using either a relativistic (DREX) or nonrelativistic (DW81) formalism. The large differences seen between the data and the two predictions, relative to the size of the uncertainties in the data, suggest the extent to which these measurements may help pinpoint weaknesses present in the underlying theory, or in approximations made in carrying out the particular calculations.

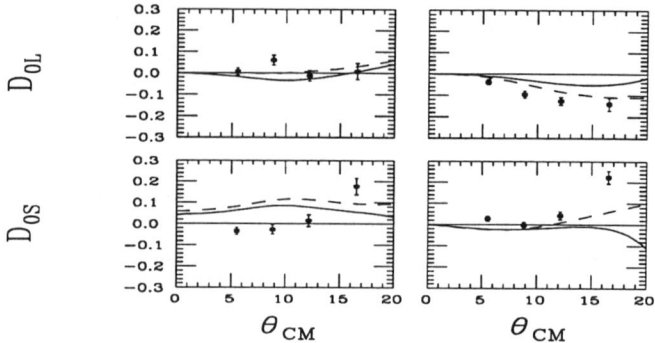

Figure 2. The four coefficients that describe the coincident asymmetry measured 45° out of the reaction plane plotted versus θ_{CM}. The physical interpretation of the four coefficients is described in the text. Also shown are the results of relativistic (DREX – solid line) and nonrelativistic (DW81 – dashed line) calculations for these observables.

Because we have made simultaneous measurements of $(\vec{p}, \vec{p}')$ and $(\vec{p}, p'\gamma)$ observables for this transition, we have, in principle, enough information to completely specify the scattering amplitude in a model-independent manner. Described in the

$$\vec{n} = \vec{p} \times \vec{p}'; \vec{K} = \vec{p} + \vec{p}'; \vec{q} = \vec{n} \times \vec{K} \tag{2}$$

coordinate system, with $\vec{p}$ ($\vec{p}'$) the incident (outgoing) beam momentum, the most general form for the amplitude for a $0^+ \to 1^+$ transition, assuming only angular momentum and parity conservation, can be written (2)

$$\begin{aligned}\hat{T}^p(\vec{p},\vec{p}\,') &= A_{n0}(\hat{\Sigma} \cdot \hat{n}) + A_{nn}(\hat{\Sigma} \cdot \hat{n})(\vec{\sigma} \cdot \hat{n}) + A_{KK}(\hat{\Sigma} \cdot \hat{K})(\vec{\sigma} \cdot \hat{K}) \\ &+ A_{Kq}(\hat{\Sigma} \cdot \hat{K})(\vec{\sigma} \cdot \hat{q}) + A_{qK}(\hat{\Sigma} \cdot \hat{q})(\vec{\sigma} \cdot \hat{K}) + A_{qq}(\hat{\Sigma} \cdot \hat{q})(\vec{\sigma} \cdot \hat{q}),\end{aligned} \tag{3}$$

where $\hat{\Sigma}$ is the polarization operator for the 1^+ nucleus, $\vec{\sigma}$ are the Pauli operators for the projectile, and the $A_{i\mu}$'s are scalar functions of energy and momentum transfer. Thus, eliminating one overall, unphysical phase, there are six amplitudes and five relative phases which must be determined for a complete specification of this amplitude. A general feature of the $A_{i\mu}$'s is that the subscript i represents the polarization component of the recoil nucleus while μ represents the polarization component of the projectile. It can be shown (2) that the singles observables restrict the relative phase information accessible to that of interference between $A_{i\mu}$'s with the same polarization components for the recoil nucleus, while interference between $A_{i\mu}$'s with different nuclear polarization components can be present in the coincident observables, provided that the photons are detected at suitably chosen angles.

Each observable, singles or coincident, can be expressed as a sum of bilinear products of the $A_{i\mu}$'s. For example, the reaction analyzing power (scaled by the singles cross section) is given by

$$\frac{d\sigma}{d\Omega_p} A_y = 2\Re(A_{n0}A_{nn}^*) + 2\Im(A_{KK}A_{Kq}^* + A_{qK}A_{qq}^*), \qquad (4)$$

while one of the coincident asymmetries (normalized to the γ-ray branching ratio to the ground state, R, and as measured if the photon is detected at an angle of 45° from the normal and in the average momentum ($\hat{K}$) plane) is given by

$$\frac{8\pi}{3R}\frac{d^2\sigma}{d\Omega_\gamma d\Omega_p}(\hat{k}) D_{0K}(\hat{k}) = -\Re(A_{n0}A_{KK}^*) + \Im(A_{nn}A_{Kq}^*), \qquad (5)$$

showing explicitly the interference between terms with nuclear polarization projections in the same (different) directions for singles (coincident) observables.

Because we have obtained more observables than there are unknowns required for a complete determination of the scattering amplitude, we are able to perform a fit to all of the data simultaneously in an attempt to determine the independent pieces, $A_{i\mu}$, of the amplitude. This involves searching the nonlinear, eleven-fold parameter space for the best values of both the real and imaginary parts of the $A_{i\mu}$'s which reproduce the measured values of the observables. We have performed a preliminary analysis using this procedure, with reasonable success at extracting the independent pieces of the amplitude. Large correlations among the $A_{i\mu}$'s and the uncertainties associated with the observables can cause ambiguities in the determination of these terms, however, and we are presently working on minimizing these effects so that a more complete and unique determination of the $A_{i\mu}$'s can be achieved.

Information about the transition has already been extracted with this procedure, however, as will be shown in the following discussion. The amplitude A_{qq} is associated with the operator combination $(\hat{\Sigma}\cdot\hat{q})(\vec{\sigma}\cdot\hat{q})$ in the transition amplitude (see Eq. 3). In free nucleon–nucleon scattering, the transition amplitude also contains a term $\delta(\vec{\sigma}_1\cdot\hat{q})(\vec{\sigma}_2\cdot\hat{q})$, where $\vec{\sigma}_1$ and $\vec{\sigma}_2$ are the Pauli operators for the two different nucleons. At low momentum transfer, where single pion exchange is expected to dominate the reaction process, the δ amplitude has been shown (6) to dominate the reaction and vary rapidly as a function of momentum transfer. Moreover, in a one-meson-exchange model, the one pion exchange term is written (7),

$$f_\pi = g_0^2 \frac{(\vec{\sigma}_1\cdot\hat{q})(\vec{\sigma}_2\cdot\hat{q})}{t - m_\pi^2}, \qquad (6)$$

where g_0 is the πNN coupling constant, m_π is the pion mass, and $t = -q^2$ is a kinematic variable. Each of the terms A_{qq}, δ, and f_π, is associated with a similar spin structure in their respective amplitudes, suggesting that the quantity A_{qq} determined in this work may be driven primarily by the pion exchange piece of the reaction mechanism.

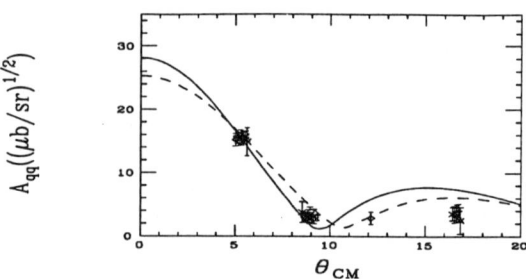

Figure 3. The empirically determined magnitude of the amplitude A_{qq} plotted versus θ_{CM}. The interpretation of the different plotting symbols is discussed in the text. Also shown are the results of the same calculations used in Fig. 2.

Shown in Fig. 3 is the magnitude of this term, as determined from the measured observables, plotted as a function of θ_{CM}, along with the results of the same calculations (5) included in Fig. 2. The different plotting symbols shown at each angle represent different combinations of the real and imaginary parts of this term which produce acceptable values of χ^2 in the fitting procedure (showing little sensitivity of this particular amplitude to some of the above mentioned ambiguities in this procedure). The abrupt change in slope seen in the calculations of this quantity as it approaches zero (near 9° for DREX and 11° for DW81) represents a sign change in this amplitude (as is verified by nearly a 180° change in the phase of this amplitude relative to the phases of other amplitudes), and seems to be reasonably well described by both DREX and DW81. It can be shown (8), in a direct-only, relativistic plane wave impulse approximation, that this amplitude is the only one which contains the pseudovector nucleon–nucleon invariant amplitude. This piece of the NN interaction is driven by π and ρ exchange, and the zero crossing in the A_{qq} amplitude suggested by the values determined from these data may reflect π–ρ interference in the reaction mechanism driving this transition.

* *Present address: Laboratory for Nuclear Science, Massachusetts Institute of Technology, Cambridge, Massachusetts 02139*

References

1. See, for example, Baghaei, H. et al., *Phys. Rev. Let.* **69**, 2054 (1992), and references therein.
2. Piekarewicz, J. et al., *Phys. Rev.* **C41**, 2277 (1990).
3. Berg, G.P.A. et al., IUCF Sci. and Tech. Rep. 220 (1993).
4. Thoennessen, M., Beene, J.R., Bertrand, F.E., and Blankenship, J.L., Oak Ridge Nat'l. Lab. Prog. Rep. (1989).
5. Piekarewicz, J. , private communication.
6. Bowyer, S.M., private communication.
7. Stoks, V.G.J. et al., *Phys. Rev.* **C46**, 792 (1993).
8. Piekarewicz, J. et al., *Phys. Rev.* **C32**, 949 (1985).

A Study of the Fermi (0^+) Transition in $^{14}C(\vec{p},n)^{14}N$ at 495 MeV

D.A. Cooper[a], S.L. Delucia[a], B.A. Luther[a], J.B. McClelland[b], D.L. Prout[b], L.J. Rybarcyk[b], E. Sugarbaker[a], T.N. Taddeucci[b]

[a] The Ohio State University, Columbus, OH 43210
[b] Los Alamos National Laboratory, Los Alamos, NM 87545

Abstract. Differential cross sections and analyzing powers have been measured for the $^{14}C(\vec{p},n)^{14}N^*$(IAS) reaction at a proton energy of 495 MeV. The data were obtained at the LAMPF Neutron Time-of-Flight Facility (NTOF) using a flight path of 400 meters over an angular distribution of $\theta_{lab} = 0°$ to $10°$ ($q = 0.0$ to $0.956 fm^{-1}$). Previous A_y results for targets with mixed Fermi ($\Delta J^\pi = 0^+$) and Gamow-Teller ($\Delta J^\pi = 1^+$) transitions are not reproduced with either DWIA or RIA calculations. The ^{14}C target offers the best opportunity to study a pure Fermi transition (2.31 MeV) resolved from the nearest Gamow-Teller (GT)strength (3.95 MeV). The results compare favorably with DWIA and RIA calculations. Thus the discrepancies for mixed Fermi and GT states most likely arise from either failure to understand nuclear structure effects or the nuclear distortion factors.

1. MOTIVATION

The nucleon charge-exchange reaction has been shown to provide an ideal way to study spin-isospin excitations of the nucleus (1). In particular, the ability to measure spin observables through the use of a polarized beam and a neutron polarimeter can provide a means of looking at specific aspects of the nucleon-nucleon interaction (2,3).

The success of the Relativistic Impulse Approximation (RIA) over the Distorted Wave Impulse Approximation (DWIA) in calculating results for elastic proton scattering (4) led to its use in inelastic proton scattering (5,6) and finally in charge exchange reactions. The (p,n) reaction is sensitive only to the isovector portions of the NN interaction, which are obscured by isoscalar contributions in the (p,p') channel.

A prior study (7) was made of the $^{15}N(\vec{p},n)^{15}O$(g.s.) reaction for two major reasons: first, the ground state of ^{15}N has a simple shell structure (primarily a single $0p_{1/2}$ proton hole in a filled $0p$ shell), and second, the large excitation energy (5.2 MeV) of the first excited state made for very modest energy resolution requirements in this initial investigation using NTOF. However, the results of this experiment show analyzing powers in significant disagreement with both RIA and DWIA calculations. The experimental analyzing powers are nearly zero for momentun transfers between 0 and 1.5 fm^{-1}, while all predictions exhibit large negative dips in the analyzing power around 0.7 fm^{-1}. Since ^{15}N suffers from a complication common to light odd-A nuclei (its ground state is

© 1995 American Institute of Physics

a "mixed" Fermi and Gamow-Teller transition) its analyzing power must be calculated as a cross section weighted average given by

$$A(\theta) = \frac{A_{GT}\sigma_{GT} + A_F\sigma_F}{\sigma_{GT} + \sigma_F}. \qquad (1)$$

Calculations show that most of the negative analyzing power seems to be coming from the Fermi part of the transition. Inability to reproduce the observed analyzing power would then indicate that either the reaction calculations do a poor job of predicting analyzing power for $(\Delta J^\pi = 0^+)$ and/or $(\Delta J^\pi = 1^+)$ transitions, or the ratio of the GT to Fermi cross section is incorrect, indicating a problem with our understanding of the nuclear structure of this transition.

In order to resolve wherein the problem lies we performed similar measurements and calculations for a nucleus in which the Fermi and Gamow-Teller transitions are distinct and separate from any other spurious states. The $^{14}C(p,n)^{14}N^*$ reaction is the best candidate for such a study. The double-differential cross section spectrum for this reaction is displayed for various lab angles in fig. 1.

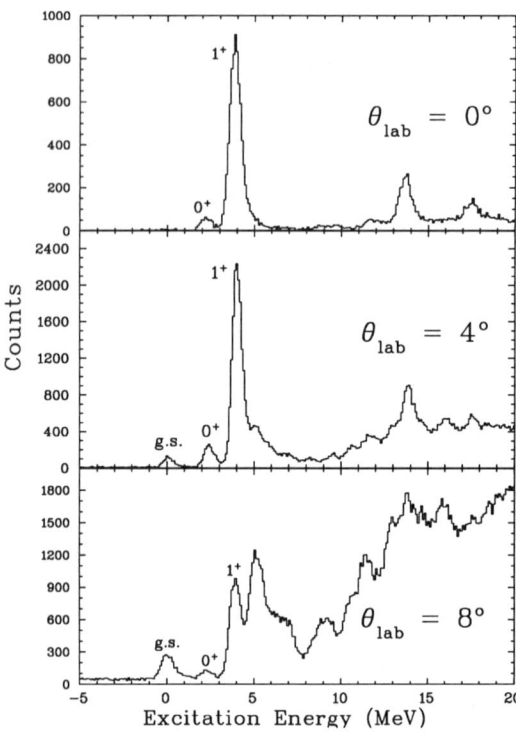

FIGURE 1. ENERGY SPECTRA FOR $^{14}C(p,n)^{14}N$ AT E_p 495 MEV. THE FERMI (0^+) AND THE GT (1^+) STATES ARE INDICATED.

2. EXPERIMENTAL METHOD

The experiment was performed using the Neutron Time-of-Flight Facility (NTOF) at the Los Alamos Meson Physics Facility (LAMPF). A pulsed polarized proton beam was provided by an optically-pumped polarized ion source (OPPIS) (8) at a typical beam current of 105 nA and a micropulse spacing of 200 ns. The polarization of the beam was cycled through normal, unpolarized, reverse, and unpolarized spin orientation at 3 min., 0.5 min., 3 min., and 0.5 min. intervals, respectively. Beam polarization was constantly monitored by two beam-line polarimeters with thin polypropylene (CH_2) targets. Average beam polarization was 60% - 63%. A set of swinger magnets was used to define different incident beam angles with respect to the fixed neutron flight path in order to obtain our angular distribution. Two sets of scintillator (two planes each) were placed at a neutron flight path of 400 m to detect outgoing neutrons. A coincidence in the sets (separated by 1.5 m) was required, so that a course velocity measurement could be used to remove events associated with slow neutrons from previous micropulses. Data were collected with typical subnanosecond time resolution that corresponded to an energy resolution of about 700 keV. The detector system is described in more detail in ref. (9).

Absolute cross section measurements were obtained by normalizing data to the yields from the ^{7}Li$(p,n)^7$Be(g.s. + 0.43 MeV) reaction, which has a known cross section at $\theta_{lab} = 0°$. Details of this procedure as well as a measurement of the ^{7}Li$(p,n)^7$Be (g.s. + 0.43 MeV) cross section can be found in ref. (10).

The ^{14}C target was the TRIUMF/LANL target (11), composed of amorphous carbon powder enriched to 89% ^{14}C and packed into a natural nickel can formed of nickel foil (0.005 cm thick) and having the dimensions of 3 cm high, 1.2 cm wide, and 0.5 cm deep. This gives a target thickness of 170 mg/cm^2. A similar, empty nickel can was used to determine that the nickel spectra did not contaminate the ^{14}C states of interest.

3. DATA & RESULTS

The time-of-flight spectra gated on either normal or reverse beam polarization were collected at $0°, 2°, 4°, 8°,$ and $10°$. A $6°$ measurement of the analyzing power has been included from data previously collected by this group. Sample spectra are shown in fig. 1. The peaks were fit to a skewed gaussian shape which reproduced the observed neutron peaks. Cross section and analyzing power results are presented in table 1. These analyzing power results are compared with theoretical calculations in figure 2. The Distorted Wave Impulse Approximation calculations (DWIA) and the Plane Wave Impulse Approximation calculations (PWIA) are from ref. (12). The Relativistic Impulse calculations are from ref. (13).

TABLE 1. Experimental Results

θ_{lab} (deg.)	$q_{c.m.}$ (fm^{-1})	E_x (MeV)	$\sigma_{c.m.}$ (mb/sr)	A_y
0	0.016	2.31	1.13 ± 0.09	0.016 ± 0.080
	0.027	3.95	17.01 ± 1.13	-0.034 ± 0.019
2	0.192	2.31	0.92 ± 0.07	-0.15 ± 0.04
	0.193	3.95	11.19 ± 0.72	0.005 ± 0.010
4	0.383	2.31	0.58 ± 0.04	-0.39 ± 0.05
	0.383	3.95	4.93 ± 0.32	0.018 ± 0.013
8	0.765	2.31	0.039 ± 0.003	-0.38 ± 0.11
	0.764	3.95	0.42 ± 0.02	-0.057 ± 0.021
10	0.955	2.31	0.019 ± 0.002	0.46 ± 0.21
	0.954	3.95	0.24 ± 0.02	0.005 ± 0.027

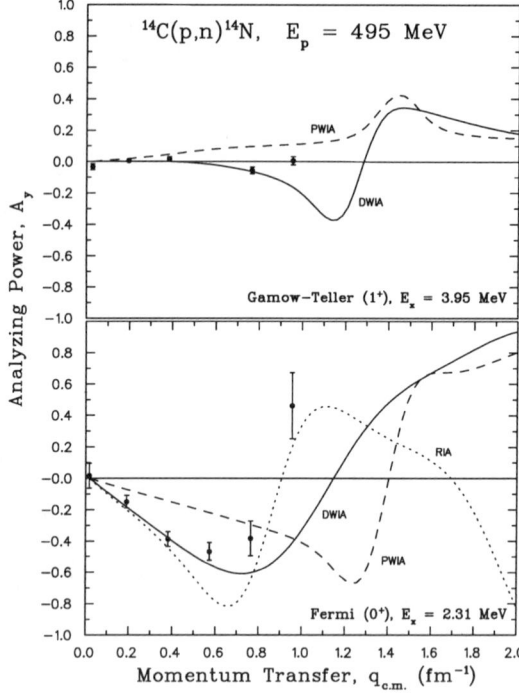

FIGURE 2. ANALYZING POWER MEASUREMENTS COMPARED TO DWIA (SOLID), PWIA (DASHED), AND RIA (DOTTED) CALCULATIONS.

4. DISCUSSION

The primary conclusion to be drawn from these results is that, although the analyzing power for the 0^+ state transition in $^{14}C(p,n)^{14}N$ does not quantitatively match either the DWIA or RIA calculations over the full range of momentum transfer measured, it does agree qualitatively. It can be seen in fig. 2 that the analyzing power for the Gamow-Teller transition remains essentially zero, as expected, while A_y for the Fermi transition has the large negative dip near q = 0.7 fm^{-1} predicted by both models. Since the calculations are quite capable of predicting the qualitative features of the analyzing power for isolated 0^+ and 1^+ states, the failure to qualitatively predict the analyzing power of the mixed state in $^{15}N(p,n)^{15}O$(g.s.) would likely be attributed to a failure to correctly model the relative contributions of GT and Fermi cross section strength. Given these results it is likely that the ratio of the 1^+ to 0^+ cross section components (σ_{GT}/σ_F) is enhanced over that predicted by existing models, leading to the discrepancy between the observed and predicted analyzing powers. A similar odd-even target discrepancy has been noted in a systematic study of the ratio of GT to Fermi unit cross sections (2).

We conclude that the A_y discrepancy at higher bombarding energies arises from a difficulty in predicting the appropriate cross sections. The good representation at the lower energies does not contradict this since the A_y distributions for GT and Fermi transitions at lower energies are very similar and thus fail to provide information about their relative contributions to a mixed transition. However, polarization transfer studies at 0° on a ^{35}Cl target (14) suggest that an enhancement of the reaction sensitivity to GT relative to Fermi strength is observed at 160 MeV as well. The source of this enhancement is still not understood, although suggestions have been made that interference between L = 0 and L = 2 amplitudes are more likely in such nuclei. (15,16)

In order to provide constraints for future theoretical work, full sets of $(\vec{p},\vec{n})$ polarization transfer observables should be measured for odd-A nuclei (such as ^{15}N or ^{13}C). It has been shown (17) that different linear combinations of the polarization transfer observables are more or less sensitive to different components of the NN interaction. Therefore, a complete set of D_{ij}'s would enable a precise diagnosis of the current problems with the interaction models. Previous work on light, odd-A targets (7,18) has shown that measurements of A_y begin to contradict theory at around 160 MeV. It would be preferable to collect data above and below this energy to provide the most stringent constraints possible on descriptions of both nuclear structure and interaction strengths related to these mixed transitions.

ACKNOWLEDGMENTS

This work is supported in part by the DOE and NSF.

REFERENCES

(1) J. Rapaport and E. Sugarbaker, Ann. Rev. Nucl. Part. Sci., to be published (1994).
(2) T.N. Taddeucci et al., Nucl. Phys. **A469**, 125 (1987).
(3) J. Rapaport et al., Phys. Rev. C **39**, 1929 (1989).
(4) J.R. Shepard, J.A. McNeil, and S.J. Wallace, Phys. Rev. Lett. **50**, 1443 (1983).
(5) E. Rost and J.R. Shepard, Phys. Rev. C **35**, 681 (1987).
(6) J.R. Shepard, E. Rost, and J.A. McNeil, Phys. Rev. C **33**, 634 (1986).
(7) D.E. Ciskowski, Ph.D. thesis, University of Texas at Austin (1989).
(8) R.L. York et al., *Prooceedings of the International Workshop on Polarized Ion Sources and Polarized Gas Jets*, Tsukuba, Japan, 1990.
(9) T.N. Taddeucci et al., Nucl. Inst. & Meth. **A241**, 448 (1985).
(10) T.N. Taddeucci et al., Phys. Rev. C **41**, 2548 (1990)
(11) W.P. Alford et al., Phys. Lett. B **179**, 20 (1986).
(12) R. Schaeffer and J. Raynal, computer code **DWBA70**, unpublished.
J.R. Comfort, revised computer code **DW81** , unpublished.
(13) E.Rost and J.R. Shepard, computer code **DREX**, unpublished and private communication.
(14) A.J. Wagner, Ph.D. thesis, The Ohio State University (1988).
(15) J.W. Watson et al., Phys. Rev. Lett. **55**, 1369 (1985).
(16) S.M. Austin, N. Anantaraman, adn W.G. Love, Phys. Rev. Lett. **73**, 30 (1994).
(17) L. Ray and J.R. Shepard, Phys. Rev. C **40**, 237 (1989).
(18) J.Rapaport et al., Phys. Rev C **36**, 500 (1987).

Measurement of the Polarization Transfer $D_{NN}(0°)$ for $(\vec{p},\vec{n})$ Reactions at 295 MeV

T. Wakasa, M. B. Greenfield*, K. Hatanaka**, S. Ishida,
N. Koori[†], H. Okamura, A. Okihana[‡], H. Otsu,
H. Sakai, N. Sakamoto, Y. Satou, and T. Uesaka

Department of Physics, University of Tokyo, Bunkyo, Tokyo 113, Japan
*International Christian University, Mitaka, Tokyo 181, Japan
**Recerch Center for Nuclear Physics, Osaka University, Ibaraki, Osaka 567, Japan
*[†]Faculty of Integrated Arts and Science, The University of Tokushima,
Tokushima 770, Japan*
[‡]Kyoto University of Education, Fushimi, Kyoto 612, Japan

Abstract. The first data of the polarization transfer coefficients $D_{NN}(0°)$ for $(\vec{p},\vec{n})$ reactions at 295 MeV by using a newly developed neutron polarimeter (NPOL) are reported. The results are compared with the distorted-wave impulse approximation (DWIA) calculations with the effective interactions parametrized by Franey and Love (FL). The FL 270 MeV interaction reproduces very well both the differential cross sections and the polarization transfer $D_{NN}(0°)$, while the FL 325 MeV interaction predicts less negative $D_{NN}(0°)$ values. It is found that the tensor part of the FL 325 MeV interaction is too strong at large momentum transfer ($Q \simeq 3.4$ fm^{-1}).

INTRODUCTION

In recent years, new information on the effective nucleon-nucleon (*NN*) interactions has been obtained through a lot of experimental and theoretical studies. Experimentally, such studies have been performed mostly by using nucleon-nucleus scatterings at intermediate energies. Among them, the isovector central interactions (V_τ, $V_{\sigma\tau}$) have been investigated extensively through the zero degree differential cross sections by using a relatively simple relationship between the differential cross section and the volume integral of the interaction. The tensor interaction has mainly been studied by using the high spin stretched states in the momentum transfer ranging approximately between 1 and 2.5 fm^{-1}, although such a study is sometimes hampered by the existence of the spin-orbit interaction. The polarization transfer coefficient D_{NN} at zero degree is another tool to study the tensor interaction.

In a framework of the plane-wave impulse approximation (PWIA), a nucleon-nucleus scattering amplitude is described as a product of the NN amplitudes and the nuclear transition density (1). The NN scattering amplitude can be represented in a standard form as

$$M(q) = A + B\sigma_{1n}\sigma_{2n} + C(\sigma_{1n} + \sigma_{2n})$$
$$+ E\sigma_{1q}\sigma_{2q} + F\sigma_{1p}\sigma_{2p}, \qquad (1)$$

or alternately as a sum of central, spin-orbit and tensor terms

$$M(q) = A + \frac{1}{3}(B + E + F)\boldsymbol{\sigma}_1 \cdot \boldsymbol{\sigma}_2 + C(\boldsymbol{\sigma}_1 + \boldsymbol{\sigma}_2) \cdot \hat{n}$$
$$+ \frac{1}{3}(E - B)S_{12}(\hat{q}) + \frac{1}{3}(F - B)S_{12}(\hat{p}), \qquad (2)$$

where

$$\hat{q} = \frac{\vec{k}_f - \vec{k}_i}{|\vec{k}_f - \vec{k}_i|}, \hat{p} = \frac{\vec{k}_i + \vec{k}_f}{|\vec{k}_i + \vec{k}_f|} \text{ and } \hat{n} = \hat{p} \times \hat{q}.$$

If a single ΔL transfer is dominant, the $D_{NN}(0°)$ for the Gamow-Teller (GT) transition is given by (2)

$$D_{NN}(0°) = \frac{-F^2}{F^2 + 2B^2}. \qquad (3)$$

If there is no exchange tensor interaction ($F = B$), then $D_{NN}(0°) = -1/3$. The experimental $D_{NN}(0°)$ value, therefore, should be a good measure to study the effective tensor interaction by observing the deviation from $-1/3$.

In this paper we report the first measurement of the polarization transfer coefficients $D_{NN}(0°)$ at 295 MeV by using a newly developed Neutron POLarimeter (NPOL) (3). We have obtained the data for (p,n) reactions on ^{6}Li, ^{7}Li, ^{11}B, ^{12}C, and ^{13}C.

EXPERIMENT

The measurement was carried out at the neutron time-of-flight (TOF) facility (4) at the Research Center for Nuclear Physics (RCNP), Osaka University. A polarized proton beam was accelerated to an energy of 295 MeV by the RCNP ring cyclotron. Beam pulses were selected to 1/4 and a separation between beam pulses was about 260 ns. The average beam intensity was 2 nA with an average polarization of 73 %. The targets were set in a beam swinger dipole magnet with thicknesses ranging from 150 to 396 mg/cm^2 and from 155

TABLE 1: Polarization transfer coefficients $D_{NN}(0°)$ for (p,n) reactions at 295 MeV.

Reaction	Transition	E_x(MeV)	$D_{NN}(0°)$	DWIA
^{6}Li$(p,n)^6$Be	GT	0.0	-0.28 ± 0.01	-0.27
^{7}Li$(p,n)^7$Be	GT + F	0.0+0.43	-0.28 ± 0.05	-0.23
^{11}B$(p,n)^{11}$C	GT + F	0.0	-0.03 ± 0.04	-0.11
	GT	2.0	-0.30 ± 0.05	-0.27
	GT	4.3+4.80	-0.21 ± 0.03	-0.27
	GT	8.10+8.43	-0.18 ± 0.04	-0.26
^{12}C$(p,n)^{12}$N	GT	0.0	-0.22 ± 0.02	-0.26
^{13}C$(p,n)^{13}$N	GT + F	0.0	-0.13 ± 0.04	-0.04
	GT	3.51	-0.27 ± 0.03	-0.28

to 703 mg/cm^2 for the differential cross section measurement and the polarization transfer $D_{NN}(0°)$ measurement, respectively. NPOL was positioned at a distance of 100 m.

NPOL consists of four planes of 2-dimensional position-sensitive scintillation detectors, each with an effective detection area of approximately 1 m^2. The first two detectors are made of the liquid scintillator BC519 and the last two detectors are made of the plastic scintillator BC408. The first two planes of liquid scintillation detectors act as scatterers and the detectors downstream of the scatterer plane can serve as catchers of doubly scattered neutrons or recoiled protons. Thin plastic scintillation detector placed in front of each detector serves to distinguish charged particles from neutrons. The neutron polarization can be determined by utilizing both the ^{1}H$(\vec{n},n)^1$H and ^{1}H$(\vec{n},p)n$ reactions. The $n + p$ events were discriminated kinematically from the $n + C$ events by using time and position information.

The effective analyzing powers ($A_{y;\text{eff}}$) of NPOL were calibrated by using polarized neutrons from the ^{2}H$(\vec{p},\vec{n})pp$ reaction at zero degree. The results are 0.290 ± 0.012 for the $(\vec{n},n)$ channel and 0.114 ± 0.003 for the $(\vec{n},p)$ channel where the uncertainties are statistical only. Details for the calibration of NPOL are described in Ref. 5.

RESULTS

The measured $D_{NN}(0°)$ values are listed in Table 1 where the uncertainties are statistical only. The systematic uncertainties are estimated to be less than 0.01 which are mainly from the uncertainties of the effective analyzing powers. The results of the GT transitions significantly deviate from $-1/3$, which is the clear evidence of the tensor interaction.

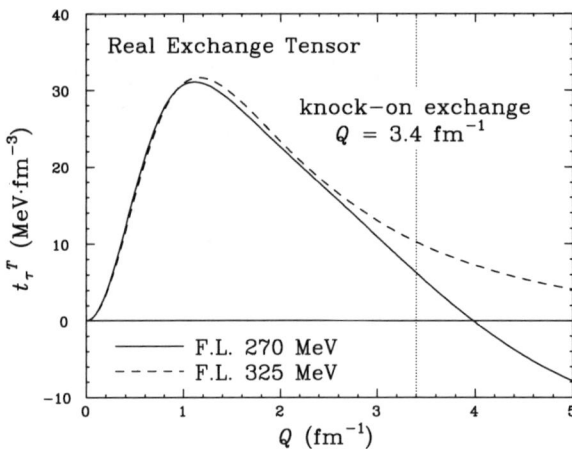

FIGURE 1: The real parts of the isovector exchange tensor interactions as a function of momentum transfer. The dotted vertical line marks the exchange momentum transfer for the ^{6}Li$(p,n)^6$Be$(g.s.)$ reaction at 295 MeV and 0 degree.

We performed the full microscopic distorted-wave impulse approximation (DWIA) calculations by using the computer code DW81 which treats the knock-on exchange amplitude exactly (6). The transition amplitudes were calculated from Cohen-Kurath wave functions (CKWF) (7) assuming harmonic oscillator radial dependence. We used the effective NN t-matrix interactions parametrized by Franey and Love (FL) at 270 MeV (8). Optical model parameters for the distorted waves were obtained from the proton elastic scattering data on ^{12}C at 318 MeV (9) and they were used for all the targets studied. The results are listed in table 1. A fairly good agreement with the experimental values are obtained for all transitions. We have examined the sensitivity of the $D_{NN}(0°)$ value to the optical potentials by using the different parameter sets obtained by the proton elastic scattering at 200 and 398 MeV (10). It was found that the choice of optical potentials causes no significant change in the $D_{NN}(0°)$ values.

DISCUSSION

The insensitivity to the optical potentials allows us to use the polarization transfer $D_{NN}(0°)$ to deduce more detailed information on the effective interactions. Let us use the $D_{NN}(0°)$ value of the ^{6}Li$(\vec{p},\vec{n})^6$Be$(g.s.)$ reaction because it has been determined most accurately. We performed the DWIA calculation with the FL 325 MeV interaction. The cross sections are not sensitive to change the effective interaction. The $D_{NN}(0°)$ value, however, is very

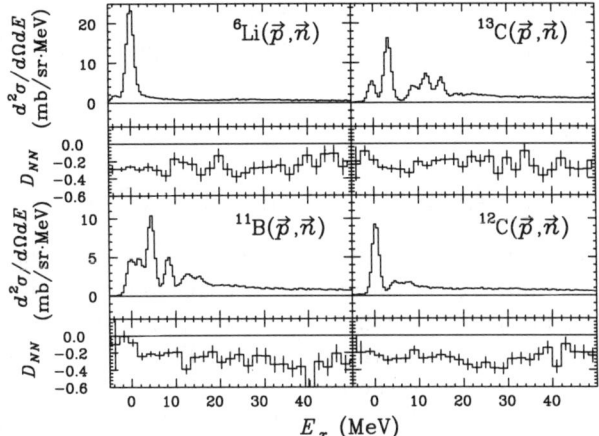

FIGURE 2: The double differential cross sections and the polarization transfer coefficients $D_{NN}(0°)$ as a function of the excitation energy.

sensitive and becomes less negative value of -0.24 compared with either the calculated value of -0.27 with the FL 270 MeV interaction or the measured value of -0.28 ± 0.01. The change of the $D_{NN}(0°)$ value is mainly due to the exchange tensor interaction and in the present situation, the relevant momentum transfer is around 3.4 fm^{-1}. The isovector tensor interaction for the exchange process is described as (11),

$$\tilde{V}_\tau^{TN} = -\frac{1}{4}(V^{TNE} + V^{TNO}), \qquad (4)$$

where V^{TNE} and V^{TNO} are the tensor-even and tensor-odd interactions, respectively. Fig.1 shows the strength of the exchange tensor interactions of the FL 270 and FL 325 MeV interactions as a function of the exchange momentum transfer (Q). Here only the real parts are plotted because the imaginary parts are small. As is clear from the figure, the tensor interaction of the FL 325 MeV interaction is about 50 % larger than that of the FL 270 MeV one at $Q \simeq 3.4$ fm^{-1}.

Fig. 2 shows the double differential cross sections and the D_{NN} values at zero degree for each reaction as a function of the excitation energy. The $D_{NN}(0°)$ data in this figure sorted into 2 MeV width to reduce statistical fluctuations. It is very astonishing to find that the continuum region shows fairly large negative $D_{NN}(0°)$ values up to an excitation energy of 50 MeV. The negative $D_{NN}(0°)$ value is a signature of the spin-flip excitations and such strong and constant spin-flip excitations are not observed at lower bombarding energies (12). This distinctive negative $D_{NN}(0°)$ value in the continuum is also observed in the heavier target nuclei such as ^{27}Al, ^{90}Zr, and ^{208}Pb (13), indicating little target dependence. Further study of this common feature may reveal useful information on the isovector spin responses in the highly excited region.

ACKNOWLEDGEMENT

We would like to thank M. Ichimura and Toru Suzuki for their valuable discussions, T. Muto for the calculation of the one body transition amplitudes, K. Tanaka for his help to obtain the ^{13}C amorphous from the target pool of KEK. This experiment was performed at RCNP under the Program Numbers E18 and E30. This work was supported financially in part by the Grant-in-Aid for Scientific Research No.6342007 and for Special Project Research on Meson Science of Ministry of Education, Science and Culture of Japan.

REFERENCES

1. Kerman, A.K., McNanus, H., and Thaler, R.M., *Ann. Phys.* **8**, 551-635, (1959).
2. Moss, J.M., *Phys. Rev. C* **26**, 727-729, (1982).
3. Sakai, H. *et al*, *Nucl. Instr. Meth.* **320**, 479-499, (1992).
4. Sakai, H., and Noro, T., *RCNP Annual Report* **1987**, 171-172, 1987.
5. Wakasa, T., Master thesis, University of Tokyo, 1994, unpublished.
6. Program DWBA70, Shaeffer, R., and Raynal, J. (unpublished); Extended version DW81 by Comfort, J.R. (unpublished).
7. Kohen, S. and Kurath, D., *Nucl. Phys.* **73**, 1-24, (1965).
8. Franey, M.A., and Love, W.G., *Phys. Rev. C* **31**, 488-498, (1985).
9. Baker, F.T., *et al.*, *Phys. Rev. C* **48**, 1106-1115, (1993).
10. Jones, K.W., *et al.*, *Phys. Rev. C* **33**, 17-21, (1986).
11. Love, W.G., and Franey, M.A., *Phys. Rev. C* **24**, 1073-1094, (1981).
12. Taddeucci, T.N., *Can. J. Phys.* **65**, 557-565, (1987).
13. Wakasa, T., *et al.*, to be published.

ISOSCALAR SPIN STRENGTH IN ^{12}C AND ^{40}Ca

C. Djalali,[a,b] M. Morlet,[b] F.T. Baker,[c] L. Bimbot,[b] J. Guillot,[b]
C. Glashausser,[e] B.N. Johnson,[a] H. Langevin-Joliot,[b] N. Marty,[b]
L. Rosier,[b] E. Tomasi-Gustafsson,[e] J. Van de Wiele,[b] and A. Willis,[b]

[a] Physics Department, University of South Carolina, Columbia, S C 29208
[b] CNRS-IN2P3, I. P. N., 91406-Orsay, France
[c] Physics Department, University of Georgia, Athens, Ga 30602
[d] Physics Department, Rutgers University, New Brunswick, N J 08903
[e] Laboratoire National Saturne, CEN Saclay, 91191 Gif/Yvette, France

Abstract. The isoscalar spin-transfer signature S_d^y has been used in inelastic scattering of 400 MeV polarized deuterons to search for $\Delta S=1$ $\Delta T=0$ strength in ^{40}Ca and ^{12}C. In ^{40}Ca, spin excitations were found in the 9 MeV region and over a broad range in the continuum with a cluster strength around 15 MeV. In ^{12}C, a previously unknown $\Delta S=1$ $\Delta T=0$ level was observed at 20.5 MeV; $\Delta S=1$ $\Delta T=0$ resonances were also found at 20 MeV and 30 MeV. Microscopic DWIA calculations agree nicely with ^{12}C data.

Spin degrees of freedom have been extensively studied, but till now very little was known about the $\Delta S=1$ $\Delta T=0$ channel. This motivated the present experimental program which has been entirely carried out at the Laboratoire National Saturne (France) using the vector and tensor polarized deuteron beam at intermediate energy. A clear signature for isoscalar spin-flip strength (S_d^y) has been obtained (1) and shown to be robust even in the presence of distortion. This has allowed us to carry out the first exploration of isoscalar spin-strength in nuclei. The relative isoscalar spin response (R_S^0) has been extracted and the $\Delta S=1$ and $\Delta S=0$ parts of the cross section have been separated in ^{12}C and ^{40}Ca (2).

SIGNATURE FOR SPIN-FLIP TRANSITIONS

For a deuteron beam, three spin-flip probabilities S_0, S_1, and S_2 may be defined; they are the probabilities for a change of 0, 1, or 2 units of <u>the spin projection along the y-axis</u>. A detailed description of the formalism can be found in references (2, 3). We will only recall the definition for the S_1 probability:

$$S_1 = \frac{1}{9}(4 - A_{yy} - P^{y'y'} - 2K_{yy}^{y'y'}) \tag{1}$$

In the plane wave approximation, $S_1=0$ for $\Delta S=0$ transitions (4). We have found that $S_1 \approx 0$ for $\Delta S=0$ transitions, even in the presence of distortion. For $\Delta S=1$ transitions S_1 is expected to be positive. However, to obtain S_1 one needs to measure the tensor polarization of the scattered deuteron which is a difficult measurement. We can get around this difficulty by considering the following observable $S_1 + 4 S_2 + \frac{1}{3}(A_{yy} - P^{y'y'})$ which does not depend on the tensor polarization of the scattered deuteron. At low momentum transfer, $A_{yy} = P^{y'y'}$ and $S_2 = 0$ even in presence of distortion, this observable called from now on S_d^y is a good approximation of $S1$:

$$S_d^y = \frac{4}{3} + \frac{2}{3} A_{yy} - 2 K_y^{y'} \tag{2}$$

Microscopic description of d-nucleus scattering

The microscopic PWIA description of deuteron-nucleus scattering which was given in Ref.(2) has been extended to include distortion effects (3). Let σ_{00}^A and σ_{10}^A be the isoscalar cross sections for a spin transfer of 0 and 1 to the nucleus. Following the method given in Ref. (5) we can define an α^A coefficient by

$$S_d^y = (\sigma_{10}^A \, \alpha^A)/(\sigma_{00}^A + \sigma_{10}^A) \tag{3}$$

For a pure $\Delta S=1$ transition ($\sigma_{00}^A = 0$), $\alpha^A = S_d^y$. By making the same approximations as in $(\vec{p}, \vec{p'})$ (5), we can replace α^A by α^{free} calculated in the free deuteron-nucleon (d-N) scattering and factorize the cross sections as

$$\sigma_{i0}^A = N_{eff} \, f_{i0} \, \sigma_{i0}^{free} \tag{4}$$

Here N_{eff} is the effective number of participating nucleons (supposed to be the same in both channels), f_{i0} the isoscalar nuclear response in the spin channel i, and σ_{i0}^{free} the d-N scattering cross section calculated for the q value of the deuteron nucleus inelastic scattering. The relative isoscalar spin response is defined as

$$R_S^0 = \frac{f_{10}}{f_{10}+f_{00}} \tag{5}$$

in analogy with the corresponding isospin-averaged quantity R_S measured in proton scattering. The $\Delta S=1$ cross-section is then given by:

$$\sigma_{10}^A = \frac{1}{\alpha^{free}} \frac{d\sigma}{d\Omega} S_d^y(\text{measured}) \tag{6}$$

where $\frac{d\sigma}{d\Omega} = \sigma_{00}^A + \sigma_{10}^A$ is the experimental cross section.

ISOSCALAR SPIN RESPONSE IN THE CONTINUUM

Recently the $\Delta S=1$ strength has been systematically investigated in the continuum of several nuclei in proton inelastic scattering (5,6). It has been found that, at an excitation energy beyond 30 MeV, the background contains up to about 80 % $\Delta S=1$ strength. This overall spin response contains both isospin components. In order to understand the role of isospin in the spin response, we have measured the isoscalar spin strength in ^{12}C and ^{40}Ca (5,6).

The ^{12}C Results

Cross-sections, analyzing powers and spin-transfer signatures have been measured from 3° to 7° and between 4 and 58 MeV of excitation energy. The spectrum of measured cross sections at 4°, the spin-flip signatures at 4° and 6°, the relative isoscalar spin-response and the $\Delta S=0$ and $\Delta S=1$ cross sections are plotted in Fig.1. The signature has high values for the 1^+ T=0 level at 12.7 MeV and the 2^- T=0 level at 18.3 MeV. A previously unknown narrow spin state is observed at 20.5 MeV. The S_d^y signature also shows concentration of spin strength around 20 (3 MeV wide) and 30 MeV (5 MeV wide) of excitation energy. For the 12.7 and 18.3 MeV levels, the angular distributions of the cross section, the vector and tensor analyzing power, the vector polarizing power and the S_d^y signature have been compared to our recent microscopic DWIA calculations. Only the differential cross sections are given in Fig.2 where the dashed (full) curve represents the plane (distorted) wave calculation using the CK wave functions (7). The dotted curves given for the 20.5 state and the wide structures centered at 20 and 30 MeV are only indicative of an L=1 shape. The distorted wave calculations are essential to reproduce the analyzing powers. The signatures on the other hand are not very sensitive to distortion effects in the present angular range. The sharp T=0 spin state at 20.5 MeV has a forward peaked angular distribution similar to that of the

12.7 MeV state, suggesting a 1^+ nature. The angular distributions of the cross-sections for the broad structures at 20 and 30 MeV have an angular distribution close to that of the 18.3 MeV level and therefore compatible with a L=1 nature. The 20 MeV structure may be the already suggested in (π, π') (8) $\Delta L=1$, $\Delta S=1$, $\Delta T=0$ resonance. The 30 MeV spin structure is observed for the first time. Beyond 35 MeV of excitation energy the isoscalar spin response drops below 0.4.

The ^{40}Ca Results

Most of these results have already been published (2) and only the main features will be given here. The strongest signature is seen in the 9 MeV region at 4° where it reaches about 0.13 ($R_S^0 \approx 0.75$). Values elsewhere range between about 0.05 and 0.10 at both angles. Except at the lowest energies, these values are close to the free values of S_d^y and they suggest the presence of a broad distribution of $\Delta S=1$, $\Delta T=0$ strength over this region. Evidence for localized isoscalar spin transitions at 8.4 and 9.2 MeV has been obtained, and a large resonance around 15 MeV contains up to 60% isoscalar spin strength (Fig. 3). The angular distribution of these isoscalar spin-flip excitations is compatible with a possible spin dipole nature. The energy position of the observed excitations is in agreement with theoretical RPA predictions (9) for the 2^- T=0 strength. In these calculations, the used Skyrme force SGII corresponds to a Landau parameter $g_0 \approx 0.25$ at half density. The comparison between these data and previous(p,p') data allowed us to get the nuclear response f_{ij} in the different spin-isospin channels (2).

Summary and conclusions

The isoscalar spin-flip signature S_d^y has been tested on known states in ^{12}C, and turned out to be a robust signature that is not affected by distortion effects at small momentum transfer ($q \leq 1 \text{fm}^{-1}$). This has allowed us to carry out the first exploration of isoscalar spin-strength in the continuum of nuclei. In ^{40}Ca, evidence for localized isoscalar spin transitions in the 9 MeV region and a large resonance around 15 MeV has been obtained. The comparison with the (p,p') results suggests a mild repulsion in the S=1, T=0 channel much weaker than in the S=1, T=1 channel, compatible with recent theoretical predictions. In ^{12}C, previously unknown spin transitions have been observed. A sharp excitation at 20.5 MeV has a forward peaked angular distribution compatible with a 1^+ nature, and a broad structure centered at 30 MeV has a possible L=1 angular distribution. A 3 MeV wide cluster centered at 20 MeV seems also to have an L=1 angular distribution in agreement with the results already obtained in (π,π'). Beyond 25 MeV of excitation energy in ^{40}Ca, the isoscalar spin response fluctuates around its Fermi gas value of 0.5, indicating a weak collectivity in this channel. In ^{12}C, the isoscalar spin response up to 35 MeV is larger than in ^{40}Ca but then becomes weaker beyond this energy.

References

1. Morlet, M., et al., *Phys. Lett.* **B247**, 228(1990).
2. Morlet, M., et al, *Phys. Rev.* **C 46**, 1008(1993).
3. Van de Wiele ,J., and Willis,A., IPN-ORS Report (1994).
4. Ohlsen, G., *Rep. Prog. Phys.* **35**, 717(1972).
5. Glashausser, C., et al, *Phys. Rev. Lett.* **58**, 2404(1987).
6. Baker, F. T., et al, *Phys. Lett.* **237B**, 337(1990).
7. Cohen, S., and Kurath, D., *Nucl. Phys.* **73**, 1(1965).
8. Bland, L., et al., *Phys. Lett.* **B144** 328 (1984).
9. Dimitriescu, T. S., and Suzuki, T., *Nucl. Phys.* **A243**, 277(1984).

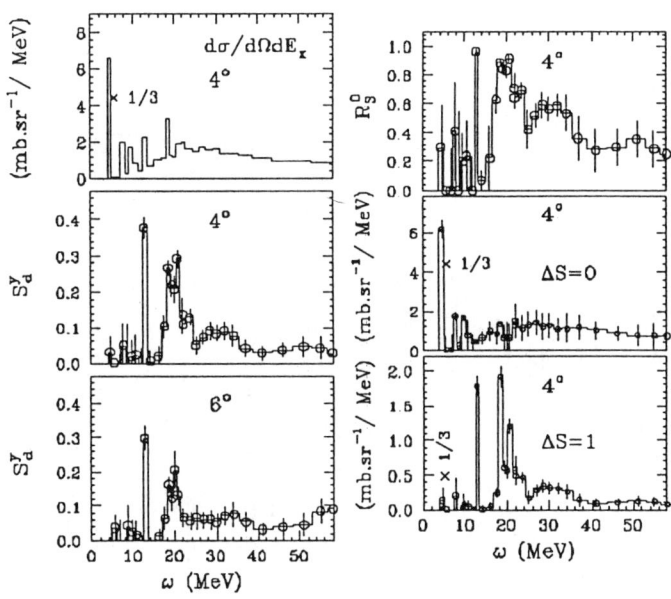

Figure 1. The ^{12}C($\vec{d},\vec{d}'$) results

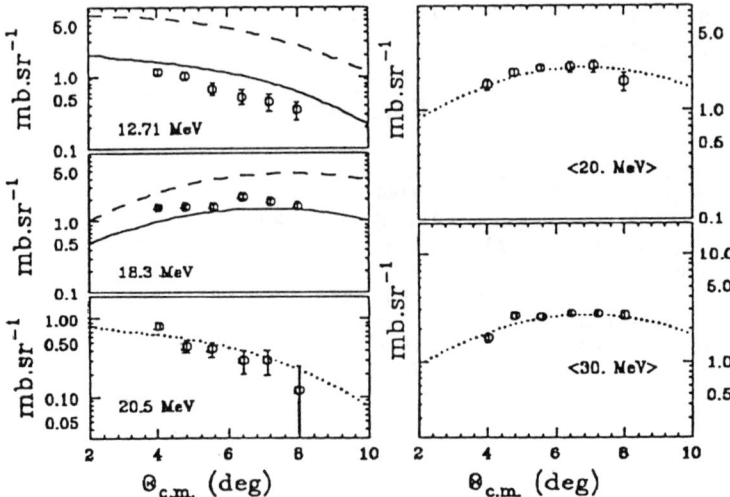

Figure 2: Angular distributions of different $\Delta S=1$ transitions in ^{12}C.

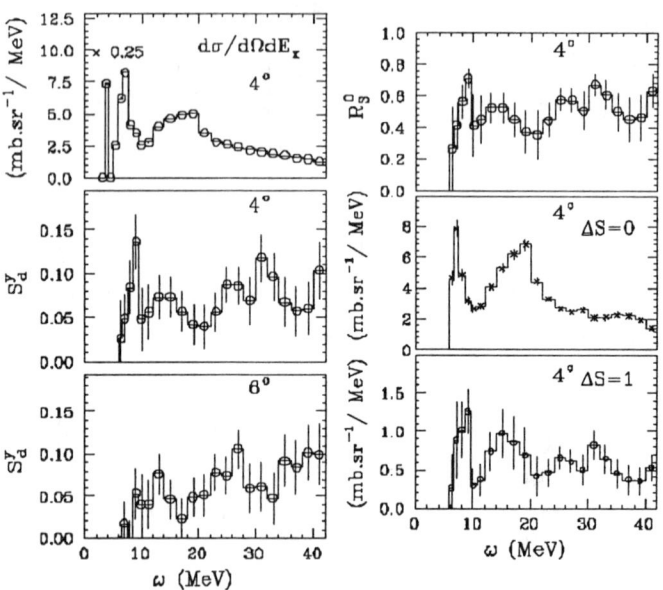

Figure 3: The ^{40}Ca($\vec{d},\vec{d'}$) results

Tensor Analyzing Power of the $^{12}C(d,{}^2He)^{12}B$ Reaction at 270 MeV

H. Okamura[a], S. Ishida[a], N. Sakamoto[a], H. Otsu[a], T. Uesaka[a],
T. Wakasa[a], Y. Satou[a], S. Fujita[a], H. Sakai[a], T. Niizeki[b],
K. Katoh[b], T. Yamashita[b], Y. Hara[b], H. Ohnuma[b], T. Ichihara[c]
and K. Hatanaka[d]

[a] *Department of Physics, University of Tokyo, Tokyo 113, Japan*
[b] *Department of Physics, Tokyo Institute of Technology, Tokyo 152, Japan*
[c] *The Institute of Physical and Chemical Research (RIKEN), Wako, Saitama 351-01, Japan*
[d] *Research Center for Nuclear Physics, Ibaraki, Osaka 567, Japan*

Abstract. The tensor analyzing power of the $(d,{}^2He)$ reaction has been measured at 270 MeV. Its usefulness for the spectroscopic study, particularly in identifying J^π of the spin-flip dipole states, is demonstrated for a ^{12}C target. The data are reasonably reproduced in terms of the DWBA theory for the three-body reaction.

INTRODUCTION

Recently the charge exchange $(d,{}^2He)$ reaction was successfully measured at intermediate energies, taking advantage of the simple one-step reaction mechanism and of the high detection efficiency of two protons.[1,2] Compared with the (n,p) reaction, it is characterized by the selective excitation of the spin-flip modes and by the relatively high count-rate and better resolutions. In addition, it is theoretically shown[3] that the tensor analyzing powers are sensitive to J^π of the residual nucleus and provide essentially the same nuclear information as the polarization transfer observables of the (n,p) reaction, the measurement of which is not feasible. The simplicity of the measurement of tensor analyzing powers should make the $(d,{}^2He)$ reaction a promising tool for the nuclear spectroscopy.

An application of interest is the study of spin-flip dipole states. The (n,p)-type reaction has the advantage that the background Gamow-Teller transition ($L=0$) is Pauli-blocked due to neutron-excess for heavy nuclei. A broad structure characterized by $L=1$ angular distributions is commonly observed in the intermediate energy (p,n) spectra for medium-heavy nuclei. Its width of about

10 MeV is believed to be caused by a superposition of the spin-dipole states (2^-, 1^-, and 0^- levels for a 0^+ target). However, the strength distribution of each J^π, which is fundamental for the understanding of the particle-hole residual-interaction, has not been clearly determined.[4] It would be fascinating to know if the 0^- collective states were enhanced through its coupling to the pionic degrees of freedom in the nucleus.

The usefulness of the tensor analyzing power in identifying J^π, however, has not been experimentally established. In this article, we present the result of the tensor analyzing powers, A_{yy} and A_{xx}, of the $(d, {}^2\text{He})$ reaction at 270 MeV on the ^{12}C target, the structure of which is well understood. This is the first measurement for well separated levels with various J^π of the residual nucleus.

EXPERIMENTAL

The experiment was performed at the RIKEN Accelerator Research Facility utilizing the newly constructed polarized ion source.[5,6] The polarization axis was rotated with the Wien filter downstream of the ion source so that it lay on the normal (sideway) direction of the scattering plane in the measurement of A_y and A_{yy} (A_{xx}). The beam polarization was monitored at 270 MeV by using the $d+p$ scattering[7] and was 70–75% of the ideal value throughout the experiment. The ^{2}He is efficiently measured by the coincidence detection of two protons emitted to close geometries. Protons are momentum-analyzed and detected at the first focal plane of the SMART spectrograph.[2,8] The $(d, {}^2\text{He})$ cross section is defined by

$$\frac{d\sigma}{d\Omega} = \frac{1}{2} \int_{4\pi} \int_{\epsilon_{min}}^{\epsilon_{max}} \frac{d^3\sigma}{d\Omega_{{}^2\text{He}} d\Omega_{pp} d\epsilon} d\Omega_{pp} d\epsilon , \tag{1}$$

where the p-p relative energy is denoted by ϵ. The factor $\frac{1}{2}$ arises from the indistinguishability of two protons. The integration limits of $\epsilon_{min}=0$ and $\epsilon_{max}=1$ MeV are set considering that the P-wave contribution is not negligible at ϵ greater than 1 MeV according to the analysis of the $^1\text{H}(d, {}^2\text{He})n$ reaction at 200 and 350 MeV.[11]

RESULTS

Figure 1 shows excitation energy spectra of the cross section and A_{xx} at the mean center-of-mass angles of 0.5° and 7°, where the Gamow-Teller and dipole states are relatively enhanced, respectively. Some discrete levels, 1_1^+ (ground), 2_1^+ (0.95 MeV), and 1_1^- (2.62 MeV), are clearly seen to be excited as well as the broad structures at 4.5 and 7.5 MeV. A striking feature is the conspicuously large A_{xx} observed at the energies corresponding to the 2_1^+ and

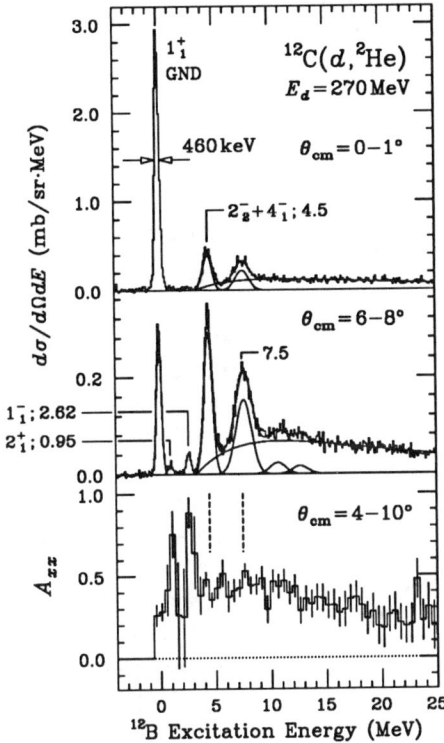

FIGURE 1. Typical excitation energy spectra of the cross section and A_{xx}.

1_1^- states. It coincides with the prediction of the PWIA theory[3] in which the A_{xx} becomes close to +1 for natural-parity states. The broad structures, on the other hand, show moderate A_{xx} consistent with the continuum at high excitation-energies. It should be noted that the isobaric-analog 1_1^- state of ^{12}N is not clearly observed in the (p, n)-type reaction since its width is broad.[9] The yield for each state has been obtained by the gaussian peak-fitting (Fig. 1). The continuum background is estimated with the semi-phenomenological parameterization of the quasi-free scattering.[10] The angular distributions of the cross section and the analyzing powers thus obtained are shown in Figs. 2 and 3, respectively. An interesting feature is the rich structures of A_{yy} and A_{xx}, while the cross sections show monotonous distributions although they are clearly characterized by L values. The uncertainty of the analyzing powers arising from the estimate of background does not have influence on the following argument since the analysing powers of the background are quite similar to those extracted for the peaks at 4.5 and 7.5 MeV.

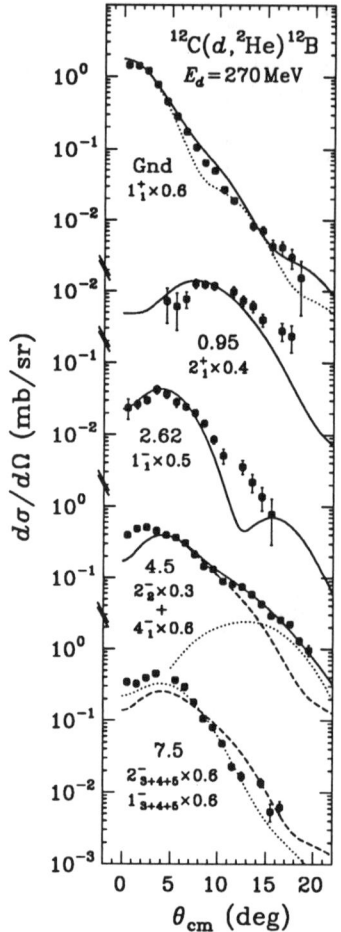

FIGURE 2. Angular distributions of the cross section. The result of DWBA analysis is also presented normalized by the factor indicated for each state. For the ground state, the dotted curve represents the calculation without the tensor part of the effective interaction. For the peak at 4.5 MeV, contributions from the 2_2^- and 4_1^- states and their incoherent sum are separately shown by dashed, dotted, and solid curves, respectively. For the peak at 7.5 MeV, the sums of the 2_3^-, 2_4^-, and 2_5^- states and of the 1_3^-, 1_4^-, and 1_5^- states are shown by dashed and dotted curves, respectively.

ANALYSIS

The data has been analyzed according to the prior-formalism of the DWBA theory for the three-body reaction.[12] The T-matrix element is expressed as

$$T^{\text{DWBA}} = \langle \chi_{^2\text{He}}^{(-)} \psi_{^2\text{He}} \Psi_{A^*} | V | \chi_d^{(+)} \varphi_d \Psi_A \rangle , \qquad (2)$$

where the projectile wave functions of the incident- and exit-channels are denoted by $\varphi_d(\boldsymbol{r})$ and $\psi_{^2\text{He}}(\epsilon; \boldsymbol{r})$, the target wave functions by $\Psi_A(\boldsymbol{x})$ and $\Psi_{A^*}(\boldsymbol{x})$, and the distorted wave functions by $\chi_d^{(+)}(\boldsymbol{R})$ and $\chi_{^2\text{He}}^{(-)}(\boldsymbol{R})$, respectively. The effective interaction between the nucleons in the projectile and in the target is denoted by $V(\boldsymbol{R}-\boldsymbol{x}+\boldsymbol{r}/2)$. The $\chi_{^2\text{He}}^{(-)}$ is assumed to be independent of $\boldsymbol{r}$ and

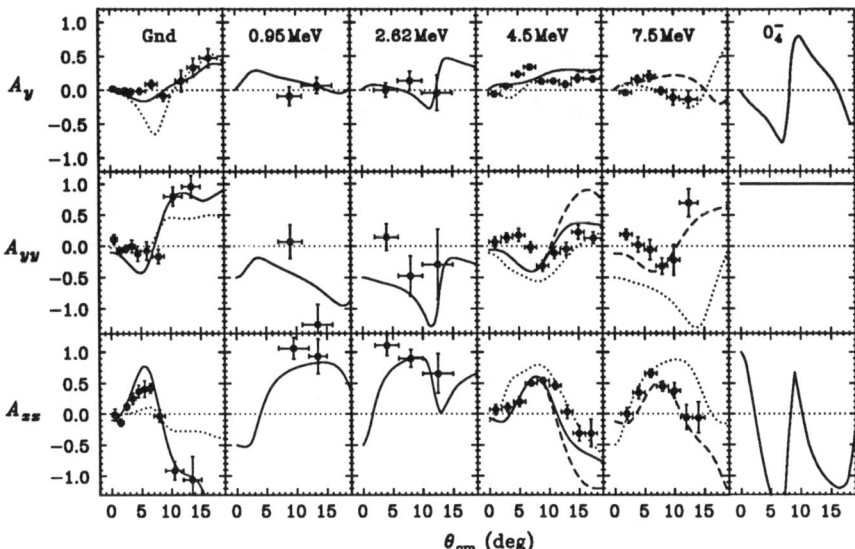

FIGURE 3. Angular distributions of the analyzing powers with the result of DWBA analysis. For the curves, see the caption of Fig. 2. The DWBA prediction for the 0_4^- state ($E_x = 14$ MeV) is also presented for comparison.

ϵ.[12] Also the φ_d and $\psi_{^2\text{He}}$ are assumed to be in the pure $^3S_1$ and $^1S_0$ states, respectively, so that the integration on r can be carried out separately. The remaining part of the T-matrix is calculated by an ordinary two-body DWBA code. The program TWOFNR[13] was used in the present calculation. The effective two-body interaction is taken from the central and tensor parts of the t-matrix parameterization at 140 MeV by Franey and Love.[14] The effect of the single-nucleon knock-on exchange is included by a short-range approximation for the central part.[15] The one-body density matrices of the target are calculated by the wave functions of Cohen and Kurath[16] and of Millener and Kurath[17] for the positive- and negative-parity final states, respectively. The optical potential for the incident-channel was determined by the preliminary data of elastic scattering measured at RIKEN.[18]

The results of calculation are shown in Figs. 2 and 3. The reasonable agreement is obtained both for the cross section and for the analyzing powers. The relative normalization factors of the cross section for the 2_1^+, 2_2^-, and 4_1^- states with respect to the ground state are consistent with those in the analysis of the intermediate energy (p, n) and (n, p) reactions.[19–21] However, the origin of the overall normalization factor, 0.6, is not understood. It may be attributed to the defect in describing the exit-channel distorted wave function.[22] In the present calculation, the parameters of ^{2}He optical-potential are chosen to be the same as those for the deuteron, except for the depth of imaginary part (W) which is modified to improve the fit of A_y. But the A_y is also sensitive to the spin-orbit part of the deuteron optical-potential and the choice of W has influence on the amplitude of the cross section by as much as 50%. The ^{2}He distorted wave should be determined by a more advanced three-body theory, such as the adiabatic approximation.[23]

In spite of this uncertainty, the DWBA calculation clearly demonstrates the usefulness of the tensor analyzing powers for the spectroscopic study. The A_{yy} and A_{xx} are weakly influenced by the choice of optical potentials, as far as the present calculation is concerned, and show the distinctively different behavior depending only on J^π. Particularly the 0^- state should be unambiguously identified since the A_{yy} is equivalent to $+1$ (Fig. 3) due to the parity conservation.[24] In the present data, however, no significant concentration of the 0^- strength has been observed up to the excitation energy of 30 MeV. It is worth noting that the large A_{xx} arises mainly from the tensor part of the effective interaction (see the dotted curve for 1_1^+ in Fig. 3). This feature is a consequence of the selection rule on spin-transfer and contrasts with ordinary deuteron-induced reactions where the large tensor analyzing powers are caused only through the higher-order effect of spin-orbit potential or the D-state of the deuteron wave function.

Finally the structure of the peak at 7.5 MeV is briefly discussed. Several shell-model states are predicted around this excitation energy.[17] According to the DWBA calculation, the sums of 2^- states and of 1^- states form the main contributions with the same order of magnitude. Although the shape of the differential cross section shows the slightly better agreement with that of the 1^- states (Fig. 2), it is largely influenced by the distorted waves and by the subtraction of the continuum background, making the estimate on the relative contribution to be ambiguous. A definite conclusion, on the other hand, is drawn from the tensor analyzing powers. Both the A_{yy} and A_{xx} clearly indicate the dominant contribution from the 2^- states rather than from the 1^- states (Fig. 3). This result disagrees with that of the intermediate energy (p,n) reactions where the dominance of the 1^- states is suggested.[20,21] But it is derived from the small difference of the shapes of the cross section and largely relies on the shell-model prediction. The discrepancy between the two reactions may be also attributed to the different selectivities on the spin-transfer. The non-spin-flip 1^- states, the strength of which is also predicted to concentrate at similar excitation energies, can be excited in the (p,n) reaction.

CONCLUSION

In summary, the ^{12}C$(d,^2$He$)^{12}$B reaction has been measured at 270 MeV with the emphasis laid on the tensor analyzing powers. The A_{yy} and A_{xx} exhibit the distinctively different behaviour depending on J^π of the residual

^{12}B. This feature is confirmed by the DWBA calculation which reproduces the data reasonably well.

The present success of the prior-form DWBA theory is encouraging. The development of the more realistic but simple description is expected, which will allow the extensive application of the $(d, ^2\text{He})$ reaction. The further measurement on heavy nuclei is currently underway and will hopefully provide the definitive understanding on the spin-dipole state.

ACKNOWLEDGEMENTS

The authors are grateful to the staff of RIKEN Accelerator Research Facility, particularly to Y. Yano, A. Goto, and M. Kase, for their invaluable assistance during the experiment. They also thank K. Ikegami, J. Fujita, N. Inabe, and T. Kubo for their work in constructing the polarized ion source.

REFERENCES

1. Ellegaard, C. et al., Phys. Rev. Lett. **59**, 974 (1987); Phys. Lett. **B231**, 365 (1989).
2. Ohnuma, H. et al., Phys. Rev. C **47**, 648 (1993).
3. Bugg, D. V. and Wilkin, C., Nucl. Phys. **A467**, 575 (1987); Wilkin, C. and Bugg, D. V., Phys. Lett. **154B**, 243 (1985).
4. Osterfeld, F., Rev. Mod. Phys. **64**, 491 (1992), and references therein.
5. Okamura, H. et al., AIP Conf. Proc. **293**, 84 (1994); contribution to this conference.
6. Okamura, H. et al., Nucl. Phys. **A577**, 89c (1994).
7. Sakamoto, N. et al., contribution to this conference.
8. Ichihara, T. et al., Nucl. Phys. **A569**, 287c (1994).
9. Sterrenburg, W. A., Harakeh, M. N., van der Werf, S. Y., and van der Woude, A., Nucl. Phys. **A405**, 109 (1983).
10. Raywood, K. J. et al., Phys. Rev. C **41**, 2836 (1990).
11. Kox, S. et al., Nucl. Phys. **A556**, 621 (1993).
12. Baur, G. and Trautmann, D., Phys. Rep. **25**, 293 (1976).
13. Igarashi, M., program TWOFNR, unpublished.
14. Franey, M. A. and Love, W. G., Phys. Rev. C **31**, 488 (1985).
15. Golin, M., Petrovich, F., and Robson, D., Phys. Lett. **64B**, 253 (1976).
16. Cohen, S. and Kurath, D., Nucl. Phys. **73**, 1 (1965).
17. Millener, D. J. and Kurath, D., Nulc. Phys. **A255**, 315 (1975).
18. Uesaka, T. et al., RIKEN Accel. Prog. Rep. **27**, 40 (1993); contribution to this conference.
19. Rapaport, J. et al., Phys. Rev. C **24**, 335 (1981).
20. Gaarde, C. et al., Nucl. Phys. **A422**, 189 (1984).
21. Yang, X. et al., Phys. Rev. C **48**, 1158 (1993).
22. Austern, N., Phys. Rev. C **30**, 1130 (1984).
23. Yahiro, M., Tostevin, J. A., and Johnson, R. C., Phys. Rev. Lett. **62**, 133 (1989).
24. Simonius, M., in *Polarization Nuclear Physics*, edited by Fick, D., Lecture Notes in Physics Vol. 30 (Springer, 1974).

Study of the ^{3,4}He($\vec{p}$,n) Reactions at $T_p = 100$ and 200 MeV

C. M. Edwards[a], M. Palarczyk[a], L. C. Bland[b],
B. D. Anderson[c], B. Brinkmöller[d], D. S. Carman[b],
D. Dehnhard[a], M. A. Espy[a], J. L. Langenbrunner[a], R. Madey[c],
Y. Wang[c], and J. W. Watson[c]

[a] University of Minnesota, [b] Indiana University Cyclotron Facility,
[c] Kent State University, [d] Universität Karlsruhe

Abstract. We report on the measurement of differential cross sections and analyzing powers for the inclusive ^{3,4}He($\vec{p}$,n) reactions at incident proton energies of $T_p = 100$ and 200 MeV. The data are compared to predictions using several different models.

There are several motivations for studying ^{3,4}He($\vec{p}$,n) reactions. First, phase shift[1] and R-matrix[2] analyses of p+^{3}He elastic scattering data found resonances in ^{4}Li to be 5–10 MeV wide. However, evidence[3] for unexpectedly narrow resonances (of 1–2 MeV width) in ^{4}Li was deduced from three-body decay spectra. The inclusive ^{4}He(p,n) reaction allows a direct measurement of the ^{4}Li spectrum at low excitation energy, thereby enabling a resolution to the controversy. Second, ^{4}He(p,n) double-differential cross sections ($d^2\sigma/d\Omega/dE$) may be used to fine-tune models of mass accretion on neutron stars[4] and particle acceleration in solar flares[5], which have relied on sparse data and semi-classical cascade model calculations. Third, broad-range neutron energy spectra enable the study of quasifree (QF) knockout of bound neutrons in the target. Such data provide a natural complement to inclusive ^{3,4}He($\vec{p}$,p') QF scattering studies[6], which show a suppression of the analyzing powers (A_y) relative to free nucleon-nucleon scattering. Fourth, measured $d^2\sigma/d\Omega/dE$ and A_y from the ^{4}He($\vec{p}$,n) reaction may be used to test Recoil-Corrected Continuum Shell Model (RCCSM) wave functions[7]. Finally, the ^{3,4}He($\vec{p}$,n) reactions are of interest to test recent theoretical models[8] purporting to understand the lack of enhancement of the spin-longitudinal relative to the spin-transverse nuclear response function. These response functions are deduced from ($\vec{p}$,$\vec{n}$)

polarization transfer observables near the QF peak, and can be constrained by $d^2\sigma/d\Omega/dE$ and A_y.

The experiment[9] was performed at the Indiana University Cyclotron Facility (IUCF). The Beam Swinger was used to vary the incident proton angle from 0° to 20° in approximately 5° steps. The time-of-flight of neutrons was measured in two detector stations, positioned along the 0° (24°) flight path at a distance of ~70 m (~50 m). For the 0° detectors, the energy resolution was 600 keV (FWHM) for the ^{12,13}C(p,n) reactions at $T_p = 200$ MeV, and 250 keV at 100 MeV. This resolution is sufficient to distinguish possibly 1–2 MeV wide states in ^{4}Li. The resolution in the 24° detector station was approximately a factor of two worse than at 0°. At these larger angles, sub-MeV resolution is not as necessary since the spectra are no longer dominated by resonant excitations. The absolute cross section normalization was obtained by comparison of measured yields from the ^{12,13}C(p,n) reactions to published cross sections[10].

A high-pressure, low-temperature gas target was developed for this experiment and provided a favorable ratio (about 5:1) of ^{3,4}He scattering centers to those of the Havar windows. The target was operated at a pressure of ~5 kTorr and near liquid nitrogen temperatures. The effective target thickness was 18 mg/cm^2 for ^{4}He and 12 mg/cm^2 for ^{3}He, providing an excellent ratio of signal to background.

Typical spectra (coarsely binned in Fig. 1, 2, and 3) are plotted as a function of missing mass (MM) with $MM = 0$ for ^{4}He(p,n) (^{3}He(p,n)) at the p+^{3}He (3p) rest mass. Within the statistical uncertainties in the ^{4}He(p,n) data there is no evidence of narrow resonances in either $d^2\sigma/d\Omega/dE$ or A_y even with the finest possible binning. Instead, the spectra near the ^{4}Li ground state (reported[2] at $MM = 4.07$ MeV), show a broad bump of at least 10-MeV width, consistent with broad $L = 1$ resonances[1,2]. Results of semi-classical cascade model calculations[4] (Fig. 1, circles) for the ^{4}He(p,n)p^{3}He reaction describe the data well at large MM and small angles, but fail near $MM = 0$ since no resonance effects are included in the model. The calculations also fall short of the absolute cross section at larger angles ($\geq 29°$), presumably because multiparticle breakup makes increasingly significant contributions to the cross sections.

Standard approaches to interpret data use the Distorted Wave Impulse Approximation (DWIA). The program THREEDEE[11] describes QF knockout reactions in the DWIA and includes distortions in the incident and outgoing channels. The calculations presented here are similar to those employed to describe the ^{3,4}He($\vec{p}$,p′) data[6]. However, it was found essential for a fit to the ^{4}He(p,n) cross section data near $MM = 0$ that an optical potential[12] was used in the outgoing p+^{3}He channel that contains a strong $L = 1$ resonance. Typical 200-MeV spectra and three curves are shown in Fig. 2: one plane-

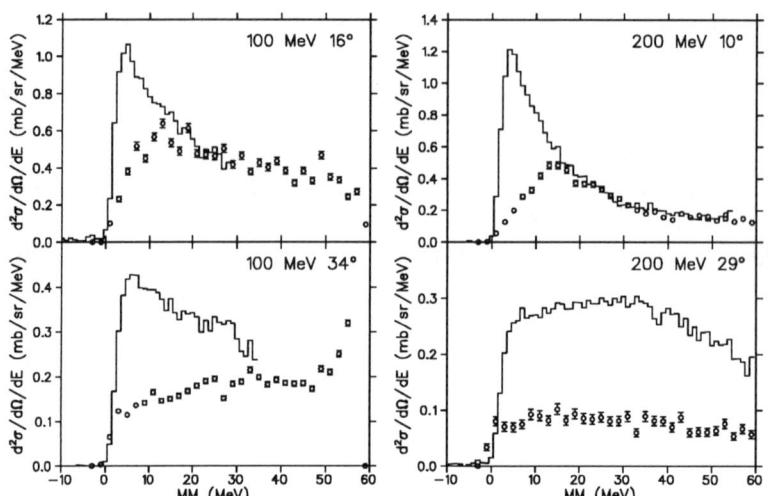

FIGURE 1. ^{4}He($\vec{p}$,n) measured $d^2\sigma/d\Omega/dE$ (bar histograms) and semi-classical cascade model calculations (circles).

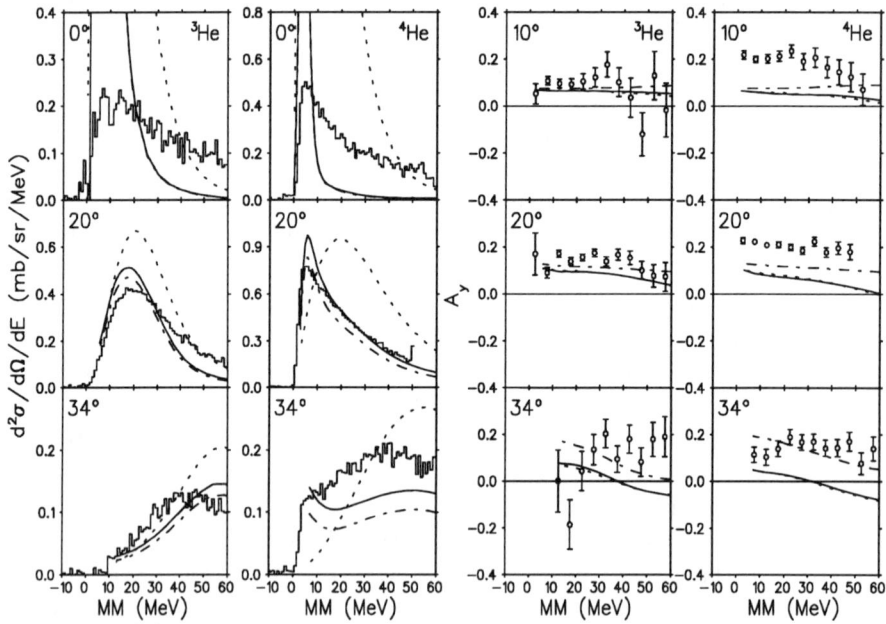

FIGURE 2. ^{3,4}He($\vec{p}$,n) 200-MeV $d^2\sigma/d\Omega/dE$ and A_y overlayed with THREEDEE[11] calculations. Dashed lines: PWIA with FEP; solid (chain-dashed) lines: DWIA with FEP (IEP) and without spin-orbit distortions.

wave (PWIA) calculation, and two distorted-wave (DWIA) calculations using two different choices (final (initial) energy prescription: FEP (IEP)) for the energy at which the two-body t-matrix is evaluated. Attempts to describe the 100-MeV spectra using THREEDEE have been unsuccessful.

At far forward angles (e.g. top panel in Fig. 2), the PWIA and DWIA calculations fail to account for the observed suppression of yield near $MM = 0$, presumably due to Pauli blocking, which is not taken into account in the THREEDEE calculations. At larger angles ($\theta_n^{lab} \geq 20°$), the PWIA-predicted spectra are close to the data for ^{3}He but bear a poor resemblance to the data for ^{4}He. Use of distortions results in a reduction of the cross section around the QF peak both for ^{3}He and ^{4}He, and a strong enhancement for ^{4}He at small MM due to the $L = 1$ resonance in the p+^{3}He channel. At 20°, the QF peak has disappeared in the DWIA prediction for ^{4}He and the measured spectrum is very well described. At 34° the DWIA shows a reduced QF peak near the MM where the data have a maximum, and a resonance enhancement for ^{4}He near $MM = 0$. The FEP and IEP calculations diverge with increasing angle, indicating that the effect of the off-shell nature of the two-body t-matrix becomes more important. The remaining difference between the ^{4}He data and the DWIA at 34° is probably due to the increased importance of multiparticle breakup, which is neglected in the DWIA calculations, and to the details of the treatment of the p+^{3}He resonance.

When spin-orbit distortions are neglected (Fig. 2), the measured A_y for ^{3}He($\vec{p}$,n) are described quite well at small angles. The ^{4}He data, however, are consistently larger than predicted, presumably due to nuclear structure effects in ^{4}Li. In general, the inclusion of spin-orbit distortions (curves not shown) worsens the qualitative agreement with the data for both ^{3}He and ^{4}He. When averaged over 0–60 MeV in MM (Fig. 4), the A_y for ^{3}He($\vec{p}$,n) is in good agreement with the free[13] n($\vec{p}$,n) values at 200 MeV, but the A_y for ^{4}He is enhanced relative to the free values. Both results are consistent with the comparisons to the THREEDEE calculations and seemingly at odds with the ($\vec{p}$,p′) findings[6].

Another set of DWIA calculations for ^{4}He($\vec{p}$,n) was obtained at both 100 and 200 MeV using the program DWBA70[14] with RCCSM wave functions[7] (Fig. 3). All RCCSM states of $J^\pi = 2^-, 1^-, 0^-, 0^+, 1^+$ and 2^+ were included. These calculations include Pauli-blocking effects, resulting in better agreement with the measured cross sections than the THREEDEE predictions at far forward angles. The RCCSM, however, also fails to describe the measured A_y.

In conclusion, we have measured inclusive neutron spectra from the ^{3,4}He($\vec{p}$,n) reactions at intermediate energies. No evidence was found for narrow states in ^{4}Li, thus resolving a recent controversy. Attempts to describe the measured cross sections using several different models were moderately

442 Study of the ^{3,4}He($\vec{p}$,n) Reactions

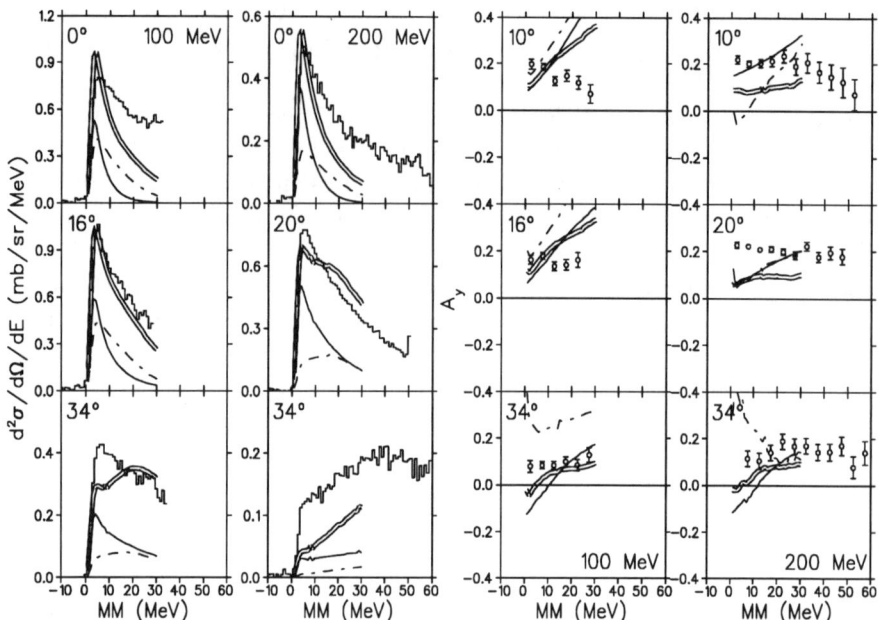

FIGURE 3. $d^2\sigma/d\Omega/dE$ and A_y for ^{4}He($\vec{p}$,n) overlayed with DWBA70 predictions. Solid lines: $J^\pi = 2^-$; chain-dashed lines: $J^\pi = 1^-$; double line: sum over all transitions. The calculations were renormalized by a factor of 1.6 at 100 MeV, and by 1.3 at 200 MeV.

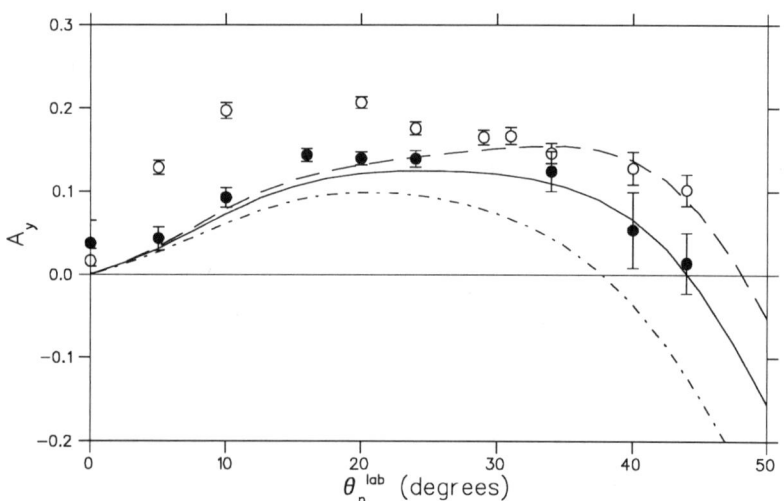

FIGURE 4. A_y for ^{3,4}He($\vec{p}$,n) at 200 MeV as a function of θ_n^{lab}. Unshaded (shaded) circles: ^{4}He($\vec{p}$,n) (^{3}He($\vec{p}$,n)) averaged over 0–60 MeV in MM. The solid, dashed, and chain-dashed lines are A_y[13] for n($\vec{p}$,n) at 200 MeV, 240 MeV, and 160 MeV, resp.

successful. The measured analyzing powers for the ^{3}He($\vec{p}$,n) reaction are consistent with the values for n($\vec{p}$,n) scattering at 200 MeV. For the ^{4}He($\vec{p}$,n) reaction at 200 MeV, however, the A_y are larger than the free values and larger than the predictions of any of the models.

The authors wish to thank Dr. L. Bildsten for the Monte Carlo calculations, and Prof. D. Halderson for the RCCSM wave functions. The help of Prof. N. S. Chant, Prof. P. G. Roos, and Dr. M. K. Jones with the code THREEDEE is greatly appreciated. Thanks also to the staff at IUCF for technical assistance and a carefully tuned beam. This work was supported by the U. S. Department of Energy and the National Science Foundation.

REFERENCES

1. T. A. Tombrello, Phys. Rev. **138**, 40 (1965).
2. D. R. Tilley et al., Nucl. Phys. **A541**, 1 (1992).
3. M. Bruno et al., Phys. Rev. C **42**, 448 (1990); B. Brinkmöller et al., Phys. Rev. C **42**, 550 (1990).
4. L. Bildsten et al., Nucl. Phys. **A516**, 77 (1990); L. Bildsten et al., Ap. J. **408**, 615 (1993); L. Bildsten, private communication.
5. R. J. Murphy et al., Ap. J. Suppl. Ser. **63**, 721 (1987).
6. J. S. Wesick et al., Phys. Rev. C **32**, 1474 (1985); L. C. Bland et al., to be published.
7. D. Halderson and R. J. Philpott, Nucl. Phys. **A321**, 295 (1979); Nucl. Phys. **A359**, 365 (1981); D. Halderson, private communication.
8. V. R. Pandharipande et al., Phys. Rev. C **49**, 789 (1994).
9. C. M. Edwards, Ph. D. Thesis, University of Minnesota, 1994.
10. J. N. Knudson et al., Phys. Rev. C **22**, 1826 (1980); G. L. Moake et al., Phys. Rev. Lett. **43**, 910 (1979); J. Rapaport et al., Phys. Rev. C **24**, 335 (1981); X. Yang et al., Phys. Rev. C **48**, 1158 (1993); T. N. Taddeucci et al., Nucl. Phys. **A469**, 125 (1987).
11. N. S. Chant and P. G. Roos, Phys. Rev. C **15**, 57 (1977); N. S. Chant et al., Phys. Rev. Lett. **43**, 495 (1979); N. S. Chant and P. G. Roos, Phys. Rev. C **27**, 1060 (1983); N. S. Chant, computer code THREEDEE (1992, unpublished).
12. B. S. Podmore and H. Sherif, "Few Body Problems in Nuclear and Particle Physics," ed. R. J. Slobodrian, B. Cujec, and K. Ramavartaram, (Laval Univ. Press, Quebec, Canada), 517 (1975).
13. R. A. Arndt et al., computer code SAID solution SM89 (1989, unpublished), and Phys. Rev. D **28**, 97 (1983).
14. R. Schaeffer and J. Raynal computer code DWBA70 (unpublished); M. A. Franey, modifications to DWBA70 (unpublished).

QUASIFREE ($\vec{p}$,Np) REACTION STUDIES AT 200 MeV

D.S. Carman[a], L.C. Bland[a], N. Chant[b], T. Gu[b], G.M. Huber[c], J. Huffman[b],
A. Klyachko[d], B.C. Markham[a], P. Roos[b], P. Schwandt[a], and K. Solberg[a]

[a] *Indiana University Cyclotron Facility, Bloomington, Indiana, 47405*
[b] *University of Maryland, College Park, Maryland, 20742*
[c] *University of Regina, Saskatchewan, Canada, S4S-0A2*
[d] *Institute of Nuclear Research, Moscow, Russia*

I. Introduction

In the study of quasifree scattering (QFS) for both inclusive (p,p') and (p,n) reactions, as well as exclusive (p,2p) and (p,np) knockout reactions, much experimental and theoretical work has been done to look for evidence of collectivity in the nuclear response as well as for possible medium modifications of the NN interaction.[1,2] The attempt to understand how the nuclear medium affects the basic NN interaction is a daunting task as one must disentangle nuclear structure effects from dynamical effects all in the presence of extremely strong final state interactions (FSI). Nevertheless, QFS represents an important process to understand as it accounts for a large fraction of the total scattering yield for nucleon-nucleus collisions, and as such, it provides a strong challenge to theoretical models. The thrust of our analysis has been to attempt to understand the main features of the QFS process through the extracted inclusive and exclusive cross sections and analyzing powers. The cross section data alone severely exercise any theoretical model which must include the appropriate physics to predict the magnitude and shape of the cross section distributions, their variation over a broad range of momentum and energy transfers, and any target dependent differences. Further restrictions on the reaction model are incurred through studies of the analyzing power. The ($\vec{p}$,p') analyzing powers in the quasifree region are strongly suppressed relative to free NN scattering independent of target or beam energy, whereas recent ($\vec{p}$,n) data show both suppression and enhancement of the analyzing powers depending on the mass number of the target.[3,4]

II. Experimental Detection Systems

The majority of the data shown in this paper was culled from IUCF experiments E358 and E336. For E336, the ejectile proton was detected by the K600 spectrometer with its associated focal plane and focal plane polarimeter detectors.[5] The ejectile nucleon detector employed for E358 (Fig.1) was placed $\sim$2 m from the primary target and subtended the angular ranges $25° \leq \Theta \leq 35°$ (in-plane) and $-4° \leq \beta \leq 4°$ (out-of-plane). Neutrons were detected by allowing them to undergo np free scattering in a series of plastic scintillators. The second scattered protons were then tracked through a stack of X and Y drift chambers and their energy was measured in an array of NaI crystals. The neutron detector efficiency was obtained from a detailed Monte Carlo calculation and was predicted to be nearly independent of energy (above the detector thresholds) having a value of $\sim 1.2 \times 10^{-3}$. The main source of background in this detector arose from target neutrons undergoing ^{12}C(n,p) scattering within the plastic scintillators. These events were removed by performing careful subtractions. When utilized as a proton detector, the plastic scintillators were removed. This detector provided an overall resolution of 1 to 2 MeV depending on its mode of operation. The central angle of the ejectile nucleon detectors for both E358 and E336 was chosen to be 30° to study the nuclear response at a momentum transfer of ~ 1.5 fm^{-1}.

The associated protons were detected in both experiments using arrays of NaI crystals and a set of multi-wire tracking chambers. These arrays spanned roughly half of the associated proton solid angle expected for single-step QFS. The acceptance, $35° \leq \Theta \leq 100°$ in-plane and $-5° \leq \beta \leq 25°$ out-of-plane, allowed for reliable integration of the exclusive yield to ascertain the extent to which the inclusive (p,p') and (p,n) reactions are dominated by single-step scattering.

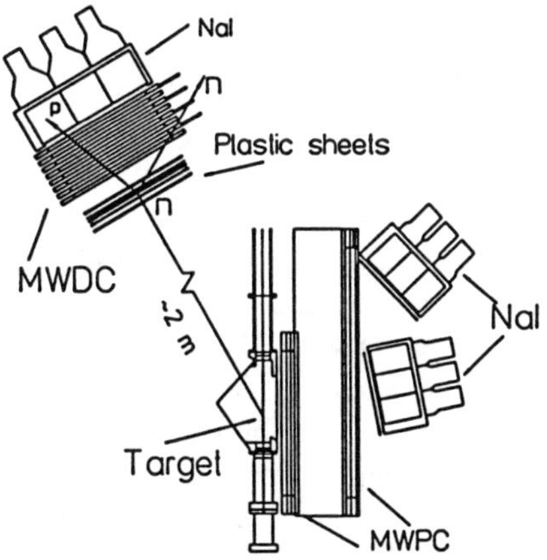

Fig.1: E358 Coincidence detection system shown at roughly 1/25 scale.

III. Theoretical Model

All data shown have been compared with either plane wave (PW) or distorted wave impulse approximation calculations employing the THREED code of Chant and Roos.[6] This calculation represents a quantum mechanical treatment of the scattering process relying on the factorization approximation and assuming the validity of the impulse approximation at 200 MeV. The strong FSI are incorporated by using scattering state wavefunctions which are solutions to the Schroedinger equation including a complex, energy-dependent optical model potential (OMP) with spin-orbit terms. The OMP parameters are generally fitted to elastic scattering data. The calculations presented in this paper use the broad energy range potentials from Nadasen[7a] (^{12}C,^{40}Ca) and van Oers[7b] (^{4}He). The ^{2}H data are compared with PW calculations. Angle-integrated observables are computed from the exclusive calculations by integrating over a Θ/β grid matched to the acceptance of the NaI for the associated proton. This angular grid was chosen to account for the rapid variations of the observables generally caused by the momentum space wavefunction of the struck nucleon. The inclusive calculations were performed in a similar spirit except that the angular integration was performed over all associated particle angles and the imaginary terms of the OMP for the associated nucleon were set to zero as there is no loss of flux if this particle is not observed.[8]

Separate calculations were performed for nucleon knockout from each relevant shell model orbital. The observables were obtained from the incoherently summed spin-dependent partial cross sections for each shell, weighted by the appropriate spectroscopic factors. The effective spectroscopic factors used in the calculations are discussed below. The calculations presented in the paper can be regarded as conventional approaches to the description of the QFS process. Deficiencies remain within these calculations and mainly center about utilizing less than ideal OMPs.

IV. Cross Section Data

A. *Exclusive Data*

A comparison is made between a sample of the preliminary E358 exclusive cross section data with the theoretical model to establish how well the exclusive QFS data is understood within this conventional approach. The independent parameters of the analysis are T_3 (the kinetic energy of the ejectile) along with Θ_5 and β_5 (the in-plane and out-of-plane angles for the associated nucleon). Variation of any of these independent variables results in a change in the bound nucleon momentum which, in the present model, is assumed to equal the missing momentum. The angles Θ_5 and β_5 are natural variables to employ for this analysis since it is necessary to integrate over them to determine the makeup of the inclusive spectra.

The ^{2}H data are compared with a PW calculation in Fig.2. The energy and angular dependence of the cross section are well reproduced by the calculation for both (p,2p) and (p,np). No renormalization was required. The ^{2}H data are expected to be reproduced by the model as the reaction closely resembles free scattering modified by the Fermi momentum of the target nucleons with only minimal FSI corrections.

The ^{12}C data (Fig.2) for $1p_{3/2}$ and $1p_{1/2}$ knockout were added together due to lack of sufficient resolution in the (p,np) data set to reliably separate them. The resolution for the (p,2p) data was adequate for the different 1p subshells to de-

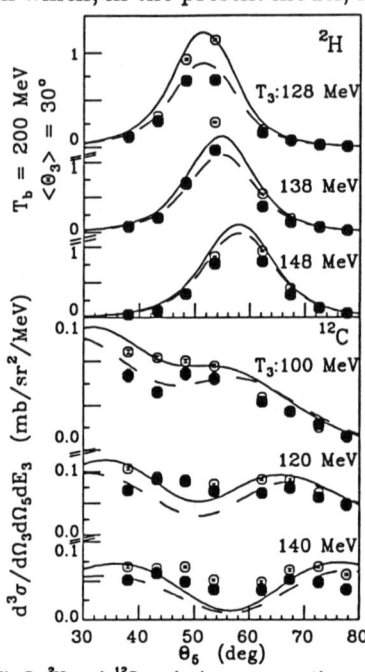

Fig.2: ^{2}H and ^{12}C exclusive cross sections. (p,2p): solid line(○), (p,np): dashed line (●)

termine the appropriate occupation numbers. We required ~30% smaller p-shell occupation numbers than from ^{12}C(e,e'p) studies. This may result since the OMP employed were fit to ^{40}Ca data and required an extrapolation to lower energies. More generally, this results from lack of detailed understanding of the strong FSI. Nevertheless the model reasonably reproduces the ^{12}C(p,Np) angular dependence over a broad range of ejectile four-momenta.

B. *Angle-Integrated Data*

The angle-integrated cross sections are necessary to provide the link between the exclusive and inclusive yields. Ignoring FSI corrections and limitations imposed by the finite angular acceptance of the BR array, this exclusive data integration was precisely the procedure followed in the calculation of the inclusive observables.

The data for p-shell knockout from ^{12}C are displayed (Fig.3) at two separate ejectile angles. The calculations required only a slight renormalization of the occupation numbers with respect to the exclusive calculations. This being the same for all bins probably indicates that the holes in our acceptance have not been properly accounted for. These holes include the gap between the BR NaI arrays, contributions from crystal escapes where less than full energy is measured, and events which deposit energy within two crystals (cross cell events) which are not included within the present analysis. As the analysis proceeds and these items are properly included, the overall normalizations of the exclusive and angle-integrated yields are expected to converge. Nevertheless, the present state of the angle-integrated data highlight that the integrand for (p,np) is reasonably understood as seen through the reliable predictions of the magnitude and energy dependence of the cross sections.

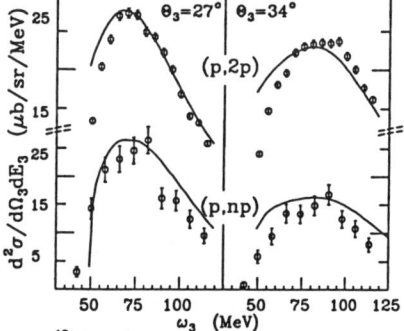

Fig.3: ^{12}C(p,Np) Angle–Integrated cross sections for two ejectile angles. T_b=200 MeV

C. *Inclusive Data*

Finally, the inclusive cross sections for ^{2}H and ^{12}C are studied. There is excellent agreement with the data and the calculations for both (p,p') and (p,n) (Fig.4). For ^{12}C, knockout for both the 1p and deeper 1s shell model orbitals have been included within the calculations. The salient point to make is that the renormalizations used for

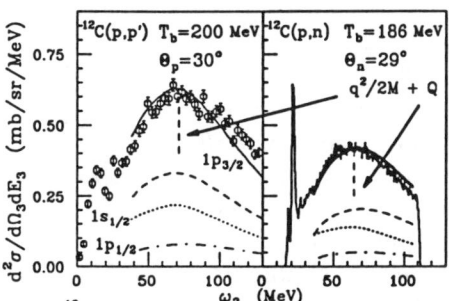

Fig.4: ^{12}C Inclusive cross sections showing individual shell contributions. ^{12}C(p,n) from ref.10

for this inclusive data agree with those as determined from the exclusive data. This suggests that the inclusive yield is composed mainly of single-step scattering. More complicated reaction dynamics, such as cluster or multinucleon knockout, do not significantly contribute. Furthermore, the agreement of the predicted peak position (at the free value of $\omega=q^2/2M$ shifted by the binding energy Q), width, and magnitude of the inclusive cross sections make it clear that there is no strong evidence for the presence of collective aspects of the nuclear response at these energies.

V. Analyzing Power Data

A. *Inclusive Data*

In a naive picture of QFS the predicted analyzing powers should agree with free space values. This picture is modified not only by the bound nucleon momentum distributions, but to some extent as well by spin dependent FSI. The predominantly spin independent FSI are not expected to effect inclusive polarization observables since they are defined by cross section ratios.

The (p,p′) analyzing powers have been measured for a range of targets at $\Theta_3=30°$ as a function of the energy transfer (Fig.5). These data are compared with the model calculations as

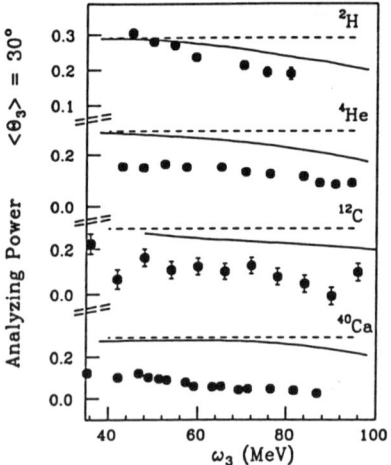

Fig.5: Inclusive (p,p) analyzing powers at $T_b=200$ MeV. Dashed line is free NN value. (^{4}He from ref.9, ^{40}Ca from ref.5)

well as the isospin-averaged free space NN values. We observe a clear suppression of the (p,p′) data with respect to the calculations. This suppression seems to follow the trend of the nuclear density and is of the same order of magnitude as that seen at higher energy.[3] This suppression has been suggested as a clear signature of a medium modification of the NN interaction within the framework of a relativistic dynamics model.[2]

B. *Angle-Integrated Data*

These data are used to ensure that the inclusive analyzing power is not affected by contributions other than single-step QFS. The observed suppression within the inclusive (p,p′) data remains for the more selective (p,2p) data. The fact that the QFS

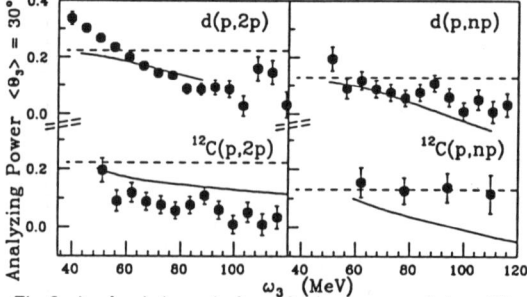

Fig.6: Angle-integrated analyzing power data with $T_b=200$ MeV. Dashed line is free NN value.

yields are reasonably understood suggests that the suppression witnessed is not caused by contributions to the inclusive reaction from other processes. This provides experimental confirmation that the analyzing power reduction is likely due to a medium modification of the NN interaction. The suppression does not appear for the (p,np) data which are more closely consistent with free scattering predictions. (Fig.6)

VI. Conclusions and Outlook

The inclusive yield at 200 MeV appears reasonably consistent with single-step scattering over a broad range of (q,ω), for a range of mass numbers, and for both (p,p') and (p,n) scattering. The reduced occupation probabilities required by the exclusive (p,Np) data indicate that a more detailed understanding of this process requires a much better understanding of the strong FSI to which the calculations are extremely sensitive.

A suppression of the inclusive (p,p') analyzing powers with respect to free scattering and the model calculations is seen and varies with nuclear density. The dependence is indicative of a medium modification. At these energies there appears to be no suppression of the (p,n) analyzing powers.

QFS studies are proceeding on a number of fronts in hopes of adding further elucidation to this important process. TRIUMF will employ its Dual Arm Spectrometer System (DASS) to study (p,2p) reactions with unprecedented energy resolution although with a limited phase space coverage, while IUCF will proceed by measuring complete sets of inclusive (p,n) polarization transfer coefficients at its nTOF facility. These new data apply further constraints on the model calculations thus allowing us to better understand not only their predictions, but how their results rely on the approximations or simplifications that are made.

VII. References

1. T.N. Taddeucci, in *Spin and Isospin in Nuclear Interactions*, edited by S. Wissink, C. Goodman, and G. Walker, 393(1991).
2. C.J. Horowitz and M.H. Iqbal, PRC 33, 2059(1986).
   C.J. Horowitz and D.P. Murdock, PRC 37, 2032(1988).
3. K. Hicks, in *Intersections Between Nuclear and Particle Physics*, edited by G.M. Bunce, 26(1988).
4. K.H. Hicks et al., PRC 47, 260(1993).
5. L.C. Bland, in *Correlations in Hadronic Systems*, edited by E. Jans, L. Lapikas, P.J. Mulders, J. Oberski, 64(1994).
6. N. Chant and P. Roos, PRC 15, 57(1977)., PRC 27, 1060(1983).
7. Nadasen et al., PRC 23, 1023(1981), van Oers et al., PRC 25, 390(1982).
8. J. Wesick et al., PRC 32, 1474(1985).
9. P. Li, Ph.D. thesis, Indiana University (1994).
10. J. Rapaport, ^{12}C(p,n) to be published.

Exclusive Measurement of $s_{1/2}$ Proton Knockout Reaction

T. Noro, M. Kawabata, M. Tanaka, K. Tamura, K. Hatanaka
N. Matsuoka, K. Takahisa, Y. Yuasa, Y. Mizuno, H. Yamazaki*
K. Sagara[†], S. Morinobu[†], M. Nakamura[††] and A. Okihana[‡]

Research Center for Nuclear Physics, Osaka University, Osaka 567, Japan
*Laboratory of Nuclear Science, Tohoku University, Sendai 982, Japan
[†]Department of Physics, Kyushu University, Fukuoka 812, Japan
[††]Department of Physics, Kyoto University, Kyoto 606, Japan
[‡]Kyoto University of Education, Kyoto 600, Japan

Abstract. A pair spectrometer system installed newly at RCNP Osaka has been made available for high resolution studies of particle-particle correlation measurements. By using this system, the analyzing power (A_y) for $1s_{1/2}$ proton knockout reaction on a ^{12}C target and $2s_{1/2}$ knockout on a ^{40}Ca have been measured with the same geometrical setting. A significant suppression of A_y has been observed in the ^{12}C case. But the suppression in the ^{40}Ca case is found much more moderate and is explained by a DWIA calculation as a distortion effect.

INTRODUCTION

Modification of the meson and nucleon properties in nuclear medium is one of the most interesting topics in current nuclear physics. This modification causes a density dependence of the nucleon-nucleon (NN) interaction and plays an important role in nucleon induced reactions at intermediate energies.

One of the most direct ways to investigate this modification is to perform an exclusive measurement of nucleon quasi-free scattering. Advantages to make exclusive measurements are as follows.

1) Kinematics of NN system can be determined by experiment. In the case of elastic and inelastic scattering or inclusive quasifree measurements, on the other hand, an ambiguity in the averaging process is inevitable. Moreover, some spin-dependent terms in the NN-amplitude is canceled out in this averaging process.

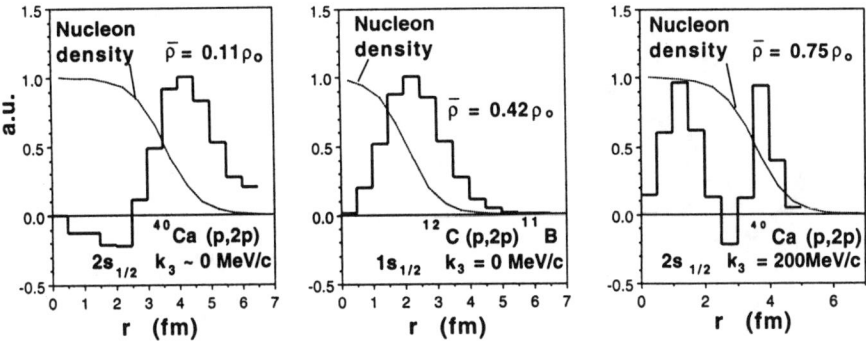

FIGURE 1: Contribution of the nuclear interior to the DWIA cross section, calculated by changing the lower cutoff parameter of the integral. $\bar{\rho}$ is averaged nucleon density weighted by this function. ρ_0 is nuclear density at the origin.

2) Contribution from the nuclear interior can be enhanced or controlled by selecting the binding energy and momentum of the knocked out nucleons. In Fig. 1, contributions of nuclear interior to DWIA cross sections are plotted for three cases of $s_{1/2}$ knockout. The averaged density weighted by these functions, which is the averaged density 'seen' by each measurement, is also shown in the figure. Note that in the right side case, the averaged density is $0.75\rho_0$, which is even higher than the averaged density of nucleus itself.

3) By choosing the $s_{1/2}$ knockout, we can expect to extract the NN spin observables in nuclei from (p,2p) observables with less ambiguity. In this case, no effective polarization effect (1) exist and spin observables of the (p,2p) reactions simply reflect those of NN-scattering, with some modification by LS potential.

At RCNP, a pair spectrometer system has been installed and we started a program on the (p,2p) reactions. Our final goal is to extract the NN-amplitude in nuclei from spin observables of (p,2p) reactions through DW calculation. Here we present the newly installed pair-spectrometer system and report the present experimental result on this program.

PAIR–SPECTROMETER SYSTEM AT RCNP

A high resolution spectrometer, named Grand Raiden, and a broad range, medium resolution spectrometer (Large Acceptance Spectrometer) were con-

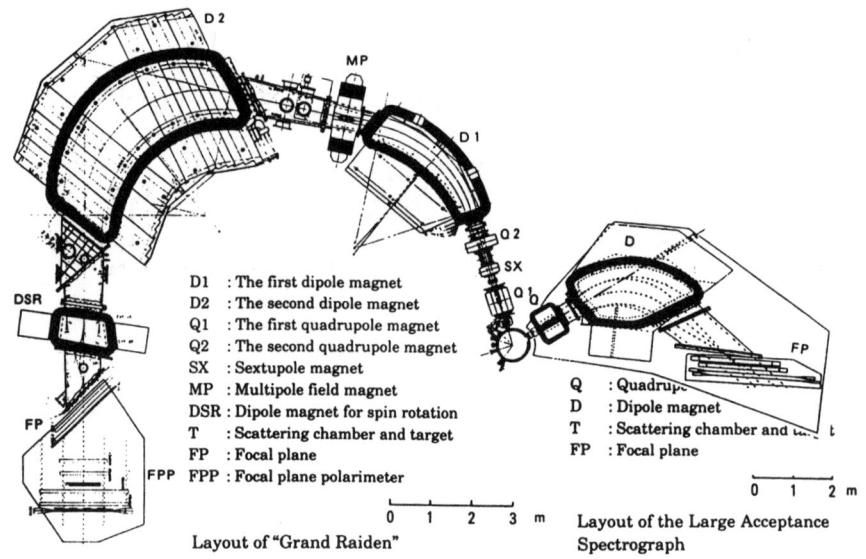

FIGURE 2: Layout of the pair spectrometer system at RCNP

structed. A schematic view of the system is shown in Fig. 2 and the specifications of these spectrometers are given briefly in Table 1.

The focal plane counter for each spectrometer system consists of two sets of Multi-Wire Drift-Chambers (MWDC's), each consists of a couple of anode planes for two-dimensional position measurement. Two planes of plastic scintillators are used in coincidence on each system as trigger counters.

A raytracing technique is utilized for higher order corrections of ion optics, particularly for LAS, and the designed energy-resolution is better than 100 keV for (p,2p) measurements at incident energy of 400MeV. However, since the dispersion matching between the beam line and the spectrometers

TABLE 1: Specifications of the spectrometers at RCNP Osaka

	Grand Raiden	LAS
Maximum magnetic rigidity ($B\rho$)	5.66 Tm	3.2 Tm
Momentum bite (P^{max}/P^{min})	1.05	1.3
Momentum resolution ($P/\Delta P$)	21000	5000
Solid angle (Ω)	5 msr	20 msr

has not been performed successfully for correlation measurements, the actual resolution reflects that of the primary beam, typical 1/1000 in energy.

At the focal plane on Grand Raiden, a focal plane polarimeter is now under construction. It is expected that polarization transfer coefficients can be measured even for (p,2p) reactions within a realistic beam time.

EXPERIMENTAL DATA AND ANALYSIS

A 392MeV polarized proton beam from the Ring Cyclotron at RCNP has been used for this measurement. The polarization of the beam was about 70%, which has been monitored periodically by p-^{12}C polarimeter on the injection beam line.

Setting condition of this measurements, the relative angle of the two spectrometers and the magnetic fields, are those corresponding to the free p-p scattering. The angular acceptances of the forward (Grand Raiden) and backward (LAS) spectrometers are about $40\text{mr}^W \times 120\text{mr}^H$ and $120\text{mr}^W \times 200\text{mr}^H$ respectively. Data are summed over the full momentum range, ±2.5%, of Grand Raiden. Since the momentum bite of LAS, ±15%, is wide enough, there are no momentum cut by backward detection both for the p-p scattering and the (p,2p) reactions.

Figure 3 shows typical spectra of separation energies on ^{40}Ca and ^{12}C target. Overall resolution is about 350keV FWHM. In the case of ^{40}Ca target, $1/2^+$ (E_x=2.5MeV) peak of ^{39}K is not separated from the adjacent $7/2^-$ level, but a DWIA estimation shows that the contribution of the latter level is less than 0.5% of the $1/2^+$ peak in this geometry.

In Figure 4, the analyzing power A_y for $1s_{1/2}$ knockout from ^{12}C target is plotted as a function of the angle of forward outgoing protons. Free p-p data from hydrogen contamination, which was simultaneously obtained with the (p,2p) data, are also plotted. It is obvious that the A_y for (p,2p) reaction is reduced to about 50% of the free value, which is consistent with the data on ^{16}O target at 500 MeV (2).

Results of PWIA and DWIA calculations, by using the code THREEDEE (3), are also plotted in the same figure. The DWIA result shows that the distortion effect is almost independent of the angle and does not explain the reduction of the slope shown in the (p,2p) data. This reduction of the slope is qualitatively explained by a PWIA calculation with t-matrix based on the relativistic approach (4). Proton effective mass of $m^*=0.82 m_N$, which corresponds to the matter density of $0.42\rho_0$, is used in this calculation.

On the other hand, A_y for $2s_{1/2}$ knockout from ^{40}Ca target shows only a little suppression from the free value as shown in Fig. 4. This small suppression is explained as a distortion effect. In this setting, as shown in Fig. 1, the

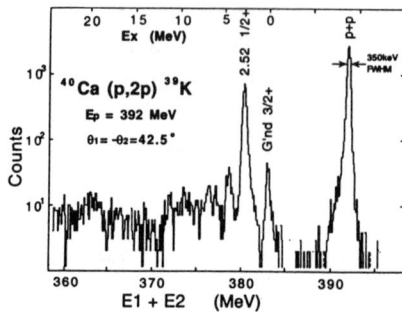 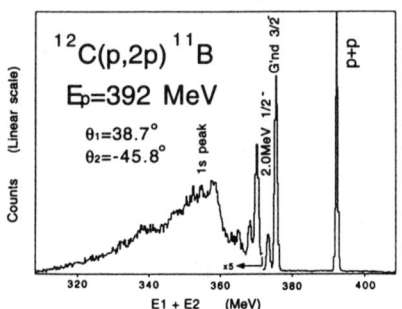

FIGURE 3: Summed energy spectra for (p,2p) reactions on ^{12}C and ^{40}C targets. Overall energy resolution is 350keV FWHM.

contributions of the inside and the surface region cancel each other and the average density is small. Therefore the result is again consistent with relativistic approach.

SUMMARY

A pair spectrometer system was installed at RCNP. By using this system, we succeeded to obtain exclusive (p,2p) data with good resolution and low background at an incident energy of 392 MeV.

For $1s_{1/2}$ proton knockout from ^{12}C, where the averaged density seen by this reaction is estimated to be $0.42\rho_0$, A_y is about 50% of the free NN value.

In the case of $2s_{1/2}$ proton knockout from ^{40}Ca, where the averaged density is estimated to be $0.1\rho_0$, the measured A_y was roughly reproduced by a DWIA calculation with the free NN matrices.

These results are qualitatively consistent with a prediction based on the relativistic approach for the nuclear reactions.

In order to pin down the origin of the A_y reduction and to investigate the medium modifications of the NN interaction, (p,2p) measurements for various averaged densities and binding energies are important. We are preparing for such experiment including an efficient measurement of the polarization transfer coefficients.

ACKNOWLEDGEMENT

The authors are pleased to thank to Prof. N.S. Chant for providing us the computer code THREEDEE. The experiment was performed under the pro-

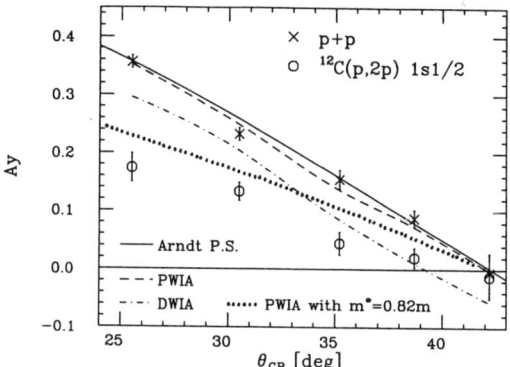

FIGURE 4: Analyzing power for ^{12}C(p,2p) reaction and free p-p scattering. PWIA and DWIA results are also shown.

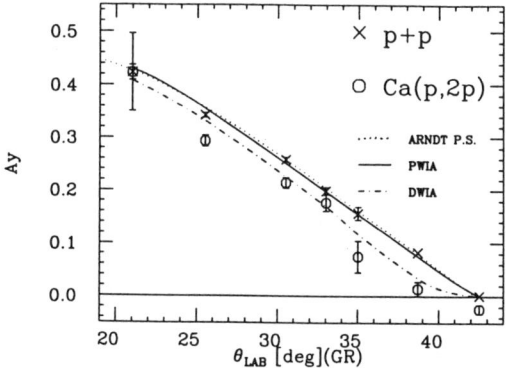

FIGURE 5: Same as Fig. 4 but for ^{40}Ca(p,2p).

gram number E42 at RCNP.

REFERENCES

1. G. Jacob et al.,Phys Lett. **45B** (1973) 181.
2. C.A. Miller et al., in "Proceedings of the 7th International Conference on Polarization Phenomena in Nuclear Physics", (Paris,1990) C6-595.
3. N.S. Chant and P.G. Roos, Phys. Rev. **C27** (1983) 1060.
4. C.J. Horowitz and M.J. Iqbal, Phys. Rev. **C33** (1986) 2059.

Study of Polarization in Quasi-Elastic Break-up Reaction ^{6}Li(p,2p)^{5}He in Complete Kinematics at 1 GeV

A.N. Prokofiev, N.P. Aleshin, S.L. Belostotski,
Yu.V. Dotsenko, V.A. Efimovykh, O.Ya. Fedorov,
A.A. Izotov, A.Yu. Kisselev, E.N. Komarov, O.V. Miklukho,
V.I. Murzin, Yu.G. Naryshkin, D.A. Prokofiev, Yu.A. Scheglov,
A.V. Shvedchikov, A.A. Zhdanov and A.A. Zhgun

Petersburg Nuclear Physics Institute, Gatchina, St.Petersburg 188350, Russia

Abstract. The polarization of recoil protons from the reaction $^6Li(p,2p)^5He$ at 1 GeV of incident proton beam energy was measured with a new two-arm magnetic spectrometer with a carbon polarimeter. The effective polarization of the P-shell protons of ^6Li nucleus was calculated and found to be in a qualitative agreement with the predictions of the cluster model rather than with those of the shell model.

INTRODUCTION

Investigation of a few-body system is of great importance and becomes more actual in connection with searching for the effects of nuclear structure. Up to now this investigation has been carried out mainly in experiments where only cross sections of various reactions were measured. The new generation of the polarization experiments where the effects of clusterization in the nuclei were studied became available recently. As one of the results of these investigations (1) the significant spin effects on ^6Li were demonstrated over the wide energy and angular ranges. None of the theoretical calculations was able to reproduce the obtained data in details. It appears that the nucleus vector and tensor analyzing powers are more sensitive to the chosen nuclear wave function than to the details of the interaction. The polarization effects found on ^7Li, ^{13}C, ^{15}N are much smaller than on ^6Li in contradiction to the theoretical predictions. It was also found out that the vector analyzing power in the reactions on very light nuclei such as deuteron (2) is large and has a tendency of increasing with the energy.

Under these circumstances it seems very interesting to measure the polarization variables in the exclusive quasi-free $(p, 2p)$ reaction on ^6Li at 1 GeV in order to reveal the effects of nucleus clusterization. With this purpose in mind we proposed to measure the polarization (P_2) of the low-energy protons produced in above mentioned reaction as well as the analyzing power (A_Z) in the reaction on polarized nuclear target. From these two observables we may obtain information on nuclear structure function $B_Z(k)$ and effective polarization of protons on nuclear shells $P_{eff}(k)$. The latter was predicted by Th. Maris et al. (3) as the result of absorption of the secondary protons and spin-orbit coupling in nuclear matter. It was shown that the predictions on the value of P_{eff} strongly correlate with the assumption on the nuclear wave function used in the calculations (4). The same is true for the nuclear structure function $B_Z(k)$.

THE EXPERIMENTAL METHOD

In the experiment with unpolarized ^6Li target we measured the polarization (P_2) of the low-energy protons knocked out in the reaction $^6Li(p,2p)^5He$ at angle $\Theta_2 = 58°$ from S- and P-shells of ^6Li nucleus. The measurements were performed at non-symmetric outgoing angles ($\Theta_1 = 15° \div 22°$, $\Theta_2 = 58°$) of the secondary protons in the conditions of a coplanar quasi-free scattering geometry and a complete reconstruction of the reaction kinematics. It is known that for such a reconstruction at a given excitation of the residual nucleus (^5He) one needs to measure five kinematical variables. In this experiment we measured the momenta (K_1, K_2) of the secondary protons and their outgoing angles (Θ_1, Θ_2, Φ_1, Φ_2). Using the obtained values of K_i, Θ_i, Φ_i and the value of beam momentum K_0 it is possible to calculate the nuclear proton separation energy ΔE and the residual nucleus momentum K_B. The latter in the impulse approximation is equal to the momentum of nuclear proton before the interaction ($\vec{K_B} = -\vec{K}$).

The experimental setup is presented in Fig.1. It consisted of proton beam magnetic channel, two-arm magnetic spectrometer, unpolarized ^6Li target (T_1), scintillation counter telescope and ionization chamber for monitoring the incoming beam parameters. The two-arm magnetic spectrometer was used for registration of the secondary protons from the reaction $^6Li(p,2p)^5He$ and for measurements of their momenta and outgoing angles. Its low-energy arm (NES-spectrometer) was installed at fixed angle $\Theta_2 = 58°$. Polarimeter

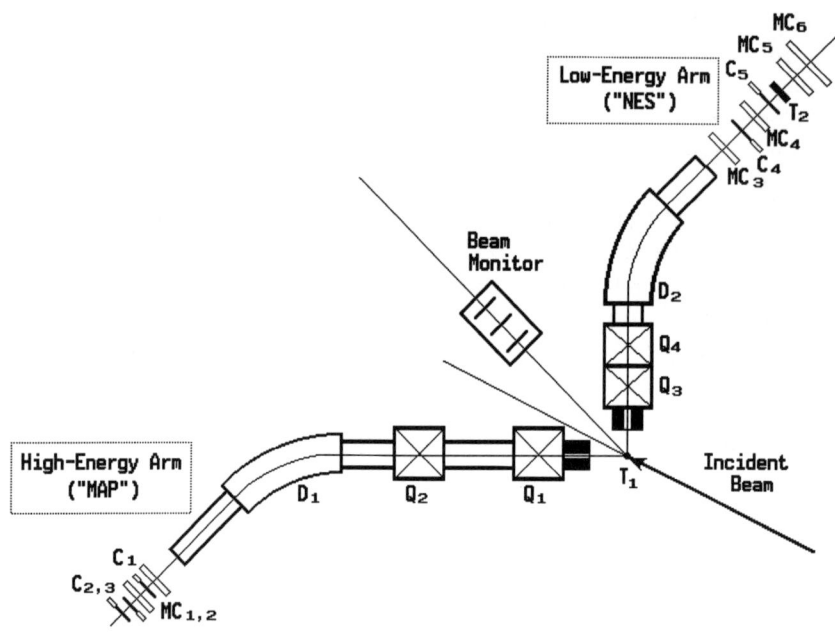

FIGURE 1. Experimental setup: $Q_1 - Q_4$ - quadrupoles; D_1, D_2 - magnets; $C_1 - C_6$ -scintillation counters; $MC_1 - MC_{10}$ - multiwire proportional chambers; T_1 - ^6Li target; T_2 - carbon analyzer of the polarimeter.

TABLE 1. Parameters of the Magnetic Spectrometers

Spectrometer	NES	MAP
Maximum particle momentum $(K/Z)^{max}$, GeV/c	1.0	1.7
Axial trajectory radius R, m	3.03	5.5
Deflection angle α, deg.	42.2	24.0
Dispersion in focal plane D, mm/%	16	22
Solid angle acceptance Ω, sr	$4.7 \cdot 10^{-3}$	$4.0 \cdot 10^{-4}$
Energy resolution (FWHM), MeV	~ 2.0	~ 1.5

located at the focal plane of the NES-spectrometer was used to measure the polarization (P_2) of the low-energy protons produced in the reaction $^6Li(p,2p)^5He$. It consisted of four proportional chambers and carbon analyzer (T_2). The main parameters of the two-arm magnetic spectrometer are listed in Table 1.

DISCUSSION OF THE RESULTS

The results of investigation of the reaction $^6Li(p,2p)^5He$ are presented in Figs.2,3. First of them represents the energy loss spectrum which allows to estimate the obtained resolution of S- and P-shells of ^6Li. The results of measurement of the polarization P_2 for P- and S-shells of ^6Li in the region of the residual nucleus momenta up to 150 MeV/c are shown in Fig.3a. The measured value of the polarization in elastic pp-scattering is shown in the figure as a point at $K_B = 0$. The solid line in Fig.3a is the result of calculations of the polarization in the reaction $^6Li(p,2p)^5He$ in the impulse approximation.

The experimental data for P-shell (Fig.3a) display a smooth decrease of the polarization P_2 as the residual nucleus momentum increases. The behavior of the polarization P_2 for S-shell is more complicated and demonstrates a deep minimum near $K_B = 100$ MeV/c. This effect apparently is not connected with any systematical errors of the measurements because the S- and P-shell data were obtained simultaneously in the same experimental conditions. The off-line data analysis was also carried out using the same routine.

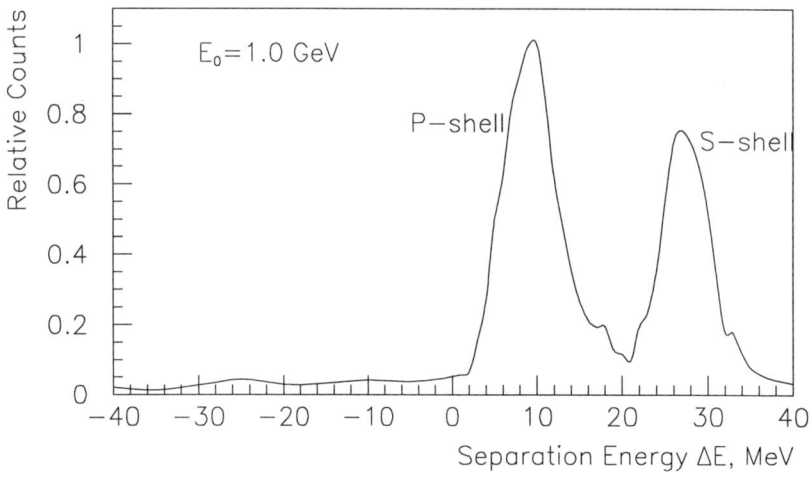

FIGURE 2. Nuclear proton separation energy spectrum for the reaction $^6Li(p,2p)^5He$ at 1 GeV.

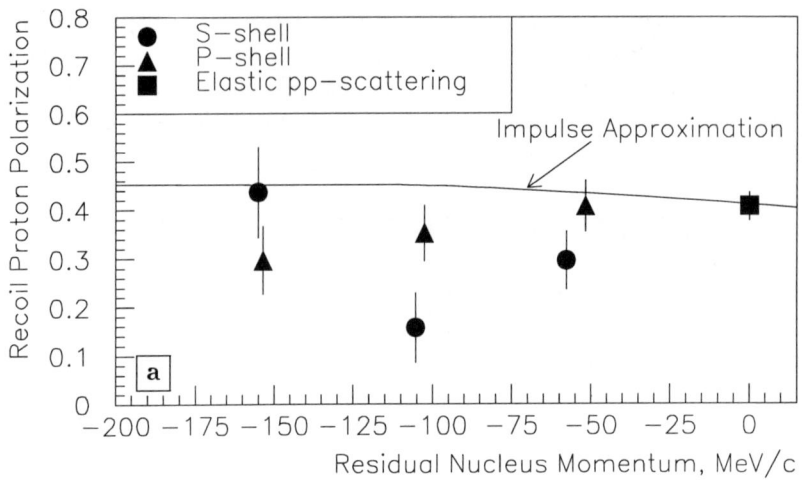

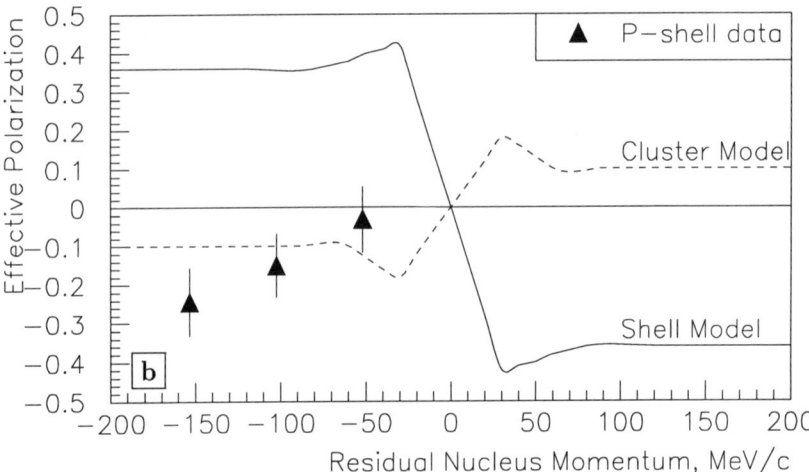

FIGURE 3. a) Polarization of the low-energy protons produced at angle $\Theta_2 = 58°$ in the reaction $^6Li(p,2p)^5He$ at 1 GeV on protons of S-shell and P-shell of ^6Li.
b) Effective polarization in the reaction $^6Li(p,2p)^5He$ for P-shell of ^6Li nucleus. Solid and dashed curves represent the results of cluster- and shell-model calculations, respectively (4). Sign of the residual nucleus momentum K_B corresponds to that of $(\vec{K}_B \cdot \vec{K}_0)$ product.

The obtained data for the effective polarization of P-shell protons are presented in Fig.3b together with the predictions calculated in (4) in the frameworks of cluster- and shell-models[1]. According to our data the mean value of the effective polarization in the region $50 \leq K_B \leq 150$ MeV/c for P-shell is equal to (5)

$$< P_{eff} > = -0.14 .$$

The obtained sign of P_{eff} is the same as predicted in (4) in the framework of primitive cluster model of ^6Li nucleus $\{^6Li - (\alpha \stackrel{s}{+} d)\}$.

The nature of the behavior of the polarization P_2 for the case of scattering on the S-shell protons (Fig.3a) is not clear yet. It is impossible to explain it in terms of P_{eff}, its value being zero for S-shell of ^6Li nucleus in the framework of pure shell model. However assuming the existence of a significant $(d-d)$ clusterization in α-core of ^6Li one can expect to find a non-zero effective polarization in the region of inner momenta close to the node of the S-state radial wave function in relative motion of the clusters due to a noticeable admixture of the D-state in $(d+d)$ cluster structure of ^4He ($11 \div 15\%$ according to theoretical predictions). Another possible explanation is the presence of $^3H + p$ component in the P-state of relative motion which can also produce an effective polarization of the same sign.

The results of this experiment are under discussion now. We plan to continue the polarization measurements in the kinematic region where the product $(\vec{K_B} \cdot \vec{K_0})$ has the positive sign (Fig.3b).

This work was partly supported by Russian Fundamental Science Foundation, Grant 93-02-16715.

REFERENCES

1. Ritt S. et al. *Phys.Rev.*, **C43**, 745-760 (1991).

2. Bazhanov N.A. et al. *Phys.Rev.*, **C47**, 395-398 (1993).

3. Jacob G. et al. *Phys.Lett.*, **45B**, 181-196 (1973).
   Nucl.Phys., **A257**, 517-532 (1976).

4. Fernandez F. et al. *Phys.Lett.*, **106B**, 15-18 (1981)

5. Aleshin N.P. et al. *Preprint PNPI-1971*, **NP-29-1994**.

---

[1]It should be mentioned that the values of effective polarization are calculated in (4) for the 320 MeV incident protons, and we cannot yet discuss the quantative agreement of our experimental results with the theoretical models. The calculation of effective polarization predictions for the reaction with 1 GeV protons is in process now.

Scattering of Polarized Protons from Polarized Targets

W. G. Love

Department of Physics and Astronomy
The University of Georgia
Athens, Georgia 30602-2451

Abstract. The inelastic scattering of polarized protons from polarized targets is considered with an emphasis on the interplay between nuclear structure and nucleon-nucleon amplitudes. Particular attention is focused on the case of spin 1/2 and spin 1 targets where explicit formulae for selected polarization observables are given in the plane wave impulse approximation where the roles of nuclear structure and nucleon-nucleon amplitudes are most transparent.

INTRODUCTION

The measurement and interpretation of spin observables obtained using polarized beams have played a crucial role in advancing our understanding of nuclear structure and nuclear scattering mechanisms. Most (1), though not all (2), of this effort has been made using unpolarized targets. For the case of polarization transfer coefficients, the plane-wave impulse approximation (PWIA) has proven to be especially useful in identifying the roles played by nuclear structure and the associated nucleon-nucleon (NN) amplitudes (3,4). The objective here is to extend this PWIA formalism to polarization observables for the case in which the target is polarized. Explicit formulae are given for some relatively simple cases.

FORMALISM

Here we consider the formulae most relevant to the present discussion of the scattering of spin 1/2 particles from polarized targets. Prior to collision the density matrix describing the beam and target may be written:

$$\rho^{(i)} = \left[\frac{1}{2} \sum_{j=0}^{3} p_j\, \sigma_j \right]_p \times \left[\frac{1}{(2I+1)} \sum_k \omega_k\, \Omega_k \right]_t, \quad \sigma_o = 1, \quad \Omega_o = 1 \qquad (1)$$

and $p_j = tr\,\rho^{(i)}\sigma_j$ and $\omega_k = tr\,\rho^{(i)}\Omega_k$ describe the initial states of polarization of the beam and target (2) and $p_o = \omega_o = 1$. σ_j is the jth component of the Pauli spin operator of the projectile and Ω_k is an analogous operator for the target of spin I. For vector polarized targets Ω_k is proportional to the kth component of the target spin; for more complicated states of the target polarization k is a composite index (5,6). The density matrix which describes the ejectile-residual nucleus system is given by $\rho^{(f)} = T\,\rho^{(i)}\,T^+$ where T is the nucleon-nucleus (NA) T-matrix. The unpolarized cross section is given by

$$\sigma_u = \frac{1}{2(2I+1)}\,tr\,T\,T^+. \qquad (2)$$

The ℓth component of polarization of the ejectile is given by:

$$p_\ell(f)\,\sigma = \sigma_u \sum_{jk} p_j(i)\,\omega_k(i)\,X(jk;\ell 0) \qquad (3)$$

where the general polarization coefficient is given by

$$X(pt;\,p't') = tr\left[T\,\sigma_p\,\Omega_t\,T^+\,\sigma_{p'}\,\Omega_{t'}\right]\Big/tr\,T\,T^+ \qquad (4)$$

and (p,t) and $(p't')$ correspond to the initial and final states of polarization of the projectile and target before and after collision respectively. σ is the polarized cross section obtained with $\ell = 0$ and $p_o = 1$. We will not consider the case in which the state of polarization of the residual nucleus is monitored. In addition, we only consider targets with at most vector polarization; polarization of higher ranks which can occur for $I > 1/2$ will be ignored. In order to work with the familiar spin operators, Eq. (3) is often rewritten (5) as

$$p_\ell(f)\sigma = \sigma_u \sum_{jk} p_j(i)\,p_k^T\,\alpha_k(I)\,X(jk;\ell 0) \qquad (5)$$

where $\alpha_o(I) = 1$ for arbitrary I and $p_k^T = tr\,\rho^{(i)}\Omega_k$ is the target polarization when α_k and Ω_k are normalized as in Table 1.

TABLE 1. Spin Parameters and Operators

	$I = 1/2$	$I > 1/2$
$\alpha_k(I)$	1 (all k)	$3/I(I+1)$ $(k \neq 0)$
Ω_k	$2S_k$	S_k

To make contact with more traditional notation, we note that:

$$X(20;00) = A_y^p, \quad X(02;00) = A_y^t, \quad \text{and} \quad X(jk;00) = C_{jk} \quad (j \neq 0, k \neq 0),$$
$$X(00;20) = P_y^p, \quad \text{and} \quad X(j0;\ell 0) = D_{j\ell} \quad (j \neq 0, \ell \neq 0) \tag{6}$$

where A_y^p and A_y^t are the analyzing powers associated with the projectile and target, C_{jk} are the usual spin correlation parameters, P_y^p is the induced polarization of the ejectile and $D_{j\ell}$ denotes the polarization transfer coefficients.

PLANE WAVE IMPULSE APPROXIMATION (PWIA)

It is in this approximation that the roles of nuclear structure and NN couplings are most transparent. The nucleon-nucleus (NA) T-matrix is

$$T_{FI} = < v', I'M' | \sum_{i=1}^{A} e^{i\vec{q}\cdot\vec{r}_i} M_{ip}(\vec{q}, E...) | v, IM > \tag{7}$$

where (v, v') and $(IM, I'M')$ denote the spin states of the projectile and target, respectively. M_{ip} is the NN scattering amplitude given by:

$$M_{ip} = \sum_{\lambda_p, \lambda_t = 0}^{3} M_{\lambda_t \lambda_p}(q) \, \sigma_{\lambda_t}(i) \, \sigma_{\lambda_p}(p) \tag{8a}$$

with

$$M_{\lambda_t \lambda_p} = M_{\lambda_t} \delta_{\lambda_t \lambda_p} + C(\delta_{0\lambda_t} \delta_{2\lambda_p} + \delta_{2\lambda_t} \delta_{0\lambda_p}), \tag{8b}$$

$$M_0 = A, \quad M_1 = F, \quad M_2 = B, \quad M_3 = E, \tag{8c}$$

and A, C, B, E and F are the Wolfenstein amplitudes (7). This gives:

$$T_{FI} = \sum_{\lambda_p \lambda_t} M_{\lambda_t \lambda_p}(q) < I'M' | \sum_i e^{i\vec{q}\cdot\vec{r}_i} \sigma_{\lambda_t}(i) | IM > < v' | \sigma_{\lambda_p} | v >. \tag{9}$$

We define a target transition "spin" operator and its four components

$$\bar{S}_t = \sum_i e^{i\vec{q}\cdot\vec{r}_i} \bar{\sigma}(i); \quad S_{\lambda_t} = \sum_i e^{i\vec{q}\cdot\vec{r}_i} \sigma_{\lambda_t}(i), \quad (\lambda_t = 0, 1, 2, 3). \tag{10}$$

The general polarization observable can then be written as:

$$\sigma_u X(pt; p't') = \sum_{[\lambda]} M_{\lambda_t \lambda_p} M^*_{\lambda'_t \lambda'_p} S_{pp'}(\lambda_p, \lambda'_p) S_{tt'}(\lambda_t, \lambda'_t) \tag{11}$$

with

$$S_{pp'}(\lambda_p,\lambda'_p) = \frac{1}{2} tr^P[\sigma_{\lambda_p} \sigma_p \sigma^{\dagger}_{\lambda'_p} \sigma_{p'}], \qquad (12a)$$

$$S_{tt'}(\lambda_t,\lambda'_t) = \frac{1}{2I+1} tr^{t'}[S_{\lambda_t}\Omega_t S^{\dagger}_{\lambda'_t}\Omega_{t'}] \qquad (12b)$$

and the sum over (λ) means sum over all λ's present; the traces are over the spin projections in the indicated subspaces. The matrices $S_{pp'}$ and $S_{tt'}$ define the "filtering" characteristics imposed by the measurement of specific polarization coefficients. Accordingly, we focus on the quantity $S_{tt'}(\lambda_t,\lambda'_t)$ and take the z-axis along the momentum transfer $\vec{q}$ (see (Eq. (8c)). The transition spin operator in Eq. (10) is a sum of terms of the *form*

$$S_\lambda = e^{i\vec{q}\cdot\vec{r}}\,\vec{\sigma}\cdot\vec{\varepsilon}_\lambda = \sum_{J,\mu}\sqrt{4\pi(2J+1)}\,(-)^{\Delta\pi+J+S(\lambda)}\,(-)^\mu\,a^\lambda_{-\mu}\,f_{\lambda\mu}\,Q^\lambda_{J\mu} \qquad (13)$$

where $S(0)=1$, $S(\lambda=1,2,3)=1$, $\Delta\pi$ is the parity change (even or odd). $a^\lambda_\mu = \bar{e}_\mu \cdot \bar{\varepsilon}_\lambda$ with $\bar{e}_\mu \cdot \bar{\varepsilon}_o = \delta_{\mu o}$, $\bar{\varepsilon}_\lambda$ is a cartesian unit vector ($\lambda=1,2,3$), $\bar{e}_\mu$ is a unit vector in a spherical basis and $Q^\lambda_{J\mu}$ are spherical tensors in the space of the target given in Table 2 for natural ($\Delta\pi = (-)^J$) and unnatural ($\Delta\pi = (-)^{J+1}$) parity amplitudes, denoted by $N\pi$ and $UN\pi$, respectively. In Table 2:

TABLE 2. $f_{\lambda\mu}$ and $Q^\lambda_{J\mu}$ of Eq. (13)

	Nπ		UNπ	
λ	$f_{\lambda\mu}$	$Q^\lambda_{J\mu}$	$f_{\lambda\mu}$	$Q^\lambda_{J\mu}$
0	$\delta_{\mu o}$	$M_{J0J,\mu}$	0	0
1	μ	$M_{J1J,\mu}/\sqrt{2}$	1	$M^t_{J\mu}$
2	μ	$M_{J1J,\mu}/\sqrt{2}$	1	$M^t_{J\mu}$
3	0	0	1	$M^\ell_{J\mu}$

$$M_{LSJ,\mu} = i^L j_L(qr)[Y_L(\hat{r})\otimes\sigma^s]^J \qquad (S=0,1). \qquad (14)$$

M^ℓ and M^t are the widely used (3) spin longitudinal and spin transverse form factors; M^t here is M^t(ref. 8)$/\sqrt{2}$. If $\vec{k}(\vec{k}')$ is the initial (final) relative momentum, we choose $\bar{y}=\bar{n}=\vec{k}\times\vec{k}'$, $\bar{z}=\bar{q}=\vec{k}-\vec{k}'$ and $\bar{x}=\bar{n}\times\bar{q}$.

POLARIZATION OBSERVABLES IN PWIA

It is instructive to consider two (relatively) simple, yet potentially realistic cases. In particular, we examine the explicit PWIA results for polarized protons scattering from polarized targets having initial spins $I=1/2$ and 1 and final spins $I'=1/2$ and 0, respectively. Typical targets of these types which have been considered are ^{13}C and ^{14}N. In Table 3 we give the formulae obtained from Eq. (11) for some of the polarization observables when there is no parity change.

TABLE 3. Polarization Observables in PWIA ($\Delta\pi$ = no)

Observable	$J^\pi = 0^+, 1^+$ $\tfrac{1}{2}+\tfrac{1}{2} \to \tfrac{1}{2}+\tfrac{1}{2}$	$J^\pi = 1^+$ $\tfrac{1}{2}+1 \to \tfrac{1}{2}+0$
σ_u	$|a|^2+|c_p|^2+|b|^2+|c_t|^2+|e|^2+|f|^2$	$|c_t|^2+|b|^2+|e|^2+|f|^2$
$\sigma_u A_y^p = \sigma_u P_y^p$	$2\operatorname{Re}(a^* c_p + b^* c_t)$	$2\operatorname{Re} b^* c_t$
$\sigma_u A_y^t$	$2\operatorname{Re}(a^* c_t + b c_p^*)$	0
$\sigma_u C_{xx}$	$2\operatorname{Re}(a^* f - b^* e)$	$-2\operatorname{Re} b^* e$
$\sigma_u C_{yy}$	$2\operatorname{Re}(a^* b + c_p^* c_t - e^* f)$	$-2\operatorname{Re} e^* f$
$\sigma_u C_{zz}$	$2\operatorname{Re}(a^* e - f^* b)$	$-2\operatorname{Re} f^* b$
$\sigma_u C_{xz}$	$2\operatorname{Im}(e^* c_p + c_t^* f)$	$2\operatorname{Im} c_t^* f$
$\sigma_u C_{zx}$	$2\operatorname{Im}(f c_p^* + c_t e^*)$	$2\operatorname{Im} c_t e^*$
$\sigma_u X(01;10)$	$2\operatorname{Re}(a^* f + b^* e)$	$2\operatorname{Re} b^* e$
$\sigma_u X(02;20)$	$2\operatorname{Re}(a^* b + c_p^* c_t + e^* f)$	$2\operatorname{Re} e^* f$
$\sigma_u X(03;30)$	$2\operatorname{Re}(a^* e + f^* b)$	$2\operatorname{Re} f^* b$

$a = A<\tilde{Q}_0^0>,\ c_p = C<\tilde{Q}_0^0>,\ f = F<\tilde{Q}_1^1>,\ b = B<\tilde{Q}_1^2>,$
$c_t = C<\tilde{Q}_1^2>,\ e = E<\tilde{Q}_1^3>$

where the superscript (subscript) denotes $\lambda_t(J)$ of Eq. (13) and $\tilde{Q}_J^\lambda = \sqrt{4\pi/(2I+1)}\, Q_J^\lambda$ of Table 2. We also note from Table 2 that $Q_J^1 = Q_J^2 = Q_J^t$ the transverse spin tensor and $Q_J^3 = Q_J^\ell$ the longitudinal spin tensor; $<\tilde{Q}_J^\lambda>$ are the reduced matrix elements (9) of Q_J^λ (times $\sqrt{4\pi/(2I+1)}$).

The relatively simple $I = 1 \to I' = 0$, $\Delta\pi$ = even case of Table 3 reflects the absence of natural parity amplitudes corresponding to a and c_p. It is instructive to make the NN amplitudes and nuclear structure factors contained in Table 3 explicit. In particular, for the $I = 1$ case,

$$\sigma_u A_y^p = 2\,\text{Re}\,B^* C |\tilde{Q}_1^t|^2, \quad \sigma_u C_{xx} = -2\,\text{Re}\,B^* E\,\tilde{Q}_1^t\,\tilde{Q}_1^\ell \qquad (15a)$$

$$\sigma_u C_{yy} = -2\,\text{Re}\,F^* E\,\tilde{Q}_1^t\,\tilde{Q}_1^\ell, \quad \sigma_u C_{zz} = -2\,\text{Re}\,F^* B |\tilde{Q}_1^t|^2 \qquad (15b)$$

where we have used the fact that the reduced matrix elements (< > omitted) $\tilde{Q}^\ell$ and $\tilde{Q}^t$ may be regarded as real (10). Using this, and assuming the NN amplitudes are known, we may solve for the quantities $\tilde{Q}^\ell \tilde{Q}^t$ and $(\tilde{Q}_1^t)^2$ from Eq. (15). Alternatively, we note, for example,

$$\sigma_u(C_{xx} \pm C_{yy}) = -2\,\text{Re}(B \pm F)^* E\,\tilde{Q}_1^t\,\tilde{Q}_1^\ell. \qquad (16)$$

The combination $C_{xx} + C_{yy}$ involves the interference between the static transverse $(\vec{\sigma}_1 \times \hat{q}) \cdot (\vec{\sigma}_2 \times \hat{q})$ $(B+F)/2$ and the longitudinal spin $(\vec{\sigma}_1 \cdot \hat{q})(\vec{\sigma}_2 \cdot \hat{q})$ E couplings (8). For isovector transitions of this spin type this observable provides information on the interference between the effective π and ρ couplings (1,8). Similarly, $C_{xx} - C_{yy}$ is sensitive to the interference between the complimentary transverse $(B-F)/2$ and the longitudinal spin couplings (8). The $B-F$ coupling dominates isoscalar spin excitations. For $1^\pm \to 0^\pm$ transitions which are dominated by $L=0$ (as is *usually* the case near $q=0$ where $\tilde{Q}_1^\ell \sim \tilde{Q}_1^t$), and for a purely central force in which $B=E=F$, we see from Eq. (15) that $C_{xx}=C_{yy}=C_{zz}$. When these conditions are met, $C_{ii}(0^o) = -2/3$ providing a signature of $J^\pi = 1^+$ transitions analogous to that provided by $D_{ii}(0^o) \cong -1/3$ under the same conditions. More generally $|\tilde{Q}_1^t|^2$ can be determined from electron scattering yielding $\sigma_u C_{zz}$ as a direct probe of the interference between the two transverse components of the NN amplitudes, B and F.

The case of an $I=1/2 \to I'=1/2$ transition is more complicated than the $I=1 \to I'=0$ case due to contributions from both natural (0^+) and unnatural (1^+) parity amplitudes. The case of $I=1/2$ has been examined earlier (11). The determination of the relative phase between the Fermi and Gamow-Teller transition amplitudes has been of particular interest. The analyzing power of the target A_y^t involves $J=0,1$ interference at $q \neq 0$. In particular,

$$\sigma_u A_y^t = 2\,\text{Re}(A^* C + BC^*)\tilde{Q}_o^o \tilde{Q}_1^t \qquad (17)$$

where $\tilde{Q}_o^o(\tilde{Q}_1^t)$ is proportional to the Fermi (Gamow-Teller) matrix element at $q=0$ (see Table 2). Additional information on the interference between different amplitudes may be obtained by combining some of the observables from Table 3. In particular,

$$\sigma_u[C_{ii} \pm X(0i;i0)] = \pm 4\,\text{Re}\begin{Bmatrix} a^*f & a^*b+c_p^*c_t & a^*e \\ b^*e & e^*f & f^*b \end{Bmatrix} \quad (18)$$

where the upper (lower) entries in brackets correspond to the +(−) sign and the three horizontal entries are for i = x,y,z, respectively. Unlike $\sigma_u A_y'$, the observables in Eq. (18) need not vanish at 0°. The (+) combinations provide information on the interference between $J = 0$ and $J = 1$ amplitudes as they enter Fermi and Gamow-Teller transitions. Complimentary information on interference between $J = S = 1$ terms may be obtained from the (−) combinations. The imaginary parts of the interfering amplitudes may also be constructed; these often require (11) 3-spin observables involving the measurement of polarization transfer from a polarized target.

In Fig. 1 we show the sensitivity of C_{NN} to distortion. Although C_{xx} and C_{zz} are somewhat more sensitive to distortion effects (DW), the PW and DW results for each of these observables are in very reasonable agreement for $\theta \leq 20°$. The closeness of agreement in each case suggests that the C_{ii} are rather robust observables for isovector transitions involving $J^\pi = 1^+$ transfers. The full 210 MeV t-matrix interaction (12) was used in the calculations.

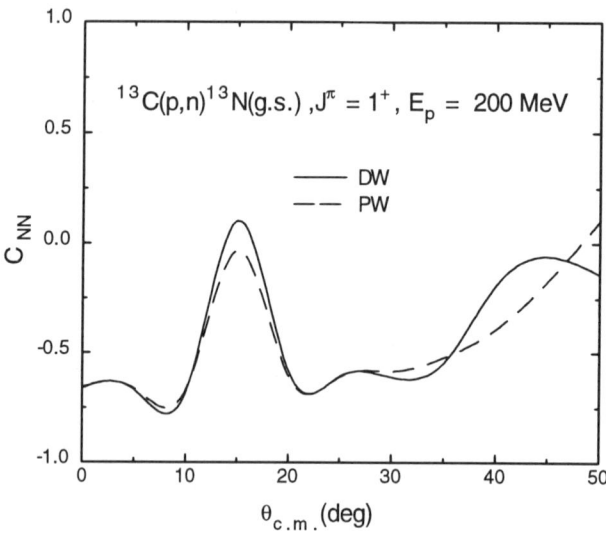

FIGURE 1. Spin-correlation coefficient $C_{NN} = C_{yy}$ in the PWIA and DWIA.

SUMMARY

Expressions for polarization observables have been given within the plane-wave impulse approximation for the inelastic scattering of polarized protons from polarized targets. The emphasis has been on making the roles of nuclear structure and the NN amplitudes explicit. Practical considerations led us to focus on spin 1/2 and spin 1 targets where detailed formulae have been given for several one and two spin observables. Generally speaking, the use of polarized targets allows interference between amplitudes which are otherwise incoherent. This, in turn, opens a new window on the interplay between nuclear structure and the NN interaction in nuclei. Although these PWIA results cannot be expected to be quantitative, they do provide insight into the new physics likely accessible using polarized targets. Moreover, preliminary $J^\pi = 1^+$ calculations suggest that the diagonal spin-correlation coefficients are rather robust with respect to the inclusion of distortion.

REFERENCES

1. Stephenson, E. J., "Elastic and Inelastic Scattering at Low and Intermediate Energies," in *Proceedings of the 7th International Conference on Polarization Phenomena in Nuclear Physics*, (Paris, France, July 9-13, 1990), pp. 85-98 and refs. therein; T. Carey, J. Phys. Soc. Jpn. **55** (1986), Suppl. pp. 172-191 and refs. therein.
2. Hoffmann, G. W. et al., *Phys. Rev. Lett.* **65**, 3096 (1990); Haüsser O., in "Spin and Isospin in Nuclear Reactions," eds. S. W. Wissink, C. D. Goodman and G. E. Walker (Plenum, N.Y., 1991), p. 381.
3. Moss, J. M., in *Spin Excitations in Nuclei*, ed. F. Petrovich et al. (Plenum, N.Y., 1984) p. 355.
4. Bleszynski, E., Bleszynski, M. and Whitten Jr., C. A., *Phys. Rev. C* **27**, 902 (1983).
5. Ohlsen, G. G., *Rep. Prog. Phys.* **35**, 717 (1972).
6. Satchler, G. R., *Direct Nuclear Reactions* (Oxford, N.Y., 1983).
7. Kerman, A. K., McManus, H. and Thaler, R. M., *Ann. Phys.* (N.Y.) **8**, 551 (1959).
8. Love, W. G., Franey, M. A. and Petrovich, F., in *Spin Excitations in Nuclei*, ed. F. Petrovich et al. (Plenum, N.Y., 1984) p. 205.
9. Bohr, A. and Mottelson, B., *Nuclear Structure* (Benjamin, N.Y., 1969) Vol. 1.
10. Brink, D. M. and Satchler, G. R., *Angular Momentum*, 2nd ed. (Clarendon, Oxford, 1971).
11. Ray, L. et al., *Phys. Rev. C* **37**, 1169 (1988); Ray, L. and Shepard, J. R., *Phys. Rev. C* **40**, 237 (1989).
12. Franey, M. A. and Love, W. G., *Phys. Rev. C* **31**, 488 (1985).

Charge exchange spin observable measurements on Pb at 795 MeV in the giant resonance region

D. Prout[1], E. Sugarbaker[1], S. Delucia[1], D. Cooper[1] B. Luther[1]
C.D. Goodman[2], B. K. Park[4], J. Rapaport[5]
L. Rybarcyk[3], T. Taddeucci[3]

1) The Ohio State University, Columbus, OH 43212 USA
2) Indiana University Cyclotron Facility, Bloomington, Indiana 47405
3) Los Alamos National Laboratory, Los Alamos, NM 87545 USA
4) New Mexico State University, Las Cruces, NM 88003 USA
5) Ohio University, Athens, OH 45701 USA

Abstract. We examine a new feature of the $0°$ (p,n) cross section spectra that appears at 35 MeV energy loss in Pb and Zr. To further investigate this structure we have measured a complete set of spin transfer observables for ^{nat}Pb at $0°$ using the $(\vec{p},\vec{n})$ reaction at 795 MeV.

Introduction

At $0°$ the dominant feature of (p,n) spectra taken at 200 MeV proton energy is the giant Gamow–Teller (GT) resonance (1). This is true of light targets such as $^{12}C(\vec{p},\vec{n})$ as well as heavy targets such as ^{90}Zr and ^{nat}Pb . A great deal of success has been attained in the theoretical description of (p,n) spectra at this energy. This success culminated in the work of Osterfeld et al. (2) who obtained excellent descriptions of 200 MeV Pb and Zr angular distributions using sophisticated DWIA–RPA calculations (3,1). Part of the success of this model relied on redistributing L=0 strength over a broad region well above the GT resonance.

At proton energies above 200 MeV a new feature appears in (p,n) spectra taken on heavy targets. At 795 MeV a significant portion of the charge-exchange cross section is concentrated in a region centered around 35–40 MeV energy loss. The target dependence of this effect is demonstrated in Fig. 1; the energy dependence is shown in Fig. 2. In the right panel of Fig. 2 is a comparison of $0°$ (p,n) Pb spectra taken at 200 MeV and 795 MeV. The 200 MeV data have been folded with a gaussian function to simulate the poorer resolution of the 795 MeV data.

There are several possible causes of this new structure. A similar feature has been observed in $0°$ cross section measurements using the $(^3He,t)$ reaction at 300 MeV/nucleon (4). In Ref. (1), this excess cross section has been attributed to the excitation of the isovector monopole.

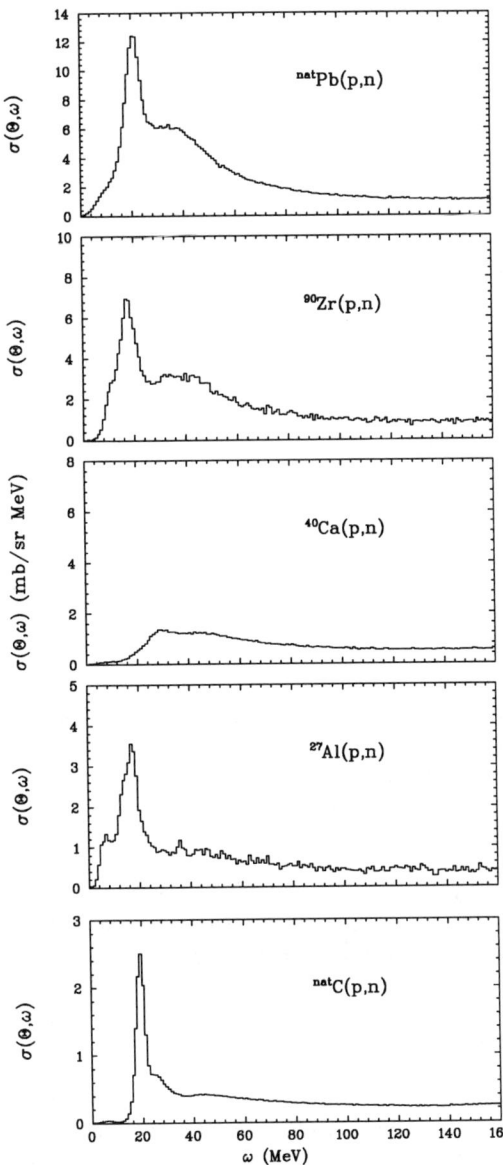

Fig. 1.— Target dependence of 0° (p,n) spectra at 795 MeV. The light targets are dominated by the GT resonance. For both Pb and Zr a bump extending to 50 MeV energy loss appears. This is absent in 200-MeV data.

472 Spin Observable Measurements on Pb

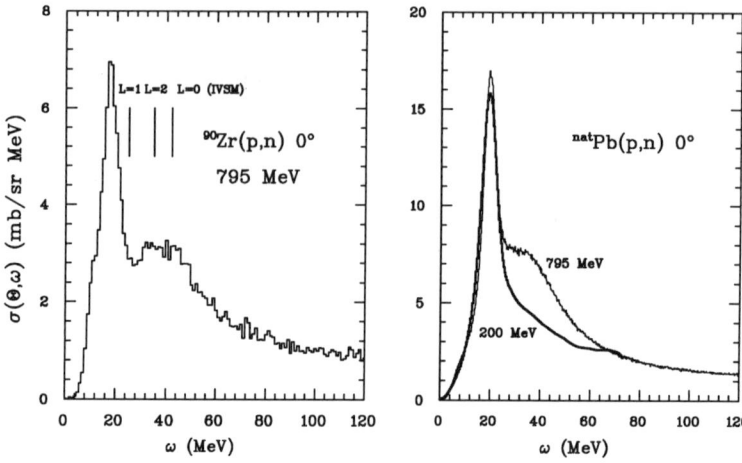

Fig. 2.— In the left panel are cross section data from the Zr(p,n) reaction at 795 MeV and 0° while the right panel shows Pb data at 200 and 795 MeV. The narrow peak in both sets of data is the GT resonance. The strength present at 40 MeV energy loss present in both sets of 795 MeV data is the region of interest.

The monopole is associated with the expansion and compression of the nucleus and occur with or withour flipping the nucleon spin. It involves no orbital angular momentum change and is thus analogous to the Fermi and GT transitions. Since the monopole is a $2\hbar\omega$ excitation it will appear at much higher energy loss than low lying Fermi or GT states: at around 35 MeV energy loss.

Charge exchange reactions at these energies have been singled out as possible means of exciting the monopole resonance. In Ref. (5) it is explained that surface peaked probes will preferentially excite the monopole. Because of the large reaction cross section at 800 MeV, the proton becomes much more strongly absorbed at 800 MeV than at 200 MeV. Thus, the (p,n) reaction becomes more surfaced peaked at this higher energy. At the same time the ratio of the spinnonflip to spinflip nucleon-nucleon (NN) cross section increases favoring S=0 type excitations. Evidence for the existance of such S=0 monopole excitations has been obtained in pion-charge-exchange experiments (6).

Another explanation for this increased cross section can be inferred from the Zr data in the left panel of Fig. 2. The expected position of three multipoles is shown above the excess cross section in Zr (7). While the L=1 excitation is at too low excitation energy to contribute, the quadrupole excitation could contribute if distortion effects were sufficiently strong (8).

A third explanation for the source of this cross section is kinematic. The high lying GT strength predicted by the DWIA model at 200 MeV, should be compressed into a smaller region of energy loss as the proton energy increases (9).

In order to further investigate this structure and to help discriminate among the above possibilities we have measured a complete set of spin transfer observables on Pb using the $(\vec{p}, \vec{n})$ reaction at 0°.

Experiment

Complete sets of 0° polarization transfer data were taken on the targets ^{nat}Pb, ^{nat}C and CD2 at the Neutron Time of Flight (NTOF) facility at the Los Alamos Meson Physics Facility (LAMPF) (10). The flight path was 200m and beam currents ranged from 20-40 na. Typical proton polarizations were 0.65. This quantity was monitored continuously by two beamline polarimeters. Absolute neutron cross sections were determined by normalizing to the 0° cross section for the ^{7}Li$(p,n)^7$Be(g.s. +0.43) reaction (27 mb/sr in center of mass). Systematic uncertainties resulting from beam current and target thickness are estimated to be < 8%. A 3% uncertainty has been associated with the absolute normalization to the ^{7}Li$(p,n)^7$Be reaction (10).

Results

The partial cross sections are defined with respect to a set of center-of-mass unit vectors and correspond to the cross section for flipping the projectile nucleon spin along each of these unit vectors. The unit vector $\hat{n}$ is normal to the scattering plane, $\hat{q}$ is in the direction of momentum transfer and $\hat{p} = \hat{n} \times \hat{p}$.

At 0° the following simple relationship holds between the spin transfer observables and the partial cross section for flipping spin along these three unit vectors:

$$\sigma_0 = \frac{1}{4}\sigma_{tot}(1 + 2D_{NN} + D_{LL}) \qquad (1)$$

$$\sigma_q = \frac{1}{4}\sigma_{tot}(1 - 2D_{NN} + D_{LL}) \qquad (2)$$

$$\sigma_{n,p} = \frac{1}{4}\sigma_{tot}(1 - D_{LL}) \qquad (3)$$

Where σ_0 is the nonspinflip cross section, σ_q is the spin-longitudinal cross section and $\sigma_{n,p}$ correspond to the two spin-transverse cross sections, which are equal at 0°. For unnatural parity transitions σ_0 is zero. We can check this with our ^{2}H data. The large cross section for the final state interaction in this target at low energy loss is entirely (L=1,S=1), since the nonspinflip interaction is Pauli blocked at 0°. Thus, this transition also provides a nominal GT benchmark against which we can compare the Pb data. For this target we have measured $D_{NN}= -0.127 \pm 0.02$ and $D_{LL}= -0.76 \pm 0.036$. Within uncertainties the value of σ_0 is zero. The difference between D_{NN} and D_{LL} is a signal of a large exchange tensor amplitude at this energy.

The results for Pb are shown in Fig. 3. The bottom panel shows the spin observables, while the top panel shows the decomposition into the spin components of expressions (1–3). In the bottom panel, the vertical and horizontal lines correspond to the transverse pieces (S=1) and the slanted lines are the longitudinal (S=1) part. The S=0 strength is shown in white.

The spin observables show little variation over the entire region. The value of D_{LL} is very close to that for ^{2}H, while D_{NN} is slightly more positive indicating there is some component of natural parity S=0 strength. The extra cross section at 35 MeV appears to be predominantly a S=1 excitation as is the GT resonance. This similarity to the GT excitation suggests this extra strength is an unnatural parity resonance. These data rule out the possibility that this extra strength is the nonspinflip isovector monopole. This cross section could still contain L=0, S=1 strength corresponding to either high lying GT or to the spin–monopole;however, in order to distinguish these possibilities from the L=2,S=1 spin–quadrupole more information is necessary. Measurements of an angular distribution of the cross section and spin observables for this reaction is planned at a lower energy. Futher analysis of these data is underway.

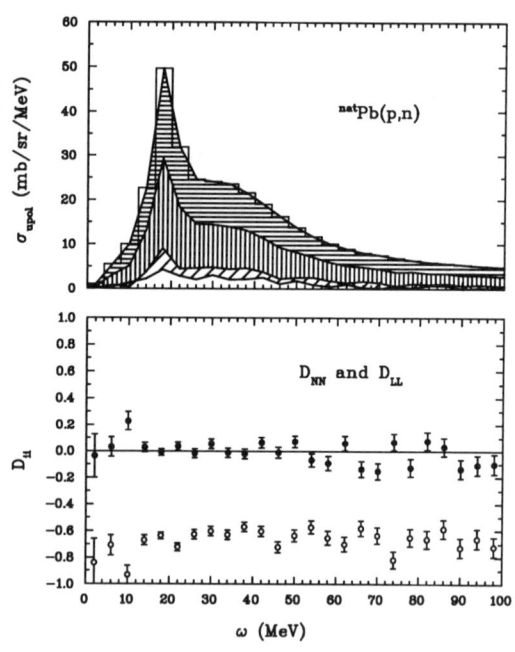

Fig. 3.— The top panel shows the ^{nat}Pb 0° cross section data from experiment 1294. The partial cross sections for the two transverse (horizontal and vertical lines), longitudinal (diagonal lines) and S=0 (white) are shown. The bottom panel shows the D_{NN} (solid) and D_{LL} (open) results that were used to decompose the cross section. This shows that the strength in the bump at 30–40 MeV energy loss is predominantly of unnatural parity.

REFERENCES

(1) F. Osterfeld, Rev. of Modern Physics **64**, 491 (1992).

(2) F. Osterfeld, D. Cha and J. Speth Phys. Rev. C **31**, 372 (1985).

(3) D. Cha, F. Osterfeld, Phys. Rev. C **39**, 694 (1989).

(4) C. Gaarde Nucl. Phys. **A369**, 258 (1981).

(5) N. Auerbach *et al.* Phys. Rev. C **28**, 280 (1983).

(6) J.D. Bowman *et al.* Phys. Rev. Lett. **50**, 1195 (1983).

(7) N. Auerbach, A. Klein Nucl. Phys. **A395**, 77 (1983).

(8) A. Brockstedt *et al.* Nucl. Phys. **A530**, 571 (1991).

(9) G. Love, Presentation at LAMPF workshop circ. 1985 (unpublished). See also G. Love, Can. J. of Physics **65** 543 (1987).

(10) X. Y. Chen *et al.*, Phys. Rev. C **47**, 2159 (1993).

Polarization Observables in $\vec{p}p \rightarrow pK\vec{Y}$ Reactions at 2.9 GeV

J. Arvieux,[d] F. Balestra,[b] Y. Bedfer,[d] R. Bertini,[b,d] L.C. Bland,[a]
S. Bossolasco,[b] F. Brochard,[d] M.P. Bussa,[b] I.V. Falomkin,[c]
L. Fava,[b] L. Ferrero,[b] R. Garfagnini,[b] D.R. Gill,[d] A. Grasso,[b]
W.W. Jacobs,[a] V.I. Lyascenko,[c] A. Maggiora,[b] D. Panzieri,[b]
G. Piragino,[b] G.B. Pontecorvo,[c] V. Serdyuk,[c] F. Tosello,[b]
V.I. Travkin,[c] S.E. Vigdor,[a] B. Zalikanov,[c] G. Zosi[b]

[a] *Indiana University and Indiana University Cyclotron Facility, USA*
[b] *Istituto di Fisica "A. Avogadro" and INFN, Torino, Italy*
[c] *JINR, Dubna, Russia*
[d] *Laboratoire National Saturne, Saclay, France*

Abstract. A program of exclusive measurements involving spin observables for $\vec{p}p \rightarrow pK\vec{Y}$ is nearing production phase. A brief description of the apparatus and its present status is given with emphasis on simulations relevant to the extraction of polarization components of $\vec{\Lambda}$ and $\vec{\Sigma}^0$.

INTRODUCTION

The Dubna-Indiana-Saclay-Torino (DISTO) collaboration is preparing to measure polarization observables in kinematically complete studies of hyperon production at the Saturne synchrotron in Saclay (1). These first <u>exclusive</u> polarization measurements at 2.9 GeV should help pin down the mechanism for associated strangeness production in proton-induced reactions within the meson-exchange regime, and may lead the way to analogous coincidence experiments at higher energies where a parton description holds sway. Our measurement of D_{NN}, simultaneously for $\vec{\Lambda}$ and $\vec{\Sigma}^0$ production, is expected to provide an especially strong constraint on various models (1,2).

DISTO APPARATUS AND PHASE SPACE SIMULATIONS

The DISTO apparatus (see Fig. 1) is designed to track the four charged products signaling Λ and Σ^0 production through a strong magnetic field: $\vec{p}p \rightarrow pK^+\vec{\Lambda}(\vec{\Sigma}^0) \rightarrow pK^+p\pi^-$. For $\vec{\Sigma}^0$ production, only the photon from the (100% branch) $\Sigma^0 \rightarrow \Lambda + \gamma$ decay is missing. A hardware (Level 1) trigger selects events with at least four charged prongs within the detector accep-

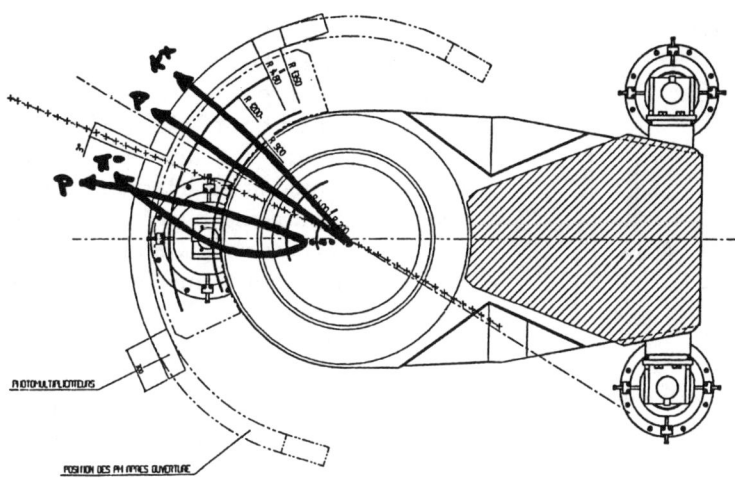

Figure 1. Dectector array for $\bar{p}p \to pK^+\vec{Y}$ shown in plan view for typical event.

tance, while a software (Level 2) trigger will search for the spatial separation between reaction and decay vertices characteristic of the weak ($\tau \sim 10^{-10}$ s) Λ decay. The tracking detectors comprise two left-right pairs of scintillating fiber chambers embedded in the 14.7 kG field ($y-u-v$ planes of 1 mm square fibers); similarly two pairs of $x-u-v$ multiwire proportional chambers will be mounted at the edge of the magnet. The detector arrays are designed to cover $\pm$ 15.5° vertically (the S170 magnet gap limit is $\pm$ 21°) and extend horizontally out to 45° on either side of the curved beam trajectory. Finally (located radially $\simeq$ 140 cm from the 2-cm thick liquid hydrogen target), a scintillator hodoscope records particle multiplicities, allows p vs K^+ time-of-flight particle identification, and provides a sample of $\bar{p}p$ scattering events to monitor the beam polarization.

We have carried out phase space simulations for both the pp $\to$ pK$^+\Lambda$ and pp $\to$ pK$^+\Sigma^0$ reactions at a bombarding energy of 2.9 GeV using the code GEANT. The subsequent re-tracking of particle trajectories was performed with HYPPO, a code written expressly for the DISTO experimental setup. At present, HYPPO is limited to analyze only events where the p and K are on one detector arm with the Λ or Σ^0 decay products on the other, resulting in reconstruction efficiencies of only about 10% (a more general version is being written). The trigger used in the simulations (and planned for the experiment) was quite general: it required at least 2 hits (summed over left + right) within the detector 0 (1st scintillating fiber chamber) acceptance, at least 4 hits at detector 1 (2nd scintillating fiber chamber), and at least 3 hits at the rear hodoscope. The results of these simulations have been used to consider several aspects of the experiment. We have found that our $\pm$ 15° vertical coverage, for example, somewhat reduces the acceptance for non-coplanar events (e.g., those falling near the center of the Dalitz plot of $M^2_{K-\Sigma^0}$

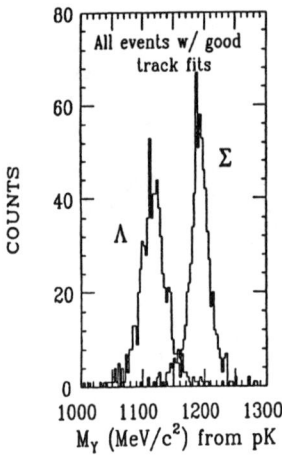

Figure 2. Missing mass.

vs. $M^2_{p-\Sigma^0}$), resulting in some loss of information concerning the reaction mechanism. On the other hand, the coverage is still complete enough to get significant out-of-plane information and to be able to clearly see the effects of baryon resonances and final state interactions between the proton and outgoing hyperon. In addition we considered hyperon transverse momentum distributions (some loss at medium p_T), and instrumental asymmetries and statistical precisions attainable in the measurement of Λ and Σ^0 polarization components (see details below). With this apparatus, particle momenta will be determined for the first time with sufficient resolution (see simulated results in Fig. 2) to separate Λ from Σ^0 production in the missing mass spectrum reconstructed at the p - K^+ vertex.

EXTRACTION OF POLARIZATION OBSERVABLES (P^N_Λ)

We have simulated event distributions relevant to the determination of the normal [parallel to $\hat{n} \equiv (\vec{k}_{inc} \times \vec{k}_Y)/|\vec{k}_{inc} \times \vec{k}_Y|$], as well as other polarization components of the Λ (the actual simulation was for Σ^0 production; results obtained for the decay Λ's produced are the same). The Λ weak decay into a proton and π^- with a 64.1% branch can be used to extract the normal polarization from the (parity-violating) asymmetry of the decay protons having emission angle θ^* with respect to $\hat{n}$ in the Λ rest frame. The differential form of this distribution is proportional to $1 + \alpha P_\Lambda cos(\theta^*)$; for $p\pi^-$, $\alpha = 0.642$. (In general, the different polarization components can be deduced from the fore-aft asymmetry in the distribution of these events vs. $cos(\theta^*)$ with respect to any spin quantization axis of interest.) For example, if we bin the data in $x \equiv |\cos\theta^*|$, then values of the polarization from bin pairs can be extracted:

$$P_{\Lambda_i} = \frac{1}{\alpha x_i} \cdot \left[\frac{\left[\frac{N(+x_i)}{N_0(+x_i)} - \frac{N(-x_i)}{N_0(-x_i)}\right]}{\left[\frac{N(+x_i)}{N_0(+x_i)} + \frac{N(-x_i)}{N_0(-x_i)}\right]} \right], \quad (1)$$

where N (N_0) represents the number of events observed in the corresponding bin with polarized (unpolarized) Λ's decaying. If there is no instrumental asymmetry, then $N_0(+x) = N_0(-x)$ for each bin.

In Fig. 3 are shown simulated results relevant to the normal Λ polarization, P^N_Λ. Our ability to determine the decay angle in the Λ rest frame is exhibited in Fig. 3a where the distribution of all reconstructed events with respect to the decay angle error (i.e., the difference between the reconstructed and generated values, $\delta\theta^*$), indicates a FWHM of $\simeq 3.0°$. This angular resolution is good enough that it should have no significant effect on the binning of the data. In Fig. 3b the distribution of the simulated (unpolarized) Λ decays is shown with respect to $\cos\theta^*$ for both the generated events (those passing the Level 1 trigger) and the reconstructed events (as mentioned earlier, only a

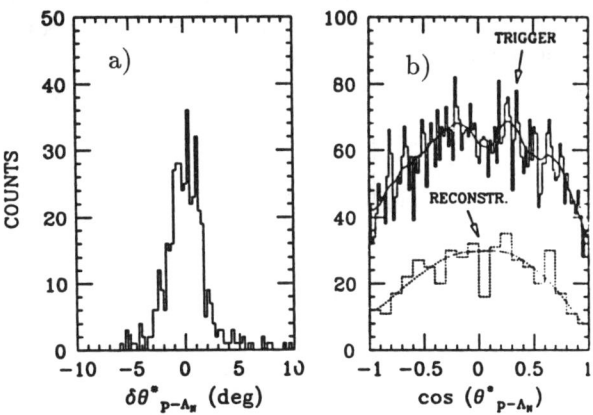

Figure 3. Simulated results for P_Λ^N; a) error $\delta\theta^*$, in decay angle of protons; b) distribution vs. $\cos\theta^*$ for trigger and reconstructed events (smoothed lines are averages of binned data).

fraction of generated events). The falloff near $\cos\theta^* = \pm 1$, most pronounced in the reconstructed sample, corresponds to a reduced probability of detecting upward or downward emitted π's with respect to a vertical Λ polarization axis. (Simulations with increased detector acceptance indicate that these latter events are the most difficult to reconstruct and thus don't effectively contribute to the figure of merit.) For our present geometry, both distributions in Fig. 3b are symmetric around $\cos\theta^*$ revealing zero instrumental asymmetry within the statistical precision of the simulation ($< \cos\theta^* > = -0.012 \pm 0.023$ for the reconstructed events). Hence, P_Λ may be determined with little systematic error from an error-weighted average of the independent P_{Λ_i} fore-aft asymmetries obtained according to the prescription in Eq. (1).

Σ^0 POLARIZATION AND NON-NORMAL COMPONENTS

The missing mass resolution we expect to obtain will allow us to separate Λ from Σ^0 production, and we have described above the procedure for extracting the Λ polarization. A measurement of the polarization for Σ^0, however, is significantly more complicated because one has to infer the polarization of the Σ^0 from a measurement of the polarization of the daughter Λ. Simulations relating to the extraction of the Σ^0 polarization normal component are displayed in Fig. 4. We note first, that the polarization transfer in the electromagnetic decay $\vec{\Sigma}^0 \rightarrow \vec{\Lambda} + \gamma$, depends (3) on the Λ's emission direction ($\hat{p}_{\Lambda-\Sigma^0}$) in the Σ^0 rest frame: $\vec{P}_\Lambda = -(\vec{P}_\Sigma^0 \cdot \hat{p}_{\Lambda-\Sigma^0})\hat{p}_{\Lambda-\Sigma^0}$. That is, the daughter

Λ is always longitudinally polarized, as viewed from the Σ^0 rest frame with a sign <u>opposite</u> and magnitude <u>reduced</u> by the projection of the parent Σ^0's polarization along the Λ emission direction. The distribution of extracted Λ emission angle errors, measured with respect to the Σ^0 spin quantization axis (e.g., parallel to $\hat{k}_{beam} \times \hat{k}_{\Sigma^0}$) is shown in Fig. 4a. In contrast to Fig. 3a, and as reflected in the wider distribution, this quantity is inferred from two separate trajectories: that of the Λ, reconstructed from p and π^-, and that of the Σ^0, reconstructed from the primary reaction products p and K$^+$. The distribution of Λ's with respect to $\hat{k}_{beam} \times \hat{k}_{\Sigma^0}$ is displayed in Fig. 4b.

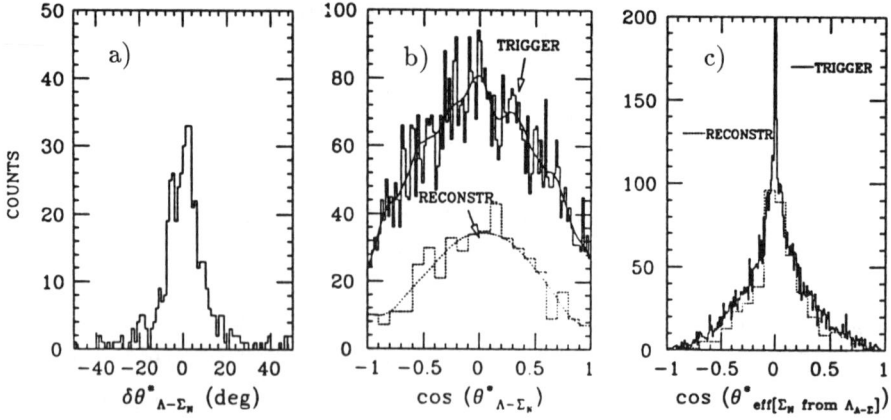

Figure 4. Simulated results for $P_{\Sigma^0}^N$; a) error $\delta\theta^*_{\Lambda-\Sigma_N^0}$ in deduced decay angle of Λ's; b) Λ distribution vs. $\cos\theta^*_{\Lambda-\Sigma_N^0}$ for trigger and reconstructed events; c) Λ decay proton distribution (lab) vs. the product of two cosines, $\cos\theta^*_{eff}$, as described in the text. Smoothed lines are averages of binned data.

In the laboratory frame, the daughter Λ is no longer purely longitudinally polarized because its momentum vector is rotated by the Lorentz transformation from the Σ^0 rest frame to the lab frame. However, the Λ decay asymmetry still depends on the cosine of the angle between the decay proton direction (in the Λ's rest frame) and the Λ's spin direction (now that of the Λ momentum in the Σ^0 rest frame). Thus, the relevant distribution of events for deducing the Σ^0 polarization is with respect to the <u>product of two independent $\cos\theta^*$ factors</u>, one describing the Λ's emission in the Σ^0 rest frame and the other the proton's emission in the Λ^0 rest frame:

$$\cos\theta^*_{eff} \equiv \cos(\theta^*_{\Lambda-\Sigma^0}) \cdot \cos(\theta^*_{p-\Lambda_{\Sigma^0}}). \tag{3}$$

The distributions with respect to $\cos\theta^*_{eff}$ displayed in Fig. 4c show an acceptance sharply peaked around zero for this product (since the two cosine factors are essentially independent, their product is sharply peaked at $\cos\theta^*_{eff} = 0$), but again, with no instrumental asymmetry. The tails of this distribution (using Eq. (1)) should allow a reasonable measurement of $P_{\Sigma^0}^N$.

The extraction of <u>non-normal</u> components of Λ or Σ^0 is also possible with DISTO, but with added difficulty due to potential sytematic errors. For example, the decay proton distributions for both longitudinal and sideways Λ polarizations (equivalent to Fig. 3b, but vs. $\cos(\theta^*_{p-\Lambda_L})$ or $\cos(\theta^*_{p-\Lambda_S})$, show strong instrumental asymmetries for generated and reconstructed events. This asymmetry is unavoidable: it reflects the important difference in momenta between pions emitted along versus opposite the Λ momentum (e.g., at $\cos(\theta^*_{p-\Lambda_L}) \simeq \pm 1$). In the case of the sideways polarization, the problem arises from trying to extract decay particle asymmetries within the bend plane of the magnet with a finite acceptance detector. It is possible, nonetheless, to extract useful information about P^L_Λ and P^S_Λ from the experiment, by "calibrating" the instrumental asymmetry with longitudinally/sideways unpolarized Λ's that must be produced (by parity conservation) by averaging over all orientations of the plane formed by the coincident proton and kaon momentum vectors. The analysis of such a "calibration sample" will give the values $N_0(\pm x)$ appearing in Eq. (1). For the two-cosine product distributions relevant to extracting polarization information of the Σ^0, the distribution for $P^S_{\Sigma^0}$ (like $P^N_{\Sigma^0}$ discussed earlier) exhibits zero instrumental asymmetry, thus simplifying its measurement. The latter results from the longitudinal preference in the Λ decay being independent of whether the Λ is emitted upward or downward (or leftward or rightward) in the Σ^0 rest frame.

DISTO TIME SCALE

The DISTO beamline, liquid hydrogen target, and scintillator hodoscope were commissioned this past year. With all SciFi chambers completed and most multi-anode photomultiplier tubes (PSPM) available this fall ('94), we will test the Sci-Fi/Hodoscope level 1 logic rates and initial level 2 software. In spring '95, completed multiwire chambers and all PSPM's on hand will enable beam tests of the complete DISTO detector assembly, level 2 trigger and track reconstuction software. Production running pp $\rightarrow$ pK$^+\vec{Y}$ (and addition of Cernekov detectors for pp $\rightarrow$ ppϕ tests) is forseen for summer '95.

From our simulations we estimate that in one day of running, we should get within each of 10 equally populated bins (e.g., subdividing the full transverse momentum transfer range) a statistical precision of ± 0.030 for the normal polarization component of direct Λ production and ± 0.097 (± 0.064) for the normal (sideways) component of Σ^0 polarization in pp $\rightarrow$ pK$^+\vec{\Sigma}^0$.

REFERENCES

1. Maggiora, A.,*et al.*, in *Workshop on Flavour and Spin in Hadronic and Electromagnetic Interactions*, Torino, Sept. 1992 (Italian Phys. Soc., Bologna, 1993) p. 111; Saturne proposal and experiment E213, R. Bertini, Spokesman.
2. Vigdor, S.E., proceedings cited in Ref. 1, pg. 317, and references therein.
3. Gatto, R., *Phys. Rev.* **109**, 610 (1957).

INTERMEDIATE ENERGY HADRON-INDUCED REACTIONS

H. Sakai
Department of Physics, University of Tokyo, Bunkyo, Tokyo 113, Japan

Abstract. The papers presented orally in the Session on "Intermediate Energy Hadron-Induced Reactions" are discussed.

1 Introduction

There were 26 contributions to this session "Intermediate Energy Hadron-Induced Reactions" 13 of which have been presented orally. Among them, about a half of the talks were given by graduate students or by just graduated young researchers. This might be taken as a good sign that this field is still attractive for young students which is a condition that we should try to maintain as long as possible.

It is certainly unfeasible to summarize or draw conclusions on all 26 contributions. Therefore I would like to discuss recent progress illustrated in a few results from 13 selected talks. I have to apologize for not being able to mention all contributions even though they were all high quality.

First of all there was only one theoretical talk given by Gary Love[1]. He presented PWIA formulae for the $\vec{p} + \vec{A}$ reaction. He showed that the spin correlation C_{ij} gives complementary and equally robust information to the polarization transfer D_{ij}. Since the progress in the technique for making polarized targets is quite impressive as was shown in several sessions during this conference, this theoretical approach should soon become increasingly valuable and useful in C_{ij} analyses.

Let me highlight some of the experimental talks by arbitrarily subdividing them into the following five categories.

2 The Polarized Deuteron as a New Spectroscopic Tool

Two talks were given on the new spectroscopic tools which exploit the characteristics of deuteron isospin and spin structure of T=0 and S=1.

2.1 Isoscalar Spinflip States via the $(\vec{d}, \vec{d'})$ Reaction

It is very difficult to get information on the isoscalar spinflip states. This is probably due to two reasons. Firstly the effective interaction V_σ is weak compared with other central interactions V_0, V_τ, and $V_{\sigma\tau}$. Secondly there has been no efficient probe. Recently a method to study the isoscalar spinflip states has been developed by employing the vector polarization transfer $K_y^{y'}$ for the $(\vec{d}, \vec{d'})$ reaction. The exact single spinflip probability can be defined as

$$S_1 = \frac{1}{9}(4 - P^{y'y'} - A_{yy} - 2K^{y'y'}_{yy}) \tag{1}$$

which is almost equivalent to S_{NN} ($=\frac{1}{2}(1-D_{NN})$) for the nucleon-nucleon scattering like (p,p') or (p,n). To extract S_1 one has to measure the tensor polarizations ($P^{y'y'}$ and $K^{y'y'}_{yy}$) of the outgoing deuteron which are very difficult to measure mainly due to the lack of efficient tensor polarimeters. Therefore a signature with properties similar to S_1 has been sought. It is proposed the quantity

$$S_d^y = \frac{4}{3} + \frac{2}{3}A_{yy} - 2K_y^{y'}, \tag{2}$$

instead of, but similar to S_1. Here one needs only the vector polarimeter aside from the tensor analyzing power A_{yy} which is easy to measure. The sensitivity of S_d^y has been tested for the ^{12}C$(\vec{d},\vec{d'})^{12}$C reaction at 400 MeV[2] by using the famous 1^+ (T=0 and S=1) state at E_x= 12.71 MeV. It has been found that $S_d^y > 0$ for S=1 and $S_d^y \simeq 0$ for S=0, respectively.

Djalali reported the S_d^y measurement for the highly excited continuum region in ^{12}C and ^{40}Ca[3]. In ^{40}Ca, isoscalar spinflip strengths were found at about 9 MeV excitation energy as a bump and at around 15 MeV as a broad bump. They found a new narrow isoscalar 1^+ state in ^{12}C at an excitation energy of 20.5 MeV, and in addition they identified two broad peaks at around 20 and 30 MeV excitation energies which have angular distributions which are characteristic of L=1. It is very surprising that such a narrow state was previously unknown in such a well studied nucleus as ^{12}C. This clearly shows the usefulness of S_d^y in the study of isoscalar spinflip states.

I would like to point out that the deuteron tensor polarimeters used at intermediate energies have been reported in a parallel session[4, 5]. Therefore the exact spinflip probability S_1 will soon be measured. Note that the exact S_1 measurement utilizing a different techniques has been reported[6].

2.2 Total Spin J Assignment via the $(\vec{d},^2\text{He})$ Reaction

The $(\vec{d},^2\text{He})$ reaction is one of the most selective reaction probes which excites exclusively the spin- and isospin-flipped states. Here the ^2He denotes the two–proton system coupled to the $^1S_0$ state. Bugg and Wilkin[7] have suggested in terms of a PWIA frame work that the tensor analyzing powers A_{xx} and A_{yy} might be useful in assigning the total spin value J. The best application may be found in the spin-dipole states whose spins are 0^-, 1^-, and 2^-. The 0^- state should have extreme values of $A_{yy} = +1$ and $A_{xx} = -2$ while the 1^- state is expected to have $A_{xx} = +1$. Thus the analyzing power measurement for the $(\vec{d},^2\text{He})$ reaction gives almost equivalent information as the polarization transfer measurement for the $(\vec{n},\vec{p})$ reaction which is not feasible at present. The former is a single scattering experiment with polarized deuteron beam and the latter is a "triple" scattering experiment, first to produce the polarized neutron, then to make the (n,p) reaction and finally to measure proton polarization.

Okamura[8] presented the first analyzing power measurement on the $^{12}\text{C}(\vec{d},^2\text{He})^{12}\text{B}$ at 270MeV at RIKEN. The overall energy resolution of 460 keV was achieved by using the spectrometer SMART. The discrete 1^- state at $E_x = 2.62$ MeV indeed shows the expected signature of $A_{xx} \simeq +1$. What is surprising is that the A_{xx} value of the broad peak at around $E_x = 7.5$ MeV is almost equivalent to that of the 2^- peak at 4.5 MeV. This result is rather amazing since the peak at 7.5 MeV was previously believed to be mostly due to the 1^- states[9], although such an assignment was mostly guided by shell model predictions[10].

3 Transverse Polarization Transfer D_{NN}

The measurement of the transverse polarization transfer coefficient D_{NN} has become a common tool for intermediate energy nuclear physics. Here we heard two talks which utilize the D_{NN} value to clarify the isospin mixing or the tensor interaction at high momentum transfer.

3.1 Isospin Mixing

Study of the isospin mixing gives information on the isospin breaking force as well as on its mixing mechanism. However such studies are rather limited because of the lack of an appropriate spectroscopic tool. Liu presented that D_{NN} could serve as a good indicator of the isospin mixing[11].

Inelastic spectra, analyzing powers, and D_{NN} by the $^{28}\text{Si}(\vec{p},\vec{p'})^{28}\text{Si}$ reaction at at 200 MeV have been measured at IUCF by using the K600 spectrometer with a focal plane polarimeter. In the excitation energy between about 15 and 19 MeV, six 6^- states have been newly identified on the basis of angular distributions and analyzing powers. Referring to the D_{NN} values for the known isospin states : $D_{NN} \sim 0.75$ for the T=0 isospin at 11.58 MeV and $D_{NN} \sim 0$ for the T=1 isospin at 14.36 MeV, the 6^- state at 17.6 MeV is identified as T=1 and other 5 states are mostly T=0 or mixed. Clear identification may not always be possible due to the angular dependence of D_{NN}.

It is thus demonstrated that D_{NN} could be a good tool for identifying the isospin. More examples are obviously needed to establish the present method as a quantitative tool to determine a mixing ratio.

3.2 Isovector Tensor Interaction

Systematic study on the $D_{NN}(0°)$ has been performed by Taddeucci et al. at IUCF for the bombarding energies between 120–200 MeV[12]. The average D_{NN} value obtained is $-\frac{1}{3}$ for the GT transitions. This is consistent with the simple PWIA prediction using only a central interaction.

Wakasa[13] reported the $D_{NN}(0°)$ measurements for the (p,n) reaction on p-shell nuclei at 300 MeV. Experiments were carried out by using the neutron polarimeter NPOL at RCNP. Measured $D_{NN}(0°)$ values for the Gamow-Teller(GT) transition are on the average about -0.28 which clearly deviates from the $-\frac{1}{3}$ which is obtained at lower bombarding energies. This indicates the contribution from the non-central interaction, namely a tensor interaction. One might wonder why the tensor interaction contributes at zero degrees. It is due to the knockon–exchange amplitudes.

The DWIA calculations with the Franey and Love (FL) 270 MeV interaction reproduce fairly well both the differential cross sections and the $D_{NN}(0°)$ value for the ^{6}Li$(\vec{p},\vec{n})^6$Li reaction, while the calculations with the FL 325 MeV interaction yields a less negative $D_{NN}(0°)$ value than the experimental value. It is attributed to the strength of the exchange interactions. They found that the exchange tensor part of the FL 325 MeV interaction is too strong at large momentum transfer (Q ~ 3.4 fm^{-1}) compared to that of the FL 270 MeV interaction. This result is very interesting since the $D_{NN}(0°)$ value might give information on the isovector tensor interaction at large momentum transfer region which is usually very difficult to assess.

4 New Isovector Spin Resonance ?

David Prout reported on the excess yield observed at around 30-40 MeV excitation energy by the (p,n) reaction at 800 MeV at zero degrees[14]. It was measured by using NTOF at LAMPF.

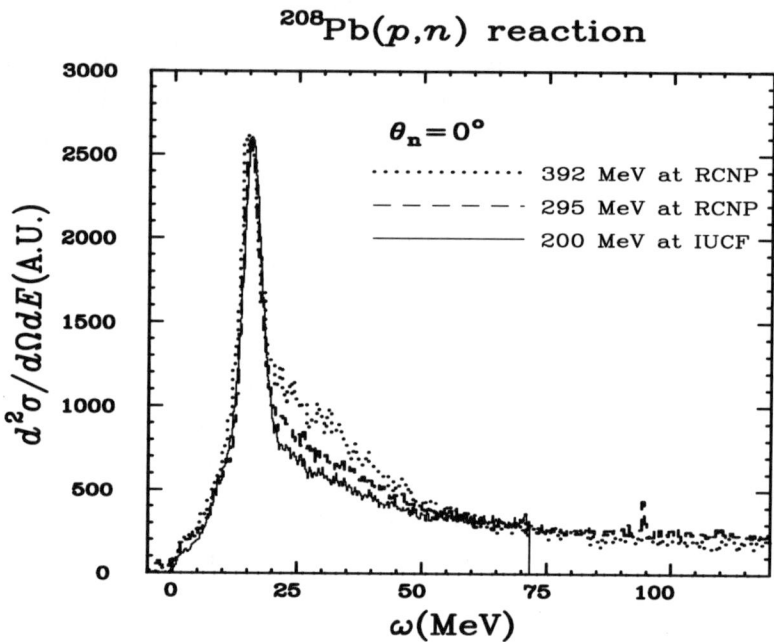

Figure 1: Energy spectra for the ^{208}Pb(p,n) reaction at 0° at 200, 300, and 400 MeV.

Extensive (p,n) measurements have been carried out using NTOF at IUCF and a lot of interesting physics on spin-isospin excitations such as the observation of Gamow-Teller giant resonances and the quenching at their strengths has come from Bloomington. Therefore one might assume that there is little to observe, particularly at small momentum transfer. Note most of this work has been done using proton energies of 120–200 MeV. Nature seems to be still kind to us since a new spin excitation mode has been observed at small momentum transfer.

The excess yield at an excitation energy region of 30–40 MeV is clearly observed in the spectra for the ^{208}Pb(p,n) reaction at 0° of 800 MeV at LAMPF[14] (Spectra are not shown here). See ref. [14]. The measurements were made on various mass targets and the strong excess yield seems to persistently exist in heavy nuclei beyond ^{90}Zr. In order to confirm whether or not this excess yield is a spin excitation, the $D_{LL}(0°)$ and $D_{NN}(0°)$ for the ^{208}Pb(p,n) reaction have been measured at 800 MeV. It is found that the excess yield is dominated by spinflip strength. Thus this excess yield is due to an unnatural parity spin-flip transition. Theoretically an isovector spin monopole(IVSM) (L=0) and an

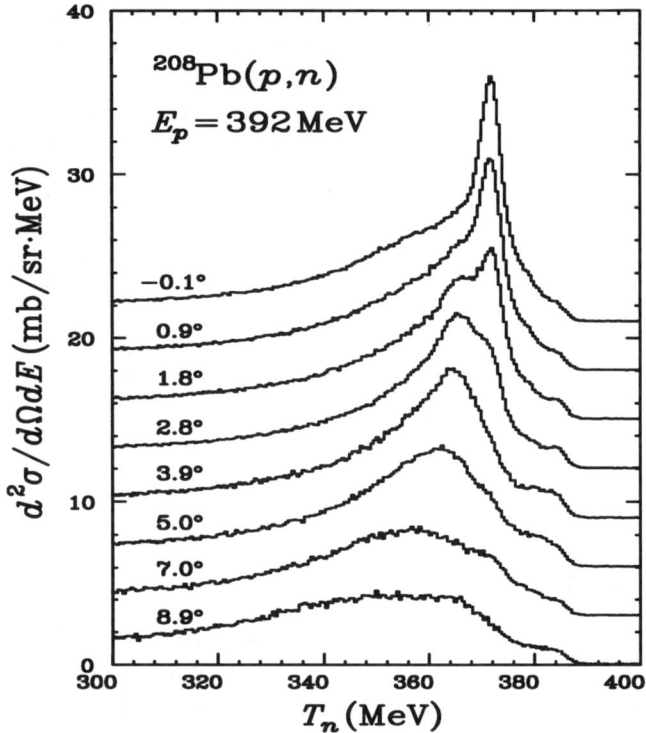

Figure 2: Angular spectra for the ^{208}Pb(p,n) reaction at 392 MeV.

isovector spin quadrupole(IVSQ) (L=2) resonances are predicted at this excitation energy region 30–40 MeV. In order to distinguish these modes an angular distribution measurement is needed.

Since the NTOF facility at LAMPF is no longer available, the measurements (Prout + Tokyo + RCNP collaboration) have been extended by using the NPOL facility at RCNP where the maximum proton energy of 400 MeV is available. The excess yield is also observed as shown in Fig.1, although it is not as prominent as that of 800 MeV. They are overlaid with different bombarding energies, 200 MeV at IUCF[12], and 295 and 392 MeV at RCNP[15]. Figure 2 shows the angular distributions for the ^{208}Pb(p,n) reaction at $E_p = 392$ MeV. It will take some time before getting a firm conclusion since a careful decomposition analysis is needed due to the overlapping resonances and also to the ambiguities of the form factors.

For a reliable analysis knowledge on the detailed spectroscopic information for the isovector spin-nonflip Fermi-type transition is also important. Available information is rather limited at intermediate energy since the (p,n) reaction populates spinflip states overwhelmingly compared with spin-nonflip states owing to the nature of the effective interactions, $|V_{\sigma\tau}| \gg |V_\tau|$. Cooper[16] reported the study in this direction : the Fermi(0^+) transition in ^{14}C$(\vec{p},n)^{14}$N at 495 MeV. It is primarily intended to resolve the failure of qualitative prediction on the analysing power of the mixed state (Fermi + Gamow-Teller) in ^{15}N$(p,n)^{15}$O(g.s.).

5 Study of Medium Modification via Polarization Observables

Although most studies in nuclear physics are related to medium modification at some level, I have selected the topics which might have to do with a relativistic treatment. Since the dramatic improvements in the quality of predictions made by the relativistic impulse approximation calculation for the spin rotation parameter in elastic scattering, a lot of experimental as well as theoretical studies have been made to clarify the relativistic effects. An effort in this direction was described by Tamii[17] who presented the spin rotation parameter measurement at 300 MeV by using a newly constructed focal plane polarimeter at RCNP.

At this time, to my knowledge, there is no consensus on whether the relativistic or non-relativistic is superior. One exception is a reduction of the analyzing power measured for the quasi-free $(\vec{p},p')$ scattering which can be explained only by the relativistic approach[18].

5.1 Study of the $(\vec{p},p'\gamma)$ Reaction

The scattering amplitudes which cannot be reached by the $(\vec{p},\vec{p'})$ reaction might be accessible through the $(\vec{p},p'\gamma)$ measurement. This is because the recoil nucleus which emits gamma-rays contains such information. A complete determination of the scattering matrix $\hat{T}$ is possible for the transition of $0^+ \to 1^+$ if we combine all $(\vec{p},p'\gamma)$ observables together with a complete set of $(\vec{p},\vec{p'})$ data[19].

Wells[20] presented the data of simultaneous measurements of $(\vec{p},\vec{p'})$ and $(\vec{p},p'\gamma)$ spin observables at 200 MeV for the 1^+, T=1 state at 15.11 MeV in ^{12}C. Measurements were carried out with the K600 spectrometer and with three large size BaF$_2$ detectors. In-plane and out-of-plane angular correlations were measured. They obtained 17 observables while 11 amplitudes are unknown in $\hat{T}$. Thus it is over-determined. They used extra observables to check internal consistency.

Measured spin observables are compared with the relativistic (DREX) and the non-relativistic (DW81) DWIA calculations. Neither calculations can describe all observables simultaneously. The coefficients B and C that characterize the in-plane p-γ angular correlation terms which are symmetric and anti-symmetric about θ_γ, respectively, show clear preference for the relativistic predictions. This tendency is also observed at an incident proton energy of 400 MeV[19].

Present measurement, in principle, is possible to determine all 11 unknown amplitudes of $\hat{T}$ in a model independent manner. This certainly puts strong constraint on both scattering and nuclear structure models. For example, amplitudes, $A_{Kq}(\hat{\Sigma} \cdot \hat{K})(\hat{\sigma} \cdot \hat{q})$ and $A_{qK}(\hat{\Sigma} \cdot \hat{q})(\hat{\sigma} \cdot \hat{K})$ are interesting since they are proportional to the composite spin-convection currents $\langle \hat{\sigma} \times \hat{J} \rangle$ and $\langle \hat{\sigma} \cdot \hat{J} \rangle$, respectively in a relativistic PWIA. $\hat{\Sigma}$ and $\hat{\sigma}$ are the polarization operator for the 1^+ recoiling nucleus and Pauli spin operator for the projectile proton, respectively. In practice, there remains still some ambiguity to fix all amplitudes. However, one of the scattering amplitudes $A_{qq}(\hat{\Sigma} \cdot \hat{q})(\hat{\sigma} \cdot \hat{q})$ has been uniquely determined. It shows a zero crossing at about $\theta_{CM} = 10°$ which is consistent with what is expected for the pion exchange amplitude. Therefore study of A_{qq} might be interesting in view of pionic correlations in nuclear medium.

5.2 Medium Modification through Quasi-Free Scattering

Edwards[21] presented inclusive quasi-free knockout measurement for the light nuclei ^{3,4}He and Carman[22] and Noro[23] talked about the exclusive quasi-free knockout measurements.

5.2.1 Inclusive Quasi-Free ^{3,4}He$(\vec{p}, n)$ Reactions

The ^{3,4}He$(\vec{p}, n)$ reactions at 200 MeV have been studied for $\theta_n = 0°$–$44°$ by using NTOF facility at IUCF by Edwards[21]. Small angle ^{3,4}He(p, n) spectra show no prominent peak indicating the Pauli blocking effect. At moderately large angles($\sim 20°$) a broad peak appears and at backward angles an energy spectrum becomes a rather flat continuum. These characteristic spectra are analyzed in terms of the Factorized DWIA (code TREEDEE by Chant and Roos[24])). It is found that the final-state-interaction(FSI) effects are important to reproduce the spectrum in respects of the reduction in cross section and of the shift in peak position in energy. The broad peaks in ^{3}He and ^{4}He are interpreted differently. That in ^{3}He is due to the quasi-free scattering and that in ^{4}He is due to the overlapping resonance states($0^-, 1^-, 2^-$). As far as the analyzing powers, the PWIA treatment seems to describe better the results than the DWIA treatment. However the spin-orbit distortions are important in explaining enhancement of A_y for ^{4}He$(\vec{p}, n)$ relative to that for $n(\vec{p}, n)$ and also of small angle A_y for ^{3}He$(\vec{p}, n)$.

5.2.2 Exclusive Quasi-Free ($\vec{p}$,Np) Reactions

Carman[22] presented very extensive measurements on the study of the quasifree ($\vec{p}$,Np) reaction on ^{2}H, ^{4}He, ^{12}C and ^{40}Ca targets at 200 MeV. Here N denotes either proton or neutron. For the $(p, 2p)$ measurement, the K600 spectrometer and the LSAA second arm detector at IUCF were used. An efficient neutron detector which utilizes n-p double scattering from the hydrogen nuclei of an organic scintillator has been developed to measure n-p coincidences for the (p, np) reaction. These detector arrangements have made it possible to cover a large portion of the phase space for the (p,Np) reactions. This study places more stress on the reaction mechanism.

Data have been compared with the DWIA(THREEDEE)[24] calculations. The triple differential cross sections for the ^{2}H,^{12}C(p,Np) reactions are generally reproduced. It is very surprising that the angle integrated cross sections for the associated particles agree with those of singles inclusive cross sections and also with calculations. This implies that the quasifree scattering yields are mainly due to the single-step scattering process at 200 MeV. Both the inclusive and angle integrated analyzing powers for the A(p, p') scattering show the suppression for A$\geq$ 4. The DWIA calculations are unable to account for this suppression. This might indicate a signature of relativistic dynamics. The angle integrated analyzing powers for the ^{2}H,^{12}C(p, np) reactions agree with those of free np scattering indicating no suppression. Note that the relativistic treatment[18] predicts no suppression for the (p, n) reaction.

There is a contribution by Otsu et al.[25] to this conference which reports the analyzing powers for the (p, n) quasifree scattering at 300 and 400 MeV. The reduction of analyzing power from the NN value is used to deduce a possible contribution from multiple scattering processes. According to that, the contribution from the multiple scattering is negligible at 400 MeV but there is a sizable contribution of 20-30% at momentum transfer $q \geq 1.8$ fm^{-1} at 300 MeV.

Noro[23] presented the high resolution study on the exclusive s$_{1/2}$ state knockout from the ^{12}C,^{40}Ca$(\vec{p}, 2p)$ reaction putting more emphasis on medium modification. The s$_{1/2}$ is chosen since there is no effective polarization effect and consequently the analyzing power can be directly compared with that of free pp-scattering although there exists small contribution by the LS potential. Measurements were made at RCNP by a high resolution spectrometer(Grand Raiden) and a broad range spectrometer(LAS) at 392 MeV under the geometrical condition of k$_3 \approx 0$ MeV/c where k$_3$ indicates the momentum of struck nucleon inside the nucleus. The overall energy resolution of about 350 keV was attained. The analyzing powers for the 1s$_{1/2}$ knockout from ^{12}C show a suppression as large as 50% of the free pp scattering. A striking contrast is seen in the analyzing powers for the 2s$_{1/2}$ knockout from ^{40}Ca. They show a very small suppression of about 10% of the free pp analyzing power values. The

THREEDEE calculations are able to reproduce the analyzing powers for ^{40}Ca but not for ^{12}C. The large suppression of ^{12}C can be explained by the relativistic PWIA calculation[18] by using the effective proton mass of $M^* = 0.82 M_p$ which corresponds the matter density of $0.42\rho_o$. Note the average density for the ^{40}Ca case is only $0.1\rho_o$ which is close to the free value.

6 Polarization in Hyperon Production

This somewhat different topic was reported by Will Jacobs[26]. He talked about the DISTO collaboration[27] which intends to measure

$$\vec{p}\, p \to p\, K\, \vec{Y} \qquad (3)$$

at 2.9 GeV. This is a kind of second generation *exclusive* experiments with more polarization observable measurements. The experiment is designed to have good separation between Λ and Σ. This experiment will tell us, for example, whether the reaction

$$\vec{p}\, p \to p\, K^+ \vec{\Lambda}(\vec{\Sigma^\circ}) \qquad (4)$$

proceeds through K^+ exchange or through π° exchange by measuring the polarization transfer coefficient D_{NN}. It can be predicted that $D_{NN} = -\frac{1}{3}$ for K^+ and $D_{NN} = 1$ for π°, respectively. The experiment is scheduled to run 1995 summer to 1996. We hope to hear the forthcoming results at the next Polarization Conference.

The author thanks M. Greenfield for reading the manuscript.

References

[1] W.G. Love, elsewhere in this proceedings.

[2] M. Morlet et al., Phys. Lett. **B247**, 228 (1990).

[3] C. Djalali et al., contribution to this conference, p111.

[4] S. Kox et al., contribution to this conference, p342.

[5] S. Ishida et al., contribution to this conference, p340.

[6] H. Sakai, Proceedings of Frontiers of High Energy Spin Physics, , ed. by T. Hasegawa et al., Universal Academy Press, Inc. Tokyo, 1993, p155. and S. Ishida et al., Phys. Lett. **B314**, 279 (1993).

[7] SD.V. Bugg and C. Wilkin, Nucl. Phys. **A467**, 575 (1987).

[8] H. Okamura et al., contribution to this conference, p139.

[9] For example, N. Olsson et al., Nucl. Phys. **A559**, 368 (1993).

[10] For example, D.J. Millener and D. Kurath, Nucl. Phys. **A255**, 315 (1975).

[11] J. Liu et al., contribution to this conference, p121.

[12] T.N. Taddeucci, Can. J. Phys. **65**, 557 (1987).

[13] T. Wakasa et al., contribution to this conference, p133.

[14] D. Prout et al., contribution to this conference, p137.

[15] D. Prout and H. Sakai et al., experiment in June, 1994 at RCNP. To be published.

[16] D.A. Cooper et al., contribution to this conference, p135.

[17] A. Tamii et al., contribution to this conference, p127.

[18] C.J. Horowitz and M.J. Roos, Phys. Rev. **C33**, 2059 (1986).

[19] J. Piekarewicz et al., Phys. Rev. **C41**, 2277 (1990).

[20] S.P. Wells et al., contribution to this conference, p115.

[21] C.M. Edwards et al., contribution to this conference, p131.

[22] D.S. Carman et al., contribution to this conference, p143.

[23] T. Noro et al., contribution to this conference, p145.

[24] N.S. Chant and P.G. Roos, Phys. Rev. **C15**, 57 (1077) and Phys. Rev. **C27**, 1060 (1983).

[25] H. Otsu et al., contribution to this conference, p149.

[26] J. Arvieux et al., contribution to this conference, p170.

[27] DISTO collaboration. E213 at SATURNE.

V. INTERMEDIATE-ENERGY ELECTROMAGNETIC INTERACTIONS

Spin Physics with an Intense CW Electron Beam in the 15–30 GeV Range: the ELFE Program

J.M. Laget

Commissariat à l'Energie Atomique
DAPNIA/SPhN, C.E.Saclay, F91191 Gif-sur-Yvette Cedex, France

INTRODUCTION

Hadronic Physics is a branch of physics which developed out of Nuclear Physics and Particle Physics. It aims at understanding the structure and the interactions of extended hadronic systems in terms of their constituents. The relevant constituents depend on the scale at which those systems are studied. At a scale of 1 fm, or larger, nuclei are made of baryons which interact by exchanging mesons. Several decades of works at energies up to 1 GeV have put this description on firm grounds. In contrast, it is known that at a scale much smaller than 0.1 fm the relevant degrees of freedom are the quarks which interact perturbatively by exchanging gluons. This has been beautifully shown by experiments performed at high energies. However the link between these two descriptions is still missing and little is known about the quark-gluon structure of hadrons. The central issue in Hadronic Physics is:

"How are hadrons and nuclei built of quarks and gluons?"

The answer to this question lies in the non-perturbative regime of QCD. One has to know how current quarks dress themselves into constituent quarks, which in turn form the observed baryons and mesons. This is a formidable challenge, which may require different and complementary approaches.

One of the most promising ways is the study of *exclusive* reactions induced, on nucleon and nuclei, by electrons and photons at high momentum transfer (in the range 10–20 GeV^2).

They select the simplest component (three valence current quarks of a baryon, a current quark-antiquark pair of a meson) in the wave function of a hadron. The corresponding hard mechanisms can be disentangled from the soft mechanisms which confine the quarks in the hadron ground or excited state. The study of many channels (elastic and inelastic nucleon form factors, Compton scattering, meson electroproduction,..) will lead to a better understanding of this simplest valence quark component of the wave function.

In the heavy quark sector (strange or charmed) they give access to the next component: the one which contains two gluons in addition to the valence quarks. This is a way to undertake the study of the gluonic content of hadronic matter. Electroproduction of J/Ψ or Φ is particularly appealing in this respect.

The study of these exclusive reactions in nuclei will allow to see how these simplest configurations evolve to the wave function of a physical hadron: how current quarks dress themselves to get hadrons. This will be revealed through the "color transparency" mechanism. At the other extreme, semi-inclusive meson electroproduction on nuclei will provide us with a *microscopic detector* to see how quark hadronization occurs in hadronic matter.

Since the momentum transfer is large, cross-sections are small: a high luminosity is mandatory. Since the reactions are exclusive, many particles are detected in coincidence: a continuous beam is mandatory. Last but not least, spin observables will be a must in being more exclusive and determining the various transition amplitudes. Those are the characteristics of the ELFE Project (1), which is currently under discussion in Europe.

THE ELFE PROJECT

ELFE is the acronym of Electron Laboratory for Europe. This facility is intended to operate a CW electron accelerator in the 15–30 GeV range and to strengthen collaborations (primarily on a european basis, but open to worldwide participations), which will use the corresponding beams. It has recently been endorsed by NUPECC (Nuclear Physics European Collaboration Committee), which found compelling the physics case and recommended to go ahead. The next couple of years will be devoted to set up the formal collaboration frame between the funding agencies of the various european countries, to optimize the design of the machine, to finalize the experimental set-up, and to enlarge the community potentially interested.

The current design assumes two stages, which could be achieved with the technology available to day. The first stage (15 GeV) of the project will be achieved by recirculating three times the electron beam in a 5 GeV supra-conducting linac. The design assumes a nominal accelerating gradient of 10 MV/m, a conservative figure consistent with the present state of the art of the technology of electromagnetic hyper-frequencies in supra-conducting cavities. The design of the recirculating arcs is made in such a way to afford the bending a 30 GeV beam, without catastrophic energy losses by synchrotron radiation. The second stage (30 GeV) will be achieved as soon as progresses in the technology of supra-conducting cavities will allow a significant increase in the accelerating gradient.

The nominal intensity will be large enough, in the range 10–50 μA, to make possible experiments with high luminosities, in the range 10^{35}–10^{38} $cm^{-2}s^{-1}$.

The large duty factor will make easier and even possible coincidence experiments.

The main physical goal of this project is to provide access to a whole class of such coincidence experiments which cannot be performed with existing facilities: the study of exclusive reactions at high momentum transfer. In those reactions, both the initial and the final state are fully determined. The high momentum transfer insures that the minimum number of participant quarks are present, at the same time, in the small interaction volume. This is the way of selecting the simplest component in the Fock expansion of the hadron wave functions: three valence current quarks in the proton wave function, for instance. It is this configuration which can be computed when solving QCD on lattice, in the quenched approximation, or which can be deduced from sum rules. The study of many exclusive channels (baryon and meson form factors, real and virtual Compton scattering, meson electroproduction, etc...) will provide many ways of probing these simple quark configurations, in order to eventually reach a mapping of the corresponding wave function.

The other originality of the project is the use of the nucleus as a "microdetector" or as a filter. Two extreme cases are worth to be considered. The first is known as "color transparency". The idea is to select one of these simplest configurations of a nucleon (or a hadron) in a nuclei and to see how it evolves toward its asymptotic wave function. The study of the interaction of the outgoing hadron with the nuclear medium, as a function of the size of the nucleus, will give us informations on the corresponding evolution. The second is the study of hadronization in nuclei. When a virtual or a real photon hits a quark and ejects it from a nucleon, this quark first propagates freely on a certain distance before dressing itself with $q\bar{q}$ pairs, as a constituent quark, and eventually resulting in a jet of hadrons. The characteristic distance over which each of these steps occurs is comparable to the size of a nucleus. Varying the size of the nucleus from which a quark is ejected is therefore a unique way to access the various steps of hadronization.

Finally, the production of heavy flavors (strangeness and charm) will open up a new way to probe hadronic matter. As in the study of all extended complex systems, the creation of an impurity (a strange or charmed quark, for instance), and the study of its propagation, will provide us with new tools.

In all these case, the determination of the various spin observables will enlarge our ability to achieve a deeper understanding of hadronic matter at short distances.

SPIN PHYSICS WITH ELECTROMAGNETIC PROBES

It is well known that spin observables are very sensitive to the details of the reaction matrix elements. Although they have been proved to provide us with a powerful way to constrain the reaction mechanisms or to get a deeper understanding of hadronic structure, considerable care has to be taken when dealing with spin observables. It makes sense to rely on them only when unpolarized data are already under control.

In addition to these features, which have been widely used with hadronic probes, spin observables in electromagnetic interactions provide us with an elegant way to disentangle the Coulomb and the Transverse parts of the hadronic current and to avoid a difficult Rosenbluth separation.

Therefore spin observables in electromagnetic interactions are expected to be extremely useful in two extreme situations. The first concerns reactions where spin observables lead to a complete reconstruction of the reaction amplitudes: multipoles, nucleon and baryonic resonance form-factors, meson form-factors, etc... This will constitute a sizable part of the CEBAF research program, at moderate momentum transfer, and will be extended at ELFE, at high momentum transfer. The second concerns reactions of which unpolarized cross-sections are already well under control and where selected spin observables will lead to specific informations. This is the case of some exclusive experiments in preparation at CEBAF, or of some semi-inclusive experiments in preparation at HERMES. They will be extended at higher momentum at ELFE.

SPIN PHYSICS WITH ELFE

Let me now give a few selected representative examples of spin physics at ELFE. This is not an exhaustive list, and I believe that new idea will emerge by the time the first beam is available.

Nucleon Form Factors

Figure 1 shows the evolution against Q^2 (the squared mass of the exchanged virtual photon) of the ratio of the Dirac form-factor of the proton, as determined in a Rosenbluth separation recently performed at SLAC. Despite the high energy available, no significant data exist above 6 GeV^2. The study of the reactions $\vec{e}p \to e'\vec{p}$ or $\vec{e}\vec{p} \to e'p$ will allow to determine this ratio up to $Q^2 \simeq 20$ GeV^2 with a good accuracy ($\simeq 2\%$ at 6 GeV^2 and better than 15% at 20 GeV^2). This is the momentum range where a perturbative treatment of the interaction of the virtual photon with the three valence quarks is expected to be valid, making possible a determination of the corresponding components of the proton wave function. Only 6 GeV^2 will be accessible at CEBAF.

To achieve this goal the luminosity will be $\mathcal{L} = 10^{35}$ $cm^{-2}s^{-1}$ when a polarized target or a 4π detector is used, and $\mathcal{L} = 10^{38}$ $cm^{-2}s^{-1}$ when two spectrometers, one being equipped with a focal plan polarimeter, are used.

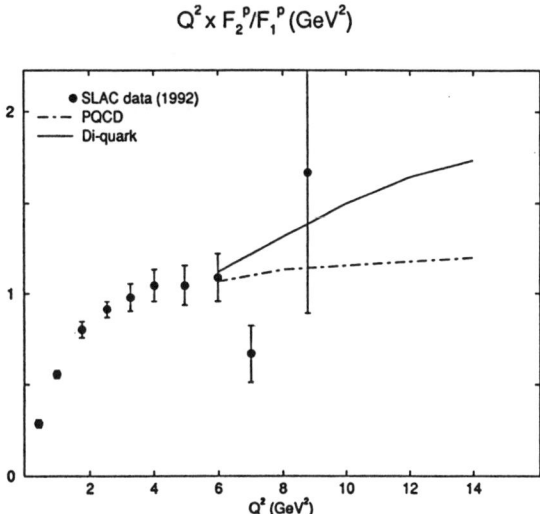

Figure 1: The ratio of the two Dirac form-factors of the proton is plotted against the squared mass of the virtual photon. The data have been obtained at SLAC. The solid line corresponds to a perturbative calculation which uses a proton wave function deduced from QCD sum rules. The dashed line corresponds to a diquark model of the proton. See Ref. 1 for details.

Virtual Compton Scattering

Compton scattering represents the simplest electromagnetic process, beyond the proton electromagnetic form-factors. In the latter, the virtual photon couples to a single quark which exchanges the minimum number of gluons with the two others. On the contrary, the incoming real or virtual photon and the outgoing real photon may couple to different quark in Compton scattering. This process is therefore more sensitive to quark correlations than form-factors, and maps out in a different way the proton wave function.

Due to the energy transferred to the intermediate three quark state, one of the quark can go on-shell during the integration upon the internal momentum of the hard diagram. This induces a logarithmic singularity and leads to a finite imaginary part in the reaction amplitude. Figure 2 shows how the electron asymmetry, in the reaction $\vec{e}p \to e'p\gamma$, is sensitive to such an imaginary part. This is a good example of the way spin observables should be used in order to pin down a specific mechanism in a channel where unpolarized cross-sections are already satisfactorily reproduced (2).

In this exclusive reaction, the four-momentum transfer t, between the incoming virtual photon and the outgoing real photon, is an other "knob" to set the size of the interaction volume, besides the squared mass Q^2 of the incoming virtual photon.

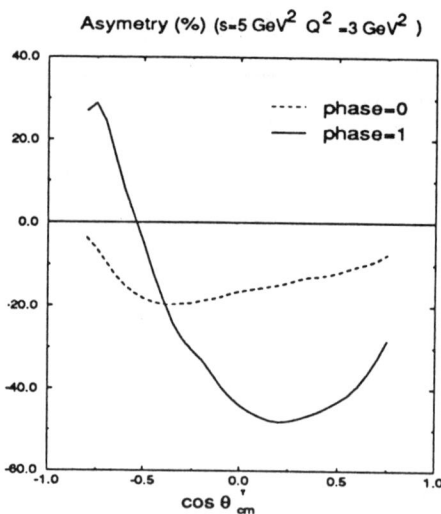

Figure 2: The electron asymmetry, in virtual Compton scattering induced by polarized electrons on protons, is plotted against the polar angle of the emitted real photon. Its azimuthal angle is 100°. The imaginary part of the hard scattering amplitude has been switched off in the dashed curve, while it has been kept in the full curve. See Ref. 1 for details.

ELFE will allow to determine the cross-section of virtual Compton scattering up to $-t \simeq 20$ GeV^2, with an accuracy of a few percent when two spectrometers are used and the experiment is performed with a luminosity of $\mathcal{L} = 10^{38}$ $cm^{-2}s^{-1}$.

Heavy Flavor Production

Figure 3 shows the variation, with the momentum transfer t, of the differential cross-section of the reaction $\gamma p \rightarrow \phi p$. Its magnitude and its t dependence are almost independent of the incoming photon energy. At low momentum transfer, where data exist, the Pomeron exchange model reproduces fairly well the existing data. At high momentum transfer only valence quark configurations with small spatial extension are selected in the vector meson and the proton. Since the ϕ is predominantly made of a $s\bar{s}$ pair, the simplest mechanism involves the exchange of two gluons: the full non-perturbative interaction between this two gluons has no time to develop into the Pomeron. In the vicinity of $-t \simeq 1$ GeV^2, the two gluon exchange model matches nicely the Pomeron exchange model. An interesting feature is the node around $-t \simeq 2.5$ GeV^2, which comes from the interference between two amplitudes: the first in which the two exchanged gluons couple to the same quark, the second where each

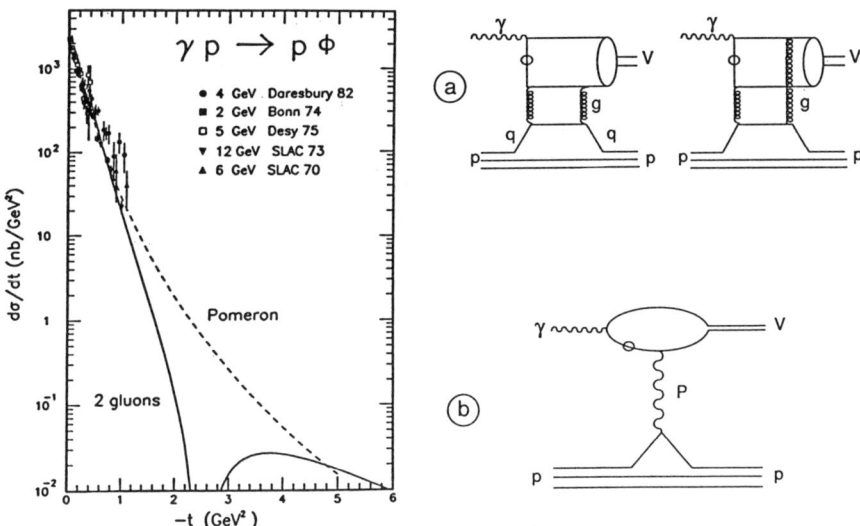

Figure 3: The cross section of the reaction $\gamma p \to \phi p$ is plotted against the four momentum transfer t between the incoming photon and the outgoing ϕ. The dashed curve is the prediction of the Pomeron exchange model, while the full curve is the prediction of the two gluon exchange model. The corresponding graphs are shown on the right hand side. See Ref. 3 for detail.

couples to a different quark. ELFE will allow to determine the shape of such a t dependence with an accuracy of a few percent up to $t \simeq 10$ GeV^2.

Now, nature does not like holes and many competing mechanisms could fill or shift it, although their probability is small. For instance, $s\bar{s}$ pairs, which may preexist in the proton wave function, may be directly knocked out by the photon and eventually recombine into a ϕ. Spin observables offer us with a possible way to disentangle such a strange component in the proton wave function from the two gluon (or the Pomeron) exchange mechanism. While the latter mechanism conserves helicity in the s-channel (S-Channel Helicity Conservation, SCHC), the former does not. The determination of the various helicity transfer coefficients, using polarized photons and determining the tensor polarization of the vector meson through the analysis of the angular distribution of its decay products, is one of ways to achieve this goal.

As an example of the relevance of this method, Figure 4 shows the behavior, against the mass of the emitted Kaon pair, of the coefficient of the $Y_{10}(\theta_k, \phi_k)$ spherical harmonic in the development into moments of the decay distribution of the ϕ meson emitted in the reaction $\gamma p \to pK^+K^-$, at 5 GeV. It has been determined just before the shutdown of the Daresbury facility (4). If only a ϕ meson were emitted, the Kaon pair would be in pure spin one state, and only spherical harmonics of rank two would survive. It turns out

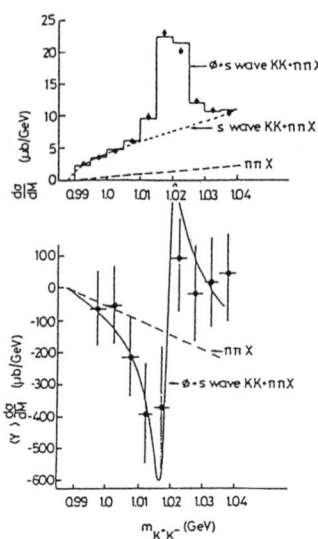

Figure 4: The mass distribution and the [10] moment of the angular distribution of the two Kaons emitted in the reaction $\gamma p \to p K^+ K^-$ at 5 GeV. Adapted from Ref. 4.

that a significant S-wave background exists below the ϕ meson peak, which would lead to spherical harmonics of rank zero, if it would be alone. The presence of spherical harmonics of rank one reveals a strong interference between S-wave and P-wave components of the two Kaon decaying system. This can occur only if the ϕ is produced with a zero helicity. This is a signature of a violation of SCHC, since the incoming real photon cannot exhibit an helicity zero component. In that experiment, performed at low energy and low momentum transfer t, many processes can be invocated to explain this feature. It should be extended at higher energy and higher momentum transfer, where the underlying mechanisms are expected to be more fundamental and more simple.

Related topics will also be addressed in the heavy quark sector. For instance, in Figure 3 the two gluons are attached to a single quark in the proton, of which the structure is taken into account through its matter form-factor (3). This is a good approximation at high t when only short distance configurations of the valence quarks are relevant. At $t = 0$ however, it can be shown (5) that the two gluon exchange cross-section becomes proportional to the gluon distribution $G(Q^2, x)$. Of course, this result is valid in the limit of large momentum transfer Q^2, in the light quark sector, or in the limit of large quark mass (charmed quark), even at low Q^2. Using a polarized target in the reactions $\vec{e}\vec{p} \to e'p\phi$ or $\vec{e}\vec{p} \to e'pJ/\Psi$ would be a way to determine the contribution of the gluonic content $\Delta G(Q^2, x)$ of the spin of the nucleon. Also the problem of the strange content of the proton can be addressed through the study of parity violation in the scattering of polarized electrons on protons, or through the study of semi-inclusive kaon production in the reaction $\vec{e}\vec{p} \to e'KX$. This brings me to the semi-inclusive part of the ELFE research program.

Semi-Inclusive Reactions

Inclusive scattering of electrons is now well under control and has been discussed at length during this conference. The status of semi-inclusive reactions is not in such a good shape. Results of the analysis of the reaction $\vec{p}(\vec{e}, e'h)$ by SMC are now available. The HERMES program will include also the study of such reactions. However the range of momentum transfer is limited due to the low luminosity of those facilities, and only the luminosity of ELFE (higher by a few order of magnitude) will allow a comprehensive study.

One of the interesting issues in this domain is the determination of the transverse spin structure function $h_1(Q^2, x)$ of the proton. I refer to the talk of Vuaridel (6) for a more comprehensive discussion. Let me briefly show how ELFE can address this problem. It has been proposed (1) to determine the spin transfer coefficients in the reaction $\vec{p}(e, e'\vec{\Lambda})$, with a transversally polarized proton target and determining the spin of the Λ through the angular distribution of its decays products. To achieve this goal, with a reasonable statistical accuracy, a luminosity of $\mathcal{L}_{eff} \simeq 10^{34} cm^{-2} s^{-1}$ should be reached. For comparison, with the SMC set-up the luminosity is $\mathcal{L}_{eff} \simeq 10^{31} cm^{-2} s^{-1}$, while with the HERA set-up it is only $\mathcal{L}_{eff} \simeq 10^{30} cm^{-2} s^{-1}$ ($\mathcal{L}_{eff} = \mathcal{L}(p_t p_b f)^2$, where p_t and p_b are respectively the polarization of the target and the beam, and f is a factor of merit).

CONCLUSION

ELFE will offer the unique opportunity to study *exclusive* reactions in a wide range of energy and momentum transfer. This will eventually lead to an accurate mapping of the simplest components of the Fock wave function of the Hadrons. In the sector of semi-inclusive reactions, ELFE will enlarge the kinematical domains already accessible. In both case spin observables will be a unique tool. As can be seen in Figure 5, none of the present facilities provide us at the same time all the conditions to successfully complete the full physics program. However, present facilities should be used in order to prepare the ELFE program. Two ways are open. Following works already done at low energy facilities, of which the duty factor has already been increased, CEBAF will offer us a significant increase of the available energy and transfer. This is the approach from the bottom, where one first increases the duty factor and then increases the energy. It allows to go from the meson sector to the

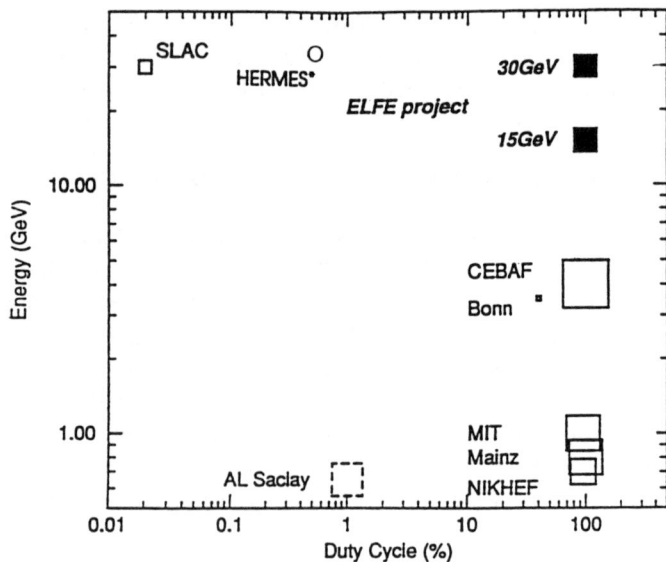

Figure 5: Comparison of ELFE with other facilities. The area of each square is proportional to the intensity of the corresponding accelerator. See Ref. 1 for more details.

constituent quark sector and eventually to the current quark sector. Following works already done at SLAC and CERN, HERMES will offer us a way to start the study of semi-inclusive reactions. This is the approach from the top, where one starts in the right energy range and then increase the duty factor and the luminosity. It allows to go from semi-inclusive to exclusive reactions.

REFERENCES

1. Arvieux, J., and De Sanctis, E., eds., "The ELFE Project", *Conference Proceedings of the Italian Physical Society* **44** (1993).
2. Kronfeld, A.S., and Nižić, *Phys. Rev.* **D44**, 3445 (1991).
3. Laget, J.-M., and Mendez Galain, R., *Nucl. Phys.* **A** (1994) in press.
4. Barber, D.P., et al., *Z. Physics* **C12**, 1 (1982).
5. Brodsky, S.J., et al., *Phys. Rev.* **D** in press.
6. Vuaridel, B., *These Proceedings*.

The measurement of the target asymmetry of the reactions $\gamma p \to \pi^+ n$ and $\gamma p \to \pi^0 p$ with the Bonn frozen spin target and the PHOENICS-detector at ELSA

Hartmut Dutz

on behalf of the PHOENICS-collaboration

Physikalisches Institut der Universität Bonn
Nussallee 12
53115 Bonn
Germany

Abstract At the Bonn Electron Stretcher Accelerator ELSA new target asymmetry data for the single pion photoproduction of the reactions $\gamma p \to \pi^+ n$ and $\gamma p \to \pi^0 p$ has been obtained. For the first time a frozen spin target has been used in combination with a tagged photon beam. A simultaneous measurement of both reaction chanels in a wide kinematical range of E_γ = 215-950 MeV and Θ_{CMS_π} = 35°-135° has been performed at the PHOENICS-detector.

Butanol (C_4H_9OH) has been used as target material. A proton polarization of more than ± 90% has been achieved in a polarizing field of 5 Tesla. The data taking has been performed during the frozen spin mode where the polarization has been maintained in a field of 0.35 Tesla. At this field and at temperatures $\leq$ 60 mK the proton relaxation time has been in the order of 5 days.

The design of the PHOENICS-detector permits a considerable kinematical overdetermination so that the nominal process could be clearly separated from the background. High quality data has been obtained. The systematical error for all data points is ΔT_{sys} = 2.5%.

1 Introduction

The analysis of the photoproduction in the resonance region has made a strong progress in the last 20 years. Quark model calculations and several partial-wave analyses, combined with the existing sets of data led to the determination of the γN couplings of the main resonances. But there are still ambiguities and the knowledge of the smaller resonances is still poor. One way to improve this situation is to perform experiments which are especially sensitive to contributions of small resonances. In this context of special interest is the value of the electric quadrupol exitation of the Δ-resonance and the contribution of the M_{1-}-amplitude.

One approach to determine the small multipol amplitudes is to measure polarization observables. They offer a higher sensitivity of the weak multipoles than the differential cross section by the occurance of terms of interference. The target asymmetry can be evaluated in terms of $cos\Theta$:

$$T(\Theta) = \frac{|\vec{q}|}{|\vec{k}|} \frac{1}{d\sigma/d\Omega} \sin\Theta(a + b\cos\Theta + c\cos^2\Theta + ...).$$

Whereby the coefficients are reflecting the contribution of the different multipoles. In case of the first resonace region ($l \leq 1$) one gets:

$$a = Im[E_{0+} 3(M_{1+}^* - E_{1+}^*)]$$
$$b = 3Im[-(M_{1+}^* - E_{1+}^*)M_{1-} + 4M_{1+}^* E_{1+}]$$

The data of a measured target asymmetry can be analysed in a multipol analysis and by taking the differential cross section into account one can determine the contribution of the multipoles and the photocoupling constants of the resonances. The Bonn electron strecher accelerator ELSA, the 'tagged photon facility' PHOENICS and new developments in the field of polarized solid state targets ('frozen spin target') permits measurements of polarization observables in the pion photoproduction with an improved quality. The target asymmetry was measured with a transversal to the production plane polarized frozen spin target and the PHOENICS-detector. We measured the reaction chanels $\gamma p \rightarrow \pi^+ n$ und $\gamma p \rightarrow \pi^\circ p$ simultaneously in a kinematic range from $E_\gamma = 215\,MeV - 950\,MeV$ and $\Theta_{CMS} = 35° - 135°$ [1, 2, 3]. The experimental set-up and the results of the target asymmetry measurement will be discussed briefly.

2 Experimental set-up

A top view of the experimental set-up of the PHOENICS-detector and the Bonn frozen spin target at ElSA is shown in fig. 1. The electron beam extracted from ELSA has an energy of 1 GeV. The bremsstrahlung photons were produced on a radiator foil of 5 μm gold and the photon beam at the target has a circular profile with a diameter of 1.1 cm (FWHM). The energy and the flux of the photon beam is determined by a tagging spectrometer. The spectrometer covers a photon energy range from E_γ 200 MeV up to $E\gamma$=950 MeV with an energy resolution of 10 MeV ($\leq$ 360 MeV) and 5 MeV at the higher energies. It is also used for a time of flight measurement of the produced secondary paricles by the photon nucleon reaction. Detailed information about the tagging system is given in [6]. The photon beam hits the frozen spin polarized solid state target which is located in the center of the PHOENICS-detector, which register the outgoing particles.

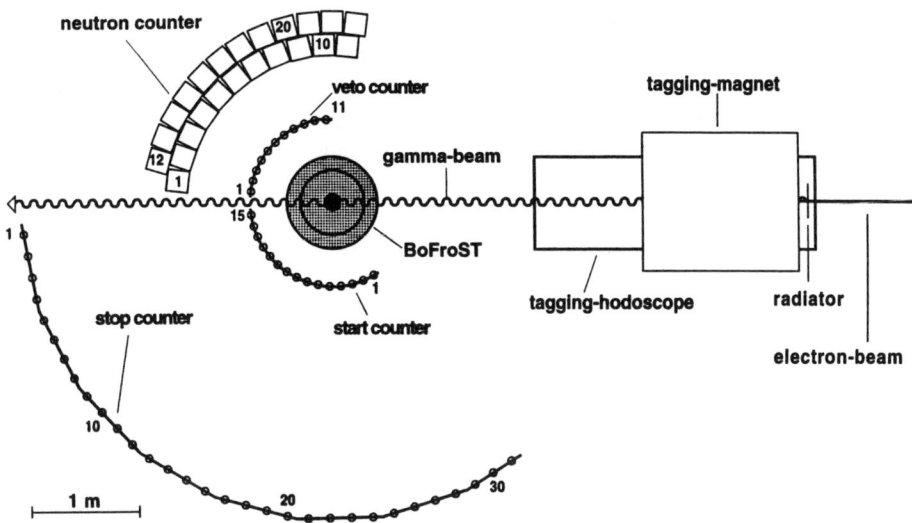

Figure 1: Top view on the experimental set-up of PHOENICS at ELSA.

2.1 The Bonn Frozen Spin Target

High nucleon polarization in a solid state sample can be achieved by means of the so - called DNP (dynamic nuclear polarization) process by which the high electron polarization of some paramagnetic impurities is transfered via microwave induced transitions to the nucleon spin system.

The set-up of the Bonn frozen spin target for this target asymmetry measurement consists of a superconducting polarizing magnet with a maximum field of 7 Tesla, and vertical superconducting 'holding magnet' with a maximum field in the target area of about 0.5 Tesla. The central part of the target is a vertical ^{3}He/^{4}He - dilution refrigerator [4]. In the frozen spin mode of the target, this setup leads to an angular acceptance for the outgoing particles in the horizontal plane of about 360° and vertical to the incoming beam of about ±20°. A schematic drawing of the frozen spin target is shown in fig. 2. The target was a 4 cm long 2.5 cm diameter cylinder filled with granules of butanol (C_4H_9OH). It was polarized by the technique of dynamic nuclear polarization in a magnetic field of 5 Tesla. During the data taking the spin orientation has been maintained by means of the vertical 'holding field' of 0.35 Tesla. At an average temperature of about 60 mK the relaxation times were in the order of $T_1 \geq 5\ days$. An average in-beam polarization P_t of 85% was measured, and a maximum polarization of $P_+ = 98\%$ (positive direction) and $P_- = 91\%$ (negative direction) was achieved. The polarization was measured by means of the standard nuclear magnetic resonance (NMR)- technique, deriving an absorption signal proportional to the nucleon polarization at the nuclear Larmor frequency. The NMR coil was connected to a phase sensitive Q-meter

(LIVERPOOL-NMR system [5]) by a $\lambda/2$ cable and was tuned to the Larmor frequency of the nucleons at the given polarizing field of 5 Tesla. The error on the polarization value is $\pm 2.5\%$ and the main source of the uncertainty is given by the determination of the so called TE-signal. Detailed information about the frozen spin target are given in ref.[1] and [4].

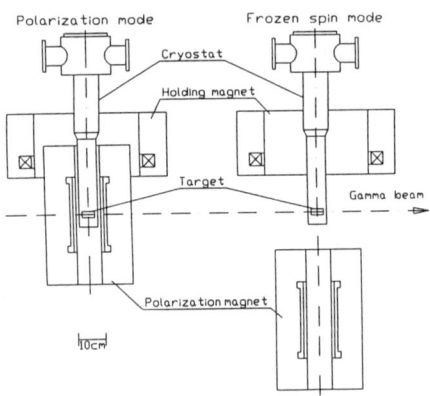

Figure 2: The frozen spin target in the polarization mode (left) and holding mode (right)

2.2 The PHOENICS-detector

The PHOENICS-detector consists of vertical oriented scintillator bars which are arranged in concentric ring segments around the target. The set-up is designed to detect 2 particles in the final state. A schematic view of the PHOENICS-detector is shown in fig. 3. Compared to the incoming photon beam the segments are placed on right and left hand side. The scintillator bars on the left side of the target (FL,CL) are designed to detect charged particles and those on the right side of the target (CR,FR) can be used to register neutral particles or charged particles respectively. For the charged pion photoproduction the π^+ was detected by the coincidence of the left side segments (CL $\wedge$ FL) and the neutron by the outer right side counters in anticoincidence with the inner right side counter ring (FR $\wedge$ $\neg$(CR)). To increase the neutron detection efficiency the FR counters were arranged in 2 rows of scintillation counters each of a thickness of 20 cm.

Each end of all scintillator bars is equipped with a photomultiplier. The anode

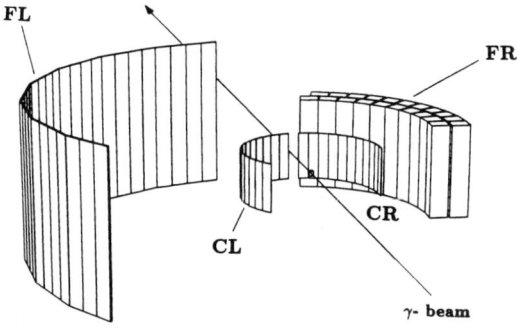

Figure 3: Schematic view of the PHOENICS-detector.

pulse of the photomultiplier is used in a charge integrating ADC and in parallel has been send to a TDC after passing a leading-edge discriminator. Thereby the time of flight of a particle is given by the sum of the time measured by the top and the bottom tube of the scintillator bar. The vertical impact position is given by the difference of the time measured by these tubes. The combination of the laboratory angular position of each counter, the time of flight and the vertical impact position of a particle the three component velocity vector β of the particle can be measured. The experimental set-up delivers seven kinematic quantities for each event of the type $\gamma p \to \pi N + X$.

A time of flight calibration of the detector components as well as the tagging system was performed before the data taking run has been started. Precise information about the detector, the calibration procedure and the time corrections are given in [7, 3].

3 Measurement and Results

We performed the measurements of the target asymmetry in the pion photoproduction in two runs of a total time of 30 days. The polarization of the target was reversed every 24 hours and the data taking time during the frozen spin mode was about 18.5 hours. We recorded 40 billion counts to estimate the target asymmetry for the reaction channels $\gamma p \to \pi^+ n$ and $\gamma p \to \pi^0 p$. The intensity of the extracted electron beam was in the order of 100 pA with a mean duty cycle of about 45%. Using a 5 μm gold radiator this beam parameter leads to a photon flux of 2-3×10^6 in the tagged energy range. The angular acceptance Θ of the detector in the laboratory system was $22.8° \leq \Theta \leq 126.4°$ for the charged particle detection and $14.5° \leq \Theta \leq 98.7°$ for the neutral particle detection respectively.

In the experiment we measured the counting rate asymmetry ϵ simultaneously both for the π^+n and the $\pi°p$ channel given by

$$\epsilon = \frac{N\uparrow - N\downarrow}{N\uparrow + N\downarrow}$$

where $N\uparrow$ and $N\downarrow$ are the counting rates for the two states of the target polarization perpendicular to the production plane. From the measured ϵ we calculated the asymmetry in the cross sections defined by

$$T = \frac{\sigma\uparrow - \sigma\downarrow}{\sigma\uparrow + \sigma\downarrow}$$

Here $\sigma\uparrow$ and $\sigma\downarrow$ represent the differential cross sections for the two directions of the proton spin. The symbol $\uparrow$ refering to the direction of $(\vec{k} \times \vec{q})$ where $\vec{k}$ and $\vec{q}$ are the momenta of the photon and the pion respectively. The counting rates $N\uparrow$ and $N\downarrow$ consists of the countig rates $n\uparrow$ and $n\downarrow$ from the 'free' polarizable protons of the target and the counting rate n_C from the bound nucleons in the target material, target containers and structure materials of the refrigerator. Because of the kinematic overdetermination of the two particle reaction and the good time resolution of the PHOENICS-detector we were able to extract the nominal πN processes of the proton $n\uparrow\downarrow$ by the difference of the πN events from the butanol sample $n_{C_4H_9OH}\uparrow\downarrow$ taken in both polarization directions and the πN-events from a carbon target n_C:

$$n\uparrow\downarrow = n_{C_4H_9OH}\uparrow\downarrow - n_C$$

This procedure leads to the following expression for the target asymmetry

$$T = \frac{n\uparrow - xn\downarrow}{xP\downarrow(n\uparrow - \alpha n_C) + xP\uparrow(n\downarrow - \beta n_C)}$$

with x the photon flux ratio of the $\uparrow$- and $\downarrow$ data set and α, β the photon flux normalisation of the carbon events of the $\uparrow$- and $\downarrow$ data set respectively. The described background substraction reduces the systematic error of the measurement onto the uncertainty of the estimation of the target polarization. The mean systematic error for all data points is $\pm 2.5\%$. The error of the background substraction is part of the statistical error of each data point. For the data analysis each recorded event was analysed under the assumption of a $\gamma p \to \pi^+ n$ or $\gamma p \to \pi° n$ event respectively. Whereby two kinematic parameters are sufficient to determine the kinematic quantities of the nominal process. As seven kinematic parameters were measured, the overdetermined parameters were used to test the assumption. This leads to the method to compare the expected value of each parameter with the measurd value event by event.

The target asymmetry was calculated from data sets of neighboring polarization runs. The granularity of 128 energy and 27 angular bins of the detection system leads to 3456 data points for the target asymmetry. To increase the statistics we sum up the data points to larger angular and energy bins. Finally we end up for the $\pi^+ n$-channel with 216 data points devided up into 22 angular distributions from 220 MeV to 393 MeV ($\Delta E_\gamma = \pm 10$ MeV) and from 425 MeV to 800 MeV ($\Delta E_\gamma = \pm 15$ MeV). The pion scattering angle varries from $35° \leq \Theta_\pi^{c.m.} \leq 135°$ ($\Delta\Theta_\pi^{c.m.} \pm 5°$). The angular distributions in the lower energy range are plotted in fig. 4. More informations are given in [8]. For the $\pi°$-production we calculated 52 data points in the energy region of the Δ-resonance from $E_\gamma = 272$ MeV to 573 MeV. The upper energy limit is given by the proton detection acceptance of the detector. The angular acceptance is similar to the π^+-channel. The angular distributions are shown in fig. 5

A comparison of the π^+-data with the multipol analysis of Arndt et al. [9] is shown in fig. 6. The agreement with our data at $E_\gamma = 345$ MeV and higher energies is good. At lower energies one can find a slightly disagreement of the calculations with our data. Here it is obvious that the multipol analysis did not take into account the E_{0+} and M_{1-} correctly. The disagreement at lower energies, from the pion threshold up to the first resonance has to be investigated more precisely.

References

[1] H. Dutz, thesis, BONN-IR-93-49.

[2] D. Krämer, thesis, BONN-IR-93-48.

[3] B. Zucht, thesis in preparation

[4] H. Dutz et al., NIM A 340 (1994) 272.

[5] G. Reicherz, thesis, BONN-IR-94-16.

[6] P. Detemple et al., NIM A 321 (1992) 479.

[7] K. Büchler et al., Nucl. Phys. A 570 (1994) 580.

[8] K.H. Althoff et al., to be published in Nucl. Phys. A.

[9] R.A. Arndt et al., Phys. Rev. C 42 (1990) 1853.

512 Measurement of the Target Asymmetry

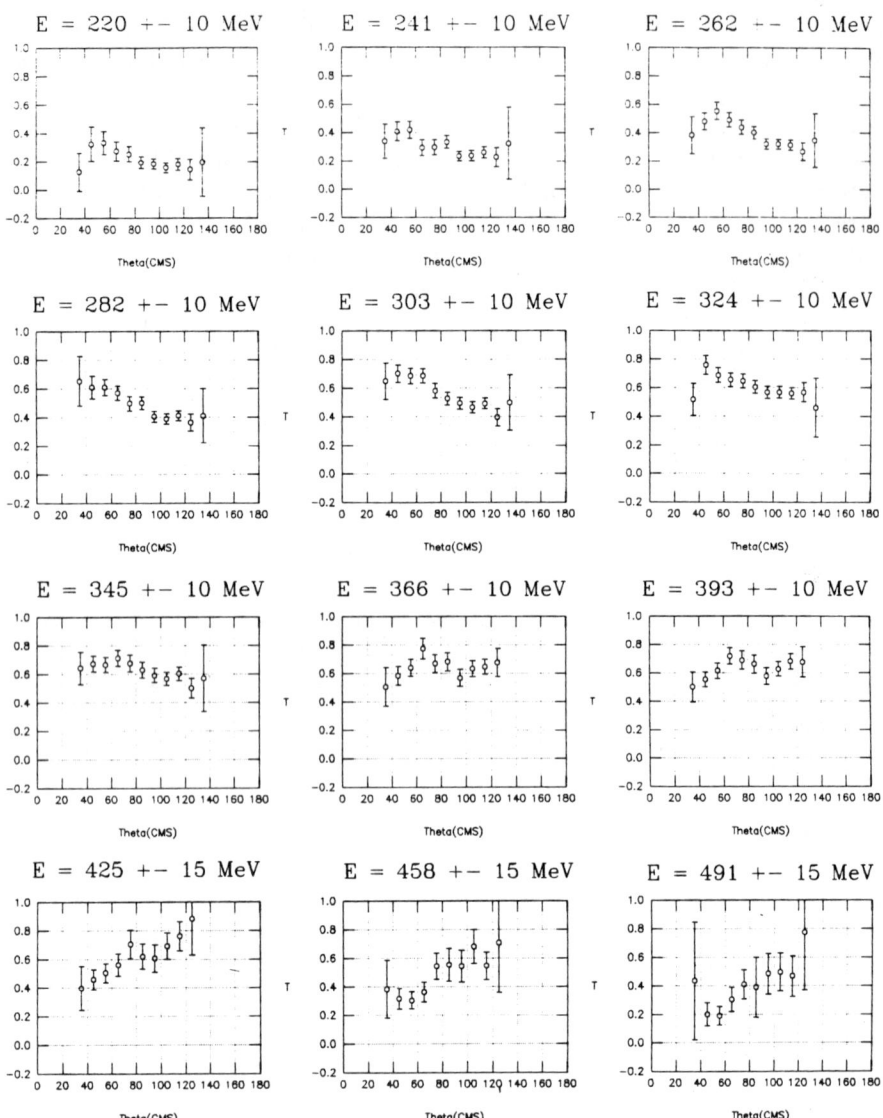

Figure 4: Measured target asymmetry of $\gamma p \to \pi^+ n$ from 220 MeV to 491 MeV

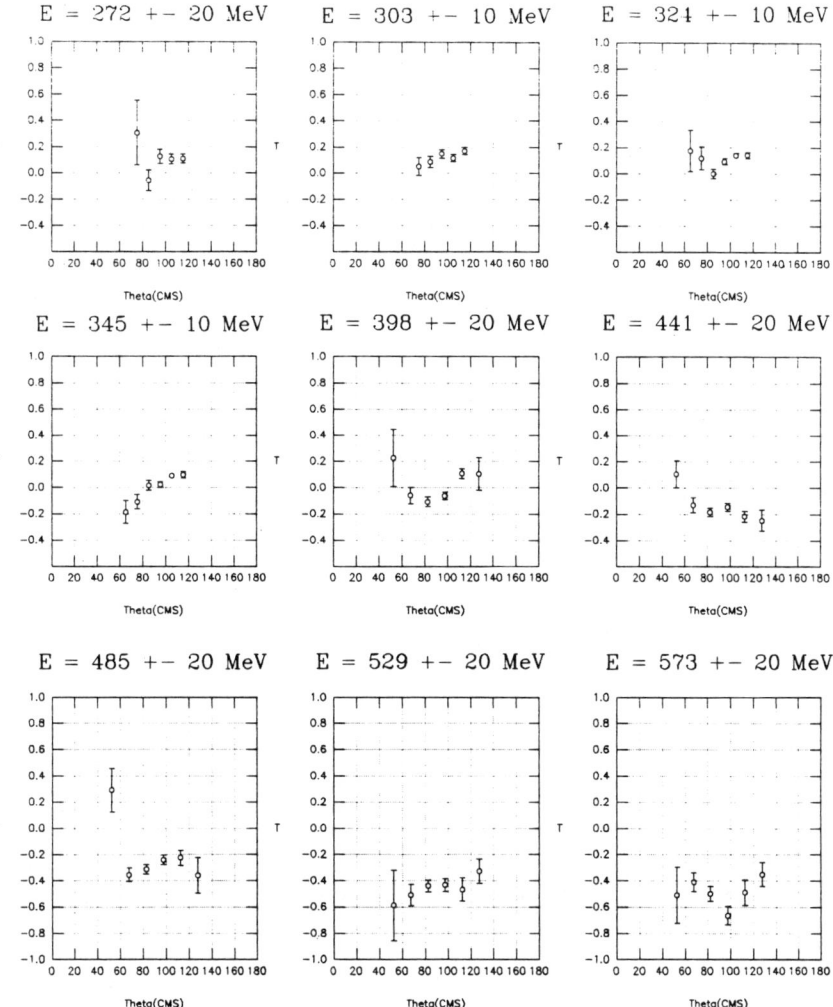

Figure 5: Measured target asymmetry of $\gamma p \to \pi^0 p$

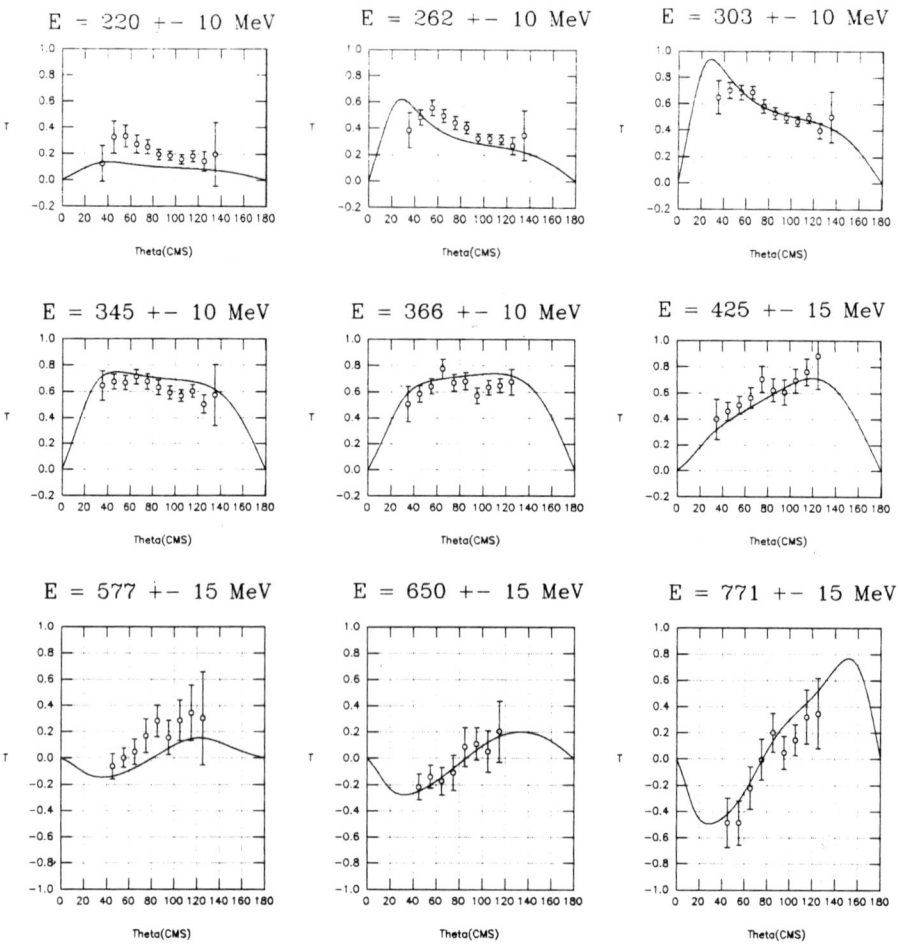

Figure 6: Target asymmetry of $\gamma p \to \pi^+ n$ compared to the result of the partial wave analysis of Arndt et al.

PHOTOPRODUCTION OF η MESONS ON THE PROTON

Marc Bouché-Pillon[a], Bijan Saghai[a], and Frank Tabakin[b]

a) Service de Physique Nucléaire, CEA/DSM/DAPNIA, Centre d'Études de Saclay
F-91191 Gif-sur-Yvette, France

b) Department of Physics and Astronomy, University of Pittsburgh,
Pittsburgh, Pennsylvania 15260

Abstract. Spin observables for the reaction $\gamma\, p \to \eta\, p$ are examined using an isobaric dynamical model and fitting the recent data from Bonn. Studying possible contributions of ten isospin 1/2 nucleonic resonances, an excellent fit is obtained by a model including only three of them: S11 (1535), D13 (1520), and F15 (1680). No evidence for a significant contribution from the Roper resonance is found. Predictions for polarization observables are shown and compare well with the very scarce and old data. General rules for such observables deduced recently by Fasano, Tabakin and Saghai are satisfied by the reported model.

All sixteen observables for the reaction $\gamma + p \to \eta + p$ have been investigated from threshold up to 1.2 GeV. The formalism is based on an isobar model [1]. In this approach electric and magnetic multipole amplitudes are expressed in terms of various isospin-1/2 nucleonic resonances plus a smooth background including S and P waves. The resonances are described by energy dependent Breit-Wigner forms. The background arises mainly from the nucleon pole diagrams, and possibly at higher energies from t-channel vector meson exchange processes [2].

The only well established information on the reaction mechanism based on old data from the 70's is the dominance of the S11(1535) resonances [1-4]. New data are becoming available from Bates [5] and preliminary differential cross sections (about 400 data points) from Bonn [6]. The highest energy for the latter data is 1.2 GeV.

In our procedure, we have fit only the recent Bonn data [6] with the MINUIT [7] least-squares minimization code implemented with our formalism. The role of the following resonances has been investigated:

S11(1535), S11(1650), P11(1440), P11(1710), P13(1720),
D15(1675), D13(1520), D13(1700), F15(1680), G17(2190).

The present results confirm the major role played by the S11(1535) resonance, while the Roper resonance is not required to reproduce the preliminary

Bonn data. Besides, among the phenomenological fits obtained, the one including only S11(1535), D13(1520), and F15(1680) resonances gives the most

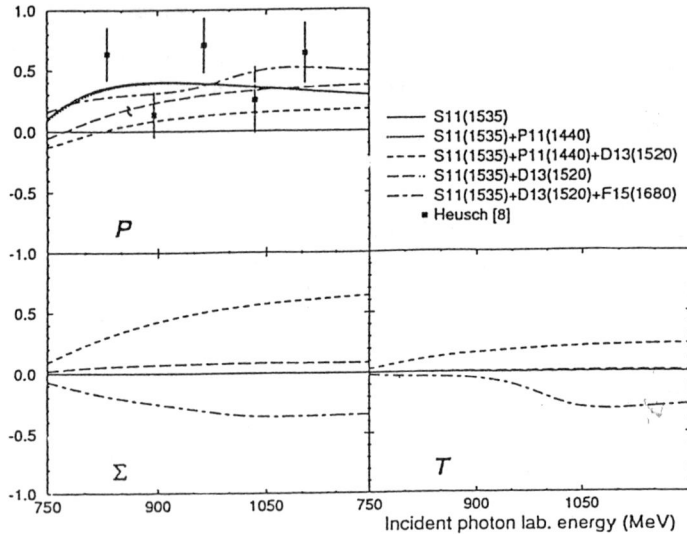

FIGURE 1. Single polarization asymmetries for $\gamma + p \to \eta + p$ at $\theta_\eta^{cm} = 90°$: recoil proton (P), Beam polarization (Σ), and polarized target asymmetries (T). The data are from Ref. [8].

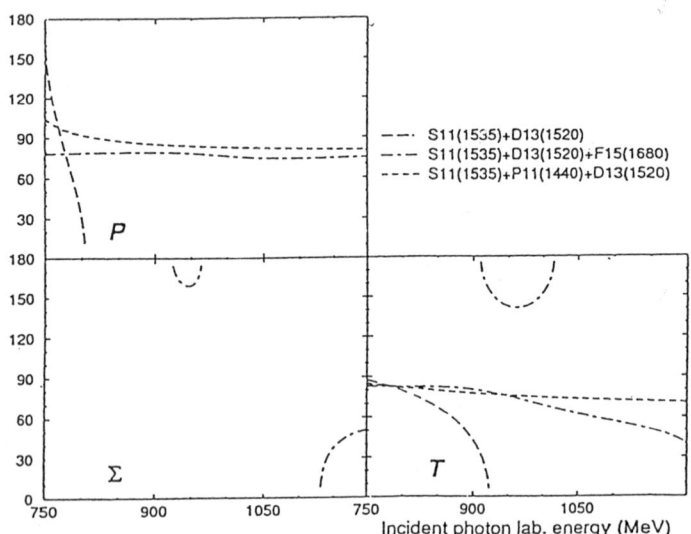

FIGURE 2. Nodal trajectories for the same observables as in Fig. 1. The locations of intervening nodes are given as a function of the photon laboratory energy.

satisfactory agreement with the Bonn data ($\chi^2_{DF} = 1.04$). The χ^2_{DF} for the other four configurations, shown in Fig. 1 are between 1.2 and 1.4. In Fig. 1, our predictions for single polarization asymmetries using five different combinations of contributing resonances are given.

The nodal structure study was found [9] to be a powerful tool in the study of pseudoscalar meson photoproduction. In Fig. 2, we show the position of the relevant nodes (angle at which the observable vanishes at a given energy) for single polarization asymmetries. Notice that all three models contain an spin 3/2 resonance (D13). According to FTS rules [9], the maximum number of nodes expected for recoil (P), beam (Σ), and target (T) asymmetries are 2, 1, and 2, respectively. Figure 2 shows that all these nodal trajectories follow nicely the predicted behavior.

We look forward to the final cross section data from Bonn and the polarization measurements planned at Bonn, CEBAF and GRAAL to help pin down the reaction mechanism.

ACKNOWLEDGMENTS

We are grateful to S. A. Dytman for his contributions at the early stage of this work. We thank M. Breuer and J.P. Didelez for having communicated the preliminary Bonn data.

REFERENCES

1. F. Tabakin, S. A. Dytman, and A. S. Rosenthal, *Proc. of Excited Baryons 88*, Troy, 4-6 August, 1988, G. Adams et al. eds., World Scientific Press, 1989 ; and references therein.
2. M. Benmerrouche and N. Mukhopadhyay, Phys. Rev. Lett. **67**, 1070 (1991).
3. C. Bennhold and H. Tanabe, Nucl. Phys. **A350**, 625 (1991).
4. M. Bouché-Pillon, DAPNIA/SPhN Report No. 93/41, 1993, (in French).
5. S. Dytman et al., Submitted for publication.
6. M. Rigney et al., "Photoproduction of Eta mesons from threshold to 1.2 GeV", *Proceedings of the 14th International Few-Body Conference*,Williamsburg, May 26-31, 1994; Ed. F. Gross , to appear ; M. Breuer, J.P. Didelez et al. private communication (1994).
7. F. James, M. Roos, MINUIT, CERN Report D506, (1981).
8. C.A. Heusch et al., Phys. Rev. Lett. **70**, 1381 (1970).
9. C.G. Fasano, F. Tabakin, and B. Saghai, Phys. Rev. C46, 2430 (1992) ; B. Saghai and F. Tabakin, this proceedings.

PHOTOPRODUCTION OF K$^+$ Σ^0 ON THE PROTON

J.C. David[a], C. Fayard[a], G.H. Lamot[a], F. Piron[b], and B. Saghai[b]

a) Institut de Physique Nucléaire de Lyon, IN2P3/CNRS,
Université Claude Bernard, F-69622 Villeurbanne Cedex, France

b) Service de Physique Nucléaire, CEA/DSM/DAPNIA, Centre d'Études de
Saclay, F-91191 Gif-sur-Yvette Cedex, France

Abstract. Investigations in progress on the associated strangeness photo- and electroproduction off the proton in the few GeV region and the negative kaon radiative capture are reported, and results for the K$^+\Sigma^0$ channel are presented.

In a thorough phenomenological investigation based on an isobaric approach [1] using a diagrammatic technique [1-3], we focus on the following reactions:

$$\gamma_{R,V} + p \to K^+ + Y, \quad K^- + p \to \gamma + Y \; ;$$

with Y=Λ, and Σ^0; γ_R, γ_V real and virtual photons, respectively.

In this contribution, given the very recent data from ELSA collaboration [4], we report on the self-analyzing K$^+$ Σ^0 final state channel from threshold up to $E_\gamma^{\text{lab}} = 2.1$ GeV.

In our approach, the first step in going beyond the existing models was to extend a recent model by Adelseck-Saghai [2], dedicated to the $\gamma + p \to K^+ + \Lambda$ reaction. This model, obtained by fitting the differential cross section data up to 1.4 GeV (reduced $\chi^2 = 1.3$), predicts well enough the total cross section and the single polarization data with the main coupling constants in agreement with the predictions of the broken SU(3)-symmetry, but it fails at higher energies. This situation was improved by introducing spin-3/2 nucleonic resonances [5]. As a next step, we generalize our formalism to include the other relevant channels with both real and virtual photons, and the effects of spin-5/2 resonances, as discussed in the following.

The present model, hereafter refered to as SALY-1 (for SAclay-LYon), is obtained [3] by fitting photo- and electroproduction cross sections (as discussed in [2]), Λ-polarization data, and the branching ratio for the radiative K$^-$ capture [6]. This model contains, besides the extended Born terms, the following resonances exchanges in the intermediate states :

K* : K*(892) , K1(1270) ;
N* : N(1440), N(1680), N(1700), N(1720) ;
Δ* : Δ (1232), Δ (1910), Δ (1920) ;
Λ* : Λ(1405), Λ (1670), Λ(1810) ;
Σ* : Σ (1660).

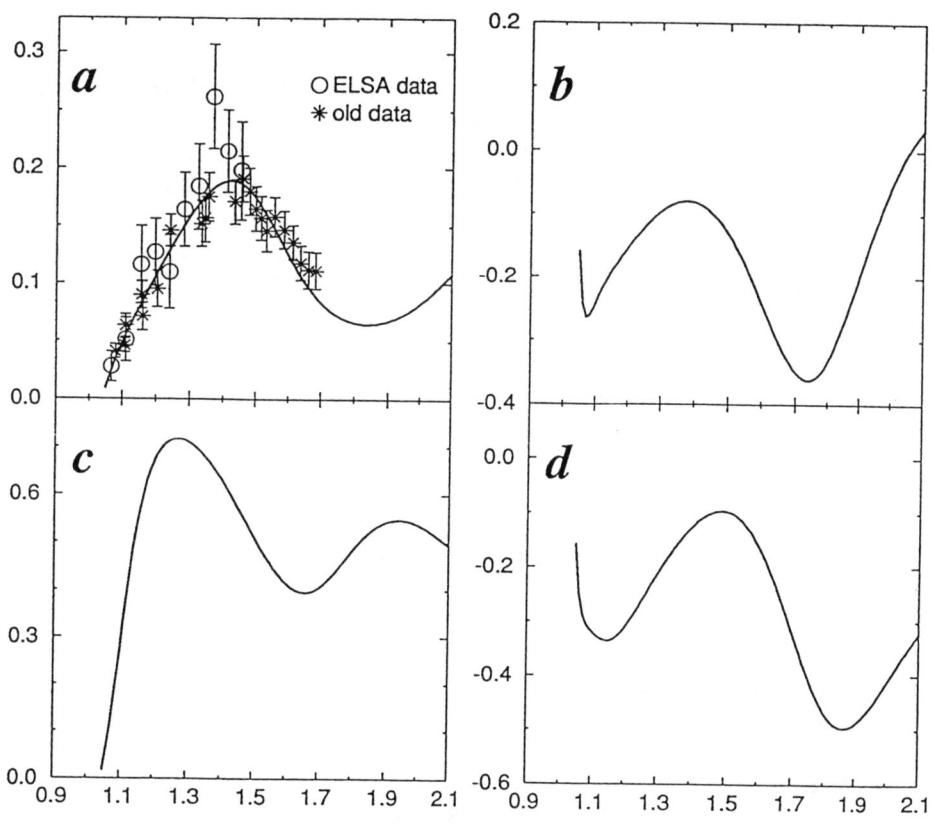

FIGURE 1. Excitation functions for the process $\gamma + p \to K^+ + \Sigma^0$ at $\Theta_K^{cm} = 90°$: a) differential cross section in the center-of-mass frame (μb/sr), b) hyperon polarization (P), c) Beam polarization (Σ), d) polarized target asymmetry (T).

The masses are given in parenthesis (MeV); the isospin-3/2 nucleonic resonances are shown separately since they contribute only to the Σ channels.

The SALY-1 model was obtained by the least-squares minimization procedure, using MINUIT [7]. The reduced χ^2 for some 500 data points is 1.4, and the two main coupling constants ($g_{K\Lambda N}/\sqrt{4\pi} = -3.2$, and $g_{K\Sigma N}/\sqrt{4\pi} = 0.8$) are in agreement with broken SU(3)-symmetry predictions as well as their extracted values from hadronic sector [2].

Figure 1(a) shows the result of our model for the differential cross section at $\Theta_K^{cm} = 90°$. In Figures 1(b) to 1(d) we produce predictions, at the same angle, for the single polarization observables which have been found [8] to be very sensitive to the dynamics of the reactions under consideration. Besides, the radiative capture branching ratios $BR = \Gamma_{(K^-p\to\gamma Y)}/\Gamma_{(K^-p\to all)}$ have been measured [6] to be $BR_{\gamma\Lambda} = (.86\pm.12)10^{-3}$ and $BR_{\gamma\Sigma} = (1.41\pm.18)10^{-3}$. The SALY-1 model reproduces well these data, giving $.95 \times 10^{-3}$ and 1.41×10^{-3}, respectively.

New data on all these observables as well as the double polarization asymmetries (expected at CEBAF, ELSA, and GRAAL) shall put strong constraints on the reaction mechanism of the elementary strangeness production process using electromagnetic probes.

REFERENCES

1. F.M. Renard and Y. Renard, *Nucl. Phys.* **B25**, 490
2. R. Adelseck and B. Saghai, *Phys. Rev.* **C42**, 108 (1990).
3. J.C. David, Ph.D. Thesis, University of Lyon, 1994 (in French); J.C. David, C. Fayard, G.H. Lamot, and B. Saghai, in preparation.
4. M. Bockhorst et al., *Z. Phys.* **C63**, 37 (1994).
5. J.C. David, C. Fayard, G.H. Lamot, and B. Saghai, "Kaon photoproduction: effect of spin 3/2 resonances", in *Proceedings of the 14th International Few-Body Conference*, Williamsburg, May 26-31, 1994; Ed. F. Gross, to appear.
6. D.A. Whitehouse et al., *Phys. Rev. Lett.* **63**, 1352 (1989).
7. F. James, M. Roos, MINUIT, CERN Report D506, (1981).
8. B. Saghai, "Spin observables in kaon photoproduction", Invited talk presented at the *6th workshop on Perspectives at Intermediate Energies*, Trieste, May 3-7, 1993, Eds. S. Boffi, C. Ciofi degli Atti, and M. Giannini, to appear.

DUALITY AND K⁺ Λ PHOTOPRODUCTION ON THE PROTON

Bijan Saghai[a], and Frank Tabakin[b]

a) Service de Physique Nucléaire, CEA/DSM/DAPNIA, Centre d'Études de Saclay
F-91191 Gif-sur-Yvette Cedex
b) Department of Physics and Astronomy, University of Pittsburgh,
Pittsburgh, Pennsylvania 15260

Abstract. Spin observables for the reaction $\gamma p \rightarrow K^+\Lambda$ are examined using three recent dynamical models and are compared to the general features of such observables deduced recently by Fasano, Tabakin and Saghai. An instructive surprise, which occur in this comparison, is then discussed showing a clear indication of the duality hypothesis in the strangeness realm.

In a recent paper [1] (FTS), we considered the general structure of the full set of 15 single and double polarization spin observables for the reaction $\gamma p \rightarrow K^+\Lambda$. Using helicity amplitudes, we deduced general rules concerning the angular structure of these observables, which were found to fall into four "Legendre classes" with four members in each class. The observables in each class have similar "nodal structure" possibilities, e.g. their values at 0° and 180° and their possible intervening nodes are of related nature. The Legendre classes of the sixteen observables, which are labeled by $\mathcal{L}_0$, $\mathcal{L}_{1a}$, $\mathcal{L}_{1b}$, and $\mathcal{L}_2$ are :

$$\mathcal{L}_0(I, \hat{E}, \hat{C}_{z'}, \hat{L}_{z'}); \quad \mathcal{L}_{1a}(\hat{P}, \hat{H}, \hat{C}_{x'}, \hat{L}_{x'}); \quad \mathcal{L}_{1b}(\hat{T}, \hat{F}, \hat{O}_{x'}, \hat{T}_{z'}); \quad \mathcal{L}_2(\hat{\Sigma}, \hat{G}, \hat{O}_{z'}, \hat{T}_{x'}).$$

In this contribution, those rules are confronted with three recent dynamical models [2-4] for the $\mathcal{L}_2$ class observables and the relationship between t-channel exchange and higher angular momentum states is discussed. In the dynamical models examined here the very limited number of resonances in s- and u-channels is suplemented by t-channel mechanisms, such as $K^*(892)$ and K1(1270) exchanges. According to the FTS rules [1], close to threshold the $\mathcal{L}_2$ class observables are expected to have either zero or ≤ 1 intervening nodes, when the total angular momenta is limited to J= 1/2 or $\leq 3/2$ states, respectively. The relevant nodal trajectories are shown in Fig. 1 for models by Adelseck-Saghai (AS) [2], Williams-Ji-Cotanch (WJC) [3], and David et al. (SALY) [4]. These curves have been calculated in the appropriate energy domain for each model, i.e. $E_\gamma \leq 1.5$ GeV for AS and $E_\gamma \leq 2.1$ GeV for WJC and SALY. All three models are based on diagrammatic technique and contain s- u- and t-channel exchanges. For our purposes, the main feature of these phenomenological models is that the first two contain only spin-1/2 baryonic resonances, while SALY also includes spin 3/2 nucleonic resonances. So, at low energy we expect no intervening nodes for the two spin-1/2 models and a maximum of 1 node for SALY. Such a behavior is verified for three $(G, O_{z'}, T_{x'})$ of the four observables depicted in Fig. 1. However, the polarized beam asymmetry Σ shows an unexpected double nodal structure for the AS model. The general form of the Σ observable is $\sin^2 \theta \, (a + b \cos \theta + c \cos^2 \theta)$. where the terms a and b arise from interference between all $J \geq 1/2$ states, while the term c arises only if $J \geq 5/2$ states contribute. Therefore to get the double nodal structure seen for the AS model near threshold, one needs a sizeable c term or, equivalently, at least J=5/2 multipoles. Such multipoles could arise from a mechanism that boosts the orbital angular momentum L to higher values. The striking of a virtual P-wave meson by a polarized incident photon (t-channel exchange) provides such a mechanism.

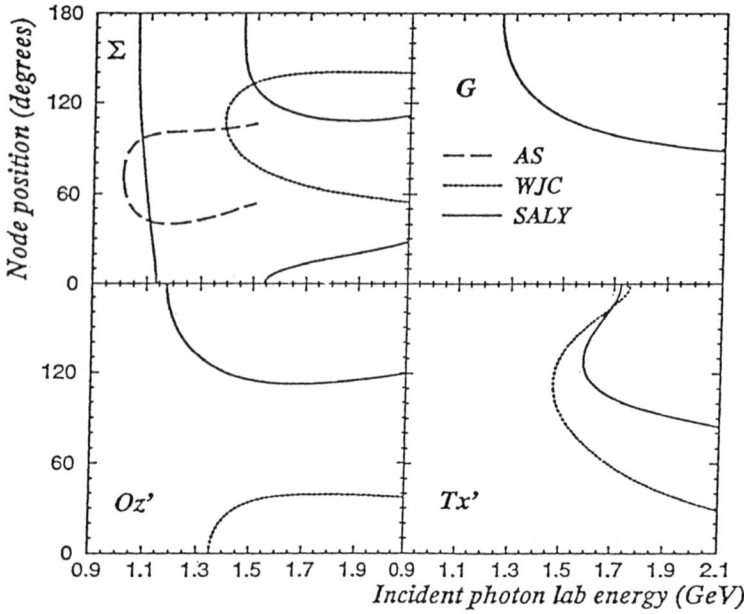

FIGURE 1. Nodal trajectories for class $\mathcal{L}_2$ polarization observable asymmetries: beam (Σ), beam-target (G), beam-recoil ($O_{z'}$), target-recoil ($T_{x'}$); with beam linearly polarized. The WJC (dotted curves) and the SALY (solid curves) satisfy the FTS rules. Only the AS model, given by the dashed curves, has double nodes near threshold for the photon asymmetry Σ observable. This indicates a role for higher J-states arising from t-channel dynamics or by duality from higher J resonances.

To understand the origins of this feature, we have performed [5] a multipole analysis for all three models. The AS model contains, besides extended Born terms, the exchange of one N*, one Λ*, and two kaonic resonances in s- (u-) and t-channels, respectively. The orbital angular momentum behavior of the multipoles is correct; namely, multipoles with smaller L's are larger. However, a configuration (fitted to the same data base as AS), containing only Born terms and the t-channel, shows a peculiar inversion: the L=2 multipoles are then more important than the L=1 multipoles. It is therefore the t-channel that introduces higher J amplitudes and gives the AS model its extra Σ nodes. In other words, a t-channel mechanism contributes to higher angular momentum amplitudes, which in a physical description corresponds to the incident photon boosting the angular momentum of a virtual t-channel meson. The magnitude of this t-channel role in contributing to higher J multipoles depends on the strength of the associated s-and u-channel resonances. That input varies from model to model. Thus we see that the nodal structure of the photon polarization asymmetry Σ provides an important constraint on the magnitude of t-channel mechanisms at low energies producing a clear manifestation of the duality hypothesis [6] in strangeness physics.

With new data forthcoming from ELSA, CEBAF, and GRAAL and more elaborate formalisms being developed which include higher spin baryonic resonances [7], the role of the t-channel, of higher J multipoles and of duality on selected spin observables as stressed in this contribution should be examined thoroughly. Here we point out that measurement of the photon polarization asymmetry can play an important role in our understanding of t-channel exchange processes.

ACKNOWLEDGMENTS

We thank F. Piron for his help in extracting nodes for the SALY model.

REFERENCES

1. C.G. Fasano, F. Tabakin, and B. Saghai, Phys. Rev. C **46**, 2430 (1992).
2. R. Adelseck and B. Saghai, Phys. Rev. C **42**, 108 (1990).
3. J.C. David, C. Fayard, G.H. Lamot, and B. Saghai, "Kaon photoproduction: effect of spin 3/2 resonances", in *Proceedings of the 14th International Few-Body Conference*, Williamsburg, May 26-31, 1994; Ed. F. Gross , to appear.
4. R. Williams, C. Ji, and S. Cotanch, Phys. Rev. C **46**, 1617 (1992).
5. B. Saghai and F. Tabakin in preparation.
6. See, e.g., S.J. Lindenbaum, *Particle-interaction physics at high energies*, Oxford, Clarendon Press, 1973, ch. 13, p. 444.
7. J.C. David, C. Fayard, G.H. Lamot, F. Piron, and B. Saghai, this proceedings.

Exclusive pion production from ^{16}O using Polarized Photons

K. Hicks[1], R. Finlay[1], J. Rapaport[1], R. Lindgren[2], V. Gladyshev[2], H. Baghaei[2], A. Cichocki[2], T. Gresko[2], B. Norum[2], R. Sealock[2], L. Smith[2], S. Thornton[2], A. Caracappa[3], S. Hoblit[3], O. Kistner[3], L. Miceli[3], A. Sandorfi[3], C. Thorn[3], M. Khandaker[4], C.S. Whisnant[5], M. Lucas[5]

[1]*Ohio U.,* [2]*U. Virginia,* [3]*BNL,* [4]*VPI,* [5] *South Carolina U.*

Abstract. The $(\vec{\gamma}, \pi^- p)$ reaction was measured for ^{16}O at photon energies of 200 to 320 MeV, and angles ranging from about 20° to 160° for the proton and 32° to 135° for the pion. The photon asymmetry data are compared with calculations in a DWIA framework. The agreement between the present data and calculations is reasonable at quasifree kinematics, but disagree as the momentum transfer to the residual nucleus becomes large.

Introduction

Data for the photoproduction of pions from hydrogen and deuterium targets at energies near the Δ resonance peak have enhanced our knowledge of the pion production vertex. In the nucleus, pion photoproduction data can be compared with data from the free nucleon to show how the observables are modified due to the presence of nuclear matter. Specifically, we hope to learn about the Δ-nucleus interaction by studying the $(\vec{\gamma}, \pi^- p)$ reaction. Photoproduction is perhaps the cleanest way to investigate pion production, since the initial interaction proceeds by the well-known electromagnetic force.

The experiment was done at the Laser-Electron Gamma Source (LEGS) at the Brookhaven National Laboratory. Distinct advantages of this facility are the > 90% polarization of the photons and the high duty cycle. The high duty cycle is ideal for coincidence measurements such as $(\vec{\gamma}, \pi^- p)$. Polarization asymmetry data are relatively free of normalization problems, and are much less sensitive to the effects of spectroscopic factors than are cross section data. The nuclear medium effects are most easily seen when both the analyzing power and the cross section are measured together and compared to predictions by theoretical models.

The experiment used an H$_2$O target, with incident photons over an energy range of 200 to 320 MeV and an angular range of 20° to 140° for the proton and 25° to 165° for the pion. The proton detectors were plastic scintillator bars stacked in pairs with thin ΔE plastic scintillator 'paddles' in front for

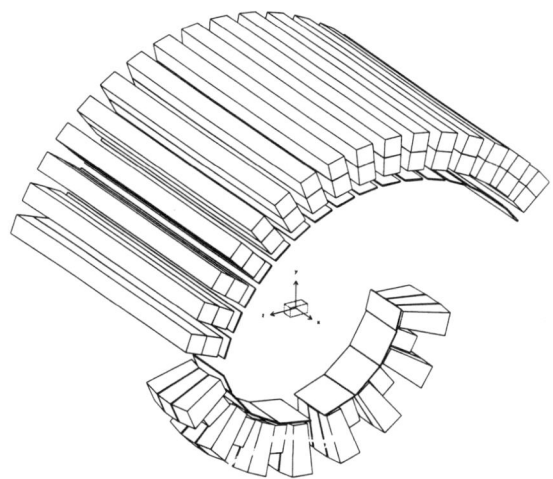

Figure 1: Isometric view of the plastic bars at LEGS.

particle identification. The pion detectors consisted of a set of CsI detectors with $\sim 1\%$ energy resolution for 100 MeV pions, and thin ΔE paddles in front. In addition, a set of thick plastic scintillators were interleaved with the CsI detectors for better solid angle coverage. This setup is shown in Fig. 1.

The current state of pion photoproduction on nuclear targets has recently been reviewed in reference [1]. This book describes the inclusive reaction (γ, π^+) which requires the nucleon at the pion vertex to remain bound in the residual nucleus. As a result, the calculations are very sensitive to the nuclear structure of the target. Uncertainties in the nuclear structure can be large, even for the p-shell targets[2, 3].

The calculations for the inclusive (γ, π^+) reaction are in good agreement with data from p-shell nuclei at incident photon energies below about 260 MeV, but at incident energies near the Δ resonance, the calculations do not agree well with the data[4]. When final-state interactions (FSI) are done in the Δ-hole model[5], calculations are in better agreement than for FSI from a conventional pion optical model; however, even the Δ-hole calculation is too low by 25% at resonance energies. This situation indicates problems for kinematics in the Δ resonance region where the nucleon must remain bound. The present experiment removes this restriction by studying the exclusive process where both the π^- and p from Δ decay are measured directly. This measurement largely removes the influence of nuclear structure and permits investigation of the basic photoproduction processes.

Previous data for the $(gamma, \pi^- p)$ reaction are sparse. A recent publication of ^{16}O data from Bates [7] shows cross sections summed over all outgoing

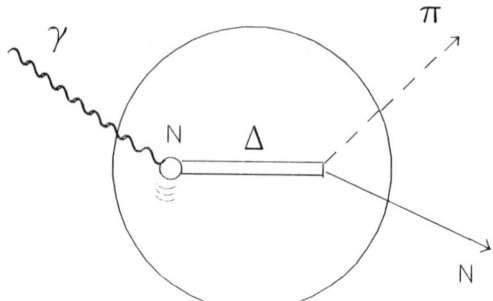

Figure 2: Impulse approximation diagram for the $(\vec{\gamma}, \pi^- p)$ reaction in the nucleus.

pion energies with statistics of only 20% or worse. An earlier data set is for a ^{12}C target from the Tomsk synchrotron[8] at only two proton angles. The low duty-cycle of these facilities is, in part, the reason for the poor statistics. These data are reviewed in references [9, 11].

$(\vec{\gamma}, \pi^- p)$ Calculations

Figure 2 is a diagramatic illustration of the DWIA: a photon enters the nucleus and interacts with a single bound nucleon to form a Δ resonance. The Δ propagates inside the nucleus and then decays into a pion and a nucleon. The outgoing pion and nucleon interact with the other nucleons while leaving the nucleus. The non-resonant Born diagrams have similar features, except that an intermediate delta is not formed.

The DWIA calculations of Lee, Wright, and Bennhold[11] use harmonic oscillator wavefunctions to describe the bound nucleon as an initial step. The full Blomqvist-Laget production operator [12] is used, which describes the elementary pion photoproduction data quite well over a wide range of energies. The pion optical potential developed by Stricker, McManus and Carr [13] was chosen, along with a global parametrization for the nucleon optical potential. The pion production operator $t_{\gamma\pi}$ depends strongly on various momenta, and thus is a nonlocal operator in coordinate space. Such nonlocalities have been found important in explaining inclusive pion photoproduction data [10]. Since nonlocalities are difficult to handle in coordinate space, the calculations have been done exactly in momentum space.

When the DWIA calculations are compared to the $(\vec{\gamma}, \pi^- p)$ cross section data of reference [7], the calculations are somewhat too high at backward pion angles, but exceeds the data at at forward pion angles by more than a factor of three [11]. The calculations include contributions from the $p_{3/2}$ and $p_{1/2}$

shells and use a spectroscopic factor of 0.6. Although basic shapes of the cross sections measured seem to be consistent with the DWIA, the magnitude of the calculations is too large. Better agreement is obtained by introducing an effective potential for the Δ resonance as it propagates through the nucleus. The Δ-nucleus potential is approximated in the calculations by reducing the Δ mass by a few percent.

Results

The *preliminary* photon spin asymmetry data $\Sigma = (\sigma_\| - \sigma_\perp)/(\sigma_\| + \sigma_\perp)$ for the present experiment are shown in Fig. 3 as a function of the pion angle, for fixed proton angles of 20° and 44°. The data in these figures has been integrated over all proton and pion energies > 10 MeV, and photon energies between 270 and 320 MeV. The asymmetry data have been corrected for accidental coincidences. The solid curve in Fig. 3 is from the model by Li and Wright described above, integrated over the same phase space. The dashed curve is the same calculation, but with the Δ mass reduced by 3%.

Data for only a selected range of angles can be shown in this paper. The general trend is that the calculations overpredict the asymmetry data at forward pion angles, but underpredict the data at backward pion angles. Near the kinematics for pion photoproduction from the deuteron (free kinematics), the calculations show good agreement with the data, within error bars.

Even though the data are still preliminary, we note that reducing the Δ mass in the calculations worsens the agreement with the asymmetry data, while at the same time reducing the cross section. The next step in the data analysis will be to extract the cross sections. It will be interesting if the cross sections for the present data are below the calculations, as one might expect from the measurements of reference [7].

Cross sections for the $D(\vec{\gamma}, pn)$ reaction were measured using the present experimental setup, using a heavy-water target. The results are in good agreement with previous results. This gives us confidence that the cross sections for the $^{16}O(\vec{\gamma}, \pi^- p)$ reaction will have small systematic errors.

Summary

The photoproduction of pions is one of the cleanest ways to test models of particle production. Calculations for the (γ, π^+) reaction fall short of the data at energies near the Δ resonance. This reaction suffers from the harsh requirement of keeping all nucleons in their ground state. The $(\vec{\gamma}, \pi^- p)$ reaction largely removes the uncertainties associated with the bound state nucleon, allowing investigation of the basic pion photoproduction process. The present high-statistics data are a considerable improvement over previous data for this

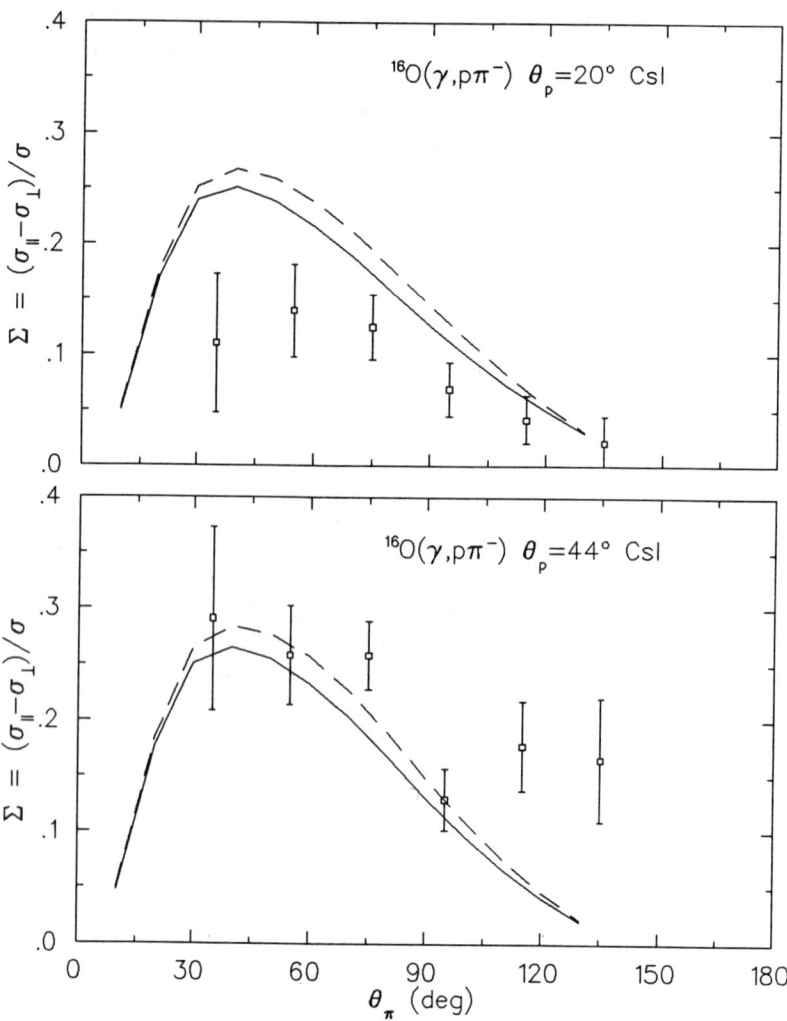

Figure 3: Photon spin asymmetry data at $E_\gamma \simeq 290$ MeV compared with calculations.

reaction, and the photon asymmetry data are much less sensitive to the normalization problems or spectroscopic factors. A full report of the data, including cross sections, is in progress. Preliminary data indicate good agreement with calculations near the free kinematics, but overpredict the asymmetries at large momentum transfer to the residual nucleus.

References

[1] A. Nagl, V. Devanathan, and H. Überall, *Nuclear Pion Production*, Springer-Verlag, Berlin, 1991.

[2] D. Koetsier, K. Amos, L. Tiator, C. Bennhold, and L.E. Wright, Phys. Rev. C45, 230 (1992).

[3] B.C. Doyle, in *Excited Baryons 1988*, World Scientific, 1989, p. 373.

[4] L. Ghedira, et al., Phys. Rev. C41, 653 (1990).

[5] R. Wittman and N.C. Mukhopadhyay, Phys. Rev. Lett. 57, 1113 (1986).

[6] L.E. Wright, private communication.

[7] L.D. Pham, et al., Phys. Rev. C 46, 621 (1992).

[8] P.S. Anan'in and I.V. Glavanakov, Sov. J. Nucl. Phys. 52, 205 (1990).

[9] G. van der Steenhoven, in *Proceedings of the 7th Annual NIKHEF conference*, Amsterdam, 1991, p. 92.

[10] L. Tiator and L. E. Wright, Phys. Rev. C 30, 989 (1984); C. Bennhold, L.Tiator and L. E. Wright, Can. J. Phys., 68, 1270 (1990).

[11] X. Li, L.E. Wright, and C. Bennhold, Phys. Rev. C 48, 816 (1993).

[12] K. I. Blomqvist and J. M. Laget, Nucl. Phys. A280, 405 (1977).

[13] K. Stricker, H. McManus, and J. A. Carr, Phys. Rev. C 19, 929 (1979); 22, 2043 (1980); 25, 952 (1982).

Status of the T_{20} Experiment at VEPP–3

S.G.Popov, S.I.Mishnev, D.M.Nikolenko, D.V.Petrov, I.A.Rachek,
A.V.Sukhanov, D.K.Toporkov, E.P.Tsentalovich[1], A.V.Volosov,
and B.B.Wojtsekhowski[2]

BINP, Novosibirsk, Russia

C.E.Jones, R.S.Kowalczyk, M.Poelker, D.H.Potterveld and
L.Young

Argonne National Laboratory, Argonne, IL, USA

R.J.Holt

University ol Illinois, Urbana-Champaign, IL, USA

R.Gilman

Rutgers University, Piscataway, NJ, USA

E.R.Kinney

Colorado University, Boulder, CO, USA

K.P.Coulter

University of Michigan, Ann Arbor, MI, USA

J.A.P.Theunissen, C.W.de Jager, H.de Vries,

NIKHEF-K, Amsterdam, The Netherlands

V.V.Nelyubin, V.V.Vikhrov

INP, St.-Petersburg, Russia

A.N. Osipov, V.N. Stibunov

INR, Tomsk, Russia

1 Introduction

It is difficult to overestimate the importance of a comprehensive analysis of the simplest nuclear system - a deuteron. Electromagnetic probes are appropriate for an experimental study of the deuteron, due to the transparent interpretation of the results. Polarization experiments are of special interest since the polarization observables, such as T_{20} in elastic $e-d$ scattering, are sensitive to very delicate aspects of nucleon-nucleon interactions, such as isoscalar meson-exchange currents, relativistic effects and quark degrees of freedom.

1.1 T_{20} experiments: finished, running and proposed

In the last decade a few experiments were performed with polarization techniques in electron–deuteron elastic scattering [1-6]. Different approaches were used: polarimeters of recoil deuterons at BATES [1, 2], a cryogenic $^{15}N\vec{D}_3$ target at Bonn [3] and an internal gaseous polarized target (first a jet target and

---
[1] Present address: MIT Bates, MA, USA
[2] Present address: CEBAF, Newport News, VA, USA

later a storage cell) at VEPP-3, Novosibirsk [4, 5, 6]. Experiments at VEPP-3 were carried out by an Argonne / Novosibirsk / St. Petersburg / Tomsk / NIKHEF collaboration. Note that as far as we know the experiments at Bates and at Bonn are finished and there are no plans to continue these experiments in order to improve the accuracy or to move to a different momentum transfer range.

A program of intensive study of spin properties of the deuteron is planned for CEBAF. The proposed experiment 93-04 is a continuation of the Bates experiment up to a momentum transfer 6.8 fm^{-1} with a new polarimeter "POLDER" [7]. Two other experiments will use the $\vec{D}(\vec{e},e'd)$ reaction with a cryogenic vector polarized $N\vec{D}_3$ target. The first experiment (spokesperson J.Mitchell) will use a magnetic spectrometer in Hall C. The second experiment (spokespersons B.Wojtsekhowski and D.Crabb) will measure the monopole deuteron form factor using the CLAS spectrometer in Hall B. Also there is discussion about doing a T_{20} experiment at HERA during the HERMES experiment. Note that the experiments at CEBAF are not approved so far [3] and the results of the measurements are expected to be obtained not earlier than 1998. In addition, these experiments are based on a different approach with different systematic uncertainties than the internal target experiment at VEPP-3 and therefore do not reduce the importance of the experiment proposed in this paper.

Fig.1 shows the world T_{20} data acquired to date.

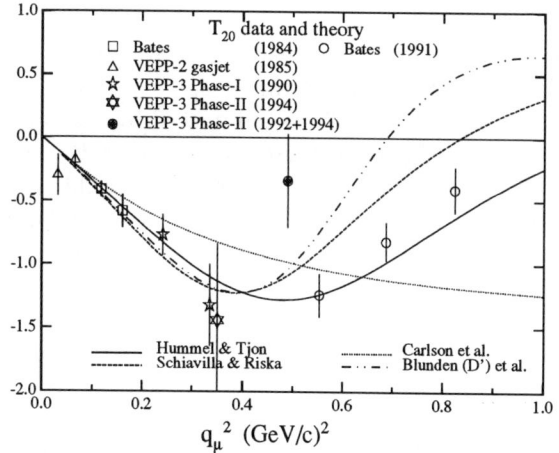

Figure 1: World T_{20} data

---

[3]The "POLDER" experiment is now approved and given a high priority

1.2 Three phases of the T_{20} experiment at VEPP-3

Started in 1987, the experiment at VEPP-3 was planned to be conducted in three phases:

   I. the use of a simple passive storage cell and a conventional atomic beam source;
   II. an active high-density storage cell with a conventional source;
   III. a laser-driven source and a high density passive storage cell.

At the final phase of this experiment it was expected to be able to make measurements of T_{20} up to a momentum transfer range 4.5 fm^{-1}.

2 Phase-I and Phase-II experience

2.1 Phase-I

The storage cell of Phase–I of the experiment was installed in the VEPP-3 straight section in February 1988. The detector set-up, the ABS and the holding field magnet were the same as for the polarized jet experiment. That is why while the total increase of the target thickness as compared with the plain jet target was a factor of about 15, the achieved gain was only 3. The main objectives of Phase-I were to show the feasibility of the storage cell concept, to investigate depolarization processes and to collect data on T_{20} in elastic scattering for 1...2.5 fm^{-1}. The latter was successfully done in a period from August 1988 to April 1989. The analysis of the elastic scattering data at the lowest momentum-transfer value allowed us to extract the properties of the storage cell internal target by normalizing to theoretical predictions: the target thickness in central part of the cell viewed by the detectors, $t_{vis} = 3 \cdot 10^{11} atoms/cm^2$; average degree of tensor polarization $P_{zz} = 0.57 \pm 0.05$. Figure 2 presents the measurements of the asymmetry at this lowest-Q value during the experiment. No substantial degradation of the polarization could be seen over a 7-months period. This demonstrated that the drifilm coating was of sufficient radiation hardness.

2.2 Phase-II

The first example of a movable storage cell was installed on the VEPP-3 straight section in May, 1989. That was a 'clam-shell' design driven by a single motor via a rotating vacuum feedthrough. First successful closing of the cell with the beam of 2 GeV electrons in VEPP-3 was achieved on July 27, 1989. The same ABS was used. A new detector was constructed and installed in February, 1991. In 1990 a long-term data-taking run has been carried out

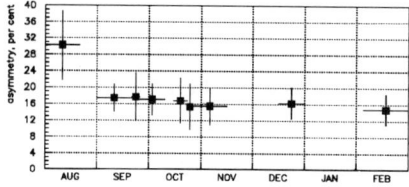
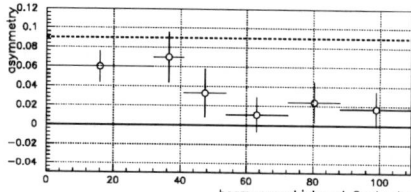

Figure 2: Asymmetry at low-Q measured during Phase–I.

Figure 3: Low-Q polarimeter asymmetry measurements during the winter-94 run. Dashed line - the predicted value for $P_{zz} = 1.0$.

to measure the averaged tensor polarization of the target with the active cell. The run showed that the polarization degree is close to zero.

In principle, the reason of low measured target polarization might be one or several of the following:

- low polarization of atoms coming from the ABS due to inefficient RF transitions;
- high level of deuterium molecules from the ABS as well as of unpolarized residual gas inside the cell;
- depolarization by the intense pulsed magnetic field of the electron beam;
- systematic errors related to non-equal luminosity integrals collected in different polarization states;
- systematic errors related to high (and unknown) level of background processes (mainly from deuteron electrodisintegration) in the data which were selected as elastic events;
- bad cell coating;

Some of these effects are known to be negligible, others can be accurately estimated and corrected for:

- During this run and all further Phase-II runs the efficiency of the RF transitions was permanently monitored by a Rabi polarimeter and was observed to be close to 100%.
- Contribution of deuterium molecules $\approx 4\%$ was obtained from the measured dissociation fraction and the known geometry of the atomic beam source;
- Contribution of the residual gas inside the cell $\approx 7\%$ was estimated from the measured vacuum in the experimental straight section;
- Depolarization by the electron beam has been studied experimentally and theoretically in Phase-I [8] and that experience was used in design of the Phase-II. This allowed us to be sure that such a depolarization is negligible.

- Polarization states (the sign of P_{zz}) were changed every 1-2 minutes, time and collected beam charge in each state were measured. It was found that integrals of beam current in each polarization state differed by less than 1% for a long-term run. Moreover, this difference is taken into account.
- Our confidence in the data analysis is based on our experience with the similar detector systems in Phase-I and in earlier experiments. Detailed discussions on the requirements for a detection system and an analysis procedure can be found in ref. [9].

The behaviour of the drifilm cell coating under real experimental conditions is the least known item. A number of reasons might cause a degradation of the cell coating:

- irradiation by synchrotron light. Its influence on drifilm properties is still not known in detail. It is obvious that this effect can be suppressed by installation of screens and light absorbers in the proper places. Unfortunately, it was found that screens placed just in front of the cell are not sufficient since there remains light coming from a rather big angle which cannot be screened. That is apparently light from the bending magnet reflected on an inner surface of the VEPP-3 vacuum chamber.

- radiation damage caused by electrons from the electron beam. Again no quantitative data on this effect is available.

- direct hitting of the cell by the electron beam. That is unlikely when the cell is stable but might occur when a movable cell is closing and the electron beam was not aligned properly.

- mechanical damage. This could occur either because of surface defects during initial drifilm covering which lead to slicing off part of the coating during installation of the cell, or due to a malfunction of the cell-moving mechanism causing an excessive strain. The latter is potentially dangerous only for a movable cell.

In the latter two cases one would expect an instantaneous degradation of the coating and, hence, of the target polarization. In the former two the degradation should be gradual. So the most probable reason of the low measured asymmetry is damage of the drifilm cell coating.

The cell has been removed from the VEPP-3 vacuum chamber in August 1990 during the summer shutdown of the ring. Damages of the drifilm coating have been observed in several places as well as dark spots on the inner surface of the cell related to an irradiation by synchrotron light.

Data were accumulated at an electron energy of 2 GeV in a period December 1991 to May 1992. The integrated charge was 156 kC. As in Phase-I we have binned the data in two q-ranges and the datum at the lowest value of Q was used to determine the average P_{zz} and the thickness of the target. We thus found that $P_{zz} = 0.68 \pm 0.22$ and the target thickness $t_{vis} = (3.0 \pm 0.5) \cdot 10^{12} atoms/cm^2$

However, during this run a serious problem was encountered – a high background rate spoiled the energy and spatial resolution of the particle detectors and enlarged the dead-time of the read-out electronics. That is why we had to work with a substantially (factor 2.5) smaller electron beam current than the VEPP-3 storage ring can potentially provide. The origin of the background was investigated and associated with electrons from the beam "halo" hitting the cell wall. To alleviate this problem, four movable collimators were installed during the summer–1992 shutdown of the ring. Further measurements have shown that the collimators did decrease the background rate by a factor 3–10 for various detectors. Besides, a new read–out electronics, based on INMOS transputers was introduced, which reduced the dead–time by an order of magnitude.

One more data-taking run at 2 GeV electron energy was performed from November 1992 to April 1993. The asymmetry at the lowest Q-value (3 ± 0.25 fm^{-1}) was found to be $a = -0.07 \pm 0.16$, while the expected value is $+0.73$ for $|P_{zz}|=1.0$, suggesting a degradation of the drifilm coating. Note that by the end of this run the total amount of beam charge passed through the cell since it was installed in 1991 was about 500 kC.

This result has shown once again that it is essential to have a fast monitor of the average tensor polarization in the target. It was decided to install an additional detector arm to detect electrons scattered over a small angle ($\approx 7°$), to be able to take data of elastic scattering at low Q (around 1 fm^{-1}) where the cross-section is high and the asymmetry is known with high accuracy. Such a detector was installed in May 1993.

A short special run with the same storage cell was carried out in June July 1993 to test the new detector and to measure the polarization of the target. In this run the holding field direction ($\theta_H = 83°$) was chosen such as to provide a maximal asymmetry for the low-Q polarimeter kinematics. An integrated beam charge of 7.3 kC was collected. It was clearly shown that such a simple "low-Q" polarimeter is an effective tool for checking the polarization i.e. a state of the cell coating. The asymmetry was measured to be $a = 0.093\pm0.030$ which corresponds to the target polarization of $P_{zz} = 0.69\pm0.22$. On the other hand, this result with the low-Q polarimeter clearly contradicts the value obtained from the normalization of the actual T_{20}–data at the low Q bin.

Despite of the positive result with the low-Q polarimeter run, it was decided to take out the storage cell during the summer–1993 shutdown of the VEPP-3 to cover the cell surface with "fresh" drifilm.

The last data-taking run has started in October 1993. However, due to various instrumentation problems only in February 1994 did we start running with the polarized target. Figure 3 shows the asymmetry measured by the low-Q polarimeter during the run. One can see that somewhere in the middle of the run the polarization decreased, implying that something disastrous happened with the cell coating. However, one has to take into account that by that time (March 1994) the integrated charge passed through the closed cell since it was installed in August 1993 was about 150 kC.

Before and after this run there were also short runs with unpolarized hydrogen and deuterium targets as well as with an 'empty' cell to calibrate the detector and to check the analysis. Such runs have been regularly performed during previous data taking runs also. These runs clearly showed that there is no false asymmetry from either the main detector or the low-Q polarimeter. The obtained energy and spatial resolution of the detector are listed in the Table 1 together with the design values. One can see that the detector parameters are slightly worse than expected. Nevertheless they are sufficient to select the events of elastic scattering at a momentum transfer up to at least 4.0 fm^{-1}. However, we are going to improve the detector performance mostly by upgrading the tracking system. This includes the installation of the new vertex chambers with an improved design and the replacing a part of θ–wire-planes in electron arms by the ones with twice smaller wire spacing.

Parameter	designed	measured
angle resolution FWHM, degrees	0.6	1.8
vertex position resolution, FWHM :		
across the beam, mm	1.0	2.5
electron energy resolution, %	7.0	8.5
hadron energy resolution, FWHM, MeV	15	14

Table 1: Detector Parameters, designed and measured

2.3 Data analysis

The layout of the particle detector is shown in Fig. 4. This detector is based on the same principle as our previous detectors [9] for internal target experiments: large angular-acceptance non-magnetic instruments detecting the scattered electron and the knocked-out hadron in coincidence. The present detector package includes a set of wire chambers for tracking, a hodoscope of plastic scintillators for a hadron detection and a layered CsI(Tl)/NaI(Tl) electron calorimeter.

The crucial task for the data analysis is the separation of elastically scattered deuterons from the much greater flux of protons from inelastic and quasielastic scattering. This separation uses constraints on the vertex position, on

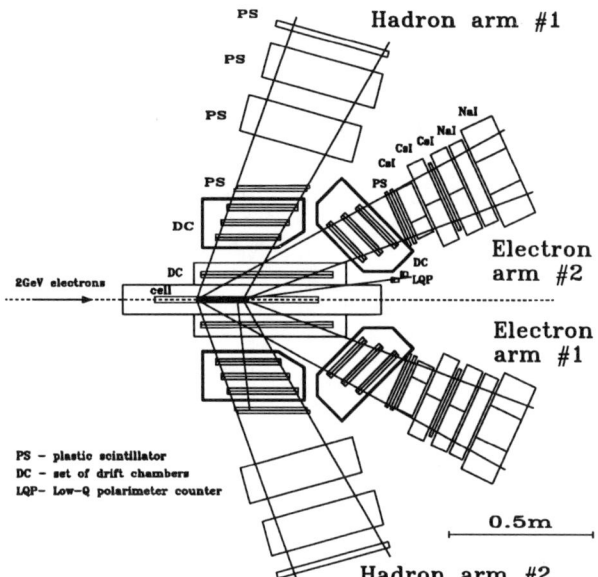

Figure 4: Layout of the detector set-up.

the electron energy deposited in the calorimeter, the kinematical correlation between the electron and deuteron scattering angles and energies and the differences in dE/dx in the scintillation counters.

The data analysis sequence comprises a number of stages:

1. Immediate rejection of events obviously not relevant to elastic scattering. The following cuts are used at this stage:

    - an energy deposition in the calorimeter exceeding ≈1.5 GeV;
    - the particle stopping in the 3rd plastic scintillator of the deuteron arm. This constrains the deuteron energy to a range from 80 to 150 MeV, corresponding to a momentum transfer range in elastic scattering of 0.55 to 0.75 GeV/c;

2. Conversion of the raw TDC and ADC data to physical parameters : angles and energies of detected particles, coordinate of escape point. Data from calibration frequently performed during the data run are used and various corrections are carried out at this stage.

3. Derivation of the correlation and identification parameters.

4. Final data analysis on a graphics workstation using PAW software package [10].

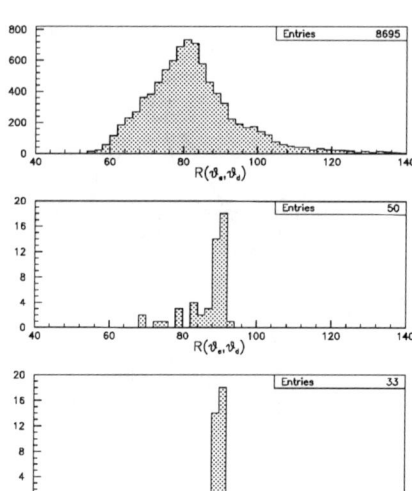

Figure 5: Histograms of the $R(\theta_e, \theta_d) = \theta_e + \theta_d$ correlation parameter. The upper histogram shows all events, the middle one the events left after applying all cuts except $R(\theta_e, \theta_d)$ and the lower one the events surviving all cuts.

The following six cuts were applied to single out the elastic scattering events when analyzing data of the winter-1994 run:

- Energy of scattered electron $\qquad$ >1500 MeV
- Vertex position transverse to
  the electron beam direction [mm] $\qquad$ -4...4
- Correlation between electron and
  deuteron azimuthal angles, $R(\phi_e, \phi_d) = \phi_e - \phi_d - \pi$ $\qquad$ -2.5°...2.5°
- Correlation between electron and
  deuteron polar angles, $R(\theta_e, \theta_d) = \theta_e + \theta_d$ $\qquad$ 88°...93°
- Correlation between deuteron energy and electron
  scattering angle, $R(E_d, \theta_e) = 10 \cdot \theta_e - A_3/12$ [a.u.] $\qquad$ 135...175
- $\Delta E - E$ hadron identification parameter
  $I(\Delta E, E) = (A_3 + 300) \cdot (A_1 + A_2 - 200)/1000$ [a.u.] $\qquad$ 170...300

Fig. 5 shows histograms of $R(\theta_e, \theta_d)$ correlation parameter before and after background rejection.

The preliminary analysis of the winter-1994 run gave the following results: the time-averaged target polarization obtained from the low-Q polarimeter data $P_{zz} = 0.52 \pm 0.08$; T_{20} values for two ranges of momentum transfer were found to be -1.44 ± 0.6 for $q^2 = 0.35(Gev/c)^2$ and -0.36 ± 0.75 for $q^2 = 0.50(Gev/c)^2$. Combining the high-Q bin value with that obtained in the earlier measurement we finally get $T_{20} = -0.34 \pm 0.37$. The results are shown in fig. 1.

2.4 Summary of Phase-I and Phase-II experience

Summing up the results and experience of Phase–I and Phase–II of the Novosibirsk T_{20} experiment we would like to stress the following.

- The storage cell concept to increase the thickness of polarized internal target proved to be feasible both for 'passive' and for 'active' cell design. However, a movable cell is more susceptible to potential damaging of the drifilm coating. This reason together with some others mentioned in next section made us choose a stable cell for the next phase of the experiment.
- The lifetime of the drifilm coating exceeds at least half a year of exposure at the interaction point of the storage ring in which **more than $\approx$150 kC** of integrated beam charge can be collected.
- The particle detectors and data acquisition system have shown an adequate performance. However a number of minor upgrades (mainly in the tracking system) will be carried out.
- Collimators installed in a proper place of the ring are effective in suppressing the high background rate produced by stray particles hitting the cell walls.
- It is very important to have a fast monitor of the average target polarization to be able to check the state of the drifilm coating during the run and to obtain an accurate value of degree of target polarization. It was shown that the Low-Q elastic scattering polarimeter should prove to be an effective tool to meet these needs.
- It is very useful to be able to change the cell quickly when the degradation of the polarization is detected. Such a design of the cell is being considered now for the next phase of the experiment.

We believe that we can now perform a successful run in the current phase, but a meaningful statistical result in a momentum transfer range $q > 3.5 fm^{-1}$ would require a long period of data-taking (about 2 years), which would interfere with the development of the Phase–III. Furthermore, in the present phase we can not replace the cell quickly, but have to wait until the shutdown of the VEPP-3 ring (usually in summer). That would further increase the time necessary to collect the required amount of data. As will be shown below, the successful development of Phase–III would provide better statistics in the same period, providing precision measurements most quickly.

3 Phase–III

A substantial increase in the luminosity will be achieved by the use of a novel source of polarized atoms which is based on the spin-exchange optical pumping technique and has been developed at Argonne National Laboratory [11].

Molecular deuterium is dissociated in an RF discharge and travels into a drifilm coated spin-exchange cell containing potassium vapor. The potassium is optically pumped and polarized by a laser beam with circular polarization in the presence of a high magnetic field. Electron polarization of potassium is transferred to deuterium atoms through spin-exchange collisions. The mixture of atoms passes through a transport tube where the potassium atoms are absorbed in a teflon sleeve. The deuterium atoms enter the storage cell after passing through an RF transition unit. The efficiency of a similar transition unit at low magnetic field and in the 20 MHz frequency region were measured to be $\varepsilon = 0.92 \pm 0.05$ [12]. The efficiency for the transitions 3→5 and 2→6 for deuterium still has to be tested. A high electron polarization of the atoms from the source (about 80%) has been measured at flow rates of $2 \cdot 10^{17}$ at/sec, while lower values were measured for flow rates exceeding $1 \cdot 10^{18}$ at/sec. The cell is a passive one with a 23×12 mm elliptical cross section and a 400 mm length manufactured from 0.1 mm thick aluminum. The cell is drifilm coated and should provide a total target thickness of about $1 \cdot 10^{14}$ at/cm^2 at a flow rate of 1×10^{17} at/sec. For these estimations a cylindrical tube with a 16 mm diameter and a 400 mm length at a 150°C wall temperature was assumed.

The presence of a potassium vapor in the flow forces us to keep most parts of the target (including the storage cell) under high temperature to prevent potassium condensation. It is obvious that a movable (like phase-II cell) and heated cell would be complicated to design. Moreover, a fixed cell can be made with a small wall thickness, thus significantly lowering the background. For these reasons we chose a fixed storage cell.

An important feature of Phase–III is compressing the electron beam near the experimental straight section through changes in the VEPP-3 electron beam optics (Fig. 6). We plan to add two new quadrupoles (DD and FD) inside the experimental straight section and to change the gradient of several existing quadrupoles (see Table 2).

Lens	D1	F1	D2	F2	DD	FD	D3	F3	D4	F4
Phase I,II	-1.67	1.67	-1.67	1.67	–	–	-1.67	1.67	-1.67	1.67
Phase III	-1.48	1.48	-1.67	2.15	-2.88	4.30	-2.15	1.67	-1.67	1.48

Table 2: Quadrupole gradients (kG/cm).

The compressed electron beam provides new possibilities:

- to use a stable storage cell with a small cross section.
- to apply an effective differential pumping scheme (two steps with high speed pumps and a tube with a conductance limiter between them) allowing acceptable vacuum conditions in the VEPP-3 ring at the high flux of atoms from an LDS (Fig. 7).

- size of DD and FD quadrupoles may be made small, which is important due to the restricted space at the straight section.

Phase	I	II	III
Polarized deuterium source type	ABS	ABS	LDS
Flux [$atoms/sec$]	$4 \cdot 10^{15}$	$4 \cdot 10^{15}$	$3 \cdot 10^{17}$
P_{zz}	≈ 1	≈ 1	$P_e \cdot K_{rf} \cdot K_{ds} =$ $.8 \cdot .9 \cdot .6 = .43$
Cell type	fixed	movable	fixed
Cell size [mm^3]	$24 \times 46 \times 940$	$9(13) \times 17(21.5) \times 569$	$13 \times 24 \times 400$
t – Visible target thickness [$atoms/cm^2$]	$3 \cdot 10^{11}$	$3 \cdot 10^{12}$	$2 \cdot 10^{14}$
Target P_{zz}	0.6	0.6	$0.8 \cdot 0.43 = 0.35$
Figure of merit $t \cdot P_{zz}^2$	10^{11}	10^{12}	$2.5 \cdot 10^{13}$
Detector system e^- scattering angle	$10° - 20°$	$20° - 30°$	
$e^- \Delta\phi \times N_{syst}$	$40° \times 4$	$60° \times 2$	
registration :			S A M E
tracking	yes	yes	
hadron energy	yes	yes	
electron energy	no	yes	
Anti-background scrapers	no	yes	yes
VEPP-3 optics	old	old	new
beam size σ_z/σ_x [mm]	0.35/1.4	0.35/1.4	0.20/0.7

Table 3: Parameters of the three phases of the T_{20} experiment at VEPP-3. P_e – atomic electron polarization, K_{rf} – efficiency of radio-frequency transition unit, K_{dc} – dissociation fraction.

The electron beam lifetime in Phase–III will be 2-3 times less than in Phase–II because of the increase in target thickness. Since it is the electrons scattered on the storage cell walls that define the singles rate of the detectors, one might expect an increase of the background. On the other hand, the smaller wall thickness of the new storage cell and the smaller size of the electron beam in the cell should reduce the background even more significantly. Therefore we expect background conditions in Phase–III to be better than in Phase–II.

The main experimental parameters of all phases are shown in Table 3.

An experimental result for the T_{20} which one can get in Phase–III is shown in Fig. 8. For this calculation the Paris potential was used, the target and

542 Status of the T_{20} Experiment

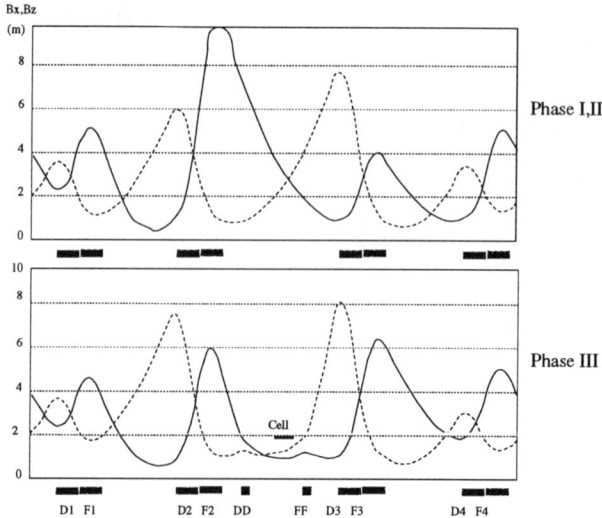

Figure 6: Radial (solid line) and vertical (dashed line) beta-functions in the experimental straight section of VEPP-3, before (upper panel) and after (lower panel) upgrade of the ring optics.

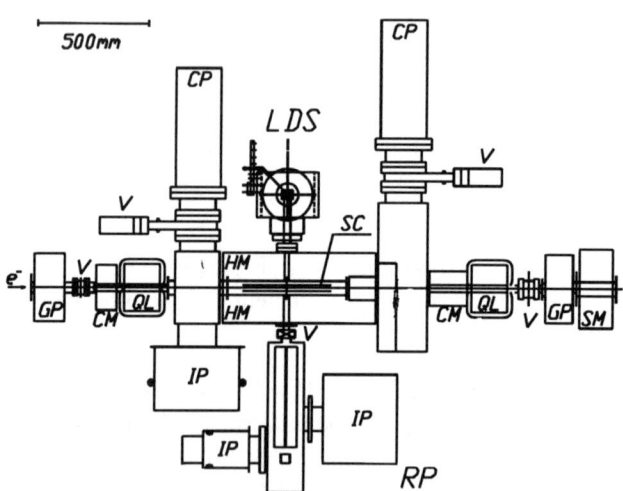

LDS-Laser-Driven Source, SC-storage cell, RP-Rabi polarimeter, HM-holding field magnet, QL-quadrupole Lense,CM-compensated magnet, SM-sextupole magnet, GP-getter pump, CP-cryopump IP-ion pump., V-valve

Figure 7: Top view of the Phase–III experimental straight section

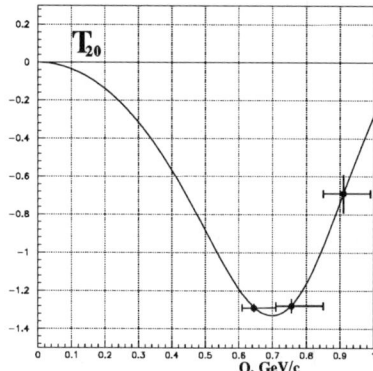

Figure 8: The expected accuracy of the T_{20} measurement in Phase–III. Calculations were performed using the Paris potential for an integrated charge of 400 kC. Target and detector parameters were taken from Table 3.

detection system parameters were taken from Table 3, an integrated electron beam charge was assumed to be 400 kC (or 100 days of beam time). The momentum transfer ranges for the three points are 3.1 - 3.6, 3.6 - 4.3 and 4.3 - 5.0 fm^{-1}. One can see that the proposed measurement will cover approximately the same momentum transfer range as in [2] but with higher accuracy.

4 Conclusion

Summing up the above we would like to note that

- experiments in Phase–I and Phase–II gave new results, besides operational experience with a storage-cell polarized target in an electron storage ring. This experience was used in the design of Phase–III of the experiment and allows us to be confident in its success;

- the new intense laser-driven source of polarized deuterium atoms, the new VEPP-3 electron beam optics and the new design of the straight section should result in a figure of merit in Phase–III about a factor of 30 larger than that in Phase–II;

- a precise measurement of T_{20} in an interesting region of momentum transfer will be performed within one or two years.

This research was supported in part by the U.S. Department of Energy, Nuclear Physics Division, under contract W-31-109-ENG-38; by the Russian Fund for Fundamental Research, under contract 123.06; by the Netherlands' Organization for Scientific Research (NWO), under contract nr. 713-119 and by the International Science Foundation, Grant N RBC000-PH1-6689-0925.

References

[1] M.E.Schultze et al., Phys. Rev. Lett. **52**(1984)597.

[2] I.The et al., Phys. Rev. Lett. **67**(1991)173.

[3] B.Boden et al., Z.Phys. **C49**(1991)175.

[4] V.F.Dmitriev et al., Phys. Lett. **157B**(1985)143.

[5] R.Gilman et al., Phys. Rev. Lett. **65**(1990)1733.

[6] E.P.Tsentalovich et al., Proc. of PAN XIII International Conf., Perugia, June 28 - July 2, 1993, p. 797, Ed. by A.Pascolini.

[7] S.Kox et al., Nucl. Instr. and Meth. **A346**(1994)527.

[8] R.Gilman et al., Nucl. Instr. and Meth. **A327**(1993)277.
K.P.Coulter, An active storage cell for a polarized gas internal target, submitted to Nucl. Instr. and Meth.

[9] L.G.Isaeva et al., Nucl. Instr. and Meth. **A325**(1993)16.
J.A.P.Theunissen et al., Nucl. Instr. and Meth. **A348**(1994)61.
E.P.Tsentalovich. Doctor Thesis (unpublished), Budker INP, Novosibirsk, 1991.
J.A.P.Theunissen. Doctor Thesis (unpublished), NIKHEF, Amsterdam, 1995.

[10] R.Brun et al. PAW - Physics Analysis Workstation. The complete Reference. CERN, Geneva, 1989.

[11] K.P.Coulter et al., Phys. Rev. Lett. **68**(1992)174.

[12] A.Zghiche et al., Proc. of Workshop on Polarized Gas Targets for Storage Ring, Heidelberg, 23-26 Sept. 1991, p.103, Ed. by H.G.Gaul, E.Steffens and K.Zapfe.

A New Deuteron Tensor Polarimeter and a Measurement of t_{20} in e-d Scattering at CEBAF.

S. Kox [a], C. Furget [a], J.S. Réal [a], L. Bimbot [b,c], C. Djalali [b,e], G.W.R. Edwards [c], M. Garçon [d], C. Glashausser [c], B.N.R. Johnson [b,e], M. Morlet [b], L. Rosier [b], E. Tomasi-Gustafsson [f], E. Voutier [a], A. Willis [b]
and the CEBAF t_{20} collaboration

[a] Institut des Sciences Nucléaires, 38026 Grenoble, France.
[b] Institut de Physique Nucléaire, 91400 Orsay, France.
[c] Rutgers University, Box 849, Piscataway, N.J. 08854, USA.
[d] DAPNIA/DPhN, CEN Saclay, 91191 Gif sur Yvette, France
[e] University of South Carolina, Columbia, S.C. 29208, USA.
[f] Laboratoire National Saturne, 91191 Gif-sur-Yvette, France.

Abstract. A new deuteron tensor polarimeter (*POLDER*), based upon the charge exchange reaction $^1\text{H}(\vec{d}, 2p)n$, has been constructed and calibrated in the energy range 175-500 MeV for all polarization signals. The large tensor figures of merit obtained make *POLDER* a unique instrument in this energy domain. Using this polarimeter at *CEBAF*, new measurements of the tensor polarization t_{20} of the deuteron produced in the e-d scattering can be performed at four-momentum transfers of $4.0 < Q < 6.8$ fm^{-1}.

In many intermediate energy experiments, the measurement of the polarization of recoil particles provides information on the mechanism involved in the primary nuclear reaction. This is deduced through the use of dedicated devices, *i.e.* polarimeters, whose functioning relies on particular nuclear reactions' producing sizeable experimental asymmetries related to the incident particles' polarization. For deuterons, the existing polarimeters have only permitted to measure recoil deuteron tensor polarization at energies lower than 250 MeV [1]. But there are many experiments which require polarimeters working at higher deuteron energies and Bugg and Wilkin [2] suggested that the $^1\text{H}(\vec{d}, 2p)n$ charge exchange reaction could be useful in this context.

Here we present a new tensor polarimeter, *POLDER*, based on this charge exchange reaction and working in the energy range 175-500 MeV. We also discuss an experiment proposed for *CEBAF*, where *POLDER* would be used to measure t_{20} of the deuteron produced in elastic e-d scattering.

POLDER

A polarization measurement requires a nuclear reaction with both large analyzing powers (T_{kq}) and large cross sections. Polarimeters are usually characterized by a figure of merit which depends on these two quantities through:

$$(F_{kq})^2 = \int (T_{kq})^2 \, \epsilon(\Omega) \, d\Omega, \qquad (1)$$

where ϵ is the ratio of the number of detected reactions to the number of incident beam particles (which depends on the cross section, the target thickness and all detection efficiencies). The quantities are integrated over the phase space covered by the polarimeter. The figure of merit allows one to compare different apparatus because it is related to the statistical error made in a polarisation measurement; the larger the figure of merit, the smaller the error for a given number of beam particles.

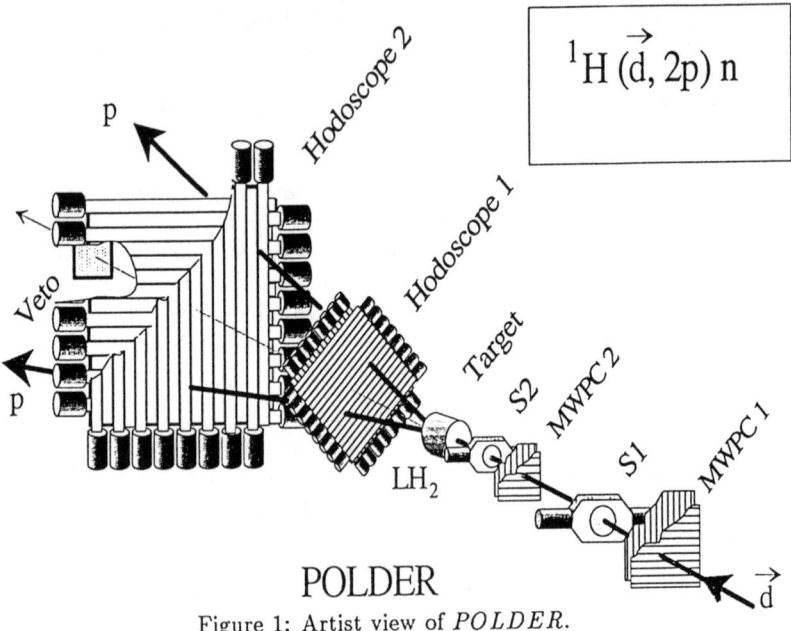

Figure 1: Artist view of *POLDER*.

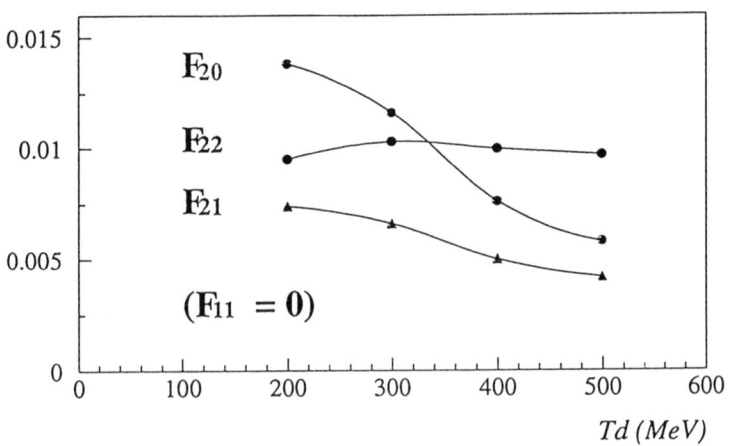

Figure 2: Figures of merit of *POLDER* as a function of the deuteron energy.

Bugg and Wilkin [2] considered the ^{1}H($\vec{d}$, $2p$)n reaction and using an impulse approximation model predicted large figures of merit for deuterons of kinetic energy 200-400 MeV, provided the two final protons have small relative energy. Experiments performed at 200 and 350 MeV [3] confirmed their predictions and the polarimeter POLDER, detailed in ref [4], was designed to work via such a measurement.

The polarimeter POLDER is composed of three parts (fig. 1). First the direction and number of incident deuterons are measured with two MWPCs and two scintillators placed upstream of a 16 cm long liquid hydrogen target. Protons produced in the ^{1}H($\vec{d}$, $2p$)n reaction are detected in two hodoscopes, composed of bars of thin plastic scintillator, located 50-70 cm and 150-210 cm downstream of the target respectively. Proton pairs with low relative momentum (k<100 MeV/c) are detected with good efficiency almost independent of beam energy in the entire region where the cross section of the charge exchange reaction is sizeable. POLDER was calibrated at SATURNE using polarized deuteron beams from 175-500 MeV. Using only time-of-flight and angle information charge exchange events were discriminated against parasitic reactions and absolute cross section measurements were performed with 1% stability. Figures of merit for all tensor moments were extracted [4] and were found (see fig. 2) to be large enough to permit good tensor polarization measurements in the 175–500 MeV deuteron energy domain. The vector analyzing power, T_{11}, was measured to be 0, as expected from the calculations [2].

Once calibrated, the polarimeter was used in a test case experiment measuring polarization observables of the ^{40}Ca($\vec{d}$,$\vec{d'}$)^{40}Ca elastic scattering. All measured observables satisfied the relations required by time reversal invariance, assuming a 1% normalization uncertainty in the measured cross section [4]. This measurement was a stringent test of the operation of the polarimeter under high background rates.

In the same experiment, POLDER was used to study an isoscalar spin flip transition in ^{12}C. In inelastic scattering induced by polarized proton beams, the spin flip probability is a good observable to measure the spin component of the nuclear excitation [5]. A polarized deuteron beam is a very efficient probe for selecting isoscalar transitions in nuclei [6]. For a deuteron beam, measuring the study of the probabilities of transferring 0, 1 or 2 units of the spin projection along the quantization axis S_0, S_1 and S_2 necessitates the use of a tensor polarimeter. However in [7] it has been shown that S_1 can be approximated by S_y^d, which requires only the measurement of the vector polarization of the scattered deuterons. This has allowed the identification of the isoscalar spin strength for different nuclei in a wide range of excitation energy using the vector polarimeter POMME [6]. However such an approximation is only valid for $\Delta S=0$ transitions and must be checked for $\Delta S=1$ transitions.

In order to test experimentally the validity of the signature S_y^d, the exact spin-flip probability S_1 has been measured with POLDER [4] for the first time at intermediate energy. The experiment was achieved on the best known isoscalar spin level, the 12.7 MeV 1^+ level in ^{12}C, excited by a 393 MeV polarized deuteron beam at SATURNE. The polarimeter was placed in the focal plane of the SPES1 spectrometer and deuterons were detected at a scattering angle of 4°.

From the measurements of the spin tensor observables of the reaction, we have deduced the S_1 signature. The result was found to be in good agreement with the approximated spin flip probability S_y^d previously measured with $POMME$.

$$S_1 = 0.43 \pm 0.05 = 0.38 \pm 0.03 = S_y^d$$

This result demonstrates that S_y^d is well suited for the study of the nuclear response in nuclei and confirms the good operation of this tensor polarimeter.

Two new experiments using $POLDER$ have been proposed. One is a study of the tensor spin observables of the $pp \rightarrow \vec{d}\pi^+$ reaction at $SATURNE$, which is useful in the theoretical understanding of the πNN system [8]. The second one will measure t_{20} of the deuteron produced in elastic e-d scattering at $CEBAF$ [9].

A MEASUREMENT OF t_{20} AT CEBAF

The electromagnetic structure of the deuteron is described by three form factors: charge monopole G_C, charge quadrupole G_Q and magnetic dipole G_M. Non-relativistically, these form factors are related to the spatial distributions of charge, quadrupole deformation and magnetization respectively. Experimentally, at least three observables of e-d scattering are needed to determine separately the three form factors. Cross section measurements at different electron angles for the same 4-momentum transfer (Q) allow the determination of the longitudinal and transverse structure functions $A(G_C^2, G_Q^2, G_M^2)$ and $B(G_M^2)$. To separate further G_C and G_Q, the measurement of another observable is required. One may either measure asymmetries induced by a tensor polarized deuterium target, or measure the tensor polarization of the recoiling deuterons (alternatively, one may use vector polarization if the electron beam is polarized). Taken together with $A(G_C^2, G_Q^2, G_M^2)$ and $B(G_M^2)$, $t_{20}(G_C, G_Q, G_M, \theta_e)$ allows the separate determination of the two charge form factors G_C and G_Q. One important feature of t_{20} is that it is nearly independent of the elementary nucleon form factors, and in particular of the poorly known neutron electric form factor.

Measurements of t_{20} have been performed [10, 11, 12, 13, 14, 15, 16] up to four-momentum transfer $Q = 4.6$ fm^{-1}. The two most recent experiments have extended the measurements into a range where theoretical models differ significantly in their predictions for t_{20}. One was performed at Bates [15], with the polarimeter $AHEAD$ [1]. The deuteron tensor polarization t_{20} was measured in the range $3.8 < Q < 4.6$ fm^{-1}, providing the first experimental evidence for a node in G_C (located around 4.4 fm^{-1}) which reflects the node in the S-state wave function of the NN system. The second experiment, performed with a polarized target in the Novosibirsk storage ring, measured the tensor analyzing power T_{20}. Additional measurements are currently underway but preliminary data [16] are available from this experiment. Where the data of the recent Bates and Novosibirsk experiments overlap, they differ significantly in the extracted value of t_{20}. These two measurements were performed using two different techniques yielding very different systematic errors. Thus there is a need for complementary measurements, and for additional precise measurements in this region of momentum transfer.

At relatively low momentum transfer ($Q < 3$ fm^{-1}), t_{20} is well determined by the non-relativistic impulse approximation (IA) [17] with small theoretical uncertainties. At high momentum transfer, various corrections to the IA become important. These include isoscalar meson-exchange currents (MEC), isobar components (IC), relativistic effects and perhaps quark degrees of freedom. Nonrelativistic models which include a $\rho\pi\gamma$ MEC [18], in order to reproduce the isoscalar form factor in the three body system [19], predict values of t_{20} which lie between the Bates and the Novosibirsk data. Relativistically covariant models [20, 21] suggest that $\rho\pi\gamma$ and $\omega\sigma\gamma$ MECs should cancel each other, resulting in good agreement with the Bates data. Beyond the region of existing data, very different predictions for the behavior of t_{20} exist from Skyrme calculations [22] and PQCD-based calculations [23, 24, 25].

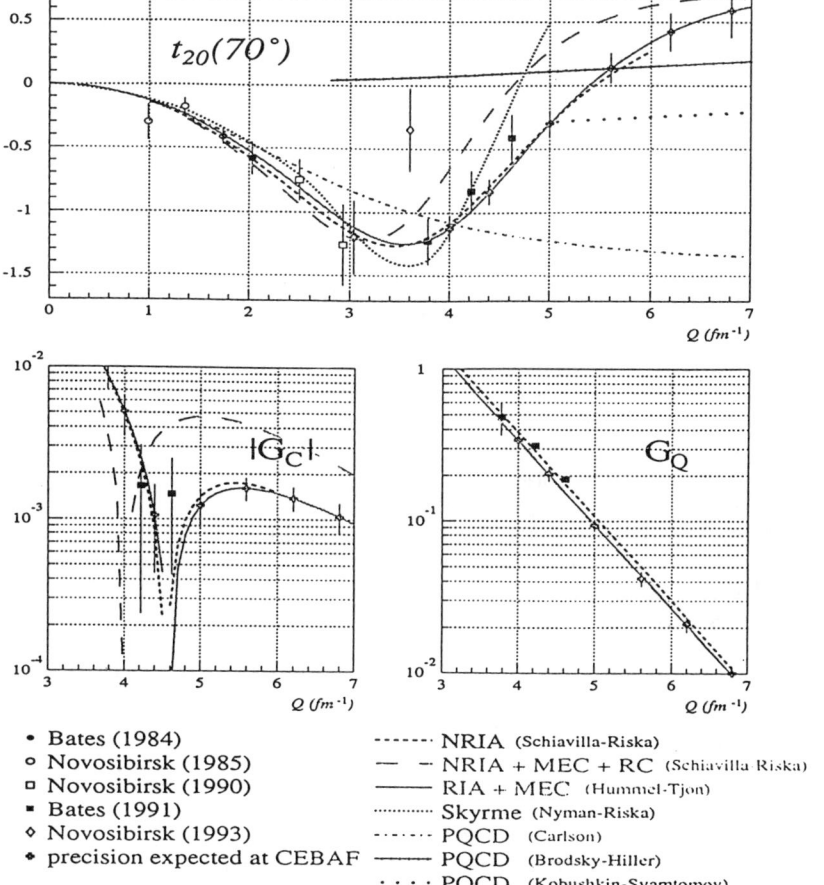

- Bates (1984)
- Novosibirsk (1985)
- Novosibirsk (1990)
- Bates (1991)
- Novosibirsk (1993)
- precision expected at CEBAF

------ NRIA (Schiavilla-Riska)
— — NRIA + MEC + RC (Schiavilla Riska)
——— RIA + MEC (Hummel-Tjon)
········ Skyrme (Nyman-Riska)
- - - - - PQCD (Carlson)
——— PQCD (Brodsky-Hiller)
· · · · PQCD (Kobushkin-Syamtomov)

Figure 3: Predictions of various theoretical models for t_{20} along with existing data. Only the precision is meaningful for the data expected at CEBAF.

The figure of merit F_{20} of POLDER is sufficiently large to permit good measurements of t_{20} in e-d at CEBAF. An experiment has thus been approved for Hall C [9]. This experiment will use the HMS spectrometer for the detection of the electron and an upgraded version of the liquid deuterium target designed for Hall C. A specific hadron channel, placed at a fixed angle, has to be designed. This will require the beam energy and electron spectrometer angles to be changed (1-4 GeV and 20-40°) to measure the data at different momentum transfers. Beam time estimates were made and it was shown [9] that with 46 days of 100 μA beam at CEBAF, six new measurements of t_{20} could be made with good precision from $4.0 < Q < 6.8$ fm^{-1} (see fig. 3). This would include new precise data in the region of the Bates and Novosibirsk measurements which should help to resolve the existing discrepancy and better determine the node of G_C. In addition, the existing data set would be extended to regions of momentum transfer where relativistic models begin to diverge and where quark degrees of freedom might begin to become manifest.

References

[1] J.M. Cameron et al., Nucl. Inst. and Meth. **A305**, 257 (1991).
[2] D.V. Bugg and C. Wilkin, Phys. Lett. **B152**, 37 (1985);
 J. Carbonell, M. Barbaro and C. Wilkin, Nucl. Phys. **A529** (1991) 653.
[3] S. Kox et al., Nucl. Phys. **A556**, 621 (1993).
[4] J.S. Réal, Thèse de l'Université de Grenoble (1994), ISN-94-003.
 S. Kox et al., Nucl. Instr. and Meth. **A346** (1994) 527.
[5] F. T. Baker, et al., Phys. Rev. C **48**, 1106(1993).
[6] M. Morlet, et al., Phys. Rev. C **46**, 1008(1993).
[7] M. Morlet, et al., Phys. Lett. **B247**, 228(1990).
[8] C. Furget et al., LNS proposal #290.
[9] S. Kox, E.J. Beise et al., CEBAF proposal PR-94-018.
[10] M.E. Schulze et al., Phys. Rev. Lett. **52**, 597 (1984).
[11] V.F. Dmitriev et al., Phys. Lett. **157B**, 143 (1985).
[12] B.B. Voitsekhovskii et al., JETP Lett. **43**, 733 (1986).
[13] R. Gilman et al., Phys. Rev. Lett. **65**, 1733 (1990).
[14] B. Boden et al., Z. Phys. **C49**, 175 (1991).
[15] I. The et al., Phys. Rev. Lett. **67**, 173 (1991);
 M. Garçon et al., Phys. Rev. C **49** (1994) 2516.
[16] C.E. Jones et al., 14th European Conference on Few Body in Physics, Amsterdam, The Netherlands, August 1993.
[17] R. Schiavilla and D.O. Riska, Phys. Rev. C **43**, 437 (1991).
[18] R. Schiavilla et al., Phys. Rev. C **41**, 309 (1990).
 H. Henning et al., Few-Body Systems, Suppl. **5**, 133 (1992).
[19] A. Amroun et al., Phys. Rev. Lett. **69**, 253 (1992).
[20] E. Hummel and J.A. Tjon, Phys. Rev. C **49**, 21 (1994).
[21] N. K. Devine and S. Wallace, Phys. Rev. C **48**, R973 (1994).
[22] E.M. Nyman and D.O. Riska, Nucl. Phys. **A468**, 473 (1987).
[23] C.E. Carlson, Nucl. Phys. **A508**, 481c (1990).
[24] S.J. Brodsky and J.R. Hiller, Phys. Rev. D **46**, 2141 (1992).
[25] A. Kobushkin and A. Syamtomov, Phys. Rev. D **49**, 1637 (1994).

Proton Knock-Out from Tensor Polarized Deuterium

E. Passchier (1), M.Ferro-Luzzi (4), Z.-L. Zhou (8), T. Botto, M. Bouwhuis,
J.F.J. van den Brand, D. Dimitroyannis, M.Doets, C.W. de Jager, J. Konijn,
D.J.J. de Lange, G.J. Nooren, N. Papadakis, I. Passchier, P. Salle, J.J.M. Steijger,
N. Vodinas, H. de Vries, C. Zegers (1); R. Alarcon, S. Choi, J. Comfort (2);
D.M. Nikolenko, S.G. Popov, I. Rachek (3); J. Lang (4), H. Arenhövel (5);
R. Ent (6); W. Leidemann (7); M. Bucholz, H.J. Bulten, M.A. Miller, J.S. Neal,
O. Unal (8)

(1) NIKHEF, P.O. Box 41882, 1009 DB Amsterdam, The Netherlands
(2) Arizona State University, U.S.A.
(3) BINP, Novosibirsk, Russia
(4) ETH, Zurich, Switserland
(5) University of Mainz, Germany
(6) CEBAF, U.S.A.
(7) Theory Centre, Trento, Italy
(8) University of Wisconsin, U.S.A.

Abstract: A status report is given on experiment 91-12, which is the first internal target experiment performed at the internal target facility of NIKHEF. The aim of this experiment is to measure the asymmetry in the (e,e'p) reaction from a tensor polarized deuterium target with unpolarized electrons. An update on the experimental setup and results of an extensive set of test measurements are presented.

1. INTRODUCTION

The first internal target experiment that will be performed at the NIKHEF-K facility is experiment 91-12. This experiment will measure tensor analyzing powers in the proton knockout from deuterium in the quasi-free scattering region (1). In this kinematical domain large asymmetries are calculated by Arenhövel (2), as a function of ϑ_{pq}^{cm} the centre of mass angle between the proton momentum and the momentum transfer q. In the region of small ϑ_{pq}^{cm} the plane wave calculation and the full calculation, which include meson-exchange currents, final-state interactions and isobar configurations, do not differ significantly. In this region the asymmetry yields information about the D-state admixture as a function of momentum. In PWIA there is a direct relation between the analyzing power and the ratio of $u_2(p)/u_0(p)$, where $u_2(p)$ and $u_0(p)$ are the deuteron wave functions for the D-state and the S-state, respectively. At higher ϑ_{pq}^{cm} ($\gtrsim 30°$) the contributions of final state interactions become large and the asymmetry becomes sensitive to NN-potential models.

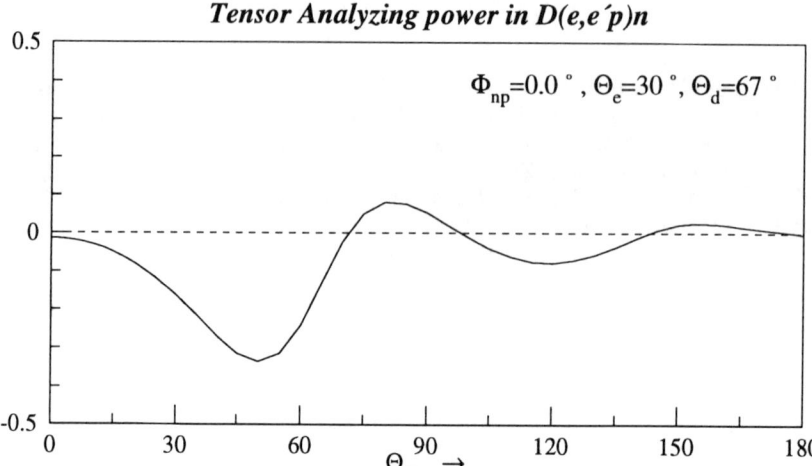

FIGURE 1. The tensor analyzing power as calculated by Arenhovel, for an electron energy of 550 MeV in the quasi elastic regime as a function of ϑ_{pq}^{cm}.

2. EXPERIMENTAL SETUP

2.1 Stored Beam

An electron beam with an zero-load energy of 700 MeV is delivered by the linear accelerator MEA, which injects into the AmPS storage/stretcher ring (3). In storage mode currents up to 150 mA have been obtained by stacking beam pulses. The 1/e lifetime exceeds 20 minutes, when no internal target or collimator is in place. The maximum energy of a stored electron beam in the ring is 525 MeV, but after installation of a 476 MHz cavity, the energy of circulating electrons can be ramped to 900 MeV. The internal target is located in a straight section at the position with minimal beta-functions (β_x = 2.0 m, β_y = 6.0 m). On the opposite side of the ring a collimator has been installed, in order to remove electrons in the tail of the beam distribution. The presented data was taken with a beam energy of 525 MeV and an average current of 25 mA.

2.2 Tensor Polarized Target

The internal target is an open-ended storage cell, through which the electron beam traverses. Into this target cell atoms are injected, through a feedtube, from an atomic beam source (ABS). The ABS consists of a dissociator, a beam-forming system, two sextupole magnets for Stern-Gerlach separation and two RF-transition units (4). The ABS produces a beam of 1×10^{16} atoms/s of tensor-polarized deuterons into two hyperfine substates. The storage cell used at present is a cylindrical aluminum tube of 40 cm length with a diameter of 15 mm and a cell-wall thickness of 25 μm, which is coated internally with teflon in order to reduce depolarization of the atoms due to cell-wall interactions. A tritium polarimeter has been built and is operational (5), yielding the polarization and atomic fraction integrated over the storage cell. The orientation of the deuteron spin is defined by a strong holding field, larger than 0.02 T, which can be directed along any direction in the scattering plane. The deflection of the stored electron beam is compensated with a small magnet, downstream of the target cell.

2.3 Detector setup

The detector setup (see fig. 1.) has been extensively described in (6), so only a brief overview will be given here. Knocked-out protons are detected in the Range Telescope (RT), consisting of a stack of 16 scintillators of 1 cm thickness.

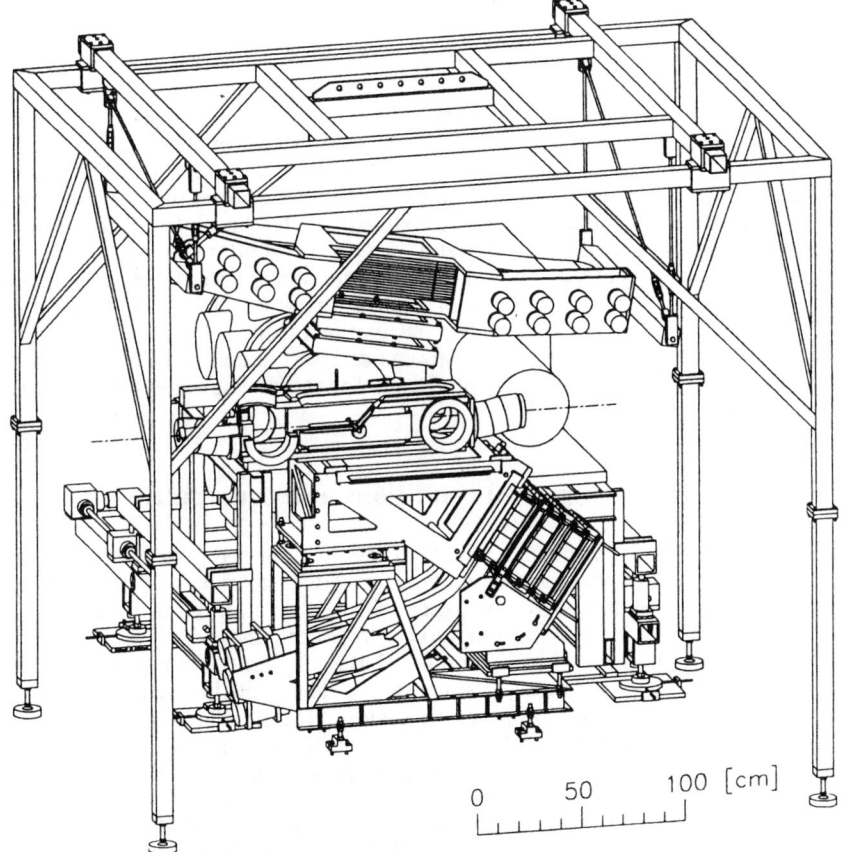

FIGURE 2. Overview of the experimental setup for experiment 91-12.

The measured energy resolution of the detector is about 3% for proton energies between 20 and 150 MeV. The first layer is only 2 mm thick, which enables the detection of elastically scattered deuterons. The track of the protons (and deuterons) is measured by two sets of wirechambers in front of the RT. The obtained angular resolution is 35 mrad. Scattered electrons are detected with a calorimeter, consisting of 6 stacks of 10 CsI(Tl) crystals of 15x6x6 cm^3 with a total thickness of 15 radiation lengths. The trigger for this detector is provided by two fast plastic scintillators, sandwiched around the first layer of CsI. The first scintillator is 5 cm thick in order to prevent Møller electrons with energies up to 10 MeV from reaching the crystals. Tracks are measured with two sets of wire chambers, each consisting of two planes, obtaining an angular resolution of 15 mrad. The angular acceptance for the proton arm is about 300 msr, for the electron arm about 180 msr, calculated from the central position in the target cell. The data acquisition system was developed at NIKHEF and is based on VME-modules and fast transputer read-out. The resolution on the time difference between the Range Telescope and the Calorimeter is 2 ns.

3. TEST RESULTS

An extensive program of test and calibration measurements was performed at NIKHEF in the last two years. In the first tests background conditions in the internal target hall were measured and the dimension and other properties of the stored electron beam were determined (6). In the next phase the complete setup was installed and the detectors were extensively tested and optimized under realistic conditions with unpolarized targets of deuterium and hydrogen. Furthermore storage cells of 20 mm and 15 mm diameter were used, and a small amount of data was taken with a tensor polarized target, from a 20 mm cell. From this last set an accurate figure for the contributions from polarized gas scattering compared to (e,e'p) events from the storage cell was deduced. Here a few of the results are presented and their implications for the measurement of the tensor analyzing power are discussed.

3.1 Background conditions and countrates

The single rates in the various elements of the detector setup was measured as a function of target thickness and beam current. After subtraction of the rates with an empty target, the data show the expected linear relation with the gas flow to the storage cell, which is proportional to the target thickness. This was confirmed for both deuterium and hydrogen. In table 1 the single rates are given, the rates for the wire chambers are those of single wires. The contributions due to the gas and due to the background are given seperately. The numbers are presented for the maximum current of 100 mA at injection, during data taking, and for a ABS-flow of 1.6×10^{16} atoms/s with the storage cell at room temperature. The ratio between single rates from the empty target and from the gas in the target cell is about 2 for the first layer of both the calorimeter and the range telescope, where for the deeper layers this ratio exceeds 5. This effect can be explained by a large amount of low-energy Møller electrons coming from the gas, which are stopped in the first detector layer.

TABLE 1. Rates in various detector elements, scaled to a maximum luminosity of 100 mA beam current and 1.6×10^{16} atoms/s flow. The total rates are separated in contributions from the gas and from the storage cell.

RANGE TELESCOPE:	Countrate Empty [kHz]	Countrate Deuterium [kHz]	Countrate total [kHz]
Plane 3	47	20	67
Plane 4	13	6	19
Scint 1	98	54	152
Scint 2	11	1.7	13
CALORIMETER:			
Plane 1y	90	39	129
Plane 1x	47	20	67
Plane 2	13	6	19
Scint 1	94	46	140
Scint 2	3.4	0.2	3.6
COINCIDENCE:			
(e,e'p) [Hz]	0.24 [Hz]	0.70 [Hz]	0.94 [Hz]
(e,e'd) [Hz]	0.0 [Hz]	0.1 [Hz]	0.1 [Hz]

For (e,e'p) coincidences between the calorimeter and the range telescope, a good track is required for both arms along with the identification of a proton in the range telescope. On this coincidence level the rate is dominated by events originating from the gas, yielding a ratio of gas to background events of 3:1. These numbers are measured with the collimator in optimal position, without the collimator this ratio is reduced to below 1:1. Cooling the cell to 100 K is predicted to improve the ratio of gas to background with a factor 1.7.

3.2 T_{20} polarimetry

The first layer of the Range Telescope has a thickness of only 2 mm, which makes it possible to measure elastically scattered deuterons. By performing timing corrections for walk of the TDCs, time of flight of deuterons, and travel time of the light through scintillators and light guides the resolution on the time difference between the proton and the electron detector is sufficient to distinguish protons from deuterons. The timing information combined with the energy signal from the first layer gives a clear separation between protons and deuterons. By aligning the spin direction of the atoms along the momentum transfer a measurement of the asymmetry in the elastic channel yields a measurement of T_{20}. At the probed kinematics ($Q^2 \approx 0.1$ (GeV/c²)²), tifferent predictions agree within 4% (7). The elastic deuteron rate of the tensor-polarized target will be about 0.1 Hz. This rate will result in a measurement of P_{zz} with an error of less than 5 % in 100 hours of beam time.

4. OUTLOOK AND CONCLUSIONS

Test measurements at the internal target setup at NIKHEF have shown the feasibility of doing an experiment with an internal tensor-polarized deuterium target and a storage cell. At present the luminosity will exceed 10^{31} atoms/cm^{2}s, when a 15 mm diameter storage cell is used, cooled to 100 K. Scraping the beam halo is essential to obtain a reasonable ratio between events from the gas and events from the cell-wall. With the cooled cell of 15 mm the dilution of the asymmetry is less than 35%, when the collimator is in optimal position. Use of non-magnetic large solid angle detectors, with proportional wire chambers for particle tracking is shown to be possible in the environment of a stored electron beam. Data taking of about 400 hours is scheduled for the end of '94 and the beginning of '95.

ACKNOWLEDGMENTS

This research was supported by the Russian Fund for Fundamental Research, under contract 123.06 and by the Netherlands' Organisation for Scientific Research (NWO), under contract 713-119.

REFERENCES

(1) NIKHEF-K electron scattering proposals 91-12 and 93-04, J.F.J. van den Brand, C.W. de Jager, spokespersons.
(2) H. Arenhövel, W. Leidemann, private communication, see also H. Arenhövel, W. Leidemann, E.L. Tomusiak, Z. Phys. **A331**, 123 (1988).
(3) Y. Wu, The optical design of AmPS, Ph.D. thesis, Technical University of Eindhoven (1991).
(4) D. Singy, P.A. Schmelzbach, W. Gruebler and W.Z. Zhang, Nucl. Instum. and Meth. **A278** 349 (1989).
(5) Z.-L. Zhou private communication, see also J.S. Price and W. Haeberli, Nucl. Instum. and Meth. **A326**, 416 (1993).
(6) C.W. de Jager, Lecture notes of 1992 RCNP KIKUCHI School, Osaka.
(7) S. G. Popov et al, Status of the t_{20} electron-deuteron scattering experiment at VEPP-3, in *Proceedings of this conference*.

Intermediate Energy Electromagnetic Interactions

M. Garçon

DAPNIA/SPhN, CEA-Saclay, 91191 Gif-sur-Yvette Cedex, France

Abstract. A brief summary of contributions to the parallel session on Intermediate Energy Electromagnetic Physics is given.

INTRODUCTION

Without being exhaustive, one can sketch the following history of the field of polarization measurements in electromagnetic interactions. Until a few years ago, very few precise measurements of spin observables were performed.

This was especially the case for experiments with lepton beams: there were pioneering experiments to study parity violation with polarized electron and muon beams, respectively at SLAC and Serpukhov; polarized targets were also used at SLAC together with the polarized electron beam. In order to determine the deuteron form factors in elastic electron-deuteron scattering, the first use of a polarized atomic beam was done inside an electron ring at Novosibirsk, and the recoil deuteron tensor polarization was first measured at Bates (MIT).

With photon beams, it has always been easier to use cryogenic solid polarized targets and consequently many measurements like photoproduction of mesons on polarized protons were performed, for example at DESY. In the study of deuteron photodisintegration, polarized photon beams were used at Kharkov and Frascati, proton recoil polarization was measured at Tokyo and Kharkov.

Due to smaller cross-sections, to limitations in the luminosity and in the degree of polarization attainable, the field of polarization has been much less rich in the study of electromagnetic processes than in the one of hadronic processes.

Technical improvements have recently led to:

   - deep inelastic scattering of polarized electrons and muons off polarized targets,

   - spin correlation and spin transfer coefficients to extract nucleon electromagnetic form factors, and

- parity violation as a tool to extract nucleon weak form factors, and these subjects have been developed in other parallel sessions at this conference.

The use of spin observables in electromagnetic processes
- to study nuclear structure and important reaction mechanisms such as relativistic effects, modification of nucleon properties in the medium,
- to elucidate properties of baryonic resonances, and
- to understand the transition from meson-nucleon dynamics to quark-gluon dynamics,

is barely starting. We anticipate a large number of new experiments addressing these questions in the next few years. Experiments and calculations reported in our session are evidence for this trend.

PHOTOPRODUCTION OF PSEUDOSCALAR MESONS ON THE PROTON

The first polarization experiment at ELSA, the stretcher ring installed at the Bonn synchrotron, was presented at this conference. Using tagged photons, a frozen spin target and the detector PHOENICS, the target asymmetry was measured in the reactions $\gamma\vec{p} \to \pi^+ n$ and $\gamma\vec{p} \to \pi^0 p$. 3150 data points were obtained for $220 < E_\gamma < 800$ MeV and $35° < \theta_\pi^{c.m.} < 135°$.

A new multipole analysis of pion photoproduction is in progress. It should allow to extract the strength of the (small!) E2 transition for this process. If this transition does occur, it could be a sign, but this is a model dependent statement that will require much investigation, of a D-wave component in the Δ resonance.

$\gamma\vec{p} \to \eta p$ will be measured soon at ELSA, and we were presented with theoretical predictions which will shed light on the role, if any, of the Roper (P_{11}) resonance in this process.

As for kaon photoproduction ($\gamma p \to K\Lambda$ and $\gamma p \to K\Sigma$), the latest calculations of the Saclay-Lyon group were presented. The present data set is well described up to $E_\gamma = 2$ GeV in a model incorporating proton resonances in the s channel, K resonances in the t channel, and hyperon resonances in the u-channel, using coupling constants in agreement with SU(3). An analysis in terms of helicity amplitudes (Saclay-Pittsburgh) allows to group the observables in classes according to their decomposition in Legendre polynomials; the number of nodes in an observable (for example the beam asymmetry) gives then an indication of the spin of intermediate resonances in the process. A conclusive test of these models will come from experiments planned at ELSA, GRAAL (tagged polarized photon beams at ESRF-Grenoble) and CEBAF.

PHOTONUCLEAR REACTIONS

In addition to the new results on deuterium and ^{3}He, and presented in the plenary session by A. Sandorfi, the LEGS collaboration at BNL has investigated $^{16}O(\vec{\gamma},p\pi^-)$ and $^{16}O(\vec{\gamma},pp)$ reactions. The first one aims at creating a Δ in the interior of the nucleus and study its propagation in the nuclear medium, while the second one addresses the question of short range two-nucleon correlations in the nucleus. For both cases, double differential cross-sections and beam asymmetries, as a function of the angles of the two outgoing particles, were extracted. The comparisons with model calculations were still preliminary and not conclusive at this stage.

DEUTERON ELECTROMAGNETIC STRUCTURE

The separate determination of the charge and quadrupole electric form factors of the deuteron necessitates the measurement of a polarization observable. Though very important for the understanding of the short-range nucleon-nucleon interaction and of the role of non-nucleonic degrees of freedom in nuclei, the measurements of such an observable are very few, for reasons given in the introduction. The observable of choice so far has been t_{20}, a measure of the alignment of the deuterons produced in the scattering. One needs either to use a tensor polarized target, or to measure the tensor polarization of the recoil deuterons.

The Novosibirsk-Argonne collaboration at VEPP-3 has been improving its measurements with an internal polarized atomic beam trapped in a cell. The target polarization is now monitored by a continuous measurement of e-d tensor asymmetry at low transfer, where it is well known. The new results presented at this conference, at $Q = 3.6$ fm^{-1}, are in agreement, (but with poor statistical significance) with earlier results from the same group shown at previous conferences, but in apparent disagreement (about 2.2 standard deviations) with the latest, and more precise, Bates data. Our understanding of isoscalar meson exchanges and of relativistic effects depends significantly on which data to believe, so that this discrepancy, if it should remain, should be clarified. The achievable luminosity should be greatly improved next year at VEPP-3 with the use of a spin-exchange optically pumped target.

The polarimeter POLDER for tensor polarized deuterons in the energy range $175 < E_d < 500$ MeV was calibrated at Saturne and is to be used at CEBAF for measurements at still higher momentum transfers ($4.0 < Q < 6.8$ fm^{-1}) of the t_{20} observable.

In quasi-elastic scattering off a tensor polarized target ($\vec{d}(e,e'p)n$), one has access, supposing the validity of the impulse approximation, to the ratio of D and S wave functions in momentum space. Such a measurement is being prepared in the stretcher ring at NIKHEF, using an internal polarized target.

There was an interesting discussion about how to deal with background rates in an electron ring.

SPIN PHYSICS AT CEBAF

There are great expectations from this new facility, which delivered its first beam on target this last July. The first polarized beam is planned for the end of 1995. The equipment that will allow the measurements of polarization observables is being implemented, either at CEBAF or by users: this includes beam polarimeters, polarized (gas or solid) targets, recoil polarimeters for protons, neutrons and deuterons. S. Nanda gave a survey of the spin physics program, and I can only refer to his written contribution for a clear overview of what to expect by the time of the next symposium.

SPIN PHYSICS AT ELFE

Looking into the more distant future, J.M. Laget presented the project ELFE (Electron Laboratory for Europe). The aim of this project is to study how nucleons and nuclei are built of quarks and gluons. A high duty cycle at energies sufficient to probe the quark structure of nucleons is needed because only through exclusive reactions performed at high momentum transfer can one select specific configurations in the nucleon wave function. We were given several examples on how polarization observables will here also play a great role. Many working groups in Europe have been studying the experimental aspects of this program and letters of intent were written. The case for a 15 to 30 GeV high duty cycle electron accelerator seems well established.

VI. LOW-ENERGY NUCLEAR REACTIONS

Cross section and polarization measurements in the ^{12}C($\vec{d}$,p) and ^{12}C(d,$\vec{n}$) reactions, and their CDCC analysis

H. Toyokawa [a] and H. Ohnuma [b]

[a] Research Center for Nuclear Physics, Osaka University, Ibaraki, Osaka 567, Japan
[b] Department of Physics, Tokyo Institute of Technology, Meguro, Tokyo 152, Japan

Abstract. The ^{12}C($\vec{d}$,p)^{13}C and ^{12}C(d,$\vec{n}$)^{13}N reactions were studied at E_d = 25 MeV. Coupled channel calculations which include continuum states of the p-n system reproduce not only cross section data but also iT_{11} and polarization data for the transitions to the ground states of final nuclei. Furthermore, cross sections for the ^{13}C(p,pn)^{12}C reactions were measured at E_p = 35 MeV as a possible more stringent test of the coupling effects between continuum states, and compared with those for the ^{13}C(p,d)^{12}C reaction and theoretical calculations.

INTRODUCTION

It has been known [1] that (d,p) and (p,d) angular distributions above incident energies of about 20 MeV can not be described well by the distorted-wave Born approximation (DWBA) theory. Johnson and his collaborators proposed the adiabatic deuteron breakup approximation (ADBA) theory [2], which turned out to be very successful in explaining (d,p) and (p,d) data. According to this theory, the incoming (or outgoing) deuteron wave contains components in which it breaks up into continuum states. Calculations based on the method of continuum discretized coupled channels (CDCC) [3] have shown that ADBA is a better description of the (d,p) reaction than DWBA but still inadequate to include effects of the neutron transfer from continuum states. Recently Masaki et al. [4] published comparisons of CDCC calculations with (d,d) and (d,p) data at 22 MeV including analyzing powers. However such comparisons are yet very scarce especially for polarization observables. Furthermore, although CDCC theory can predict transfer cross sections from/to continuum states themselves, no experimental data to be compared with such calculations exist. We have measured not only cross sections but also vector analyzing powers for the (d,p) reaction and neutron polarizations for the (d,n) reaction on ^{12}C. In addition, cross sections for the reactions ^{13}C(p,d) and ^{13}C(p,pn) were measured. The last reaction was studied with an experimental setup in which detection of the $^1S_0$ p-n pair was favored. Obtained experimental data were compared with CDCC calculations.

EXPERIMENT

The ($\vec{d}$,p), (p,d) and (p,pn) experiments were carried out at the Institute for Nuclear Study, University of Tokyo, with a 25 MeV vector-polarized deuteron beam and a 35 MeV proton beam. Angular distributions for the ($\vec{d}$,p) and (p,d) reactions were measured with a magnetic spectrometer [5] and a focal plane detector system [6]. In the (p,pn) experiment protons and neutrons emitted at the same angle were measured in coincidence. In this case protons and neutrons were detected by three sets of Si counter telescopes and by NE213 liquid scintillation counters placed behind them, respectively. Such collinear geometry enhanced detection of p-n pairs in the $^1S_0$ state. The contribution from the $^3S_1$ state was estimated from Monte Carlo calculations based on the Watson-Migdal approximation [7] to be about 18 - 23 % depending on the angle, and subtracted in the cross section calculation. Effective solid angles [8] for the (p,pn($^1S_0$)) reaction were obtained also from the Watson-Migdal formalism, treating this reaction as if it were a reaction with a two-body final state.

The (d,$\vec{n}$) experiment was carried out at the Cyclotron and Radioisotope Center, Tohoku University, with a 25 MeV deuteron beam and the time-of-flight facilities. Neutron polarizations were measured using a liquid-^{4}He polarimeter [9].

Cross sections and iT$_{11}$ for the ^{12}C(d,p)^{13}C reactions leading to the ground state (1/2$^-$), 3.09 MeV state (1/2$^+$), 3.68 MeV state (3/2$^-$) and the 3.85 MeV state (5/2$^+$) of ^{13}C are shown in Fig. 1. Figure 2 shows the cross section and polarization for the ^{12}C(d,n)^{13}N(g.s., 1/2$^-$) reaction. Cross sections for the ^{13}C(p,d)^{12}C and ^{13}C(p,pn)^{12}C reactions leading to the ground state (0$^+$) and the 4.44 MeV state (2$^+$) of ^{12}C are shown in Fig. 3.

ANALYSIS

First, standard DWBA and ADBA analyses were made by using an exact-finite-range (EFR) DWBA code TWOFNR [10]. Optical potential parameters given in Ref. [11] were used for proton and neutron channels throughout the present analysis. The well-depth description was used to generate the form factor of the transferred neutron (or proton) with parameters $r_0 = 1.25$ fm, $a = 0.65$ fm and V$_{so}$ = 6.5 MeV. The deuteron optical potential used in the EFR-DWBA calculation was taken from [12]. The real and imaginary parts of the adiabatic potentials [2] used in the present ADBA analysis were calculated from nucleon optical potentials obtained from Ref. [11] for the p-^{12}C and n-^{12}C systems at a half of the deuteron energy and the Reid soft-core potential [13]. The L·S potential was taken to be the same as for the nucleon optical potential.

As an attempt to include the deuteron breakup effect more explicitly, we also performed coupled-channel Born approximation (CCBA) calculations with CDCC wave functions. The codes CDC2RT and HICALST [4] were used in the CDCC calculation to construct base wave functions, solve coupled equations, and generate distorted waves. Solutions of these coupled equations were then fed into TWOFNR to calculate cross sections and iT_{11} (or polarization) treating the transfer channel in the first-order Born approximation. The base wave functions for the bound and unbound states of the p-n system were eigenfunctions of the Reid soft-core potential [13]. The ^3D components were mixed in the $^3S_1$ channels in the case of the (d,p), (d,n) and (p,d) reactions, while the coupling is only between the $^1S_0$ channels in the case of the (p,pn) reaction. The wave number k_{pn} for the relative p-n motion was divided into discrete bins to take into account of coupling between continuum states and effects of the transfer from (or pickup into) continuum states. The coupling terms between the bound state and continuum states and between continuum states were calculated for each bin by using the Reid soft-core potential and p- and n-^{12}C optical potentials [11] to obtain the CDCC equations [3].

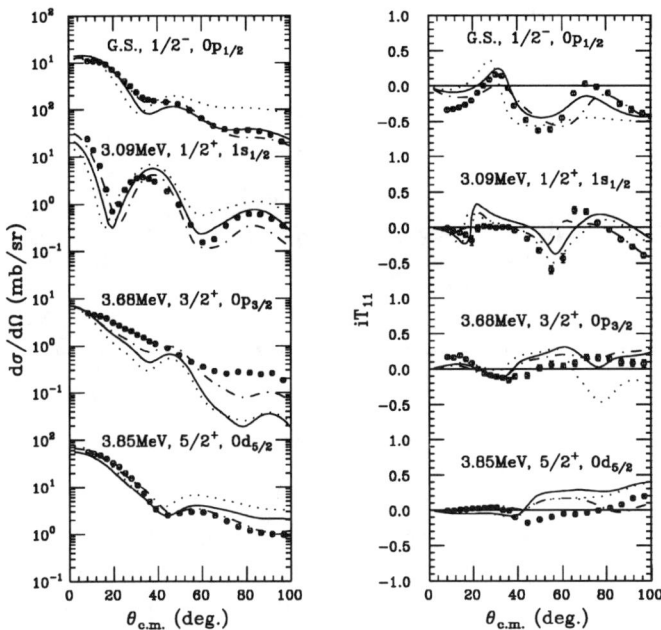

FIGURE 1. Experimental and calculated differential cross sections and iT_{11} for the $^{12}C(d,p)^{13}C$ reaction

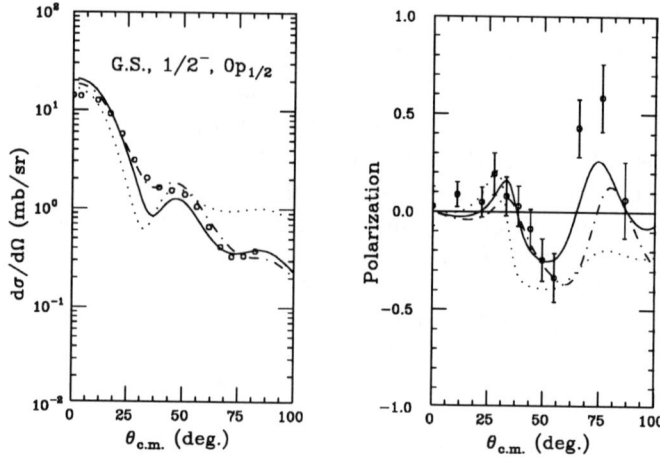

FIGURE 2. Experimental and calculated differential cross sections and polarizations for the $^{12}C(d,n)^{13}N$ reaction

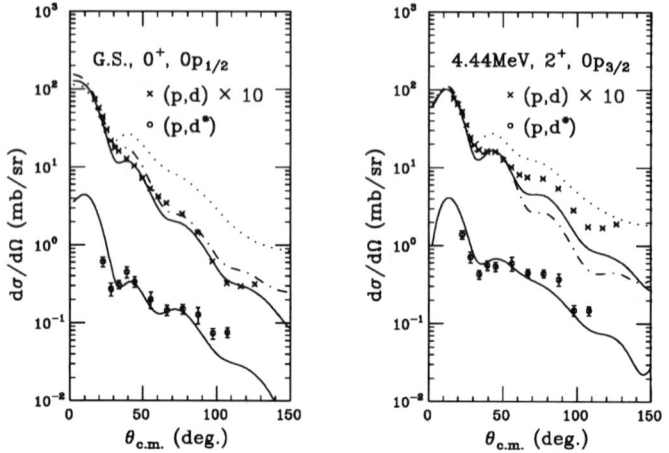

FIGURE 3. Experimental and calculated differential cross sections for the $^{13}C(p,d)^{12}C$ and $^{13}C(p,pn)^{12}C$ reactions

The EFR-DWBA results are shown by the dotted lines, the ADBA results by the dot-dashed lines, and the CDCC-CCBA results by the solid lines in Figs. 1-3. Calculated cross sections were normalized to the data at forward angles to obtain spectroscopic factors. Additional density-of-state factors are incorporated in the (p,pn) cross sections, but no other normalizations are introduced. The spectroscopic factors thus obtained were in good agreement with those from the Cohen-Kurath shell model wave functions [14].

DISCUSSION

Although the $^{12}\text{C}(\vec{d},p)$ and $(d,\vec{n})$ reactions leading to the ground states of ^{13}C and ^{13}N are considered to be good single-particle transitions, DWBA calculations can not describe these data very well. They give not only much slower falloff of the cross section angular distributions, but also give poor descriptions of the iT_{11} and polarization data for these transitions. On the other hand, considerable improvements are observed both in cross sections and in iT_{11}/polarization with ADBA and CDCC-CCBA calculations, indicating that deuteron breakup effects are essential in these reactions. Coupling of the transferred neutron with inelastically excited ^{12}C core has been shown to be important for the $^{12}\text{C}(d,p)$ transitions to higher lying states, especially for those to the positive parity states [15]. Maybe this is responsible for less satisfactory results obtained with ADBA and CDCC-CCBA calculations for these transitions.

The $^{13}\text{C}(p,d)$ angular distributions are also very poorly described by DWBA calculations, while ADBA and CDCC-CCBA calculations reproduce the experimental angular distribution shapes very well and give reasonable spectroscopic factors. The $^{13}\text{C}(p,pn)^{12}\text{C}$ cross sections decrease more slowly at large angles than the $^{13}\text{C}(p,d)^{12}\text{C}$ cross sections. Although there seem to exist slight shifts to larger angles, the present CDCC-CCBA calculations also reproduce overall features of the experimental data for the (p,pn) reaction including the cross section magnitude.

The CDCC-CCBA results for the (d,p), (d,n) and (p,d) reaction did not depend much on the maximum value of k_{pn} or on the bin size. On the other hand, it was found necessary to take as large k_{pn} values as possible for the convergence of (p,pn) results. This may be due to the fact that we are looking at continuum states themselves in the (p,pn) reaction while well defined bound deuterons are coming in or going out in the other reactions. Nevertheless, inclusion of neutron pickup/transfer to/from continuum states modifies calculated (d,p), (d,n) and (p,d) angular distribution shapes to some extent.

SUMMARY

We have measured cross section and analyzing power/polarization angular distributions for the $^{12}\text{C}(\vec{d},p)^{13}\text{C}$ and $^{12}\text{C}(d,\vec{n})^{13}\text{N}$ reactions at an incident deuteron energy of 25 MeV. Cross sections for the $^{13}\text{C}(p,d)^{12}\text{C}$ and $^{13}\text{C}(p,pn(^1S_0))^{12}\text{C}$ reactions were also measured at an incident proton energy of 35 MeV. DWBA calculations can not reproduce the data at all, and inclusion of the breakup effect, either by using an approximate treatment such as ADBA or by using the CDCC method, is essential to explain the data. The CDCC-CCBA calculation is found to give not only good descriptions of the cross section data, but also iT_{11} and polarization data.

REFERENCES

1. J. L. Yntema and H. Ohnuma, Phys. Rev. Let. **19** (1967) 1341; H. Ohnuma et al., J. Phys. Soc. Jpn. **36** (1974) 1236.
2. R. C. Johnson and P. J. R. Soper, Phys. Rev. **C1** (1970) 976; J. D. Harvey and R. C. Johnson, Phys. Rev. **C3** (1971) 636.
3. Y. Iseri et al., Prog. Theor. Phys. **69** (1983) 1038; M. Yahiro et al., Prog. Theor. Phys. Suppl. **89** (1986) 32; M. Kawai et al., Prog. Theor. Phys. Suppl. **89** (1986) 118.
4. M. Masaki et al., Nucl. Phys. **A573** (1994) 1.
5. S. Kato et al., Nucl. Instrum. Methods **154** (1978) 19.
6. M. H. Tanaka et al., Nucl. Instrum. Methods **195** (1982) 509.
7. K. M. Watson, Phys. Rev. **88** (1952) 1163; A. B. Migdal, Sov. Phys. JETP **1** (1955) 2.
8. T. Motobayashi et al., Nucl. Instrum. Methods **A271** (1988) 491.
9. K. Ieki et al., Nucl. Instrum. Methods **A262** (1987) 323.
10. M. Igarashi, computer code TWOFNR, private communication.
11. B. A. Watson, P. P. Singh and R. E. Segel, Phys. Rev. **182** (1969) 977.
12. G. Perrin et al., Nucl. Phys. **A193** (1972) 215.
13. R. V. Reid, Annals of Phys. **50** (1968) 411.
14. S. Cohen and D. Kurath, Nucl. Phys. **A101** (1967) 1.
15. H. Ohnuma et al., Nucl. Phys. **A448** (1986) 205.

Spectroscopy of ^{88}Y by means of the ^{91}Zr$(\vec{p}, \alpha)^{88}$Y Reaction at 22MeV

P.Guazzoni*, U.Atzrott[†], G.Cata-Danil*, G.Graw[‡],
R.Hertenberger[‡], D.Hofer[‡], M.Jaskola•, P.Schiemenz[‡],
G.Staudt[†], J.Tropilo•, E.Zanotti-Müller[‡] and L.Zetta*

*Dipartimento di Fisica dell'Università and I.N.F.N., I-20133 Milano, Italy
[†]Physikalisches Institut Universität, D-72076 Tübingen, Germany
*Institutul de Fizica Atomica, Bucarest, Romania
[‡]Sektion Physik der Universität München, D-85748 Garching, Germany
•Soltan Institute of Nuclear Studies, Swierk, Poland

Abstract. This contribution presents high resolution results for the ^{91}Zr$(\vec{p}, \alpha)^{88}$Y. The homology concept allows to find a correspondence between the observed multiplets of highly excited states in ^{88}Y and the appertaining parent states in ^{87}Y. Therefore spin and parity values can be attributed to the states of the multiplets through comparison of different experimental results.

In low resolution experiments (1), it has been shown that the spectra of α-particles emitted in (p, α) reactions, induced on target nuclei having one unpaired nucleon outside a magic shell, display distinctive features which indicate that this last, slightly bound nucleon, acts as a spectator in the process. These features are: 1) weak population of residual nucleus levels below an excitation energy, strictly related to the energy gap, in the nucleon state spacing at the filling of the magic shell 2) excitation of homologous states (i.e. of states with a close structural relationship) of residual nuclei from (p, α) on adjacent target nuclei, one m a g i c with a magic neutron and/or proton shell (parent), and the other, n e a r - m a g i c, with one more nucleon outside the magic shell (son).

To study the spectator role of the unpaired nucleon in different mass regions, ^{208}Pb$(\vec{p}, \alpha)^{205}$Tl (2), ^{209}Bi$(\vec{p}, \alpha)^{206}$Pb (3) and ^{90,91}Zr$(\vec{p}, \alpha)^{87,88}$Y reactions were studied at 22 MeV with the Munich Tandem, using the polarized ion source, the Q3D magnetic spectrograph and the position and angle resolving light ion detector. In fig.1 the 20° energy spectra for the reactions ^{90}Zr$(\vec{p}, \alpha)^{87}$Y up to E_x= 3 MeV and of ^{91}Zr$(\vec{p}, \alpha)^{88}$Y up to E_x=5.2 MeV are reported. The ^{88}Y states, shown in the middle, correspond to the coupling of the 51st $2d_{\frac{5}{2}}$ neutron in ^{91}Zr with the configurations excited in ^{87}Y.

In the case of weak coupling between the parent state and the spectator proton or neutron it is expected: a) the angular distributions of cross sections

570 Spectroscopy of ^{88}Y

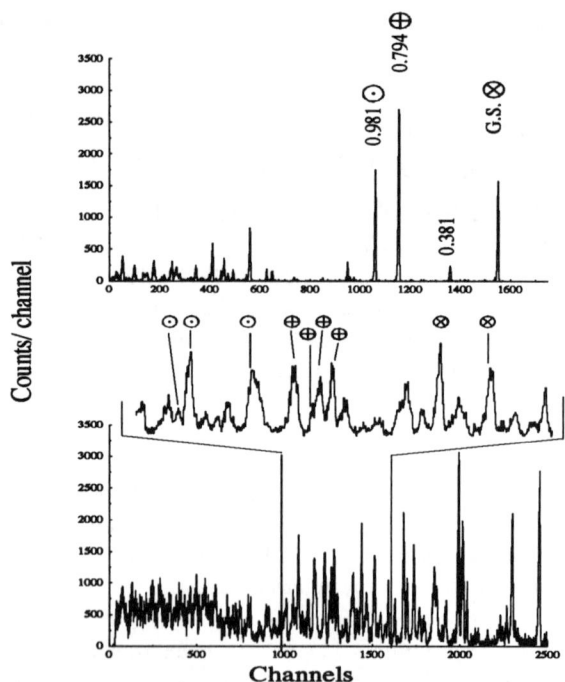

FIGURE 1: Energy spectra of α particles measured at 20° for the ^{90}Zr(p,α)^{87}Y (up) and ^{91}Zr(p,α)^{88}Y (down) reactions. In the middle levels of ^{88}Y homologous to the lowest energy states of ^{87}Y are presented in enlarged scale.

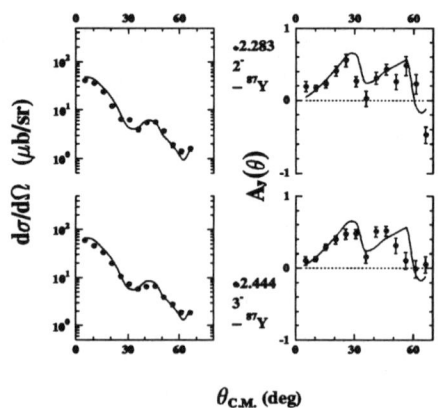

FIGURE 2: Comparison of the experimental cross sections and analyzing powers for population of the doublet of levels of ^{88}Y (solid points) homologous to the 1/2$^-$ g.s. of ^{87}Y with the experimental cross sections and analyzing powers for population of the parent state (solid lines).

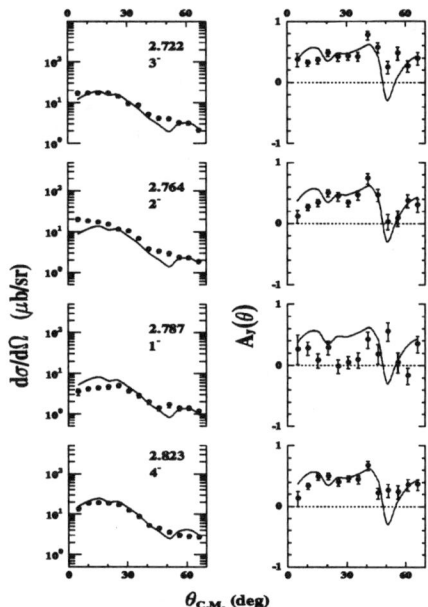

FIGURE 3: Comparison of the experimental cross sections and analyzing powers for population of four levels of ^{88}Y (solid points) considered homologous to the $5/2^-$ 0.794 MeV of ^{87}Y level with the experimental cross sections and analyzing powers for population of the parent state (solid lines).

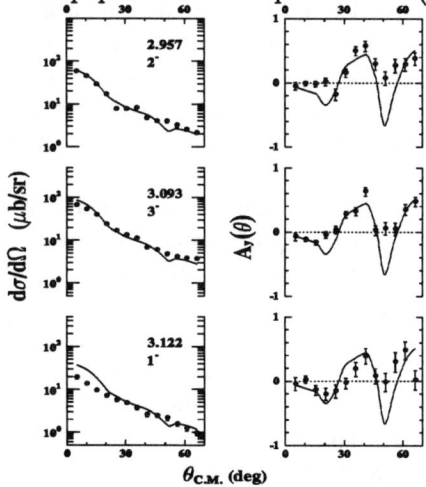

FIGURE 4: Comparison of the experimental cross sections and analyzing powers for population of three levels of ^{88}Y (solid points) considered homologous to the $3/2^-$ 0.982 MeV of ^{87}Y level with the experimental cross sections and analyzing powers for population of the parent state (solid lines).

TABLE 1: Energies, spin and parity of homologous states

^{87}Y		^{88}Y		
E_{exc}	J^π	E_{exc}	J^π	J^π_{ad}
g.s.	$\frac{1}{2}^-$	2.283	2^-	
		2.244	3^-	(10^+)
0.794	$\frac{5}{2}^-$	2.722	3^-	$3^+,4^+,5^+$
		2.764	2^-	
		2.787	1^-	
		2.823	4^-	
		3.150	$(^-)$	
		3.207	$(^-)$	
0.982	$\frac{3}{2}^-$	2.957	2^-	
		2.997	$(^-)$	
		3.023	$(^-)$	
		3.049	$(^-)$	
		3.093	3^-	
		3.122	1^-	

and analyzing powers for transitions to homologous states to be very similar in shape, because the processes exciting these states are essentially the same; b) the differential cross section for the population of a parent state to be approximately equal in magnitude to the sum of the cross sections of the transition to the multiplet of homologous son states which corresponds to the given parent state; c) the relative cross section for the population of a homologous son state with spin J in a given multiplet to be proportional to $(2J+1)$.

In fig.s 2,3,4 the comparison between the measured $\sigma(\theta)$ and $A_y(\theta)$ for the transitions to the various multiplets of homologous states in ^{88}Y, scaled for each level i by the factor $\frac{2J_i+1}{\Sigma_i(2J_i+1)}$ and the angular distributions of the cross sections and asymmetries for the transitions to the corresponding parent levels in ^{87}Y are shown.

In Table 1 multiplets of homologous states we identified are reported. The levels, for which only parity is indicated, were identified on the hypothesis of a possible fragmentation of highest spin levels.

The cumulative cross section and asymmetry for the doublet of levels homologous to the $\frac{1}{2}^-$ g.s. of ^{87}Y are in perfect agreement with those relative to the g.s. of ^{87}Y either in absolute value, and in shape.

In the case of the $\frac{5}{2}^-$ 0.794 MeV parent level, we have identified only a quartet, instead of the sextet of son states : are missing the 0^- and 5^- levels.

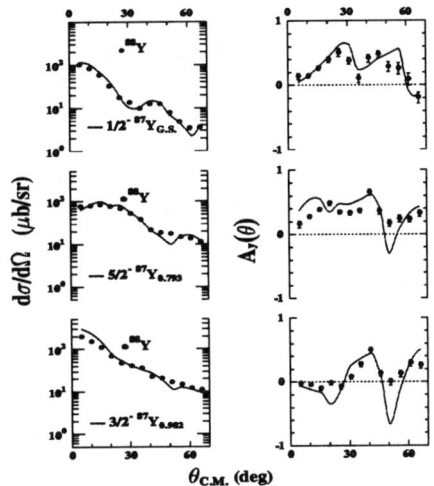

FIGURE 5: Cumulative cross sections and analyzing powers for population of multiplets of states of ^{88}Y (solid points) homologous to the low-lying parent states of ^{87}Y, including the fragmentated level contributions, compared with the cross sections and analyzing powers for population of these ^{88}Y states (solid lines).

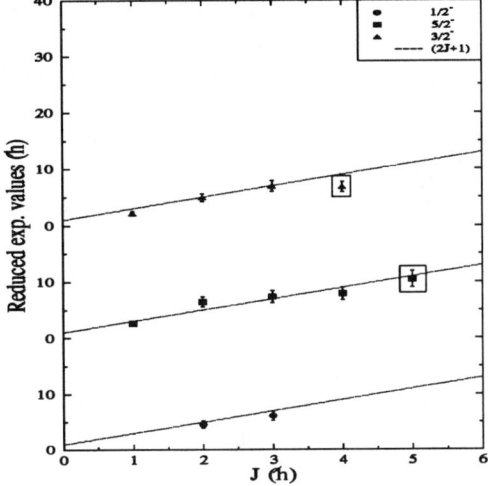

FIGURE 6: Reduced experimental values $\frac{\sigma(J_i)_{88Y}}{\sigma_{87Y}}\Sigma_i(2J_i+1)$ for the multiplets of states (solid points) homologous to the ^{87}Y parent states reported as a function of J and compared with the straight line $(2J+1)$ predicted by the weak coupling model (dashed lines). The two surrounded points correspond to the sum of the different branched levels.

The cross section of the 0^- member of the multiplet is expected to be very small and consequently its observation is difficult. On the contrary the cross section of the 5^- member of the same multiplet is expected to be about 30% of the cumulative cross section of the multiplet of states homologous to the $\frac{5}{2}^-$ 0.794 MeV parent level of ^{87}Y. There is no experimental evidence for this high cross section, whereas for the levels of ^{88}Y at 3.150 MeV and 3.207 MeV the shape of the differential cross sections and of the asymmetries fits very well the corresponding ones of the parent state.

In the case of the $\frac{3}{2}^-$ 0.982 MeV parent level, we have identified only a triplet, instead of the quartet of son states. The 4^- level of this multiplet is missing: its cross section is expected to be about 38% of the cumulative cross section for this multiplet. There is no experimental evidence for this high cross section, while for the levels of ^{88}Y at 2.997MeV, 3.023 MeV and 3.049 MeV the shape of the angular cross sections and of the asymmetries is reproduced by the differential cross section and asymmetry for the excitation of the 0.982 MeV parent state in ^{87}Y. If the previously discussed levels are included in the cumulative cross section of the proper multiplet, the parent state angular distributions of cross section and asymmetry are practically reproduced (fig.5). So it seems that the highest spin homologous son levels are fragmented.

In fig.6 the quantity $\frac{\sigma(J_i)_{88Y}}{\sigma_{87Y}}\Sigma_i(2J_i+1)$ for each multiplet of ^{88}Y states homologous to the ^{87}Y parent states we have identified are reported as a function of J_i, together with the straight $(2J+1)$ line. The satisfactory agreement between the experimental data and the prediction of the weak coupling model supports the spin and parity assignement of the ^{88}Y homologous levels.

We want to stress that the homology concept allows us to single out the dominant configuration of a given transition simply by comparison with experimental results of the parent nucleus, without applying complex shell model calculation.

REFERENCES

1. E.Gadioli, E.Gadioli Erba, R.Gaggini, P.Guazzoni, P.Michelato, A.Moroni and L.Zetta - *Z. Phys.* **A310**, 43-54 (1983)
2. E.Gadioli, P.Guazzoni, S. Mattioli, L.Zetta, G.Graw, R.Hertenberger, D.Hofer, H.Kader, P.Schiemenz, R.Neu, H. Abele and G.Staudt - *Phys. Rev.* **C43**, 2572-2585 (1991)
3. E.Gadioli, P.Guazzoni, M.Jaskola, L.Zetta, G.Colombo, G.Graw, R.Hertenberger, D.Hofer, H.Kader, P.Schiemenz, R.Neu and G.Staudt - *Phys. Rev.* **C47**, 1129-1142 (1993) and P.Guazzoni, M.Jaskola, L.Zetta, G.Graw, R.Hertenberger, D.Hofer, H.Kader, P.Schiemenz, U. Atzrott, R.Neu and G.Staudt - *Phys. Rev.* **C49**, 2784-2787 (1994)

General Formulae of Cross Sections and Analyzing Powers in Low Energy Limit
– Application to ^{2}H$(\vec{d},p)^3$H Reactions –

M. Tanifuji* and H. Kameyama†

Department of Physics, Hosei University, Tokyo 102, Japan
†*Chiba-Keizai College, Chiba 263, Japan*

Abstract. Data of cross sections and analyzing powers by polarized beam in ^{2}H$(\vec{d},p)^3$H reactions at E_d=30 and 70keV are analyzed by the use of the general formulae of cross sections and analyzing powers which are derived in a model-independent way. The calculations by the formulae reproduce the data very well suggesting their high validity in this energy region.

INTRODUCTION

Because of interest in fusion reactor applications, several measurements have been performed on analyzing powers by polarized beam in ^{2}H$(\vec{d},p)^3$H reactions at extremely low incident energies[1-3]. By the use of the invariant-amplitude method[4], we have derived in a model-independent way the general formulae of cross sections and analyzing powers[5] for $A(\vec{a},b)B$ reactions when spin-dependent interactions are weak. Since the formulae have high validity in the low energy limit, we will apply the formulae to the data on the ^{2}H$(\vec{d},p)^3$H reaction at E_d=30keV and 70keV.

INVARIANT-AMPLITUDE METHOD

The T matrix M for the reaction can be expanded into spin space tensors $S_{K\kappa}$

$$M = \sum_{K\kappa}(-)^\kappa S_{K\kappa} R_{K-\kappa}, \qquad (1)$$

where $R_{K-\kappa}$ is the counter part of $S_{K\kappa}$, the tensor in the ordinary space, and K and κ are the rank of the tensor and its z component, respectively. Designating the initial and final states by the z components ν of spins s of the related particles and the incident and final momenta $\vec{k}_i$ and $\vec{k}_f$, the invariant-amplitude method[4] gives the matrix element of the (K,κ) component of M as

$$<\nu_b\nu_B; \vec{k}_f|(-)^\kappa S_{K\kappa}R_{K-\kappa}|\nu_a\nu_A; \vec{k}_i>$$
$$= \sum_{s_is_f}(s_a\nu_a s_A\nu_A|s_i\nu_i)(s_b\nu_b s_B\nu_B|s_f\nu_f)(-)^{s_f-\nu_f}\sum_K(s_i\nu_i s_f-\nu_f|K\kappa)$$
$$\times \sum_{\ell_i=\bar{K}-K}^{K}[C_{\ell_i}(\Omega_i)\otimes C_{\ell_f}(\Omega_f)]_\kappa^K F(s_is_fK\ell_i;\cos\theta), \qquad (2)$$

where $s_i(s_f)$ is the channel spin in the initial(final) state, and $\Omega_i(\Omega_f)$ is the solid angle of $\vec{k}_i(\vec{k}_f)$. Here $\bar{K}=K$ for $K=even$ and $K+1$ for $K=odd$ when the total parity is conserved. In the following we will treat only this case, because the reaction is assumed to take place predominantly through central interactions. The quantity $F(s_is_fK\ell_i;\cos\theta)$ is the invariant amplitude and a function of $\cos\theta$, where θ is the relative angle between $\vec{k}_i$ and $\vec{k}_f$. The characteristic of $F(s_is_fK\ell_i;\cos\theta)$ is the designation by K, the rank of spin-space tensor associated. For example, $F(s_is_f0\ell_i;\cos\theta)$, $F(s_is_f1\ell_i;\cos\theta)$ and $F(s_is_f2\ell_i;\cos\theta)$ are respectively the reaction amplitudes due to interactions which are scalar, vector and second rank tensor in the spin space.

CROSS SECTIONS AND ANALYZING POWERS

When spin-dependent interactions are weak, it is sufficient to keep the spin-dependent amplitudes($K\neq 0$) up to their linear terms in observables. This approximation will be applicable to reactions in the low energy limit, because of very weak velocity-dependent interactions. Using this approximation and applying Eq.(2) to the transition amplitude, we get the analyzing powers T_{kq} in the coordinate axes, $z//\vec{k}_i$ and $y//\vec{k}_i\times\vec{k}_f$,

$$T_{kq} = \frac{2k_f}{k_i\hat{s}_a\hat{s}_A^2\sigma}\binom{Re}{iIm}\sum_{s_is_f}\hat{s}_i W(s_fs_Aks_a;s_as_i)$$
$$\times \sum_{\ell_i}(\ell_i 0\ell_f q|kq)C_{\ell_f q}(\theta,\varphi=0)F(s_is_fk\ell_i;\cos\theta)F^*(s_fs_f 00;\cos\theta) \qquad (3)$$

with

$$\sigma = \frac{k_f}{k_i\hat{s}_a^2\hat{s}_A^2}\sum_{s_f}|F(s_fs_f 00;\cos\theta)|^2. \qquad (4)$$

Specifying k and q, the explicit forms of iT_{11} and T_{2q} are obtained from Eq.(3),

$$iT_{11} = \frac{\beta}{\sigma}\sin\theta, \tag{5}$$

$$T_{20} = -\sqrt{\frac{2}{3}}\frac{\alpha}{\sigma}[(3\cos^2\theta - 1) + \varepsilon_1\cos\theta + \varepsilon_2], \tag{6}$$

$$T_{21} = 2\frac{\alpha}{\sigma}(\cos\theta\sin\theta + \frac{1}{4}\varepsilon_1\sin\theta), \tag{7}$$

$$T_{22} = -\frac{\alpha}{\sigma}\sin^2\theta \tag{8}$$

or

$$A_{yy} = -\sqrt{3}T_{22} - \frac{1}{\sqrt{2}}T_{20} = \frac{1}{\sqrt{3}}\frac{\alpha}{\sigma}(2 + \varepsilon_1\cos\theta + \varepsilon_2), \tag{9}$$

where α, ε_1 and ε_2 are composed of $F(s_is_f2\ell_i;\cos\theta)$ and $F(s_fs_f00;\cos\theta)$ and β is of $F(s_is_f1\ell_i;\cos\theta)$ and $F(s_fs_f00;\cos\theta)$. The θ-dependence of the invariant amplitudes is described by the Legendre polynomial expansion,

$$F(s_is_fK\ell_i;\cos\theta) = F_0(s_is_fK\ell_i)\sum_{\ell=0}^{\ell_{max}}\gamma_\ell^K(s_is_f\ell_i)P_\ell(\cos\theta), \tag{10}$$

where $F_0(s_is_fK\ell_i)$ and γ_ℓ^K are the the theoretical parameters, except for $\gamma_0^K = \sqrt{2}$.

APPLICATION TO ^{2}H($\vec{\text{d}}$,p)^{3}H REACTIONS

Let us take account fully the equivalence between the projectile and the target, which has been treated in an incomplete way in the previous work[6]. Further, to reduce the number of the adjustable parameters, we assume the θ-dependence of the invariant amplitude to be common for each class of K and all parameters to be real. Numerical searches indicate that the scalar amplitudes converge at $\ell_{max} = 2$ and the vector and tensor amplitudes at $\ell_{max} = 1$. Then, one can write

$$\sigma = \sigma_0\left\{[1 + \gamma_2^0 P_2(\cos\theta)]^2 + [\xi\gamma_1^0 P_1(\cos\theta)]^2\right\}, \tag{11}$$

$$\alpha = \alpha_0\left\{[1 + \gamma_2^0 P_2(\cos\theta)] + \frac{\sqrt{3}\xi\gamma_1^0}{2\zeta_1}P_1(\cos\theta) - \frac{\xi\gamma_1^0\zeta_2\gamma_1^2}{2}[P_1(\cos\theta)]^2\right\}, \tag{12}$$

$$\beta = \beta_0\left\{[1 + \gamma_2^0 P_2(\cos\theta)] + \frac{\xi\gamma_1^0}{2\eta_2}P_1(\cos\theta) + \frac{\xi\gamma_1^0\gamma_1^1}{\sqrt{3}}(\frac{\sqrt{5}}{2}\eta_1 - \eta_3)[P_1(\cos\theta)]^2\right\}, \tag{13}$$

where $\xi = F_0(1100)/F_0(0000)$, $\eta_1 = F_0(2111)/F_0(1111)$, $\eta_2 = F_0(1011)/F_0(1111)$, $\eta_3 = F_0(0111)/F_0(1011)$, $\zeta_1 = F_0(2020)/F_0(2120)$ and $\zeta_2 = F_0(1120)/F_0(2020)$.

The comparison of the present formulae with the experimental data[2] is made in two ways. One is for the simplified version of the calculation where ε_1, ε_2 and the θ-dependence of $F(s_i s_f K\ell_i; \cos\theta)$ are neglected. In this case, the angular distributions of iT_{11} and $T_{2q}(q=0,1,2)$ are described by $P_1^1(\cos\theta)$ and $P_2^q(\cos\theta)$, respectively, and hence A_{yy} is a θ-independent constant and their magnitudes are determined by one parameter β_0/σ_0 and α_0/σ_0 for the vector and tensor analyzing powers, respectively. Another way, the refined version, takes account into the calculation ε_1 and ε_2 as θ-independent constants and the θ-dependencies of $F(s_i s_f K\ell_i; \cos\theta)$ for all amplitudes.

Numerical Result and Reaction Mechanism

The numerical results are shown in Figs.1 and 2 for $E_d = 30$keV and 70keV, respectively, where the simplified version and the refined version are represented by the dashed lines and the solid lines, respectively, the parameters for which are given in Table 1. The dashed lines describe the characteristic features of the angular distributions of iT_{11} and $T_{2q}(q=0,1,2)$; that is, their gross structures are represented by $P_1^1(\cos\theta)$ and $P_2^q(\cos\theta)$, respectively.

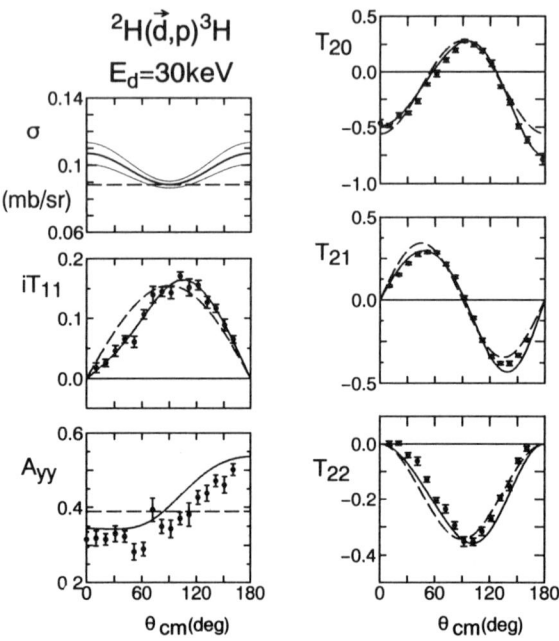

FIGURE 1. Comparison between calculations and experimental data in $^2\text{H}(\vec{d},p)^3\text{H}$ reactions at $E_d = 30$keV. The dashed and solid lines are for the simplified version and the refined version, respectively. The parameters are in Table 1 and the analyzing power data are taken from Ref.2. The cross section data[7] are between the two thin solid lines.

On the other hand, the simplified version fails in describing the angular distribution shapes of σ and A_{yy}. The solid lines reproduce all of the data very well though A_{yy} at large angles are systematically overestimated at E_d =30keV.

Since $\ell_i(\ell_f)$ is the orbital angular momentum in the initial(final) state in the simplified version, the above results indicate that the main reaction mechanism consists of the transitions from the S wave(initial) to the S wave(final) by the scalar interaction, those from the S wave to the D wave by the tensor interaction and those from the P wave to the P wave by the vector interaction. The calculations in the refined version introduce some additional mechanisms. For example, the non-zero γ_1^0 and γ_2^0 induce the P wave and the D wave for both of the incident and final channels in the scalar amplitudes, because, from Eqs.(2) and (10), the orbital angular momentum is given by $\vec{\ell_i} + \vec{\ell}$ and $\vec{\ell_f} + \vec{\ell}$ in the initial and final channels, respectively. That means the additional transition takes place from the P wave to the P wave and from the D wave to the D wave by the scalar interactions. As is seen in the figures, the corrections due to such additional reaction mechanisms are indispensable to reproduce the data of σ, A_{yy} and the details of other analyzing powers.

The success of the theory at energies both of 30keV and 70keV indicates the present formulae to be useful in such an energy region. This encourages us in further applications of the formulae.

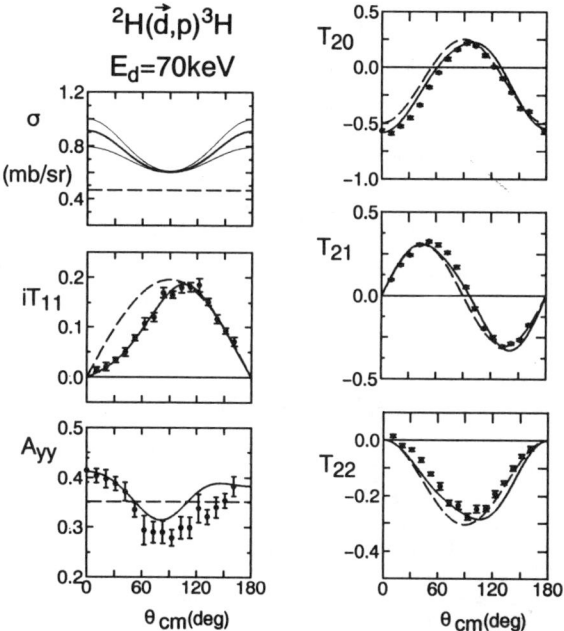

FIGURE 2. Comparison between calculations and experimental data in $^2\text{H}(\vec{d},p)^3\text{H}$ reactions at E_d =70keV. For further details, the captions in Fig.1 are referred to.

TABLE 1. Parameters for Calculations.

	30keV	70keV
σ_0(mb/sr)	0.089	0.47
α_0(mb/sr)	0.030	0.14
β_0 (mb/sr)	0.014	0.091
ε_1	0.18	0.50
ε_2	0.034	0.064
$\xi\gamma_1^0$	0.37	0.98
$1/\eta_2$	-0.51	-1.1
$1/\zeta_1$	-0.97	-0.45
$\gamma_1^1(\sqrt{5}\eta_1/2 - \eta_3)$	0	0.49
$\zeta_2\gamma_1^2$	-1.3	-2.1
γ_2^0	0	-0.18

Finally, the suppression of the total cross section due to polarizations is estimated when both of the projectile and the target are polarized in the z direction. Neglecting the contribution from the vector amplitudes, which is small, the cross section ratio $\sigma_{\rightrightarrows}^t/\sigma^t$ is given by the present theory in the sense of the simplified version,

$$\frac{\sigma_{\rightrightarrows}^t}{\sigma^t} = \frac{9}{5}(1 + \frac{1}{\zeta_1^2})\frac{|F_0(2020)|^2}{|F_0(0000)|^2} \ . \tag{14}$$

When $F(0000)$ and $F(2020)$ are real

$$\frac{|F_0(2020)|}{|F_0(0000)|} = \sqrt{\frac{2}{3}\frac{\alpha_0}{\sigma_0}} \ . \tag{15}$$

Using the values of σ_0, α_0 and ζ_1 in Table 1, we get for E_d =30keV

$$\frac{\sigma_{\rightrightarrows}^t}{\sigma^t} = 0.3 \ . \tag{16}$$

This is a rough estimation and the imaginary parts of $F_0(0000)$ and $F_0(2020)$ may modify the number. However, the cross section suppression due to polarizations seems to be not so large.

REFERENCES

1. H.Paetz gen. Schieck, B.Becker, R.Randermann, S.Lemaître, P.Niessen, R.Reckenfelderbäumer and L.Sydow, Phys.Lett. **B276**, 290 (1992).
2. Y.Tagishi, N.Nakamoto, K.Katoh, J.Togawa, T.Hisamune, T.Yoshida and Y.Aoki, Phys.Rev.C **46**, R1155 (1992).
3. K.A.Fletcher, Z.Ayer, T.C.Black, R.K.Das, H.J.Karwowski, E.J.Ludwig and G.M.Hale, Phys.Rev.C **49**, 2305 (1994).
4. M.Tanifuji and K.Yazaki, Prog.Theor.Phys. **40**, 1023 (1968).
5. M.Tanifuji and H.Kameyama, to be published.
6. M.Tanifuji, Phys.Lett. **B289**, 233 (1992).
7. R.E.Brown and N.Jarmie, Phys.Rev.C **41**, 1391(1990).

Inelastic Scattering of Low- and Intermediate-Energy Polarized Protons from Various Light Nuclei

A. Plavko, M. Onegin, and V. Kudriashov

St. Petersburg State Technical University, St. Petersburg 195251, Russia

C. Olmer, A. Bacher, P. Schwandt, and E. Stephenson

Indiana University Cyclotron Facility, Bloomington, Indiana 47408, USA

In this paper we investigate similarities and differences in the inelastic scattering of polarized protons from various light nuclei for both low and intermediate bombarding energies E_p. In this respect, the present paper is an extension of our recent work (Ref. 1). There we found that the distributions of analyzing power A(q) and differential cross section $\sigma(q)$ in $(\vec{p},p')$ scattering at E_p=135 MeV for the excitation of 2_1^+ states are very similar in shape for ^{28}Si and ^{20}Ne, respectively (see Fig. 1), as well as for ^{16}O (see Ref. 1).

At low beam energies E_p, the behavior of the analyzing power $A(\theta)$ is different for the 2_1^+ states of light nuclei. For example, although the angular distributions $A(\theta)$ remain similar for ^{28}Si and ^{20}Ne, they are significantly different from $A(\theta)$ for ^{16}O. The latter is also different from $A(\theta)$ for ^{22}Ne, ^{24}Mg, ^{26}Mg and even ^{18}O (Ref. 2).

In Fig. 1 the angular distributions of $A(\theta)$ and $\sigma(\theta)$, respectively, are also compared for ^{28}Si and ^{20}Ne in the case of the 4_1^+ and 1_1^- states. The $A(\theta)$ behavior is still the same as for the 2_1^+ excitation, in that the $A(\theta)$ distributions are practically identical for the two nuclei both for the excitation of the 4_1^+ and 1_1^- states. At

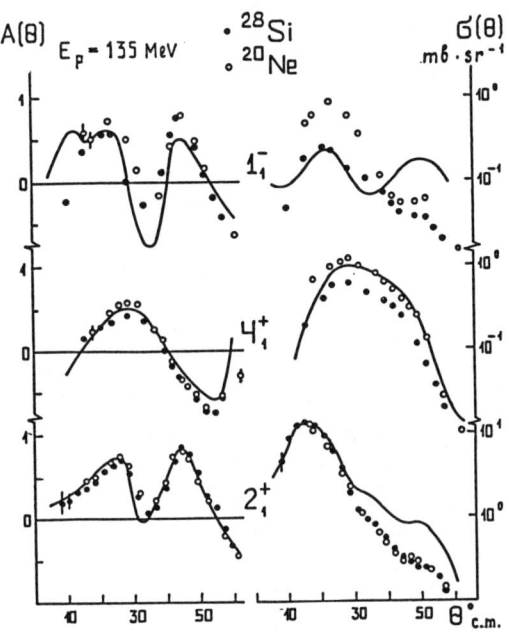

FIGURE 1. $A(\Theta)$ and $\sigma(\Theta)$ for $1^-, 2^+, 4^+$ excitations in ^{20}Ne and ^{28}Si at E_p = 135 MeV. Curves in this and all subsequent Figs. are calculations based on effective NN interactions.

low E_p, the $A(\theta)$ behavior for the 4_1^+ state was again different (see, e.g., systematic data for the 4^+ states in various light nuclei in Ref. 3).

At low E_p, the $A(\theta)$ and $\sigma(\theta)$ variations for the 4^+ excited states are usually explained by the change in central hexadecapole deformations. The main characteristics of the $A(\theta)$ variations in the case of the 2^+ excited states can be attributed to spin-orbit quadrupole deformations, as has been described in numerous papers. A similar view was taken in the study of ^{20}Ne($\vec{p}$,p$'$) scattering at E_p=135 MeV (Ref. 4).

However, ($\vec{p}$,p$'$) scattering at E_p=135 MeV can also be studied from a different point of view. For example, the experimental $A(\theta)$ distributions, as shown in Fig. 1, do not exhibit any differences between the two light nuclei, and therefore do not differentiate between any (central or spin-orbit) collective deformations. Is similar $A(\theta)$ behavior possible in ($\vec{p}$,p$'$) scattering at low E_p? It turns out that it *is* possible for certain states in light nuclei as described in Ref. 2, namely for negative-parity states such as 1^-, 2^-, 3^-, etc. As is known from numerous phenomenological analyses, the $A(\theta)$ distributions for those states are least affected by collective spin-orbit deformations.

Refs. 1 and 5 describe the possibility of using the concept of an effective NN interaction at low E_p for the study of various nuclear states. The most difficult task in this microscopic approach appears to be the description of $A(\theta)$ for the 4_1^+ state in ^{28}Si (as well as other light nuclei). For this state it was found that the $A(\theta)$ shape strongly depends on the central hexadecapole deformation when the collective rotational model was used in coupled-channels calculations. It also seems unrealistic to apply the concept of an NN interaction in a microscopic framework at low E_p to the description of $A(\theta)$ for the 2_1^+ state in ^{28}Si and other s-d shell nuclei, where collective (vibrational or rotational) models previously established the necessity of introducing strong spin-orbit quadrupole deformations. However, in Ref. 6 it is shown that in those cases where spin-orbit deformation can be ignored, the description of both $A(\theta)$ and the product $A(\theta)\sigma(\theta)$ based on an effective NN interaction appears to be quite good. This approach is also acceptable for the description of $\sigma(\theta)$ at all E_p (Refs. 1, 5) since cross-section calculations, as is well-known, primarily reflect the shape of the transition form factor, whereas analyzing powers are generally more sensitive to the details of the effective NN interaction.

Fig. 2 demonstrates fairly well that experimental $\sigma(\theta)$ distributions follow the form factors obtained from (e,e$'$) scattering both at E_p=135 MeV and E_p=26.3 MeV (the latter experimental data have been taken from Ref. 7) in those cases where the momentum transfer is small. Here the concept of the effective NN interaction allowed us to apply experimental charge transition densities extracted from (e,e$'$) scattering to (p,p$'$) scattering. It is impossible

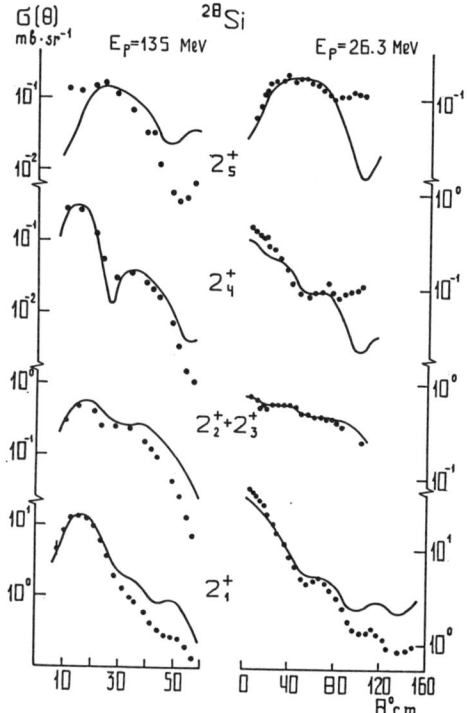 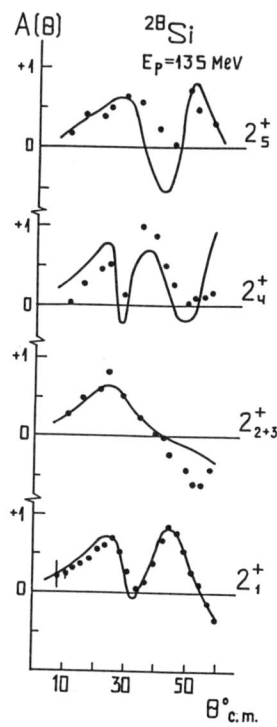

FIGURE 2. $\sigma(\Theta)$ for 2^+ excitations in ^{28}Si at $E_p = 135$ MeV (left) and 26.3 MeV (right).

FIGURE 3. $A(\Theta)$ for 2^+ excitations in ^{28}Si at $E_p = 135$ MeV.

to achieve that directly using any collective model. The present calculations are of the type used in Ref. 8 and based on the program LEA by Kelly (Ref. 9). In the case of E_p=135 MeV we have used the Paris-Hamburg (PH) effective interaction (Ref. 10), whereas at low E_p the effective NN has been obtained empirically, using a parametrized model (Ref. 11) guided by nuclear matter theory.

The $A(\theta)$ distributions at E_p=135 MeV are shown in Fig. 3 and are reasonably well described by the calculations. It has been shown that not only the transition form factors but also the effective interaction appear to be quite realistic. As was mentioned above and shown in Ref. 1, the description of $A(\theta)$ at low E_p is much less successful, but it should be noted that the available experimental data at low energies are also incomplete.

On the whole, a faily good description of the main features of the angular distributions, primarily for $\sigma(\theta)$ at not too high momentum transfers, can be achieved on the basis of the effective NN interaction and charge transition densities. However, it is difficult to describe the $\sigma(\theta)$ data on the basis of a collective band-like character of these states (except for 2_1^+) or using the shell model.

As for the results for ^{28}Si, calculated using an effective NN interaction and shown in Fig. 1, it should be noted that data fits of a similar quality (or, in the case of $\sigma(\theta)$ for the 1_1^- state, of better quality) can be obtained using a collective model for the corresponding states of ^{20}Ne (Ref. 4). So we have to decide what is preferable: the use of empirical transition densities extracted from (e,e') scattering or the parametrization of a band-like character of the excited states.

A similar problem arises in the case of excitation of the 4_1^+ and 5_1^- states of ^{28}Si. The A(q) distributions for these states are presented in Fig. 4, showing systematic data taken from Refs. 8, 12-15. On the one hand, changes in experimental A(q) distributions with E_p are of the same type for the 4_1^+ and 5_1^- states. This has been confirmed by our calculations (solid curves) based on the effective NN interaction of the PH type and transition densities extracted from (e,e') scattering which are quite similar in shape for these two states. On the other hand, such a description for the 4_1^+ state of ^{28}Si becomes qualitative only already at E_p=65 MeV, and deteriorates further with decreasing E_p (Ref. 5), whereas the use of the rotational model, in contrast, leads to better agreement with the experimental $(\vec{p},p')$ scattering data (Ref. 16).

Unfortunately, the collective-band concept turned out to be unacceptable for the description of $A(\theta)$ and $\sigma(\theta)$ in the case of the 5_1^- state of ^{28}Si at E_p=65 MeV (Ref. 15). Moreover, although the use of macroscopic transition densities for the description of $A(\theta)$ in the case of the 4_1^+ state of ^{28}Si was successful at E_p=25 MeV (Ref. 16), at lower proton energies this procedure becomes very problematic (Ref. 17).

Although the application of the effective NN interaction concept does not yield very good agreement with certain experimental data at low E_p, it appears to be more universal. For cases of non-collective excitations or collective excitations of a not-well-understood nature, this concept is absolutely indispensable.

An extremely important property of the concept of the NN interaction is that at E_p=100-200 MeV (and also at a certain range of higher E_p) for normal-parity excitations the effective NN interaction depends primarily upon the local density and is almost independent of the nucleus, the excitation state, or deformation. This is demonstrated very clearly in Fig. 5 (at E_p=135 MeV) where new results for the $2_1^+ - 4_1^+$ rotational band in ^{28}Si are compared to the well-known data of Kelly et al. for the same band in ^{16}O, also based on the ground state. Our calculations using the PH interaction and charge transition densities for ^{28}Si are also in a good agreement with the

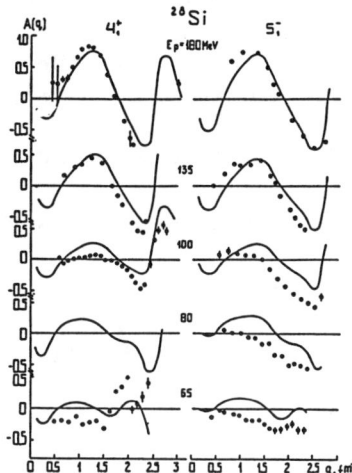 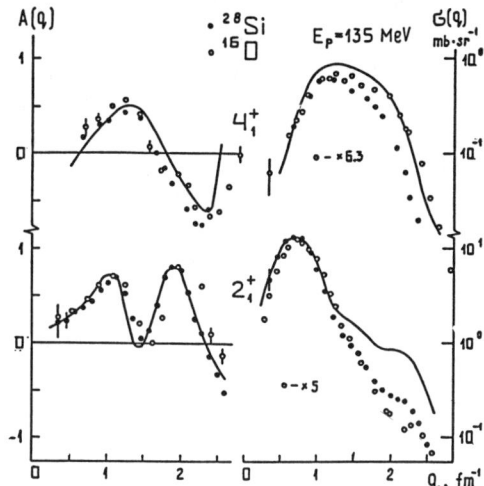

FIGURE 4. A(q) for 4^+ (left) and 5^- (right) excitations in ^{28}Si for E_p = 65–180 MeV.

FIGURE 5. A(q) and $\sigma(q)$ for $2^+, 4^+$ excitations in ^{16}O and ^{28}Si at E_p = 135 MeV.

experimental A(q) and $\sigma(q)$ distributions for ^{16}O. The ratios of cross sections at the maxima shown in Fig. 5 are in a good agreement with the ratios of the corresponding charge transition densities, even in a simple plane-wave approximation.

The present paper demonstrates that, despite the somewhat more complex character of low-energy nuclear dynamics, the physics of inelastic scattering at low (non-resonance) energies and that at intermediate energies have more common than different properties.

ACKNOWLEDGMENTS

This work was financed by the Russian Foundation for Fundamental Research under Grant No. 93-02-3325.

REFERENCES

1. A.V. Plavko, M.S. Onegin, V.I. Kudriashov, C. Olmer and P. Schwandt, Bulletin of the Russian Academy of Sciences, Physics Series **58**, 97 (1994).
2. A.V. Plavko in *Program of Experimental Research at the Meson Facility of the Institute of Nuclear Research of the USSR Academy of Sciences*

(Proced. of the 3rd All-Union Seminar, 23-27 April 1983, Zvenigorod) Moscow, 1984, p.280.
3. A.V. Plavko, V.I. Kudriashov, R.M. Lombard and I.-L. Escudie, Report of the Institute of Nuclear Physics of Leningrad **N424**, July 1978, Leningrad.
4. M.S. Munro, G.G. Shute and B.M. Spicer *et al.*, IUCF Scientific and Technical Report 1988-1989, p.24.
5. A.V. Plavko, M.S. Onegin and V.I. Kudriashov, Bulletin of the Russian Academy of Sciences, Physics Series (to be published).
6. A.V. Plavko, Abstracts of Contributed Papers, 7th Int. Conf. on Polarization Phenomena in Nucl. Phys. (Paris, 9-13 July 1990), p.58 B.
7. J.J.A. Zalmstra, M.N. Harakeh and J.F.A. van Hienen, Nucl. Phys. **A526**, 59 (1991).
8. Q. Chen, J.J. Kelly, P.P. Singh *et al.*, Phys. Rev. C **41**, 2514 (1990).
9. J.J. Kelly, Computer Program LEA (unpublished).
10. H.V. Geramb in *The Interaction Between Medium Energy Nucleons in Nuclei* (Bloomington, Indiana, 1982), AIP Conf. Proc. **97** (AIP, New York, 1982), p.44.
11. J.J. Kelly, Phys. Rev. C **39**, 2120 (1989).
12. C. Olmer, A.D. Bacher, G.T. Emery *et al.*, Phys. Rev. C **29**, 361 (1984).
13. A.V. Plavko, C. Olmer and P. Schwandt in *Nuclear Spectroscopy and Structure of the Atomic Nucleus*, Proceed. Intern. Sympos. (Minsk, 16-19 April 1991), St.Petersburg, Nauka, 1991, p.267.
14. A.V. Plavko, M.S. Onegin, C. Olmer and P. Schwandt in *Nuclear Spectroscopy and Structure of the Atomic Nucleus*, Proceed. Internat. Sympos. (Dubna, 20-23 April 1993), St.Petersburg, 1993, p.206.
15. S. Kato, K. Okada, M. Kondo *et al.*, Phys. Rev. C **31**, 1616 (1985).
16. B.J. Verhaar in *Lecture Notes in Physics, Polarization in Nuclear Physics* (Proceed. of meeting held at Ebermannstadt, 1973), Springer-Verlag, Berlin-Heidelberg-New York, 1974, p.268.
17. M. Haller, J. Kiener, W. Kretschmer *et al.*, Proceed. Sixth Int. Symp. Polar. Phenom. in Nucl. Phys. (Osaka, 1985), J. Phys. Soc. Jpn. **55** Suppl., 590 (1986).

Singlet-State Contributions to Deuteron Elastic Scattering

Y. Iseri* and M. Tanifuji[†]

*Department of Physics, Chiba-Keizai College, Chiba 263, Japan
[†]Department of Physics, Hosei University, Tokyo 102, Japan

Abstract. In elastic scattering of deuterons, effects of virtual breakup of the deuteron to spin-singlet states are investigated with the CDCC method. The tensor analyzing power A_{yy} is seriously affected by the singlet breakup, while the cross section and the vector analyzing power A_y are little. The analysis with the spin-space tensor amplitudes introduced by the invariant-amplitude method clarifies that the singlet-breakup effect considered here plays as an effective T_L-type tensor interaction.

Contributions of virtual breakup of deuterons have been extensively investigated on deuteron elastic scatterings. Many of the investigations until now, however, have treated only the breakup to spin-triplet states in the p-n continuum. The transition between the triplet and the singlet states is caused by a part of spin-orbit interactions in the nucleon optical potentials. The central force cannot produce the spin flip process, such as the transition from the triplet to the singlet and its reverse. Because the central part is stronger than the spin-orbit part in most cases and the deuteron ground state is triplet, it has been believed that the triplet breakup gives much larger contribution to the deuteron elastic scattering than the singlet breakup. It turns out to be true for the cross section, but not for some polarization observables. Furthermore the singlet-breakup effect is expected to become large at some high incident energies, because effective strength of the spin-orbit coupling increases with the deuteron incident energy.

In this work we treat the singlet-breakup process at the deuteron incident energy $T_d=$ 400 and 80 MeV with the coupled-discretized-continuum channels (CDCC) method[1,2] based on the p-n-target three-body model. The p-target and n-target interactions are taken to be the respective optical potential at half the deuteron incident energy. The Reid soft-core potential is used for the p-n interaction. The model space of the p-n system is composed of the $^3S_1$, $^3D_1$, $^3D_2$ and $^3D_3$ states for the triplet and the $^1P_1$ state for the singlet. The breakup states of every partial wave are truncated with the maximum value of the wave number for the p-n relative motion $k_{max} = 1.0$ fm^{-1}. These truncated continuum states are discretized into two bins at $T_d = 400$ MeV and into four bins at $T_d = 80$ MeV.

In the CDCC method we transform the variables of the input nucleon-target potentials, r_p and r_n, into the p-n relative coordinate ρ and the center of mass coordinate to the target R. The spin-orbit potentials are transformed as

$$U_{pA}^{SO}(r_p)\ell_p \cdot \sigma_p + U_{nA}^{SO}(r_n)\ell_n \cdot \sigma_n \tag{1}$$
$$= \tfrac{1}{2}U_+ (L+\ell)\cdot S_+ + \tfrac{1}{i}U_- (R\times\nabla_\rho + \tfrac{1}{4}\rho\times\nabla_R)\cdot S_+$$
$$+ \tfrac{1}{2}U_- (L+\ell)\cdot S_- + \tfrac{1}{i}U_+ (R\times\nabla_\rho + \tfrac{1}{4}\rho\times\nabla_R)\cdot S_-,$$

where

$$U_\pm = U_\pm(R,\rho) = U_{pA}^{SO}(r_p) \pm U_{nA}^{SO}(r_n), \tag{2}$$
$$S_\pm = \tfrac{1}{2}(\sigma_p \pm \sigma_n). \tag{3}$$

The coupling between the triplet and the singlet state is produced by the terms including S_- in the interaction (1). We take account of only the $L \cdot S_-$ term among those S_- terms because its effect is expected to be largest. In the S_+ terms we neglect the term including ∇_R.

In the following calculation for $T_d = 400$ MeV we use nucleon optical potentials which are derived by the Dirac phenomenology in the effective Schrödinger form. In the case of $T_d = 80$ MeV standard Woods-Saxon shaped potentials are used.[2]

Figure 1 shows the comparison between the experimental data and the results of the calculations for the differential cross section, σ, and the vector and tensor analyzing powers, A_y and A_{yy}, in the scattering from ^{58}Ni at $T_d = 400$ MeV. The triplet breakup causes much change in the cross section while it yields little effect in the analyzing powers. On the other hand, the singlet breakup gives remarkable change in the tensor analyzing power A_{yy}, while its effect is negligible on the cross section and very small on the vector analyzing power. The singlet-breakup contribution to A_{yy} improves the fit to the data in the region of $20° \lesssim \theta_{cm} \lesssim 30°$.

Recently Al-Khalili et al. have evaluated the singlet-breakup contribution to the scattering amplitude by the two-step model.[3] In this model the effect of the multi-step process is ignored, while it is included in the CDCC calculation. Their results for the features of the singlet-breakup effect agree with the CDCC results fairly well. It indicates that the two-step model is fairly good in this energy region.

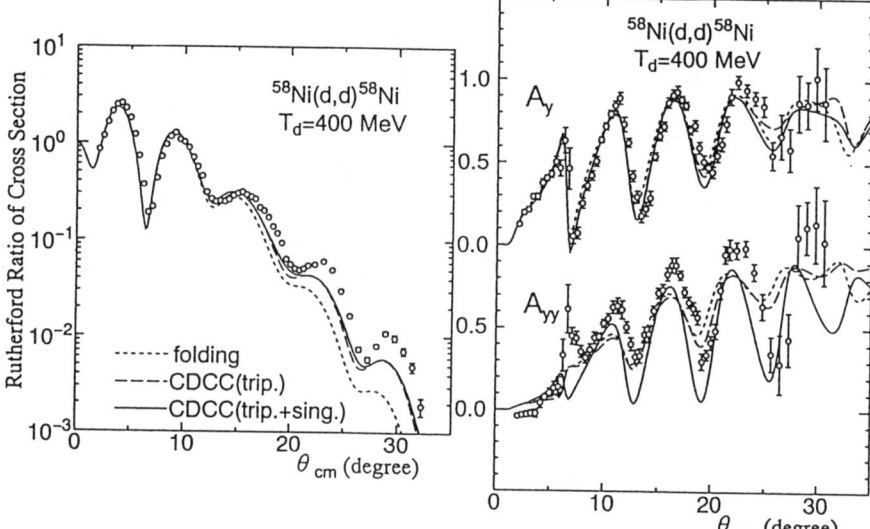

FIGURE.1. Cross section and analyzing powers in ^{58}Ni$(\vec{d},d)^{58}$Ni at $T_d = 400$ MeV. The dotted lines are the simple folding model without the breakup effect; the dashed lines are the CDCC calculation with the triplet breakup; the solid lines are the CDCC calcultion with the triplet and the singlet breakup. The experimental data are taken from Ref.[4].

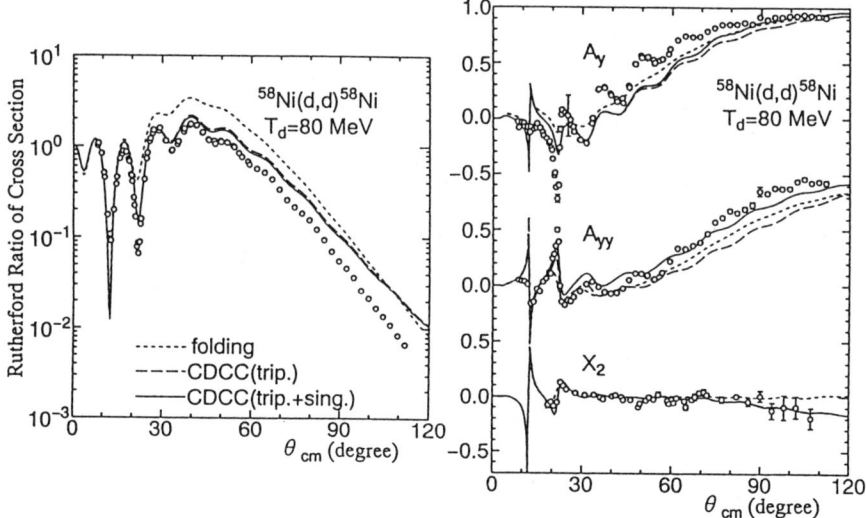

FIGURE.2. Cross section and analyzing powers in ^{58}Ni$(\vec{d},d)^{58}$Ni at $T_d = 80$ MeV. The meanings of the lines are same as Fig.1. The experimental data are taken from Ref.[5].

The results for $T_d = 80$ MeV are shown in Fig. 2. The sensitivity of the observables to the breakup effects is almost same as that in the case of $T_d = 400$ MeV. The singlet breakup effect is remarkable in A_{yy} but is not in σ and A_y.

In order to understand above observable dependence of the breakup effects, we introduce the spin-space tensor amplitudes, U, S, T_α and T_β, by the invariant-amplitude method. These amplitudes are defined as a linear combination of the scattering amplitudes and characterize a specific spin-dependent interaction.[6] The U is a scalar amplitude in the spin space and describes the scattering amplitude due to the central (spin-independent) interaction in the sense of the effective interaction. The S is a vector amplitude and describes the one due to the spin-orbit interaction. The T_α and the T_β are tensor amplitudes and characterize the T_R-type and the T_L-type tensor interaction, respectively. Among these amplitudes, the magnitude of U is expected to be the largest because the central interaction is usually stronger than spin-dependent interactions.

In Fig.3 the magnitudes of the spin-space tensor amplitudes at $T_d = 400$ MeV are displayed. One finds in this fugure that $|U|$ and $|S|$ are affected by the triplet breakup in the region $\theta \gtrsim 15°$ but not by the singlet breakup. Contrary to this $|T_\beta|$ is seriously affected by the singlet breakup in the whole angular region. This result implys that the singlet breakup process produces considerable contribution to the effective T_L-type tensor interaction. It is also

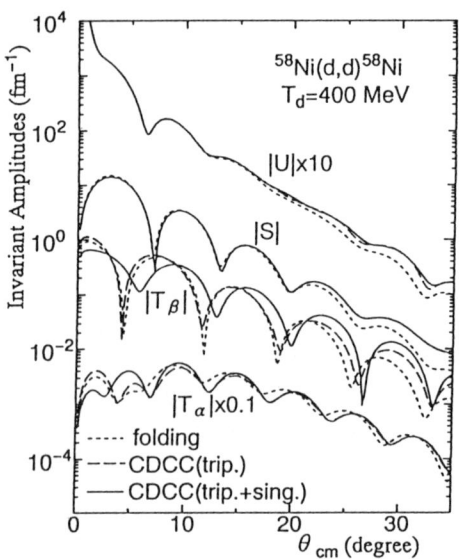

FIGURE.3. Magnitudes of the spin-space tensor amplitudes in $^{58}\text{Ni}(\vec{d}, d)^{58}\text{Ni}$ at $T_d = 400$ MeV. The meanings of the lines are same as Fig.1.

implied that the triplet-breakup effects produce considerable changes in the effective central and spin-orbit interactions.

The cross section and analyzing powers can be described by U, S, T_α and T_β. The approximate formulae for these observables up to the first order of the spin-dependent terms are given by

$$\sigma \simeq \frac{1}{9}|U|^2, \tag{4}$$

$$A_y \simeq \frac{2\sqrt{2}}{9\sigma}\mathrm{Im}(US^*), \tag{5}$$

$$A_{yy} \simeq \frac{1}{9\sigma}\left\{-8\mathrm{Re}(UT_\beta^*) + \frac{2\sqrt{2}}{\sin\theta}\mathrm{Re}(UT_\alpha^*)\right\}. \tag{6}$$

The singlet-breakup effect which behaves as the effective T_L-type tensor interaction contributes mainly to the amplitude T_β. The T_β is included in A_{yy} but not in σ and A_y up to the first order of the spin-dependent terms. The tensor analyzing power A_{yy} is, therefore, more sensitive to the singlet-breakup effect than σ and A_y. Similarly, the change in the cross section due to the triplet-breakup effect is explained as the change in U through the effective central interaction generated from the triplet-breakup process.

The reason why the virtual singlet-breakup process produce the effective T_L-type tensor interaction is explained by the two-step model[6]. At these incident energies the two-step model is expected to be qualitatively good. In the first step the deuteron is excited from the ground state to the singlet-breakup state by the interaction $U_- \boldsymbol{L} \cdot \boldsymbol{S}_-$. The same interaction is used in the second step where the deuteron is deexcited from the breakup state to the ground state. This two-step process is characterized by the operator $(\boldsymbol{L} \cdot \boldsymbol{S}_-)^2$ and

$$(\boldsymbol{L} \cdot \boldsymbol{S}_-)^2 = -[S_+ \otimes S_+]_2 \cdot [L \otimes L]_2 - \tfrac{1}{2}\boldsymbol{L} \cdot \boldsymbol{S}_+ + \tfrac{1}{3}\boldsymbol{L}^2 \tag{7}$$

The first term in the right-hand side of (7) is just the T_L-type tensor operator. Thus the second-rank tensor interaction generated from the virtual breakup to the singlet states by the $\boldsymbol{L} \cdot \boldsymbol{S}_-$ interaction can be treated as if it is the usual T_L-type tensor interaction.

We attempt to express the effective interaction due to the singlet breakup as a local potential.[7] The two-step model is adopted with adiabatic and closure approximations. The effective local potential obtained by the numerical estimation reproduces well the result of the corresponding CDCC calculation.[8] This indicates a possibility of describing the singlet-breakup effect by the local potential.

REFERENCES

1. Iseri,Y., Kameyama,H., Kamimura,M., Yahiro,M. and Tanifuji,M., *Nucl.Phys.* **A490**,383(1988);
   Iseri,Y., Tanifuji,M., Aoki,Y. and Kawai,M., *Phys.Lett.* **B265**,207(1991).

2. Yahiro,M., Iseri,Y., Kameyama,H., Kamimura,M. and Kawai,M., *Prog.Theor.Phys.Suppl.* **89**,32(1986).

3. Al-Khalili,J.S., Tostevin,J.A. and Johnson,R.C., *Phys.Rev.* **C41**,R806(1990); *Nucl.Phys.* **A514**,649(1990).

4. Sen,N.van. et al., *Phys.Lett.* **156B**,185(1985) and *private communications*.

5. Stephenson,E.J. et al., *Phys.Rev.* **C28**,134(1983) and *private communications*.

6. Iseri,Y., Tanifuji,M., Kameyama,H., Kamimura,M. and Yahiro,M., *Nucl.Phys.* **A533**,574(1991).

7. Tanifuji,M. and Iseri,Y., *Prog.Theor.Phys.* **87**,247(1992).

8. Iseri,Y. and Tanifuji,M., *to be published*.

A Constrained Dispersive Optical Model for the neutron-nucleus interaction from -80 to +80 MeV for the mass region $27 \leq A \leq 32$

M. A. Al-Ohali
King Fahd University of Petroleum and Minerals, Dhahran, Saudi Arabia 31261
and Triangle Universities Nuclear Laboratory, Durham, North Carolina 27707

C. R. Howell, W. Tornow and R. L. Walter
Duke University and Triangle Universities Nuclear Laboratory
Durham, North Carolina 27707

Abstract. A Constrained Dispersive Optical Model (CDOM) analysis was performed for the neutron-nucleus interaction in the energy domain from -80 to 80 MeV for the three nuclei in the center of the 2s-1d shell nuclei. The CDOM incorporates the dispersion relation which connects the real and imaginary parts of the nuclear mean field. Parameters for the model were derived by fitting the neutron differential elastic cross-section, the total cross-section and the analyzing power data for ^{27}Al, ^{28}Si and ^{32}S. The parameters were also adjusted slightly to improve overall agreement to single-particle bound-state energies.

INTRODUCTION

The nucleon-nucleus Optical Model potential has played an important role in the interpretation of data for many types of nuclear reactions. In particular, the OM potential has had great success in describing the average behavior of elastic scattering observables (total and differential cross-section and analyzing powers). However, increasingly accurate experimental data made it clear that a standard formulation of the OM potential was no longer adequate to describe the experimental data over a wide energy range. In addition, it is not possible to extend the conventional OM derived from scattering data into the negative energy region (that is, to connect the OM to the shell model potential) in a physically meaningful way. During recent years, a great deal of effort was devoted to apply the Dispersion Relation (DR) for nucleon-nucleus scattering. The DR is an integral relation that connects the real and imaginary parts of a complex-valued analytical function. The main advantage of the Dispersive Optical Model (DOM) model is that, in a natural way it accounts for the energy dependence of the strength and the radius parameter of the central real potential. Also, it allows one to use the wealth of nucleon-nucleus scattering data that is available in the literature to characterize the extension of the Nuclear Mean Field (NMF) into the shell-model region in a consistent manner. Mahaux and coworkers (1) have laid much of the groundwork for applying the DR to nucleon-nucleus problems. The current reported here builds upon this foundation.

The nuclei in the 2s-1d shell provide an attractive ground for nuclear structure studies. In addition, nuclei near the center of the shell are known to have excited states with a highly collective nature. This feature requires that core polarization effects for the unbound nucleon be considered in the analysis of the nucleon interactions with such systems (2).

Global optical models of the conventional type exist in literature for neutron scattering from heavier nuclei, and typically for relatively narrow energy regions. However, the global parameters for these models are less successful in describing the scattering data for nuclei with A < 40 than with A > 40. In this work, we applied the DR to investigate the nature of neutron scattering from nuclei in the center of the 2s-1d shell. From that result, a constrained set of DOM parameters were developed to provide a description of the observables for neutrons scattered elastically from nuclei in the mass region $27 \leq A \leq 32$. Having used input from measured bound-state energies in addition to the scattering data, our model applies to the energy range from -80 MeV to +80 MeV.

THE NUCLEAR MEAN FIELD AND *CDOM* FORMALISM

The nuclear mean field M is written with a real and an imaginary part as follows:

$$M(r,E) = V(r,E) + i\, W(r,E) \qquad (1)$$

$$V(r,E) = V_{HF}(r,E) + \Delta V(r,E) \qquad (2)$$

$$V_{HF}(r,E) = V(0) \cdot f(r) \cdot \{ \exp[-\alpha \cdot E / V(0)] \} \qquad (3)$$

$$\Delta V(r,E) = (P/\pi) \int_{-\infty}^{\infty} [W(r,E')/(E'-E)]\, dE' \qquad (4)$$

where $V(r,E)$ is the real potential, and $W(r,E)$ is conventional absorptive potential that contains surface and volume contributions. The $V_{HF}(r,E)$ is the Hartree-Fock contribution. In equation (3), the $V(0)$ is the Hartree-Fock potential depth at Fermi energy, the $f(r)$ is the conventional Woods-Saxon form and the α can be considered as the slope of the Hartree-Fock potential depth. The dispersion relation is given by equation (4) and it produces a contribution $\Delta V(r,E)$ to the real potential that depends on the strength and shape of the imaginary potential. The P in the DR denotes a principal-value integral. The energy dependences of the volume and the surface absorption are represented in a form similar to the expression in Jeukenne and Mahaux (3). In the present case, we chose the energy dependence that is symmetric with respect to Fermi energy E_F. Figure 1 shows the strengths W_S of the surface absorption term and W_V of the volume absorption term for n-^{28}Si.

THE *CDOM* DEVELOPMENT

In order to apply the DOM approach, it is important to have a fairly wide energy range of data in the positive and negative energy regions. In the positive energy region, the data base consisted of $\sigma(\theta)$, $A_y(\theta)$ and σ_T data from

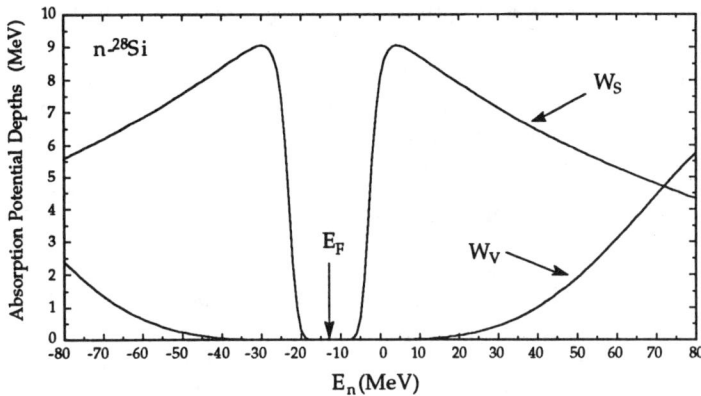

Figure 1. The dispersive OM surface and volume absorption terms for n - ^{28}Si.

experiments done at TUNL and several other labs. Table 1 lists the database for the positive energy region for the three nuclei of the present study: ^{27}Al, ^{28}Si and ^{32}S. For the negative energy region, the binding energies of the single-particle and single-hole states are used. This information is obtained from the one-neutron transfer reactions: stripping, pickup, and knock-out reactions. The empirical information for the bound states was only obtained for ^{28}Si and ^{32}S.

TABLE 1. Database used to develop the CDOM

		^{27}Al	^{28}Si	^{32}S
$\sigma(\theta)$	# of data sets	14 sets	16 sets	16 sets
	Energy range	2-25 MeV	2-40 MeV	2-40 MeV
$A_y(\theta)$	# of data sets	3 sets	4 sets	4 sets
	Energy range	14-17 MeV	10-17 MeV	10-17 MeV
σ_T	# of data	25	25	25
	Energy range	0.3-80 MeV	0.3-80 MeV	0.3-80 MeV

METHOD OF ANALYSIS, RESULTS AND DISCUSSION

The first step to develop the CDOM was to generate individual DOM parameters for each system utilizing the empirical information. Searches were performed simultaneously on the large database of the scattering region to obtain optimum values for the DOM parameters to fit all the scattering data. These optimum values (determined from chi-squared minimization) were then adjusted slightly to improve the agreement with the empirical bound-state binding energies. The process was repeated until a reasonable fit was obtained to the scattering data as well as to the binding energies of the bound states for the nucleus. For developing the interaction for ^{27}Al, the single-particle bound-state energies for ^{28}Si were used to provide a modest control.

For the CDOM development, we started with the average values of the individual DOM parameters resulting from the three solutions above. Using average values for the three DOMs, we were able to obtain a fairly good description of the scattering observables. The most obvious problems were the calculations of the differential cross section distributions. It was clear that for ^{27}Al the absorption strengths needed to be reduced, while for the ^{28}Si and the ^{32}S they needed to be increased. These adjustments indicate that nuclear structure or correlation effects must be treated differently for these nuclei. Therefore, we allowed the strengths of the volume and surface absorption to differ for all three nuclei. Reducing the constraint on the absorption strengths allowed the total chi-squared to drop by 18% and yielded better fits of the scattering data for all three systems. The final CDOM has about 30% less volume absorption and 10% less surface absorption for ^{27}Al than for ^{28}Si and ^{32}S. No attempt to further optimize the single-particle binding energies was made while developing the final CDOM.

Samples of the data and calculations are plotted in Figs. 2 and 3. Figure 2 shows the CDOM prediction for the n- ^{28}Si total cross section up to 80 MeV. Figure 3 shows the CDOM calculations for the differential cross sections for n-^{32}S and the analyzing powers for n-^{27}Al.

The binding energies for bound single-particle states were calculated from the final CDOM. For n-^{28}Si and n-^{32}S, the two systems for which empirical data exist, the calculations compare reasonably well with the data, all the way down to the $1s_{1/2}$ states near -70 MeV. The next stage in studying the CDOM is to calculate spectroscopic factors, spectral functions and occupation probabilities for the bound states and to compare these quantities to empirical results obtained from neutron transfer reactions and (e,e'n) reactions.

In comparison to the conventional spherical optical model derived for the energy range from 14 to 40 MeV for the same systems (see Ref. 4), our CDOM exhibits better fits to the scattering data. This success is most likely tied to the complicated energy dependence of the central potential that is generated by the addition of the dispersion-relation contributions; that is, the energy dependence of the true real potential arises more naturally in the DOM and therefore the potential is more able to describe the data over the wide energy range -80 to + 80 MeV.

ACKNOWLEDGMENT

This work was supported in part by the U.S. Dept. of Energy, the Office of High Energy and Nuclear Physics, under grant # DEFG 05-01-ER4000619. The first author acknowledges the support of King Fahd University of Petroleum and Minerals (KFUPM) Dhahran, Saudi Arabia, during the period of conducting this work at TUNL.

REFERENCES

1. Mahaux, C., and Sartor, R., *Adv. Nucl. Phys.* **20**, 1 (1991), and references therein.
2. Winfield, J. S., Austin, S. M., DeVito, R., Berg, U., Chen, Z., and Sterrenburg, W., *Phys. Rev.* **C33**, 1 (1986).
3. Jeukenne, J.P., and Mahaux, C., *Nucl. Phys.* **A394**, 445 (1983).
4. Martin, Ph., *Nucl. Phys.* **A466**, 119 (1987).

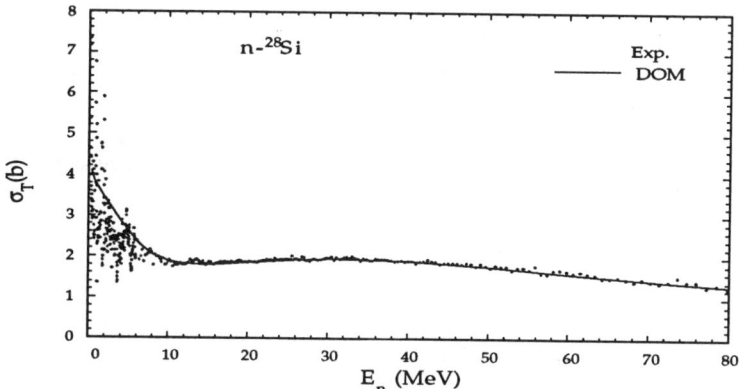

Figure 2. The CDOM calculation (curve) of the total cross section for n-^{28}Si compared to the experimental values.

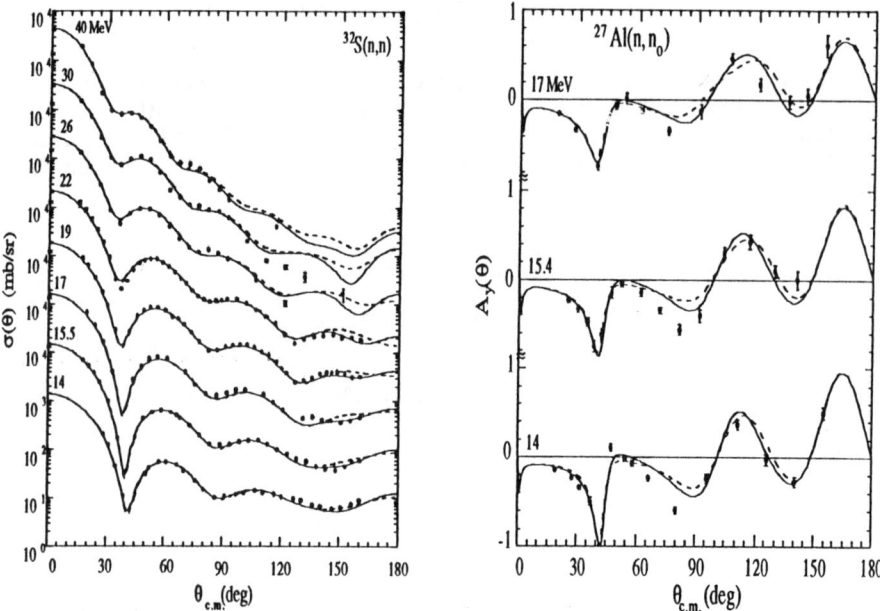

Figure 3. Cross sections for n-^{32}S and analyzing powers for n-^{27}Al. *Solid curve*: individual DOM calculations. *Dashed curves*: constrained CDOM calculations. *Dots*: experimental data.

Analyzing Powers for $^6\vec{\text{Li}}+^{12}\text{C}$ Scattering at 30 and 50 MeV

P.L. Kerr, E.L. Reber, P.V. Green, K.W. Kemper, A.J. Mendez
K. Mohajeri, E.G. Myers, and B.G. Schmidt

Department of Physics, Florida State University
Tallahassee, FL 32306-3016 USA

V. Hnizdo

Department of Physics, University of Witwatersrand
Johannesburg, 2050 South Africa

Comparison between elastic analyzing powers at 30 and 50 MeV show no decrease in their magnitude at the higher energy. A combined optical model analysis of both energies shows T_{21} to arise from the tensor interaction, T_{20} to be an interference between tensor and J dependent interactions and iT_{11} to be the most complicated, arising from an explicit spin-orbit, tensor and J dependent interactions. The inelastic ^{12}C vector analyzing powers require an explicit spin-orbit interaction to reproduce the magnitude of the oscillations.

The widely held theoretical belief that spin-dependent effects would not be large in heavy-ion scattering was shown to be in error by a Hamburg-Heidelberg collaboration who measured large vector analyzing powers for polarized ^{6}Li scattering (1). Since then ^{6}Li vector analyzing powers have been observed on numerous targets and at bombarding energies up to 70 MeV (2). While scattering studies with tensor polarized ^{6}Li beams are more limited, they also have produced reasonably large analyzing powers. Complete sets of elastic scattering analyzing powers have now been published for $^6\vec{\text{Li}}+^{12}\text{C}$ (3), ^{26}Mg (4) and ^{120}Sn (5).

The present work reports a complete set of analyzing powers for 50 MeV $^6\vec{\text{Li}}+^{12}\text{C}$ elastic scattering which when combined with earlier 30 MeV data (3) allow for the energy-dependence of the analyzing powers to be studied in detail. A laser pumped ^{6}Li ion source, based on the very successful Heidelberg-Marburg source (6,7), has been combined with the FSU tandem/linac accelerator to produce the 50 MeV $^6\vec{\text{Li}}$ beam. Typical beam currents on target were 50 enA.

Elastic scattering data for $^6\vec{\text{Li}}+^{12}\text{C}$ at the two energies are shown in Fig. 1. These energies are 4 and 6.7 times the classical Coulomb barrier height and so are dominated by nuclear scattering. The major feature of the data is

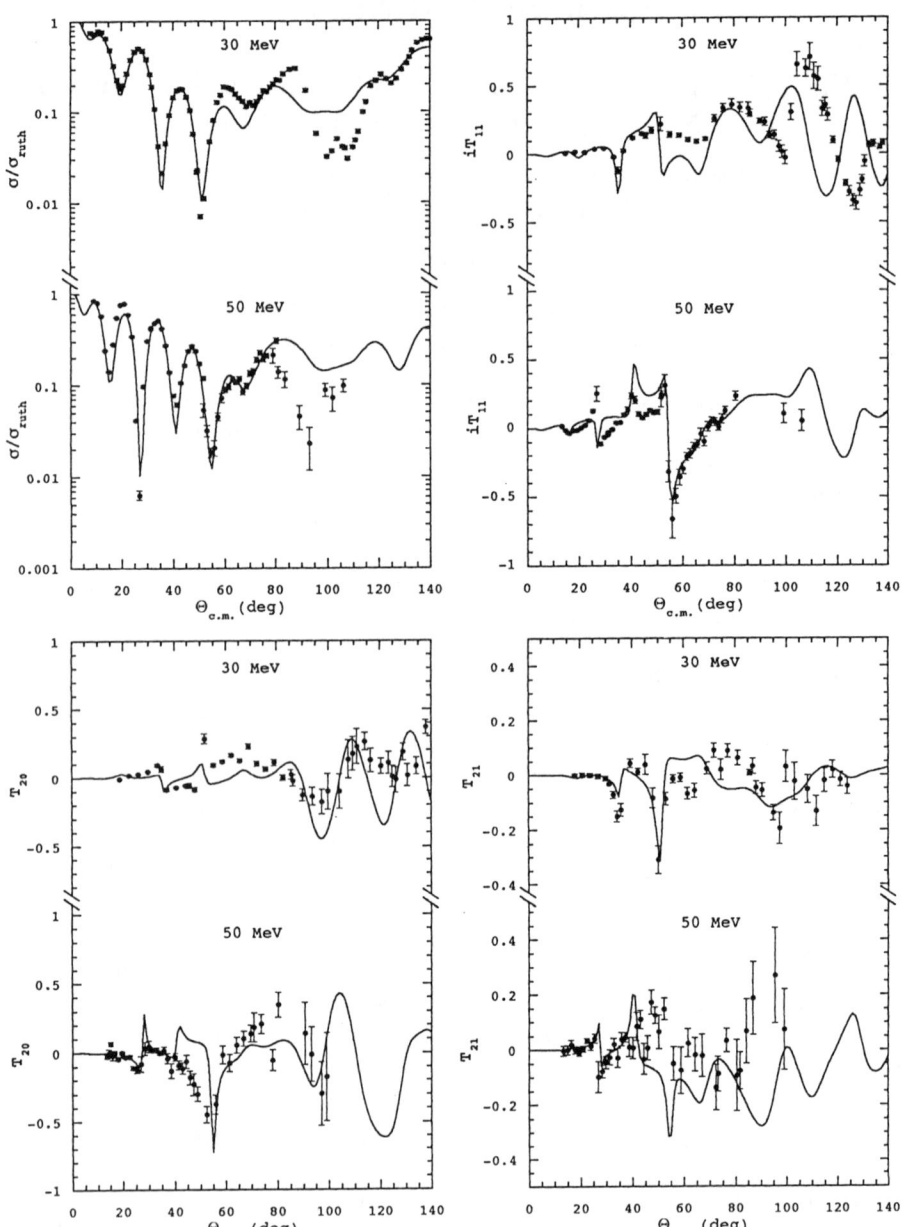

Figure 1. Cross sections and analyzing powers for $\vec{^6\text{Li}} + {}^{12}\text{C}$. Note that the analyzing powers iT_{11} and T_{20} are as large or larger at 50 MeV as at 30 MeV, except for T_{21} which is smaller at 50 MeV.

that there is no decrease in the magnitude of the analyzing powers except for T_{21} at the higher energy. As expected, the major structure in the analyzing powers occurs in the region of sharp interference minima in the elastic scattering, but not every minimum produces structure in the analyzing power.

An optical model analysis of the data at the two energies with standard Woods-Saxon real and imaginary potentials that included real spin-orbit and tensor potentials and an angular momentum (J) dependent absorption was carried out for the 30 and 50 MeV data. The J dependent absorption has been shown to provide a reasonable simulation of coupled channels calculations by comparing results of these calculations with those using a semi-microscopic interaction for the 30 MeV data. Starting parameters were taken from the 30 MeV analysis (3). The philosophy in the calculations was to obtain the best description of the data at both energies with as few parameter differences as possible. The calculations are shown in Fig. 1. Aside from the expected increase in the imaginary potential at the higher energy, the major difference is in the real tensor potential where the diffuseness must be greatly reduced at 50 MeV to provide a reasonable description of T_{20} and T_{21}. The parameters for 50 MeV are given here, with the changes to them for the 30 MeV calculation given in parentheses beside them. The parameters used in the calculations are: V=347 MeV, r_r=0.83 fm, a_r=0.77(0.80) fm, W=9.5(7.5) MeV, r_i=2.4 fm, a_i=0.85(1.0) fm, $V_{\ell s}$=2.70 MeV, $r_{\ell s}$=0.98 fm, $a_{\ell s}$=0.53 fm, V_T=2.0 MeV, r_T=1.1 fm, a_T=0.16(0.80) fm, J_c=15(12), ΔJ=2.50 with r_c=2.24 fm. A description of the potential is given in (3).

That the ^{6}Li analyzing powers arise from an interference between spin-orbit (SO), J dependent absorption and tensor (T) interactions can be seen from Fig. 2, where calculations are presented that have the central potential plus the SO, J dependence and T terms turned on one at a time. From the present analysis it appears that T_{21} arises solely from the tensor interaction, that T_{20} arises from an interference between the tensor and spin-orbit interactions and that iT_{11} is the most complicated analyzing power, with all three terms contributing. Extensive parameter searches showed that it was possible to describe iT_{11} with only a spin-orbit interaction but then the description of σ/σ_R was lost.

The analyzing powers for the inelastic excitation of the 2^+ and 3^- states in ^{12}C are structured with maximum amplitudes of 0.4. Coupled channels calculations that make use of the semi-microscopic M3Y interaction have been carried out to learn the origin of these analyzing powers. The details of these calculations are given in Ref. 8. These calculations include reorientation of the ^{6}Li(1^+)g.s., which is able to describe the elastic T_{21}. It is in the inelastic iT_{11} that the clearest evidence for an explicit spin-orbit potential

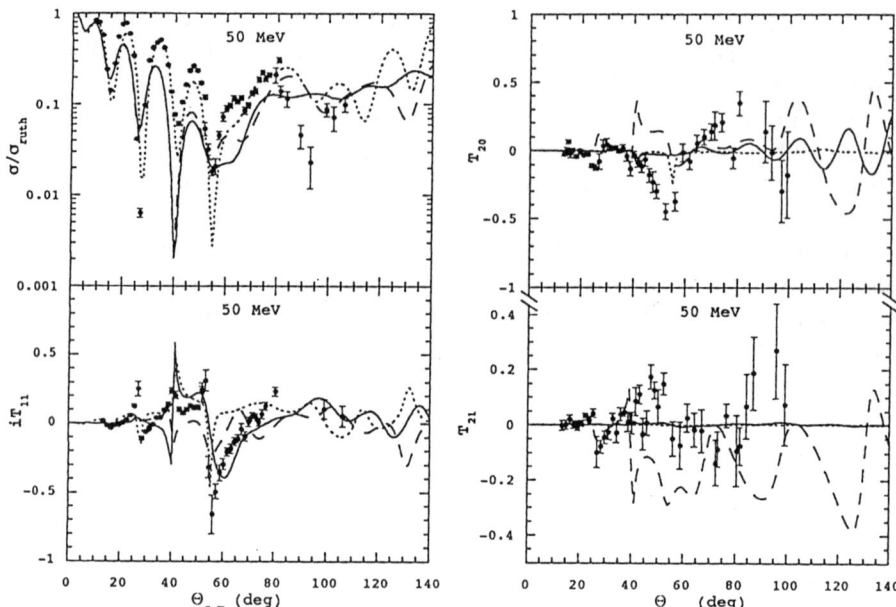

Figure 2. Differential cross-section and analyzing powers for elastic scattering of $\vec{^6\text{Li}} + {}^{12}\text{C}$ at $E_{lab} = 50$ MeV. The curves are optical model calculations which include central potentials and the addition of only a spin-orbit (solid), only a J dependent (dotted), or only a tensor potential (dashed).

is found. While the elastic iT_{11} is a complicated interference between coupled channel, and spin-orbit contributions with each component yielding large analyzing powers, channel coupling alone gives virtually zero iT_{11} for both the 2^+ and 3^- states. Only when an explicit spin-orbit is included in the calculations is the magnitude of the oscillations reproduced. This result has been found also in earlier analyses (4,5,8) of inelastic scattering data. Figure 3 shows coupled channels calculations carried out for the 2^+ state iT_{11} with and without inclusion of a spin-orbit interaction.

The greatest failure of the calculations was their inability to produce any sizable tensor analyzing powers for the ${}^{12}\text{C}(3^-)$ state data. Figure 4 shows the coupled channels calculations that produced the largest tensor analyzing powers. These calculations did not include a tensor interaction because when one was included it greatly overpredicted the ground state T_{21}.

In summary, modern laser pumped alkali ion sources produce intense beams so that high quality analyzing power data can be measured. These data allow detailed studies on the interplay between the different interactions contributing to these analyzing powers to be carried out.

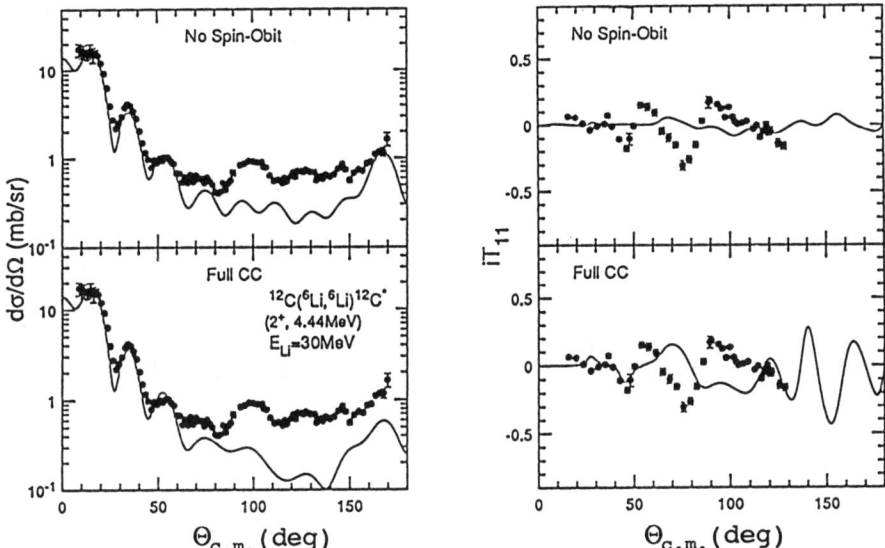

Figure 3. Coupled channels iT_{11} calculations for the $^{12}C(2^+)$ state with and without an explicit spin orbit interaction.

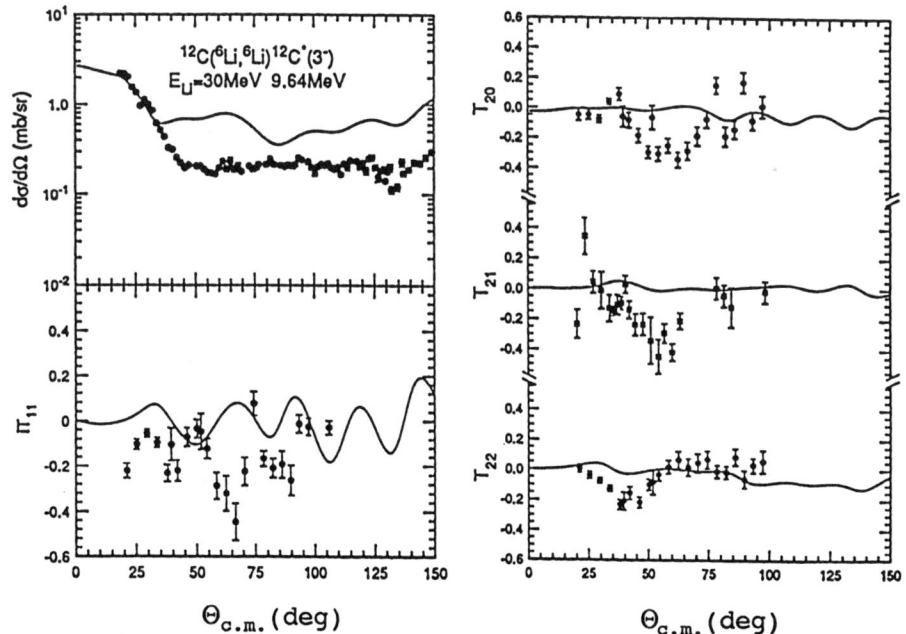

Figure 4. Coupled channels calculations for the $^{12}C(3^-)$ state. These calculations show the maximum tensor analyzing powers obtained while still describing the elastic data.

This work was supported by the National Science Foundation and the State of Florida.

REFERENCES

1. W. Weiss *et al.*, Phys. Lett. **61B**, 237 (1976).
2. P.R. Dee *et al.*, Daresbury Annual Report 1992-1993.
3. E.L. Reber *et al.*, Phys. Rev. **C4**, R1 (1994).
4. K. Rusek *et al.*, Nucl. Phys. **A503**, 223 (1989).
5. K. Becker *et al.*, Nucl. Phys. **A535**, 189 (1991).
6. H. Jänsch *et al.*, Nucl. Instrum. and Meth. **A254**, 7 (1987).
7. H. Reich and H.-J. Jänsch, Nucl. Instrum. and Meth. **A288**, 349 (1990).
8. S.P. Van Verst *et al.*, Phys. Rev. **C39**, 853 (1989).

Semi-Classical Analysis of Scattering of Deformed Heavy-Ions below the Coulomb Barrier

R C Johnson and E J Roberts
Department of Physics, University of Surrey, Guildford, GU2 5XH

C V Sukumar
Theoretical Physics, University of Oxford, 1 Keble Road, Oxford, OX1 3NP

D M Brink
European Centre for Theoretical Studies in Nuclear Physics and Related Areas,
Facoltà di Scienze, 38050 Pove, Trento, Italy.

Abstract. Polarization observables for the sub-Coulomb elastic scattering of a quadrupole deformed projectile of spin $^3/_2$ from a spinless spherical target are evaluated using a new semi-classical method based on a path-integral formalism. Analytic expressions are obtained which agree well with coupled channels calculations and which predict definite deviations from the 'shape-effect' relations for tensor analyzing powers.

1. INTRODUCTION

An important general problem in quantum mechanics is the study of systems with several degrees of freedom where a sub-set of these can be treated classically to a good approximation but others have to be treated quantum mechanically. A well-known example in nuclear physics is the Coulomb excitation of heavy ions where the relative motion of the colliding nuclei is well described as a Rutherford orbit but the treatment of multipole excitation of their internal degrees of freedom must be treated by quantum mechanics. In the well known Alder-Winther (AW) theory (1) which corresponds to this picture, the excitation process is described by a time dependent Hamiltonian in which the time dependence appears through the relative co-ordinate of the two nuclei as they traverse a Coulomb trajectory. The transition amplitude from internal state α to internal state β for the classical trajectory $\underline{r}_c(t)$ is given by

$$T_{\beta\alpha}(\underline{r}_c(t)) = U_{\beta\alpha}(+\infty, -\infty),\quad (1)$$

where $U_{\beta\alpha}(t,t_0)$ satisfies

$$i\hbar\frac{\partial U(t,t_0)}{\partial t} = h(\underline{r}_c(t),\xi)U \quad (2a)$$

$$U(t_0,t_0) = 1, \quad (2b)$$

and $h(\underline{r},\xi)$ is given by

$$h(\underline{r},\xi) = H_0(\xi) + V(\underline{r},\xi), \quad (3)$$

where $H_0(\xi)$ is the internal Hamiltonian of the nuclei and $V(\underline{r},\xi)$ is the coupling between the internal and relative motion. The complete Hamiltonian of the system is

$$H = \frac{p^2}{2m} + V_0(r) + h(\underline{r},\xi), \quad (4)$$

where m is the reduced mass, $\underline{p}$ is the momentum conjugate to $\underline{r}$, and $V_0(r)$ is the monopole nucleus-nucleus Coulomb interaction.

In the AW approach the differential cross-section for a transition from α to β is given by

$$\frac{d\sigma_{\beta\alpha}}{d\Omega} = \frac{v_\beta}{v_\alpha}\left(\frac{d\sigma}{d\Omega}\right)_c |T_{\beta\alpha}|^2 \quad (5)$$

where v_β and v_α are the relative velocities in the entrance and exit channels and $(d\sigma/d\Omega)_c$ is the Rutherford cross-section for elastic scattering by $V_0(r)$.

An interesting question for theory is how should one make corrections to the AW theory in a systematic way. One answer is, of course, to use one of the modern multi-channel computer programmes which are now available, but these calculations are formidable and themselves involve subtle questions of convergence and accuracy. In addition, such an approach involves little insight into the physics of the problem. Polarization studies have a special role to play because, as we will see, in certain special cases of experimental interest, polarization effects arise entirely from processes ignored in the AW theory and hence provide a testing ground for any approach which claims to identify the leading corrections.

2. THE THEORY OF SUKUMAR AND BRINK

For definiteness we refer to the case of Coulomb excitation of a projectile by an inert target. Pechukas (2) has shown how in this case the propagator between initial and final states of relative motion of the two nuclei and internal states α

and β of the projectile (described by $H_0(\xi)$ in eq.(3)) can be expressed as a path integral over all paths $\underline{r}(t)$ and associated transition amplitude $U_{\beta\alpha}(\underline{r}(t))$ satisfying eq.(2). Sukumar and Brink (3) - (6) have developed a systematic expansion of this expression about the classical path associated with the potential $V_0(r)$ i.e. the relevant Coulomb trajectory in the present case.

The leading term in this expansion leads to the AW theory, eqs.(1) - (5). Corrections arise from quantum fluctuations about the Rutherford orbit which modify the orbit itself and correctly take into account the modification of the coupling between the relative motion and the internal degrees of freedom produced by these fluctuations. For the case of Coulomb excitation the expansion is an asymptotic one in powers of $1/\eta$ for fixed q_2 and ξ, where η is the usual Sommerfeld parameter, ξ is the usual adiabaticity parameter and q_2 is a measure of the strength of the quadrupole coupling. These key dimensionless parameters are given by

$$\eta = Z_1 Z_2 e^2 / \hbar v, \quad \xi = \eta \Delta E / m v^2 \tag{6a}$$

$$q_2 = \eta \chi, \tag{6b}$$

where v is initial relative velocity and χ is the ratio of the strength of the quadrupole and monopole interactions at the distance of closest approach.

$$\chi = \frac{Q}{8 Z_1 a_c^2}, \quad a_c = Z_1 Z_2 e^2 / m v^2. \tag{7}$$

Here Q and Z_1, are suitably defined quadrupole and monopole charge moments of the projectile and Z_2 is the charge of the target.

The Sukumar-Brink expansion is expected to be useful for large enough values of η for any value of Q. It has been shown (5) to given an excellent account of corrections to AW results for excitation probabilities for spin zero target and projectile grand states in cases where $\eta \approx 20$, $q_2 \approx 1$ and $\chi \approx 0.05$.

3. RELEVANCE OF ANALYSING POWER DATA

The dramatic failure of the AW method for polarisation calculations is best illustrated by considering the elastic scattering of a projectile of spin I, spectroscopic quadrupole moment Q, by a spin zero target in a situation in which projectile excitation can be ignored. This is a good starting point for many cases where experimental data exist. The only effect of the quadrupole coupling in the interaction $V(\underline{r},\xi)$ of eq.(3) is to couple different values of the projection M of I along the z-axis i.e. re-orientation, and β and α in eqs.(1) and (5) refer to possible values of M. It is easy to show (6) that for real interactions V_0 and V (no absorption) the matrix U satisfies $U^\dagger U = 1$ and hence that eq.(5) predicts that the cross-section for an unobserved final spin state is independent of the initial spin state:

$$\sum_{\beta} \frac{d\sigma_{\beta\alpha}}{d\Omega} = \left(\frac{d\sigma}{d\Omega}\right)_c \sum_{\beta} |T_{\beta\alpha}|^2 = \left(\frac{d\sigma}{d\Omega}\right)_c . \qquad (8)$$

All analyzing powers are therefore predicted to vanish, in strong disagreement with relevant experiments, quantum coupled channel calculations and other semi-classical approaches (3), (8), (9), (10).

In the calculations referred to at the end of Section 2 the Sukumar-Brink expansion was applied to cross-sections for which the zeroth order AW theory already gave large non-vanishing results. In the present application the whole of the theoretical predictions for analyzing powers arise from the leading order $1/\eta$ correction terms in the Sukumar-Brink expansion. We apply our formalism specifically to the case of polarised ^{7}Li ($I=^3/_2$) scattering on ^{58}Ni at 10MeV for which data exist (7). For this reaction and a ^{7}Li spectropic quadrupole movement of $Q=-3.666$ fm^2 the parameter χ of eq.(7) is -0.003325 and $\eta=11.07$. Therefore $q_2=\eta\chi\approx-0.03$ and is very small and we can simplify the Sukumar-Brink expressions by retaining only terms of first order in q_2.

4. ANALYZING POWERS IN THE SUKUMAR-BRINK THEORY

According to (6) the cross-section for a transition from spin-projection M to spin projection M' is

$$\left(\frac{d\sigma}{d\Omega}\right)_{M'M} = \left(\frac{d\sigma}{d\Omega}\right)_c \{|T_{M'M}|^2 + 2Re(T_{M'M}\delta T^*_{M'M})\} \qquad (9)$$

where T is given by eq.(1) and δT is expressible entirely in terms of partial derivatives of T with respect to parameters which characterize an arbitrary orbit, the derivatives being evaluated on that Rutherford orbit which corresponds to the scattering event of interest.

Analyzing powers are given by (standard Madison notation)

$$T_{kq} = \frac{Trace(\Delta\tau_{kq})}{Trace(1+\Delta)} \qquad (10)$$

where

$$\Delta = (T^+\delta T + \delta T^+ T) \qquad (11)$$

and the result $T^+T=1$ has been used.

Sukumar and Brink (6) work in a co-ordinate system SB with z-axis along the direction of momentum transfer, and x-axis along $\mathbf{k}_i \wedge \mathbf{k}_f$. They express Δ as a sum of seven terms. For re-orientation calculations only 3 of these survive (11). To first order in q_2 we obtain (11)

$$\Delta = \frac{i\hbar}{2L}\left[\frac{\partial}{\partial\theta_0} + L\left(\frac{\partial\theta_0}{\partial L}\right)_{E0}\right]\left(\frac{\partial^2}{\partial\theta_0^2} - \frac{\partial^2}{\partial\alpha^2} + \frac{1}{\tan\theta_0}\frac{\partial^2}{\partial\gamma^2} - \tan\theta_0\frac{\partial^2}{\partial\beta^2}\right)T^+, \qquad (12)$$

where L is angular momentum for the Rutherford orbit, E_0 the incident centre of mass kinetic energy and the $\theta_0 = (\pi - \theta)/2$ where θ is the scattering angle. The angles α, β, γ specify the orientation in space of an arbitrary orbit with respect to the Rutherford orbit and are defined in (11). The derivatives in eq.(12) are evaluated at $\alpha = \beta = \gamma = 0$, i.e. on the Rutherford orbit.

The detailed evaluation of these expressions for $I = 3/2$ is described in (11) for the case of a quadrupole re-orientation interaction of the form

$$h = \frac{Z_2 e^2 Q}{r^3} \sqrt{\frac{\pi}{5}} \sum_q \tau_{2q}(I) Y_{2q}^*(\hat{r}) \tag{13}$$

The results for tensor analyzing powers in the SB frame are

$$T_{20}^{SB}(\theta_0) = -\chi f_1(\theta_0), \tag{14a}$$

$$T_{22}^{SB}(\theta_0) = T_{2-2}^{SB}(\theta_0) = \sqrt{\frac{3}{2}} \chi f_2(\theta_0), \tag{14b}$$

$$T_{21}^{SB}(\theta_0) = T_{2-1}^{SB}(\theta_0) = 0 \tag{14c}$$

where the 2 universal functions f_1 and f_2 are given by

$$f_1(\theta_0) = 2\cos^2\theta_0 + 6\cot^2\theta_0 + 9\cot^4\theta_0 \left(1 - \frac{2\theta_0}{\sin 2\theta_0}\right) \tag{15a}$$

$$f_2(\theta_0) = \frac{5}{9} f_1(\theta_0) - \frac{16}{9} \cos^2\theta_0 \tag{15b}$$

Eq.(14c) is a general result to first order in Q (12). For a quadrupole interaction all components of the rank 1 and 3 analyzing powers are predicted to vanish. The corresponding components in the Madison co-ordinate systems are found to be

$$T_{20}(\theta_0) = -\chi \left[\frac{(3\cos^2\theta_0 - 1)}{2} f_1 + \frac{3}{2}\sin^2\theta_0 f_2\right] \tag{16a}$$

$$T_{21}(\theta_0) = -\sqrt{\frac{3}{2}} \chi \sin\theta_0 \cos\theta_0 [f_1 - f_2], \tag{16b}$$

$$T_{22} = -\sqrt{\frac{3}{8}} \chi [\sin^2\theta_0 f_1 + (1 + \cos^2\theta_0) f_2] \tag{16c}$$

Expression (16a) for T_{20} agrees with that given by Grawert and Derner (10). Expressions for T_{21} and T_{22} were not given in (10). Note that these expressions are independent of $\hbar$. This is a consequence of retaining only terms of first order in q_2. Contributions to the leading order $1/\eta$ Sukumar-Brink expansion from terms of higher order in q_2 ($= \eta \chi$) will involve $\hbar$.

5. DISCUSSION

The results of section 4 have been applied to ^{7}Li scattering on ^{58}Ni at 10 MeV and compared with quantum mechanical coupled channels calculations using the code FRESCO (13) for the same Hamiltonian. We refer to reference (11) for the details. Excellent agreement (within 0.7%) is obtained. A detailed comparison with experiment requires additional consideration, including excitation effects (9), (20), but with the parameters given at the end of section 3, qualitative agreement is obtained with measured values of T_{20} (8).

It should be noted that the expressions (14) and (16) predict definite deviations from the shape-effect relations (14). In the SB frame the latter imply $T_{22}^{SB} = 0$ for all θ, which is definitely not the case. Another way of saying this is that the analyzing power for re-orientation involves 2 universal functions $f_1(\theta)$ and $f_2(\theta)$. The shape-effect relations are obtained by putting $f_2 = 0$ in eqs.(14) or (16). The predicted T_{20} then differs from that given by Grawert and Derner (10). The shape-effect relations were originally derived (14) on the basis of a semi-classical picture of the effect of deformation on elastic scattering. The Sukumar-Brink theory shows that even to first order in the quadrupole moment Q there are corrections to this picture which are independent of Planck's constant. It would be interesting to know how far these predicted deviations are in agreement with experiment.

Finally we emphasise that the expressions (14) and (16) are valid only to first order in q_2, the quadrupole coupling parameter defined in eqs. (6) and (7). The Sukumar-Brink theory makes predictions for any value of q_2, and only requires that the Sommerfeld parameter be sufficiently large. It would be of interest to test the predictions for analyzing powers in a case where q_2 is not small and for which semiclassical methods based on DWBA ideas (10) will surely break-down.

1. Alder K and Winther A, *Coulomb Excitation* (Academic Press, New York and London, 1966)
2. Pechukas P, *Phys Rev*, **181**, 174 (1969)
3. Sukumar C V and Brink D M, *Nucl Phys*, **A404**, 121 (1983)
4. dos Aidos F D and Brink D M, *J Phys G:Nucl Phys*, **11**, 249 (1985)
5. dos Aidos F D, Sukumar C V and Brink D M, *Nucl Phys*, **A448**, 333 (1986)
6. Sukumar C V and Brink D M, *Nucl Phys*, **A560**, 863 (1993)
7. Fick D, Grawert G, Turkiewicz I M, *Phys Rep*, **214**, 1-111 (1992)
8. Weller A et al, *Phys Rev Letts*, **55**, 480 (1985)
9. Voelk H-G and Fick D, *Nucl Phys*, **A530**, 475 (1991)
10. Grawert G and Derner J Chr, *Nucl Phys*, **A496**, 165 (1989)
11. Roberts E, Sukumar C V, Johnson R C and Brink D M, accepted for publication in *Nucl Phys A*, September 1994
12. Hooton D J and Johnson R C, *Nucl Phys*, **A175**, 583 (1971)
13. Thompson I J, *Comp Phys Rep* 7, 1 - 7 (1988)
14. Tungate G and Fick D, *Lecture Notes in Physics 89*, ed H V Geramb (Springer, Heidelberg 1979) 404.

Polarized ^{6}Li Studies at the Nuclear Structure Facility, Daresbury

R.P. Ward,[a] C.O. Blyth,[a] H.D. Choi,[a] N.M. Clarke,[a]
K.A. Connell,[b] N.J. Davis,[c] P.R. Dee,[a] S.J. Hall,[a] O. Karban,[a]
K.I. Pearce,[d] C.N. Pinder,[d] S. Roman,[a] K. Rusek,[e]
D.B. Steski,[b] and G. Tungate[a]

[a] *School of Physics and Space Research, University of Birmingham, Edgbaston, Birmingham B15 2TT, England*
[b] *Science and Engineering Research Council, Daresbury Laboratory, Warrington WA4 4AD, England*
[c] *Department of Physics, University of Edinburgh, Mayfield Road, Edinburgh EH9 3JZ, Scotland*
[d] *Wheatstone Laboratory, King's College London, Strand, London WC2R 2LS, England*
[e] *Soltan Institute for Nuclear Studies, Zaklad 1, Hoza 69, 00 681 Warsaw, Poland*

Abstract. Data have been obtained for the elastic and inelastic scattering of polarized ^{6}Li by ^{26}Mg and ^{58}Ni at energies of 60 MeV and 70.5 MeV, respectively. These data have been compared with the results of coupled-channels (CC) and continuum-discretized coupled-channels (CDCC) calculations performed with interaction potentials generated by the cluster-folding (CF) technique. CDCC calculations, including couplings to the resonant and nonresonant excited states of ^{6}Li and target excitation, have been found to reproduce data for ^{6}Li +^{26}Mg elastic and inelastic scattering without renormalization of the CF potentials. CC calculations of tensor analyzing powers for the elastic scattering of ^{6}Li by ^{58}Ni are shown to be highly sensitive to the D-state component of the ^{6}Li ground state wavefunction.

INTRODUCTION

Studies of polarized ^{7}Li interactions performed at the Nuclear Structure Facility (NSF), Daresbury were reported to the previous polarization conference (1). In the final year of NSF operations, the polarized heavy ion programme was extended to include ^{6}Li studies. Studies of polarized ^{6}Li scattering by ^{26}Mg and ^{58}Ni (at 60 and 70.5 MeV, respectively) were performed before the closure of the accelerator.

It is widely recognised that the breakup of ^{6}Li into $\alpha + d$ plays a crucial role in the elastic scattering of ^{6}Li. The ^{6}Li $+^{26}$Mg system has been studied previously at 44 MeV (2,3), where the data were found to be described by CC calculations including couplings between the ground state of ^{6}Li and the $T = 0$ triplet of resonant excited states at 2.18, 4.31 and 5.65 MeV. However, to achieve this description, two artificial parameters—renormalization factors N_R and N_I for the real and imaginary potentials, respectively—had to be introduced into the calculations. A careful study of this anomalous renormalization of the CF potentials (4) by the inclusion of couplings to nonresonant excited states failed to find its origin. However, it was deduced that N_R is energy dependent and that analyses of data at ≥ 10 MeV per nucleon should be free of the renormalization.

The D-state components of the ground state wavefunctions of light nuclei are of considerable interest. In an $\alpha + d$ model of the ^{6}Li ground state the spectroscopic amplitude b of the D-state admixture to the ^{6}Li ground state wavefunction is related to the quadrupole moment of ^{6}Li. This amplitude is also related to the second-rank tensor potential between ^{6}Li and a target nucleus, enabling study of the ^{6}Li D-state by measurement of tensor analyzing powers for ^{6}Li elastic scattering. However, since the cluster model tensor potential has two components, one from the $d +$ target second-rank tensor potential and the other proportional to b, precise knowledge of the $d +$ target second-rank tensor potential at the relevant energy is essential.

The data presented here were obtained in a series of experiments using the NSF polarized heavy ion source (5) and the Charissa scattering chamber. Details of one such experiment have been given by Ward et al. (6). The analyses described here used the coupled-channels computer code FRESCO (7) in conjunction with diagonal and off-diagonal potentials generated by cluster-folding.

THE $^6\vec{\text{Li}} + {}^{26}$Mg SYSTEM

Differential cross-sections and vector analyzing powers were measured for the elastic and inelastic scattering of 60 MeV ^{6}Li by ^{26}Mg (8). In common with previous analyses of this system at 44 MeV (2,3), the data were found to be described by CC calculations including couplings between the ground state of ^{6}Li and the $T = 0$ triplet of resonant excited states. A CDCC analysis of the data has also been performed (9), using potentials generated from $\alpha + ^{26}$Mg (10) and $d + {}^{26}$Mg (11,12) potentials by the CF technique. The $d + {}^{26}$Mg tensor potential was omitted, as no appropriate study exists. These calculations took into account not only couplings between the ^{6}Li ground state and the resonant excited states but also couplings to $L = 0, 1, 2$ nonresonant continuum states above the breakup threshold at 1.47 MeV. Extensive test calculations were first performed and it was found that the continuum could be divided into

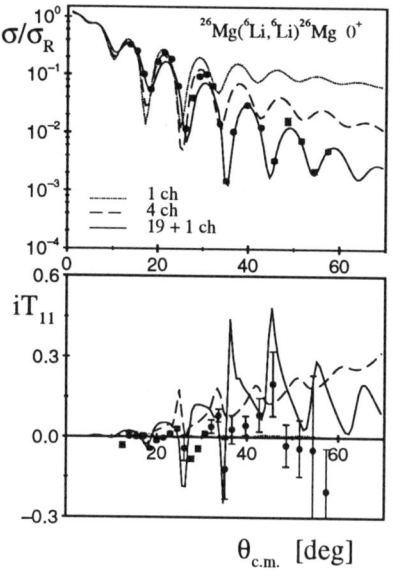 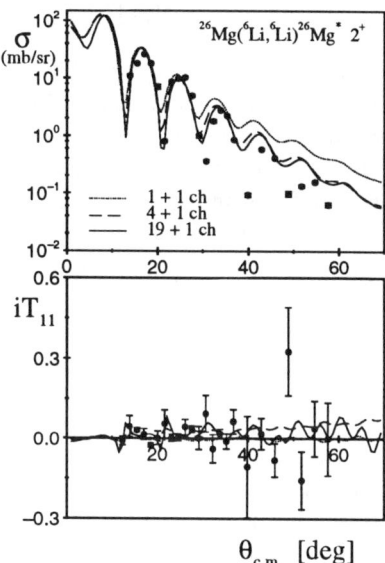

Figure 1: Differential cross-sections and vector analyzing powers for the elastic and inelastic scattering of ^{6}Li at 60 MeV. The predictions are described in the text.

$\Delta k = 0.2\,\mathrm{fm}^{-1}$ bins with $0.2\,\mathrm{fm}^{-1} \leq k \leq 0.8\,\mathrm{fm}^{-1}$. This range corresponds to excitation energies between 2.1 MeV and 11.5 MeV relative to the ^{6}Li ground state.

Fig. 1 shows the results of 1-channel (optical model), 4-channel (resonant states) and 20-channel (resonant states, nonresonant states and target excitation) calculations without renormalization of the interaction potentials. The 1-channel calculation overpredicts the differential cross-section and yields a very small vector analyzing power. The inclusion of couplings between the ground state of ^{6}Li and the resonant states improves the description of the differential cross-section data and generates a dynamic spin-orbit potential, giving rise to oscillatory structure in the predicted vector analyzing power. The inclusion of the nonresonant excited states and target excitation results in an excellent description of the differential cross-section data. The effects of target excitation were found to be weaker than those of projectile excitation. Data for inelastic scattering populating the first excited state of ^{26}Mg are simultaneously described by the 20-channel calculation, as also shown in Fig. 1. Clearly, the effect of the nonresonant states is much less profound for target excitation than for elastic scattering. This is the first analysis to describe polarized ^{6}Li scattering data without renormalization of the CF potentials.

THE $^6\vec{\text{Li}} + ^{58}$Ni SYSTEM

Angular distributions of differential cross-section, iT_{11}, T_{20} and $^T T_{20}$ were measured for the elastic scattering of polarized ^{6}Li by ^{58}Ni at 70.5 MeV (13). The data have been compared with CC calculations with all diagonal and off-diagonal potentials calculated from empirical $\alpha + {}^{58}$Ni and $d + {}^{58}$Ni potentials (14,15) using the CF method. Here, the second-rank tensor potential was included in the calculations. With unrenormalized interaction potentials and no couplings to excited states the differential cross-section data were overpredicted. The dotted lines in Fig. 2 are the results of a CC calculation including couplings between the ground and the three resonant excited states of the projectile. The dashed lines include couplings to the resonant excited states of the projectile and the first two excited states of the target. The inclusion of couplings to the resonant excited states of the projectile and the first two excited states of the target reduced the predicted differential cross-section for elastic scattering. However, even with these couplings, a phase difference was seen to exist between data and predictions. This was removed by renormalization of the potentials, the values $N_R = 0.7$ and $N_I = 0.8$ being optimal; the corresponding prediction is the solid line in Fig. 2.

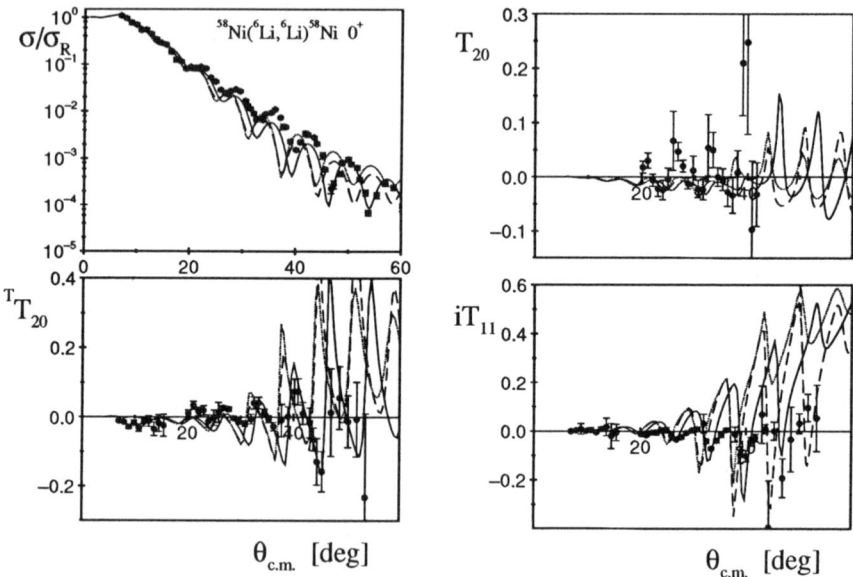

Figure 2: Differential cross-sections and analyzing powers for the elastic scattering of ^{6}Li by ^{58}Ni. The predictions are described in the text.

Fig. 3 demonstrates that calculations of the analyzing power $^T T_{20}$ are highly sensitive to the magnitude and sign of the D-state spectroscopic am-

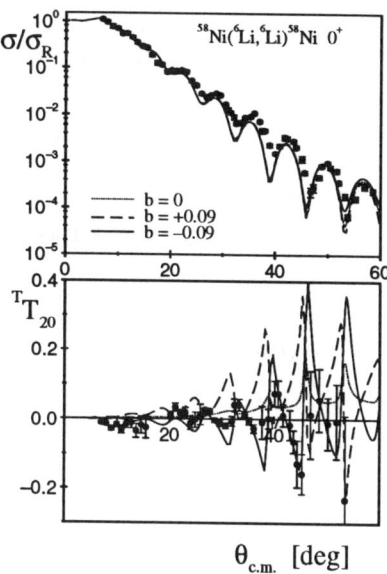

Figure 3: Differential cross-sections and $^T T_{20}$ for the elastic scattering of ^{6}Li by ^{58}Ni at 70.5 MeV. The predictions are described in the text.

plitude b. The dotted line in Fig. 3 is a 4-channel calculation with $b = 0$, while the dashed and solid lines are similar calculations with $b = +0.09$ and $b = -0.09$, respectively. Clearly, the data are best described by calculations with $b = -0.09$, in agreement with the value obtained from the quadrupole moment of ^{6}Li (16) and from $(\vec{d},^6\mathrm{Li})$ studies (17) but in contradiction to the values obtained from polarized ^{6}Li fragmentation measurements (18) and three body calculations (19). A CDCC analysis of these data is planned.

ACKNOWLEDGMENTS

This work was supported in part by the Science and Engineering Research Council of the United Kingdom.

REFERENCES

1. O. Karban, G. Kuburas, C.O. Blyth, H.D. Choi, N.J. Davis, S.J. Hall, S. Roman, G. Tungate, and I.M. Turkiewicz, in *Proceedings of the 7th International Conference on Polarization Phenomena in Nuclear Physics*, edited by A. Boudard and Y. Terrien (Les Éditions de Physique, Paris, 1990), pp. 435.

2. K. Rusek, J. Giroux, H.J. Jänsch, H. Vogt, K. Becker, K. Blatt, A. Gerlach, W. Korsch, H. Leucker, K. Luck, H. Reich, H.-G. Völk, and D. Fick, Nucl. Phys. **A503**, 223 (1989).

3. Y. Hirabayashi and Y. Sakuragi, Nucl. Phys. **A536**, 375 (1992).

4. Y. Hirabayashi, Phys. Rev. C **44**, 1581 (1991).

5. O. Karban, W.C. Hardy, K.A. Connell, S.E. Darden, C.O. Blyth, H.D. Choi, S.J. Hall, S. Roman, and G. Tungate, Nucl. Instrum. Methods A **274**, 4 (1989).

6. R.P. Ward, N.M. Clarke, C.N. Pinder, K.I. Pearce, C.O. Blyth, H.D. Choi, P.R. Dee, S. Roman, G. Tungate, and N.J. Davis, Phys. Rev. C **48**, 2366 (1993).

7. I.J. Thompson, Comput. Phys. Rep. **7**, 167 (1988).

8. R.P. Ward, N.M. Clarke, K.I. Pearce, C.N. Pinder, C.O. Blyth, H.D. Choi, P.R. Dee, S. Roman, G. Tungate, and N.J. Davis, Phys. Rev. C **50**, 918 (1994).

9. K. Rusek, N.M. Clarke, and R.P. Ward, Phys. Rev. C **50**, (1994) (in press).

10. P.P. Singh, R.E. Malmin, M. High, and D.W. Devins, Phys. Rev. Lett. **23**, 1124 (1968).

11. J.M. Lohr and W. Haeberli, Nucl. Phys. **A232**, 381 (1974).

12. W.W. Daehnick, J.D. Childs, and Z. Vrcelj, Phys. Rev. C **21**, 2253 (1980).

13. P.R. Dee, C.O. Blyth, N.M. Clarke, S.J. Hall, O. Karban, S. Roman, K. Rusek, G. Tungate, R.P. Ward, K.A. Connell, D.B. Steski, and N.J. Davis, in preparation for submission to Phys. Rev. C.

14. H. Chang, B. Ridley, T. Braid, T. Conlon, E. Gibson, and N. King, Nucl. Phys. **A270**, 413 (1976).

15. M. Takei, Y. Aoki, Y. Tagishi, and K. Yagi, Nucl. Phys. **A472**, 41 (1987).

16. H. Nishioka, J.A. Tostevin, and R.C. Johnson, Phys. Lett. B **124**, 17 (1983).

17. J.E. Bowsher, T.B. Clegg, H.J. Karwowski, E.J. Ludwig, W.J. Thompson, and J.A. Tostevin, Phys. Rev. C **45**, 2824 (1992).

18. V. Punjabi, C.F. Perdrisat, E. Cheung, J. Yonnet, M. Boivin, E. Tomasi-Gustafsson, R. Siebert, R. Frascaria, E. Warde, S. Belostotsky, O. Miklucho, V. Sulimov, R. Abegg, and D.R. Lehman, Phys. Rev. C **46**, 984 (1992).

19. D.R. Lehman and W.C. Parke, Phys. Rev. C **31**, 1920 (1985) and Phys. Rev. C **37**, E2266 (1988).

Fusion of a polarized projectile with a polarized target

J.A. Christley, R.C. Johnson and I.J. Thompson

Department of Physics, University of Surrey,
Guildford, Surrey GU2 5XH, U.K.

Abstract. The fusion cross sections for a polarized target with both unpolarized and polarized projectiles are studied. Expressions for the observables are given for the case when both nuclei are polarized. Calculations for fusion of an aligned ^{165}Ho target with ^{16}O and polarized ^{7}Li beams are presented.

INTRODUCTION

The fusion of heavy ions at energies close to the Coulomb barrier is sensitive to intrinsic degrees of freedom of the interacting nuclei. Experiments with polarized projectiles provide additional information [1] to help identify these degrees of freedom. It is predicted [2] that fusion of a polarized ^{165}Ho target depends strongly on the target alignment. In this paper we consider the additional information available if both the target and projectile are polarized. We will briefly summarize the predictions [2] for $\vec{^{16}\text{O}} + {^{165}\vec{\text{Ho}}}$ and present new calculations for a simple model of $\vec{^7\text{Li}} + {^{165}\vec{\text{Ho}}}$.

FUSION POLARIZATION OBSERVABLES

When one nucleus is polarized the fusion cross section for the incident density matrix ρ may be expressed in terms of the cross sections calculated assuming pure initial m-substates (with the quantization axis in the beam direction):

$$\sigma_F(\rho) = \sum_m <m|\rho|m> \sigma_F(m) \qquad (1)$$

or as a product of the unpolarized cross section and a tensor expansion

$$\sigma_F(\rho) = \sigma_F(\text{unpol.}) \left[1 + \sum_{k=2,4,\ldots} t_{k0}^* T_{k0}^{\text{fus}}\right] \qquad (2)$$

where t_{k0} is a tensor coefficient of the incident density matrix and T_{k0}^{fus} is the fusion tensor analysing power describing the sensitivity of the fusion cross section to polarization [1,3].

When both the projectile (spin I_a with projection m_a) and target (I_A, m_A) are polarized the fusion cross section is dependent on off-diagonal elements of the incident density matrix (with the quantization axis in the beam direction)

$$\sigma_F(\rho) = \sum_{m_a m_A m'_a m'_A} <m_a m_A|\rho|m'_a m'_A> T^{\rm F}(m_a m_A; m'_a m'_A), \qquad (3)$$

where $T^{\rm F}(m_a m_A; m'_a m'_A)$ is the contribution to fusion including interference between initial spin projections m_a and m_A, and m'_a and m'_A in the incident channel α. It is calculated from the difference between the total cross section (calculated from the optical theorem) and the sum of cross sections for the N direct reaction channels β included in our model:

$$T^{\rm F}(m_a m_A; m'_a m'_A) = \frac{2i\pi}{k_\alpha} \left(f^{\alpha\alpha\,*}_{m_a m_A; m'_a m'_A}(\theta=0) - f^{\alpha\alpha}_{m'_a m'_A; m_a m_A}(\theta=0) \right)$$
$$- \sum_{\beta=1}^{N} \int d\Omega_\beta \frac{v_\beta}{v_\alpha} \left(f^{\beta\alpha}_{m_b m_B; m_a m_A}(\Omega_\beta) f^{\beta\alpha\,*}_{m_b m_B; m'_a m'_A}(\Omega_\beta) \right). \qquad (4)$$

To express the total cross section we assume a screened Coulomb potential but the terms dependent on the screening cancel in the expression for the fusion cross section [4].

The fusion cross section can be expressed as a tensor expansion

$$\sigma_F(\rho^{\rm inc}) = \sigma_F({\rm unpol.}) \left[1 + \sum_{\substack{k_a, k_A, K \\ (k_a+k_A>0)}} \rho_K(k_a, k_A) \, T^{\rm F}_K(k_a, k_A) \right] \qquad (5)$$

where $\rho_K(k_a, k_A)$ is a coefficient in the tensor expansion of the incident density matrix

$$\rho_K(k_a, k_A) = (-i)^{k_a+k_A-K} [\, t_{k_a q}(I_a) \times t_{k_A -q}(I_A) \,]_{K0} \qquad (6)$$

and the rank-K tensor coefficients $T^{\rm F}_K(k_a, k_A)$ parameterize how the fusion cross section depends on the initial spin states of both the projectile and the target

$$T^{\rm F}_K(k_a, k_A) \times \sigma_F({\rm unpol.}) = i^{k_a+k_A-K} \sum_q <k_a q k_A -q|K0> \qquad (7)$$
$$\sum_{m_a m_A m'_a m'_A} \frac{T^{\rm F}(m_a m_A; m'_a m'_A)}{(2I_a+1)(2I_A+1)} <m'_a|\tau_{k_a q}(I_a)|m_a><m'_A|\tau_{k_A -q}(I_A)|m_A>$$

Symmetry under rotation and parity conservation dictate that $T_K^F(k_a, k_A)$ vanishes for odd values of K and only terms coupling to total projection $Q = 0$ contribute. We have included complex phases so the observables $T_K^F(k_a, k_A)$ are purely real.

The new notation includes the usual tensor analysing powers dependent on the polarization of one nucleus:

$$\rho_{k_a}(k_a, 0) = t_{k_a 0}(I_a), \qquad \rho_{k_A}(0, k_A) = t_{k_A 0}(I_A),$$
$$T_{k_a}^F(k_a, 0) = T_{k_a 0}^{\text{fus}}, \qquad T_{k_A}^F(0, k_A) = T_{k_A 0}^{\text{fus}},$$

and additional terms dependent on the polarization of both nuclei. For example

$$T_0^F(2,2), \quad T_2^F(2,2), \quad T_4^F(2,2) \tag{8}$$

quantify the sensitivity to the coupling of the quadrupole polarization of both nuclei to zero, two and four respectively.

Partial wave expressions for $T_K^F(k_a, k_A)$ can be obtained by substituting for the scattering amplitudes. For example in the channel spin coupling scheme [3] the fusion tensor observables can be written as:

$$T_K^F(k_a, k_A) \times \sigma_F(\text{unpol.}) = \frac{\pi}{k_\alpha^2} \sum_{L_\alpha, S_\alpha, L'_\alpha, S'_\alpha, J_T} \frac{(2J_T+1) A_{L_\alpha, S_\alpha, L'_\alpha, S'_\alpha}^{k_a, k_A, K}(J_T)}{(2I_a+1)(2I_A+1)}$$

$$\left(\delta_{L_\alpha L'_\alpha} \delta_{S_\alpha S'_\alpha} - \sum_{\beta L_\beta S_\beta} \frac{v_\beta}{v_\alpha} e^{i(\sigma_{L_\alpha}(\alpha) - \sigma_{L'_\alpha}(\alpha))} S_{L_\beta S_\beta \beta; L'_\alpha S'_\alpha \alpha}^{J_T *} S_{L_\beta S_\beta \beta; L_\alpha S_\alpha \alpha}^{J_T} \right) \tag{9}$$

where we have factored out angular momentum terms in a similar manner to reference [5]

$$A_{L_\alpha, S_\alpha, L'_\alpha, S'_\alpha}^{k_a, k_A, K}(J_T) = \hat{L}_\alpha \hat{L}'_\alpha \hat{k}_a \hat{k}_A \hat{S}_\alpha \hat{S}'_\alpha \hat{I}_A \hat{I}_a (-)^{J_T - S'_\alpha - k_a - k_A}$$

$$i^{k_a + k_A - K} W(L_\alpha S_\alpha L'_\alpha S'_\alpha; J_T K) < L'_\alpha 0 L_\alpha 0 | K 0 > \begin{Bmatrix} S_\alpha & S'_\alpha & K \\ I_a & I_a & k_a \\ I_A & I_A & k_A \end{Bmatrix} \tag{10}$$

Similar expressions have previously been published for the total cross section for polarized neutrons (eg. [5] and references therein). Gould et al. [5] have shown that for the total cross section terms $T_K^T(k_a, k_A)$ with $k_a + k_A + K$ odd must vanish if the S-matrix is symmetric under time reversal. The same is not true for the fusion cross section. We are still working on how to interpret the new observables.

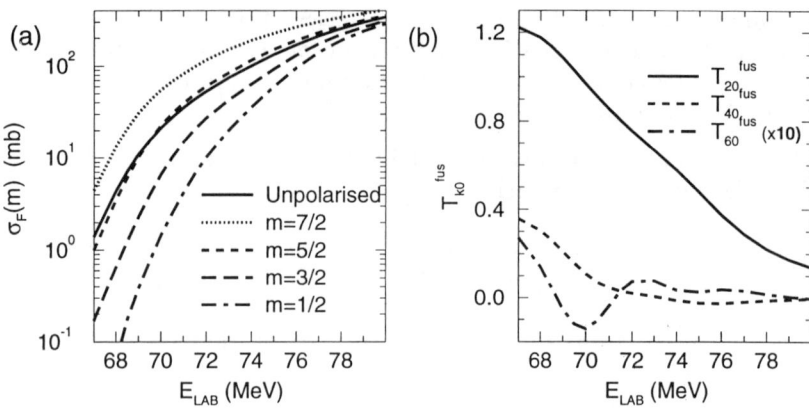

FIGURE 1: Fusion cross sections (a) and analysing powers (b) for $^{16}O + ^{165}Ho$ from reference [2]. (Note change of scale for K=6).

EXAMPLE 1: FUSION OF $^{16}O + ^{165}Ho$

We will present the results of coupled channels calculations using the code FRESCO [2,6] for two reactions. In both examples a ^{165}Ho target is chosen because it has a large static deformation ($\beta_2 \approx 0.33$) and can be aligned relatively easily. The low lying states are well described by a rotational band based on a $I_0 = 7/2$ ground state.

In our first example we summarize the results of coupled channels calculations for $^{16}O + ^{165}Ho$ [2]. These calculations include nuclear and Coulomb excitation and reorientation of the target ground state rotational band. Two previous studies of fusion for this system gave very different predictions. Stokstad and Gross [7] predicted that aligning the target would change the fusion cross section by up to two orders of magnitude, whereas Jacobs and Smilansky [8] predict the maximum change in cross section to be 40%.

In figure 1 we show the fusion cross section and analysing powers from coupled channel calculations [2]. The fusion cross section for an unpolarized target and for pure initial m-substates of the target (with the quantization axis in the beam direction) are shown in figure 1a. The cross sections differ by more than an order of magnitude at low energies. The largest cross section is obtained for $m = 7/2$ where the nuclear overlap is largest assuming a head-on collision. Our predictions for an experiment are that an unpolarized cross section of 10 mb decreases by 40% when the target is aligned perpendicular to the beam direction and is enhanced by 175% when the target is aligned in the beam direction. We have shown [2] that Stokstad and Gross

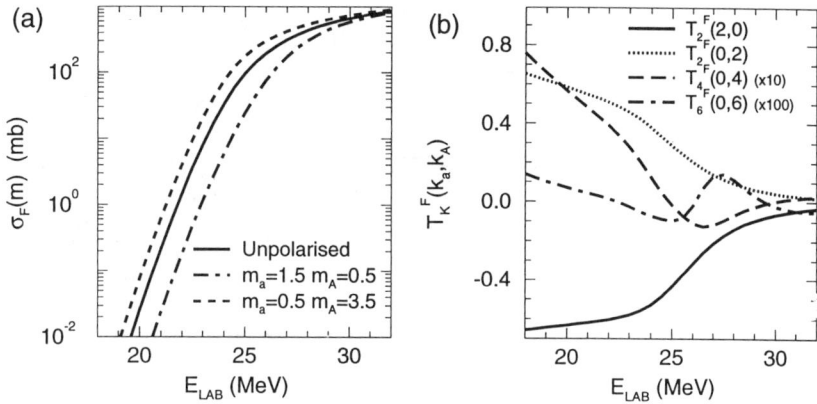

FIGURE 2: (a) Fusion cross sections and (b) $T_K^F(k_a, k_A)$ for ^{7}Li + ^{165}Ho. The tensor observables shown depend on the polarization of only one of the nuclei and are equivalent to the tensor analysing powers. (Note change of scale for $K = 4, 6$).

[7] overestimate the dependence of fusion on alignment mainly due to an unrealistic representation of the target alignment, and Jacobs and Smilansky [8] may have underestimated the effect of alignment due to an overcorrection for the isocentrifugal approximation used in their method.

The fusion tensor analysing powers [2] are shown in figure 1b. These are more sensitive than the unpolarized cross section to channel couplings and may be used to separate the effects of different multipole components in the coupling. For example the large positive T_{20}^{fus} at low energy indicates a large prolate quadrupole deformation and T_{40}^{fus} gives information on the sign and magnitude of the hexadecapole deformation.

EXAMPLE 2: FUSION OF ^{7}Li + ^{165}Ho

In our second example we consider fusion of an aligned ^{165}Ho target with a polarized ^{7}Li beam ($I_a = 3/2$, $\beta_2 \approx 1.45$). A simple coupled channels model is used which includes only nuclear quadrupole reorientation of the ground states of both nuclei (ignoring simultaneous reorientation).

Figure 2a shows the fusion cross section for an unpolarized experiment and for pure initial m-substates (with the quantization axis in the incident beam direction). We have only shown the results for the m-substate combinations that give the largest deviations from the unpolarized result. The largest cross section corresponds to the combination that gives the largest nuclear overlap for a head-on collision.

In figure 2b we show the tensor observables $T_2^F(2,0), T_2^F(0,2), T_4^F(0,4)$ and $T_6^F(0,6)$ (which are equivalent to T_{20}^{fus} for the projectile and $T_{20}^{\text{fus}}, T_{40}^{\text{fus}}, T_{60}^{\text{fus}}$

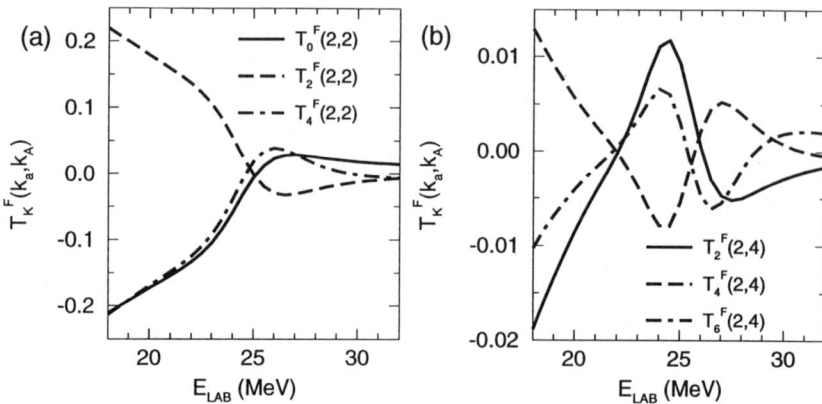

FIGURE 3: Fusion observables (a) $T_K^F(2,2)$ and (b) $T_K^F(2,4)$ for ^{7}Li + ^{165}Ho. These observables are sensitive to the polarization of both nuclei and are not accessible in a reaction where one nucleus is unpolarized.

for the target). The target analysing powers show the same behaviour as in example 1. The negative $T_2^F(2,0)$ indicates that the projectile is oblate.

Two sets of the new observables dependent on the polarization of both nuclei are shown in figure 3. $T_K^F(2,2)$ for $K = 0, 2, 4$ describe the sensitivity to the quadrupole polarization of both nuclei and the set $T_K^F(2,4)$ describe the sensitivity to the coupling of the projectile quadrupole and target hexadecapole polarization.

These results are for a simple model which ignores inelastic excitation and Coulomb reorientation. However they do demonstrate that the new observables may contain additional information on the potentials and the reaction mechanism that is not otherwise accessible. Further work is in progress on the interpretation of this new information.

This research was supported by SERC grants GR/H 53648 and GR/H 53556.

REFERENCES

1. Fick D., Grawert G. and Turkiewicz I.M., *Phys. Rep.* **214** 1 (1992)
2. Christley J.A., Johnson R.C. and Thompson I.J., *J. Phys.* **G20** 169 (1994).
3. Satchler G.R., *Direct Nuclear Reactions*, (OUP, Oxford, 1983), p. 175-178.
4. Christley J.A., Johnson R.C. and Thompson I.J., to be published.
5. Gould C.R., Haase D.G., Roberson N.R., Postma H. and Bowman J.D., *Int. J. Mod. Phys.* **A5** 2181 (1990).
6. Thompson I.J., *Comp. Phys. Comm.* **7** 167 (1988).
7. Stokstad R.G. and Gross E.E., *Phys. Rev.* **C 23** 281 (1981).
8. Jacobs P. and Smilansky U., *Nucl. Phys.* **A 438** 536 (1985).

Spin Polarization of ^{23}Mg in ^{24}Mg + Au, Cu and Al Collisions at 91 A MeV

K. Matsuta,[a,c] S. Fukuda,[a] T. Izumikawa,[a] M. Tanigaki,[a]
M. Fukuda,[a] M. Nakazato,[a] M. Mihara,[a] T. Onishi,[a] T. Yamaguchi,[a]
T. Miyake,[a] M. Sasaki,[a] A. Harada,[a] T. Ohtsubo,[a] Y. Nojiri,[a]
T. Minamisono,[a] K. Yoshida,[b] A. Ozawa,[b] T. Kobayashi,[b]
I. Tanihata,[b] J.R. Alonso,[c] G.F. Krebs,[c] and T.J.M. Symons[c]

a *Osaka University, Toyonaka, Osaka 560, Japan*
b *RIKEN, Wako, Saitama 351-01, Japan*
c *Lawrence Berkeley Laboratory, Berkeley, California 94720, USA*

Abstract. Spin polarization of beta-emitting fragment ^{23}Mg(I^π=3/2$^+$, $T_{1/2}$=11.3 s) produced through the projectile fragmentation process in ^{24}Mg + Au, Cu and Al collisions has been observed at 91 AMeV. General trend in the observed momentum dependence of polarization is reproduced well qualitatively by a simple fragmentation model based on the participant-spectator picture, for heavy and light targets. However the polarization behavior differs from this model in terms of zero crossing momentum, which become prominent in the case of Cu target, where the polarization is not monotone function of the fragment momentum.

INTRODUCTION

The technique of polarized radioactive nuclear beams is very useful for the study of nuclear structure. It has been applied for the measurement of nuclear moments(1-4). This technique can also be applied to the study of fundamental interactions. Precise measurement of the asymmetry parameter of ^{23}Mg beta decay, which discloses validity of Cabbibo universality(5), can be a good example of such a new application. In order to expand applicability of this technique, however, clear understanding of the polarization mechanism in heavy ion collisions in the intermediate energy region is crucial. Although the polarization mechanism has been studied using ^{14}N, ^{18}O, ^{40}Ca, and ^{46}Ti projectiles(1-4,6-9), the middle mass region in the sd shell has been missing. In the present experiment, spin polarization of beta-emitting fragment ^{23}Mg(I^π=3/2$^+$, $T_{1/2}$=11.3 s) produced through the projectile fragmentation process of ^{24}Mg has been observed at 91 A MeV.

EXPERIMENTAL PROCEDURE

The ^{24}Mg beam of 100 A MeV extracted from the RIKEN's K=540 ring cyclotron was used to bombard three kinds of targets (Au, Cu and Al). Energy loss of the beam in the target was 18 A MeV, so that the effective energy was 91 A MeV. ^{23}Mg fragments emerging from the target at a certain deflection angle were selected by a defining slit. They were then separated out from various fragments and were momentum analyzed by the RIPS (RIken Projectile fragment Separator). The separated ^{23}Mg nuclei were then implanted in a 50μm-thick Pt catcher placed in a strong magnetic field(4.0 kOe or 2.7 kOe) to maintain the polarization produced in the collision. The Pt foil was cooled to a low temperature(28K or 13K) to increase the relaxation time of the polarization. Beta rays emitted from the stopped nuclei were detected by two sets of plastic scintillation counter telescopes placed above and below the catcher relative to the polarization axis. The polarization was measured by means of counting asymmetry in these counters. For background rejection, a pulsed beam method was used, where the beam was on for 10 s in every 45 s and beta rays were observed in the following 35s of counting time. To obtain a reliable polarization, geometrical counter asymmetry was canceled by flipping the spin ensemble using the AFP(adiabatic fast passage) method in NMR in every other beam-count cycle.

RESULTS AND DISCUSSION

Spin polarization of ^{23}Mg was measured as a function of fragment momentum at a reaction angle of 2.0°. Fig. 1 shows the observed polarization together with the momentum distribution of ^{23}Mg for three targets. As expected from the participant-spectator picture of the collision at the present energy region, the observed fragment momentum shows a Gaussian distribution centered near the momentum corresponding to the beam velocity with a width close to that caused from the Fermi motion of a nucleon(10).

The polarization observed for the Au target is a monotone increasing function of momentum(Fig. 1-a), while the polarization for Al target is a decreasing function(Fig. 1-c), following the same trend observed in previous studies(6,7,9). As is shown by solid lines in Fig. 1, these momentum dependencies are reproduced well qualitatively by a simple fragmentation model(7, 11) based on the participant-spectator picture of the high energy heavy ion collision and the orbital deflection of projectiles caused by the combined Coulomb-nuclear potential to either positive(Au target) or negative (Al target) angles.

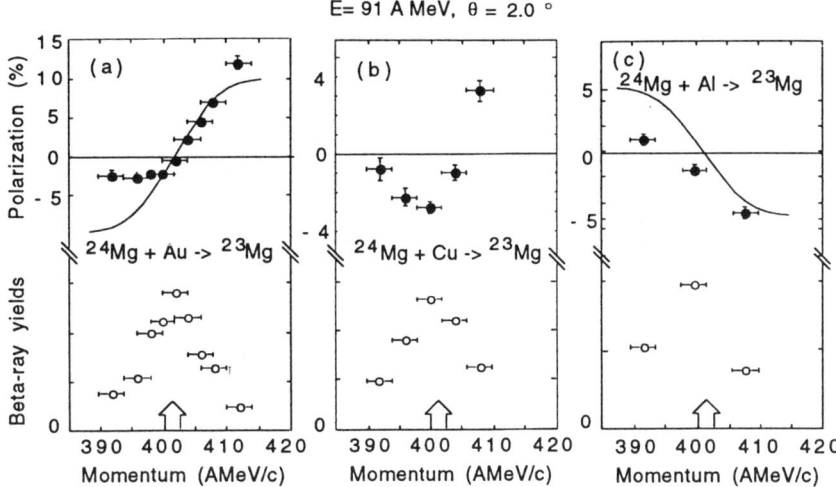

FIGURE 1. Momentum dependence of the polarization of ^{23}Mg.
Closed and open circles are the polarizations and beta-ray yields, respectively. Arrows indicate the momentum corresponding to the beam velocity. Solid lines are the predictions from a simple polarization model, scaled down by 1/10 and 1/20 for Au and Al targets, respectively.

Although the general trend is well reproduced, details of the observed polarization differs from the above mentioned simple model. The predicted polarization is a monotone function crossing the momentum axis near the beam velocity and is anti-symmetric relative to the zero-crossing momentum. For the Au target the absolute polarization on the lower momentum side is significantly smaller than that on the higher momentum side; this effect has been seen in previous works. This trend may be caused from the unknown fraction of deep inelastic collisions (DIC) in the lower momentum region. The contribution from DIC to the polarization is not yet clear. For the Al target, the zero-crossing momentum is significantly lower than the beam velocity. This behavior may be understood by introducing a deceleration mechanism such as a nuclear frictional force. However, such a mechanism causes additional polarization.

In contrast to the polarization for heavy and light targets, the polarization for the Cu target has a minimum near the central momentum(Fig. 1-b) as was reported for light projectiles in ref. 2 and ref. 8. This trend is supposed to be a result of the mixing of deflection angles of both signs. However, this momentum dependence can not be reproduced by a simple extension of the above mentioned simple model.

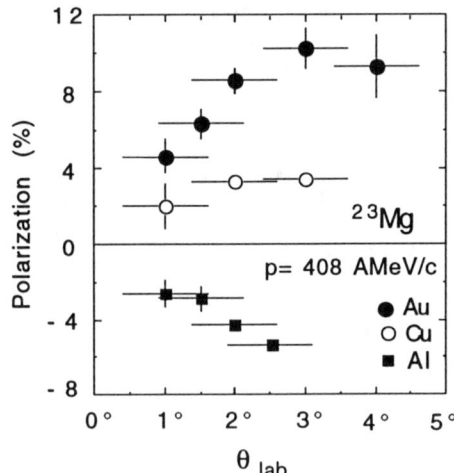

FIGURE 2. Reaction angle dependence of the polarization of ^{23}Mg. Closed circles, open circles and closed squares are the polarizations for Au, Cu and Al targets, respectively.

It can be reproduced only if there is some deceleration mechanism for the negatively deflected component. Instead of the deceleration mechanism, Okuno et al. recently introduced an idea of recession of the abrasion point(8). This mechanism introduces additional negative polarization for relatively large angles, resulting in the lowering of the zero-crossing momentum in the case of light targets. Although the recession necessary for explaining the data in ref. 8 seems too large compared with that expected from the finite mean free path of the abraded nucleons as is already discussed in the reference, it is worth trying to apply to the present data.

Spin polarization was also measured as a function of reaction angle at a momentum 1.7% higher than the beam velocity, as shown in Fig 2. For Au and Cu targets, the polarization increases as the angle and reaches a saturated value around 2°, which is explained well by the beam divergence due to the original beam emittance and multiple scattering of the beam in the target. On the other hand, the absolute polarization from the Al target keeps increasing in the observed range of reaction angle. Since multiple scattering of the beam is smaller in the light target, this angle dependence reflects the polarization mechanism. Thus the present behavior in angle dependence allow one to check the validity of the idea of recession of the abrasion point, since the additional polarization introduced by this new model changes its sign from positive to negative as the reaction angle increases.

There is another difficulty in understanding the polarization mechanism, i.e., the quenching of the absolute degree of polarization from the predicted polarization. This quenching, which is seen in all the present cases and in the previous works, can not be explained well by the depolarization in the implantation process, by the depolarization due to gamma transitions of the exited fragment nuclei, or by the mixing of the coulomb breakup together with the fragmentation process. This quenching may be caused by the coexistence of the positively and negatively deflected components even for both extreme cases where either of the components become dominant.

Once the polarization technique was established in ^{23}Mg, the technique can be applied to the study of beta decay. For further study, polarized ^{23}Mg nuclei produced at the angle of 2° at 1.7%-higher momentum from the beam velocity with Au target was used to maximize the figure of merit which is the yield rate times the square of the polarization. Study of the weak vector coupling constant G_V for mixed transitions like ^{23}Mg beta decay provides a sharp test for CVC(Conserved Vector Current) theory. G_V for the mixed transition is obtained from the asymmetry parameter as well as the half life of the transition. In the present experiment, a preliminary result for the asymmetry parameter of ^{23}Mg beta decay was obtained. A thin Pt catcher foil was further cooled down to 13 K. Polarization effects were measured by means of asymmetric beta-ray emission for both the ground state transition and the transition to the first excited state. The transition to the excited state was tagged by the 440 keV gamma rays.

Fig. 3 shows the obtained beta-ray asymmetries $\mathcal{A}P$ for beta-ray singles and beta-gamma coincidence events. Since the value $\mathcal{A}_{ex}$ is known to be -0.6, the ratio of two asymmetries, $\mathcal{A}_{ex}/\mathcal{A}_0'=0.90\pm0.19$, gives the asymmetry parameter for the ground transition as $\mathcal{A}_0=-0.65(13)$, which is in good agreement with the value -0.55 predicted from CVC. A more precise measurement is being planned in order to obtain the coupling constant G_V with an accuracy of about 1%.

ACKNOWLEDGMENTS

The present work was partially supported by the Grant in Aid for Scientific Research, and the Monbusho International Science Research Program of the ministry of Education, Culture and Science, Japan. Support was also given by the USA-Japan Collaborative Research, funded by both the Japan Scientific Foundation, and the National Science Foundation, USA. It was also supported in part by the US Department of Energy, under contract No. DE-AC03-76SF0098, and by the Kurata Research Grant from the Kurata Foundation.

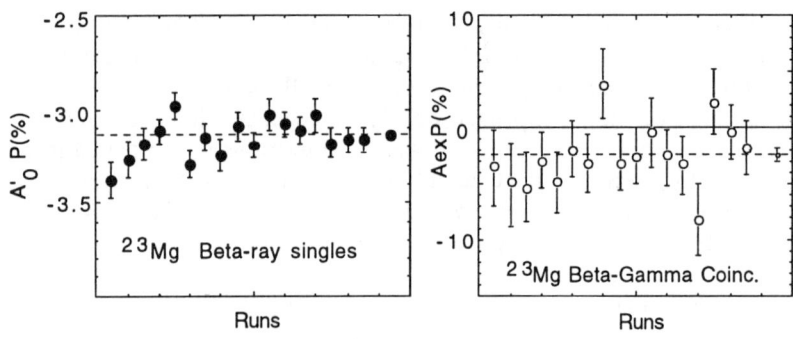

FIGURE 3. Beta-ray Asymmetries.
The last data points and the broken lines indicate average asymmetries.

REFERENCES

1. K. Matsuta et al., Hyperfine Interactions **78**, 123 (1993).
2. K. Asahi et al., Proc. 2nd Int. Conf. on Radioactive Nuclear Beams, Louvain- la Neuve, Belgium, 1991, ed. Th Delbar (IOP Publishing, London 1992) 155.
3. H. Okuno et al., Hyperfine Interactions **78**, 97 (1993).
4. H. Izumi, Proc. of Symp. on Science with Short Lived Nuclear Beams '93, Dec. 16-18, 1993, Tokyo Japan, ed S. Kubono and T. Ohtsuka :(JHP-Suppl.-15, 1994) 49.
5. J.C. Hardy and I.S. Towner, Phys. Lett. **58B**, 261 (1975).
6. K. Matsuta et al., Phys. Lett. **B281**, 214 (1992);
   Proc. 2nd Int. Conf. on Radioactive Nuclear Beams, Louvain-la Neuve, Belgium, 1991, ed. Th Delbar (IOP Publishing, London 1992) 361;
   Hyperfine Interactions **78**, 127 (1993).
7. K. Asahi et al., Phys. Lett. **B251,** 488 (1990);
   Proc. Symp. on Nuclear Collective Motion and Nuclear Reaction Dynamics, Wako, Japan, 1989, ed. K.-I.Kubo, M. Ichimura, M. Ishihara, S. Yamaji, (World Scientific, Singapore 1990) 239.
8. H. Okuno et al., Phys. Lett. **B335**, 29 (1994).
9. M. Ishihara, Colloque de Physque C6, Suppl. au n°22, Tome 51, C6-231 (1990);
   Nucl. Phys. **A538**, 309c (1992).
10. A.S. Goldhaber, Phys. Lett. **53B**, 306 (1974).
11. K. Asahi and M. Ishihara, RIKEN Accel. Prog. Rep. **20**, 21 (1986).

SPINFLIP PROBABILITY VIA THE ^{26}Mg $(^{3}\text{He}, t\gamma)$ ^{26}Al REACTION

H. Sakai[a], T. Aoyama[a], M.N. Harakeh[b], K. Kubota[a],
H. Okamura[a], H. Otsu[a], Y. Satou[a], M. Tanaka[c], T. Uesaka[a],
and T. Wakasa[a]

a Department of Physics, University of Tokyo, Bunkyo, Tokyo 113, Japan
b KVI, Zernikelaan 25, 9747 AA Groningen, The Netherlands
c INS, University of Tokyo, Tanashi, Tokyo 188, Japan

Abstract. The spinflip probability for the reaction ^{26}Mg$(^{3}\text{He},t\gamma)^{26}$Al ($1^+; 1.06$MeV) has been measured at E(^{3}He)= 53 MeV and $\theta_{lab} = 10°$. The extracted value is $40.6 \pm 3.6\%$. A microscopic DWBA calculation with empirically determined effective interactions reproduces the observed cross sections fairly well but the spinflip probability poorly.

Introduction

The charge-exchange (^{3}He,t) reaction is considered to be a powerful tool for investigating spin-isospin excitations in nuclei. Since this reaction populates both spinflip and non-spinflip states, measuring the spinflip probability constitutes an excellent method to discriminate between the two types of excitation. However, the spinflip probability for the (^{3}He,t) reaction, to our knowledge, has never been measured. This is partly due to the difficulty in preparing a polarized ^{3}He beam and a triton polarimeter for the polarization-transfer measurement which is commonly used in the nucleon-induced reactions like $(\vec{p},\vec{p}')$ or $(\vec{p},\vec{n})$. The (^{3}He,$t\gamma$) reaction, in which γ-decay is measured in coincidence with the triton ejectile, offers an unique method to extract the spinflip probability. The method was first utilized by Schmidt et al. [1] to investigate the spinflip probability in inelastic proton scattering. Because the technique used in the experiment is simple, it is frequently used for the $(p,p'\gamma)$ reaction. The same technique has been extended recently to the (d,d') reaction [2] which gives exact single spinflip probability for the spin-one particle, i.e. deuteron.

We measured, for the first time, the spinflip probability for the charge-exchange reaction

$$^{26}\text{Mg}\,(^{3}\text{He}, t\gamma)\,^{26}\text{Al}\;(1^+; 1.06\text{MeV} \to 0^+; 0.23\text{MeV})$$

© 1995 American Institute of Physics

at $E_{^3He} = 53$ MeV and $\theta_{lab} = 10°$, by applying the method of Schmidt et al. This can be viewed as a benchmark test since the 1^+ state at an excitation energy of 1.058 MeV in ^{26}Al is a spinflip state.

Experiment and Result

The present experiment was performed with a ^{3}He beam of 53 MeV provided by the SF Cyclotron at the Institute for Nuclear Study, University of Tokyo. The target was a 2.85 mg/cm^2 thick metallic foil of ^{26}Mg enriched to > 95 %. The tritons were detected by a magnetic spectrograph with a solid angle of 5 msr. The γ-rays were measured in coincidence with the tritons in a GSO(Ce) (Gd_2SiO_5 doped by Ce) scintillator of dimensions 3.5×3.5×1.0 cm^3 placed perpendicular to the reaction plane at a distance of 10 cm from the target. The ^{3}He beam had to be stopped inside the scattering chamber so that a large amount of lead blocks was set inside the scattering chamber to shield the GSO scintillator from the γ-ray background produced in the Faraday cup material. The singles count rates of the GSO scintillator and the spectrograph with a beam current of 5 nA were 25000 c/s and 100 c/s, respectively. The coincident rate was about 1 c/s.

Figure 1 shows the position spectrum of the triton. The 1^+ state at

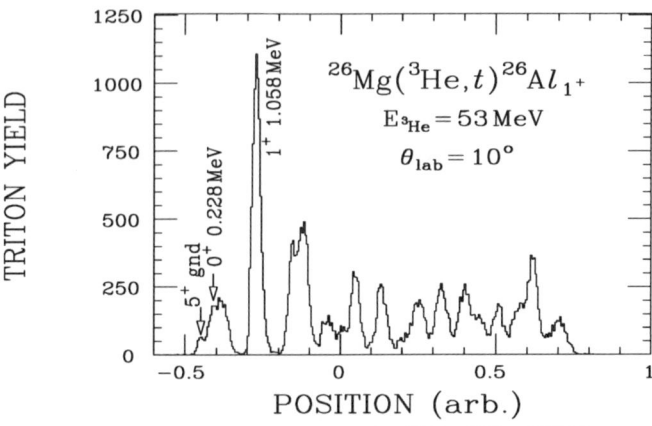

Figure 1: The triton position spectrum for the ^{26}Mg(^{3}He,t)^{26}Al reaction at 53 MeV and $\theta_{lab} = 10°$.

$E_x = 1.058$ MeV in ^{26}Al in which we are interested is clearly observed. It is well separated from nearby states. Figure 2 shows the angular distribution of this 1^+ state.

For the coincidence measurement we set the spectrograph at $\theta_{lab} = 10°$ where the cross section is maximum (see Fig. 2).

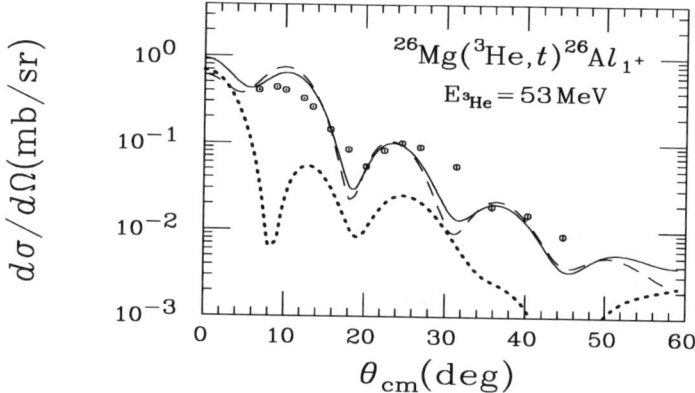

Figure 2: Differential cross section of the 1^+ state at $E_x =1.058$ MeV in ^{26}Al.

The true coincidence γ-ray spectrum is obtained after gating on the transition to the 1^+ state in the triton spectrum. This is shown in Fig. 3. The accidental coincidences have been corrected for.

Spinflip Probability

By applying the Bohr theorem [3], one can show that the spinflip probability S becomes equal to P_1 for the 1^+ state. Here, P_m is the magnetic m-substate population probability. Note that the final state populated by the γ-decay must have J = 0 to apply the Bohr theorem. Therefore, if a deexciting γ-ray emitted in the direction perpendicular to the reaction plane is detected in coincidence with an ejectile triton, the S value can be obtained by using the radiation pattern of γ-rays as

$$S = \frac{8\pi}{3}\frac{d^2\sigma_{coin}}{d\Omega_\gamma d\Omega_t}/\frac{d\sigma_t}{d\Omega_t} = \frac{8\pi}{3}\frac{1}{\Delta\Omega_\gamma \epsilon_\gamma}\frac{Y_{coin}}{Y_t}, \quad (1)$$

where $\frac{d\sigma_t}{d\Omega_t}$ is the singles differential cross section, $\frac{d^2\sigma_{coin}}{d\Omega_\gamma d\Omega_t}$ is the coincidence double-differential cross section, Y_t is the number of singles events of tritons populating the 1^+ state (1.06 MeV), Y_{coin} is the number of true coincidence events where γ-rays ($1^+ \to 0^+$) are detected at around 830 keV and $\Delta\Omega_\gamma \epsilon_\gamma$ is the total detection efficiency of the GSO(Ce) scintillator.

The γ-ray detection efficiency for $E_\gamma = 0.83$ MeV was calibrated by using standard γ-ray sources, ^{60}Co and ^{137}Cs, and was found

$$\Delta\Omega_\gamma \epsilon_\gamma = 4\pi \times 1.05 \times 10^{-3} \text{ sr}. \quad (2)$$

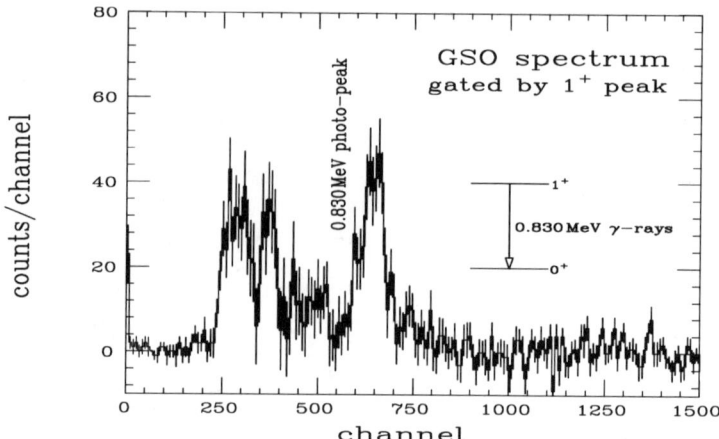

Figure 3: The GSO γ-ray spectrum gated by the transition to the 1^+ state in the triton spectrum. The photo-peak counts of 0.830 MeV γ-rays ($1^+ \to 0^+$) give the coincidence counts.

The effect due to the finite size of the GSO(Ce) detector has been corrected for.

The spin-flip probability, thus extracted, is

$$S = 40.6 \pm 3.6\%,$$

and is plotted in Fig. 4.

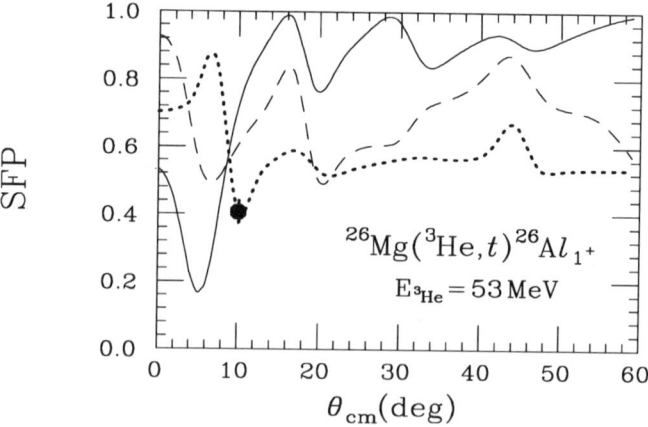

Figure 4: Spin-flip probability measured for the 1^+ state at $E_x = 1.058$ MeV in ^{26}Al.

Microscopic DWBA Calculation and Discussion

A microscopic distorted-wave Born approximation (DWBA) calculation was performed with the code DW81 [4]. The effective interactions used in the calculation are those of Van der Werf et al. [5] which are derived phenomenologically by fitting differential cross sections of various states for targets ranging from ^{12}C to ^{90}Zr at $E_{^3He} = 66 - 90$ MeV. The effective interactions derived in terms of single Yukawa potentials ($Y(r)$) are

$$V_{eff}(r) = \{V_{\sigma\tau}(\vec{\sigma_1}\cdot\vec{\sigma_2})Y(r/R_{\sigma\tau}) + V_{T\tau}S_{12}r^2Y(r/R_{T\tau})\}(\vec{\tau_1}\cdot\vec{\tau_2}) \quad (3)$$

where $\quad V_{\sigma\tau} = -3.0$ MeV and $V_{T\tau} = -6.50$ MeV·fm^{-2} \quad (4)

$\quad\quad R_{\sigma\tau} = 1.415$ fm and $R_{T\tau} = 0.878$ fm. \quad (5)

Here, S_{12} is the usual tensor operator.

Recently these effective interactions have been successfully applied in the analysis of the ^{24}Mg(^{3}He,t)^{24}Al reaction at 81 MeV [6]. The wave functions are those of Brown and Wildenthal [7]. The optical potential parameters are taken from ref. [5]. The calculated results are shown in Figs. 1 and 2. The effect of the angular acceptance of the spectrograph has been corrected for. The dotted and dashed curves are the calculations with central $V_{\sigma\tau}$ and tensor $V_{T\tau}$ interactions, respectively. The solid curve represents the calculation with the full interaction ($V_{\sigma\tau} + V_{T\tau}$). The differential cross sections are reproduced rather well by the calculation while the spinflip probability is overestimated by a factor of almost two. The cross sections are entirely dominated by the tensor interaction. In other words the $V_{\sigma\tau}$ interaction plays a minor role.

The effective interaction by Schaeffer [8], which has been derived for the low energy ($\sim$ 30 MeV) (^{3}He,t) reaction, has stronger potentials, $V_{\sigma\tau} = -6.6$ MeV and $V_{T\tau} = -8.5$ MeV.fm^{-2}. If we calculate the spinflip probability S by using the Schaeffer's effective interaction, we get a larger S value of 0.80 at around 10° compared to that obtained with the Van der Werf et al. interaction, and the cross section becomes about 50% larger.

S also does not depend on the wave function. If we calculate S by assuming a simple configuration such as $1d_{\frac{5}{2}} \rightarrow 1d_{\frac{5}{2}}$, the S value changes by less than 0.003 from the original S obtained with the Brown-Wildenthal wave function.

The ^{3}He optical potential for ^{26}Mg used in the analyses has no spin-orbit term. The ^{3}He spin-orbit potential for ^{26}Mg has been determined by Cohler et al. [9] at 33.4 MeV. If we calculate S by using the optical potential of Cohler et al., we get $S = 0.703$ which differs only 0.017 from the original value at $\theta = 10°$ obtained with the optical potential of ref. [5].

These examples clearly demonstrate the stability of the spinflip probability against changes in the effective interaction, wave function and optical potential. Thus it is very difficult to reproduce the observed cross sections and spinflip probability simultaneously by varying these parameters within reasonable ranges.

The (^{3}He,t) reaction may have contributions from two-step processes. It has been shown [9] that two-step contributions may affect the absolute values of the cross sections, but tend to have angular distributions with shapes similar to the corresponding one-step process. Therefore it is worthwhile to study whether the contribution of the two-step process is able to explain the difference in the spinflip probability. It is also interesting to extend the present experiment to higher bombarding energies where the one-step process becomes dominant.

Acknowledgement

The authors wish to thank S.Y. van der Werf for his support in the DWBA calculation. One of the authors (MNH) acknowledges the financial support by the JSPS.

References

1. F.H. Schmidt et al., Nucl. Phys. **52**, 353(1964).
2. S. Ishida et al., Phys. Lett. **B314**, 279(1993).
3. A. Bohr, Nucl. Phys. **10**, 486(1959).
4. J.R. Comfort extended version of the program DWBA70 by R. Schaeffer and J. Raynal (unpublished).
5. S.Y. van der Werf et al., Nucl. Phys. **A496**, 305(1989).
6. M.B. Greenfield et al., Nucl. Phys. **A524**, 228(1991).
7. B.A. Brown and B.H. Wildenthal, At. Data Nucl. Data Tables **33**, 347(1985).
8. R. Schaeffer, Nucl. Phys. **A164**, 145(1971).
9. M.D. Cohler et al., J. Phys. G **2**, L151(1976).

Low Energy Nuclear Reactions

D. Fick

Philipps-Universität, Fachbereich Physik, D-35032 Marburg, Germany

Abstract The contributions to the „Low Energy Nuclear Reaction" sessions are reviewed. Emphasis is solely put to contributions which deal with the coupling to the nonresonant continuum.

Most of the contributions to the low energy nuclear reactions sessions reflect topics which had been already at the forefront of interest at many of the previous Polarization Conferences. These contributions can be looked up in the present Proceedings and will not be discussed here further. Rather this rapporteur talk intends to concentrate on one topic, which albeit being present since quite some time nevertheless gained importance only during the previous decayed: coupling to the nonresonant continuum in the description of elastic scattering and / or direct reaction data. In fact it was widely discussed during the Polarization Conference at Osaka in 1980, but it attracted only little attention at the following Conference in Paris. It is an important topic for the loosely bound projectiles as deuterons, ^{6}Li etc. which were and are widely used in investigations of elastic scattering and direct nuclear reactions. Even though by obvious physical reasons coupling to the nonresonant continuum certainly had to be emphasized always as important its treatment was for long abandoned mainly by technical reasons. In fact, if treated rigorously one would face a three body problem with all its complications as well conceptually as computerwise. Up to now it is not solvable rigorously. That is probably why the problem was inhibited for a long time along a reasoning : "What cannot be treated realisticly is just not present".

However, almost ten years ago with increasing available computer power a realistic approach to the problem was achieved via the "continuum discretized coupled" channel (CDCC) method (1,2). In this approach the continuum above the energetically lowest breakup threshold is discretized in a set of momentum bins with respect to the momentum $\hbar k$ of the relativ motion of the breakup nuclei. That is e.g. n-p for the deuteron and d-α for ^{6}Li etc. The cluster wave functions $\psi(r;k) = f(k)\chi(r;k)$ in a momentum bin are averaged over the bin width Δk and normalized to unity assuming for continuum state wave functions $f(k)=1$. Each

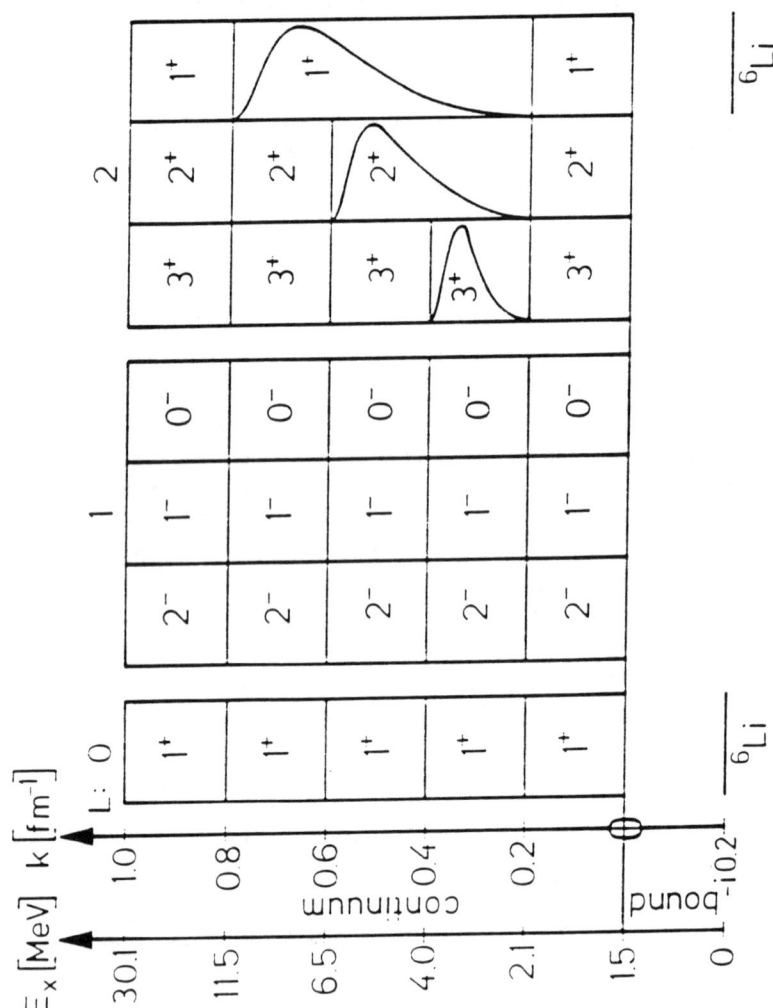

FIGURE. 1 Example of a model space of ^{6}Li → d+α breakup states. The D-wave resonances are only schematically displayed

bin is treated as an excited state of the projectile represented by a way fonction ψ (r) at an energy corresponding to the mean energy of the bin. To it the total spin $\vec{J} = \vec{L} + \vec{S}$ and parity $\pi = (-1)^L$ is attributed, L being the relativ angular momentum of the breakup nuclei and S the channels spin. The question of the size and the number of the momentum bins and the maximum angular momentum to be taken into account has to be answered individually for each problem relying mainly on its physics but very often also on computer space available. It is a pitty that in most of the recent publications a proper description of this part cannot be found in enough detail.

If resonances are present in the continuum, as for ^{6}Li, the cluster wave function varies "resonance like" over a certain momentum interval. This can be taken into account by adding to the nonresonant continuum a resonant part setting for the specific values of L and J of the resonance (1,2).

$$f(k) = \frac{i\Gamma/2}{\varepsilon(k) - \varepsilon_{res} + i\Gamma/2}.$$

A coupling scheme for ^{6}Li, which has some resemblance with an example discussed below (3,4) may look as displayed in Fig. 1. In addition to the ground state of ^{6}Li ($J^\pi=1^+$) consisting of a S- and D-state 3 D-wave resonances with $J^\pi = 3^+, 2^+, 1^+$ are present. For the nonresonant part bins with $\Delta k=0.2 fm^{-1}$ in k-space have been assumed for the relativ angular momenta L = 0,1,2 of the d-α relativ motion, if not already covered with a resonance. Alltogether this represents a 66-channel problem to which channels from target excitation have to be added. Doubtless, even with present days computer capabilities, this is still beyond the present possibility. Thus, truncations have to be made. Nevertheless great progress has been reported in three papers submitted to the conference (3,5,6). However, what seems even more important to me : the work in two of these papers (3,6) was now nicely performed by experimentalists in detail after theorists prepared the case in general. Even more important, one contribution (3) used an existing and rather easy accessible coupled channel (CC) code (FRESCO) (7) onto which the CDCC-method was implemented.

The first contribution to the conference to be mentioned (5) dealt among others with the elastic scattering of 400 MeV deuterons on ^{58}Ni. (see their fig.1). A double folding potential (Reid soft core) which took into account properly the spin orbit parts and in which as well the S- as the D-wave part of the deuteron groundstate was considered underestimates the differential cross-section beyond 50° by far if inserted into a conventional optical model code. Furthermore it has problems to describe A_{yy} over the whole angular range. Including coupling to the continuum as well in the channel spin S=1 (triplet) as in the S=0 (singulet) channel rises the cross-section but not yet to the observed values. Also A_{yy} is now better described. But the first pronounced maximum (here around 7°) is still failed completely, as in other calculations which deal with the elastic scattering of hundreds of MeV deuterons.

A further contribution (6) dealt with the description of an almost classical example, the ^{12}C(d,p)^{13}C reaction initiated by 25 MeV deuterons. DWBA calculations with double folding potentials (Reid soft core) as input overestimate beyond the first diffraction minimum the cross section by far and iT_{11} is described poorly as a whole (see their figure). As far as the cross section is concerned an adiabatic treatment of the continuum (8) does almost the job. However, a few minor discrepancies are straightend out by applying CDCC calculations. They were however restricted to the triplet S-wave continuum. It has to be seen whether the still remaining discrepancies in describing the iT_{11} data can be related to this severe truncation of the model space.

The final contribution to be discussed dealt with the elastic scattering of 60 MeV ^{6}Li on ^{26}Mg using a d-α cluster double folding potential (S-state of ^{6}Li only) as input into FRESCO (7). The differential cross section beyond 30° is overestimated by far (see their fig.1). The folding potential has to be renormalized to half of its value (4) to account in the calculations for the observed experimental data. Moreover, as it is wellknown since long (9) such calculations yield values for iT_{11} very close to zero. It is, however, common wisdom since years (10,1) that breakup effects of the weakly bound ^{6}Li projectile are responsible for this famous anomaly in the normalization of cluster folding potentials of weakly bound projectiles. It appears also if instead of a cluster folding potential a M3Y double folding potential is used (11). In fact, it was shown by a potential inversion method that the breakup effects can be interpreted as an additional effective repulsive real potential (12). Including the D-wave resonances in ^{6}Li - in this specific contribution by a weak binding energy approximation - reduces the discrepancy in the cross section to half of its previous size and generates also sizeable values of iT_{11} (9) which, however, in the structure of their angular distributions, have no real resemblance to the structure of the measured ones. However, the famous puzzle is resolved if into the coupled channels calculations as in ref.(3) additionally to the D-wave resonances the nonresonant S-, P- and D-wave continuum inbetween k=0.2fm^{-1} and 0.8fm^{-1} in bins of Δk=0.2fm^{-1} are included (see fig.1) for those parts in k-space not already covered by a D-wave resonance. Without any renormalization of the cluster folding potential the cross section is described almost to perfection and also the predictions for iT_{11} show now similarities to the data. To achieve this in the final 20 channel calculations the coupling to the first excited state of ^{26}Mg has to be treated as well. However, the omission of the D-state of the ^{6}Li groundstate, which inclusion would have doubled the number of channels to just 40 may be blamed for the remaining discrepancies. They certainly will show up more severely in a proper description of second rank tensor analyzing powers which are quite sensitive to the D-state of ^{6}Li (9).

To achieve this results on a "alpha"-workstation the widely known CC code FRESCO (7) had been used into which the CDCC method has been implemented. This example shows that coupling to the continuum is now also accessible to experimental groups. This is to my opinion the real big achievement shown by the group from Warsaw and Birmingham, in particular since faster workstations, which soon will be on hand to experimentalist, certainly will allow in future for a further extension of the model space.

REFERENCES

1. Sakuragi, Y., *Phys. Rev.* **C35**, 2161 (1987) and references therein.
2. Hirabayashi, Y., *Phys. Rev.* **C44**, 1584 (1991) and references therein.
3. Ward, R. P., et al., contribution to this Conference.
4. Rusek, K., Clarke, N. M., and Ward, P. P., *Phys. Rev.* **C50**, in press.
5. Iseri, Y., and Tanifuji, M., contribution to this Conference.
6. Toyokawa, H., and Ohnuma, H., contribution to this Conference.
7. Thompson, I. J., *Comp. Phys. Rep.* **7**, 167 (1988).
8. Johnson, R. C., and Soper, P. J. R., *Phys. Rev.* **C1**, 976 (1970).
   Harvey, J. D., and Johnson, R. C., *Phys. Rev.* **C3**, 636 (1971).
9. Fick, D., Grawert, G., and Turkiewicz, I. M., *Phys. Rep.* **214**, 1 (1992) and references therein.
10. Thompson, I. J., and Nagarajan, M. A., *Phys. Lett.* **106B**, 163 (1981).
11. Satchler, G. R:, and Love, W. G., *Phys. Rep.* **55**, 183 (1979).
12. Mackintosh, R. S., and Kobos, A. M., *Phys. Lett.* **116B**, 95 (1982).
    Ioannides, A. A., and Mackintosh, R. S., *Phys. Lett.* **169B**, 113 (1986).
13. van Sen, N., et al., *Phys. Rev.* **C28**, 134 (1983).

VII. POLARIZED SOURCES AND TARGETS

Study of a Polarized Hydrogen Ion Source with Deuterium Plasma Ionizer

A.S. Belov*, G.E. Derevyankin†, V.G. Dudnikov†,
V.S. Klenov*, L.P. Nechaeva*, Yu.V. Plohinsky*,
G.A. Vasil'ev* and V.P. Yakushev*

* *Institute for Nuclear Research of the Russian Academy of Sciences, Moscow, 117312, Russian Federation*
† *Budker Institute for Nuclear Physics of the Russian Academy of Sciences, Novosibirsk, 630090, Russian Federation*

The atomic beam type source of polarized hydrogen ions developed at the INR Moscow [1,2] produces pulsed beams (pulse duration 100 μsec) of polarized protons (H^+) with peak currents up to 6 mA, polarization 85%, normalized emittance 2 π mm mrad, and polarized negative hydrogen ions (H^-) with peak currents up to 200 μA and normalized emittance 2.2 π mm mrad.

The source is shown schematically in Fig. 1. A beam of polarized neutral hydrogen atoms is produced by a conventional atomic beam apparatus which consists of an rf dissociator with pulsed gas and rf power supply, an atomic beam separating and focusing system composed of two sextupole electromagnets with pole tip magnetic field of about 9 kG, and a weak-field rf transition unit.

Resonant charge-exchange between polarized hydrogen atoms and unpolarized deuterium ions in a deuterium plasma is used for production of polarized protons or negative hydrogen ions.

The deuterium plasma ionizer works as follows. The deuterium plasma is produced by a pulsed arc-discharge plasma source. A surface plasma converter (not shown in Fig. 1) is installed downstream of the plasma source in order to produce a deuterium plasma enriched in negative deuterium ions. The deuterium plasma is injected into the ionization region in a direction opposite to the polarized atomic hydrogen beam direction. In the ionization region both beams intersect and polarized hydrogen ions are formed by the charge exchange reaction between polarized hydrogen atoms and deuterium ions. The polarized ions formed are confined in the radial direction by a longitudinal magnetic field (1.2 kG) created in the ionization region by a solenoid, and move along magnetic field lines to the three-electrode ion-optical system in which polarized ions, together with unpolarized plasma

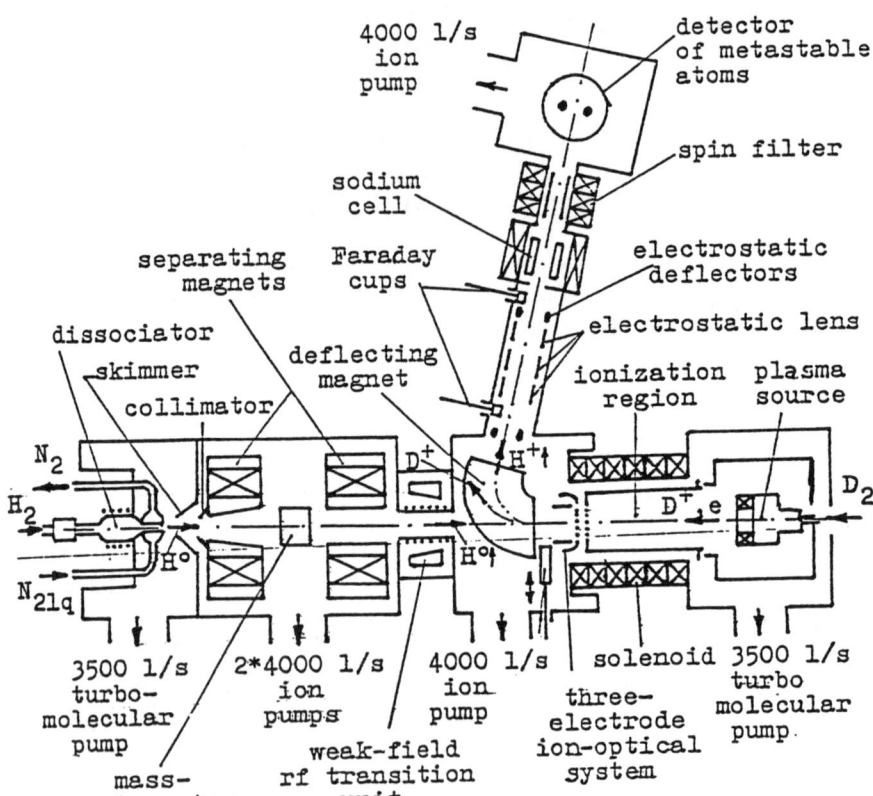

FIGURE 1. Schematic diagram of the atomic-beam type polarized hydrogen ion source.

ions, are accelerated to an energy of 20 keV. The polarized hydrogen ion beam is then separated from the deuterium ions in the deflecting magnet and extracted from the source. The polarization of the H^+ beam has been measured by a low energy polarimeter based on properties of hydrogen atoms in the metastable 2S state. This polarimeter, also shown in Fig. 1, consists of a sodium cell, spin filter and detector of metastable atoms. Its principle and operation are described in detail in Ref. [1]. We will use this polarimeter also for polarization measurements of H^- beams. For this application we have developed a pulsed gas target in a strong longitudinal magnetic field in order to convert the polarized H^- beam into polarized H^+ without loss of polarization by double stripping of H^- when the beam passes through the target.

The SPIN Collaboration on acceleration of polarized proton beams in the Fermilab Tevatron accelerator [3] is considering a polarized H^- source of the type described here as a possible candidate for this project. For this source to qualify it is necessary to increase the intensity of the polarized H^- beam generated by the source to the 1 mA level. We plan to achieve this by making several possible improvements to the present source :

1. Development of a deuterium plasma source which should produce more negative deuterium ions in the deuterium plasma injected into the ionizer.
2. Increase of the polarized atomic hydrogen beam intensity by use of stronger sextupole magnets, with pole-tip magnetic fields up to 4.5 T achievable with superconducting coils.
3. Increase of the polarized atomic hydrogen density in the ionizing region by means of storage of hydrogen atoms in a storage cell similar to those developed for internal polarized gas targets.

The existing deuterium plasma source produces up to 1.5 mA of D^- beam measured after beam extraction from the ionizer. It has been described in detail in Ref. [2]. A shortcoming of this plasma source is that it produces a plasma outside the ionizer solenoid in a weak magnetic field, so a fraction of the plasma and D^- ions generated is lost during injection of the plasma into the ionizer solenoid. A plasma source which can work in a strong longitudinal magnetic field would be better suited for injection of a deuterium plasma with higher D^- ion concentration into the ionizer solenoid. We fabricated and tested a different plasma source, similar to one described in Ref. [4], namely a ring-magnetron source which works in a longitudinal magnetic field of about 700 G. The initial tests showed that this source actually produces less D^- ion current in comparison with our previous plasma source (only few microamperes after the deflecting magnet). Apparently, one of the reasons for the low D^- output is connected with relatively high density of atomic deuterium gas inside the magnetron and correspondingly high probability for charge-exchange of D^- with D^0 inside the magnetron. This leads to a loss of energy of D^- ions formed on the magnetron cathode and prevents their leaving the source. A second reason could be connected with the low longitudinal velocity of extracted D^- ions, leading to destruction on the magnetron walls.

Subsequently, a new version of the plasma converter has been fabricated in which a longitudinal magnetic field is created in the area where D^- ions are formed in order to decrease the loss of D^- ions during plasma injection into the solenoid. The deuterium plasma source with this version of the converter is shown schematically in Fig. 2.

The deuterium plasma is generated by an arc-discharge between the cathode and the anode of the plasma source. The plasma jet is then injected

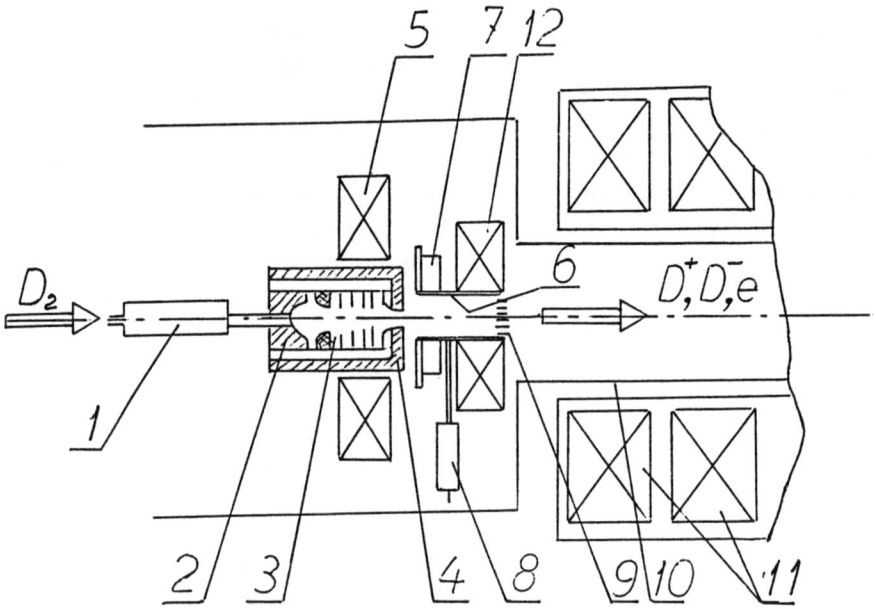

FIGURE 2. Schematic of the plasma source with plasma converter. 1: electromagnetic valve, 2: cathode, 3: diaphragms, 4: anode, 5: coil, 6: converter cylinder, 7: permanent magnets, 8: cesium oven, 9: molybdenum sheets, 10: electrostatic screen, 11: solenoid coils, 12: converter coil.

into the plasma converter in which D^- ions are formed by conversion of D^+ ions and neutral atoms into D^- on cesium-covered walls of the converter. The longitudinal magnetic field is created by the converter coil (element 12 in Fig. 2). The plasma source with this converter is now undergoing bench tests.

It is well-known that the acceptance angle of separating/focusing sextupole magnets is directly proportional to the pole tip magnetic field of the sextupoles. Increasing the pole tip magnetic field from 9 kG, which is typical for an electromagnet sextupole, to 45 kG, which can be achieved by a special design of a sextupole with superconducting coils, leads to a five-fold increase of the acceptance angle of the sextupole. The gain in density of the atomic hydrogen beam focused into the ionization region may be less than this factor because of non-ideal matching of the atomic beam emittance to the acceptance of the ionization region, and also because of chromatic aberration in the focusing of atomic beams with large velocity dispersion.

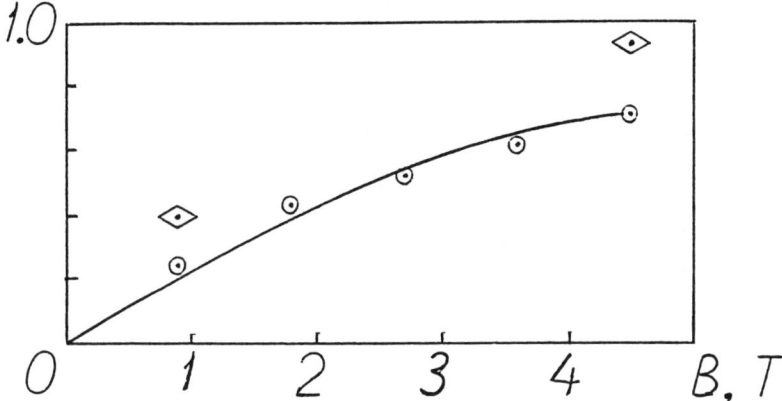

FIGURE 3. Polarized atomic hydrogen beam density in the ionization region, averaged over a 15 mm diameter, 25 cm long ionization region, as function of sextupole magnet pole-tip field. Circles: single sextupole, rhombi: system of two sextupoles.

We have made calculations of the focusing of an atomic beam by high-field sextupoles for an atomic beam having a most probable atom velocity of 2×10^5 cm/s and velocity spread of 0.9×10^5 cm/s, corresponding to atoms obtained from a liquid nitrogen cooled dissociator and having Mach number M=4 [5].

Some results of the calculations are shown in Fig. 3. Here a polarized atomic hydrogen beam density (in relative units) is shown vs. magnetic field at the sextupole pole tips. The atomic beam density was averaged over an ionization volume of 1.5 cm diameter and 25 cm length. The center of the ionization region was chosen 50 cm downstream the sextupole exit. The parameters of the sextupole magnets, such as their length and distance between opposite pole tips, were optimized in the calculation for each magnetic field value. The results of these calculations show that the average density of the atomic beam focused into the ionizer can be increased 2.4 times over that obtained with the existing focussing sextupoles.

The parameters of the optimized sextupole magnet system are given below:

	Magnet 1	Magnet 2
Distance between opposing pole tips	3.6 cm	5.6 cm
Length	20 cm	15 cm
Pole-tip field	4.5 T	4.5 T
Distance between the magnets	44 cm	

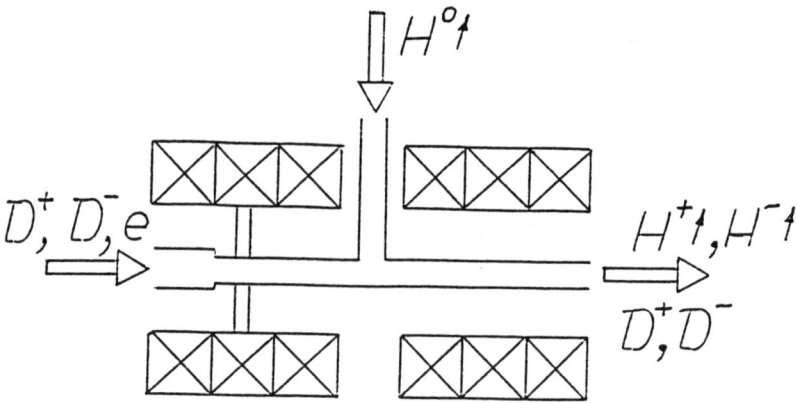

FIGURE 4. Scheme for using a storage of polarized hydrogen atoms in the deuterium plasma ionizer.

Next, the possibility of increasing the intensity of polarized hydrogen ion beams by storing polarized atomic hydrogen in the ionization region is considered. The storage of polarized atoms in cells of appropriate materials or wall surface coatings is possible due to small depolarization of atoms during their collisions with the cell walls. This effect has been used in hydrogen masers [6]. More recently, it has been proposed for storing polarized hydrogen atoms in storage cells used as internal gas targets in particle accelerators [7].

Development of storage cell techniques for polarized atomic hydrogen has resulted in polarized atomic hydrogen gas targets with thickness of 1×10^{14} atoms/cm^2 in continuous mode of operation [8]. Recently, very interesting results were obtained showing that in strong magnetic fields depolarization of atomic hydrogen is small even for storage cell walls made of uncoated metals, for example, aluminum alloys [9]. This is important for use of the storage cell in the deuterium plasma ionizer because the deuterium plasma can cause detoriation of wall coatings.

A possible scheme for a deuterium plasma ionizer incorporating storage of polarized hydrogen atoms is shown in Fig. 4.

It is important to note that for **pulsed** polarized ion sources the atomic beam intensity is 2-3 times greater than for sources operating in continuous mode. Taking into account the gain factor of 2.5 which can be obtained by use of high-field sextupoles, and a gain factor of 2 for pulsed-mode operation, one can expect polarized atomic hydrogen target thicknesses of 5×10^{14} atoms/cm^2 for a deuterium plasma ionizer using the storage cell technique.

The cross-section for the resonant charge-exchange reaction between D$^+$ ions and H atoms at an ion energy of about 10 eV is 5×10^{-15} cm^2, and this value for D^- ions and H atoms. Hence essentially complete neutralization of D^+ or D^- ions passing through the storage-cell target would be possible, and polarized beams of H^+ or H^- ions with intensities comparable to those of unpolarized D^+ or D^- beams could be extracted from the deuterium plasma ionizer.

ACKNOWLEDGMENTS

We would like to thank P. Schwandt and T.B. Clegg for useful discussions. The research described in this publication was made possible in part by Grant N M2T000 from the International Science Foundation.

REFERENCES

1. A.S. Belov, S.K. Esin, S.A. Kubalov, V.E. Kuzik and V.P. Yakushev, Nucl. Instr. and Meth. in Phys. Res. **A255**, 442 (1987).

2. A.S. Belov, V.G. Dudnikov, V.E. Kuzik, Yu.V. Plohinsky and V.P. Yakushev. Nucl. Instr. and Meth. in Phys. Res., **A333**, 256 (1993).

3. Progress report on Acceleration of Polarized Protons in Fermilab Tevatron Accelerator, SPIN Collaboration, 1994 (unpublished).

4. J.G. Alessi, Proc. of Intern. Workshop on Polarized Ion Sources and Targets, Helv. Phys. Acta **59**, 547 (1986).

5. A.S. Belov, S.A. Kubalov, V.E. Kuzik and V.P. Yakushev, Nucl. Instr. and Meth. in Phys. Res. **A239**, 443 (1985).

6. D. Kleppner, H.M. Goldenberg, N.F. Ramsey, Phys. Rev. **126**, 603 (1962).

7. W. Haeberli, Proc. Workshop on Nucl. Phys. with Stored, Cooled Beams, Indiana 1984, AIP Conf. Proc. **128**, 251 (1985).

8. K. Zapfe, Proc. Workshop on Polarized Ion Sources and Polarized Gas Targets. Madison, WI, 1993, AIP Conf. Proc. **293**, 3 (1993).

9. J.S. Price and W. Haeberli, Proc. Workshop on Polarized Ion Sources and Polarized Gas Targets. Madison, WI, 1993, AIP Conf. Proc. **293** (1993), p. 18.

Spin-exchange Polarization Study at the TRIUMF OPPIS

A.N. Zelenski*, C.D.P. Levy, W.T.H. van Oers, P.W. Schmor, J. Welz, G.W. Wight

TRIUMF, 4004 Wesbrook Mall, Vancouver, B.C. Canada V6T 2A3

Abstract. The TRIUMF optically pumped polarized ion source (OPPIS) now produces 56 μA H$^-$ ion current and 85% polarization for routine operation and 120 μA and 78% polarization in high current mode. This is a result of the latest ECR primary proton source improvements and optimization of optical pumping efficiency. It has been shown that in a pulsed mode suitable for high energy accelerator applications, it is possible to produce 1 – 2 mA of polarized H$^-$ current in a charge-exchange OPPIS based on the INR-scheme(1). A possible INR-scheme upgrade to the 10 mA range is discussed. The experimental results on spin-exchange polarization presented in this paper hold promise of polarized H$^-$ current increasing to the 10 mA range.

INTRODUCTION

Understanding is growing of the importance of polarization experiments with fixed targets or colliding beams in the TeV energy range(2,3). The development of polarization facilities for 500 GeV pp collisions at RHIC has succeeded recently with the partial "siberian snake" application for polarization preservation in the AGS synchrotron. The studies undertaken by the SPIN collaboration have shown the feasibility of polarized beam acceleration in the FNAL Main injector and Tevatron-Collider(4). An essential experimental facility is a high current source of polarized H$^-$ ions. A pulsed current of 1.2 mA is required to produce a polarized proton-antiproton collision luminosity of about 10^{32} cm^{-2} s^{-1}, equal to the projected unpolarized value at FNAL(4). The development of such a source is vital to the future polarization program, because of competition with experiments using unpolarized beam. Ideally, the polarized ion source should deliver polarized H$^-$ current similar to unpolarized beam (i.e. in excess of 10 mA). Here will be examined limitations on maximum current in different types of polarized ion source

---

*Visitor from Institute for Nuclear Research, Moscow, Russia

and then the latest results of the spin-exchange polarization technique development will be presented, which promises that a pulsed, optically-pumped polarized ion source will meet the above requirements. It is instructive to consider the two competing techniques – the atomic beam source (ABS) and the optically-pumped polarized ion source – in the same way, since they have many common features. The general polarization scheme is basically the same. In the first step, electron-spin polarized atoms are produced; in the second, electron polarization is transferred to nuclei by means of hyperfine interactions, and in the last step atoms are ionized. The difference is the energy of the atomic H beam.

a)   Polarization of a low energy beam (temperatures below a few hundred K). In the ABS a separated sextupole magnet selects hydrogen atoms in one particular spin state. The acceptance of separated magnets is limited by the maximum sextupole field and minimum beam velocity. Intrabeam scattering limits the usefulness of any temperature decrease below 30 K. It appears that parameters have been well-optimized and 10^{17} atoms/sec in a ionizer acceptance cannot be substantially improved. The most optimistic extrapolation based on the use of a superconducting sextupole gives only a factor of 3 increase of atomic flux.

Scattering is an inevitable limit only for the conventional ABS technique with the separating magnet. There is a possibility of H polarization in spin-exchange collisions with optically-pumped alkali-metal vapour in a cell. A polarized atomic H flux in excess of 10^{18} sec^{-1} has been obtained by the use of K–H spin-exchange collisions. With a proper choice of cell wall material and temperature regime, recombination can be reduced to 20–30%. Selective ionization of atomic hydrogen must be used to preserve polarization. Ionizers using a Cs beam or a D-enriched plasma are selective but the presently obtained ionization efficiency is quite low (0.3–0.5%). Nonetheless, with a 10^{18} sec^{-1} atomic H flux up to 1.0 mA polarized H$^-$ current would perhaps be produced.

b) Polarization of a fast 0.5 – 5.0 keV beam. An advantage of the fast atomic H beam is the ease of conversion to a negative H$^-$ ion beam. In the sodium ionizer cell the equilibrium yield is about 9% at 1 – 5 keV beam energies. At 0.5 – 1.0 keV energies the yield is 16% in a Rb ionizer cell. The electron polarization of a fast H beam can be obtained either in a charge-exchange process, in which primary protons capture polarized electrons from optically-pumped alkali-metal vapour, or in spin-exchange collisions. All present operational OPPIS's are based on charge-exchange polarization. The currents from OPPIS using ECR primary proton sources are limited by the ECR plasma temperature and emittance degradation, which is introduced by the beam extraction system operating in a high magnetic field.

In the INR OPPIS this limitation has been overcome by using a high-brightness atomic H beam injector which is situated outside the magnetic field(5). In this injector the high density plasma is initially expanded in vacuum and efficiently cooled down to 0.2 eV. A four-grid extraction system is used for extraction of a low divergence beam with up to 0.4 A/cm^2 current density. The proton beam is focussed by a solenoidal magnetic lens and then neutralized in a pulsed H_2 gas cell. As a result an atomic H beam intensity up to 3×10^{18} atoms/sec within the ionizer acceptance was obtained(6). This is a 30 times higher flux than for the best ABS. In the INR OPPIS the atomic H beam is injected into a high magnetic field solenoid where it is ionized in a pulsed He gas cell. In effect the He ionizer cell serves as a proton source in a high magnetic field, like the ECR source, but with higher current density and much less beam divergence. The Novosibirsk source is capable of producing a few hundred mA of equivalent atomic H beam within the ionizer acceptance, which would allow a few tens of mA of polarized H$^-$ beam to be obtained(6). However, it is expected that space-charge of the proton beam and the Rb ions produced in the charge-exchange polarization process will increase the beam divergence and reduce the current at some point.

In the INR scheme both injected and outgoing beams are neutral so space-charge is compensated. The equilibrium fraction of neutral H beam after the He ionizer is 20% and without separation the unpolarized atoms will reduce the final proton polarization. A bias voltage applied to the He ionizer gives the polarized and unpolarized beam energy separation but destroys space-charge compensation. It is possible to reduce the fraction of unpolarized atoms produced i H$^+$ - He collisions by using a mixture of He and optically-pumped Rb vap   r in the same 100 cm-long cell. The Rb vapour density must be near t'   radiation trapping limit of 10^{13} atoms/cm^3 and the He density about $\cdot$ 10^{14} atoms/cm^3. In this case the full He thickness will be enough to pro· le a high ionization efficiency but neutralization will mainly be due to polarized electron pick-up from Rb atoms, since the average neutralization length on Rb: $(\sigma_{+0}N)^{-1}_{Rb}$ is 12 cm and on He: $(\sigma_{+0}N)^{-1}_{He}$ is 200 cm. Simultaneously He – Rb buffering collisions will increase the Rb polarization relaxation time, and spin-exchange collisions in a thick 10^{15} atoms/cm^2 Rb cell will further help increase polarization. We believe that this scheme will avoid space-charge current limitations and will probably compete with the spin-exchange technique in the 10 mA polarized H$^-$ current range.

EXPERIMENTAL STUDIES OF SPIN-EXCHANGE POLARIZATION

The experiments were done at the operational TRIUMF OPPIS setup, which was modified in the following way (see Fig. 1). A duoplasmatron followed by a Rb neutralizing cell was used as an injector of atomic H beam. The OPPIS ECR source was removed and the magnetic field in the solenoid was adjusted to a flat shape. A longer 60 cm optically-pumped Rb cell was installed (inner liner diameter 12 mm). The Sona transition, ionizer and polarization diagnostics were exactly the same as for the well-studied charge-exchange OPPIS, so that all polarization measurements could be compared with well-known reference points.

The proton polarization was measured at 300 keV in a polarimeter based on the $\vec{p}(^6Li, {}^3He)^4He$ reaction. This polarimeter was calibrated by comparison with the reference 200 MeV polarimeters. In all experiments fast spin-reversal was used with a 40 Hz repetition rate and a synchronous detection technique for noise reduction. As a result, the proton polarization measurement accuracy of about ±1.5% was routinely obtained in a 5-minute integration time. For Rb thickness and polarization measurements the well-developed technique of Faraday rotation was used.

The results of the H^- polarization measurements are presented in Fig. 2. The polarization grows quickly with Rb thickness up to a thickness of 3×10^{14} atoms/cm². A wide plateau is the result of competition between spin-exchange

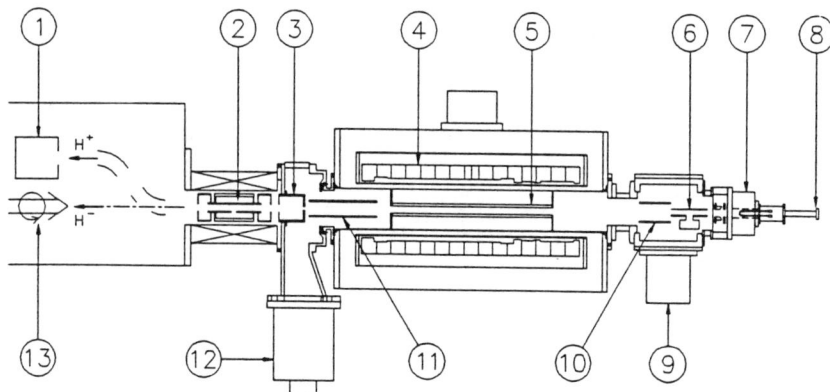

Figure 1. Setup for spin-exchange polarization measurements. 1) Lyman-alpha polarimeter, 2) ionizer Rb cell, 3) Sona transition region, 4) 10 kG S.C. solenoid magnet, 5) 60 cm optically-pumped Rb cell, 6) neutralizer Rb cel, 7) duoplasmatron, 8) window for probe laser, 9) turbopump, 10-11) deflection plates, 12) cryopump, 13) pump laser light.

collisions, which push polarization up, and radiation trapping, which pulls Rb and hence proton polarization down. The maximum thickness of highly polarized Rb is limited by available laser power in the dc mode of operation. Nonetheless, the maximum 48% H⁻ polarization is very satisfactory for a dc source without any dryfilm coating. In the pulsed mode suitable for high energy accelerators the density and thickness could be at least doubled by using high power pulsed lasers. The proton polarization would then increase to 80% (see Ref. (7)).

The energy dependence of spin-exchange cross sections has been studied in the range of 0.5 – 2.0 keV (see Fig. 3). The cross sections are close to calculated values but a substantial shift in the position of the maximum was

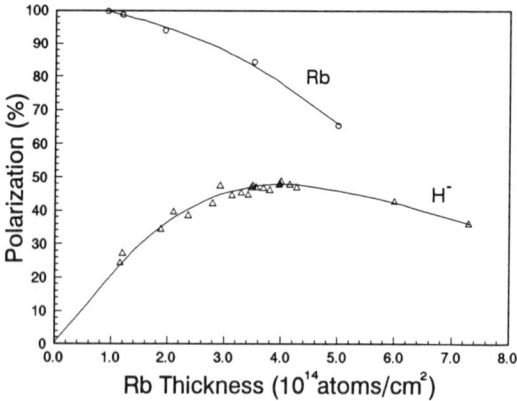

Figure 2. Rb and H⁻ polarization dependence on Rb thickness, using an uncoated 60 cm long cell.

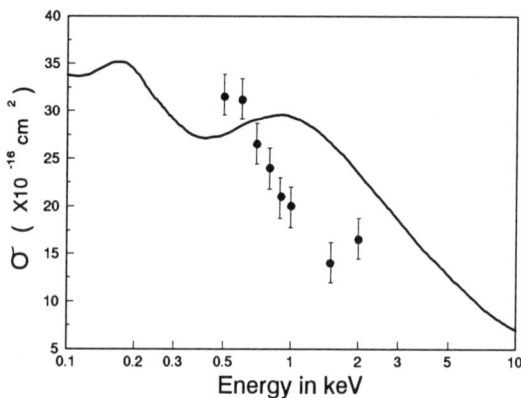

Figure 3. H-Rb spin-exchange cross section dependence on atomic H beam energy. Data points, present work; solid line, calculated value.(8)

observed. A polarized H⁻ current of about 10 μA was obtained with a duoplasmatron from the old TRIUMF Lamb-shift source without any special optimization. It is planned to develop a high brightness atomic hydrogen injector in collaboration with BINP Novosibirsk to replace the duoplasmatron. With a new atomic beam injector one expects the polarized H⁻ ion current to be in the 10 mA range(9).

CONCLUSIONS

Recent development of polarized ion sources and polarization preservation techniques make feasible high energy polarized beam facilities. The use of the optical pumping technique has remarkably improved the polarized target and polarized ion source parameters. At present the TRIUMF dc OPPIS produces polarized H⁻ current of 120 μA. The pulsed INR-type OPPIS could produce at least 1 mA H⁻ current; therefore the TRIUMF-INR-KEK-BINP team has proposed source development in the frame work of the SPIN collaboration for experiments at FNAL. The results of the present work are very promising for the future development of a 10 mA pulsed spin-exchange OPPIS.

ACKNOWLEDGEMENTS

We would like to thank G. Dutto, D. Swenson, Y. Mori and D. Tupa for helpful discussions. We wish to acknowledge the help of the engineering and technical staff of the Ion Source Injector group for assistance in OPPIS development and especially the contributions of M. McDonald and R. Ruegg. We would like to thank A.D. Krisch for his encouragement and support of high-current OPPIS development.

REFERENCES

1. Zelenski, A.N., et al., AIP Conf. Proc. No. 293, 175, (1994).
2. Bourelly, C., et al., Phys. Rep. **177**, 319, (1989).
3. Taxil, P., Proc. 10th Intern. Symp. on High Energy Spin Physics, Universal Academy Press, Inc., Tokyo, Japan, 215, (1992).
4. 1994 SPIN progress report, Univ. of Michigan, spokesman A.D. Krisch.
5. Zelenski, A.N., et al., Nucl. Instrum. Methods, **A245**, 223, (1986).
6. Belchenko, Ya. I., et al., Rev. Sci. Instrum. **61**, 378 (1990).
7. Zelenski, A.N., et al., "Development of a spin-exchange polarized H⁻ ion source for high energy accelerators", presented at the European Particle Accelerator Conference, London, UK, June 27–July 1, 1994.
8. Swenson, D.R., et al., J. Phys. B., **18**, 4433 (1985).
9. Zelenski, A.N., Kochanovski, S.A., AIP Conf. Proc., No. 293, 164, (1994).

A Dual-Optically-Pumped Polarized Negative Deuterium Ion Source

M. Kinsho*, Y. Mori, K. Ikegami, and A. Takagi

*Research Fellow of the Japan Society for the Promotion of Science
Oho 1-1, Tsukuba-shi, Ibaraki-ken 305, JAPAN
National Laboratory for High Energy Physics (KEK)
Oho 1-1, Tsukuba-shi, Ibaraki-ken 305, JAPAN

Abstract. To obtain highly vector polarized negative deuterium ions, we have been developed a dual-optically-pumped polarized negative deuterium ion source. It is possible to select a pure nuclear-spin with this scheme and negative deuterium ions with a 100 % nuclear-spin vector polarization can be produced in principle. We have obtained a 70 % of nuclear-spin vector polarized negative deuterium ion beam so far.

INTRODUCTION

Polarized protons have been successfully accelerated in the KEK 12-GeV proton synchrotron (KEK-PS) since 1985[1]. Many experiments have been carried out with polarized proton beams so far. A deuteron beam was also accelerated in the KEK-PS since 1991. Following this success, it has been strongly requested to accelerate polarized deuteron beams.

An optically pumped polarized ion source (OPPIS) has been used for acceleration of polarized proton beams in the KEK-PS so far. The idea of this type of polarized ion source was proposed by Anderson[2] and the first operational ion source has been successfully developed at KEK[3]. Afterwards, various institutes have developed an OPPIS for their accelerators[4][5][6]. It has been believed that this type of polarized ion source is not useful to produce a highly nuclear-spin (vector and tensor) polarized deuterium ions. In 1988, Schneider and Clegg[7] proposed a new nuclear-spin state selection method based on the dual optical pumping scheme. In spite of this possibility of making a highly polarized deuteron beam by optical pumping, they concluded eventually in their paper that this dual optical pumping scheme might be not practical because efficient optical pumping of the thick target in the ionizer is difficult due to radiation trapping. However, we have re-examined the dual-pumped scheme in detail and found that radiation trapping was not a serious problem and highly polarized deuterons could be obtained with the dual-pumped scheme. In the preliminary experiments, we found that the dual-optical pumping scheme for generating a highly polarized negative deuterium ion worked well[8].

Recently, we increase the magnetic field strength at the neutralizer with a 2.7 T super conducting solenoid. By this modification, 70 % nuclear-spin vector polarized negative deuterium ion beam was obtained.

OPPIS FOR DEUTERONS

A block diagram of the dual-optically-pumped polarized ion source is shown in Fig.1 schematically. The idea of this scheme is as follows: After picking up the polarized electrons from optically pumped alkali atoms, for example, deuterium atoms are electron-spin polarized in the state of $m_j = +1/2$ as shown in Fig.2. These electron-spin polarized deuterium atoms equally populate three hyperfine sub-levels $I_Z=+1$, 0, and -1 in a high magnetic field, which are labeled the states 1, 2, and 3, respectively in the figure. Using the Sona transition, the state 1 ($m_j = +1/2$, $I_Z=+1$) goes to the state 1' ($m_j = -1/2$, $I_Z=-1$), the state 2 ($m_j = +1/2$, $I_Z=0$) goes to the state 2' ($m_j = +1/2$, $I_Z=-1$), and the state 3 goes to the state 3' ($m_j = +1/2$, $I_Z=0$), respectively. Therefore, the deuterium atoms with only the hyperfine level of $I_Z=-1$ (state 1' in Fig. 2) has an opposite electron-spin state, $m_j = -1/2$, of the other two sub-levels (2' and 3') after Sona transition. When the alkali atoms in the ionizer are also optically pumped and their electrons are to be spin polarized in the $m_j = +1/2$ state, only deuterium atoms with the electron-spin state of $m_j = -1/2$ (state 1') can form negative ions because of the Pauli exclusion principle. This process is shown in Fig.3, schematically. The nuclear-spin state of the negative deuterium ions in this case is $I_Z=-1$, the nuclear vector polarization becomes -1. The nuclear tensor polarization is, in this case, -1. Using a proper rf transition simultaneously, a pure nuclear tensor polarization of -2 may become possible.

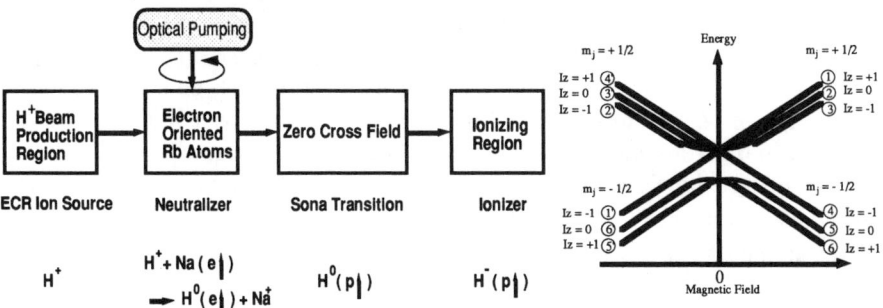

FIGURE 1. Block diagram of the dual optically pumped polarized negative deuterium ion source

FIGURE2. Hyperfine sublevels of deuterium atom in Sona transition

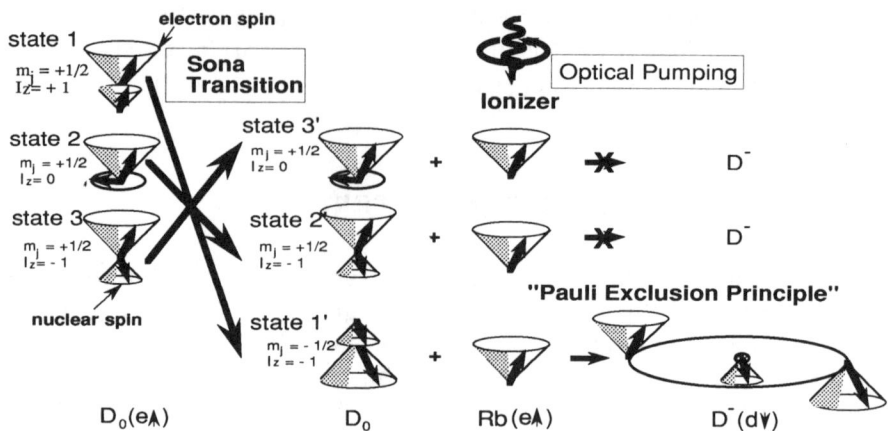

FIGURE 3. Principle of the dual-optically pumped polarized negative deuterium ion source

EXPERIMENT OF DUAL OPTICAL PUMPING

In order to check this dual-optically pumped scheme experimentally, it was possible to use a following simple method. The beam intensity of negative deuterium ions after the ionizer can be written in following equation.

$$I = \frac{I_{D^o} N_0}{6}(3 - P_{D^o} P_e^i)\sigma_0 \, l. \tag{1}$$

Here, I_{D0} is the intensity of electron-spin polarized deuteron atoms in the neutralizer, N_0 the density of the alkali atoms in the ionizer, P_{D0} the electron-spin polarization of deuteron atoms in the neutralizer, P_e^i the electron-spin polarization of the alkali atoms in the ionizer, σ_0 is the spin independent electron pick-up cross section to form negative deuterium ions and l the ionizer length, respectively. As can be clearly seen from this equation, the intensity of electron polarized deuterium ions is determined by both electron-spin polarization of the deuteron atoms after the neutralizer and the alkali atoms in the ionizer.

We define ε as follows,

$$\varepsilon \equiv \frac{I_{off} - I_{on}}{I_{off}}. \tag{2}$$

Here I_{off} and I_{on} are the beam intensities when the pumping laser for the ionizer is off and on, respectively. Thus, from eq.(1) ε can be expressed in the following form.

$$\varepsilon = \frac{1}{3} P_{D^0} P_e^{\ i}. \tag{3}$$

It is clearly seen from this equation that ε depends only on P_{D0} and P_e^i. When the electron-spin polarization directions of the alkali atoms in both neutralizer and ionizer are parallel, ε becomes positive. On the other hand, when the electron-spin polarization directions of the alkali atoms in each cell are anti-parallel, ε becomes negative.

Figure 4 shows the changes of the beam intensities of negative deuterium ions in each case that the electron-spin polarization directions of the alkali atoms in the neutralizer and the ionizer are parallel and anti-parallel, respectively. Rubidium atom was used for alkali atom in both neutralizer and ionizer. The upper trace shows the beam intensity of negative deuterium ions when the rubidium atoms in the ionizer are electron-spin polarized whose polarization direction is parallel to the polarization direction of the electron-spin polarized deuterium atoms after the neutralizer. The lower one corresponds to that when the polarization directions of rubidium atoms in the ionizer and deuterium atoms after neutralizer are opposite. The middle trace shows the beam intensity when the optical pumping of the ionizer is off. Apparently seen from this figure, the negative deuterium intensity is increased or decreased according to the polarization of rubidium atoms in the ionizer as expected from eq.(3).

From these experimental results, it was confirmed that this dual optical pumping scheme for generating highly polarized negative deuterium ions worked well.

From definition, the deuteron vector polarization (P_D^-) is written as follows,

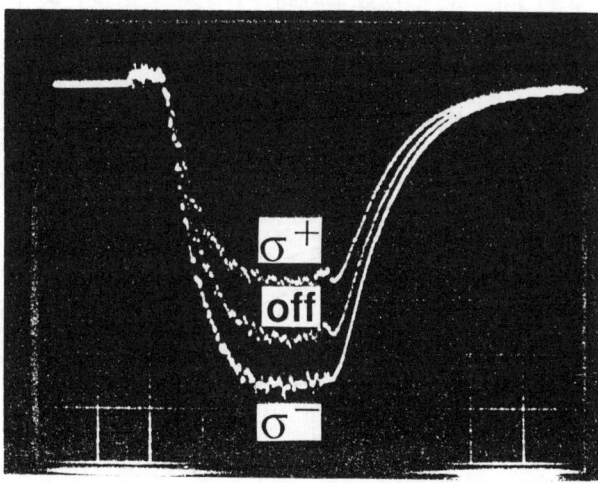

FIGURE 4. The changes of the beam intensities of negative deuterium ions

$$P_{D^-} = \frac{I_+ - I_-}{I_+ + I_0 + I_-}. \quad (4)$$

Here I_+ is the beam fraction of the deuteron nuclear spin component of $m_I=+1$, I_- the fraction of $m_I=-1$ and I_0 the fraction of $m_I=0$, respectively. With eq.(3), P_{D^-} is given by the following equation

$$P_{D^-} = \frac{-2\varepsilon}{P_e^i (1-\varepsilon)}. \quad (5)$$

The relationship between P_{D^-}, P_e^i and ε are shown in Fig. 5. With the measurements of P_e^i and ε, we can estimate the deuteron vector polarization of negative deuterium ions (P_{D^-}) from eq. (5).

The result of the experiment is also shown in Fig. 5. The closed circle in the figure shows the experimental result. The electron-spin polarization of alkali atoms in the ionizer (P_e^i) was measured with a Faraday rotation method. The error shown in the figure was measurement error of Faraday rotation.

We obtained $P_{D^-} = -0.70 \pm 0.05$ in the experiment. In the present apparatus, we have used a super conducting solenoid at the neutralizer where the magnetic field strength was 2.7 T. In the previous apparatus which used a normal conducting solenoid whose magnetic field strength was 1.2 T at the neutralizer, the obtained

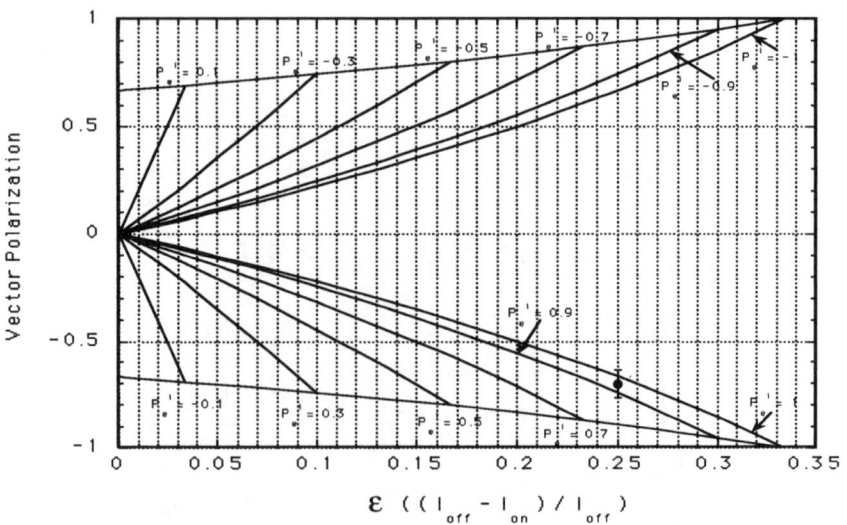

FIGURE 5. Relation between P_{D^-} and ε. The closed circle in the figure shows the experimental result.

nuclear-spin vector polarization was 55 %[8]. This is because the spin-orbit coupling in neutral deuterium atoms, which are created by picking up polarized electrons from the optically pumped alkali atoms in the neutralizer, reduces the polarization. When the magnetic field strength at the neutralizer is 1.2 T, the theoretical maximum polarization is limited to less than 80 %.

CONCLUSION

We are developing the dual optically pumped polarized negative ion source. Through the series of experiments, the difference of ion currents was measured by switching optical pumping on and off, and we obtained the following results;
(1) It was proved that the dual optical pumping method worked well as expected theoretically.
(2) 70 % of vector polarization of negative deuterium ion beam was obtained in the present experiment.

It was previously thought that the optically pumped polarized ion source was not useful for producing highly polarized deuterons. Our result open up a new possibilities for the optically pumped polarized ion source.

REFERENCES

1. S.Hiramatsu, et.al., AIP Proc., Protvino(1984)
2. W.L.Anderson., Nucl. Inst. and Meth. 167,363(1979).
3. Y.Mori, K.Ikegami, Z.Igarashi, A.Takagi,and S.Fukumoto., AIP Proc. 117, New York(1984)123.
4. R.L.York, O.B. Van Dyck, D.R.Swenson and D.Tupa: Proc. of the Int. Workshop on Polarized Ion Sources and Polarized Gas Jets, KEK Report 90-15(1990), 142.
5. L.Buchmann,C.D.P.Levy, M.McDonald, R.Ruegg, and P.W.Schmor: ibid 161.
6. A. Zelenskii, S.A.Kokhanovskii, V.G.Polushkin and K.N. Vishnevskii:ibid, 154.
7. M.B.Schneider and T.B.Clegg: Nucl. Inst.. and Meth.,A254,630(1987).
8. M.Kinsho, Y.Mori, K.Ikegami, and A.Takagi : Rev. Sci. Instrum., Vol. 65, No. 4, April (1994) 1388

Polarized Ion Source Operation at IUCF

V. Derenchuk, A. Belov[*], R. Brown, J. Collins, J. Sowinski,
E. Stephenson and M. Wedekind

Indiana University Cyclotron Facility, Bloomington, IN 47408, U.S.A.
[*]*Institute for Nuclear Research of the Russian Academy of Sciences,
Moscow, 117312, Russian Federation.*

Abstract. The IUCF high intensity polarized ion source (HIPIOS), based on the source in operation at TUNL (1) and employing cold (~30 K) atomic beam technology with an electron cyclotron resonance ionizer, has recently delivered beam to the first users. The results of the development work required to make the source operate reliably, with reasonable beam parameters are described. Methods used to measure the polarization and possible sources of unpolarized background are also discussed.

INTRODUCTION

During 1994, HIPIOS (Fig. 1) has been used to deliver beam to the first users. Source reliability at the highest intensities has improved considerably over this time period. No source maintenance is required for periods of at least two weeks with 200 µA delivered, after mass analysis, to the entrance of 600 kV acceleration column. The transition units and source tune are now being optimized to give higher polarization. Most recently, minor changes in the ionizer and extraction region design have resulted in the greatest improvements in source reliability. HIPIOS and its nominal operating parameters have been described in detail elsewhere (2).

ATOMIC BEAM SECTION

Atomic beam profile measurements, taken with a retractable compression tube mounted 10 cm upstream of the ECR ionizer entrance, have been analysed (Fig. 2). This data shows that the atomic beam is totally contained within a diameter of 22 mm and has an atomic beam density of $3.3*10^{11}$ atoms/cm^3.

Long term, reliable operation with constant intensity has been a major goal in development of the atomic beam section (ABS). In order to achieve this goal, we have used a time-of-flight (TOF) mass spectrometer (Fig. 3), designed by A. Belov and built at INR in Russia, to study the dissociated fraction of the hydrogen beam from the dissociator. During normal operation, the dissociation fraction was measured to be 88.5±0.5% (Fig. 4). As a result of these tests, we determined that

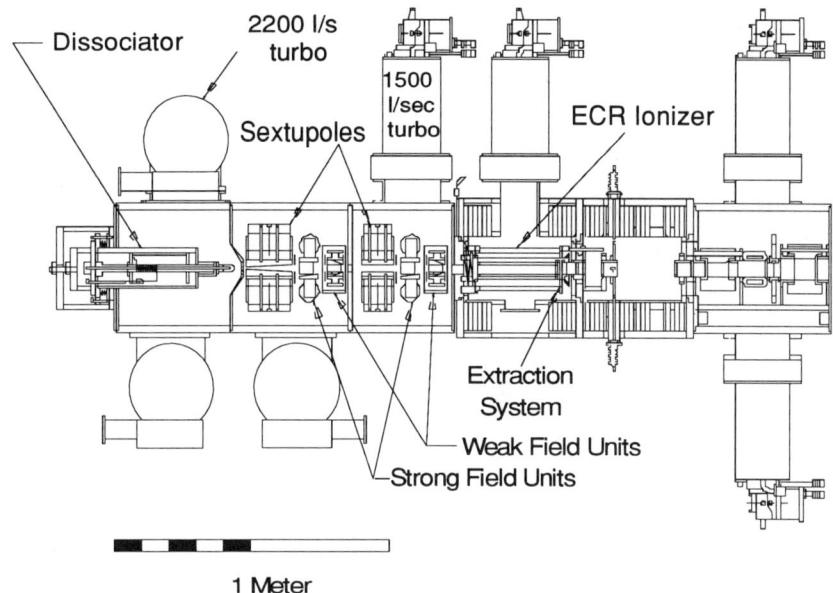

FIGURE 1. Schematic layout of HIPIOS.

the dissociated fraction (d.f.) can be correlated to the heat required to maintain the cold nozzle temperature of 30K and the total flux as measured in the compression tube. When the d.f. is low, the cold nozzle requires less heater power to maintain a constant temperature, most likely due to recombination of H^0 (an exothermic reaction) on the surface of the nozzle. We now use the heater power as a diagnostic to maintain a high dissociation fraction.

Due to improvements made during the past year, the ABS has operated for two weeks with a constant optimized intensity. We will be testing further modifications to the cold nozzle design, in hopes of increasing the peak performance lifetime.

ECR DEVELOPMENT

A gridded extraction system and quartz tube liner were added to the ECR ionizer in 1993. Initial results were presented at the Madison workshop (3). Since then, we have gained considerable operating experience with the extraction system operating in the accel-accel mode.

Initial extraction system tests were made with a mass analysis magnet followed by a pepper pot emittance measuring device. With an atomic beam intensity of $2.0*10^{16}$ atoms/sec at the compression tube, a beam current of 0.68 mA was

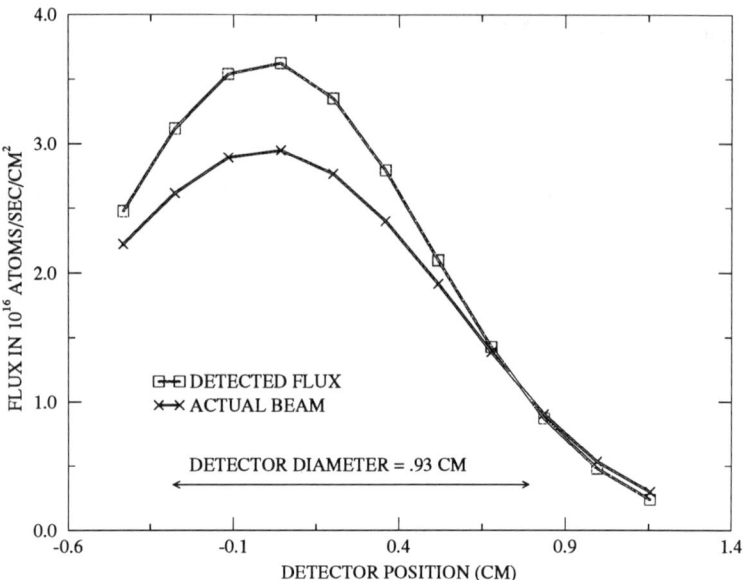

FIGURE 2. Flux measured 10 cm upstream of the ECR ionizer.

measured upstream of the mass analysis magnet with the sextupoles on and 0.18 mA with the sextupoles off. Mass analyzed beam currents between 200 and 250 µA were typical. Depending on the ionizer tune and ABS performance, the measured normalized emittance varied from 0.8 π-mm-mrad to 2.0 π-mm-mrad. After installation of the Terminal C beam transport line, over 200 µA has been measured at the entrance of the 600 kV acceleration column with a source extraction potential of 15.5 kV.

We have also been studying the long term reliability of the gridded extraction system. As suggested by T. Clegg, we are using a D_2 buffer gas for the ECR in order to reduce the sputtering rate of the molybdenum grids. It is expected that the life of the grids will determine the required maintenance period for the ionizer. The third grid had a lifetime of about two weeks before grid damage and plating of the insulators on which the grids are mounted causes sparking and beam instability. It has now been replaced by a tube element with no decrease in analyzed beam intensity. Beam optics calculations of a single gridded extraction system have been made (4) and preliminary results indicate that there should be no loss in source performance. Since the beam energy is defined by the first grid, sputtering is very slow.

The use of D_2 as a buffer gas was also suggested independently by A. Belov.

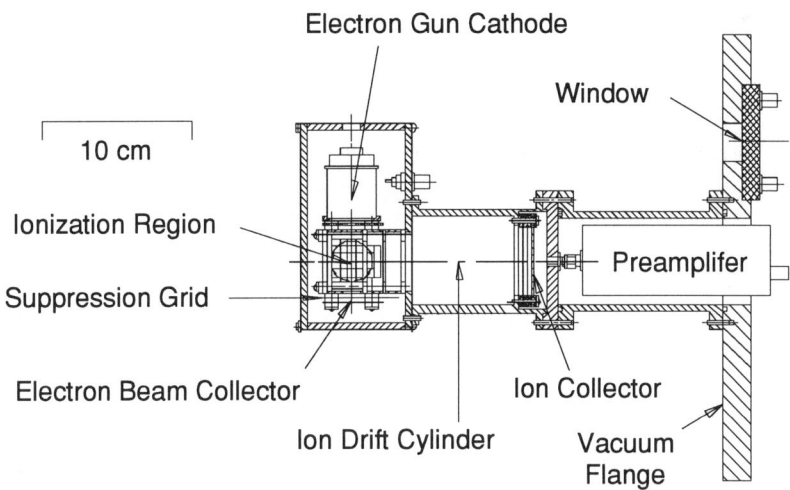

FIGURE 3. Time of flight mass spectrometer designed and built by A.Belov at INR, Moscow for use in HIPIOS. The atomic beam travels through the ionization region (into the page) and is ionized by a colliding electron beam. A voltage pulse accelerates the ions which are subsequently collected on the ion collector. Installed at the entrance of the ECR, it was used to measure the dissociation fraction of the atomic beam.

He proposed that the deuterium would increase the ionization efficiency by collisional charge exchange with atomic hydrogen. During recent operating experience, we have observed a 30% improvement in beam intensity delivered to the end of the Terminal C beam transport line. Also, kHz beam noise was reduced and short term stability improved by a factor of two when compared with N_2 buffer gas. This effect could be explained by the increase in supply of electrons to the plasma from the buffer gas.

Testing of the single-gap, ramp-waveform buncher in November 1993, led us to the conclusion that, due to the space charge forces of the beam being bunched ($\approx$ 700 μA of total beam), bunching was required after mass analysis and at a higher beam energy. A sinusoidal buncher has been installed near the entrance to the acceleration column and the ramp-waveform buncher removed.

POLARIZATION MEASUREMENTS

A nuclear spin filter type polarimeter is currently being commissioned in Terminal C. T. Clegg loaned IUCF the Nuclear Spin Filter used at TUNL for polarimetry of the TUNL polarized ion source beam (5). Advantages of using this polarimeter, as opposed to a Lyman α type low energy polarimeter, is its ability to measure both proton and deuteron polarization with an accuracy of better

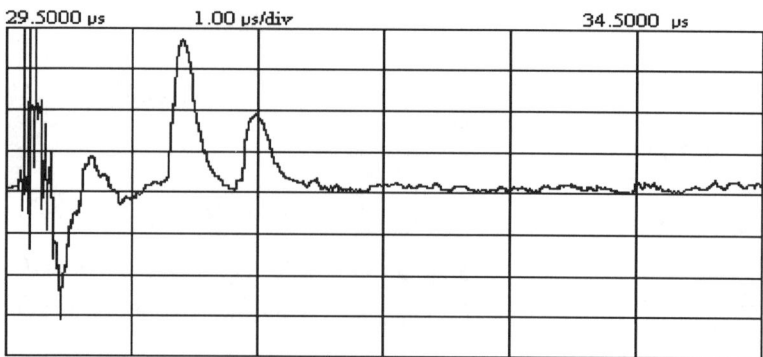

FIGURE 4. Oscilloscope trace of the time of flight spectrum from A. Belov's mass spectrum. The TOF was mounted immediately upstream of the HIPIOS ECR ionizer. The first large peak is the ionized H^0 and the peak to the right is H_2. The dissociation fraction was measured to be 76.0% in this case and 88.5% during long term optimal running conditions.

than a few percent. Both polarimeters are able to measure polarization 'real time'.

A lock-in amplifier technique, first used at PSI by P. Schmelzbach, was used to observe a small .01% change in beam intensity with the 1→3 and the 2→4 transition units. To see such a small change, we turned the units on and off at a rate of 80 Hz and fed the Stop 5 beam signal into a lock-in amplifier. With about 120 µA on the stop we observed a 5 nA to 10 nA change in the beam intensity when the units were tuned optimally. This technique may be used to tune the transition units before acceleration.

Tensor polarization measurements of a HIPIOS 386 keV deuteron beam have been made using a D(d,p)T polarimeter (6) in BL1C. This polarimeter was useful for tuning transition units that produced states with significant tensor polarization. Absolute measurements were extremely difficult as the results seemed to depend strongly on beam position and tune.

Deuteron beam from HIPIOS was also accelerated through the Injector cyclotron. The BL2 polarimeter was used to make absolute measurements of the efficiencies of transition units. It was found that both weak field units and the strong field 3→5 transition unit produced polarizations of about 65% to 75% of the maximum possible value. The strong field 3→6 transition after sextupole 2 produced an unexpected 3→5 transition. We did not observe a 3→6 transition and expect that the 3→6 electrodes are not excited correctly.

Proton beam has been accelerated through both cyclotrons and delivered to a user for the first time. One shift of development on the BL2 polarimeter was used to set and measure the polarization from both transition units. The polarization measured with the 1→3 transition unit was about 74%. The maximum polarization observed with the 2→4 transition unit was only 45%. The operation of the weak field 1→3 transition is very encouraging, especially for a first time measurement. We have discovered a RF power delivery problem for the strong field transition and anticipate that with further development it will perform equally well.

A second method of flipping the spin of the proton beam with a period of about 10 s was tested. Keeping the weak field 1→3 transition unit on, the spin of the 15 keV beam was rotated $180°$ by reversing the polarity of the spin rotation solenoid. Although measurements were made on the BL2 polarimeter using this spin reversal method, it created motion and beam intensity fluctuations at the experimenters target. Further development of this method is required before it can become useful for most users.

REFERENCES

1. Clegg, T.B., et al., "A New Atomic Beam Polarized Source for the Triangle Universities National Laboratory: Overview, Operating Experience and Performance," NIM, 1994.

2. Derenchuk, V., Wedekind, M., Brown, R., Ellison, T., Friesel, D., Hicks, J., Jenner, D., Pei, A., Petri, H., Schwandt, P., Sowinski, J., "Performance and Status of the IUCF High Intensity Polarized Ion Source", in *Proceedings from the 13th International Conference on Cyclotrons and Their Applications,* Vancouver, Canada, 1992, pp. 330-333.

3. Derenchuk,V.P., Brown,R., Wedekind,M., "Polarized Ion Source Development at IUCF," in *AIP Conference Proceedings 293, Polarized Ion Sources and Polarized Gas Targets,* Madison, WI, 1993, pp.72-75.

4. Schwandt, P., Private communication.

5. Lemieux, S.K., Clegg, T.B., Karwowski, H.J., Thompson, W.J., and Crosson, E.R., "A Nuclear Spin Filter Polarimeter", *NIM A 333*, p. 434 (1993).

6. Arview et al, *NIM A 273,* 1988, p. 48.

The Polarized Ion Source for COSY*

P.D. Eversheim[a], R. Gebel[a], M. Altmeier[a], O. Felden[a], C. Heimann[a],
M. Kammermann[a], W. Kretschmer[b], R. Weidmann[b], K. Mümmler[b],
B. Aumüller[b], A. Glombik[b], H. Paetz gen. Schieck[c], S. Lemaitre[c],
R. Reckenfelderbäumer[c], M. Eggert[c], O. Suttorp[c]

[a] *Institut für Strahlen- und Kernphysik, Universität Bonn, D-53115 Bonn, Germany*
[b] *Physikalisches Institut, Universität Erlangen, D-91058 Erlangen, Germany*
[c] *Institut für Kernphysik, Universität Köln, D-50937 Köln, Germany*

Abstract. The polarized ion source for the cooler synchrotron COSY-Jülich has been set in operation. The source produces pulsed $\vec{H}^-$ or $\vec{D}^-$ beams in an charge-exchange process. The working scheme of this colliding-beams source, its master equations, which control the performance of the source, and first results are discussed.

INTRODUCTION

With view to the experience and the intention of the collaboration to provide a polarized beam that compares with the available unpolarized beam intensity in COSY, a colliding-beams source (CBS) was built by three university groups from Bonn, Erlangen and Köln. This type of source produces polarized H^- or D^- ions that allow for a non Liouville stripping injection into the COSY ring. This injection scheme is about an order of magnitude more efficient compared to the alternative stacking injection of H^+ or D^+. In addition, the charge-exchange reaction $\vec{H}^\circ + Cs^\circ \to \vec{H}^- + Cs^+$ of the CBS works selective for atoms, so that some unpolarized molecular background can not reduce the carefully prepared polarization of the atomic beam. The source comprises three major components: the atomic beam source, the Cs-beam source, and the charge-exchange region (cf. Fig. 1).

THE ATOMIC BEAM SOURCE

The atomic-beam source produces an intense neutral polarized atomic hydrogen-beam. At first, gas molecules are dissociated in an inductively coupled 300-400 W RF discharge. The atoms are cooled to about 30 K, while passing the aluminum nozzle of 20mm length and 3 mm diameter. Therefore, the atoms are considerably

---

*This work was supported by the Forschungszentrum-Jülich and the BMFT, Germany

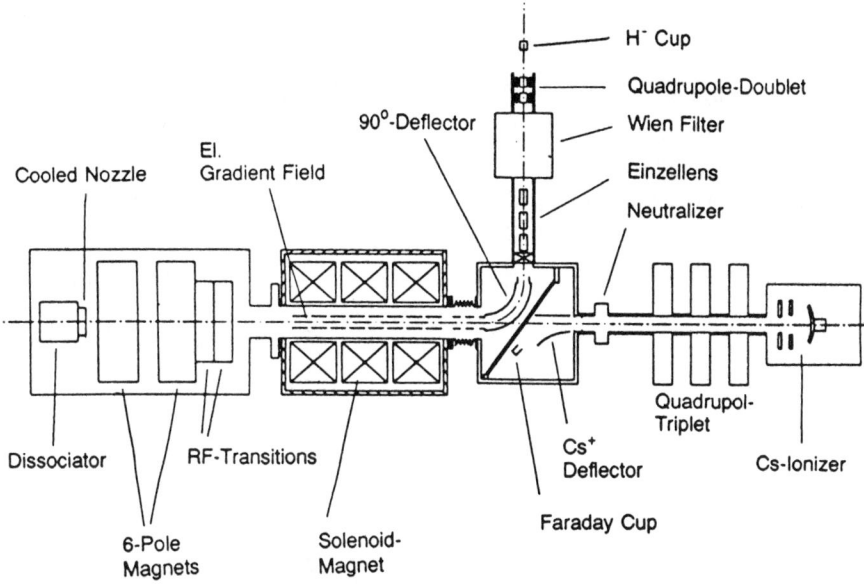

Fig. 1 The polarized ion-source for COSY

slowed down (1) with the consequences that i) shorter 6-pole magnets can be used, ii) the first tapered 6-pole magnet accepts an increased solid angle and iii) the dwell time of the atoms in the charge-exchange region is increased in proportion to the decrease of the beam velocity. These beneficial effects are reduced by gas scattering in- and outside the nozzle (2).

In order to optimize the transport into the charge-exchange region, the supersonic beam-velocity distribution offered to the 6-pole magnets has to be measured. In a next step these distributions are used as an input to a beam-transport simulation, which in turn helps to answer the following questions:
- How well is the transmission through the 6-pole magnets matched to the velocity distribution of the atomic beam (Fig. 2)?
- What is the best position of the second, the so called compressor magnet (Fig. 3)?
- Does the magnet design allow to adjust the maximum of the atomic beam density in the charge-exchange region (Fig. 4)?

At the position of the charge-exchange region an hydrogen intensity I_{H^0} of $I_{H^0} > 4 \cdot 10^{16}$ atoms/sec has been verified.

The RF transitions have been tested in the Bonn polarized ion source. Efficiencies in excess of 90% have been measured (3).

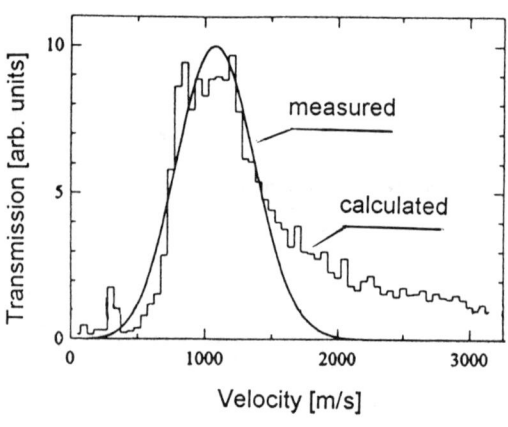

Fig. 2 Transmission through the 6-pole magnets for a "white" velocity distribution (calculated) and the velocity distribution for a nozzle temperature of 27 K (measured).

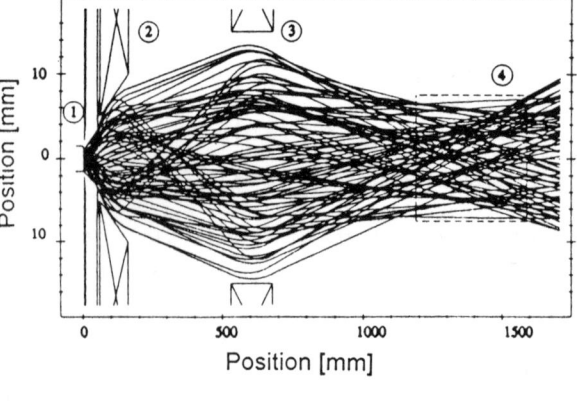

Fig. 3 Monte-Carlo Simulation. ① Nozzle, ② and ③ 6-pole magnets, ④ Charge-exchange region

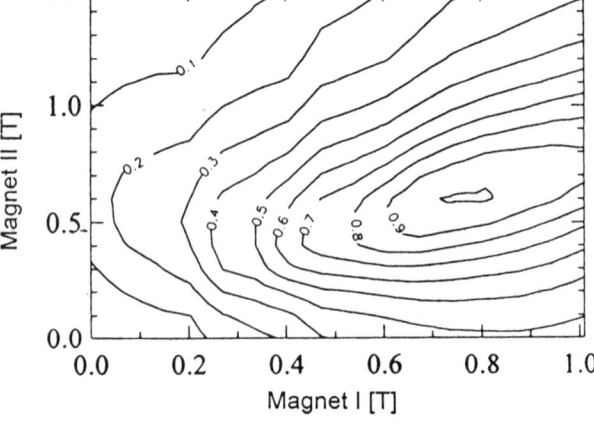

Fig. 4 Contour plot of H⁰ density in the charge-exchange region for 6-pole magnet I and II excitation

THE CESIUM BEAM SOURCE

The fast, neutral Cs^0-beam for the charge-exchange reaction in the solenoid is produced in two steps. First, Cs vapour is thermally ionized on a hot (1200 °C) porous W surface at an appropriate beam potential of about 40-60 kV. Since it turned out to be difficult (4) to transport this Cs^+-beam further than about 450 mm, this beam has to be focused into the charge-exchange solenoid by magnetic quadrupols (5). After having passed the solenoid, the Cs^+ intensity can be measured by a calorimeter. The Cs^+-beam emittance ϵ_{Cs} was measured to be less than $\epsilon_{Cs} \leq 65\ \pi$ mm mrad. According to Fig. 5, the Cs-beam source has a well defined perveance $1/\alpha$ for an optimum beam in the calorimeter. Therefore, the Cs^+ intensity I_{Cs^+} is related to the ionizer potential U_{But} by:

$$U_{But} = \alpha \cdot (I_{Cs^+})^{2/3} \qquad (1)$$

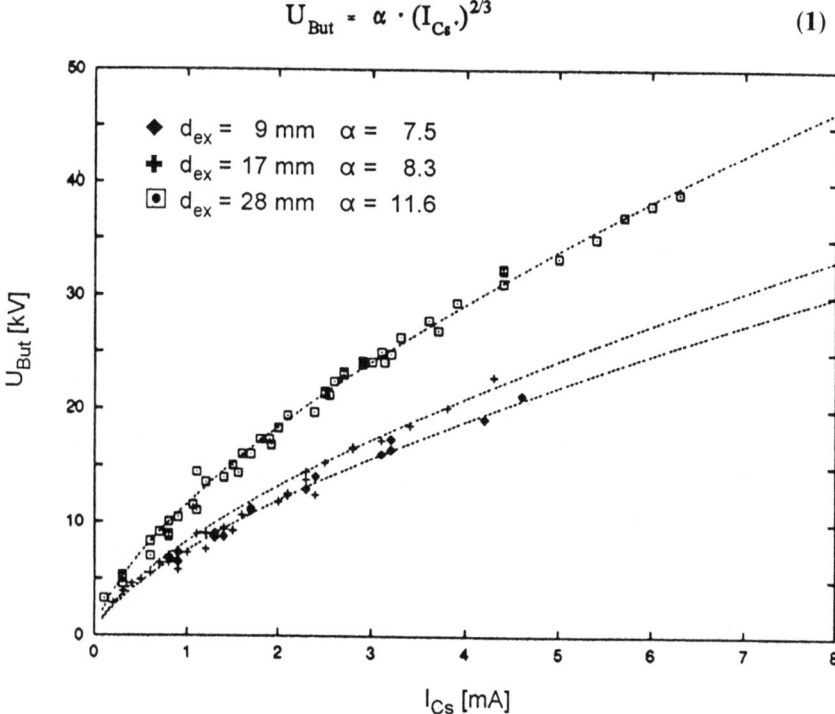

Fig. 5 Perveance $1/\alpha$ from $U_{But} = \alpha \cdot (Ics+)^{2/3}$ for optimum beam in the calorimeter and various extraction gaps d_{ex}

Usually, the charged Cs^+ beam is electrically deflected in front of the solenoid into a Faraday cup. Only just before the injection into the COSY ring is due, the neutralizer placed between the quadrupoles and the solenoid is filled with Cs vapour. The charged, fast Cs^+ beam becomes neutralized, and enteres the 400 mm long charge-exchange region inside the solenoid through an orifice in the 90°-deflector.

The neutralizer comprises a Cs oven, a cell that is filled with Cs vapour, and a magnetically driven flapper valve between the oven and the cell. With the help of a Langmuir-Taylor detector it could be shown that i) it takes about 150 ms after the flapper has been actuated until the Cs-vapour jet has filled the cell, and that ii) the width of this Cs jet is only 30 mm. The neutralizer efficiency η was measured to exceed 90 %.

THE CHARGE-EXCHANGE REGION

In the charge-exchange region of length L = 400 mm the neutral polarized hydrogen-beam with the velocity v and intensity I_{H^0} meets the fast neutral Cs beam with the intensity I_{Cs^0} in a mean cross section area F. With a given cross section σ for the charge-exchange reaction $\vec{H}^0 + Cs^0 \rightarrow \vec{H}^- + Cs^+$ the intensity of I_{H^-} the extracted H^- beam can be calculated:

$$I_{H^-} = I_{H^0} \cdot \frac{L \cdot \sigma}{F \cdot v} \cdot I_{Cs^0} = n_{H^0} \cdot L \cdot \sigma \cdot I_{Cs^0} \qquad (2)$$

In the charge-exchange solenoid the polarization is preserved by the longitudinal field $\vec{B}$ with its vector potential $\vec{A}$. Table 1 shows how much polarization can be expected for a given solenoidal field.

On the other hand the solenoid field defines the emittance of the beam. As a consequence of the conservation of total angular momentum $\vec{M}$ the "spacial part" $\vec{L}$ of the total angular momentum will grow to the extent that the H^- ions leave the solenoid field and the vector potential $\vec{A}$ becomes zero.

$$\vec{M} = e\vec{A}r + \vec{L} \qquad (3)$$

At this point the "magnetic part" $e\vec{A}r$ of the total angular momentum has been converted to a macroscopic spacial angular momentum of the beam. The resulting transversal emittance for protons is given by:

Table 1. Calculated maximum polarization for two states

Solenoid field [Gauss]	Max. polarization [%]
0	50
500	85
1000	94.5

$$\epsilon = 1.1 \cdot 10^{-2} \cdot \pi \cdot B\, r^2 \cdot 1/\sqrt{U_{ex}} \qquad (4)$$

[Gauss]; r [mm]; U_{ex} [kV]; ϵ_{H^-} [mm mrad]

For the nominal acceptance of the injector cyclotron $\epsilon_{H^-}\cdot\pi$ = 500 mm mrad U_{ex} = 4.5 kV, and r = 5.5 mm, B must not exceed 1000 Gauss. By varying the solenoid field, the transversal emittance ϵ_{H^-} can be traded for polarization.

The longitudinal phase space -or the energy spread of the beam- is tuned by the electrical drift field inside the solenoid. This field accelerates the $\vec{H}^-$ ions toward the extraction end. This effect can be increased by rising the magnetic field at one end of the solenoid, thus providing a magnetic mirror.

STATUS AND FIRST RESULTS

The polarized ion source for COSY has been set to operation. The H⁻ beam has been extracted and measured with an H⁻ Faraday cup behind the Wien filter. The Wien filter was tuned to select the H⁻ atoms and suppress the electrons. With an Cs ionizer potential of U_{ex} = 45 kV, the Cs source was operated to deliver 0.7 mA Cs⁺ within an area of 15 mm cross section. At the H⁻ cup 1.8 µA have be measured with 45 nA background. The RF transitions had been installed between the 6-pole magnets. Thus, when switching these transitions on- and off, an H⁻ beam polarization P of more than P = 90 % could be derived from an intensity modulation at the H⁻ cup. In addition the emittance ϵ_{H^-} of the H⁻ beam (at low solenoid field) turned out to be about $\epsilon_{H^-} \sim 3.6\ \pi$ mm mrad (MeV)$^{1/2}$.

REFERENCES

1) H.G. Mathews; Ph.D. Thesis, University of Bonn, 1979
   H.G. Mathews, A. Kruger, S. Penselin, A. Weinig; Nucl. Instr. Meth. **213** (1983) 155
2) W. Haeberli; Helv. Phys. Acta **59** (1986) 513
3) P.D. Eversheim et al.; AIP **293** (1993) 92
4) T. Wise, W. Haeberli; Nucl. Instr. Meth. **B6** (1985) 566
5) H. Paetz gen. Schieck et al.; AIP **293** (1993) 97

The HERMES-FILTEX target source for polarized hydrogen and deuterium [1]

F. Stock for the HERMES target group [2]

MPI für Kernphysik, Postfach 10 39 80, 69029 Heidelberg, Germany

Abstract: The performance of the FILTEX atomic beam source in the FILTEX test experiment is reported with final results on maximum target polarization $P_{max} = 0.84(0.02)$ and density $n_{max} = 1.09(0.04) \cdot 10^{14}$ atoms/cm^2. The modifications for the HERMES experiment are explained with their implications for further development. The new sixpole magnet system and a new set of rf-transitions for the HERMES atomic beam source are described. The rf-transitions show efficiencies $\geq 98\,\%$ at an input power of 2 W.

THE FILTEX ATOMIC BEAM SOURCE

The FILTEX [1] atomic beam source (ABS) provides a beam of nuclear polarized hydrogen atoms designed to feed an internal storage cell target. A thermal atomic beam is produced using a rf-dissociator followed by a cooled aluminum nozzle and two skimmers. Polarization is achieved by a combination of Stern-Gerlach permanent sixpole magnets for electron spin separation and a radio frequency transition for exchange of the occupation numbers of different hyperfine substates. The FILTEX ABS was described before in publications [2, 3]. The high output flow with two substates of $8.1(0.2) \cdot 10^{16}$ $\vec{H}$/s into the feed tube of the storage cell made the successful performance of the FILTEX test experiment at the Heidelberg Test Storage Ring possible. In αp-scattering at $E_\alpha = 27$ MeV a maximum areal target density in the storage cell (250 mm long, 11 mm diameter) of $n_{max} = 1.09(0.04) \cdot 10^{14}$ atoms/cm^2 with two substates and a maximum target polarization of $P_{max} = 0.84(0.02)$ with one substate were measured. At the point of the optimum figure of merit $P^2 \cdot n$ the polarization was $P = 0.80(0.02)$ with a target density of $n = 5.5(0.2) \cdot 10^{13}$ atoms/cm^2 at a storage cell temperature of $T_{cell} = 115(5)$ K. Polarization build up in the stored beam due to spin dependent attenuation by the polarized target was observed [4, 5].

---

[1] Supported under various contracts by the Bundesministerium für Forschung und Technologie, Bonn.
[2] D. Fick, U. Funk, G. Graw, B. Lorentz, B. Povh, M. Rall, F. Rathmann, K. Reinmüller, P. Schiemenz, E. Steffens, E. Wittmann

THE HERMES TARGET SOURCE FOR HYDROGEN AND DEUTERIUM

HERMES [6] will study the spin structure of the nucleons by deep inelastic scattering of longitudinally polarized electrons at about 30 GeV in the HERA electron ring from transversally or longitudinally polarized H, D or ^{3}He atoms. Like in the FILTEX test experiment, a storage cell target will be used. The hydrogen and deuterium will be provided by the FILTEX ABS which has been modified to meet the new requirements for the HERMES target. The ^{3}He atoms will be supplied from a laser driven source [7].

Basic Concept

For each target gas, vector and – for deuterium – tensor polarization with maximum values at both signs are essential for high precision data. Depolarizing interactions between the electron beam and the target atoms imply the use of a strong target holding field on the order of 0.1 T to preserve the target polarization [8]. At such fields the spins of electron and nucleus of the target atoms are decoupled. Therefore it is possible to use two hyperfine substates for each required target polarization. On the other hand, the strength of the holding field makes its quick reversal impossible. Hence, a reversal of the target polarization must be accomplished by the injection of different hyperfine substates. Table 1 lists the six different preparations of the atomic beam and the maximum possible target polarizations. The polarizations are achieved by suitable combinations of weak field (WF), medium field (MF) and strong field (SF) rf-transitions. This concept is chosen very close to a proposal of the Wisconsin group [9]. However, it differs in the use of a SF(3-5) transition for deuterium. By using this transition at most two transitions are needed at one time.

Table 1: The different preparations of the atomic beam for the required target polarizations at HERMES. The various rf transitions are called the strong field (SF), the medium field (MF) and the weak field (WF) transition.

states	hydrogen (states 1–4)		deuterium (states 1–6)			
	vector		vector		tensor	
after 1st set of sixpoles	1, 2	1, 2	1, 2, 3	1, 2, 3	1, 2, 3	1, 2, 3
1st transition	—	—	MF (3-4)	MF (3-4)	MF (1-4)	MF (1-4)
states in beam	1, 2	1, 2	1, 2, 4	1, 2, 4	2, 3, 4	2, 3, 4
after 2nd set of sixpoles	1, 2	1, 2	1, 2	1, 2	2, 3	2, 3
2nd transition	—	WF (1-3)	—	WF	—	—
states in beam	1, 2	2, 3	1, 2	3, 4	2, 3	2, 3
3rd transition	SF (2-4)	—	SF (2-6)	—	SF (2-6)	SF (3-5)
states in beam	1, 4	2, 3	1, 6	3, 4	3, 6	2, 5
P_z	+1	−1	+1	−1	0	0
P_{zz}	—	—	+1	+1	+1	−2

The Structure of the HERMES Target Source

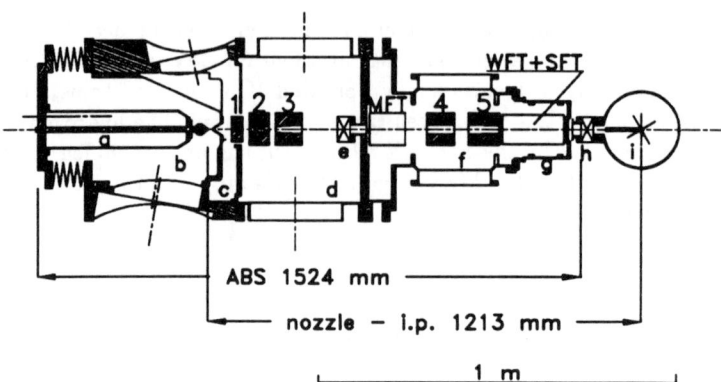

Figure 1: Vertical cut through the HERMES ABS along the atomic beam. The target chamber is also sketched. Details to be seen are the dissociator (a), the dissociator chamber (b), the skimmer chamber (c), the first sixpole chamber (d), the second sixpole chamber (f) with the extension (g) and the target chamber (i) with the storage cell. There are two valves to seal off the different chambers (e, h). The view is rotated by 30° into the horizontal plane.

Figure 1 shows a vertical cut through the HERMES ABS. The vacuum system with four pump stages is nearly identical to the FILTEX ABS. The second sixpole chamber had to be modified and is equipped with an extension, in order to house all rf-transitions and the second part of the sixpole system. The sixpole magnets are numbered 1 through 5, the rf-transitions (MFT and WFT + SFT) are indicated by rectangular boxes. The ABS is about 1.5 m long, the distance between nozzle and interaction point is about 1.2 m. Like the FILTEX ABS, the HERMES ABS is inclined by 30° out of the horizontal plane. For better visibility this inclination is omitted in figure 1.

New Developments for the HERMES Target Source

The modifications for the HERMES ABS emphasized the development of a new sixpole magnet system and a new set of rf-transitions for hydrogen and deuterium.

The sixpole system of the HERMES Target Source

The presence of a rf-transition between the last sixpole magnet and the feed tube of the storage cell is different from the FILTEX system. Because of the larger focal distance, a completely new design of the magnet system was necessary. For the design of the sixpole system a tracking code is employed which needs the velocity distribution of the atomic beam as an input data. Since there were no data on the velocity distribution of a deuterium atomic beam existing for the nozzle type in use, precise measurements for both deuterium and hydrogen were carried out. They resulted in parametrizations for the drift velocity and the atomic beam temperature as a function of the nozzle temperature:

$$v_{drift} = (1351 + 6.1 \cdot T_{nozzle[K]}) \text{ (m/s)}$$
$$T_{beam} = 0.290 \cdot T_{nozzle[K]} \text{ (K)} \quad \text{(hydrogen)} \quad (1)$$

$$v_{drift} = (1070 + 3.45 \cdot T_{nozzle[K]}) \text{ (m/s)}$$
$$T_{beam} = 0.267 \cdot T_{nozzle[K]} \text{ (K)} \quad \text{(deuterium)} \quad (2)$$

These parametrizations are valid for a dissociator throughput of $Q_{diss} = 1$ mbar·l/s, a nozzle temperature of 80 K $\leq T_{nozzle} \leq$ 180 K and a nozzle diameter of 2 mm.

An iterative algorithm was used to find an optimum sixpole system. The most important criteria of the design were a good focussing of deuterium state 1 atoms as well as the deflection of those atoms which are focussed in the first part of the sixpole system and then deflected after spin flip in the medium field transition. Table 2 lists the geometry of the HERMES atomic beam system, especially the dimensions of the new sixpole magnets.

The RF-Transitions for the HERMES Target Source

The new development for the rf-transitions was to create a very compact design with a very small extension in atomic beam direction in order to have a small distance between the last sixpole and the feed tube of the storage cell.

The rf-transitions consist of an electrical magnet for the holding field and a rf-resonator. The magnets are designed as window frame magnets with two pairs of

Table 2: The beam forming system and the sixpole system for the HERMES atomic beam source. The medium field transition is located between sixpoles 3 and 4, the weak field and the strong field transition are located between sixpole 5 and the feed tube of the storage cell.

	outer diameter [mm]	inner diameter [mm]	length [mm]	distance [mm]
nozzle	–	2	–	15
skimmer	–	5	–	35
collimator	–	6.4	–	13
1. sixpole	60	8.6 – 10.4	30.0	20
2. sixpole	80	12.0 – 16.0	55.0	20
3. sixpole	80	16.7 – 25.0	75.0	350
4. sixpole	80	26.6	75.0	40
5. sixpole	80	26.6 – 22.0	85.0	300
feed tube	–	10	100	

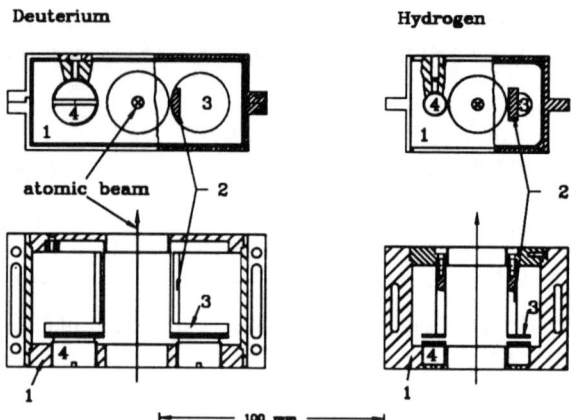

Figure 2: To scale drawing of the strong field resonators for hydrogen (right) and deuterium (left). Views along (above) and perpendicular (below) to the atomic beam are shown. The resonator rods (2) are bolted and soldered into the frame (1). The rods end in a condensator plate (3). The opposite plate (4) can be continuously adjusted for matching.

coils. One pair of coils produces a homogenious field on which is superimposed a gradient field provided by the second pair of coils. The magnet for the MF-transition is 100 mm long. A second magnet is 170 mm long and houses the WF- and the SF-transition. WF- and MF-transitions use copper coils, for the SF-transitions quarterwave resonators are used. The length of the resonator for hydrogen is 37 mm inside and 53 mm in total. For deuterium it is 44 mm and 58 mm. The opening for the atomic beam is 25 mm in diameter for both resonators. All components are mounted inside the vacuum. Figure 2 shows the quarterwave resonators for hydrogen and deuterium.

Table 3: Results of the efficiency measurements with the rf transitions of the HERMES–ABS. The error for the efficiencies is ± 1 %. The input power is in all cases 2 W with a reflected power of ≤ 0.05 W.

transition			signal ratio on/off		efficiency [%]	frequency [MHz]
			calculated	measured		
H	SF	(2-4)	0.623	0.634	98.2	1425.2
	WF	(1-3)	0.597	0.606	98.5	14.0
D	SF	(2-6)	0.728	0.733	99.2	341.5
	SF	(3-5)	0.742	0.741	100.2	341.5
	WF		0.754	0.750	100.5	6.0
	MF	(3-4)	0.683	0.694	98.5	22.3
	MF	(1-4)	0.683	0.694	98.5	22.3

Test of the HERMES RF-Transitions

To test the rf-transitions, the atomic beam of the HERMES ABS was analyzed with a 200 mm long sixpole magnet and the atomic flow was measured with a compression tube. The measured signal ratios of the pressure in the compression tube were compared with the calculated ratios yielded by the tracking code. In this test, a provisional sixpole system was used. The results of the efficiency measurements are listed in table 3. All transitions show efficiencies better than 98 % at an input power of 2 W and a reflected power of ≤ 0.05 W.

CONCLUSION AND FUTURE ACTIVITIES

During the FILTEX test experiment, the HERMES-FILTEX ABS has proved to be very reliable under longterm operation. The new sixpole system in combination with the new set of rf-transitions can provide the maximum possible variety of vector and tensor polarization for the HERMES target. For the HERMES system a hydrogen flow into the feed tube of about $6.6 \cdot 10^{16}$ atoms/s with a polarization of about 0.98 is expected. The ABS was taken to DESY at the end of August 1994 and the sixpole system was delivered in September 1994. Tests of the ABS including a new setup of the nozzle cooling for temperatures lower than liquid nitrogen temperature are scheduled for October 1994.

ACKNOWLEDGEMENTS

The developments at the HERMES ABS are to a large extent the work of Bernd Lorentz for the velocity distribution measurement and the design of the sixpole system, Eva Wittmann for the magnets of the rf-transitions and Martin Rall for the strong field resonators of the rf-transitions. Thanks are due to the whole HERMES target group, especially F. Rathmann and E. Steffens. Last but not least the support and inventiveness of the workshops at the MPI with V. Mallinger, K. Hahn and H. Fuchs is highly appreciated.

REFERENCES

1. H. Döbbeling *et al.*; Proposal CERN/PSSC/85-80 (1985) and Addendum (1986).
2. F. Stock; Proc. Workshop on Pol. Ion Sources and Pol. Gas Targets, Madison 1993, eds. L.W. Anderson und W. Haeberli; AIP 293 (1994) 22.
3. F. Stock *et al.*; Nucl. Instr. Meths. A 343 (1994) 334.
4. F. Rathmann *et al.*; Phys. Rev. Lett. 71 (1993) 1379.
5. F. Rathmann; contribution to this conference.
6. HERMES-Collaboration; *Proposal*; DESY - PRC 90/91 (1990).
7. K. Lee; Phys. Rev. C70 (1993) 738.
8. HERMES-Collaboration; *Technical Design Report*; DESY - PRC 93/06 (1993).
9. A.D. Roberts *et al.*; Nucl. Instr. Meths. A 322 (1992) 6.

POLARIZED INTERNAL GAS TARGET FOR HYDROGEN[*] AND DEUTERIUM AT THE IUCF COOLER RING

T. Wise, W. Haeberli, B. Lorentz, F. Rathmann, and M. A. Ross
University of Wisconsin, Madison, Wisconsin, USA
W. A. Dezarn, J. Doskow, J. G. Hardie, H. O. Meyer, R. E. Pollock, B. von Przewoski, T. Rinckel, and F. Sperisen
Indiana University and Indiana University Cyclotron Facility, Bloomington, Indiana, USA
P. V. Pancella
Western Michigan University, Kalamazoo, Michigan, USA

Abstract. A polarized internal H gas storage cell target has been successfully operated in the IUCF cooler ring. A target thickness for atomic hydrogen has been measured to be $3.5 \pm 0.3 \times 10^{13}/cm^2$ in a single spin state. Target polarization (0.75 ± 0.01) showed no sign of deterioration after two weeks of operation with an average beam current of 100 µA. Detection of p-p elastic events at 200 MeV bombarding energy has been demonstrated by use of coincidences between silicon detectors near the beam and a forward detector array. This target is presently being used to measure p-p elastic scattering spin correlation coefficients at 200 MeV.

INTRODUCTION

We describe the characteristics of an internal polarized hydrogen gas target that has been installed and tested in the IUCF storage ring. Figure 1 shows a schematic of a typical storage cell. Spin polarized atoms enter the cell through a fill tube and execute multiple wall collisions before their escape. These collisions enhance the target thickness by a factor of several hundred over a gas jet target. The storage cell principle was first tested in Madison in 1980 [1,2].

ATOMIC BEAM SOURCE

The Wisconsin atomic beam source [3] provides the spin polarized hydrogen beam that is injected into the cell. Figure 2 shows the arrangement of the source. The beam is formed by passing cooled atomic gas through permanent six-pole separation magnets with 1.5 T pole tip field. The separation magnets focus states 1 and 2 and eliminate states 3 and 4 from the beam. For hydrogen, the beam intensity that enters the fill tube has been measured to be 6.7×10^{16} atoms/s with hyperfine states 1 and 2 included in the beam. Three features of the permanent magnets lead to the observed high beam intensity: the high pole tip field strength, the taper in the bore of the first four elements and segmentation into short individual elements. The segmentation improves the vacuum in the interior of the magnet bore thereby reducing scattering losses.

---

[*] Work supported in part by the U. S. National Science Foundation and the U. S. Department of Energy.

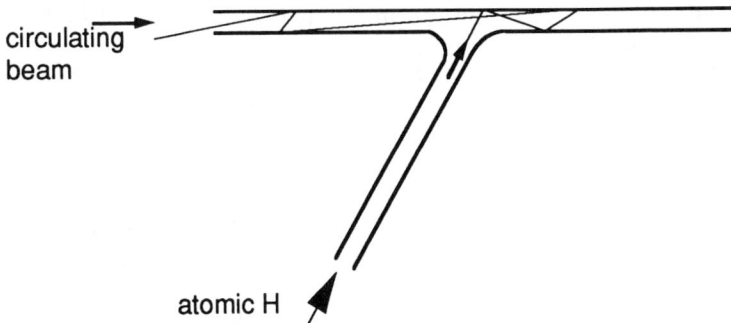

FIGURE 1. Schematic of typical storage cell. One possible path for a stored hydrogen atom is illustrated.

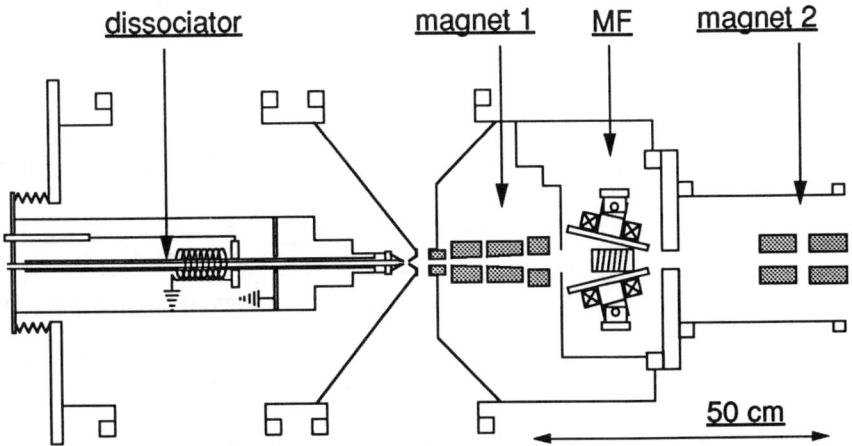

FIGURE 2. Schematic of the Wisconsin atomic beam source.

For present purposes, free choice of the polarization direction is required, which makes it necessary to use a weak guide field (a few Gauss) to avoid interference with ring operation. In that case the polarization of state 2 atoms approaches zero. A medium field 2↔3 RF transition [4] has been developed to remove state 2 atoms from the atomic beam so that the beam will consist mainly of atoms in state 1 after passing through the second group of six-pole magnets. State 1 atoms have nuclear polarization 1 independent of guide field strength.

In this case the expected atomic beam polarization will depend on the efficiency of the medium field transition, on the rejection of state 3 atoms by the second magnet group and on the magnitude of the guide field over the target region. The beam polarization is given by the relation $P = \frac{1-\varepsilon a - (1-\varepsilon a)R}{1+e+(1-\varepsilon)R} = 0.87$ where $(1-\varepsilon) = 0.96$ is the measured efficiency of the medium field RF transition, $R=0.05$ is the fraction of state 3 atoms remaining after passing the second magnet group as calculated by a transport code and a=0.02 is the polarization of state 2 atoms in a 1mT guide field. In the absence of depolarizing effects from interaction of atoms with the walls of the target or with the circulating beam the value of $P = 0.87$ is an upper limit to the target polarization.

At the same time the atomic beam entering the cell will be reduced by the factor $0.5(1+\varepsilon+R(1-\varepsilon))$ yielding a flux of 3.6×10^{16} atoms/s into the feed tube.

STORAGE CELL

An end view of the target cell assembly is shown schematically in Fig. 3. The cell is formed by assembly of four 25 cm long quadrants. Each quadrant has a 25 cm long 5 micron thick Teflon foil stretched across two aluminum knife edges separated by 8 mm. When assembled, the four foils enclose a volume with 8 mm square cross-section, open to the beam at each end. The foils also function as windows for low energy recoil particles. To allow for horizontal mounting of the atomic beam the windows are mounted 45 degrees from the vertical. To allow for passage of the atomic beam into the interior of the cell two adjacent knife edges are flared in the center to form part of a 10 mm diameter, 130 mm long fill tube, not shown in Fig. 3. In addition, the fill tube is pointed 30 degrees upstream from a line normal to the circulating beam to provide room for detection of forward particles. For studies of systematic errors, a Teflon capillary tube has been inserted into the fill tube near its intersection with the cell. The capillary is connected to a variable gas leak to make an unpolarized target of controlled thickness. More detail on the construction and performance of this cell can be found in Ref. 5.

Recoil particles are detected by eight silicon strip detectors, two detectors per window. Each detector is 5 cm from the beam axis and has an azimuthal acceptance of 40°. The detectors are divided into 28 strips each. A resistive type position readout permits vertex reconstruction with about 2mm resoluton along the beam direction.

Teflon was chosen for the cell windows for its known small depolarization of atomic hydrogen [2,6,7]. The foils are cut from commercially available rolls of 5μm PTFE Teflon [8]. The foil thickness was chosen to permit passage of low energy recoils through the Teflon and into the silicon strip detectors. Recently the

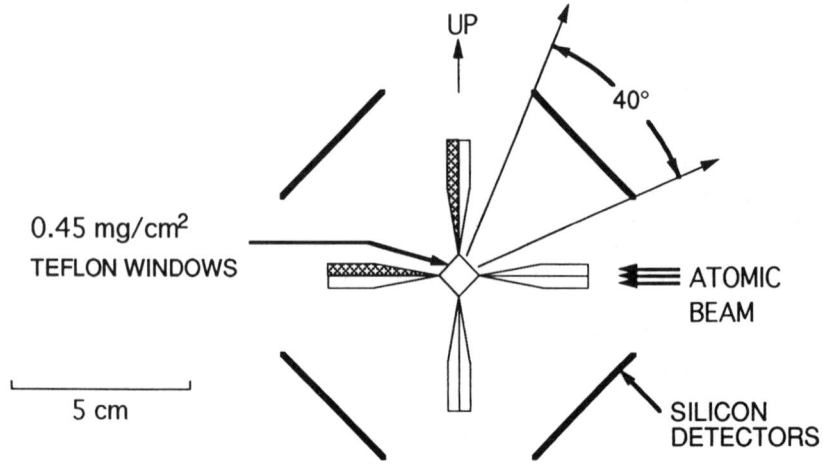

FIGURE 3. End view schematic of target cell construction. One quadrant is shaded.

foil material has been rolled to a thickness of 2 μm and successfully installed onto a target cell. For the case of p-p elastic scattering at 4° forward angle the low energy recoil particles are incident on the Teflon window at 0.97 MeV and have 0.88 MeV remaining after passing through a wall of this thickness. P-P elastic events are distinguished from background by requiring a coincidence between the forward particle detector and a silicon detector. Forward particles at four degrees are at the acceptance limit of the current downstream detector array. The thickness of the exit windows does not limit elastic scattering detection at this time.

TARGET THICKNESS

The target thickness can be calculated from gas conductance calculations and the atomic beam flux. Using the H flux in a single spin state and the room temperature cell conductance for an ideal geometry we calculate a conductance for mass 1 atoms of 11.6 liter/s. In practice the gas conductance of the cell is approximately 10% higher than this due to the required flaring at the center of the cell. The calculated target thickness is then $3.5 \times 10^{13}/cm^2$. The calculated density at the center of the cell is $2.8 \times 10^{12}/cm^3$ and varies linearly to zero at the ends.

The target density has been measured directly by comparison with an H_2 target of known thickness. Observed count rates were corrected for differences in the circulating beam current. The measured thickness is $3.5 \pm 0.3 \times 10^{13}/cm^2$ in agreement with the calculated value. The error is due to an uncertainty in the circulating beam current and an uncertainty in the H_2 target thickness that arises from the injection geometry of the H_2 gas.

GUIDE FIELD

The direction of the target polarization is defined entirely by the direction of a weak guide field over the cell. For current and proposed experiments it is necessary to have free choice of the polarization direction. For this purpose coils have been constructed to provide fields that can be independently directed along three orthogonal directions --vertically, horizontally or along the proton beam. The coils are located outside the vacuum vessel to avoid interference with the acceptance of the detectors. Because of geometrical constraints the field strength varies by a factor of nearly two along the cell but the field direction is uniform along the cell axis provided the coils are accurately aligned. The guide fields required to obtain maximum polarization are about 0.5 mT averaged along the cell. The horizontal coils are shown in Fig. 4.

The vertical and horizontal guide fields distort orbit of the stored beam. Such distortions are undesirable because they may introduce systematic errors in phase with changes in the target polarization. The perturbation to the ring orbit has been minimized by the addition of compensation coils up and downstream of the target. The guide field extends into the region of the ABS and can affect the efficiency of the MF transition by changing the field strength at the RF coils. Careful measurement of the relative target thickness for opposite signs of the horizontal holding field showed a modulation of the target thickness in phase with a sign change of the coil current. This effect was eliminated by shunting a fraction of the horizontal field current into auxiliary windings on the MF magnet to cancel the field modulation. Finally, stray fields from nearby equipment were canceled to first order to permit use of weaker guide fields.

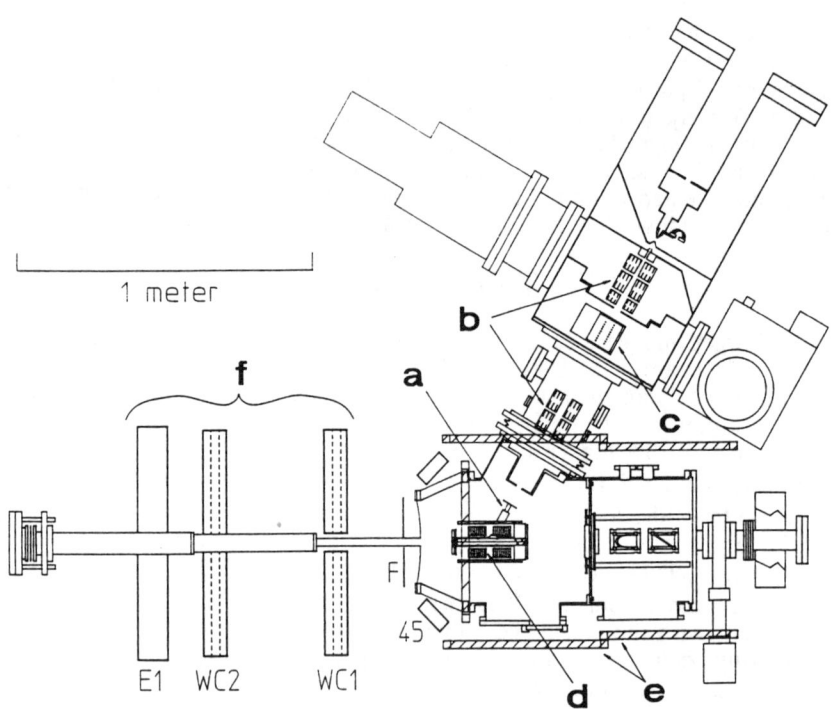

FIGURE 4. Target cell, atomic beam source, and detectors installed in a straight section of the IUCF cooler ring. a) cell feed pipe, b) separation six-pole magnets, c) medium field RF transition, d) silicon strip detectors, e) horizontal holding field coils, f) forward particle detector array.

POLARIZATION

Target polarization was measured by making use of the known pp analyzing power at 200 MeV averaged over 6°-18° and weighted by cross-section. A coincidence was required between the silicon detectors (low energy recoils) and the forward particle detector array. The measured target polarization is 0.75 ±0.01 and 0.73 ±0.01 for vertical and horizontal guide fields respectively. Target polarization showed no sign of deterioration after ten days of operation with beam currents of up to 100 µA. A peak current of 500 µA has passed through the cell.

DEUTERIUM

A polarized deuterium gas target may also be produced with this hardware. To obtain large vector and tensor polarization it is necessary to insert RF transitions in the drift region between the last 6-pole magnet and the storage cell fill tube. As with hydrogen it is necessary to operate with a weak guide field over the cell. A proposed scheme of RF transitions is shown in Table 1.

DEUTERIUM	VECTOR		TENSOR	
states populated after 1st sixpole	1+2+3	1+2+3	1+2+3	1+2+3
transition #1 (MF)	MF $3\leftrightarrow4$	MF $3\leftrightarrow4$	MF $3\leftrightarrow4$ & $2\leftrightarrow3$	MF $3\leftrightarrow4$ & $2\leftrightarrow3$
states populated after transition #1	1+2+4	1+2+4	1+3+4	1+3+4
states populated after 2nd sixpole	1+2	1+2	1+3	1+3
transition #2 (WF)	OFF	WF	OFF	OFF
states populated after transition #2	1+2	3+4	1+3	1+3
transition #3 (MF)	OFF	OFF	MF $3\leftrightarrow4$	MF $3\leftrightarrow4$
final states populated	1+2	3+4	1+4	2+3
P_z	+2/3	-2/3	0	0
P_{zz}	0	0	+1	-1

TABLE 1. RF transition arrangement to produce a vector or tensor polarized deuterium target in a weak guide field.

REFERENCES

1. M. D.. Barker et al. in "Polarization Phenomena in Nuclear Physics." AIP Conf. Proc. **69**, 931 (1981).
2. W. Haeberli in "Nuclear Physics with Cooled Stored Beams." AIP Conf. Proc. **128**, 251 (1984).
3. T. Wise, A. .D. Roberts, and W. Haeberli, *Nucl. Inst. Meth. A* **336** (1993).
4. A. D.. Roberts, P. Elmer, M. A. Ross, T. Wise, and W. Haeberli, *Nucl. Inst. Meth. A* **332** (1992).
5. M. A. Ross, A. D. Roberts, T. Wise, W. Haeberli, W. A. Dezarn, J. Doskow, H. O. Meyer, R. E. Pollock, B. v. Przewoski, T. Rinckel, F. Sperisen, P. V. Pancella, *Nucl. Inst. Meth. A* **344** (1994).
6. W. Haeberli, in *Frontiers of High Energy Spin Physics*, T. Hasegawa et al. eds., (Universal Academy press, Inc., 1993) p.335.
7. J. S. Price, and W. Haeberli, *Nucl. Inst. Meth. A* **349**, 321 (1994).
8. Supplied by Goodfellow Corporation.

A high Density polarized Hydrogen Gas Target for Storage Rings*

Kirsten Zapfe†, B. Braun‡, H.-G. Gaul, M. Grieser, B. Povh,
M. Rall, E. Steffens, F. Stock, J. Tonhäuser

Max-Planck-Institut für Kernphysik, 69029 Heidelberg, Germany

C. Montag†, F. Rathmann§, D. Fick

Philipps-Universität, Fachbereich Physik, 35032 Marburg, Germany

W. Haeberli

Department of Physics, University of Wisconsin, Madison 53706, WI, USA

Abstract. A storage cell target of polarized hydrogen gas was installed in a storage ring to study the target characteristics (nuclear polarization, target thickness, radiation resistance). The density of the target gas was enhanced by cooling the cell. Using a weak transverse guide field (5 G), the nuclear polarization of the target was measured to be 0.80 ± 0.02 when atoms in a single hyperfine state were selected. The areal density of the target under these conditions was $5.5 \cdot 10^{13}$ $\vec{H}/cm^2$ ($T = 115$ K), while for two spin states (applicable to experiments in high energy rings, where a strong magnetic field can be applied to the target) the target thickness was found to be $8.2 \cdot 10^{13}$ $\vec{H}/cm^2$.

Introduction

Over the last few years intense interest has developed in the use of polarized gases as internal storage cell targets in particle storage rings in nuclear and high-energy physics (1). Experiments using this novel technique have advantages over corresponding experiments with conventional solid polarized targets: gas targets of H and D atoms can be prepared in chemically and isotopically pure form, and thus avoid the background from the large fraction of inert nucleons such as the background of N in $N\vec{H}_3$ targets. If the gas target is produced by the atomic-beam method, the sign of the target polarization can be reversed in a fraction of a second by switching RF transition units which change the population of the hyperfine states.

---

*Supported by the Bundesministerium für Forschung und Technologie (Germany) under various contracts.
†now at Deutsches Elektronen-Synchrotron DESY, 22603 Hamburg, Germany.
‡now at Sektion Physik der Universität München, 85748 Garching, Germany.
§now at Department of Physics, University of Wisconsin, Madison 53706, WI, USA.

Similarly, for deuterons, rapid change between vector and tensor-polarized target can be achieved. In addition gas targets are not subject to the kind of radiation damage that limits the permitted beam dose on solid targets. Furthermore they are thin enough to cause negligible energy loss for beam or reaction products, so that even low energy reaction products can be detected.

One example of the use of internal polarized gas targets is the HERMES experiment at DESY, which is designed to study the spin structure of the nucleon in a complete and precise way (2). It will measure spin dependent deep inelastic scattering off the proton and neutron using the longitudinally polarized electron or positron beam of the HERA storage ring at a beam energy of 27.5 GeV. Polarized internal gas targets of hydrogen, deuterium and ^{3}He will be used with both longitudinal and transverse polarization. The physics program will first focus on precise measurements of the spin structure functions $g_1(x,Q^2)$ and $g_2(x,Q^2)$ of the proton and the neutron with high statistical and systematic accuracy. This allows a precise determination of the Bjorken sum rule, the Ellis-Jaffe sum rule for the proton and the neutron and the Burkhardt-Cottingham sum rule. Using deuterium as a spin-1 target, the new spin structure functions $b_1(x,Q^2)$ and $\Delta(x,Q^2)$ are accessible. A second focus of HERMES are the different flavour contributions to the nucleon spin which can be separated using semi-inclusive asymmetries with final state hadron detection. Additionally, several measurements with unpolarized targets have been proposed, e.g. measuring the flavor asymmetry of the quark-sea, the flavor distribution of the valence quarks etc. The HERMES experiment is presently assembled and will start data taking in spring 1995.

The polarized Hydrogen Target

In the following tests of a high-density polarized hydrogen gas target in the low energy heavy ion storage ring (test storage ring TSR) at the Max-Planck-Institut for Nuclear Physics in Heidelberg are reported (3). The results were an important step towards the final approval of the HERMES experiment where the same source for polarized hydrogen and deuterium will be used. α-particles were used to determine the target thickness and the polarization of the target protons because $p\alpha$ elastic scattering has a large and well known proton analyzing power (4). Further, the feasibility of the filter method, i.e. polarizing a stored beam by the interaction with the internal target was demonstrated the first time using a proton beam as reported in Ref. (5).

In the present experiment, polarized hydrogen atoms produced by an atomic-beam source (6) were injected through a 10 mm diameter and 100 mm long entrance tube into a target tube of 250 mm length and 11 mm diameter. The storage cell was coated with a thin layer of teflon to reduce depolarization of atoms by wall collisions (7) and was placed in one of the straight sections of the storage ring. A circulating beam of 27 MeV α-particles was stored in the ring and cooled by electron cooling in order to compensate for emittance blow up and energy loss of the beam caused by the target. The density of the target gas was enhanced by cooling the cell walls to temperatures as low as 80 K. A weak guide field (5 G) over the target defined the orientation of the target spin. Recoil protons, after penetrating the 0.2 mm thick Al wall of the cell and a 0.125 mm thick Kapton window, were detected in scintillation counter telescopes located at $\Theta_{lab} = 21°$. The analyzing power at this angle, averaged over the acceptance of the detector, is $A = 0.84 \pm 0.02$ (4). A detailed description of the scattering chamber and the detector system is found in Ref. (8).

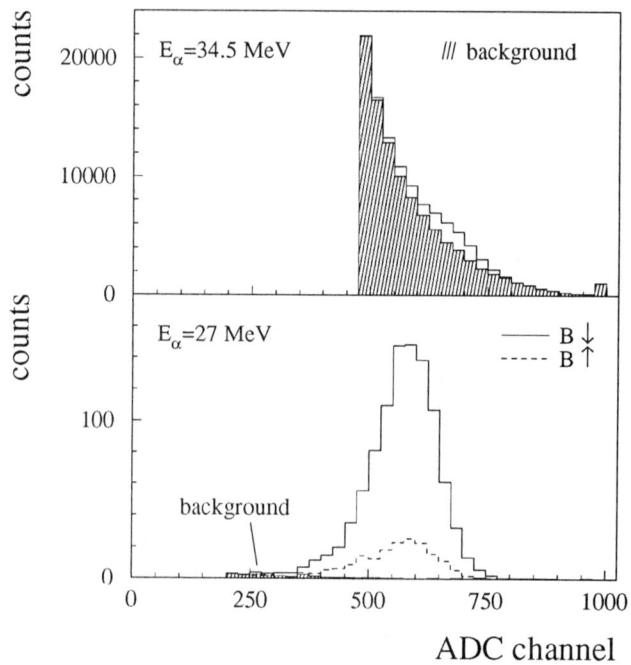

Figure 1: Illustration of the improved beam quality and background conditions resulting from use of an internal target in the storage ring. The top graph shows the pulse height distribution of recoil protons when an external α-beam from the tandem accelerator was used to bombard the target. In the bottom graph the same target is used as an internal target in the storage ring. No background is visible. Additionally the graph shows the large change in count rate from reversal of the target polarization.

Results

A pulse height spectrum of recoil protons in the left detector for the magnetic guide field up and down is shown in the bottom part of Figure 1 for one of the recorded runs. Almost no low-energy background is visible in the pulse height spectrum. This is in striking contrast to tests done earlier with the same storage cell using the external beam of the tandem accelerator instead of the beam stored in the ring (top part of Figure 1) (8). Only after great effort in beam collimation was it possible to detect the recoil protons from the gas target among the large background of other particles. Thus illustrates the great improvement in beam quality that results from storing and cooling the α-particle beam.

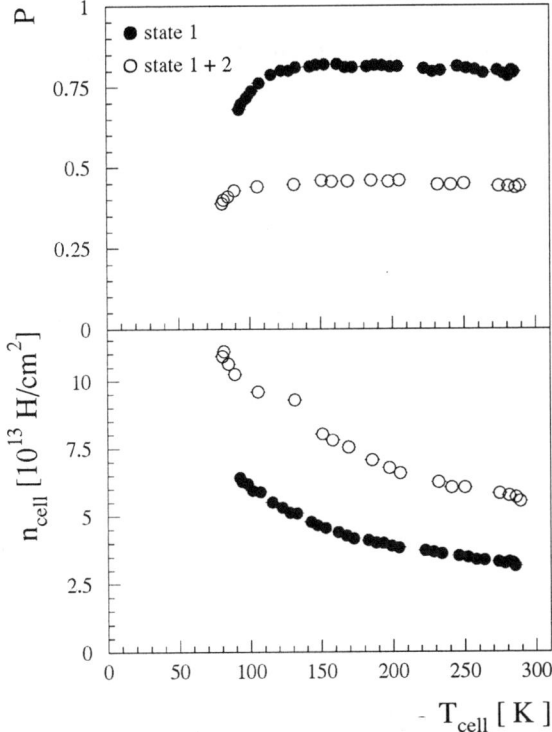

Figure 2: Target polarization (top) and areal density (bottom) of the polarized hydrogen target as function of the cell wall temperature in a weak magnetic field (5 G). The solid dots refer to the condition when an RF transition is turned on to reject one of the hyperfine states. The open circles refer to measurements with two hyperfine states.

Measurements of target polarization and target thickness as a function of wall temperature are shown in Figure 2. The open circles refer to measurements in which the atoms produced by the atomic-beam source were in two hyperfine states (states 1 and 2 in the usual notation), which is the result if no RF transitions are used. Whereas state 1 is a pure state, whose nuclear polarization $P = 1$ is independent of the magnetic field, state 2 is a mixed state whose polarization changes from $P = 0$ in a weak magnetic field to $P = 1$ in a strong field. Correspondingly the maximum nuclear polarization produced with state 1 and 2 in a weak magnetic field is $P = 0.5$. The solid dots refer to measurements in which atoms in state 2 were removed by turning on an RF transition unit located between two sets of sixpole magnets in the atomic-beam source. Ideally only state-1 atoms remain, whose polarization is $P = 1$. The results show a polarization $\geq 80\%$ of the maximum possible value, virtually independent of wall temperature down to 120 K. The error bars are statistical errors only. There is an additional overall scale factor of ±2 % primarily due to uncertainties in the $p\alpha$ analyzing power. The target polarization was found to be independent of beam current in the range investigated (0.02 to 0.3 mA).

The target thicknesses shown in Figure 2 were determined from the count rate and the known $p\alpha$ cross section, detector geometry and beam current. For comparison, the target thickness was also deduced from the small energy loss (0.05 eV) which the beam suffers in the target. The energy loss was determined from the change in revolution frequency of the beam when the electron cooling was turned off. The expected target thickness is also readily calculated from the known atomic-beam intensity and the geometry (gas conductance) of the storage cell. The output flow of polarized atoms in state 1 and 2 from the atomic-beam source ($8.1 \cdot 10^{16}$ H/s (6)) injected into the 10 mm diameter entrance tube of the storage cell is deduced from the pressure rise in a so-called compression tube. The expected target thickness and the two measurements mentioned above all agree with each other within about 10 %.

For two hyperfine states, a target thickness of $(8.2 \pm 0.3) \cdot 10^{13}$ $\vec{H}/cm^2$ and a target polarization of 0.46 ± 0.01 is observed at 115 K. It should be emphasized, that for experiments at higher beam energies, where a strong magnetic field (e.g. 2 kG) can be applied like HERMES, high target thickness and high polarization can be obtained simultaneously by inducing RF transitions between the atomic-beam source and the storage cell. In the present tests the low beam energy prevented use of a strong magnetic field.

One concern about the use of storage cell targets is the possibility that radiation damage may alter the wall coatings and lead to depolarization in wall collisions. This question was addressed by repeating the measurements shown in Figure 2 after a proton beam of intensity up to 1 mA had passed though the cell over a four week period. No change in target polarization (≤ 1.5 %) was detected.

Application at HERMES

Obviously, the design of an internal target dependends on the particular storage ring in which the target will be used. At HERMES the polarized hydrogen and deuterium gas target will be installed in the electron ring of HERA. The main differences to the low energy storage ring in Heidelberg are the high energy ($E \approx 27$ GeV) and the high peak intensity of the electron beam bunches.

Without any special means about 100 W of synchrotron radiation would be incident upon the storage cell, heating and irradiating the cell surface. About 10^{13} - 10^{14} Hz of soft synchrotron photons would be Compton scattered into the acceptance of the detector and would overwhelm the wire chambers. A special collimator system has been designed and successfully tested to shield the cell and the detector properly against synchrotron radiation (9).

The generation of wake-fields (i.e excitations due to electromagnetic interaction of the electron bunches with the target region) results in power deposition into the storage cell or other components close to the beam. To minimize heating of these elements thin-walled meshes will be used to make smooth transitions between different cross sections along the electron beam axis but still allow for pumping of the gas diffusing out of the cell.

Beam induced depolarization of the target atoms by the intense magnetic fields of the HERA electron beam bunches can be overcome by the presence of a holding field up to 0.35 T for hydrogen and deuterium. Using this strong magnetic field, high target thickness and high polarization will be obtained at the same time as explained above.

Conclusions

The results reported here show the excellent performance of an internal polarized hydrogen gas target based on the storage cell technique for storage rings. The high density and high polarization were stable over three months of beam time. No deterioration of target polarization after prolonged exposure to a circulating beam of up to 1 mA has been observed. The polarized gas target for the HERMES experiment will be installed by the end of '94 and data taking will start in spring '95.

References

1. See, for instance, Proceedings of the 10th International Symposion on High Energy Spin Physics, Nagoya, 1992, edited by T. Hasegawa et al. (Univ. Acad. Press, Tokyo, 1993).
2. H.E. Jackson, *HERMES*, contribution to this conference; HERMES-Collaboration: Technical Design Report, DESY-PRC 93/06 (1993).
3. K. Zapfe et al., *A high Density polarized Hydrogen Gas Target for Storage Rings*, accept. for publ. in Rev. Sci. Instrum.
4. P. Schwandt, T.B. Clegg and W. Haeberli, Nucl. Instr. Meth. **A163** (1971) 432.
5. F. Rathmann, *Polarizing a stored, cooled Proton Beam by spin-dependent Interaction with a polarized Hydrogen Gas Target*, contribution to this conference; F. Rathmann et al., Phys. Rev. Lett. **71** (1993) 1379.
6. F. Stock, *The HERMES-FILTEX Target Source for polarized Hydrogen and Deuterium*, contribution to this conference; F. Stock et al., Nucl. Instr. Meth. **A343** (1994) 334.
7. J.S. Price and W. Haeberli, *Measurement of Cell Wall Depolarization of polarized Hydrogen Gas Targets in a weak Magnetic Field*, accept. for publ. in Nucl. Instr. Meth. **A**.
8. M. Düren et al., Nucl. Instr. Meth. **A322** (1992) 13.
9. S.F. Pate, *More Details on HERMES*, contribution to this conference.

An Internal Polarized ^{3}He Target for Electron Storage Rings

L.H. Kramer, D. DeSchepper, R.G. Milner, S.F. Pate, and T. Shin

MIT-Bates Linear Accelerator Center and Laboratory for Nuclear Science, Massachusetts Institute of Technology, Cambridge, Massachusetts 02139

> **Abstract:** We describe an internal polarized ^{3}He target under construction which will be used in several electron storage ring experiments. The target is based on the technique of metastability exchange laser optical pumping, where the polarized atoms flow into a cryogenically-cooled storage cell. This novel technique allows for high precision measurements where the beam interacts with the pure atomic species. Both the HERMES experiment at DESY and the BLAST detector at the MIT Bates Laboratory will use the polarized ^{3}He target in their measurements. Details of the target system, including the provisions needed to incorporate the target into the electron storage ring, are presented.

An internal polarized ^{3}He target is currently being constructed for use in several electron storage ring experiments. Interest in polarized ^{3}He targets originates in their use as an effective neutron target in nuclear(1) and particle(2) physics. In a polarized internal gas target a windowless storage cell is fed by a polarized source, as illustrated in Figure 1. Polarized atoms are injected into a T-shaped storage cell which confines the atoms to a region near the beam axis, increasing the areal target density. Advantages of these targets include high polarization, no dilution of asymmetry due to interactions with unpolarized nucleons in the storage cell, relatively rapid polarization reversal, and low thickness at high beam luminosity. When combined with a high current electron storage ring beam, the polarized internal target provides an excellent platform for high precision studies.

One target currently under construction will be used for the HERMES experiment(3), which begins data taking in the HERA ring at DESY in the spring of 1995. HERMES will measure spin-dependent deep inelastic scattering from the neutron and proton by the scattering of 27 GeV longitudinally polarized electrons from polarized hydrogen, deuterium, and ^{3}He. The ^{3}He target will be used to determine the neutron spin-dependent structure functions over a large range in x to high precision. The integral over x will test several sum rules, including the Bjorken and Ellis-Jaffe sum rules. In addition, leading hadrons from the spin dependent deep inelastic scattering will be measured, leading to information on the relative contribution of valance *vs.* sea quark distribution.

Another target to be used in the BLAST detector(4) is in the preliminary design stage. BLAST is the Bates Large Acceptance Spectrometer Toroid which is a non-

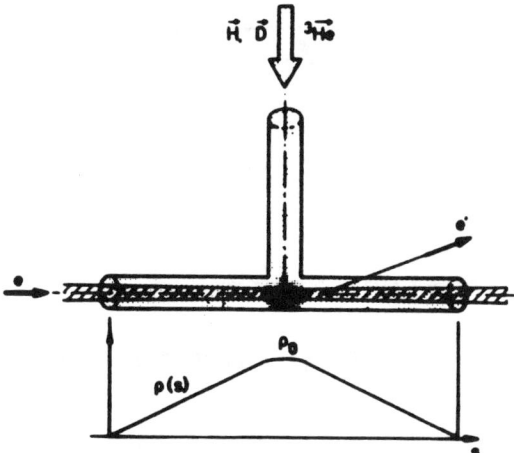

FIGURE 1. Principle of the internal target storage cell. $\rho(s)$ is the volume density of the stored gas.

focusing magnetic spectrometer with eight coils arranged in a toroidal configuration. The initial detector package will instrument two of the eight sectors with the ability to detect electrons and an optional number of hadrons in coincidence. The focus of the BLAST effort is to study several nuclear effects including a broad investigation of the spin-dependent electromagnetic response of nuclei below $Q^2 \leq 1$ (GeV/c)^2. The spin-dependent asymmetry for both inclusive and exclusive (including (e,e'N) and (e,e'$\pi^{\pm}$)) channels will be measured in the quasielastic, 'dip,' and resonance regions for proton, deuteron, and ^{3}He targets.

The target is based on the technique of metastability exchange laser optical pumping(5). To understand this scheme, consider the relevant energy levels of the ^{3}He atom as shown in Figure 2. If a weak electric discharge is maintained in a low pressure ^{3}He gas, a small fraction of the atoms ($\approx 10^{-6}$) will be in the long-lived 2^3S_1 metastable state. Circularly polarized pumping light incident upon the sample along a weak applied magnetic field excites transitions between the $^3S_1$ and $^3P_0$ states. Angular momentum is thus transferred from the pumping light to the metastable atoms, and the metastable atoms become polarized. Transfer of polarization to the ground-state atoms is achieved through metastability exchange collisions.

To create an internal target, ^{3}He atoms flow through a glass pumping cell of volume V at a rate of F atoms/sec and into a storage cell. As shown in the schematic diagram of Figure 3, the ^{3}He gas has an input density of ρ_i and traverses an input conductance C_1. The polarized gas in the pumping cell has a density of ρ_p and exits through a conductance C_2 to the storage cell. The average residence time

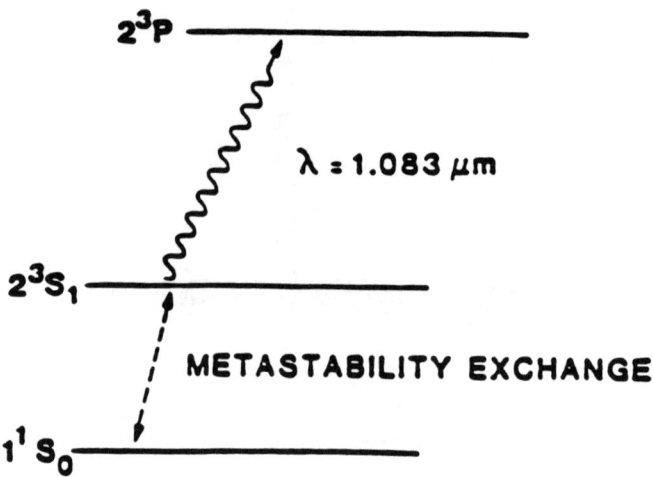

FIGURE 2. The important energy levels of a ^{3}He atom in an external magnetic field. The metastable 2^3S_1 state is polarized by optical pumping to the 2^3P_0 state with circularly polarized light at wave length 1.083 μm.

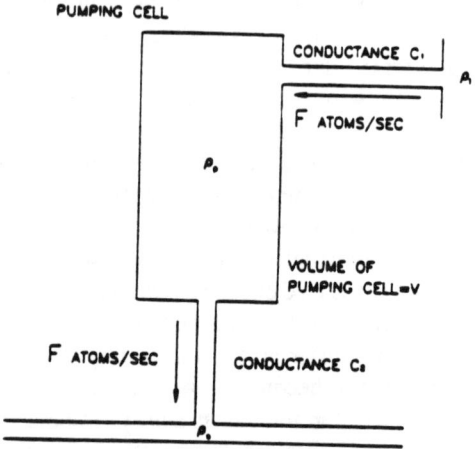

FIGURE 3. A schematic diagram of the internal target. Typically $\rho_i \approx 5$ torr, $\rho_p \approx 0.5$ torr, and $\rho_0 \approx 0$.

of an atom in the pumping cell, t_r, is given by

$$t_r = \frac{\rho_p V}{F}. \tag{1}$$

In an equilibrium configuration

$$F = C_1(\rho_i - \rho_p) = \rho_p C_2, \tag{2}$$

and so

$$t_r = \frac{V}{C_2}. \tag{3}$$

Thus, the residence time of an atom in the pumping cell depends only on the quantities V and C_2. Further, from (2) we have

$$\frac{\rho_p}{\rho_i} = \frac{C_1}{C_1 + C_2}, \tag{4}$$

and so a measurement of ρ_p as a function of ρ_i measures the ratio of conductances C_1 and C_2. Note that in the intermediate flow region C_1 and C_2 are functions of ρ_p.

In the pumping cell the atoms are polarized by absorbtion of the pumping light at the helium transition from the laser. If we consider a sample of ^{3}He atoms in a sealed cell it will be polarized to an equilibrium polarization P_0^s with a pump-up time-constant of t_p as

$$P(t) = P_0^s(1 - e^{-\frac{t}{t_p}}). \tag{5}$$

Similarly, the atoms in an identical pumping cell, where we now flow at a rate F atoms/sec, will be polarized to an equilibrium polarization P_0^f with a pump-up time constant t_f where

$$\frac{1}{t_f} = \frac{1}{t_r} + \frac{1}{t_p}. \tag{6}$$

The equilibrium polarization obtained with a flow-through system, P_0^f, and the equilibrium polarization obtained on a sealed copy of the pumping cell, P_0^s, are related by

$$P_0^f = P_0^s \frac{t_r}{t_r + t_p}, \tag{7}$$

where the difference between sealed and flow-through systems is the presence of a polarization relaxation with time constant t_r due to atoms exiting the pumping cell through C_2. From (7) we see that for high polarization in the flow through system we require that $t_r \gg t_p$, i.e. the pump-up time must be much shorter than the sitting time in the pumping cell.

A similar target was built for the CE-25 experiment (see Figure 4) (7), which ran at the Indiana University Cyclotron Facility Cooler Ring in 1992 and 1993. The experiment studied the spin structure of the ^{3}He ground state through the

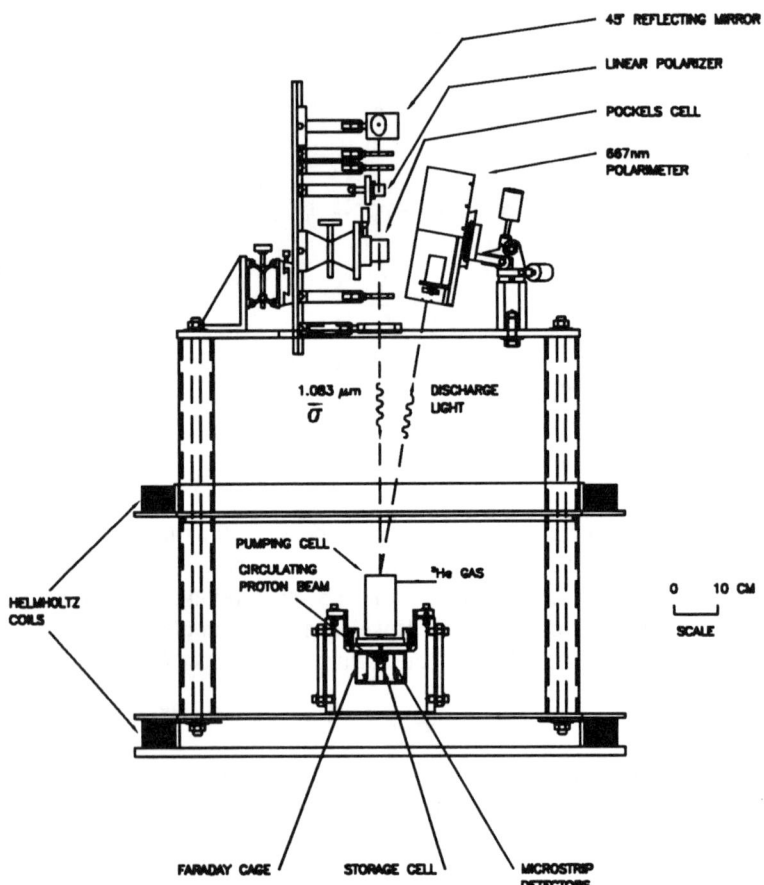

FIGURE 4. A schematic diagram of the CE-25 polarized ^{3}He internal gas target.

measurement of the spin dependent cross section for both ^{3}He(p,2p) and ^{3}He(p,pn) quasielastic scattering over a large kinematic range. The polarized target was combined with a polarized proton beam in the storage ring and a large acceptance non-magnetic spectrometer to carry out the measurements. The target proved to be highly reliable during the measurements. Experimental techniques are reported in (8) and the first results are given in (9) and (10). This target has served as a foundation for the current targets, although several modifications are necessary for use in an electron storage ring.

Several factors must be considered when interfacing the target system into the storage ring. The target must not interfere with the operation of the stored beam. This is accomplished by arranging the storage cell to be well outside the phase-space of the beam and minimizing the gas load into the machine. For the HERMES target, the cell will have an elliptical cross section of 9.8×29.0 mm^2. This corresponds to a $\pm 20\sigma$ beam clearance and should allow for a maximum of 60 mA of circulating electron current. The gas load is minimized with a set of five turbo pump stations, where the pressure beyond the pump stations is below 10^{-8} torr. Additionally, heating generated in the target region due to wakefields created by the time structure of the stored beam must be minimized. This is accomplished by arranging smooth transitions between beamline components.

Experiments which measure electromagnetic interactions further constrain the target system. To achieve the higher densities needed for these interactions, the storage cell is cryogenically cooled to ~ 15K(11), increasing the target thickness by a factor of five. For HERMES measurement of the $g_1(x,Q^2)$ and $g_2(x,Q^2)$ structure functions requires the target to be polarized along two different directions. To minimize systematic uncertainties, it is necessary to be able to randomly switch between all polarization directions quickly, on the order of every 30-60 seconds. To accomplish this two sets of holding field coils along with a cubical optical pumping cell are employed. The BLAST target also faces several technical design challenges, including minimizing the magnetic gradients from the toroid in the target region, which lead to depolarization, and assembling the target within the confines of the detector.

When combined with an electron storage ring, a polarized internal target provides a unique opportunity to study spin dependent effects in nuclei. In the case of the ^{3}He, measurement of asymmetries will lead to knowledge of the spin structure and form factors of the neutron. The targets under construction will play an important role in fundamental measurements over the next few years. A large kinematic range will be explored, from from quasielastic to deep inelastic scattering. Typical figures of merit for a ^{3}He target with a flow rate of 10^{17} atoms/sec are densities of 10^{15} atoms/cm^2, polarizations over 50%.

---

1. Blankleider, B. and Woloshyn, R.M., *Phys. Rev.* **C 29**, 538 (1984).
2. Milner, R.G., Proceedings of the Workshop on Polarized ^{3}He beams and targets, Princeton, NJ, October 1984, p.186 (AIP Conference Proceedings No. 131).
3. HERMES proposal to DESY, DESY-PRC-90-01, unpublished, 1990.
4. Bates Large Acceptance Spectrometer Toroid (BLAST) proposal, 1992.
5. Colegrove, F.D., Schearer, L.D., and Walters, G.K., *Phys. Rev.* **132**, 2561 (1963).
6. Daniels, J.M., Schearer, L.D., Leduc, M., and Nacher, P.J., *Jour. Opt. Soc. Amer.* **4**, 1133 (1987)
7. Lee, K., et al., *Nucl. Instr. and Meth.* **A 333**, 194 (1993).
8. Block, C., et al., submitted to *Nucl. Instr. and Meth. A*, (1994).
9. Lee, K., et al., *Nucl. Phys. Rev. Lett.* **70**, 738 (1993).
10. Miller, M.A., et al., submitted to *Phys. Rev. Lett.*, (1993).
11. Kramer, L.H., et al., to be published.

Status on the Michigan-MIT Ultra-Cold Polarized Hydrogen Jet Target

V. G. Luppov, B. B. Blinov, J. A. Bywater, S. Chin, V. V. Churakov*,
G. R. Court[†], W. A. Kaufman, D. Kleppner[‡], A. D. Krisch, Yu. M. Melnik*,
J. B. Muldavin, T. S. Nurushev, J. S. Price, A. F. Prudkoglyad*,
R. S. Raymond, V. B. Shutov**, J. A. Stewart

Randall Lab of Physics, University of Michigan, Ann Arbor, 48109-1120, USA
*Inst. for High Energy Physics, RU-142284, Protvino, Russia
[†]Physics Department, Liverpool University, P.O. Box 142, Liverpool, L693BX, Great Britain
[‡]Department of Physics, Massachusetts Institute of Technology, Cambridge, Massachusetts 02139, USA
**Joint Institute for Nuclear Research, RU-141980, Dubna, Russia

Abstract. Progress on the Mark-II ultra-cold polarized atomic hydrogen gas Jet target for the experiments NEPTUN-A and NEPTUN at UNK is presented. We describe the performance and the present status of different components of the jet.

To study of spin effects in high energy p-p collisions in the NEPTUN-A (1) and NEPTUN (2) experiments we are developing an ultra-cold high density jet target of proton-spin polarized hydrogen atoms. This method uses an ultra-cold separation cell coated with superfluid helium-4 and a high magnetic field to produce an electron-spin polarized atomic hydrogen beam (3,4).

The Michigan ultra-cold prototype Jet (4) produced an electron-spin polarized atomic hydrogen beam with dc flow of $3.7 \cdot 10^{15}$ H/s, which corresponds to a density of $3 \cdot 10^{11}$ H/cm^3. With our Mark-II Jet we expect to reach nuclear-spin polarized hydrogen atom density of about $5 \cdot 10^{12}$ H/cm^3, corresponding to the target thickness of 10^{13} H/cm^2. We plan to have a 2 cm jet width along the accelerator beam.

A schematic diagram of the Mark-II Jet is shown in Figure 1. The atomic hydrogen is produced in a room temperature rf dissociator and guided to an ultra-cold stabilization cell coated with superfluid helium-4. The double walls of the cell form the mixing chamber of the dilution refrigerator. The cell's entrance and exit apertures are respectively located at about 95% and 60% of the central field of the 12T superconducting solenoid. After the hydrogen atoms are thermalized by collisions with the cell surface, the magnetic field gradient physically separates the atoms according to their electron-spin state. The atoms in the two lowest hyperfine states ($|3>, |4>$) are attracted toward the high field region and escape from the cell. They recombine on bare surfaces and are cryopumped. The atoms in the two higher hyperfine states ($|1>, |2>$) are repelled toward the low field region and effuse from the exit aperture, forming the electon-spin polarized beam.

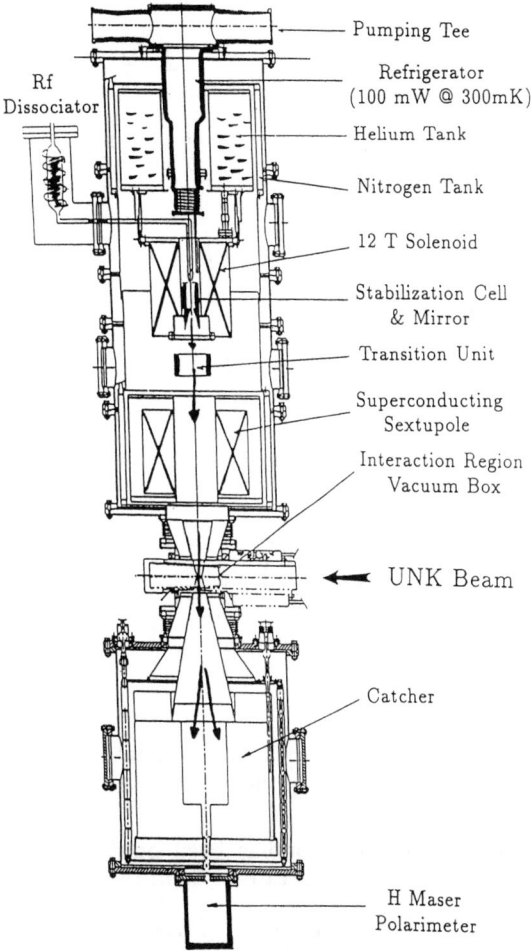

FIGURE 1. Schematic diagram of the Mark-II Ultra-Cold Jet.

After an rf transition unit, which interchanges atoms in states $|2>$ and $|4>$ we have a superconducting sextupole. The sextupole selects atoms in electron spin state $+1/2$ by focusing atoms in state $|1>$ into the interaction region and defocusing atoms in state $|4>$, which are then cryopumped. The nuclear-spin polarized beam that passes through the interaction region is caught by a cryopumping catcher. A maser polarimeter below the catcher monitors the beam proton polarization.

Most of the vacuum jackets, nitrogen tanks and the main helium reservoir have been built and tested. The 12 T solenoid has been wound and successfully tested. It consists of an inner Nb_3Sn, outer NbTi coils and a NbTi bucking coil. Due to the bucking coil, located downstream of the solenoid the axial field falls off with about a 1 T/cm gradient and has a very short tail, which lightens the rf unit magnetic shielding problem.

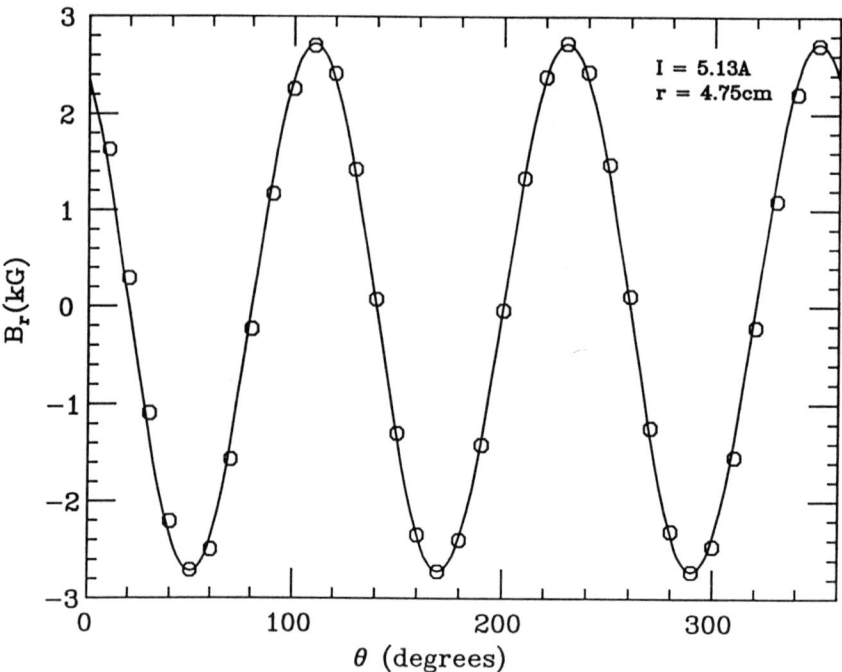

FIGURE 2. Azimuthal scan of the sextupole radial magnetic field near pole tips. o measurements, solid line - sinusoidal fit to the data.

The 100 mW dilution refrigerator has been built and is now undergoing cryogenic tests. It is designed to operate at 300 mK and circulate 30 mmol/s of helium-3.

We plan to use a helium-film coated quasi-parabolic mirror to better focus the atomic hydrogen effusing from the stabilization cell (5). Simulations indicate that with no mirror approximately 25% of the beam can be focused to the 10mm x 20mm interaction spot. We expect that a specially designed mirrow will significantly increase that fraction. We are now working on a mirror shape taking into account the axial and radial gradients of the solenoid field.

The $|2>$ to $|4>$ rf unit uses the adiabatic passage transition method. A ring dielectric resonator is used to accomodate the 6 cm beam diameter. The preliminary design and some tests on a prototype unit have been done (6).

The focusing sextupole magnet with superconducting coils and iron poles has been built and tested. To increase the Mark-II acceptance the 20 cm long sextupole has a bore diameter of 11 cm. Simulations tailored to the Mark II geometry show that an optimum pole tip magnetic field is about 3.2 kG. This number corresponds to a coil current of 6 A. One of the sextupole test results is shown in Figure 2.

The catcher consists of 13 m^2 of copper cryocondensation fins cooled to 3 K. It will keep the background pressure in the interaction region chamber at a level of 10^{-9} Torr. All parts of the catcher, except the upper shielding, have been fabricated and are now under assembly.

The MIT hydrogen maser polarimeter has been built and is now under test at Michigan. To monitor the state populations it employs a room temperature hydrogen maser that is operated in a transient, sub-threshold mode (7).

ACKNOWLEDGEMENTS

This work is supported by the U.S. Department of Energy.

REFERENCES

1. Krisch, A. D., "Polarization in High $P_\perp^2$ $p-p$ Elastic Scattering at UNK" in Proceedings of the Workshop on Physics at UNK, Protvino, Russia, 1989, pp. 152-174.
2. Solovianov, V. L., "Study of Spin Effects with Jet Target on the UNK Internal Beam Facility at 0.4-3.0 TeV/c"in Proceedings of the Workshop on the Experimental Program at UNK, Protvino, Russia, 1987, pp. 191 - 213
3. Mertig, M., et al., "Accumulation of Hydrogen Atoms in a Low Temperature Storage Cell of a Polarized Hydrogen Gas Jet Source "in Proceedings of the 9-th Intern. Symposium on High Energy Physics, Bonn, 1990, pp. 164-167.
4. Kaufman, W. A., Roser, T., and Vuaridel, B., Nuclear Instruments and Methods A 335, 17-29 (1993).
5. Luppov, V. G., et al., Physical Review Letters 71, 2045-2048 (1993).
6. Muldavin, J. B., and Price, J. S., "Progress in Development of a 4K RF Transition Unit for Use with an Ultra-Cold Polarized Hydrogen Beam", in Bulletin of the American Physical Society, Vol.39, No. 2, p.1109 (1994)
7. Kleppner, D., Goldenberg, H. M., and Ramsey, N. F., Physical Review 126, 603-615 (1962)

High Intensity Polarized Ion Source at RCNP

K. Hatanaka, K. Takahisa, H. Tamura, H. Kaneko*
and I. Miura

Research Center for Nuclear Physics, Osaka University, Osaka 567, Japan
*Sumitomo Heavy Industry Accelerator Service, Shinagawa-ku 141, Tokyo

Abstract. A new polarized ion source has been constructed at RCNP. It employs cold(~35 K) atomic beam technology and an electron cyclotron resonance ionizer. The source was assembled and installed on the shielding roof of the AVF cyclotron in Feb. 1994. Performance tests of the atomic beam section was started in March. Protons were extracted from the ionizer in April, and the optimization of the source was performed after this. A new axial injection line to the AVF cyclotron was installed in this summer. The new system should push experimental programs with polarized beams accelerated by the K=400 ring cyclotron. The design of the system and results from operation of the source will be described.

INTRODUCTION

At Research Center for Nuclear Physics, Osaka University, the first polarized ion source was constructed in 1975 and extensive researches have been performed since then with polarized protons and deuterons accelerated by the RCNP AVF cyclotron. In 1992, a K=400 ring cyclotron was completed as the post accelerator. It requires the high quality beam to be injected from the AVF cyclotron. The emittance of the beam extracted from the AVF cyclotron is reduced in both the transverse and longitudinal phase spaces to match the acceptance of the ring cyclotron. Consequently, the intensity of the 400 MeV proton beam is restriced to around 10 nA on target in the usual operation. A large fraction of the experimental program at RCNP is concentrated on studies of spin degrees of freedom. In order to enhance the opportunities in spin physics research using the ring cyclotron, the construction of a new high intensity polarized ion source was proposed as a two years project in 1993.

The new source built at RCNP is schematically illustrated in Fig. 1. Its design is based on sources in operation at PSI[1], TUNL[2], IUCF[3] and

RIKEN[4], which employ cold (~30 K) atomic beam technology and an electron cyclotron resonance ionizer. The source produces in excess of 100 μA DC, H^+ and D^+ ion beams, and is coupled by a high-efficiency bunching system and a high-transmission injection line to the AVF cyclotron. The first beam injection trial to the AVF cyclotron is scheduled in this fall.

FEATURES OF THE SOURCE

The beam of H or D atoms is produced from H_2 or D_2 gas in a 13.6 MHz, 50~200 W discharge contained in a pyrex tube of 20 mm inner diameter. The discharge is cooled by water flowing between the dissociator tube and the second, surrounding pyrex tube of larger diameter, 30 mm. Atoms emerging from the discharge pass through a Macor section at the end of the tube and into a copper or an aluminum nozzle. The Macor serves to isolate the nozzle thermally at ~35 K from the high temperature of the discharge. The nozzle is cooled by conduction to the cold head of a 9 W, closed-cycle helium refrigerator. Cooled atoms emerge as a directed jet from a 3 mm diameter nozzle orifice into the first vacuum chamber which is evacuated by two 2800 l/sec turbomolecular pumps with a magnetic suspension. With 15 std-cc/min of H_2 flowing into the dissociator, the pressure in the first chamber is $< 2.1 \times 10^{-3}$ Pa.

An atomic beam is formed at the entrance to the second vacuum chamber when the beam passes through a 6 mm diameter skimmer aperture placed 25 mm from the end of the nozzle. This skimmer separates the first and second

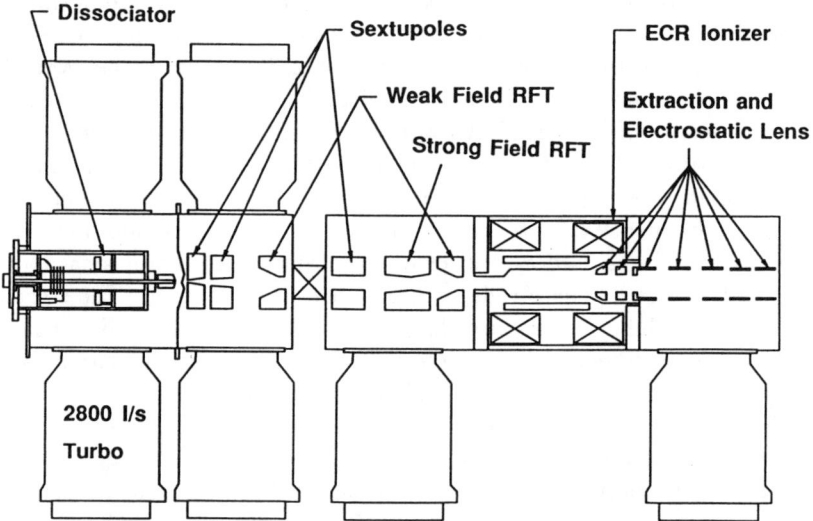

FIGURE 1. The schematic layout of the RCNP polarized ion source.

chambers. With other two 2800 l/sec turbomolecular pumps on the second chamber, the pressure there is $< 1.2 \times 10^{-4}$ Pa. The third vacuum chamber follows where another 2800 l/sec turbomolecular pump holds the pressure during operation $< 3 \times 10^{-5}$ Pa.

The atomic beam enters the first sextupole magnet system 25 mm beyond the skimmer orifice. The system consists of two sextupole magnets which are separated by 30 mm in order to efficiently evacuate the beam path. The first magnet is 50 mm long and has axially tapered pole tips with inner diameters of 16 mm at its entrance and 22 mm at its exit. The second one is 70 mm long and has inner diameters of 22 mm at its entrance and 30 mm at its exit. The atomic beam enters the third sextupole 272 mm beyond. The last sextupole is 100 mm long and has straight pole tips with a constant 30 mm tip-to-tip diameter. The sextupole magnet is segmented in 24 pieces[5] of oriented Nd-Fe-B material, NEOMAX-35H, and the outer diameter is 120 mm. Each magnet is put in a vacuum tight container whose outer diameter is 170 mm. The remanent field B_r in the direction of the easy axis of the material was measured to be 1.245 T, and the maximum energy product $(BH)_{max}$ 295 kJ/m^3. The size of sextupole magnets and their positions were optimized from calculations using a Monte-Carlo code for the atomic beam transport which was written at the University of Wisconsin[6]. In the calculation, the radius of the effective ionizing region was taken to be 10 mm.

The ECR ionizer was designed following Schmelzbach[1]. The entrance of the ionizer is 375 mm downstream from the exit of the last sextupole magnet. The inner diameter of solenoid coils is 180 mm. The current in the entrance and exit solenoid coils can be adjusted independently in order to optimize the shape of the axial mirror field. At the entrance of the ionizer, the solenoid coil has a 50 mm thick iron yoke with a hole of 36 mm in diameter to shield the rf transition units against the stray field from the ionizer. The sextupole magnet consists of six NEOMAX-35H bars which are 20 mm wide, 15 mm high and 210 mm long. They are magnetized in the width direction, and are assembled in a sextupole magnet with a 110 mm tip-to-tip diameter. The microwave frequency is 2.45 GHz at present, and the maximum power is 200 W. A pyrex tube is used as a discharge container, and the plasma potential is defined by the first electrode with an opening of 50 mm. The solenoidal coils have the ability to produce B_{min} of 230 mT for the axial mirror field. If the low magnetic field 87.5 mT deteriorates the proton polarization, we can increase the microwave frequency higher than 6 GHz.

The desired states of nuclear polarization for H_0 and D_0 beams are produced by three sextupole magnets with one weak field radio-frequency transition (RFT) unit between the second and the third magnet, and a second weak field and one (1400 MHz for H_0) or two (455 MHz and 331 MHz for D_0) strong field units following the last sextupole. With this configuration, both the pure vector and the pure tensor states are produced for deuterons.

INJECTION LINE TO THE AVF CYCLOTRON

A new injection line for the AVF cyclotron was designed. So far, polarized protons and deuterons have been externally injected into the AVF cyclotron. On the other hand, unpolarized ions including light heavy ions have been

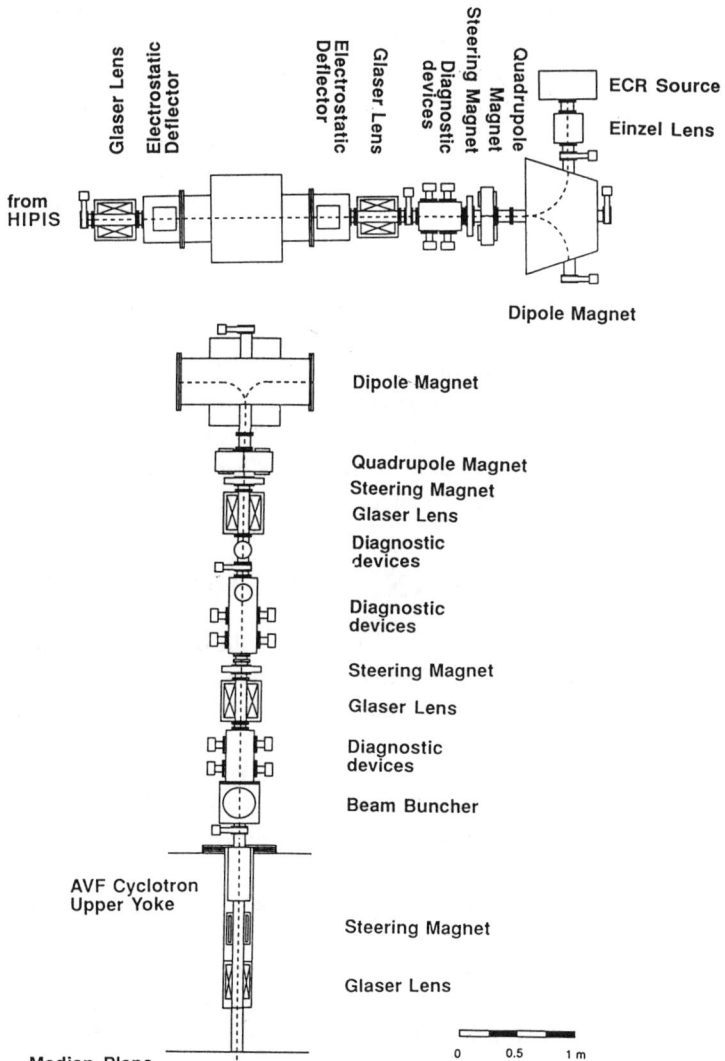

FIGURE 2. The schematic layout of the injection line.

produced by an internal ion source. When ions are externally injected with a mirror system, structures in the central region of the cyclotron have to be changed from those used for the internal source. The vacuum chamber of the cyclotron is opened in each time when accelerated ions are changed from polarized protons/deuterons to unpolarized ions, and vice versa. In the new system, both polarized ions and unpolarized ions produced with a ECR source, NEOMAFIOS built by R. Geller, are injected through the common vertical injection line and the common inflection system. A schematic layout of the new injection line is shown in Fig. 2. At the exit of the source, protons and deuterons are longitudinally polarized. Protons are deflected by 96.7 degrees with a dipole magnet, and deuterons by 105 degrees, so that the beam remains longitudinally polarized after the 90 degrees deflection into the vertical beam line. For this purpose, small parts of pole pieces are added at the magnet entrance when deuterons are injected to the cyclotron. An electrostatic deflector is used to compensate the beam trajectory shifts. The same method will be applied, when polarized ^{3}He^{2+} ions are injected in future. In the system, electromagnetic lenses are used instead of electrostatic ones in the old system. They ensure an efficient focusing and a good vacuum through the line. A beam buncher is placed at 2.5 m upstream from the median plane of the cyclotron. It consists of two parallel mesh plates forming a single gap, and is excited by an RF voltage with a sawtooth-like waveform generated by combining RF sine waves with the first three higher harmonics. At present a mirror system is used to inflect ions into the median plane at the center of the cyclotron. A design work is now in progress to replace it by a spiral inflector.

STATUS AND RESULTS

The source was assembled and installed next to the old one on the shielding roof of the AVF cyclotron in Feb. 1994. Performance tests of the atomic beam section were started in March. Protons were extracted from the ionizer at the end of April, and the optimization of the source was performed after this. At the end of July, we started to disassemble and remove the old source and injection line. The new source was connected with the cyclotron by the new beam line in the beginning of September. The source could not be operated for one and a half months.

Although the available time was limited, efforts were made to check the performance of the source, especially its long term stability. The atomic beam intensity was measured with a compression tube which was installed in place of the strong field RF transition units. The entrance of the compression tube is 9.3 mm in diameter with a length of 40 mm. In usual operation, pressure differences of $(1.6 \sim 2.4) \times 10^{-3}$ Pa were observed at a gas flux of 15 std-cc/min and a microwave power of 80 W with the compression tube which has not been calibrated yet. N_2 gas was fed at the rate of 0.02 std-cc/min in near the nozzle through the Macor section. Initially the dissociator nozzle was made of copper, and the atomic beam intensity was observed to rapidly decrease after a 4 day operation. A heat cycle procedure, heating the nozzle to 300 K and then cooling down to 35 K, could not recover the source. The accumulation of green stuffs was observed on the surface of the nozzle. It was supposed due to a small contamination of water in the hydrogen gas line. The material of the nozzle was changed from copper to aluminum. In the beginning of the operation, the aluminum nozzle produced the atomic beam of the same intensity as the copper nozzle. After a 4 day operation, the atomic beam began to decrease. In this case, however, the heat cycle procedure could recover the nozzle. It was found that the aluminum nozzle can be recycled every four days and the source works for longer than two weeks keeping good performance.

The ionized protons in the ECR plasma were analyzed with a 90 degrees bending magnet. Protons of 80-150 μA were ordinally extracted at the 12 kV extraction voltage and with the 20 W ECR microwave power. About 20 % of these protons remained even if the dissociator microwave was turned off. The ionizer steadily works with the addition of the N_2 buffer gas at the rate of around 0.05 std-cc/min.

The first trial to inject the beam into the AVF cyclotron is scheduled at the end of this September, and the beam polarization will be measured in the beginning of October.

ACKNOWLEDGEMENT

The authors like to thank H. Okamura, T. Clegg, P. Schmelzbach, Y. Mori(KEK), M. Wedekind(IUCF), and their colleagues for providing information on their sources, W. Haeberli and A. Roberts for making available their computer code to calculate the atomic beam transport.

REFERENCES

1. P. A. Schmelzbach, AIP Conference Proceedings 293(1994) 65.
2. T. B. Clegg, AIP Conference Proceedings 187(1989) 1227.
3. V. Derenchuk et al., AIP Conference Proceedings 293(1994) 72.
4. H. Okamura et al., AIP Conference Proceedings 293(1994) 84.
5. K. Halbach, Nuclear Instruments and Methods 169(1980) 1.
6. A. Roberts, private communication.

Polarizing Stored Beams by Interaction with Polarized Electrons

C.J. Horowitz and H.O. Meyer

Department of Physics, Indiana University, Bloomington, IN, USA

Abstract. A polarized, internal electron target gradually polarizes a proton beam in a storage ring. We have derived the spin-transfer cross section for $\vec{e}(p,\vec{p})e$ scattering. A recent measurement of the polarizing effect of a polarized atomic hydrogen target is explained when the effect of the atomic electrons is included. We also consider the interaction of a stored beam with a pure electron target which can be realized either by a comoving electron beam or by trapping of electrons in a potential well. In the future, this could provide a practical way to polarize antiprotons.

INTRODUCTION

It has been shown by experiment that a polarized, internal hydrogen target polarizes an initially unpolarized orbiting beam [1]. Originally, this polarization effect has been attributed to the strong interaction between the two nucleons. However, it turns out that the polarized atomic target electrons are also important. Here, we discuss spin dependence in $\vec{e}(p,\vec{p})e$ scattering and show that there is a surprisingly large, electromagnetic polarization transfer, due to the interference of the hyperfine interaction with the Coulomb amplitude which is large at small scattering angles. A more detailed account of this work has recently been published [2].

SPIN-TRANSFER CROSS SECTION FOR $\vec{e}(p,\vec{p})e$ SCATTERING

Plane-Wave Born Approximation

Our calculation is carried out in the electron rest frame. The initial electron polarization is along $\hat{i}$, the scattered proton is polarized along $\hat{j}$, where $\hat{i},\hat{j}$ are unit vectors expressed in terms of the basis vectors $\hat{\ell}$ (longitudinal), $\hat{m}$ (sideways) and $\hat{n}$ (normal to the scattering plane), as defined in eq.(2.2) of ref.[3]. We write m_p, **p** and **k** for the mass, the initial and the final momentum of the proton, and, respectively, m_e, $\mathbf{p}_e$ and $\mathbf{k}_e$ for the electron. Then, according to eq.(B.1) of ref.[4], we write

$$\frac{d\sigma}{d\Omega} K_{j00i} \approx \frac{m_p^2}{4\pi^2} |M_{ij}|^2 \qquad (1)$$

where K_{j00i} is the coefficient for polarization transfer from the target to the projectile (formally defined in ref.[3]), and M_{ij} is the invariant matrix between the appropriate spin states in the initial and final state. The latter, in plane-wave Born approximation (PWBA), is given by

$$M_{ij} = \frac{4\pi\alpha}{q_\mu^2} U_p' (\gamma_\mu + \lambda_p i \sigma_{\mu\nu} \frac{q^\nu}{2M}) U_p \cdot U_e' \gamma^\mu U_e , \qquad (2)$$

where $q_\mu = k_p - p_p$ is the momentum transfer to the proton, and $\lambda_p = 1.793$, the anomalous moment of the proton. The square of eq.(2), summed over the spin of the final electron and averaged over the spin of the initial proton is evaluated, and the resulting spin transfer cross section for $\hat{i} = \hat{j} = \hat{n}$ (both normal to the scattering plane) is

$$\frac{d\sigma}{d\Omega} K_{n00n} = - \frac{\alpha^2 (1+\lambda_p) m_e}{2 p_e^2 m_p \sin^2(\theta/2)} . \qquad (3)$$

Carrying out the calculation for other spin directions, one finds the following relations between various polarization observables: $K_{l00l} = K_{n00n}$, $K_{m00m} = 0$, $K_{j00i} = 0$, for $j \ne i$, and $A_{00ji} = K_{j00i}$ (spin-transfer equals spin-correlation coefficient).

The maximum scattering angle of a proton from an electron at rest is $\theta_{max} = m_e/m_p = 0.54$ mrad. This is well within the acceptance angle of any storage ring. Thus, the scattered protons stay in the ring, and we are interested in the *total* cross section

$$\sigma_{ij} \equiv \int_{\theta_{min}}^{\theta_{max}} \frac{d\sigma}{d\Omega} K_{j00i} d\Omega . \qquad (4)$$

At large impact parameter, the Coulomb field of the target electron is screened and there is no interaction. This can be taken into account by cutting off the integration at a minimum angle θ_{min}. We define θ_{min} so that the minimum momentum transfer is the inverse of the screening distance Λ, or $2p_e \sin(\theta_{min}/2) = 1/\Lambda$.

Including Coulomb Distortion

The use of Coulomb-distorted waves in the Born approximation modifies the plane-wave result in two ways. First, the matrix element of the short-ranged hyperfine interaction scales with the square of the wavefunction at the origin which is modified by the Coulomb interaction by approximately

$$C_0^2 = \frac{2\pi\eta}{e^{2\pi\eta}-1}, \quad \eta = -\frac{z\alpha}{v} , \qquad (5)$$

where z is the beam charge and v is the relative velocity. Second, there is an angle-dependent relative phase between the Coulomb and the hyperfine amplitudes. Together, these effects lead to an extra factor in the expression for σ_{ij} as it would be obtained from eqs.(3) and (4). The result for $\hat{i} = \hat{j} = \hat{n}$ is

$$\sigma_{nn} = -\left[\frac{4\pi\alpha^2 (1+\lambda_p) m_e}{p_e^2 m_p}\right] C_0^2 (\tfrac{v}{2\alpha}) \sin\left[\tfrac{2\alpha}{v} \ln(2p_e\Lambda)\right] . \tag{6}$$

As a stored beam circulates through an internal polarized electron target the polarization P_B of the beam changes according to [5],

$$\frac{dP_B}{dt} = (1-P_B^2)\, f_R\, d\, P_e\, \hat{\sigma} . \tag{7}$$

Here f is the orbit frequency of the beam (of order 10^6 Hz), d is the target thickness (particles per unit area), P_e the electron polarization, and $\hat{\sigma}$ the spin-dependent cross section, averaged over the orientation of the scattering plane. For an electron target polarized transverse to the beam we have $\hat{\sigma} = \tfrac{1}{2}(\sigma_{nn}+\sigma_{mm}) = \sigma_{nn}/2$, while for longitudinal electron polarization $\hat{\sigma} = \sigma_{ll} = \sigma_{nn}$, where σ_{nn} is given by eq.(6).

POLARIZED ELECTRON TARGETS

Atomic electrons

Let us now consider polarized target electrons as they occur in a polarized internal hydrogen target. Such a situation was realized recently at the TSR in Heidelberg [1], where a 23 MeV proton beam interacted with an internal target of 6×10^{13} polarized hydrogen atoms per cm^2. The screening distance in an atom, given by the Bohr radius (5.29×10^{-4} fm), determines θ_{min}. Eq.(6) for this case yields $\sigma_{nn} = -140$ mb, or $\hat{\sigma} = -70$ mb. This result is also shown as a function of proton bombarding energy in Fig.1. This represents an electronic contribution to the beam polarization of a magnitude comparable to the experimental result [1], but of *opposite* sign. When this result is combined with the effect of the polarized target protons which contribute by selective spin state removal and by Coulomb-nuclear interference scattering (see [5]), excellent agreement with the TSR measurement is obtained. Thus, the effect of a polarized electron target on the polarization of a stored beam seems to be established experimentally.

In an atomic, polarized target one has to contend with beam loss due to NN scattering. The beam lifetime is characterized by the so-called loss cross section [6]. After one lifetime, 1/e of the original beam intensity remains, and the value of the beam polarization is determined by the ratio between the polarizing cross section $\hat{\sigma}$ and the loss cross section σ_L. At low energy, where σ_L is dominated by Rutherford scattering, this ratio increases with beam energy. However, above about 1 GeV, the major contribution to σ_L is from the weakly energy-dependent strong interaction, while $\hat{\sigma}$ (due only to the polarized electrons) continues to decrease. This poses a *fundamental* upper limit of a few percent for the attainable polarization at the end of a beam lifetime.

Pure Electron Targets

In a pure electron target, we approximate the screening length by the Debye length Λ_D which, for the targets discussed below, is much larger than in an atomic target. For instance, for an electron gas of density $n_e=10^{10}$ cm^{-3} and temperature $kT=0.1$ eV, $\Lambda_D=10^{10}$ fm. This larger screening length raises the spin cross section by about a factor of three (see Fig.1). The absence of a nuclear target leads to long beam lifetimes, bounded only by the quality of the ring vacuum. The task of providing a dense, pure electron target is a technical problem. Aside from this there are no fundamental limits to being able to polarize a stored beam.

Electron beams

A pure electron target could be provided by a comoving electron beam, which may be similar to an electron cooling beam. The design constraints, however, are different, since low beam temperature may be traded for intensity. Clearly, such an arrangement requires a high-intensity, high-duty-factor, polarized electron source, but the recent, rapid development in this field indicates that this technical limitation may be overcome. One advantage of a comoving beam is the possibility to choose the best relative energy between the protons and the electrons (about 5 MeV, see Fig.1), while keeping the energy of the stored beam high to ensure a long lifetime. For example, if it were possible to prepare an electron-beam target with $P_e=0.5$ and of the same number density as the *atomic* target discussed above, the resulting polarization rate could be as high as 7.5% per hour.

Trapped Electrons

Trapping of electrons in a Penning trap has been used as a diagnostic tool in the Indiana Cooler. A realistic extrapolation of the performance of such a device yields a target thickness of 10^{12} electrons per cm^2 [7]. This is still less than the electron thickness of an internal hydrogen target. But the task of developing a high-density electron trap for use in nuclear physics is new and one has to look to future advances in this area, perhaps borrowing from techniques developed for plasma confinement in nuclear fusion. For a trap in which electrons can be *accumulated*, the requirement on the source of polarized electrons is less demanding. If a polarized trapped-electron target of thickness 10^{13} cm^{-2} were available in the TSR ring with its present performance [1], 5% proton polarization would be achieved at the end of one lifetime.

POLARIZING ANTIPROTONS

Another, more speculative, consequence of electron-induced polarization is its possible use in polarizing stored antiprotons. For antiproton-electron scattering, C_0^2 in eq.(5) is simply evaluated for a negative beam charge or a *positive* η parameter. The rest of the calculation stays the same. The spin transfer cross section σ_{nn}, eq.(6), for scattering of antiprotons from transversely polarized electrons versus the laboratory kinetic energy is shown as dashed lines in Fig.1. Polarized antiproton beams are of great interest in nuclear and particle physics, but none of the many methods that have been proposed in the past [8] has proven practical. With a reasonably optimistic projection of future advances in the technology of internal, polarized electron targets, the method described here would be the answer to this old problem.

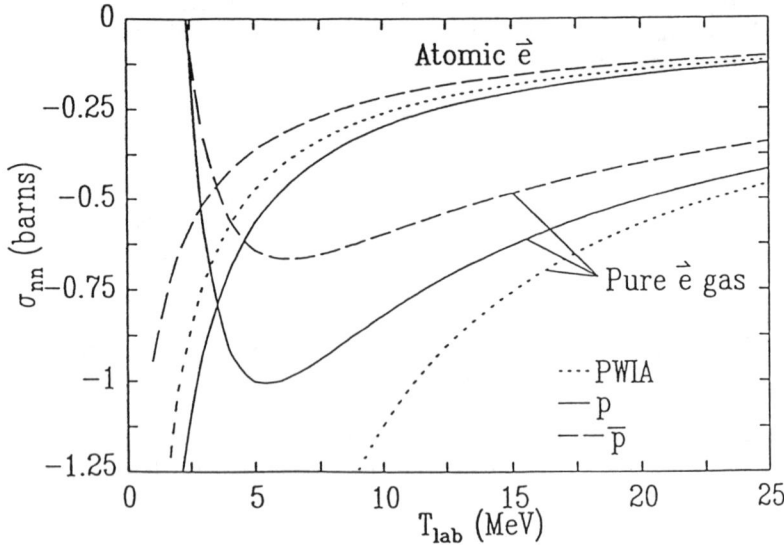

FIGURE 1. Spin transfer cross section σ_{nn} for $\vec{pe}$ scattering versus beam kinetic energy T_{lab}. The dotted curves are the plane wave Born approximation results, eq.(3) and (4), which are the same for protons and antiprotons, while the solid and dashed curves are approximate distorted-wave Born approximation calculations, eq.(6), for protons and antiprotons, respectively. The upper three curves give the electronic contribution for an atomic hydrogen target while the lower curves are for a pure electron target.

REFERENCES

[1] F. Rathmann et al., Phys. Rev. Lett. **71**, 1379 (1993).
[2] C.J. Horowitz and H.O. Meyer, Phys. Rev. Lett. **72**, 3981 (1994).
[3] J. Bystricky, F. Lehar and P. Winternitz, J. de Physique **39**, 1 (1978).
[4] J.D. Bjorken and S.D. Drell, Relativistic Quantum Mechanics, Mc Graw-Hill, New York, 1964.
[5] H.O. Meyer, Phys. Rev. E**50**, 1485 (1994).
[6] R.E. Pollock, Ann. Rev. Nucl. Sc. **41**, 357 (1991).
[7] J. H. Malmberg et al, "Experiments with Pure Electron Plasmas", in Non-Neutral Plasma Physics, eds. C. W. Roberson, C. F. Driscoll, Am. Inst. of Physics, N.Y., 1988, p.28.
[8] Proc. of Workshop on Polarized Antiprotons, eds. A. D. Krisch and O. Chamberlain, Bodega Bay 1985, AIP Conf. Proc. **145**, Am. Inst. of Physics, N.Y., 1986.

A Possible Method to Produce a Polarized Antiproton Beam at Intermediate Energies

H. Spinka, E.W. Vaandering,[a] and J.S. Hofmann[b]

High Energy Physics Division
Argonne National Laboratory, Argonne, Illinois 60439, USA

The lack of existence of polarized antiproton beams has severely limited the detailed understanding of the antinucleon-nucleon ($\overline{N}N$) interaction at low and intermediate energies. In particular, it would be desirable to have improved knowledge of the $\overline{N}N$ short range force and annihilation. A complete description of $\overline{N}N \rightarrow \overline{N}N$ scattering requires a determination of five complex scattering amplitudes for each of two isospins. At least 19 different spin observables must be measured at each c.m. angle and energy to reconstruct these amplitudes in a model independent fashion. However, only a few different types of spin observables for the $\overline{p}p \rightarrow \overline{p}p$ and $\overline{p}p \rightarrow \overline{n}n$ reactions have been measured. This situation is similar to the status of the NN interactions in the early 1960's. Furthermore, the $\overline{N}N$ interaction is highly inelastic because of annihilation channels, even at the lowest beam momenta, whereas NN scattering is essentially elastic up to about 800 MeV/c. A polarized antiproton beam incident on a polarized proton or deuterium target would permit about a dozen new spin observables to be measured in $\overline{N}N \rightarrow \overline{N}N$ scattering, as well as many new spin observables for $\overline{N}N \rightarrow \pi\pi, KK$, etc. reactions, making an enormous impact on the understanding of the $\overline{N}N$ interaction.

Several methods have been discussed to produce polarized $\overline{p}$ beams at momenta up to a few GeV/c [1-5]. Three of these methods rely on dedicated use of the accelerator or a storage ring. The antiprotons are produced with a beam of high energy protons, collected, and finally stored in the accelerator or storage ring for hours while the $\overline{p}$'s are slowly polarized. These methods involve formation of $\overline{H}$ atoms [1,2], Stern-Gerlach separation of $\overline{p}$ spins in an inhomogeneous magnetic field [3], and differences in $\overline{p}p$ total cross sections for parallel and antiparallel spins [4-6]. A fourth method would scatter an unpolarized $\overline{p}$ beam from carbon, but the resulting $\overline{p}$ beam polarization

a. Present address: University of Colorado, Boulder, CO 80309.
b. Present address: Visix Software, Inc., Reston, VA.

would be quite small [7]. This paper presents the conceptual design of a still different method, not requiring dedicated running in an accelerator such as LEAR. It would be ideal for a kaon factory beam line. Only one polarized antiproton beam has been successfully built [8], but its operation is based on different principles, and it operated at much higher energies (~ 200 GeV).

The method to produce a polarized $\bar{p}$ beam at intermediate energies is shown schematically in Fig. 1. A secondary beam of unpolarized antiprotons strikes a liquid hydrogen (LH_2) target. Elastically scattered $\bar{p}$'s are refocussed at the experimental target, for example a polarized proton target (PPT), with a toroidal magnet which accepts a range of scattering angles and azimuthal angles, ϕ. The scattered $\bar{p}$'s have a known polarization [9-13] which is perpendicular to the scattering plane. If there is no bending magnet after the toroid, the $\bar{p}$'s will have transverse polarizations at the experimental target. Note that reversing the magnetic field direction would permit polarized protons to be collected and focussed on the target from an incident unpolarized proton beam. This would allow for tests of systematic errors with the well-studied pp elastic scattering reaction.

In the ideal case, no $\bar{p}$ spin precession would occur for an incident beam with a small spot size, since the toroidal magnetic field would always be antiparallel to the $\bar{p}$ spin direction. A realistic toroid design will limit the acceptance for antiprotons in ϕ due to the presence of conductor windings, and will cause small spin precessions due to a nonideal toroidal field. Small spin precessions will also occur when the ratio of the incident beam spot size to the toroid radius is sizeable.

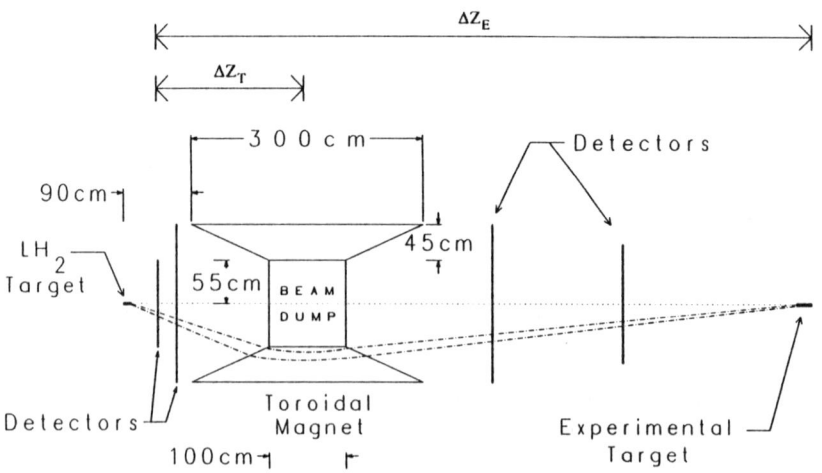

FIGURE 1. Drawing of a possible design for a polarized $\bar{p}$ beam line at 1089 MeV/c. The incident unpolarized $\bar{p}$'s enter from the left and strike the LH_2 target. Two examples of scattered $\bar{p}$'s are shown, along with detectors to measure the particle trajectories.

There are a number of important considerations for the detailed design of (1) the toroid, (2) the choice of distance between the LH$_2$ target and the toroid (ΔZ_T - see Fig. 1), and (3) the distance from the LH$_2$ target to the nominal beam focus at the experimental target (ΔZ_E). The distance ΔZ_T should be kept small in order to minimize the size and cost of the toroidal magnet. The beam line design should permit operation over a wide range of polarized beam momenta. The spot size at the focus should be minimized, since it is costly to construct a large diameter and volume polarized target. The ratio of distances $\Delta Z_E/\Delta Z_T$ should be large to reduce the beam divergence at the beam focus. The intensity of the polarized antiproton beam should be maximized to allow higher statistics experiments and/or to reduce running time per measurement. One way to achieve higher intensity is to design the beam line to accept sizeable divergence, momentum spread, and spot size for the unpolarized $\bar{p}$ beam incident on the LH$_2$ target. For example, a set of correction coils in addition to the main toroidal coils will allow a smaller focus for the polarized antiprotons at the experimental target over a wide range of momentum and other beam parameters. A realistic design will be a compromise of the considerations above and other factors, such as cost and available space.

The trajectories of the $\bar{p}$'s scattered in the LH$_2$ target must be measured with several sets of position sensitive detectors, such as multiwire proportional chambers. One reason is to verify that a $\bar{p}p$ elastic scattering occurred in the LH$_2$ target, based on the observed scattering angle and particle bend in the toroid. A well separated $\bar{p}$ beam and/or a threshold Cerenkov to identify π^- and K^- particles in the polarized beam will be highly desirable, since pions and kaons will scatter at forward angles in the LH$_2$ target with kinematics similar to $\bar{p}p$ elastic scattering. Another reason to measure the trajectories of the scattered antiprotons is to tag the $\bar{p}$ polarization direction and magnitude on a particle-by-particle basis, since these will vary with the beam phase space. Corrections to the antiproton spin direction due to a nonideal toroidal field can also be included. A third reason is to measure the $\bar{p}$ angle and position at the experimental target, so that the scattering angle and interaction point can be determined. This beam line design gives sizeable beam divergences, especially at low momenta.

Detailed calculations [14] were performed to obtain estimated intensities and polarizations for the polarized antiproton beam design of Fig. 1. A Monte Carlo computer program was written to perform these calculations. The incident unpolarized $\bar{p}$ beam was assumed to be a secondary beam with $\pm 5\%$ momentum spread, ± 5.0 mrad divergence, and ± 1.0 cm spot size at the LH$_2$ target. The target length was taken to be 10 cm.

In the program, antiprotons were scattered in the LH$_2$ target with a cross section and polarization calculated from Legendre polynomial fits [11-13] to the experimental data from 0.4 to 1.7 GeV/c [9-13]. The square of the statistical uncertainty in spin observable measurements is proportional to $Q^{-1} = (P^2 d\sigma/d\Omega_{c.m.})^{-1}$, where P and $d\sigma/d\Omega$ are the $\bar{p}p$ elastic analyzing power and differential cross section. The quantity Q can be used as a "figure of merit" to compare scattering at different angles, and it is desired to maximize Q. Fig. 2 shows typical data, fits, and estimates of Q for one beam momentum as a function of 4-momentum transfer squared, t, from Ref. 11. The results of the fits suggest Q is maximized for $-t \simeq 0.1 - 0.15 (GeV/c)^2$ for beam momenta above about 0.6 GeV/c. The LH_2 target to toroid distance ΔZ_T was chosen to give a central value of $t \simeq -0.12 (GeV/c)^2$ accepted by the beam line for the Monte Carlo calculations.

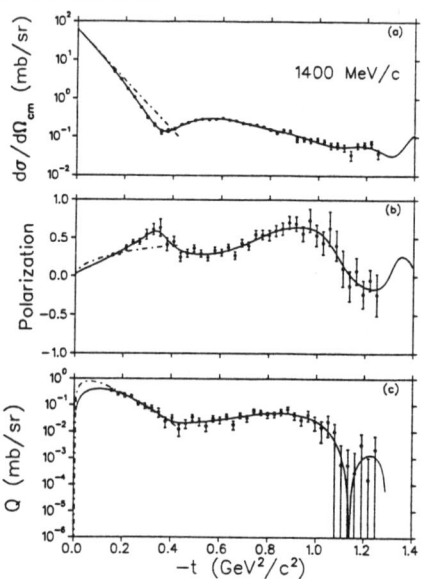

FIGURE 2. Plots of the c.m. differential cross section, polarization, and figure of merit, Q, as functions of the 4-momentum transfer squared, t. The laboratory momentum is 1400 MeV/c and the data are from Ref. 11. The Legendre polynomial fits are shown as solid lines.

For simplicity, an idealized toroidal field was assumed. In cylindrical coordinates centered on the toroid, the magnitude of the magnetic field was assumed to vary as r^{-1}. The magnetic field cross section was taken to be trapezoidal with values chosen to give the best beam line performance among a number of cases considered ($|z| \leq 0.5$ m at r = 0.55 m and $|z| \leq 1.5$ m at r = 1.0 m). It was assumed that the toroid coils subtended half the ϕ acceptance at the LH_2 target, and no correction coils were included. The maximum magnetic field was 1.3 T at r = 0.55 m.

The scattered antiproton trajectories were numerically integrated through the toroidal magnetic field in the Monte Carlo program. Particle trajectories were considered good if the $\bar{p}$'s passed through a 4-cm diameter, 10-cm long cylinder representing the experimental target; other trajectories were rejected. Polarized targets with similar dimensions have been constructed and used in experiments. The value of ΔZ_E was varied in order to approximately maximize the fraction of good trajectories.

The results of these calculations suggest that the $\bar{p}$ beam polarization and intensity will be typically 0.2 and 2×10^{-4} per incident unpolarized antiproton, respectively, in the momentum range 0.5 to 2.5 GeV/c. Several possibilities to increase the intensity exist, such as increasing the experimental target volume or the LH_2 target length, or adding correction coils to the toroid. The intensity was also found to be sensitive to the incident unpolarized beam momentum spread, but nearly independent of divergence or spot size up to ±20 mrad and ±1.5 cm, respectively.

Reversal of polarization direction by 180° is a common way to reduce systematic errors in spin experiments. For the transverse spins in the beam line in Fig. 1, this can be accomplished by adding a solenoid after the toroid to precess the spins by ±90°. The change in solenoid current from +90° to −90° spin precession can occur in minutes, even for superconducting solenoids. With this option for reversal of the beam spin direction, experiments on unpolarized targets become feasible, such as for studies of polarized $\bar{p}$ scattering from nuclear targets.

One problem is that the polarized $\bar{p}$ beam in Fig. 1 contains no longitudinal polarization. This can be solved by the addition of three dipole magnets as in Fig. 3 and Ref. 15. This solution keeps the average beam position and direction fixed at the experimental target, so that the detectors after the beam focus would not need to move as a function of beam momentum. The bend in the final dipole, $\theta = 90° \cdot m/[E \cdot (g/2 - 1)]$, is momentum dependent and is determined by the requirement that the $\bar{p}$ spin precess by 90° in the horizontal plane. The $\bar{p}$ mass and energy in the expression above are m and E. Knowledge of the field map of the final dipole magnet and of the $\bar{p}$ trajectory will enable the spin precession to be accurately computed for each particle, even for cases with large beam divergence at low momenta.

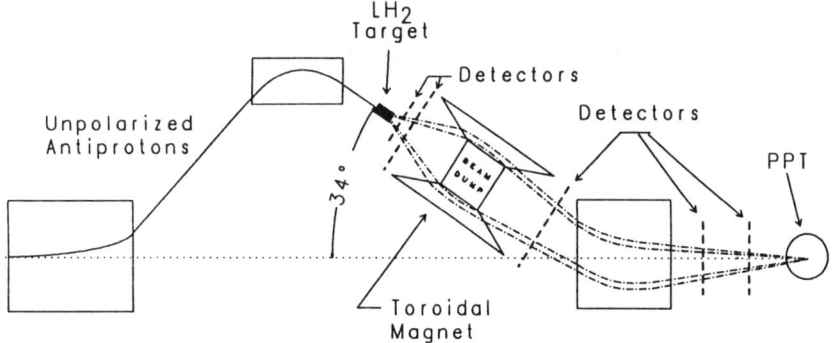

FIGURE 3. Drawing of a possible design for a ~1.0 GeV/c transversely and longitudinally polarized $\bar{p}$ beam line. The bending magnets are required to keep the average angle of the polarized antiprotons at the experimental target fixed as a function of momentum.

A feasible and conservative design for a medium energy polarized antiproton beam has been presented. The design requires an intense beam of unpolarized antiprotons ($\geq 10^7$/sec) from a typical secondary beam line in order to achieve reasonable $\bar{p}p$ elastic scattering count rates. All three beam spin directions can be achieved. Methods were discussed to reverse the spin directions in modest times, and to change to a polarized proton beam if desired. It is expected that experiments with such a beam would have a profound effect on the understanding of the $\overline{N}N$ interaction at intermediate energies.

We wish to express our gratitude to our colleagues at Valparaiso University, CEA Saclay, France, and Argonne National Laboratory for helpful suggestions during the course of this work. We are also thankful for help with some of the figures by D. Lopiano. This work was supported in part by the U.S. Department of Energy, Division of High Energy Physics, Contract W-31-109-ENG-38.

References

1. K. Imai, Proc. 6th Int. Symp. on Polarization Phenomena in Nuclear Physics, Osaka, Japan, Eds. M. Kondo et al., Suppl. Jour. Roy. Soc. Japan **55**, 1136-1139 (1986).
2. H. Poth, Proc. Conf. on Intersections Between Particle and Nuclear Physics, Lake Louise, Canada, Amer. Inst. Phys. Conf. Proc. **150**, 480-489 (1986).
3. Y. Onel, A. Penzo, and R. Rossmanith, Proc. Conf. on Intersections Between Particle and Nuclear Physics, Lake Louise, Canada, Amer. Inst. Phys. Conf. Proc. **150**, 1229-1231 (1986); T.O. Niinikoski and R. Rossmanith, Nucl. Instr. Meth. **A255**, 460-465 (1987).
4. P.L. Csonka, Nucl. Instr. Meth. **63**, 247-252 (1968).
5. G. Graw, Physics With Polarized Beams on Polarized Targets, eds. J. Sowinski and S.E. Vigdor (World Scientific, Singapore, 1990) pp. 328-348.
6. F. Rathmann et al., Phys. Rev. Lett. **71**, 1379-1382 (1993).
7. R. Birsa et al., Phys. Lett. **155B**, 437-441 (1985).
8. D.P. Grosnick et al., Nucl. Instr. Meth. **A290**, 269-292 (1990).
9. C. Daum et al., Nucl. Phys. **B6**, 617-627 (1968).
10. M.G. Albrow et al., Nucl. Phys. **B37**, 349-363 (1972).
11. R.A. Kunne et al., Nucl. Phys. **B323**, 1-36 (1989).
12. R. Bertini et al., Phys. Lett. **228B**, 531-535 (1989).
13. F. Perrot-Kunne et al., Phys. Lett. **261B**, 188-190 (1991).
14. E.W. Vaandering, H.M. Spinka, and J.S. Hofmann, accepted for publ. in Nucl. Instr. Meth. A.
15. E. Colton et al., Nucl. Instr. Meth. **151**, 85-88 (1978); E.P. Colton, IEEE Trans. Nucl. Sci. **NS-26**, 3206-3208 (1979); H. Spinka et al., Nucl. Instr. Meth. **211**, 239-261 (1983).

VIII. CONCLUDING PRESENTATION

Spin Physics in the Next Decade

J.M. Moss

Physics Division, Los Alamos National Laboratory
Los Alamos, NM 87545

Abstract. I discuss several areas of polarized structure function physics that will be feasible in the next 5 to 10 years at polarized storage ring facilities, operating either in the collider mode or with polarized internal jet targets.

INTRODUCTION

After agreeing to this assignment, I decided to review the Proceedings of the 6th International Polarization Conference[1] to see whether anyone had agreed to prognosticate about the next ten years back in 1985. Unfortunately no-one had, at least in print. In spite of the lack of guidance I will proceed with little trepidation, since it seems clear that a few areas of polarized experimental technology, currently under development, will drive the field for at least the next 5-10 years. My personal opinion is that the fascinating field of nonperturbative QCD will be led by polarized beam experiments. With these statements of faith made I will focus on two developments that will surely produce a wealth of new physics within the next ten years. These are, A) polarized colliding beams, and B) polarized jet targets in conjunction with polarized beams in storage rings. In order to make the overview manageable, I will cover only selected areas where the physics goals are the elucidation of nonperturbative aspects of QCD applied to nucleons and nuclei.

A major resurgence of interest in the spin structure of the nucleon followed the 1988 publication[2] of small-x measurements of $g_1^p(x)$ by the European Muon Collaboration (EMC). Analysis of the EMC data and larger-x deep-inelastic scattering (DIS) data from SLAC[3] gave rise to the so-called "spin crisis" -- the observation that the integral contribution of the up and down quarks to the proton's helicity was small. The past six years has seen an enormous effort to acquire more precise DIS data using polarized electron and muon beams. New measurements[4] of $g_1^p(x)$ confirm earlier EMC results, and new data on the neutron from DIS on polarized ^{2}H[5] and polarized ^{3}He[6] add significantly to our knowledge of quark polarization of the nucleon system.

THE HERMES EXPERIMENT

The HERMES experiment[7] will employ polarized gas jet targets in the HERA 30 GeV electron ring to make a qualitative advance in the experimental science of DIS. The targets are jets of ^{1}H, ^{2}H, and ^{3}He flowing though an active cell region of ~40 cm length yielding target densities in the range $1\text{-}10 * 10^{14}$ nucleons cm^{-2}. Combined with electron currents of 30 mA, this gives luminosities of $1.8\text{-}50*10^{31}$ cm^{-2}sec^{-1}, values very competitive with fixed-target electron experiments. What is much better than just competitive is, of course, the nearly pure target of polarized light ions, free from the dominant background of cryogenic polarized targets where the desired target material is little more than an impurity. The result is that HERMES will be able to measure $g_1^p(x)$ in the large x region with better errors than are currently feasible in the best polarized (e,e') experiments. HERMES will also be able to make significantly better measurements of the $g_1^n(x)$ structure function. In my opinion the most interesting feature of the HERMES experiment will be its capability to detect hadrons produced in coincidence with deeply inelastic scattered electrons. Although exclusive DIS is not as theoretically clean, reactions such as (e,e' K^+) may yield information about anti-quark polarization.

Of course a large advantage of polarized experiments using electron storage rings is that the polarization comes almost free (This characterization would certainly be disputed by accelerator physicists.). The well-known Sokolov-Ternov effect[8], a collective effect of synchrotron radiation, builds up a polarization transverse to the plane of the ring; the electron spin must then be rotated into the longitudinal orientation at the HERMES spectrometer. The capability to produce and maintain this polarization (avoiding depolarizing resonances)in the HERA electron ring has been convincingly demonstrated recently[7] as is seen in Fig. 1. HERMES is scheduled to begin data taking in 1995. The current plan is to run ~3000 hours per year.

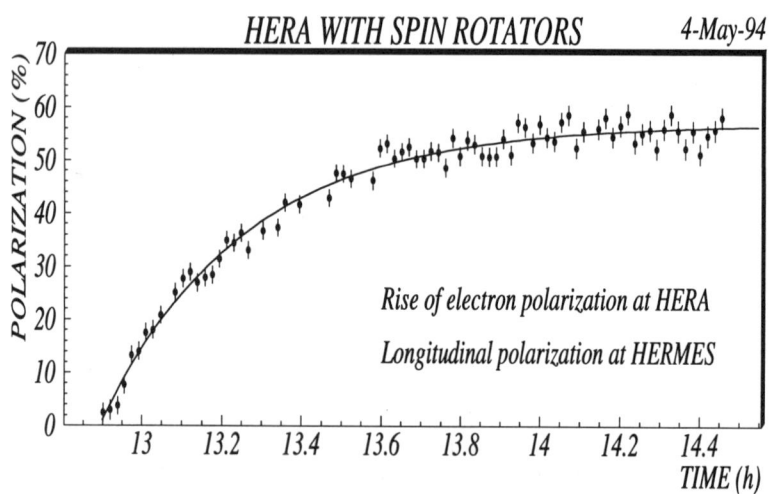

Figure 1. Production of electron polarization in HERA[7].

There is an exceedingly interesting and important HERMES-like experiment that could be carried out with an internal target in a hypothetical polarized-proton storage ring. I will return to this at the end of my talk.

POLARIZED HADRON COLLIDER PHYSICS

In spite of the large body of new data and the promise of even better experiments in the near future, polarized DIS measurements have major limitations. They do not directly provide any information about anti-quark or gluon helicity distributions. Neither can they probe the potentially equally interesting, and completely unknown, chiral-odd quark structure functions[9]. Hence there is great interest in polarized hard-hadronic processes that could offer new physics insight, complimentary to polarized DIS.

At present it looks very likely that the Relativistic Heavy-Ion Collider (RHIC) at Brookhaven National Laboratory will be successful in building the capability to provide polarized-proton collisions from $\sqrt{s}$ = 50-500 GeV; luminosities would be proportional to $\sqrt{s}$ with L = 2 10^{32} sec^{-1}cm^{-2} at $\sqrt{s}$= 500 GeV. This will presumably occur sometime after the scheduled turn-on of RHIC in 1999. The polarized physics program at RHIC is being advanced by the RHIC spin collaboration (RSC), an active and broad-based group composed of experimentalist, theorists, and accelerator physicists. An overview of many of the areas of physics to be investigated at RHIC has been published recently by the RSC[10]. Unlike electron polarization in storage rings which comes almost free, one has to be quite deliberate about producing high-energy polarized ions. Fortunately there now exists proven technology to preserve polarization in synchrotrons and storage rings, even while the beam passes through many potentially depolarizing resonances. The method involves "Siberian snakes", sets of (usually) dipole magnets which rotate the spin direction by 180º while preserving the geometry of the particle's orbit. The Indiana University Cooler has been the site of several important tests of the Siberian snake concept[11].

At present the future of the polarized-collider program looks very promising. A consortium of Japanese physicists under the leadership of the Institute for Chemical and Physical Research (RIKEN) will probably play a major role creating the polarized-beam capability at RHIC. In addition RIKEN will make a substantial contribution to upgrading the PHENIX detector for polarized physics.

Before discussing the polarized program at RHIC, I should also mention the strong interest at KEK in building a somewhat lower energy collider[12] which would be have the capability of both heavy-ion and polarized-beam collisions at somewhat lower energies than RHIC; for polarized-proton operation one would have $\sqrt{s}$ = 38 GeV. As I will discuss later, this energy range is very well suited to polarized structure function experiments, and very much complimentary to the energy range of the RHIC program. At the moment the future of the KEK PS collider is uncertain because of the laboratory's decision to proceed with the construction of a B factory.

Polarized Structure Function Physics

A polarized proton at high energies is a complex object. For example, a longitudinally polarized proton generally contains quarks aligned and anti-aligned with the proton's helicity. Within the parton model hadron polarization observables are related to the corresponding quantities at the parton level by a convolution equation of the type shown below for A_{LL}.

$$A_{LL}\sigma(x_1,x_2) = \sum_{i,j} \Delta f_i(x_1)\Delta f_j(x_2)\hat{a}^{ij}_{LL}(\hat{s},\hat{t},\hat{u})\sigma^{ij}(\hat{s},\hat{t},\hat{u}), \quad (1)$$

with,

$$\sigma(x_1,x_2) = \sum_{i,j} f_i(x_1)f_j(x_2)\sigma^{ij}(\hat{s},\hat{t},\hat{u}),$$
$$\Delta f_i(x) = f_i^+(x) - f_i^-(x),$$
$$f_i(x) = f_i^+(x) + f_i^-(x).$$

Where $\hat{s},\hat{t},\hat{u}$ are the usual Mandelstam variables at the parton level. The superscripts +(-) refer to spin projections parallel (antiparallel) to the hadron's spin. The summation is over all parton fusion processes which may contribute to the measured final state.

Polarized Drell-Yan Process

Before describing the PHENIX detector, I want to give a brief overview of the Drell-Yan[13] (DY) process, a well-understood electromagnetic reaction that shows much promise as a source of new information about polarized quark and antiquark structure. The process, denoted schematically by,

$$q(x_1, hadron1) + \bar{q}(x2, hadron2) \xrightarrow{\gamma^*} l^+l^-$$

has been studied for over twenty years, largely by measuring muon pairs in fixed-target experiments at Fermilab and CERN[14]. With the development of next-to-leading-order corrections QCD, it can be characterized as well understood theoretically. Phenomenological evaluations of parton structure functions now routinely use DY as well as DIS data.

The polarized DY process has received a great deal of publicity in part because of the well-understood nature of the reaction, and part because qualitatively new structure function physics can be obtained. As an example, a chiral-add structure function, the transversity[9], can be measured via transverse spin correlation in the DY process; the transversity is not accessible in DIS. In spite of the interest in the transverse spin correlation, the longitudinal spin correlation in the DY process is much more amenable to discussion because of its close relation to DIS. The equation for A_{LL} in the DY process reads,

$$A_{LL}^{DY}(x_1,x_2) = \frac{\sum_i e_i^2 \Delta f_i(x_1)\Delta \bar{f}_i(x_2) + (2 \leftrightarrow 1)}{\sum_i e_i^2 f_i(x_1)\bar{f}_i(x_2) + (2 \leftrightarrow 1)}. \quad (2)$$

For pp collisions in the kinematic range $x_F > 0.2$, a considerable simplification results from the dominant $4/9 \cdot u(x_1)\bar{u}(x_2)$ term in the denominator of Eq. 2. One finds

$$A_{LL}^{DY}(x_1, x_2) \approx \frac{\Delta u(x_1)}{u(x_1)} \times \frac{\Delta \bar{u}(x_1)}{\bar{u}(x_1)} + \frac{1}{4} \frac{\Delta d(x_1)}{u(x_1)} \times \frac{\Delta \bar{d}(x_1)}{\bar{u}(x_1)}. \tag{3}$$

Here, the notation $f_u(x) \equiv u(x)$ has been introduced. Neglecting small sea-quark effects for $x \geq 0.2$ one can express the structure functions of DIS as,

$$F_1^p \approx 1/9 \cdot u(x),$$
$$\Delta u(x) = 18/15(4g_1^p(x) - g_1^n(x)), \tag{4}$$
$$\Delta u(x) = 18/15(-g_1^p(x) + 4g_1^n(x)).$$

The relations and the measured values of $g_1^{p,n}$ shown in Figs. 2 and 3 allow even further simplification. For $x \geq 0.2$ one sees that $g_1^n \approx 0$, and with the above relations Eq. 3 becomes

$$A_{LL}^{DY}(x_1, x_2) \approx A_{LL}^{pDIS}(x_1) \left[\frac{16}{15} \frac{\Delta \bar{u}}{\bar{u}}(x_2) - \frac{1}{15} \frac{\Delta \bar{d}}{\bar{u}}(x_2) \right],$$
$$\approx A_{LL}^{pDIS}(x_1) \bullet \frac{\Delta \bar{u}}{\bar{u}}(x_2). \tag{5}$$

Thus for the DY process with $x_F > 0.2$ a longitudinally polarized proton can be thought of as a beam of polarized quarks with $p \approx A_{LL}^{pDIS}(x_1)$. It is apparent from the large measured values of that longitudinal spin correlation in the DY process will be a sensitive measure of the polarization of the $\bar{u}$ quark distribution of the proton.

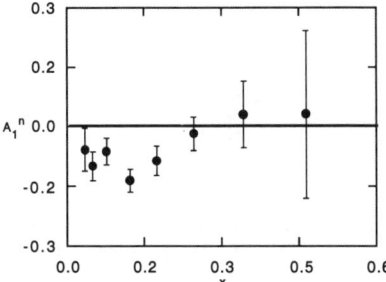
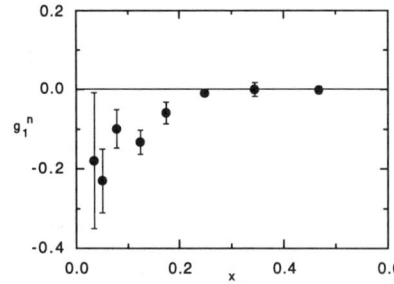

Figure 2. Longitudinal asymmetry and $g_1(x)$ structure function for deep-inelastic lepton scattering from the proton from CERN[2,4] and SLAC[3].

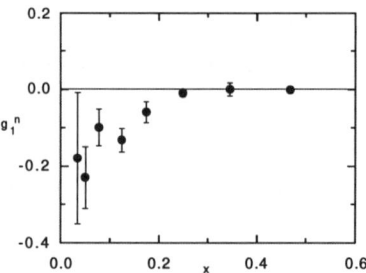

Figure 3. The g_1 structure function of the neutron from deep-inelastic electron scattering from ^{3}He[6].

The PHENIX Detector at RHIC

The baseline version of the PHENIX detector is shown in Fig. 4. I will discuss only one aspect of the PHENIX program, the polarized DY process. Recently the PHENIX Collaboration has expanded its spin activities considerably through an apparently successful collaboration with the RIKEN Institute.

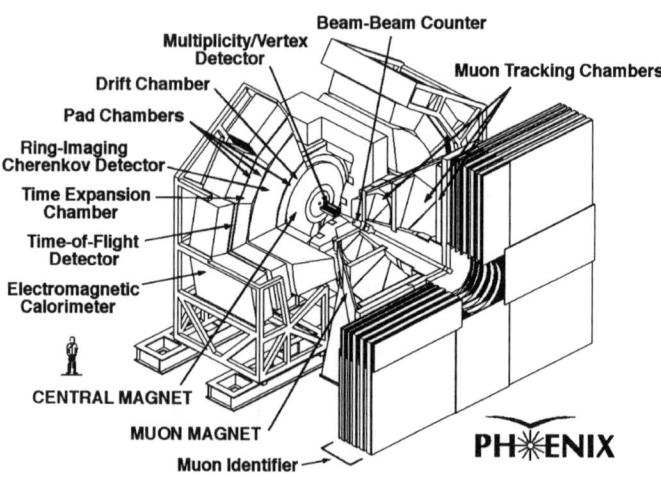

Figure 4. Baseline version of the PHENIX detector showing the first muon endcap. If the RIKEN/PHENIX spin initiative is successful the detector will have a 2nd muon endcap on the opposite side of the central magnet.

If successful, a major investment by the RIKEN Spin Group will make feasible a significant enhancement of the polarized collider program of the PHENIX Detector using the muon detection capability. The crucial addition of a second muon endcap and muon identification in the central rapidity region will yield a detector with large acceptance for high-mass muon pairs. The qualitative improvement of the upgraded PHENIX muon system is shown clearly in Fig. 5.

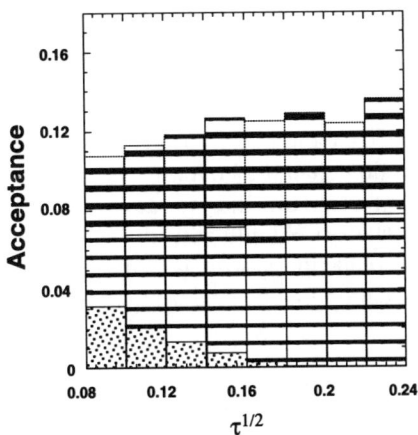

Figure 5. Acceptance of three versions of the PHENIX muon subsystem for Drell-Yan pairs with $M > 4$ GeV. Calculations were performed at $\sqrt{s} = 50$ GeV but are approximately valid at higher energies as well. The largest acceptance is for two endcaps and muon identification in the central spectrometer; the second largest for two endcaps only, and the smallest for the baseline version of PHENIX with only one endcap.

The PHENIX/Spin collaboration has recently made the case that the polarized DY process is best carried out at relatively low colliding energies, e. g. $\sqrt{s} = 50$ GeV. First it is likely that antiquark polarizations are largest at relatively large values of x_2; this leads to the requirement that $\tau = M^2/s = x_1 x_2$ be large. An additional feature dependent only on parton-model scaling is that the cross section scales as,

$$d^2\sigma / d\sqrt{\tau} dx_F \propto \frac{1}{s} F(\tau, x_F), \qquad (6)$$

carrying the strong message to minimize the energy required to get to the scale defined by $\tau = x_1 x_2$. When combined with the roughly linear increase in the luminosity of RHIC with energy, one finds the rate for the DY process scales as, $R \approx 1/\sqrt{s}$. A colliding-beam energy of 50 GeV appears to be optimum for determination of antiquark polarization in the range $x_2 = 0.04 - 0.2$. An additional incentive for relatively low-energy operation for DY physics has to do with the most serious source of background, semileptonic decay of charmed hadrons. The charm production cross section will increase by 6-10 in going from $\sqrt{s} = 50$ GeV to $\sqrt{s} = 200$ GeV.

Based on current luminosity projections, the PHENIX/Spin collaboration estimates that it can collect ~10K DY events in the course of a running period at $\sqrt{s} = 50$ GeV; this will permit significant measurements to be made of antiquark helicity, and both quark and antiquark transversity.

The STAR Detector at RHIC

The STAR (Solenoidal Tracker at RHIC) detector is based around a large solenoid magnet and time projection chamber covering the central rapidity region, $|\eta| \leq 1$. The STAR collaboration would like to add to the base detector a large electromagnetic calorimeter (EM) and shower-maximum detector in the central rapidity region, and an endcap calorimeter covering the range, $1 \leq |\eta| \leq 2$. A schematic drawing of the proposed EM calorimeter is shown in Fig. 6

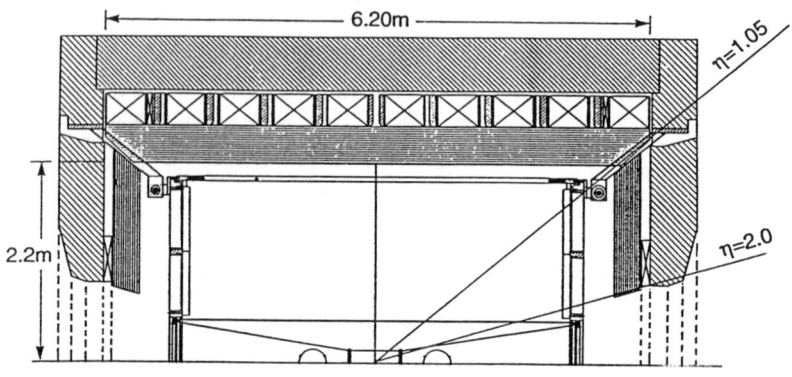

Figure 6. Schematic diagram of the proposed STAR EM calorimeter.

The STAR physics program emphasizes gluon polarization observed through jet and direct-photon production, and quark/antiquark polarization from $W^{\pm}$ and Z^0 production and decay into electrons[15]. I will discuss only one of the many processes of interest to the STAR collaboration -- direct-photon plus jet production.

Direct-photon production has been touted as a tool for studying the gluon structure function for many years. Only recently with the advent of high quality data from CDF and D0 at the Tevatron collider has quality of the experimental data lived up to this expectation. The process of interest to STAR is the coincident detection of a photon and associated jet via the gluon Compton scattering process,

$$g + q(\bar{q}) \to \gamma + jet$$

The idea is to isolate the kinematic regime where this process is dominant over the competing annihilation diagram. It is essential to concentrate on the forward rapidity region where largely polarized quarks scatter from modest x gluons -- much the same philosophy as described previously for the DY process. The parton-level spin correlation for gluon Compton scattering is peaked at forward angles resulting in a large gain in the figure of merit for the experiment if STAR has at least one endcap EM calorimeter. An endcap calorimeter has the additional advantage also yields valuable acceptance in the kinematic region where the parton-level asymmetry for gluon Compton scattering is largest[15]. Running at $\sqrt{s}$ = 200 GeV, the polarization of the gluon structure function could be measured with reasonable statistics in the range 0.02 to 0.2. This would address a very important and currently completely unresolved problem having the do with the axial current anomaly and its effect on spin sum rules[16].

Physics with a Jet Target in a Polarized-Proton Storage Ring

Although it is not feasible today because there are no polarized-proton storage rings, one can hope that in the next 5-10 years such facilities will be a reality. The idea is simple: put a polarized jet cell of ~10^{14} /cm² in a circulating current of ~100 mA of polarized protons at energies from 100 GeV to 1000 GeV. This would yield L=6.25*10^{31} cm⁻²sec.⁻¹, very respectable by collider standards, and large enough to collect several times 10^4 Drell-Yan events in a typical 1/3 year running period. The beam lifetimes at this luminosity look acceptable on paper (several hours), but of course much groundwork would have to be done to establish realistic rates. The spectrometer required is unspecified at the moment, but might look similar to the existing HERMES detector. As has been argued previously, the energy range $14 \leq \sqrt{s} \leq 43$ GeV is about right for polarized DY physics, the lower side of this range would definitely complement the polarized collider experiments planned at RHIC.

One experiment that would be quite feasible in a proton storage ring, and problematical for collider operation, is the measurement of the polarization of the antidown quark of the proton. This would involve a stored polarized-proton beam and a polarized neutron target, such as is available in the form of polarized ³He. Applying charge symmetry to equation 5, one has for a p+n experiment,

$$A_{LL}^{DY}(x_1, x_2) \approx A_{LL}^{pDIS}(x_1) \bullet \frac{\Delta \bar{d}}{\bar{d}}(x_2). \qquad (7)$$

I will close with the observation that polarized hadron physics (with both hadrons polarized) has sufficient promise that several facilities should be able to share the glory.

References

1. Proceedings of the XI International Symposium on Polarization Phenomena in Nuclear Physics. Osaka, Japan, 1985, Physical Society of Japan, Komiyama Printing Co., Tokyo, Japan.

2. J. Ashman et al., Phys. Lett **B206**, 364(1988); Nucl. Phys. **B328**, 1(1989).

3. M.J. Alguard et al., Phys. rev. Lett. **37**, 1261(1978); **41**, 70(1978); G. Baum et al., Phys. Rev. Lett. **51**, 1135(1983).

4. D. Adams et al., Phys. Lett. **B329**, 399(1994).

5. B. Adeva et al., Phys. Lett. **B302**, 533(1993).

6. P.L. Anthony et al., Phys. Rev. Lett. **71**, 959(1993).

7. The HERMES Experiment at the Deutsches Elektronen Synchrotron (DESY), DESY World Wide Web Server home page.

8. A.A. Sokolov and I.M. Ternov, Sov. Phys. Dokl. **8**, 1203(1964).

9. R.L. Jaffe and Xiangdong Xi, Phys. Rev. Lett. **67**, 552(1991), and references therein.

10. G. Bunce et al., Particle World **3**, 1(1992).

11. A.D. Krisch et al., Phys. Rev. Lett. **63**, 1137(1989).

12. Proc. of the Workshop on the PS-Spin Collider, Ed. by A. Mori, Tsukuba, Japan 1992, KEK Proceedings 93-5 (unpublished).

13. S.D. Drell and T-M. Yan, Phys. Rev. Lett. **25**, 316(1971).

14. W.J. Stirling and M.R. Whalley, Journal of Phys. **G 19**, D1(1993).

15. STAR electromagnetic calorimeter conceptual design report, LBL PUB-5380.

16. R.L. Jaffe and A. Manohar, Nucl. Phys. **B337**, 509(1991); and references therein.

# APPENDICES

SPIN'94

Conference Schedule and Plenary Session Talks

start   talk   author, institution
time length title

FRIDAY MORNING, SEPTEMBER 16
[Alumni Hall]

OPENING

8:30   15   *George Walker, Indiana University Vice-President*
               Opening remarks

NUCLEAR: Quarks to Hadrons to Nuclear Matter Chair: B.D. Serot, Indiana

8:45   30   *A.W. Thomas, Adelaide*
               Non-perturbative aspects of the spin structure functions of nucleons and nuclei
9:25   30   *T.N. Taddeucci, Los Alamos*
               The nuclear spin-isospin response to quasifree nucleon scattering

10:05 20   **COFFEE BREAK**

HIGH ENERGY Chair: A. Masaike, Tokyo

10:25 30   *S.E. Vigdor, Indiana*
               LISS: Planning for spin physics with multi-GeV nucleon beams at IUCF
11:05 30   *T. Devlin, Rutgers*
               Discovery of hyperon polarization at Fermilab
11:45 30   *S. Nurushev, Protvino*
               Summary of SPIN'93 International Workshop at Protvino

FRIDAY AFTERNOON, SEPTEMBER 16

PARALLEL SESSIONS

	NUCLEAR	HIGH ENERGY
14:00	Hadron Form Factors I [Frangipani]	Lepton Accelerators, Beams and Polarimeters I [Maple]
14:00	Polarized Sources and Targets I [Oak]	Polarized Solid Targets I [Sassafras]
14:00	Low Energy Nuclear Reactions I [Persimmon]	Hadron Accelerators and Beams I [Whittenberger]

15:45 30   **COFFEE BREAK**

16:15	Hadron Form Factors II [Frangipani]	Strong Interactions at High Energy I [Walnut]
16:15	Polarized Sources and Targets II [Oak]	Polarized Electron Sources [Maple]
16:15	Low Energy Nuclear Reactions II [Persimmon]	Hadron Accelerators and Beams II [Whittenberger]

SATURDAY MORNING, SEPTEMBER 17
[Alumni Hall]

HIGH ENERGY Chair: A. Efremov, Dubna

8:30   30   *M. Placidi, CERN*
Polarization at LEP

9:10   30   *L. Pondrom, Wisconsin*
Consequences of hyperon polarization

9:50   25   **COFFEE BREAK**

NUCLEAR: Symmetry Tests in Complex Systems Chair: M. Simonius, Zurich

10:15   30   *B. Heckel, U. Washington*
Limit on the electric dipole moment of the ^{199}Hg atom

10:55   30   *J. Deutsch, Louvain*
Search for incomplete parity violation in leptonic and semi-leptonic processes: status and perspectives

11:35   30   *Y.-F. Yen, Los Alamos*
Study of parity and time-reversal violation in neutron-nucleus interactions

SATURDAY AFTERNOON, SEPTEMBER 17
[Alumni Hall]

HIGH ENERGY/NUCLEAR PLENARY SESSION: Nucleon Spin Structure Functions Chair: V. Hughes, Yale

14:00   30   *R. Windmolders, CERN*
SMC results for nucleon spin structure functions

14:40   30   *D. Day, Virginia*
E143 results from SLAC

15:20   30   *H. Jackson, Argonne*
HERMES

16:00   30   *J. Soffer, Marseille*
Use of weak interactions to measure spin distributions

16:45–17:00   **Buses leave for IUCF reception and tour**
17:00–20:00   **IUCF reception and tour**

SUNDAY MORNING, SEPTEMBER 18

9:00   **Buses depart from Union for all-day excursion and Belle of Louisville cruise**

MONDAY MORNING, SEPTEMBER 19
[Alumni Hall]

NUCLEAR: Polarization Effects in Strange Systems   Chair: H. Fearing, TRIUMF

8:30	30	H. Ejiri, Osaka Polarization and weak decays of hypernuclei
9:10	30	K. Kilian, Jülich Polarization effects in $\bar{p}p \rightarrow \bar{\Lambda}\Lambda$
9:50	25	**COFFEE BREAK**

HIGH ENERGY   Chair: G. Fidecaro, CERN

10:15	30	M. Woods, SLAC Polarization at SLAC
10:55	30	G. Voss, DESY Polarization at HERA
11:35	30	I. Ternov, Moscow State The spin of relativistic nuclei in external magnetic fields

MONDAY AFTERNOON, SEPTEMBER 19

NUCLEAR PLENARY SESSION: Few Body Studies   Chair: C. Elster, Ohio
[Whittenberger]

14:00	30	A. Sandorfi, Brookhaven Few-body photodisintegration with polarized photons
14:40	30	R. Milner, MIT The spin structure of ^{3}He
15:20	20	*COFFEE BREAK*
15:40	30	W. Tornow, Duke New results in nucleon-nucleon scattering at low energies
16:20	30	H. Spinka, Argonne New results in nucleon-nucleon scattering at intermediate energies

HIGH ENERGY PARALLEL SESSIONS

14:00	Strong Interactions at High Energy II [Walnut]
14:00	Electroweak Interactions I [Frangipani]
14:00	Solid Polarized Targets II [Maple]
14:00	Hadron Accelerators and Beams III [Oak]

15:45	30	**COFFEE BREAK**
16:15		Solid Polarized Targets III [Maple]
16:15		Nucleon Spin Structure Functions I [Frangipani]
16:15		Hadron Accelerators and Beams IV [Oak]

20:00–22:00 **CONCERT and RECEPTION** [Waldron Arts Center]

TUESDAY MORNING, SEPTEMBER 20
[Alumni Hall]

HIGH ENERGY Chair: E. Courant, Brookhaven/Michigan

8:30	30	*L.W. Anderson, Wisconsin* Report on the Madison workshop on sources and targets
9:05	30	*W. Meyer, Bonn* **RAPPORTEUR:** Solid polarized targets
9:40	30	*S. Ozaki, Brookhaven* Spin at RHIC

10:15	20	**COFFEE BREAK**

NUCLEAR: Nucleon Form Factors Chair: R. Jaffe, MIT

10:35	30	*E. Reichert, Mainz* Determination of the neutron electric form factor in quasielastic collisions of polarized electrons with ^{3}He and ^{2}H
11:15	30	*I. Stancu, Riverside* Measurements of the proton's axial form factor via neutrino-proton scattering
11:55	20	*E. Beise, Maryland* **RAPPORTEUR:** Hadron form factors
12:15	20	*D. Fick, Marburg* **RAPPORTEUR:** Low energy nuclear reactions

TUESDAY AFTERNOON, SEPTEMBER 20

PARALLEL SESSIONS

	NUCLEAR	HIGH ENERGY
14:00	Symmetries I [Walnut]	Lepton Accelerators, Beams and Polarimeters II [Maple]
14:00	Few-Body Systems I [Persimmon]	Nucleon Spin Structure Functions II [Whittenberger]
14:00	Intermediate-Energy Hadron-Induced Reactions I [Frangipani]	Strong Interactions at High Energy III [Oak]

15:45	30	**COFFEE BREAK**

16:15	Symmetries II [Walnut]	Lepton Accelerators, Beams and Polarimeters III [Maple]
16:15	Few-Body Systems II [Persimmon]	Strong Interactions at High Energy IV [Oak]
16:15	Intermediate-Energy Electromagnetic Interactions I [Frangipani]	Electroweak Interactions II [Whittenberger]

18:30–20:00 **RECEPTION** [Solarium]
20:00 **BANQUET** [Alumni Hall]

WEDNESDAY MORNING, SEPTEMBER 21
[Alumni Hall]

NUCLEAR Chair: H.-O. Meyer, Indiana

8:30	30	W. Haeberli, Wisconsin Proton-proton interactions with polarized internal targets in storage rings
9:10	20	A. Eiró, Lisbon **RAPPORTEUR:** Few Body Systems
9:30	20	S. Page, Manitoba **RAPPORTEUR:** Symmetries
9:50	25	**COFFEE BREAK**

HIGH ENERGY Chair: W. Happer, Princeton

10:15	30	M. Anselmino, Torino **RAPPORTEUR:** Strong interactions at high energy
10:55	30	H. Steiner, Berkeley **RAPPORTEUR:** Electroweak interactions
11:35	30	T. Roser, Brookhaven **RAPPORTEUR:** Hadron beams and accelerators

WEDNESDAY AFTERNOON, SEPTEMBER 21

PARALLEL SESSIONS

	NUCLEAR	HIGH ENERGY
14:00	Intermediate-Energy Hadron-Induced Reactions II [Frangipani]	Lepton Accelerators, Beams and Polarimeters IV [Maple]
14:00	Intermediate-Energy Electromagnetic Interactions II [Oak]	Nucleon Spin Structure Functions III [Whittenberger]
14:00	Polarized Sources and Targets III [Persimmon]	

15:45–17:00 **COFFEE + POSTER SESSION**

17:30 **PRE-OPERA BUFFET DINNER** [Tudor Room]
18:45 **OPERA DRESS REHEARSAL** [Musical Arts Center]

THURSDAY MORNING, SEPTEMBER 22
[Alumni Hall]

HIGH ENERGY Chair: L. Soloviev, Protvino

8:30	30	D. Barber, DESY
		RAPPORTEUR: Polarized beams and polarimeters at lepton accelerators
9:05	30	R. Voss, CERN
		RAPPORTEUR: Hadron Structure

NUCLEAR Chair: J. Arvieux, Saclay

9:35	20	Y. Mori, KEK
		RAPPORTEUR: Polarized sources and gaseous targets
9:55	20	M. Garçon, Saclay
		RAPPORTEUR: Intermediate energy electromagnetic interactions
10:15	20	H. Sakai, Tokyo
		RAPPORTEUR: Intermediate energy hadron-induced reactions
10:35	25	**COFFEE BREAK**

CONCLUDING TALKS Chair: J. Arvieux, Saclay

HIGH ENERGY

11:00	45	A. Krisch, Michigan
		Highlights and future prospects

NUCLEAR

11:45	45	J. Moss, Los Alamos
		Spin physics in the next decade
12:30		**END OF CONFERENCE**

SPIN'94
Parallel Sessions

| start time | talk length | author, institution title |

FRIDAY, SEPTEMBER 16

NUCLEAR

HADRON FORM FACTORS I Chair: C. Horowitz, Indiana
[Frangipani Room]

2:00 20 *J. Mitchell, CEBAF*
       Measurement of the neutron electric form factor with a polarized deuterium target at CEBAF

2:25 20 *A. Lai, Kent State*
       A new neutron polarimeter for measurements of G_E^n from the $d(\vec{e},e'\vec{n})p$ reaction

2:50 20 *J.O. Hansen, MIT*
       Measurement of G_E^n with a polarized ^{3}He target

3:15 20 *M. Pitt, Caltech*
       The SAMPLE experiment: parity violating electron scattering from the proton and deuteron

POLARIZED SOURCES AND TARGETS I: Ion Sources Chair: P. Schmor, Manitoba [Oak Room]

2:00 15 *A. Belov, Russian Academy of Sciences*
       A study of the source of polarized negative hydrogen ions with a deuterium plasma ionizer

2:18 15 *A.N. Zelenski, TRIUMF*
       Spin exchange polarization studies at the TRIUMF OPPIS

2:36 15 *M. Kinsho, KEK*
       A dual-optically-pumped polarized negative deuterium ion source

2:54 15 *V. Derenchuk, IUCF*
       IUCF high intensity polarized ion source operation

3:12 15 *A.J. Mendez, TUNL*
       Investigation of ion beam extraction system performance for the TUNL polarized ion source

3:30 15 *P.D. Eversheim, Bonn*
       The polarized ion source for COSY

LOW ENERGY NUCLEAR REACTIONS I: Light ions Chair: R.C. Johnson, Surrey [Persimmon Room]

2:00	15	H. Toyokawa, Osaka Cross section and iT_{11} for the ^{12}C(d,p) reaction and coupling between the $^3S_1$ states
2:18	15	L. Zetta, Milano Spectroscopy of ^{88}Y by means of the ^{91}Zr(p,α)^{88}Y reaction at 22 MeV
2:36	15	M. Tanifuji, Hosei Analyzing power formulae for nuclear reactions in the low energy limit, application to ^{2}H(d,p)^{3}H reactions
2:54	15	A. Plavko, St. Petersburg Technical Inelastic scattering of polarized protons at low energies from various medium and light-weight nuclei
3:12	15	Y. Iseri, Chiba Singlet state contributions to deuteron elastic scattering
3:30	15	M.A. Al-Ohali, Duke A dispersive optical model for the neutron-nucleus interaction from -80 to $+80$ MeV for nuclei in the mass region $27 \leq A \leq 32$

HIGH ENERGY

LEPTON ACCELERATORS, BEAMS AND POLARIMETERS I: Lepton Polarimeters Chair: M. Woods, SLAC [Maple Room]

2:00	15	H.R. Band, Wisconsin Möller polarimeters for SLAC
2:20	15	A. Afanasev, CEBAF Atomic electron binding effects for Möller polarimetry at electron beams
2:40	15	C. Cavata, Saclay The Compton polarimeter for CEBAF
3:00	15	F. Zetsche, DESY/Hamburg Experience with the fast polarimeter at HERA
3:20	15	G. Kezerashvili, Novosibirsk The fast polarimeter for the VEPP–4M collider
3:40	15	Y. Derbenev, IUCF/Michigan RF-resonance beam polarimeter

POLARIZED SOLID TARGETS I Chair: G. Court, Liverpool [Sassafras Room]

2:00	25	A. Honig, Syracuse Large, accessible, highly polarized frozen-spin solid HD targets
2:30	15	S. Whisnant, South Carolina SPHICE: a strongly polarized H and D target
2:50	15	D. Haase, North Carolina State A statically polarized solid ^{3}He target

| 3:10 | 20 | S. Penttila, Los Alamos
LAMPF neutron spin filter |
| 3:35 | 15 | S. Bultmann, Bielefeld
^{13}C cross relaxation measurements |
| 3:55 | 15 | I. Plis, Dubna
Target with frozen nuclear polarization for experiments at low energies |

HADRON ACCELERATORS AND BEAMS I: Polarized beam preservation and manipulation Chair: H. Sato, KEK [Whittenberger Auditorium]

| 2:00 | 20 | E.D. Courant, Brookhaven/Michigan
Prospects of RHIC spin and Tevatron spin projects |
| 2:25 | 10 | D.D. Caussyn, Michigan
First partial Siberian snake test during acceleration |
| 2:40 | 10 | H. Huang, Indiana/Brookhaven
Preservation of proton polarization by a partial Siberian snake |
| 2:55 | 10 | T. Toyama, KEK
Prospect for polarized beam acceleration at the KEK PS |
| 3:10 | 10 | F. Rathmann, Marburg
Polarizing a stored, cooled proton beam by spin-dependent interaction with a polarized hydrogen gas target |
| 3:25 | 20 | H.-O. Meyer, Indiana
Towards longitudinal beam polarization in the IUCF Cooler
Beam depolarization with an internal target
Effect of a polarized hydrogen target on the polarization of a stored proton beam |

| 3:45 | 30 | **COFFEE BREAK** |

NUCLEAR

HADRON FORM FACTORS II Chair: W. Turchinetz, MIT
[Frangipani Room]

| 4:15 | 20 | D. Beck, U. Illinois
Measurement of the g_0 proton form factor by parity-violating electron scattering at CEBAF |
| 4:40 | 20 | S. Pate, MIT
More details on HERMES |
| 5:05 | 20 | A. Yu. Umnikov, Alberta
The deuteron spin structure function in a Bethe-Salpeter approach and the extraction of the neutron spin-dependent structure function |
| 5:30 | 20 | B. Vuaridel, Geneva
Transverse polarization in deep inelastic scattering |

POLARIZED SOURCES AND TARGETS II: Gaseous targets Chair: R.J. Holt, Illinois [Oak Room]

4:15	15	F. Stock, Heidelberg The HERMES-FILTEX target source for polarized hydrogen and deuterium
4:33	15	T. Wise, Wisconsin Spin-polarized internal gas target for hydrogen and deuterium in the IUCF Cooler Ring
4:51	15	D.K. Toporkov, Novosibirsk Laser-driven internal polarized deuterium target for the VEPP-3 electron storage ring
5:09	15	K. Zapfe, Munich/DESY High density polarized hydrogen gas target for storage rings
5:27	15	L. Kramer, MIT An internal polarized ^{3}He target for electron storage rings
5:45	15	V.G. Luppov, Michigan Status of the Michigan-MIT ultra cold polarized hydrogen jet target

LOW ENERGY NUCLEAR REACTIONS II: Heavy Ions Chair: K. Kemper, Florida State [Persimmon Room]

4:15	15	E.L. Reber, Florida State Analyzing powers for elastic and inelastic scattering of polarized ^{6}Li from ^{12}C at 30 MeV
4:33	15	R.C. Johnson, Surrey Semi-classical analysis of scattering of deformed heavy-ions below the Coulomb barrier
4:51	15	R.P. Ward, Birmingham Polarized ^{6}Li studies at the Nuclear Structure Facility, Daresbury
5:09	15	J.A. Christley, Surrey Fusion of a polarized projectile with a polarized target
5:27	15	K. Matsuta, Osaka Spin polarization of ^{23}Mg in ^{24}Mg+Au, Cu and Al collisions at 91 A MeV
5:45	15	H. Sakai, Tokyo Spin-flip probability via the ^{26}Mg(^{3}He,tγ)^{26}Al reaction

HIGH ENERGY

STRONG INTERACTIONS AT HIGH ENERGY I: Elastic Scattering at Small and Large Angles Chair: M. Anselmino, Torino [Walnut Room]

4:15	15	D. Grosnick, Argonne Measurement of $\Delta\sigma_L$(pp) and $\Delta\sigma_L(\bar{p}p)$ at 200 GeV/c
4:35	15	S. Troshin, Protvino Theoretical aspects of single-spin studies

4:55	15	*G.P. Ramsey, Loyola* Polarization and N–N elastic scattering amplitudes
5:15	15	*P. Draper, Texas at Arlington* Proton-proton elastic scattering experiment at RHIC
5:35	15	*V. Solovianov, Protvino* Experiment NEPTUN – physics and status
5:55	15	*A.M.T. Lin, Michigan* NEPTUN – a spectrometer for measuring the spin analyzing power in p–p elastic scattering at large $p_\perp^2$ at 400 GeV (and 3 TeV) at UNK

POLARIZED ELECTRON SOURCES Chair: R. Prepost, Wisconsin
[Maple Room]

4:15	20	*C. Prescott, SLAC* Review of the SLAC and Les Houches Workshops
4:40	20	*H. Aoyagi, Nagoya* Recent progress on cathode development from Nagoya and KEK
5:05	20	*T. Maruyama, SLAC* Recent progress on cathode development from SLAC/Wisconsin
5:30	15	*S. Cohen, Orsay* Results from the Orsay polarized electron source
5:50	15	*H. Tang, SLAC* Prospects for a DC-gun based polarized electron source for the next generation of linear colliders

HADRON ACCELERATORS AND BEAMS II Chair: H. Sato, KEK
[Whittenberger Auditorium]

4:15	15	*D. Underwood, Argonne* Polarimeters for high energy polarized beams
4:35	10	*R.A. Phelps, Michigan* Spin flipping a stored polarized proton beam
4:50	10	*Y. Derbenev, IUCF/Michigan* A concept for Stern-Gerlach polarization in storage rings
5:05	10	*H. Okamura, RIKEN* Technique for rotating the spin direction at RIKEN
5:20	10	*S.B. Nurushev, Protvino* Extraction and transformation of polarization at RHIC at 250 GeV/c
5:35	10	*V.I. Ptitsin, Novosibirsk* Helical spin rotators and snakes

MONDAY, SEPTEMBER 19

HIGH ENERGY

STRONG INTERACTIONS AT HIGH ENERGY II: Single Spin Asymmetry
Chair: G. Bunce, Brookhaven [Walnut Room]

2:00	15	N. Saito, Kyoto Measurement of single spin asymmetry for direct photon production in pp collisions at 200 GeV/c
2:20	15	W. Nowak, DESY Single spin asymmetries in proton-proton and proton-neutron scattering at 820 GeV
2:40	15	S. Timm, Carnegie-Mellon Energy, p_T, and x_F dependence of the polarization of Σ^+ hyperons produced by 800 GeV/c protons
3:00	15	K.A. Johns, Arizona Ξ and Ω^- polarizations produced by high energy polarized and unpolarized neutral beams
3:20	15	S.B. Nurushev, Protvino Single spin asymmetries and invariant cross sections of high p_T inclusive π^0 production in $\bar{p}$–p interactions
3:40	15	G.J. Musulmanbekov, Dubna Simulation of one-spin meson asymmetry in $\vec{p}$–p and $\vec{p}$–A collisions at high energy

ELECTROWEAK INTERACTIONS I Chair: H. Steiner, Berkeley
[Frangipani Room]

2:00	20	K.T. Pitts, Fermilab Electroweak coupling measurements from polarized Bhabha scattering at the Z resonance
2:25	20	M. Woods, SLAC Precise measurement of the left-right cross section asymmetry in Z production by e^+e^- collisions
2:50	15	O. Adriani, INFN Firenze Measurement of the tau polarization with the L3 detector
3:10	15	M. Maolinbay, Zürich Study of the reaction $e^+e^- \to \gamma\gamma(\gamma)$ at the Z^0 pole with the L3 detector at LEP
3:30	15	M. Jezabek, Krakow Polarization in top pair production and decay near threshold

SOLID POLARIZED TARGETS II Chair: S. Pentilla, Los Alamos
[Maple Room]

2:00	15	J. Ball, Saclay Polarized target for nucleon-nucleon experiments at Saturne

2:20	20	*J. Kyynarainen, CERN* The new SMC dilution refrigerator
2:45	15	*P. Delheij, TRIUMF* The CHAOS polarized target
3:05	15	*S. Goertz, Bonn* Irradiated lithium hydrides as polarized target materials
3:25	15	*M. Iinuma, Kyoto* Proton polarization at room temperature
3:45	15	*N. Piskunov, Dubna* The new Dubna movable target

HADRON ACCELERATORS AND BEAMS III: Polarimeters Chair: L. Ratner, Brookhaven/Michigan [Oak Room]

2:00	10	*C.D. Roper, TUNL* Static and electromagnetic field requirements for a Lamb-shift spin filter polarimeter
2:10	10	*A.J. Mendez, TUNL* Modeling the hyperfine state selectivity of a short Lamb-shift spin-filter polarimeter
2:20	10	*V. Derenchuk, IUCF* The IUCF/TUNL spin filter polarimeter
2:30	10	*Discussion of first 3 papers*
2:40	10	*R. Gilman, CEBAF/Rutgers* Physics with a focal plane polarimeter for Hall A at CEBAF
2:50	10	*S.M. Bowyer, IUCF* A calibration of the K600 focal plane polarimeter
3:00	10	*M. Yosoi, Kyoto* Focal plane polarimeter for the GRAND RAIDEN at RCNP
3:10	10	*H. Sakai, Tokyo* Facility for the (p,n) polarization transfer measurement at RCNP
3:20	10	*Y. Tagishi, Tsukuba* A proton polarimeter using a liquid helium target
3:30	10	*B. Braun, München* The Breit-Rabi polarimeter for polarized internal H/D targets
3:40	15	*Discussion of above papers*
3:45	30	**COFFEE BREAK**

HIGH ENERGY

SOLID POLARIZED TARGETS III Chair: D. Crabb, Virginia [Maple Room]

4:15	25	*T. Averett, Virginia* Operation of a polarized target with ammonia in a high intensity electron beam

4:45	25	*H. Dutz, Bonn* The Bonn polarized target facility
5:15	20	*D. Kramer, Bielefeld* The SMC polarized target – systems and operations
5:40	20	*G. Court, Liverpool* Review of non-linear corrections in CW Q-meter target polarization measurements

NUCLEON SPIN STRUCTURE FUNCTIONS I Chair: R. Voss, CERN [Frangipani Room]

4:15	20	*A. Zanetti, Trieste* More details on SMC experiments
4:40	20	*A. Feltham, Basel* More details on SLAC E–143
5:05	20	*J. Ralston, Kansas* How gluons spin in the proton
5:30	20	*B. Li, Kentucky* Quark spin and quark angular momentum contents of the proton

HADRON ACCELERATORS AND BEAMS IV: Polarimeters Chair: L. Ratner, Brookhaven/Michigan [Oak Room]

4:15	10	*E.J. Ludwig, North Carolina* The use of the ^{3}He(d,p)^{4}He reaction for low-energy polarimetry
4:25	10	*S. Ishida, Tokyo* Construction of the deuteron polarimeter at RIKEN
4:35	10	*E. Tomasi-Gustafsson, Saturne* A deuteron vector and tensor polarimeter up to 2 GeV
4:45	20	*Discussion of first 3 papers*
5:05	10	*S. Nurushev, Protvino* Proton polarization determination by elastic p–e scattering
5:15	10	*T. Wakasa, Tokyo* Effective analyzing powers of NPOL at 295 and 384 MeV
5:25	10	*J.W. Watson, Kent* The Kent State "2π" neutron polarimeter
5:35	15	*Discussion of above papers*

Parallel Sessions 747

TUESDAY, SEPTEMBER 20

NUCLEAR

SYMMETRIES I: Parity Violation Chair: P.D. Eversheim
[Walnut Room]

2:00	18	J. Birchall, TRIUMF	
		Parity violation in p–p scattering at TRIUMF	
2:22	15	M. Shmatikov, Moscow	
		Theoretical overview of parity violation in p-p scattering	
2:40	16	Y. Masuda, KEK	
		Neutron spin rotation and P and T violations	
3:00	16	M. Leuschner, IUCF	
		Parity nonconservation in ^{207}Pb	
3:20	12	O. Yilmaz, Indiana	
		Relativistic effects on parity violation in nuclei	
3:35	12	K. Kimura, Nagasaki	
		Search for parity nonconservation in the compound nuclear reaction via an isobaric analog resonance of ^{90}Zr + p	

FEW BODY SYSTEMS I: Two Baryons Chair: H. von Geramb, Hamburg
[Persimmon Room]

2:00	15	R.T. Braun, TUNL	
		Neutron-proton analyzing power at 12 MeV and the charged πNN coupling constant	
2:18	15	S.M. Bowyer, IUCF	
		A measurement of the spin transfer observable D_{NN} for p+p elastic scattering at T_p=200 MeV	
2:36	15	C.A. Davis, TRIUMF	
		Zero-crossing angle of the np analyzing power below 300 MeV	
2:54	15	B. Vuaridel, Geneva	
		Spin observables in neutron-proton elastic scattering	
3:12	15	Y.D. Kim, KEK/Tsukuba	
		Hyperon-nucleon scattering experiment at KEK	
3:30	15	P. Heimberg, Northwestern	
		Differential cross section and analyzing power of $p(\vec{p}, \pi^+)d$ near threshold	

INTERMEDIATE ENERGY HADRON-INDUCED REACTIONS I
Chair: W.G. Love, Georgia [Frangipani Room]

2:00	12	A. Tamii, Kyoto	
		Test measurement of spin rotation parameters in proton elastic scattering from ^{58}Ni at E_p=300 MeV	
2:15	12	J. Liu, IUCF	
		Fragmentation of "stretched" 6^- strength in ^{28}Si$(\vec{p}, \vec{p}')^{28}$Si	

748 Parallel Sessions

2:30	12	S.P. Wells, IUCF
		Simultaneous measurement of $(\vec{p},\vec{p}')$ and $(\vec{p},p'\gamma)$ observables for the 15.11 MeV, 1^+, T=1 state in ^{12}C at 200 MeV
2:45	12	D.A. Cooper, Ohio State
		A study of the Fermi (0^+) transition in ^{14}C(p,n)^{14}N at 495 MeV
3:00	12	T. Wakasa, Tokyo
		Measurement of the polarization transfer $D_{NN}(0°)$ for (p,n) reactions at 300 MeV
3:15	12	C. Djalali, South Carolina
		Isoscalar spin strength in ^{12}C and ^{40}Ca
3:30	12	H. Okamura, Tokyo
		Tensor analyzing power of the ^{12}C(d,^{2}He)^{12}B reaction at 270 MeV

HIGH ENERGY

LEPTON ACCELERATORS, BEAMS AND POLARIMETERS II
Chair: M. Minty, SLAC [Maple Room]

2:00	20	D. Barber, DESY
		Spin decoherence in electron storage rings
2:25	20	M. Düren, Erlangen/DESY
		Transverse and longitudinal electron polarization at HERA
2:50	20	M. Böge, DESY/Hamburg
		Optimization of spin polarization in the HERA electron ring using beam-based alignment procedures
3:15	20	Y. Shatunov, Novisibirsk
		Spin control system for the South Hall Ring at Bates Linear Accelerator Center
3:40	20	Y. Eidelman, Novosibirsk
		Developments in the computer code SPINLIE

NUCLEON SPIN STRUCTURE FUNCTIONS II Chair: R. Voss, CERN
[Whittenberger Auditorium]

2:00	20	N. Kochelev, Dubna
		Vacuum QCD and new information on nucleon structure functions
2:25	20	H. Borel, Saclay
		More details on SLAC E-143
2:50	20	X. Ji, MIT
		Using PQCD to probe non-perturbative QCD
3:15	20	J. Qiu, Iowa
		Twist-3 and structure functions
3:40	20	B. Kamal, McGill
		Direct γ production to measure the gluon structure function

STRONG INTERACTIONS AT HIGH ENERGY III: Inclusive processes, QCD, and other topics Chair: M. Anselmino, Torino [Oak Room]

2:00	15	W. Tang, SLAC Polarization as a probe to the production mechanism of charmonium in π–N collisions
2:20	15	F. Murgia, Cagliari Single spin asymmetry in inclusive pion production
2:40	15	T. Meng, Berlin Inclusive meson and lepton-pair production in single-spin hadron-hadron collisions
3:00	15	A. Brandenburg, SLAC Angular distributions in the Drell-Yan process: a closer look at higher twist effects
3:20	15	A.P. Contagouris, McGill/Athens Higher-order QCD corrections to processes with longitudinally and transversely polarized particles
3:40	15	O. Teryaev, Dubna On the twist–3 single and double asymmetries
3:45	30	**COFFEE BREAK**

NUCLEAR

SYMMETRIES II Chair: H.E. Conzett, Berkeley
[Walnut Room]

4:15	20	W.T.H. van Oers, TRIUMF Measurement of charge symmetry breaking in np elastic scattering at 350 MeV
4:40	20	G.A. Miller, U. Washington Theoretical overview of charge symmetry violation
5:05	16	P.R. Huffman, Duke An experiment to test P-even time reversal invariance with MeV neutrons
5:25	16	P.D. Eversheim, Bonn Test of time reversal invariance in proton-deuteron scattering
5:45	12	B.M.K. Nefkens, UCLA Polarization and tests of discrete symmetries

FEW BODY SYSTEMS II: Three and four nucleons Chair: F.D. Santos, Lisboa
[Persimmon Room]

4:15	15	L.D. Knutson, Wisconsin Polarization observables for p–d breakup and the nuclear three-body force
4:33	15	I.M. Sitnik, Dubna Status and future of polarization phenomena investigations in backward elastic deuteron-proton scattering

4:51 15 Z. Ayer, North Carolina
        Determination of the asymptotic D- to S-state ratio for the triton and ^{3}He via $(\vec{d},t)$ and $(\vec{d},^3\text{He})$ reactions
5:09 15 M. Miller, Wisconsin
        Measurement of spin observables in quasielastic scattering of polarized protons from polarized ^{3}He: $^3\vec{\text{He}}(\vec{p},2p)$, $^3\vec{\text{He}}(\vec{p},pn)$ and $^3\vec{\text{He}}(\vec{p},pd)$
5:27 15 W.J. Cummings, TRIUMF
        Elastic π^+ scattering on polarized ^{3}He
5:45 15 W. Kretschmer, Erlangen
        Polarization transfer in p–d scattering at 22.7 MeV

INTERMEDIATE ENERGY ELECTROMAGNETIC INTERACTIONS I
Chair: K. de Jager, NIKHEF [Frangipani Room]

4:15 30 J.M. Laget, Saclay
        Spin physics with an intense CW electron beam in the 15-30 GeV range
4:50 15 H. Dutz, Bonn
        Target asymmetry measurements of $\gamma p \to \pi^+ n$ and $\gamma p \to \pi^0 p$ with PHOENICS and ELSA
5:10 20 B. Saghai, Saclay
        Pseudoscalar meson photoproduction on the proton
5:35 15 K.H. Hicks, Ohio
        Exclusive pion and two proton photoproduction from ^{16}O using polarized photons

HIGH ENERGY

LEPTON ACCELERATORS, BEAMS AND POLARIMETERS III
Chair: M. Minty, SLAC [Maple Room]

4:15 15 R. Assmann, CERN
        Deterministic harmonic spin matching in LEP
4:35 15 B. Dehning, CERN
        Energy calibration with resonant depolarization at LEP in 1993
4:55 15 H. Grote, CERN
        A Richter-Schwitters test spin rotator for LEP
5:15 15 Y. Bashmakov, RAS
        Radiation and spin separation of high energy positrons by a bent crystal
5:35 15 Y. Shatunov, Novisibirsk
        Spin flip by rf field at storage rings with Siberian snakes

STRONG INTERACTIONS AT HIGH ENERGY IV: Inclusive processes, QCD, and other topics Chair: G. Bunce, Brookhaven [Oak Room]

4:15 15 V. Rykov, Wayne State and IHEP, Protvino
        Study of W and Z production processes at RHIC

4:35	15	P.M. Border, Minnesota
		A precise mesurement of the Ω^- magnetic moment
4:55	15	R. Rylko, London
		Spin effects in production of charmed hadrons
5:15	15	M. Chavleishvili, Dubna
		Spin phenomena in two-body processes

ELECTROWEAK INTERACTIONS II Chair: C. Prescott, SLAC
[Whittenberger Auditorium]

4:15	15	W. Kretschmer, Erlangen
		Neutrino physics with KARMEN
4:35	15	P. Gumplinger, British Columbia
		Measuring the muon polarization in $K \to \pi\mu\nu$ with the E-246 detector at KEK
4:55	15	D. Ciampa, Minnesota
		A precise measurement of non-leptonic decay parameters α and β/γ for $\Omega^- \to \Lambda K^-$ decays
5:15	15	E.C. Dukes, Virginia
		A new experiment to search for CP violation in hyperon decays
5:35	15	H. Noumi, KEK
		Study of polarized $^5_\Lambda$He weak decay via the (π^+, K^+) reaction on ^{6}Li

WEDNESDAY, SEPTEMBER 21

NUCLEAR

INTERMEDIATE ENERGY HADRON-INDUCED REACTIONS II
Chair: N.S. Chant [Frangipani Room]

2:00	12	C.M. Edwards, Minnesota
		Study of the ^{3,4}He(p,n) reactions at T_p=100 and 200 MeV
2:15	12	D.S. Carman, IUCF
		Quasifree (p,Np) reaction studies from ^{2}H and ^{12}C at 200 MeV
2:30	12	T. Noro, RCNP
		Exclusive measurement of $s_{1/2}$ proton knockout reaction
2:45	12	W.G. Love, Georgia
		Scattering of polarized protons from polarized targets
3:00	12	D. Prout, Ohio State
		Charge exchange spin observable measurements on Pb at 795 MeV in the giant resonance region
3:15	12	W.W. Jacobs, IUCF
		Polarization observables in $\vec{p}p \to pK^+\vec{Y}$ reactions at 2.9 GeV

INTERMEDIATE ENERGY ELECTROMAGNETIC INTERACTIONS II
Chair: C. Glashausser, Rutgers [Oak Room]

2:00	30	S. Nanda, CEBAF CEBAF spin physics
2:35	15	S. Popov, Novosibirsk Status of the t_{20} electron–deuteron scattering experiment at VEPP-3
2:53	15	S. Kox, Grenoble The new deuteron polarimeter POLDER and the t_{20} experiment at CEBAF
3:11	15	A. Afanasev, CEBAF Induced nucleon polarization in the d(e,e′N) reaction near threshold
3:29	15	E. Passchier, NIKHEF Proton knockout from tensor-polarized deuterium

POLARIZED SOURCES AND TARGETS III Chair: K. Zapfe, DESY
[Persimmon Room]

2:00	15	K. Hatanaka, RCNP High intensity polarized ion source at RCNP
2:20	15	W.J. Cummings, TRIUMF Application of high power GaAlAs diode laser arrays for optically pumped spin exchange polarized ^{3}He targets
2:40	15	K.P. Coulter, Michigan Advances in alkali spin-exchange pumped ^{3}He target technology
3:00	15	C.J. Horowitz, Indiana Polarized stored beams by interaction with polarized electrons
3:20	15	H. Spinka, Argonne A possible method to produce a polarized antiproton beam at intermediate energies

HIGH ENERGY

LEPTON ACCELERATORS, BEAMS, AND POLARIMETERS IV
Chair: M. Minty, SLAC [Maple Room]

2:00	15	M. Berz, Michigan State Description and normal form analysis of spin dynamics using differential algebra
2:20	15	R. Lieu, U.C. Berkeley Synchrotron radiation: inverse Compton effect
2:40	15	Y. Derbenev, IUCF/Michigan RF-intrinsic spin flipper
3:00	15	G. Bunce, Brookhaven The new muon $g-2$ experiment at BNL
3:20	15	F. Feinstein, Saclay Muon polarimeters at SMC

NUCLEON SPIN STRUCTURE FUNCTIONS III Chair: R. Voss, CERN
[Whittenberger Auditorium]

2:00	20	*L. Tkatchev, Dubna*
		Search for jet handedness in hadronic Z^0 decays at DELPHI
2:25	20	*A. Efremov, Dubna*
		Puzzling correlation of handedness in $Z \to 2$ jet decay
2:50	20	*A. Dorokhov, Moscow*
		Instanton-induced helicity and flavor asymmetries in the light quark sea of the nucleon
3:15	20	*A. Yokosawa, Argonne*
		RHIC valence, sea, and gluon spin structure function measurements

SPIN'94 PARTICIPANTS LIST

Nuclear

Andrei Afanasev
CEBAF
12000 Jefferson Ave
Newport News, VA 23606
USA
afanas@cebaf.gov
(tel.) 1-804-249-7011
(fax) 1-804-249-7363

Nural Akchurin
University of Iowa
Dept. of Physics & Astronomy
Iowa City, IA 52242
USA
akchurin@iaquark.physics.uiowa.edu
(tel.) 1-319-335-1941
(fax) 1-319-335-1753

Mohammad Al-Ohali
KFUPM
Dept. of Physics
P.O. Box 2015
Dhahran 31261
SAUDI ARABIA

Antonio Amorim
Universidade de Lisboa
Centro de Fisica Nuclear
Av. Gama Pinto 2
1699-Lisboa Codex
PORTUGAL
faamorim@skull.cc.fc.ul.pt
(tel.) 351-1-7950790
(fax) 351-1-7956289

L. Wilmer Anderson
University of Wisconsin
Dept. of Physics
1150 University Ave.
Madison, WI 53706
USA
lwanderson@uwnuc0.physics.wisc.edu
(tel.) 1-608-262-8962
(fax) 1-608-262-3598

Jacques Arvieux
Laboratoire National Saturne
C.E. Saclay
F-91191 Gif-sur-Yvette
Cedex
FRANCE
arvieux@frcpn11.in2p3.fr
(tel.) 33-1-69082203
(fax) 33-1-69082970

Ralph Assmann
Kreuzbergstr. 46
D-53127 Bonn
GERMANY
(tel.) 49-228-25-32-74

Ulrich Atzrott
Universität Tübingen
Physikalisches Institut
Auf der Morgenstelle 14
D-72076 Tübingen
GERMANY
atzrott@pit.physik.uni-teubingen.de
(tel.) 49-7071-293432
(fax) 49-7071-296296

Angela Betker
Indiana University Cyclotron Facility
2401 Milo B. Sampson Lane
Bloomington, IN 47408
USA
angie@iucf.indiana.edu
(tel.) 1-812-855-9365
(fax) 1-812-855-6645

Jim Birchall
University of Manitoba
Department of Physics
Winnipeg
Manitoba R3T 2N2
CANADA
birchall@physics.umanitoba.ca
(tel.) 1-204-474-6205
(fax) 1-204-269-8489

Les Bland
Indiana Univ. Cyclotron Fac.
2401 Milo B. Sampson Lane
Bloomington, IN 47408
USA
bland@iucf.indiana.edu
(tel.) 1-812-855-6051
(fax) 1-812-855-6645

Herve Borel
C. E. Saclay
DAPNIA/SPhN
Orme des Merisiers
F-91191 Gif-sur-Yvette
FRANCE
borel@phnx7.saclay.cea.fr
(tel.) 33-1-69-08-75-09
(fax) 33-1-69-08-75-84

Sonya M. Bowyer
Indiana University Cyclotron Facility
2401 Milo B. Sampson Lane
Bloomington, IN 47408
USA
smbowyer@iucf.indiana.edu
(tel.) 1-812-855-9365
(fax) 1-812-855-6645

Bernd Braun
Universität München
Sektion Physik, MPI für Kernphysik
Postfach 10 39 80
69029 Heidelberg
GERMANY
bbr@dxnhd5.mpi-hd.mpg.de
(tel.) 49-6221-516-432
(fax) 49-6221-516-540

Richard Braun
Duke University
422 TUNL
Durham, NC 27708-0308
USA
braun@tunl.duke.edu
(tel.) 1-919-660-2635
(fax) 1-919-660-2634

Stephen Bueltmann
CERN
CH-1211
Genève 23
SWITZERLAND
stephen@na47sun05.cern.ch
(tel.) 41-22-767-3453
(fax) 41-22-785-0672

John M. Cameron
Indiana University Cyclotron Facility
2401 Milo B. Sampson Lane
Bloomington, In 47408
USA
cameron@iucf.indiana.edu
(tel.) 1-812-855-9365
(fax) 1-812-855-6645

Peter Cameron
BNL
Bldg. 830
Upton, NY 11973
USA
cameron@bnlux1.bnl.gov
(tel.) 1-516-282-7657
(fax) 1-516-282-3079

Christian Cavata
DAPNIA/SPhN CEN Saclay
F-91191 Gif sur Yvette
Cedex
FRANCE
cavata@phnx7.saclay.cea.fr
(tel.) 33-1-6908-3237
(fax) 33-1-6908-7584

Shining Chang
Indiana University Cyclotron Facility
2401 Milo B. Sampson Lane
Bloomington, IN 47408
USA
sxchang@iucf.indiana.edu
(tel.) 1-812-855-0923
(fax) 1-812-855-6645

Nicholas S. Chant
University of Maryland
Dept. of Physics
College Park, MD 20742-4111
USA
chant@enp.umd.edu
(tel.) 1-301-405-6104
(fax) 1-301-314-9525

Michael P. Chavleishvili
Joint Inst. for Nuclear Rsch.
141980, Dubna
Moscow region
RUSSIA
chavlei@theor.jinrc.dubna.su
(tel.) 7-09621-6-2156
(fax) 7-09621-6-5084

James Christley
University of Surrey
Department of Physics
Guildford
Surrey GU2 5XH
UNITED KINGDOM
j.christley@ph.surrey.ac.uk
(tel.) 44-483259405
(fax) 44-483259501

Thomas B. Clegg
University of North Carolina
Dept. of Physics & Astronomy
Phillips Hall
Chapel Hill, NC 27599-3255
USA
clegg@tunl.duke.edu
(tel.) 1-919-962-2079
(fax) 1-919-962-0480

Samuel Cohen
Institut de Physique Nucléaire
91406 Orsay Cedex
FRANCE
cohen_s@ipncls.in2p3.fr
(tel.) 33-1-6941-7778
(fax) 33-1-6941-6470

Andreas P. Contogouris
McGill University
Dept. of Physics
Montreal, P.Q.
CANADA H3A 2T8
secretariat@hep.physics.mcgill.ca
(tel.) 1-514-398-6520
(fax) 1-514-398-3733

Homer E. Conzett
Lawrence Berkeley Laboratory
One Cyclotron Road
Berkeley, CA 94720
USA
heconzett@lbl.gov
(tel.) 1-510-486-7813
(fax) 1-510-486-7983

Daniel A. Cooper
Ohio State University
Van de Graaf Laboratory
1302 Kinnear Road
Columbus, OH 43212
USA
dcooper@mps.ohio-state.edu
(tel.) 1-614-292-4775
(fax) 1-614-292-4833

Donald G. Crabb
University of Virginia
Dept. of Physics
McCormick Road
Charlottesville, VA 22901
USA
dgc3q@virginia.edu
(tel.) 1-804-924-6790
(fax) 1-804-924-4576

Raquel Crespo
Universidade Técnica de Lisboa
Departmento do Física
do Instituto Superior Técnico (IST)
Avenida Rovisco Pais
P-1095 Lisboa
PORTUGAL
(tel.) 351-1-7584128

William J. Cummings
TRIUMF
4004 Wesbrook Mall
Vancouver, B.C., V6T 2A3
CANADA
cummings@erich.triumf.ca
(tel.) 1-604-222-1047
(fax) 1-604-222-1074

Daniel Dale
University of Kentucky
Dept. of Physics
Lexington, KY 40506
USA
dale@zeppo.pa.uky.edu

Charles A. Davis
TRIUMF
4004 Wesbrook Mall
Vancouver, B.C.
CANADA V6T 2A3
cymru@erich.triumf.ca
(tel.) 1-604-222-1047
(fax) 1-604-222-1074

Donal Day
University of Virginia
Department of Physics
McCormick Road
Charlottesville, VA 22901
USA
dbd@virginia.edu
(tel.) 1-804-924-6566
(fax) 1-804-924-4576

Kees de Jager
NIKHEF-K
P.O. Box 41882
1009 DB Amsterdam
THE NETHERLANDS
kees@paramount.nikhefk.nikhef.nl
(tel.) 31-20-5922143
(fax) 31-20-5922165

Paul Delheij
TRIUMF
4004 Wesbrook Mall
Vancouver, B.C., V6T 2A3
CANADA
delh@triumf.ca
(tel.) 1-604-222-1047
(fax) 1-604-222-1074

Vladimir P. Derenchuk
Indiana University Cyclotron Facility
2401 Milo B. Sampson Lane
Bloomington, IN 47408
USA
laddie@iucf.indiana.edu
(tel.) 1-812-855-9365
(fax) 1-812-855-6645

Jules Deutsch
Univ. Catholique de Louvain
Inst. de Physique
Chemin du Cyclotron 2
Louvain-la-Neuve, B1348
BELGIUM
wins%"deutsch@fynu.ucl.ac.be"
(tel.) 32-10-473273
(fax) 32-10-452183

William De Zarn
Indiana University Cyclotron Facility
2401 Milo B. Sampson Lane
Bloomington, IN 47408
USA
dezarnwa@iucf.indiana.edu
(tel.) 1-812-855-3613
(fax) 1-812-855-6645

Chaden Djalali
University of S. Carolina
Dept. of Physics
Columbia, SC 29208
USA
djalali@nuc002.psc.scarolina.edu
(tel.) 1-803-777-4318
(fax) 1-803-777-3065

Fraser Duncan
University of Maryland
Department of Physics
College Park, MD 20742
USA
duncan@enp.umd.edu
(tel.) 1-301-405-6105
(fax) 1-301-314-9525

Michael Düren
DESY
HERMES Collaboration
Notkestrasse 85
D-22603 Hamburg
GERMANY
dueren@vxdsya.desy.de
(tel.) 49-40-89982089
(fax) 49-40-89983438

Hartmut Dutz
Universität Bonn
Physikalisches Institut
Nussallee 12
D-53115 Bonn
GERMANY
dutz@pib1.physik.uni-bonn.de
(tel.) 49-228-733610
(fax) 49-228-737869

Carla M. Edwards
c/o University of Minnesota Group
Mail Stop H846
Los Alamos National Laboratory
Los Alamos, NM 87545
USA
carla@stpaul.lampf.lanl.gov
(tel.) 1-505-665-7695
(fax) 1-505-665-7920

Ana Maria Eiró
Universidade de Lisboa
Centro de Física Nuclear
Av. Gama Pinto 2
1699 Lisboa Codex
PORTUGAL
eiro@a1f3.cc.fc.ul.pt
(tel.) 351-1-7950790
(fax) 351-1-7954288

Hiroyasu Ejiri
Osaka University
Research Center for Nuclear Physics
10-1 Mihogaoka, Ibaraki
Osaka 567
JAPAN
ejiri@rcnpvx.rcnp.osaka-u.ac.jp
(tel.) 81-68798929
(fax) 81-68798899

Charlotte Elster
Ohio University
Department of Physics and Astronomy
Clippinger Research Laboratories
Athens, OH 45701-2979
USA
elster@stringray.phy.ohiou.edu
(tel.) 1-614-593-1697
(fax) 1-614-593-0433

Dieter Eversheim
Universität Bonn
Institut für Strahlen und Kernphysik
Nussallee 14-16
D-53115 Bonn
GERMANY
evershei@servax.iskp.uni-bonn.de
(tel.) 49-228-735299
(fax) 49-228-733728

Willie R. Falk
University of Manitoba
Dept. of Physics
Winnipeg
CANADA R3T 2N2
falk@physics.umanitoba.ca
(tel.) 1-204-474-9856
(fax) 1-204-269-8489

Harold W. Fearing
TRIUMF
4004 Wesbrook Mall
Vancouver BC V6T 2A3
CANADA
fearing@triumf.ca
(tel.) 1-604-222-1047
(fax) 1-604-222-1074

Andrew Feltham
Inst. für Physik der Universität Basel
Experimental Kernphysik
Klingelbergstrasse 82
4056 Basel
SWITZERLAND
feltham@urz.unibas.ch
(tel.) 41-61-267-3728
(fax) 41-61-267-3784

Dieter Fick
Philipps Universität-Marburg
Fachbereich Physik
D35032 Marburg
GERMANY
fick@mv13a.physik.uni-marburg.de
(tel.) 49-06421-282017
(fax) 49-06421-287033

Giuseppe Fidecaro
CERN
CH1211 Geneva 23
SWITZERLAND
fde@cernvm.cern.ch
(tel.) 41-22-767-2686
(fax) 41-22-783-0672

Wilbur Franklin
Indiana University Cyclotron Facility
Milo B. Sampson Lane
USA
Bloomington, IN 47408
franklin@iucf.indiana.edu
(tel.) 1-812-855-9365
(fax) 1-812-855-6645

Michel Garcon
DAPNIA/SPhN, Bat. 703
CEN-Saclay
F-91191 Gif-sur-Yvette
Cedex
FRANCE
garcon@phnx7.saclay.cea.fr
(tel.) 33-1-6908-8623
(fax) 33-1-6908-7584

Susan V. Gardner
Indiana University
Nuclear Theory Center
2401 Milo B. Sampson Lane
Bloomington, IN 47408
USA
gardner@iucf.indiana.edu
(tel.) 1-812-855-9365
(fax) 1-812-855-6645

Shalev Gilad
MIT
77 Massachusetts Avenue
Bldg. 26-449
Cambridge, MA 02139
USA
gilad@pierre.mit.edu
(tel.) 1-617-253-7785
(fax) 1-617-258-5440

Ron Gilman
Rutgers University
Department of Physics and Astronomy
P.O. Box 849
Piscataway, NJ 08855-0849
USA
gilman@ruthep.rutgers.edu
(tel.) 1-908-445-5489
(fax) 1-908-445-4343

Charles Glashausser
Rutgers University
Serin Physics Laboratory
Box 849
Piscataway, NJ 08854
USA
glashausser@ruthep.rutgers.edu
(tel.) 1-908-445-2526
(fax) 1-908-445-4343

Charles Goodman
Indiana University Cyclotron Facility
2401 Milo B. Sampson Lane
Bloomington, IN 47408
USA
goodman@iucf.indiana.edu
(tel.) 1-812-855-9365
(fax) 1-812-855-6645

Christoper R. Gould
North Carolina State Univ.
Dept. of Physics
Box 8202
Raleigh, NC 27695-8202
USA
chris_gould@ncsu.edu
(tel.) 1-919-515-3380
(fax) 1-919-660-2634

Gerhard Graw
Universität München
Sektion Physik der Ludwig-Maximilians
Am Goulombwall 1
D-85748 Garching
GERMANY
graw@physik.uni-muenchen.de
(tel.) 49-89-32096155
(fax) 49-89-8544469

David Grosnick
Argonne National Labratory
HEP362
9700 South Cass Ave.
Argonne, IL 60439
USA
dpg@hep.anl.gov
(tel.) 1-708-252-7529
(fax) 1-708-252-5782

Paolo Guazzoni
Univ. di Milano
Dipartimento di Fisica
Via Celoria 16
I-20133 Milano
ITALY
guazzoni@milano.infn.it
(tel.) 39-2-2392249
(fax) 39-2-2392297

David G. Haase
North Carolina State Univ.
Dept. of Physics
Box 8202
Raleigh, NC 27695-8202
USA
david_haase@ncsu.edu
(tel.) 1-919-515-6118
(fax) 1-919-515-7545

Ole Hansen
MIT
Lab for Nuclear Science
Room 26-650B
77 Massachusetts Avenue
Cambridge, MA 02139-4307
USA
ole@mitlns.mit.edu
(tel.) 1-617-253-4798
(fax) 1-617-258-5440

William Happer
Princeton University
Department of Physics
Princeton, NJ 08544
USA
happer@pupgg.princeton.edu
(tel.) 1-609-258-4382
(fax) 1-609-258-2496

Ryan Hartman
Indiana University Cyclotron Facility
2401 Milo B. Sampson Lane
Bloomington, IN 47401
USA
hartman@iucf.indiana.edu
(tel.) 1-812-855-9365
(fax) 1-812-855-6645

762 SPIN'94 Participants

Takeo Hasegawa
Miyazaki University
Faculty of Engineering
1-1 Gakuen-Kibanadai-Nishi-
Miyazaki-shi, 889-21
JAPAN
hastake@kekvax.kek.jp
(tel.) 81-0985-58-2811X4321
(fax) 81-0985-58-1647

Kichiji Hatanaka
Osaka University
RCNP
10-1 Mihogaoka, Ibaraki
Osaka 567
JAPAN
hatanaka@rcnpvx.rcnp.osaka-u.ac.jp
(tel.) 81-6-879-8934
(fax) 81-6-879-8899

Blayne R. Heckel
University of Washington
Physics FM-15
Seattle, WA 98195
USA
heckel@phys.washington.edu
(tel.) 1-206-685-2401
(fax) 1-206-685-0635

Peter Heimberg
Northwestern University
Department of Physics & Astronomy
Evanston, IL 60208
USA
heimberg@nuhepb.phys.nwu.edu
(tel.) 1-708-491-8607
(fax) 1-708-491-9982

Bill Hersman
University of New Hampshire
Dept. of Physics
Durham, NH 03824
USA
(tel.) 1-603-862-3512
(fax) 1-603-862-2998

K. H. Hicks
Dept. of Physics
Ohio University
Athens, OH 45701
USA
khicks1@ohiou.edu
(tel.) 1-614-593-1981
(fax) 1-614-593-1436

Gregory C. Hillhouse
University of the Western Cape
Dept. of Physics
Private Bag X17
Bellville 7535
SOUTH AFRICA
hillhouse@nacdh4.nac.ac.za
(tel.) 27-021-959-2556
(fax) 27-021-959-2266

Yorck Holler
DESY, Dpt. MEA
Postfach
D-22603 Hamburg
GERMANY
meahol@dsyibm.desy.de
(tel.) 49-40-8998-3743
(fax) 49-40-8998-3438

Roy J. Holt
University of Illinois
Dept. of Physics
1110 W. Green Street
Urbana, IL 61801
USA
holt@uinpla.npl.uiuc.edu
(tel.) 1-217-244-6039
(fax) 1-217-333-1215

Arnold Honig
Syracuse University
Dept. of Physics
201 Physics Building
Syracuse, NY 13244-1130
USA
honig@suhep.phy.syr.edu
(tel.) 1-315-443-3888
(fax) 1-315-443-9103

Haixin Huang
Brookhaven National Laboratory
AGS Department
Building 911B
Upton, NY 11973
USA
huang@bnldag.bnl.gov
(tel.) 1-516-282-5446
(fax) 1-516-282-5954

Paul Huffman
Duke University
Triangle Univesities Nuclear Laboratory
P.O. Box 90308
Durham, NC 27708-0308
USA
huffman@tunl.tunl.duke.edu
(tel.) 1-919-660-2639
(fax) 1-919-660-2634

Vernon W. Hughes
Yale University
Dept. of Physics, 465 J. W. Gibbs
P.O. Box 208121
New Haven, CT 06520-8121
USA
hughes@yalph2.physics.yale.edu
(tel.) 1-203-432-3819
(fax) 1-203-432-3804

Masataka Iinuma
Kyoto University
Department of Physics
Kitashirakawa
Kyoto, 606-1
JAPAN
iinuma%kytax1.dnet@kekux.kek.jp
(tel.) 81-75-7533871
(fax) 81-75-7115175

Yasunori Iseri
Chiba-Keizai College
Todoroki-cho, 4-3-30
Inage-ku
Chiba 263
JAPAN
(tel.) 81-43-255-3451
(fax) 81-43-252-6050

Satoru Ishida
University of Tokyo
Department of Physics
Hongo 7-3-1, Bunkyo
Tokyo 113
JAPAN
sishida@rikvax.riken.go.jp
(tel.) 81-3-5689-7343
(fax) 81-3-3811-0960

Harold E. Jackson, Jr.
Argonne National Laboratory
Bldg. 203 B-237
Argonne, IL 60439
USA
hal@anl.gov
(tel.) 1-708-252-4013
(fax) 1-708-252-3903

William W. Jacobs
Indiana University Cyclotron Facility
2401 Milo B. Sampson Lane
Bloomington, IN 47408
USA
jacobs@iucf.indiana.edu
(tel.) 1-812-855-8873
(fax) 1-812-339-6645

Robert L. Jaffe
6-311 MIT
77 Massachusetts Avenue
Cambridge, MA 02139
USA
jaffe@mitlns.mit.edu
(tel.) 1-617-253-4858
(fax) 1-617-253-8674

Brajesh K. Jain
Bhabha Atomic Research Centre
Nuclear Physics Division, BARC
Bombay 400 085
INDIA
bkjain@magnum.barct1.ernet.in
(tel.) 91-22-555-6071
(fax) 91-22-556-0750

764   SPIN'94 Participants

Ronald C. Johnson
University of Surrey
Dept. of Physics
Guildford
Surrey GU2 5XH
ENGLAND
(tel.) 44-483-259375
(fax) 44-483-259501

Kyungseon Joo
MIT
77 Massachusetts Avenue
Bldg. 26-648
Cambridge, MA 02139
USA
kjoo@mitlns.mit.edu
(tel.) 1-617-253-4230
(fax) 1-617-258-5440

Basim Kamal
McGill University
Physics Department
Montreal, Quebec, H3A 2T8
CANADA
cxbk@musica.mcgill.ca
(tel.) 1-514-398-6502

Hugon Karwowski
University of N. Carolina
CB #3290
Chapel Hill, NC 27599
USA
hugon@tunl.duke.edu
(tel.) 1-919-962-7206

Kirby W. Kemper
Florida State University
Dept. of Physics, B-159
Tallaahassee, FL 32306-3016
USA
kirby@fsulcd.physics.fsu.edu
(tel.) 1-904-644-2585
(fax) 1-904-644-9848

Phil Kerr
Florida State University
Dept. of Physics B-159
Tallahassee, FL 32304
USA
kerr@nucmar.physics.fsu.edu
(tel.) 1-904-574-3731
(fax) 1-904-644-9848

G. Ya. Kezerashvili
Budker Institute of Nuclear Physics
630090 Novosibirsk
RUSSIA
guramkez@inp.nsk.su
(tel.) 7-3832-359420
(fax) 7-3832-352163

I. Khudeir
A E C of Syria
Nuclear Physics Division
P. O. Box 6091
Damascus
SYRIA
(tel.) 963-11-6668114
(fax) 963-11-6620317

Kurt Kilian
Forschungszentrum Jülich GmbH
Institut für Kernphysik
Briefpost: 52425 Jülich
Fracht/Paketpost: 52428 Jülich
GERMANY
kph001@djukfa11
(tel.) 49-2461-61-5943
(fax) 49-2461-61-3930

Kikuo Kimura
Nagasaki Inst. of Applied Sci.
536 Aba-machi
Nagasaki 851-01
JAPAN
kimura@nias.ac.jp
(tel.) 81-958-39-3111
(fax) 81-958-30-1126

Michikazu Kinsho
National Lab for High En. Phys.
(KEK), 1-1 Oho-machi
Tsukuba-shi
Ibaraki 305
JAPAN
kinsho@kekvax.kek.jp
(tel.) 81-298-64-5215
(fax) 81-298-64-3182

Lynn D. Knutson
University of Wisconsin
Dept. of Physics
1150 University Avenue
Madison, WI 53706
USA
knutson@uwnuc0.physics.wisc.edu
(tel.) 1-608-262-3096
(fax) 1-608-262-3598

Alexander Komives
Indiana University Cyclotron Facility
2401 Milo B. Sampson Ln.
Bloomington, IN 47408
USA
komives@iucf.indiana.edu
(tel.) 1-812-855-9365
(fax) 1-812-855-6645

Serge Kox
Institut des Sci. Nucleaires
53 Avenue des Martyrs
38026 Grenoble Cedex
FRANCE
kox@frcpn11.in2p3.fr
(tel.) 33-76-28-4155
(fax) 33-76-28-4004

Dirk Kramer
CERN
PPE Division
1211 Geneva 23
SWITZERLAND
kraemer@uxnhd.cern.ch
(tel.) 41-22-785-6428

Laird Kramer
MIT-Lab for Nuclear Science
Building 26-533
Cambridge, MA 02139
USA
kramer@mitlns.mit.edu
(tel.) 1-612-253-3761
(fax) 1-612-258-5440

Wolfgang Kretschmer
Universität Erlangen-Nürnberg
Physikalisches Institut,
Erwin-Rommel-Str. 1
D-91058 Erlangen
GERMANY
pi4kret@pkvx1.physik.uni-erlangen.de
(tel.) 49-9131-857075
(fax) 49-9131-15249

Alan D. Krisch
University of Michigan
Randall Lab of Physics
Ann Arbor, MI 48109-1120
USA
krisch@umiphys
(tel.) 1-313-936-1027
(fax) 1-313-936-0794

Ronald Kunne
Laboratoire National Saturne
CE Saclay
F-91191 Gif-sur-Yvette
Cedex
FRANCE
kunne@frcpn11.in2p3.fr
(tel.) 31-1-6908-3358
(fax) 31-1-6908-2970

J. M. Laget
CEA (French Atomic Energy Com.)
Dapnia/SPhn
CE. Saclay, F91191-Gif-sur-Yvette
Cedex
FRANCE
laget@phnx7.saclay.cea.fr
(tel.) 33-1-6928-7554
(fax) 33-1-6928-7584

Mark Leuschner
Indiana University Cyclotron Facility
2401 Milo B. Sampson Lane
Bloomington, IN 47408
USA
leuschner@iucf.indiana.edu
(tel.) 1-812-855-0189
(fax) 1-812-855-6645

C.D. Philip Levy
TRIUMF
4004 Wesbrook Mall
Vancouver B.C.
CANADA V6T 2A3
levy@triumf.ca
(tel.) 1-604-222-1047
(fax) 1-604-222-1074

Derun Li
Indiana University Cyclotron Facility
2401 Milo B. Sampson Lane
Bloomington, IN 47408
USA
deli@iucf.indiana.edu
(tel.) 1-812-855-9365
(fax) 1-812-855-6645

John T. Londergan
Indiana University
Dept. of Physics
Swain Hall W. 117
Bloomington, IN 47405
USA
londergan@iucf.indiana.edu
(tel.) 1-812-855-1247
(fax) 1-812-855-5533

Bernd Lorentz
Univ. of Wisconson-Madison
Nuclear Physics
1150 University Ave.
Madison, WI 53706-1390
USA
lorentz@uwnuc0.physics.wisc.edu
(tel.) 1-608-262-6555
(fax) 1-608-262-3598

W. Gary Love
University of Georgia
Department of Physics & Astronomy
Athens, GA 30602
USA
wglove@sb.dcs.uga.edu
(tel.) 1-706-542-3083
(fax) 1-706-542-2492

Edward J. Ludwig
University of North Carolina
Dept. of Physics & Astronomy
Chapel Hill, NC 27599-3255
USA
ludwig@tunl.duke.edu
(tel.) 1-919-962-5001
(fax) 1-919-962-0480

Malcolm MacFarlane
Indiana University
Nuclear Theory Center
2401 Milo B. Sampson Lane
Bloomington, IN 47408
USA
macfarlane@iucf.indiana.edu
(tel.) 1-812-855-2953
(fax) 1-812-855-6645

Yousef I. Makdisi
Brookhaven National Laboratory
RHIC Project, Bldg. 510C
P.O. Box 5000
Upton, NY 11973-5000
USA
makdisi@bnldag.ags.bnl.gov
(tel.) 1-516-282-4932
(fax) 1-516-282-2532

Yasuhiro Masuda
KEK
1-1 Oho
Tsukuba-shi
Ibaraki-ken 305
JAPAN
masuda@kekvax.kek.jp
(tel.) 81-298-641171
(fax) 81-298-643202

Kensaku Matsuta
Lawrence Berkeley Laboratory
One Cyclotron Road
MS 64-121
Berkeley, CA 94720
USA
matsuta@lbl.gov
(tel.) 1-510-486-4145
(fax) 1-510-486-4565

Tony Mendez
Univ. of North Carolina
Triangle Univ. Nuclear Laboratory
TUNL/Duke University
Box 90308
Durham, NC 27708-0308
USA
mendez@tunl.duke.edu
(tel.) 1-919-660-2626
(fax) 1-919-660-2634

Hans-Otto Meyer
Indiana University
Dept. of Physics
Swain Hall West
Bloomington, IN 47405
USA
meyer@iucf.indiana.edu
(tel.) 1-812-855-9365
(fax) 1-812-855-6645

Werner Meyer
Physikalisches Institut-Bonn
Nussallee 12
D-53115 Bonn
GERMANY
meyer@pib1.physik.uni-bonn.de
(tel.) 49-228-732230
(fax) 49-228-737869

Dan Miller
Indiana University Cyclotron Facility
Bloomington, IN 47405
USA
miller@iucf.indiana.edu
(tel.) 1-812-855-9365
(fax) 1-812-855-6645

Gerald A. Miller
University of Washington
Dept. of Physics, FM-15
Seattle, WA 98195
USA
miller@alpher.np1.washington.edu
(tel.) 1-206-543-2995
(fax) 1-206-685-0635

Michael A. Miller
University of Wisconsin
1150 University Avenue
Madison, WI 53706
USA
miller@uwnuc1.physics.wisc.edu
(tel.) 1-608-262-3091
(fax) 1-608-262-3598

Richard G. Milner
MIT
26-447, Dept. of Physics
Cambridge, MA 02139
USA
milner@mitlns.mit.edu
(tel.) 1-617-258-5439
(fax) 1-617-258-5440

Joseph H. Mitchell
CEBAF
MS 12H
12000 Jefferson Avenue
Newport News, VA 23606
USA
mitchell%micro@cebaf.gov
(tel.) 1-804-249-7851
(fax) 1-804-249-5000

Yoshiharu Mori
National Lab for High En. Phys
(KEK), Oho 1-1
Tsukuba-shi
Ibaraki-ken 305
JAPAN
moriy@kekvax.kek.ac.jp
(tel.) 81-298-64-1171
(fax) 81-298-64-3182

Toshiyuki Morii
Kobe University
Faculty of Human Development
Tsurukabuto, Nada
Kobe 657
JAPAN
morii@cphys.cla.kobe-u.ac.jp
(tel.) 81-78-803-0917
(fax) 81-78-803-0831

Joel M. Moss
Los Alamos National Laboratory
P2-Mail Stop D456
LANL
Los Alamos, NM 87545
USA
jmm@lanl.gov
(tel.) 1-505-667-1029
(fax) 1-505-665-4986

Francesco Murgia
Istituto Nazionale Di Fisica Nucleare
Sezione Di Cagliari
Via Ada Negri 18
I-09127 Cagliari
ITALY
murgia@vaxca.ca.infn.it
(tel.) 39-70670834
(fax) 39-70657823

Genis Musulmanbekov
Joint Inst. for Nuc. Research
LCTA
Head Post Office P.O. Box 79
101000 Moscow
RUSSIA
genis@vsdl28.jinr.dubna.su
(fax) 7-096-21-6-5145

Sergey I. Nagorny
Kharkov Inst. of Phys. & Tech.
National Science Center, Theoretic Division
Academicheskaya St., 1
Kharkov 310108
UKRAINE
kfti%kfti.kharkov.ua@relay.ussr.eu.net
(tel.) 7-057-235-6024
(fax) 7-057-235-1738

Sirish Nanda
CEBAF
12000 Jefferson Ave.
Newport News, VA 23606
USA
nanda@cebaf.gov

Hermann Nann
Indiana University Cyclotron Facility
2401 Milo B. Sampson Lane
Bloomington, IN 47408
USA
nann@iucf.indiana.edu
(tel.) 1-812-855-2884
(fax) 1-812-855-6645

Bernard Nefkens
UCLA
Dept. of Physics
405 Hilgard Avenue
Los Angeles, CA 90024
USA
bnefkens@uclapp.physics.ucla.edu
(tel.) 1-310-825-4970
(fax) 1-310-206-4397

Edwin Norbeck
University of Iowa
Dept. of Physics and Astronomy
Iowa City, IA 52242
USA
norbeck@iowa.physics.uiowa.edu
(tel.) 1-319-335-0903
(fax) 1-319-335-1753

Tetsuo Noro
Osaka University
Research Center for Nuclear Physics
10-1 Mihogaoka, Ibaraki-567
JAPAN
noro@rcnpvx.rcnp.osaka-u.ac.jp
(tel.) 81-6-879-8933
(fax) 81-6-879-8899

Hiroyuki Noumi
KEK
National Laboratory for High Energy Physics
1-1 OHO
Tsukuba, Ibaraki, 305
JAPAN
noumi@kekvax.kek.jp
(tel.) 81-298-64-1171
(fax) 81-298-64-7831

Wolf-Dieter Nowak
DESY-IfH Zeuthen
D-15735 Zeuthen
GERMANY
wdn@znher2.ifh.de
(tel.) 49-33762-77349
(fax) 49-33762-77330

Sandibek B. Nurushev
Inst. for High Energy Physics
142284 Protvino
Moscow Region
RUSSIA
nurushev@mx.ihep.su
(tel.) 7-095-230-3228
(fax) 7-095-230-2337

Hideaki Ohgaki
Electrotechnical Laboratory
1-1-4 Umezono
Tsukuba-shi, Ibaraki-305
JAPAN
e8904@etlcom3.etl.go.jp
(tel.) 81-298-58-5643
(fax) 81-298-52-7944

Hajime Ohnuma
Tokyo Institute of Technology
Dept. of Physics
2-12-1 Oh-Okayama, Meguro-KU
Tokyo
JAPAN
ohnuma@rikvax.riken.go.jp
(tel.) 81-3-5734-2080
(fax) 81-3-5734-2742

Hiroyuki Okamura
University of Tokyo
Dept. of Physics
Hongo 7-3-1
Bunkyo-ku
Tokyo 113
JAPAN
okamura@rikvax.riken.go.jp
(tel.) 81-3-3812-2111
(fax) 81-3-3811-0960

Hideaki Otsu
University of Tokyo
Dept. of Physics
7-3-1, Hongo, Bunkyo-ku
Tokyo 113
JAPAN
otsu@rikvax.riken.go.jp
(tel.) 81-48-462-1111x4131
(fax) 81-48-461-5301

H. Paetz gen Schieck
Universität Köln
Institut für Kernphysik
Zuelpicher Strasse 77
D-50937 Köln
GERMANY
schieck@lucie.ikp.uni-koeln.de
(tel.) 49-221-470-3620
(fax) 49-221-470-5168

Shelley Page
University of Manitoba
c/o TRIUMF
4004 Wesbrook Mall
Vancouver, BC V6T 2A3
CANADA
shelly@erich.triumf.ca
(tel.) 1-604-222-1047
(fax) 1-604-222-1074

Paul V. Pancella
Western Michigan University
Dept. of Physics
Kalamazoo, MI 49008-5151
USA
pancella@wmich.edu
(tel.) 1-616-387-4962

Erik Passchier
NIKHEF-K
Postbus 41882
1009 DB Amsterdam
THE NETHERLANDS
erik@nikhefk.nikhef.nl
(tel.) 31-20-5922147
(fax) 31-20-5922165

Stephen Pate
MIT
Building 26-405
77 Massachusetts Avenue
Cambridge, MA 02139
USA
pate@marie.mit.edu
(tel.) 1-617-253-4868
(fax) 1-617-258-5440

Seppo I. Penttila
Los Alamos National Laboratory
P-ll, MS-H846
Los Alamos, NM 87545
USA
penttila@lampf.lanl.gov
(tel.) 1-505-665-0641
(fax) 1-505-665-7920

Nikolai Piskunov
Joint Institute for Nuc. Rsch.
LHE JINR
Dubna
Moscow Region
141980 RUSSIA
piskunov@lhe06.jinr.dubna.su
(tel.) 7-096-2163023
(fax) 7-096-2165889

Mark Pitt
Caltech 106-38
Pasadena, CA 91125
USA
pitt@almach.caltech.edu
(tel.) 1-818-395-4587
(fax) 1-818-564-8708

Kevin Pitts
Fermilab
P.O. Box 500
M.S. #318
Batavia, IL 60510
USA
kpitts@fnald.fnal.gov
(tel.) 1-708-840-8718
(fax) 1-708-840-2968

Anatoli Plavko
St. Petersburg State Technical Univ.
40-3-9 Thorez Prospect
St. Petersburg
194223 RUSSIA
root@fmf.stu.spb.su
(tel.) 7-812-552-4751
(fax) 7-812-552-6086

Iouri A. Plis
Joint Inst. for Nuclear Rsch.
Laboratory of Nuclear Problems
141980 Dubna
Moscow Region
RUSSIA
plis@main1.jinr.dubna.su
(tel.) 7-096-2162757
(fax) 7-096-2166666

Stanislav G. Popov
Budker Inst. of Nuc. Physics
Prosp. Lavrent'eva 11
630090 Novosibirsk
RUSSIA
stas@nikhefk.nikhef.nl
(tel.) 7-383-235-9714
(fax) 7-383-235-2163

Scott Price
University of Michigan
Randall Laboratory of Physics
Ann Arbor, MI 48109-1010
USA
price@mich1.physics.lsa.umich.edu
(tel.) 1-313-764-5114
(fax) 1-313-763-9027

David Prout
Ohio State University
1302 Kinnear Road
Columbus, OH 43212
USA
prout@ohstpy.mps.ohio-state.edu
(tel.) 1-614-292-4775
(fax) 1-614-292-4833

Modesto Pusterla
Padova University
Dept. of Physics
Via Marzolo
Padova 35131
ITALY
pusterla@padova.infn.it
(tel.) 39-49-831767
(fax) 39-49-844245

Jack Rapaport
Indiana University Cyclotron Facility
2401 Milo B. Sampson Lane
Bloomington, IN 47408
USA
rapaport@iucf.indiana.edu
(tel.) 1-812-855-9365
(fax) 1-812-855-6645

Lazarus G. Ratner
Two Canterbury Ct.
E. Setauket, NY 11733
USA
ratner@bnldag.ags.bnl.gov
(tel.) 1-516-473-6585
(fax) 1-516-282-5954

Richard Raymond
University of Michigan
Dept. of Physics
Ann Arbor, MI 48109
USA
raymond@mich.physics.lsa.umich.edu
(tel.) 1-313-764-5113
(fax) 1-313-763-9027

Erwin Reichert
Institüt für Physik
der Johannes Gutenberg Universität
D-55099 Mainz
GERMANY
reichert@vipmza.physik.uni-mainz.de
(tel.) 49-61-31-392729
(fax) 49-61-31-392991

Tom Rinckel
Indiana University Cyclotron Facility
2401 Milo B. Sampson Lane
Bloomington, IN 47408
USA
rinckel@iucf.indiana.edu
(tel.) 1-812-855-0095
(fax) 1-812-855-6645

N. Russell Roberson
Duke University
Triangle Universities Nuclear Laboratory
Box 90308
Durham, NC 27708-0308
USA
roberson@tunl.duke.edu
(tel.) 1-919-660-2600
(fax) 1-919-660-2634

Renato Roncaglia
Indiana University
Nuclear Theory Center
2401 Milo B. Sampson Lane
Bloomington, IN 47408
USA
renato@iucf.indiana.edu
(tel.) 1-812-855-9365
(fax) 1-812-855-6645

Chris Roper
Triangle Universities Nuclear Laboratory
Duke University
Box 90308
Durham, NC 27708-0308
USA
roper@tunl.duke.edu
(tel.) 1-919-660-2635
(fax) 1-919-660-2634

Thomas Roser
Brookhaven National Laboratory
Bldg. 911B, AGS Dept.
Upton, NY 11973-5000
USA
roser@bnl.gov
(tel.) 1-516-282-7084
(fax) 1-516-282-5954

Robert Rylko
University of London
Queen Mary & Westfield College
Mile End Road
London E1 4NS
ENGLAND
rylko@v1.ph.qmw.ac.uk
(tel.) 44-71-9755555x4003
(fax) 44-81-9819465

Bijan Saghai
Centre d'Etudes de Saclay
Service de Physique Nucleaire
CE-Saclay, F-91191
Gif-sur-Yvette CEDEX
FRANCE
saghai@phnx7.saclay.cea.fr
(tel.) 33-1-69087561
(fax) 33-1-69087584

Naohito Saito
Kyoto University
Department of Physics
Kitashirakawa
Kyoto 606-01
JAPAN
saito@kytax.scphys.kyoto-u.ac.jp
(tel.) 81-75-753-3842
(fax) 81-75-711-5175

Takeji Sakae
Kyushu University
6-10-1, Hakozaki, Higashi-ku
Fukuoka 812
JAPAN
sakaetne@mbox.nc.kyushu-u.ac.jp
(tel.) 81-092-6411101X5821
(fax) 81-092-6417098

Hide Sakai
University of Tokyo
Dept. of Physics, Fac. of Science
Hongo 7-3-1, Bunkyo
Tokyo 113
JAPAN
sakai@tkyvax.phys.s.u-tokyo.ac.jp
(tel.) 81-3-5689-7343
(fax) 81-3-3811-0960

Naruhiko Sakamoto
University of Tokyo
Department of Physics
7-3-1, Hongo, Bunkyo-ku
Tokyo 113
JAPAN
nsakamoto@rikvax.riken.go.jp
(tel.) 81-3-5689-7343
(fax) 81-3-3811-0960

Andrew M. Sandorfi
Brookhaven National Laboratory
Dept. of Physics
Building 510A
Upton, NY 11973
USA
sandorfi@bnlcl1.bnl.gov
(tel.) 1-516-282-7951
(fax) 1-516-282-5568

Filipe D. Santos
Universidade de Lisboa
Centro de Física Nuclear DA
Av. Gama Pinto 2
1699 Lisboa Codex
PORTUGAL
santos@alf4.cc.fc.ul.pt
(tel.) 351-1-7597713
(fax) 351-1-7597716

Georgios Savopulos
Indiana University Cyclotron Facility
2401 Milo B. Sampson Lane
Bloomington, IN 47408
USA
gsavopul@venus.iucf.indiana.edu
(tel.) 1-812-855-3613
(fax) 1-812-855-6645

Pierre A. Schmelzbach
Paul Scherrer Institute
F1, Accelerator Division
CH-5232
Villigen-PSI
SWITZERLAND
schmelzbach@cvax.psi.ch
(tel.) 41-1-992111
(fax) 41-1-993383

Bill Schmitt
Indiana University Cyclotron Facility
2401 Milo B. Sampson Lane
Bloomington, IN 47408
USA
schmitt@iucf.indiana.edu
(tel.) 1-812-855-9365
(fax) 1-812-855-6645

Paul W. Schmor
TRIUMF
4004 Wesbrook Mall
Vancouver, B.C.
V6T 2A3 CANADA
schmor@triumf.ca
(tel.) 1-604-222-1047
(fax) 1-604-222-1074

Peter Schwandt
Indiana University
Dept. of Physics
Bloomington, IN 47408
USA
schwandt@iucf.indiana.edu
(tel.) 1-812-855-9365
(fax) 1-812-855-6645

Benjamin Shahbazian
Joint Inst. for Nuclear Rsch.
Dubna
Moscow Region 141980
RUSSIA
shahbazi@lhe22.jinr.dubna.su
(tel.) 7-096-216-2985
(fax) 7-095-975-2381

Michael Shmatikov
Kurchatov Institute
Russian Research Center
B. Chermushkinskaya 25
117259 Moscow
RUSSIA
msh@ofpnp.kiae.su
(tel.) 7-095-196-7736
(fax) 7-095-123-6584

Markus Simonius
Institute for Particle Physics
ETH-Hoenggerberg
CH-8093 Zürich
SWITZERLAND
simonius@imp.phys.ethz.ch
(tel.) 41-1-633-2038
(fax) 41-1-633-1067

Klaus Sinram
DESY, Dpt. MEA
Notkestr. 85, Postfach
D22603 Hamburg
GERMANY
measin@dsyibm.desy.de
(tel.) 49-40-8998-3714
(fax) 49-40-8998-3438

Igor M. Sitnik
Joint Inst. for Nuc. Research
Laboratory for High Energy
141980 Dubna
RUSSIA
sitnik@main1.jinr.dubna.su
(tel.) 7-096-2163023

Dennis Sivers
Portland Physics Institute
4780 SW Macadam #101
Portland, OR 97201
USA
sivers@anlhep
(tel.) 1-503-223-2680
(fax) 1-503-223-2750

Morag Smith
University of Wisconsin
1150 University Ave.
Madison, WI 53706
USA
mksmith@uwnuc0.physics.wisc.edu
(tel.) 1-608-263-2263
(fax) 1-608-262-3598

Todd B. Smith
University of Michigan
4063 Randall Lab
Ann Arbor, MI 48109
USA
smith@mich.physics.lsa.umich.edu
(tel.) 1-313-763-5981
(fax) 1-313-763-9694

W. Mike Snow
Indiana University Cyclotron Facility
2401 Milo B. Sampson Lane
Bloomington, IN 47408
USA
snow@iucf.indiana.edu
(tel.) 1-812-855-7914
(fax) 1-812-855-6645

Franz Sperisen
Indiana University Cyclotron Facility
2401 Milo B. Sampson Lane
Bloomington, IN 47408
USA
sperisen@iucf.indiana.edu
(tel.) 1-812-855-2948
(fax) 1-812-855-6645

Harold Spinka
Argonne National Laboratory
High Energy Physics Division
Building 362, 9700 S. Cass Avenue
Argonne, IL 60439
USA
hms@hep.anl.gov
(tel.) 1-708-252-6317
(fax) 1-708-252-5782

Mark Spraker
Indiana University Cyclotron Facility
Bloomington, IN 47405
spraker@iucf.indiana.edu
(tel.) 1-812-855-9365
(fax) 1-812-855-6645

Keith Stantz
Indiana University Cyclotron Facility
2401 Milo B. Samspon Lane
Bloomington, IN 47408
USA
stantz@iucf.indiana.edu
(tel.) 1-812-855-9365
(fax) 1-812-855-6645

Edward J. Stephenson
Indiana University Cyclotron Facility
2401 Milo B. Sampson Lane
Bloomington, IN 47408
USA
stephenson@iucf.indiana.edu
(tel.) 1-812-855-9365
(fax) 1-812-855-6645

Friedemann Stock
MPI Kernphysik
Postfach 10 39 80
69029 Heidelberg
GERMANY
frs@dxnhd1.mpi-hd.mpg.de
(tel.) 49-6221-516339
(fax) 49-6221-516540

Igor Strakovsky
Virginia Polytechnic Institute
Dept. of Physics
Blacksburg, VA 24061-0435
USA
igor@vtinte.phys.vt.edu
(tel.) 1-703-231-7410
(fax) 1-703-231-7511

Leonid Strunov
JINR
Head Post Office
PO Box 79
101000 Moscow
RUSSIA
strunov@sunhe.jinr.dubna.su
(tel.) 7-09621-62885
(fax) 7-095-9752381

Evan R. Sugarbaker
Ohio State University
Dept. of Physics
174 West 18th Avenue
Columbus, OH 43210-1106
USA
sugarbak@mps.ohio-state.edu
(tel.) 1-614-292-4775
(fax) 1-614-292-4833

John Szymanski
Indiana University Cyclotron Facility
2401 Milo B. Sampson Lane
Bloomington, IN 47408
szymanski@iucf.indiana.edu
(tel.) 1-812-855-2882
(fax) 1-812-855-6645

Terry N. Taddeucci
Los Alamos National Lab
P-17, MS-H803
Los Alamos, NM 87545
USA
taddeucci@lampf.lanl.gov
(tel.) 1-505-665-3114
(fax) 1-505-665-4121

Yoshihiro Tagishi
University of Tsukuba
Institute of Physics
Ibaraki 305
JAPAN
tagishi@tsukuba.ac.jp
(tel.) 81-298-53-2567
(fax) 81-298-53-2565

Mauro Taiuti
Istituto Nazionale di Fisica Nucleare (INFN)
via Dodecanneso 33
I-16100, Genova
ITALY
taiuti@genova.infn.it
(tel.) 39-10-3536336
(fax) 39-10-313358

Atsushi Tamii
Kyoto University
Department of Physics
Kitashirakawa Oiwaki-cho
Sakyo-ku, Kyoto, 606
JAPAN
tamii@kytvs1.scphys.kyoto-u.ac.jp
(tel.) 81-075-753-3866
(fax) 81-075-753-3887

Masayoshi Tanaka
Kobe Tokiwa Jr. College
2-6-2 Ohtani-cho, Nagata-ku
Kobe 653
JAPAN
tanaka@rcnpvx.rcnp.osaka-u.ac.jp
(tel.) 078-611-1821
(fax) 078-643-4361

Makoto Tanifuji
Hosei University
Dept. of Physics
2-17-1, Fujimi, Chiyoda
Tokyo 102
JAPAN
(tel.) 81-03-3264-9431
(fax) 81-03-3264-9326

Michael J. Tannenbaum
Brookhaven National Laboratory
Dept. of Physics, 510C
P.O. Box 5000
Upton, NY 11973-5000
USA
sapin@bnldag.ags.bnl.gov
(tel.) 1-516-282-3722
(fax) 1-516-282-3253

Anthony W. Thomas
University of Adelaide
Dept. of Physics
P.O. Box 498
Adelaide SA 5005
AUSTRALIA
athomas@physics.adelaide.edu.au
(tel.) 61-8-303-5113
(fax) 61-8-303-4380

Alan K. Thompson
National Institute of Standards & Technology
Bldg. 235/Mail Stop A106
Gaithersburg, MD 20899
USA
akt@rrdstrad.nist.gov
(tel.) 1-301-975-4666
(fax) 1-301-921-9847

Ian J. Thompson
University of Notre Dame
Physics Department
Notre Dame, IN 46556-5670
USA
ithompso@doc.helios.nd.edu
(tel.) 1-219-631-7262
(fax) 1-219-631-5952

Egle Tomasi-Gustafsson
CE-Saclay
LNS/DIR
F-91191 Gif-sur-Yvette Cedex
FRANCE
tomasi@lnsbm2.saclay.cea.fr
(tel.) 33-1-6-908-3459
(fax) 33-1-6-908-9011

Dmitri Toporkov
Sib. Div. of the Academy of Science
Budker Institute of Nuclear Physics
630090 Novosibirsk
RUSSIA
tdm@inp.nsk.su
(tel.) 7-3832-359910
(fax) 7-3832-352163

Werner Tornow
Duke University
Dept. of Physics
Box 90308
Durham, NC 27706
USA
tornow@tunl.tunl.duke.edu
(tel.) 1-919-660-2637
(fax) 1-919-660-2634

Takeshi Toyama
Nat. Lab. for High En. Physics (KEK)
1-1 Oho-machi
Tsukuba-shi
Ibaraki-ken, 305
JAPAN
toyama@kekvax.kek.jp
(tel.) 81-298-64-5277
(fax) 81-298-64-3182

Hidenori Toyokawa
Research Center for Nuclear Physics
Osaka University
10-1, Mihogaoka, Ibaraki
Osaka 567
JAPAN
toyokawa@rcnpvx.rcnp.osaka-u.ac.jp
(tel.) 81-6-879-8939
(fax) 81-6-879-8899

Sergei M. Troshin
Institute for High En. Physics
142284 Protvino
Moscow Region
RUSSIA
troshin@mx.ihep.su
(tel.) 7-095-289-2732
(fax) 7-095-230-2337

Garry Tungate
University of Birmingham
School of Physics & Space Research
Birmingham B152TT
ENGLAND
psi@np.ph.birmingham.ac.uk
(tel.) 44-021-414-4685
(fax) 44-021-414-4719

William Turchinetz
MIT, Bates Lab
P.O. Box 846
Middleton, MA 01949
USA
billt@bates.mit.edu
(tel.) 1-617-253-9214
(fax) 1-617-253-9599

Tamotsu Ueda
Ehime University
Department of Physics, Faculty of Science
Bunkyo 2-5
Ehime 790
JAPAN
(tel.) 81-899-247111x3541
(fax) 81-899-232545

Tomohiro Uesaka
University of Tokyo
Dept. of Physics
Hongo 7-3-1
Bunkyo-ku
Tokyo 113
JAPAN
uesaka@rikvax.riken.go.jp
(tel.) 81-3-5689-7343
(fax) 81-3-3811-0960

Alex Umnikov
University of Alberta
Theoretical Physics Institute
Dept. of Physics, Edmonton
Alberta T6G 2J1
CANADA
umnikov@phys.ualberta.ca
(tel.) 1-403-492-5575
(fax) 1-403-492-3408

Orhan Unal
Univ. of Wisconson-Madison
Dept. of Physics
1150 University Ave.
Madison, WI 53706
USA
unal@uwnuc0.physics.wisc.edu
(tel.) 1-608-262-6555
(fax) 1-608-262-3598

Jacques Van DeWiele
Universite Paris-sud
Institut de Physique Nucleaire
BP n°1
91406 Orsay, Cedex
FRANCE
vandewi@ipncls.in2p3.fr
(tel.) 33-1-6941-7328
(fax) 33-1-6941-6470

Willem T. H. Van Oers
University of Manitoba, TRIUMF
4004 Wesbrook Mall
Vancouver BC V6T 2A3
CANADA
vanoers@triumf.ca
(tel.) 1-604-222-1047
(fax) 1-604-222-1074

Steven E. Vigdor
Indiana University
Dept. of Physics
Bloomington, IN 47405
USA
vigdor@iucf.indiana.edu
(tel.) 1-812-855-9365
(fax) 1-812-855-6645

Nikolaos Vodinas
NIKHEF
Postbus 41882
1009 DB Amsterdam
THE NETHERLANDS
vodinas@nikhefk.nikhef.nl
(tel.) 31-20-5922089
(fax) 31-20-5922165

Heinrich V. von Geramb
Universität Hamburg
Theoretische Kernphysik
Luruper Chaussee 149
D-22761 Hamburg
GERMANY
i04ger@dsyibm.desy.de
(tel.) 49-040-8998-2131
(fax) 49-040-8998-2143

Barbara von Przewoski
Indiana University Cyclotron Facility
2401 Milo B. Sampson Ln.
Bloomington, IN 47405
USA
przewoski@iucf.indiana.edu
(tel.) 1-812-855-9365
(fax) 1-812-855-6645

Rudiger Voss
CERN
PPE Division
CH-1211
Geneva, 23
SWITZERLAND
rvoss@cernvm.cern.ch
(tel.) 41-22-767-6447
(fax) 41-22-785-0672

Tomotsugu Wakasa
University of Tokyo
Department of Physics
Hongo 7-3-1, Bunkyo-ku
Tokyo 113
JAPAN
wakasa@rikvax.riken.go.jp
(tel.) 81-3-5689-7343
(fax) 81-3-3811-0960

Richard Walter
DUKE
Dept. of Physics
Durham, NC 27708-0305
USA
walter@tunl.duke.edu
(tel.) 1-919-660-2629
(fax) 1-919-660-2634

Roger P. Ward
University of Birmingham
School of Physics & Space Research
Edgbaston
Birmingham B15 2TT
UNITED KINGDOM
r.p.ward@bham.ac.uk
(tel.) 44-021-414-4674
(fax) 44-021-414-4719

L. Kieffer Warman
Indiana University Cyvlotron Facility
2401 Milo B. Sampson Ln.
Bloomington, IN 47408
USA
kieffer@iucf.indiana.edu
(tel.) 1-812-855-9365
(fax) 1-812-855-6645

Glen Warren
MIT
Rm 26-648
77 Massachusetts Avenue
Cambridge, MA 02139
USA
gwarren@mitlns.mit.edu
(tel.) 1-617-253-7977
(fax) 1-617-258-5440

John W. Watson
Kent State University
Dept. of Physics
Kent, OH 44242
USA
watson@ksuvxd.kent.edu
(tel.) 1-216-672-2771
(fax) 1-216-672-2938

C. Steven Whisnant
University of South Carolina
Department of Physics & Astronomy
Columbia, SC 29208
USA
whisnant@nuc003.psc.scarolina.edu
(tel.) 1-803-777-9025
(fax) 1-803-777-3065

W. Scott Wilburn
Duke University at TUNL
Box 90308
Duke University
Durham, NC 27708-0308
USA
wilburn@tunl.duke.edu
(tel.) 1-919-660-2624
(fax) 1-919-660-2634

Thomas Wise
Univ. of Wisconsin-Madison
Nuclear Physics
1150 University Ave
Madison, WI 53706-1390
USA
wise@uwnuc0.physics.wisc.edu
(tel.) 1-608-262-6555
(fax) 1-608-262-3845

Scott W. Wissink
Indiana University Cyclotron Facility
2401 Milo B. Sampson Lane
Bloomington, IN 47408
USA
wissink@iucf.indiana.edu
(tel.) 1-812-855-5192
(fax) 1-812-855-6645

Teruya Yamanishi
Kobe University
Graduate School of Science & Technology
1-1 Rokkodai-machi, Nada-ku
Kobe-shi 657
JAPAN
yamanisi@cphys.c1a.kobe-u.ac.jp
(tel.) 81-78-881-1212
(fax) 81-78-803-0831

Yi-Fen Yen
Los Alamos National Laboratory
MS H846
Los Alamos, NM 87545
USA
yen@lampf.lanl.gov
(tel.) 1-505-665-8322
(fax) 1-505-665-7920

Osman Yilmaz
Middle East Technical Univ.
Department of Physics
06531 Ankara
TURKEY
a10268@vm.cc.metu.edu.tr
(tel.) 90-312-2101000
(fax) 90-312-2101281

Masaru Yosoi
Kyoto University
Department of Physics
Oiwake-cho, Kitashirakawa, Sakyo-ku
Kyoto 606
JAPAN
yosoi@kytvs1.scphys.kyoto-u.ac.jp
(tel.) 81-075-753-3832
(fax) 81-075-753-3887

Anna Marie Zanetti
INFN - Area di Ricerca
Palazzina L3
Padriciano 99
34012 Trieste
ITALY
zanetti@trieste.infn.it
(tel.) 39-40-3756227
(fax) 39-40-3756258

Kirsten Zapfe
DESY/HERMES Collaboration
Notkestrasse 85
D-2260 Hamburg, 3
GERMANY
zapfe@vxdesy.desy.de
(tel.) 49-40-8998-3743
(fax) 49-40-8998-3438

Anatoli Zelenski
TRIUMF
4004 Wesbrook Mall
Vancouver B.C., V6T 2A3
CANADA
zelenski@erich.triumf.ca
(tel.) 1-604-222-1047
(fax) 1-604-222-1074

Luisa Zetta
Istituto Nazionale di Fisica Nucleare
Dipartimento di Fisica
via Celoria 16
20133 Milano
ITALY
zetta@milano.infn.it
(tel.) 39-2-2392290
(fax) 39-2-2392297

Yong Zhou
Univ. of Wisconsin-Madison
1150 University Ave.
Madison, WI 53706
USA
yong@uwnuc0.physics.wisc.edu
(tel.) 1-608-263-2263
(fax) 1-608-262-3598

Pawel Zupranski
Soltan Inst. for Nuc. Studies
Hoza 69
00-68 Warsaw, 9
POLAND
zupran@apollo.fuw.edu.pl
(tel.) 48-2-6213829
(fax) 48-2-6213829

AUTHOR INDEX

A

Abegg, R., 162, 302
Ackerstaff, K., 319
Ahmidouch, A., 308
Ahn, J.K., 314
Akimune, H., 395
Al-Ohali, M.A., 593
Alarcon, R., 551
Aleshin, N.P., 456
Alonso, J.R., 623
Altmeier, M., 668
Andalkar, A., 156
Anderson, B.D., 438
Andresen, H.G., 18
Annand, J.R.M., 18
Aoyama, T., 629
Arenhövel, H., 551
Arnold, J., 308
Arrington, J., 63
Arvieux, J., 476
Atzrott, U., 569
Aulenbacher, K., 18
Aumüller, B., 668
Ayer, Z., 337

B

Bacher,, A.D., 296, 401, 407, 581
Baghaei, H., 524
Baker, F.T., 425
Balestra, F., 476
Bassalleck, B., 314
Beck, D., 63
Becker, J., 18
Bedfer, Y., 476
Beene, J., 407
Beise, E.J., 63, 95
Belostotski, S.L., 456
Belov, A.S., 643, 662
Bent, R.D., 319
Berdoz, A.R., 136, 162, 302
Berg, G.P.A., 407
Bertini, R., 476
Bertrand, F., 407
Betker, A., 407
Bimbot, L., 425, 545
Birchall, J., 136, 162, 302

Bisplinghoff, J., 191
Blanchard, S.P., 349
Bland, L.C., 438, 444, 476
Blinov, B.B., 698
Bloch, C., 343
Blomgren, J., 319
Blume-Werry, J., 18
Blyth, C.O., 611
Bossolasco, S., 476
Botto, T., 551
Bouché-Pillon, Marc, 515
Bouwhuis, M., 551
Bowman, J.D., 120, 136, 156
Bowyer, S.M., 296, 401, 407
Bowyer, T.W., 296
Brash, J.E., 349
Braun, B., 686
Braun, R.T., 290
Brink, D.M., 605
Brinkmöller, B., 349, 438
Brochard, F., 476
Brown, J.D., 319
Brown, R., 662
Bucholz, M., 551
Bugg, D.V., 296
Bulten, H.J., 343, 551
Burleson, G.R., 349
Bussa, M.P., 476
Bywater, J.A., 698

C

Cain, B., 156
Campbell, J.R., 136, 162, 302
Candell, E., 63
Caracappa, A., 524
Cardman, L., 63
Carman, D.S., 438, 444
Carr, R., 63
Cata-Danil, G., 569
Chang, S., 296, 401, 407
Chant, N., 444
Chen, F-J., 319
Chin, S., 698
Choi, H.D., 611
Choi, S., 551
Christley, J.A., 617
Chung, M.S., 314

Chung, W.M., 314
Churakov, V.V., 698
Cichocki, A., 524
Clarke, N.M., 611
Collins, J., 662
Comfort, J., 551
Connell, K.A., 611
Conzett, H.E., 191
Cooper, D.A., 413, 470
Coulter, K.P., 530
Court, G.R., 698
Crawford, B.E., 120
Cummings, W.J., 349

D

Daito, I., 395
Das, R.K., 337
Daum, M., 308
David, J.C., 518
Davis, B.J., 349
Davis, C.A., 136, 162, 302
Davis, N.J., 611
de Jager, C.W., 530, 551
de Lange, D.J.J., 551
de Vries, H., 530, 551
Dee, P.R., 611
Dehnhard, D., 349, 438
Delheij, P.P.J., 120, 162, 349
Delucia, S.L., 413, 470
Demierre, Ph., 308
Derenchuk, V., 662
Derevyankin, G.E., 643
DeSchepper, D., 343, 692
Deutsch, Jules, 112
Dezarn, W.A., 680
Dimitroyannis, D., 551
Djalali, C., 425, 545
Dodson, G., 63
Doets, M., 551
Dombo, Th., 18
Doskow, J., 680
Dotsenko, Yu.V., 456
Dow, K., 63
Drescher, P., 18
Drevenak, R., 308
Ducret, J.E., 18
Dudnikov, V.G., 643
Duncan, F., 63
Dutz, H., 505

E

Eden, T., 47
Edwards, C.M., 349, 438
Edwards, G.W.R., 545
Efimovykh, V.A., 456
Eggert, M., 668
Eiró, Ana M., 361
Ejiri, H., 386
Ent, R., 343, 551
Enyo, H., 314
Ernst, J., 191
Espy, M.A., 349, 438
Eversheim, P.D., 191, 668
Eyl, D., 18

F

Falomkin, I.V., 476
Farkhondeh, M., 63
Fava, L., 476
Fayard, C., 518
Fedorov, O.Ya., 456
Felden, O., 668
Ferrero, L., 476
Ferro-Luzzi, M., 551
Fick, D., 635, 686
Filippone, B., 63
Finger, M., 308
Finger, Jr., M., 308
Finlay, R., 524
Fisher, H., 18
Forest, T., 63
Foster, C., 407
Frankle, C.M., 120
Franklin, W., 296, 407
Franz, J., 308
Frey, A., 18
Fujita, S., 431
Fujiwara, M., 395
Fukuda, K., 120
Fukuda, M., 623
Fukuda, S., 623
Fukuda, T., 314
Funahashi, H., 314
Furget, C., 545

G

Gan, L., 162, 302

Gao, H., 63
Garçon, M., 545, 557
Garfagnini, R., 476
Gaul, H.-G., 686
Gebel, R., 668
Gill, D.R., 476
Gilman, R., 530
Girit, I.C., 156
Gladyshev, V., 524
Glashausser, C., 425, 545
Glombik, A., 668
González Trotter, D.E., 290
Goodman, C.D., 343, 470
Goto, Y., 314
Goujon, N., 308
Gould, C.R., 120, 185
Grabmayr, P., 18
Grasso, A., 476
Graw, G., 569
Green, A.A., 120, 136
Green, P.V., 599
Green, P.W., 136, 162, 302
Greenfield, M.B., 419
Greeniaus, L.G., 162, 302
Gresko, T., 524
Grieser, M., 686
Gu, T., 444
Guazzoni, P., 569
Guillot, J., 425

H

Haase, D.G. 120, 185
Haeberli, W., 213, 680, 686
Hajdas, W., 308
Halbert, M., 407
Hall, S., 18
Hall, S.J., 611
Hamian, A.A., 136
Hansen, J.-O., 55, 343
Hara, Y., 431
Harada, A., 623
Harakeh, M.N., 629
Hardie, G., 319
Hardie, J.G., 680
Hartmann, P., 18
Hatanaka, K., 395, 419, 431, 450, 702
Häusser, O., 349
Hautle, P., 308
Healey, D.C., 136, 162

Heckel, Blayne R., 107
Heer, E., 308
Hehl, T., 18
Heil, W., 18
Heimann, C., 668
Heimberg, P., 319
Helmer, R., 136, 162, 302
Henderson, R., 349
Hertenberger, R., 569
Hess, R., 308
Hicks, K., 407, 524
Higashi, A., 314
Hinterberger, F., 191
Hoblit, S., 524
Hofer, D., 569
Hoffmann, J., 18
Hofmann, J.S., 713
Holt, R.J., 530
Horen, D., 407
Horowitz, C.J., 708
Hosono, K., 395
Howell, C.R. 290, 593
Huber, G.M., 444
Huffman, P.R., 185
Huffman, J., 444

I

Ichihara, T., 431
Ieiri, M., 314
Iinuma, M., 120, 314
Ikegami, K., 656
Imai, K., 314
Inomata, T., 395
Iseri, Y., 587
Ishida, S., 419, 431
Itow, Y., 314
Izotov, A.A., 456
Izumikawa, T., 623

J

Jacobs, W.W., 343, 476
Jacobsen, E., 319
Jahn, R., 191
Janout Jr., Z., 308
Jaskola, M., 569
Jennings, B.K., 349
Johnson, B.N., 425, 545
Johnson, R.C., 605, 617

Jones, C.E., 343, 530
Jones, M.K., 349

K

Kameyama, H., 575
Kammermann, M., 668
Kaneko, H., 702
Kaptari, L.P., 79
Karban, O., 611
Karwowski, H.J., 337
Katoh, K., 431
Kaufman, W.A., 698
Kawabata, M., 450
Kazakov, K.Yu., 79
Keith, C.D., 185
Kellie, J.D., 18
Kemper, K.W., 599
Kerr, P.L., 599
Khandaker, M., 524
Khanna, F.C., 79
Kim, G.D., 314
Kim, Y.D., 314
Kinney, E.R., 530
Kinsho, M., 656
Kisselev, A.Yu., 456
Kistner, O., 524
Klein, F., 18
Klenov, V.S., 643
Kleppner, D., 698
Klyachko, A., 444
Knott, J.E., 156
Knudson, J.N., 120
Knutson, L.D., 325
Kobayashi, T., 623
Koger, R., 308
Kolb, N., 162
Komarov, E.N., 456
Komives, A., 156
Konijn, J., 551
Konter, J.A., 308
Koori, N., 419
Korkmaz, E., 136, 162, 302
Korsch, W., 63, 343
Kowalczyk, R.S., 530
Kowalski, S., 63
Kox, S., 545
Kozlowska, B., 337
Kramer, L.H., 343, 692
Krebs, G.F., 623

Kretschmer, W., 191, 355, 668
Krisch, A.D., 698
Kubota, K., 629
Kudriashov, V., 581

L

Lacker, H., 308
Ladygin, V.P., 331
Laget, J.M., 495
Lai, A., 47
Lail, B.A., 349
Lamot, G.H., 518
Lang, J., 551
Langenbrunner, J.L., 349, 438
Langevin-Joliot, H., 425
Larson, B., 349
Lechanoine-LeLuc, C., 308
Leduc, M., 18
Lee, J.M., 314
Lee, K., 343
Lee, L.R., 136, 162
Lehar, F., 308
Leidemann, W., 551
Lemaitre, S., 668
Leuschner, M., 156, 343
Levy, C.D.P., 136, 162, 650
Li, J., 162, 302
Lindgren, R., 524
Lisantti, J., 407
Liu, J., 296, 401, 407
Lorentz, B., 680
Lorenzon, W., 343, 349
Love, W.G., 462
Lowie, L.Y., 120
Lucas, M., 524
Ludwig, E.J., 337
Lung, A., 63
Luppov, V.G., 698
Luther, B.A., 413, 470
Lyascenko, V.I., 476

M

Machleidt, R., 290
Madey, R., 47, 438
Maeda, K., 349
Maggiora, A., 476
Makins, N.C.R., 343
Mango, S., 308

Marchlenski, D., 343
Markham, B.C., 444
Marty, N., 425
Masaike, A., 120, 314
Mascarini, Ch., 308
Masuda, Yasuhiro, 120, 148
Matsuda, Y., 120, 314
Matsuoka, N., 450
Matsuta, K., 623
McClelland, J.B., 413
McKeown, R., 63
Meierhoff, M., 18
Melnik, Yu.M., 698
Mendez, A.J., 599
Meyer, H.-O., 319, 343, 680, 708
Meyers, E.G., 599
Miceli, L., 524
Mihara, S., 314
Mihara, M., 623
Miklukho, O.V., 456
Miller, C.A., 162, 302
Miller, Gerald A., 172
Miller, M.A., 343, 551
Milner, Richard G., 245, 343, 692
Minamisono, T., 623
Mischke, R.E., 136
Mishnev, S.I., 530
Mitchell, G.E., 120
Miura, I., 702
Miyake, T., 623
Mizuno, Y., 450
Mohajeri, K., 599
Mohring, R., 63
Möller, H., 18
Montag, C., 686
Mori, Y., 656
Morinobu, S., 450
Morlet, M., 425, 545
Morris, C.L., 349
Moss, J.M., 721
Mueller, B., 63
Mueller, P., 407
Muldavin, J.B., 698
Mümmler, K., 668
Murzin, V.I., 456

N

Nachtigall, Ch., 18
Nakamura, M., 395, 450
Nakazato, M., 623
Nann, H., 319
Napolitano, J., 63
Naryshkin, Yu.G., 456
Neal, J.S., 343, 551
Nechaeva, L.P., 643
Nefkens, B.M.K., 197
Nelson, B., 349
Nelyubin, V.V., 530
Niizeki, T., 431
Nikolenko, D.M., 530, 551
Nojiri, Y., 623
Nooren, G.J., 551
Noro, T., 395, 450
Norum, B., 524
Nurushev, T.S., 698

O

O'Donnell, J.M., 349
Ohnuma, H., 431, 563
Ohtsubo, T., 623
Okada, K., 314
Okamura, H., 419, 431, 629
Okihana, A., 419, 450
Olive, D., 407
Olmer, C., 401, 581
Onegin, M., 581
Onishi, T., 623
Opper, A.K., 162, 302
Osipov, A.N., 530
Ostrick, M., 18
Otsu, H., 419, 431, 629
Otten, E.W., 18
Owens, R.O., 18
Ozawa, A., 623

P

Paetz gen. Schieck, H., 191, 668
Page, S.A., 136, 162, 202, 302
Palarczyk, M., 349, 438
Pancella, P.V., 319, 343, 680
Panzieri, D., 476
Papadakis, N., 551
Park, B.K., 349, 470
Park, I.S., 314
Park, Y.M., 314
Passchier, E., 551
Passchier, I., 551

Pate, S.F., 71, 343, 692
Pearce, K.I., 611
Penttilä, S.I., 120, 349
Petrov, D.V., 530
Petrov, M., 156
Pickar, M.A., 319
Pinder, C.N., 611
Piragino, G., 476
Piron, F., 518
Pitt, M., 63
Pitts, W.K., 296, 343
Plavko, A., 581
Plohinsky, Yu.V., 643
Plützer, S., 18
Poelker, M., 530
Pollock, R.E., 680
Pontecorvo, G.B., 476
Popov, S.G., 530, 551
Postma, H., 120, 162
Potterveld, D.H., 530
Povh, B., 686
Price, J.S., 698
Prokofiev, A.N., 456
Prokofiev, D.A., 456
Prout, D.L., 413, 470
Prudkoglyad, A.F., 698

R

Rachek, I.A., 530, 551
Rall, M., 686
Ramsay, W.D., 136, 162, 302
Rapaport, J., 470, 524
Rapin, D., 308
Rathmann, F., 680, 686
Raymond, R.S., 698
Réal, J.S., 545
Reber, E.L., 599
Reckenfelderbäumer, R., 668
Reichert, E., 18
Reitzner, S.D., 136
Rinckel, T., 319, 343, 680
Roberson, N.R., 120, 185
Roberts, E.J., 605
Rohe, D., 18
Roman, S., 611
Roos, P., 444
Roper, C.D., 290
Rosier, L., 425, 545
Ross, M.A., 680

Rössle, E., 308
Roy, G., 136
Rusek, K., 611
Rybarcyk, L.J., 413, 470

S

Sagara, K., 450
Saghai, Bijan, 515, 518, 521
Saito, N., 314
Sakaguchi, H., 395
Sakai, H., 419, 431, 482, 629
Sakamoto, N., 419, 431
Salinas, F., 290
Salle, P., 551
Sandorfi, A.M., 230, 524
Sasaki, M., 623
Satou, Y., 419, 431, 629
Savopulos, G., 343
Schäfer, M., 18
Shearer, L.D., 18
Scheglov, Yu.A., 456
Schiemenz, P., 569
Schmelzbach, P.A., 308
Schmidt, B.G., 599
Schmieden, H., 18
Schmitt, H., 308
Schmitt, W., 407
Schmor, P.W., 136, 650
Schwandt, P., 444, 581
Sealock, R., 524
Seely, M.L., 185
Seestrom, S.J., 120
Segel, R.E., 319
Sekulovich, A.M., 136, 302
Serdyuk, V., 476
Sereni, P., 308
Setze, H.R., 290
Sharapov, E.I., 120
Shimizu, H.M., 120
Shin, T., 692
Shin, Y.M., 314
Shmatikov, M., 142
Shutov, V.B., 698
Shvedchikov, A.V., 456
Sim, K.S., 314
Šimičević, N., 63
Sitnik, I.M., 331
Slunecka, M., 308
Smith, A., 343

Smith, L., 524
Snow, W.M., 156
Solberg, K., 444
Soukup, J., 136, 162
Sowinski, J., 296, 343, 662
Sperisen, F., 343, 680
Spinka, H., 275, 713
Stachetzki, R., 308
Stancu, I., 32
Staudt, G., 569
Steffens, E., 686
Steffens, K., 18
Steijger, J.J.M., 551
Stephenson, E.J., 296, 401, 407, 581, 662
Stephenson, S.L., 120
Steski, D.B., 611
Stewart, J.A., 698
Stibunov, V.N., 530
Stinson, G.M., 136, 162
Stock, F., 674, 686
Stocki, T., 136
Stracener, D., 407
Sugarbaker, E.R., 343, 413, 470
Sukhanov, A.V., 530
Sukumar, C.V., 605
Sum, V., 136, 302
Surkau, R., 18
Susukita, R., 314
Suttorp, O., 668
Swenson, D.R., 349
Symons, T.J.M., 623
Szymanski, J.J., 156

T

Tabakin, Frank, 515, 521
Taddeucci, T.N., 371, 413, 470
Takagi, A., 656
Takahisa, K., 450, 702
Takashima, R., 314
Takeuchi, F., 314
Tamii, A., 395
Tamura, H., 702
Tamura, K., 450
Tanaka, M., 450, 629
Tanifuji, M., 575, 587
Tanigaki, M., 623
Tanihata, I., 623
Teglia, A., 308
Terburg, B., 63
Theunissen, J.A.P., 530

Thiessen, D., 349
Thomas, A.W., 3
Thompson, I.J., 617
Thorn, C., 524
Thornton, S., 524
Titov, N.A., 136
Tlustý, P., 314
Tomasi-Gustafsson, E., 425, 545
Tonhäuser, J., 686
Toporkov, D.K., 530
Tornow, Werner, 260, 290, 593
Tosello, F., 476
Toyama, S., 395
Toyokawa, H., 563
Travkin, V.I., 476
Tropilo, J., 569
Tschalär, C., 343
Tsentalovich, E.P., 530
Tungate, G., 611
Tupa, D., 349

U

Uesaka, T., 419, 431, 629
Umnikov, A.Yu., 79
Unal, O., 343, 551

V

Vaandering, E.W., 713
van den Brand, J.F.J., 343, 551
van den Brandt, B., 308
van Oers, W.T.H., 136, 162, 302, 650
Van de Wiele, J., 431
Varner, R., 407
Vasil'ev, G.A., 643
Vigdor, S.E., 476
Vikhrov, V.V., 530
Vodinas, N., 551
Volosov, A.V., 530
von Przewoski, B., 319, 343, 680
Voutier, E., 545
Vuaridel, B., 87, 308

W

Wakasa, T., 419, 431, 629
Walcher, Th., 18
Walter, R.L., 290, 593
Wang, Y., 438

Ward, R.P., 611
Watson, J.W., 438
Wedekind, M., 662
Weidmann, R., 668
Welch, T.P., 343
Wells, S.P., 296, 401, 407
Welz, J., 650
Whisnant, C.S., 524
Wight, G.W., 650
Wilburn, W.S., 185
Willis, A., 425, 545
Wise, T., 680
Wissink, S.W., 296, 401, 407
Witkowski, M., 63
Wojtsekhowski, B.B., 530

Y

Yakushev, V.P., 643
Yamagoshi, M., 395
Yamaguchi, T., 623
Yamashita, S., 314
Yamashita, T., 431
Yamazaki, H., 450
Yen, Yi-Fen, 120
Yokkaichi, S., 314
Yoshida, K., 623
Yoshida, M., 314
Yoshimura, M., 395
Yosoi, M., 395
Young, L., 530
Yuan, V.W., 120
Yuasa, Y., 450

Z

Zalikanov, B., 476
Zanotti-Müller, E., 569
Zapfe, Kirsten, 686
Zegers, C., 551
Zelenski, A.N., 136, 162, 650
Zetta, L., 569
Zhao, J., 162, 302
Zhao, Q., 349
Zhdanov, A.A., 456
Zhgun, A.A., 456
Zhou, Z-L., 343, 551
Zhuralev, A., 319
Zosi, G., 476

AIP Conference Proceedings

		L.C. Number	ISBN18-307-2
No. 189	Relativistic, Quantum Electrodynamic, and Weak Interaction Effects in Atoms (Santa Barbara, CA, 1988)	89-84431	0-88318-389-7
No. 190	Radio-frequency Power in Plasmas (Irvine, CA, 1989)	89-45805	0-88318-397-8
No. 191	Advances in Laser Science—IV (Atlanta, GA, 1988)	89-85595	0-88318-391-9
No. 192	Vacuum Mechatronics (First International Workshop) (Santa Barbara, CA, 1989)	89-45905	0-88318-394-3
No. 193	Advanced Accelerator Concepts (Lake Arrowhead, CA, 1989)	89-45914	0-88318-393-5
No. 194	Quantum Fluids and Solids—1989 (Gainesville, FL, 1989)	89-81079	0-88318-395-1
No. 195	Dense Z-Pinches (Laguna Beach, CA, 1989)	89-46212	0-88318-396-X
No. 196	Heavy Quark Physics (Ithaca, NY, 1989)	89-81583	0-88318-644-6
No. 197	Drops and Bubbles (Monterey, CA, 1988)	89-46360	0-88318-392-7
No. 198	Astrophysics in Antarctica (Newark, DE, 1989)	89-46421	0-88318-398-6
No. 199	Surface Conditioning of Vacuum Systems (Los Angeles, CA, 1989)	89-82542	0-88318-756-6
No. 200	High T_c Superconducting Thin Films: Processing, Characterization, and Applications (Boston, MA, 1989)	90-80006	0-88318-759-0
No. 201	QED Structure Functions (Ann Arbor, MI, 1989)	90-80229	0-88318-671-3
No. 202	NASA Workshop on Physics From a Lunar Base (Stanford, CA, 1989)	90-55073	0-88318-646-2
No. 203	Particle Astrophysics: The NASA Cosmic Ray Program for the 1990s and Beyond (Greenbelt, MD, 1989)	90-55077	0-88318-763-9
No. 204	Aspects of Electron-Molecule Scattering and Photoionization (New Haven, CT, 1989)	90-55175	0-88318-764-7
No. 205	The Physics of Electronic and Atomic Collisions (XVI International Conference) (New York, NY, 1989)	90-53183	0-88318-390-0

No. 206	Atomic Processes in Plasmas (Gaithersburg, MD, 1989)	90-55265	0-88318-769-8
No. 207	Astrophysics from the Moon (Annapolis, MD, 1990)	90-55582	0-88318-770-1
No. 208	Current Topics in Shock Waves (Bethlehem, PA, 1989)	90-55617	0-88318-776-0
No. 209	Computing for High Luminosity and High Intensity Facilities (Santa Fe, NM, 1990)	90-55634	0-88318-786-8
No. 210	Production and Neutralization of Negative Ions and Beams (Brookhaven, NY, 1990)	90-55316	0-88318-786-8
No. 211	High-Energy Astrophysics in the 21st Century (Taos, NM, 1989)	90-55644	0-88318-803-1
No. 212	Accelerator Instrumentation (Brookhaven, NY, 1989)	90-55838	0-88318-645-4
No. 213	Frontiers in Condensed Matter Theory (New York, NY, 1989)	90-6421	0-88318-771-X 0-88318-772-8 (pbk.)
No. 214	Beam Dynamics Issues of High-Luminosity Asymmetric Collider Rings (Berkeley, CA, 1990)	90-55857	0-88318-767-1
No. 215	X-Ray and Inner-Shell Processes (Knoxville, TN, 1990)	90-84700	0-88318-790-6
No. 216	Spectral Line Shapes, Vol. 6 (Austin, TX, 1990)	90-06278	0-88318-791-4
No. 217	Space Nuclear Power Systems (Albuquerque, NM, 1991)	90-56220	0-88318-838-4
No. 218	Positron Beams for Solids and Surfaces (London, Canada, 1990)	90-56407	0-88318-842-2
No. 219	Superconductivity and Its Applications (Buffalo, NY, 1990)	91-55020	0-88318-835-X
No. 220	High Energy Gamma-Ray Astronomy (Ann Arbor, MI, 1990)	91-70876	0-88318-812-0
No. 221	Particle Production Near Threshold (Nashville, IN, 1990)	91-55134	0-88318-829-5
No. 222	After the First Three Minutes (College Park, MD, 1990)	91-55214	0-88318-828-7
No. 223	Polarized Collider Workshop (University Park, PA, 1990)	91-71303	0-88318-826-0
No. 224	LAMPF Workshop on (π, K) Physics (Los Alamos, NM, 1990)	91-71304	0-88318-825-2

No. 225	Half Collision Resonance Phenomena in Molecules (Caracas, Venezuela, 1990)	91-55210	0-88318-840-6
No. 226	The Living Cell in Four Dimensions (Gif sur Yvette, France, 1990)	91-55209	0-88318-794-9
No. 227	Advanced Processing and Characterization Technologies (Clearwater, FL, 1991)	91-55194	0-88318-910-0
No. 228	Anomalous Nuclear Effects in Deuterium/ Solid Systems (Provo, UT, 1990)	91-55245	0-88318-833-3
No. 229	Accelerator Instrumentation (Batavia, IL, 1990)	91-55347	0-88318-832-1
No. 230	Nonlinear Dynamics and Particle Acceleration (Tsukuba, Japan, 1990)	91-55348	0-88318-824-4
No. 231	Boron-Rich Solids (Albuquerque, NM, 1990)	91-53024	0-88318-793-4
No. 232	Gamma-Ray Line Astrophysics (Paris-Saclay, France, 1990)	91-55492	0-88318-875-9
No. 233	Atomic Physics 12 (Ann Arbor, MI, 1990)	91-55595	088318-811-2
No. 234	Amorphous Silicon Materials and Solar Cells (Denver, CO, 1991)	91-55575	088318-831-7
No. 235	Physics and Chemistry of MCT and Novel IR Detector Materials (San Francisco, CA, 1990)	91-55493	0-88318-931-3
No. 236	Vacuum Design of Synchrotron Light Sources (Argonne, IL, 1990)	91-55527	0-88318-873-2
No. 237	Kent M. Terwilliger Memorial Symposium (Ann Arbor, MI, 1989)	91-55576	0-88318-788-4
No. 238	Capture Gamma-Ray Spectroscopy (Pacific Grove, CA, 1990)	91-57923	0-88318-830-9
No. 239	Advances in Biomolecular Simulations (Obernai, France, 1991)	91-58106	0-88318-940-2
No. 240	Joint Soviet-American Workshop on the Physics of Semiconductor Lasers (Leningrad, USSR, 1991)	91-58537	0-88318-936-4
No. 241	Scanned Probe Microscopy (Santa Barbara, CA, 1991)	91-76758	0-88318-816-3
No. 242	Strong, Weak, and Electromagnetic Interactions in Nuclei, Atoms, and Astrophysics: A Workshop in Honor of Stewart D. Bloom's Retirement (Livermore, CA, 1991)	91-76876	0-88318-943-7

No. 243	Intersections Between Particle and Nuclear Physics (Tucson, AZ, 1991)	91-77580	0-88318-950-X
No. 244	Radio Frequency Power in Plasmas (Charleston, SC, 1991)	91-77853	0-88318-937-2
No. 245	Basic Space Science (Bangalore, India, 1991)	91-78379	0-88318-951-8
No. 246	Space Nuclear Power Systems (Albuquerque, NM, 1992)	91-58793	1-56396-027-3 1-56396-026-5 (pbk.)
No. 247	Global Warming: Physics and Facts (Washington, DC, 1991)	91-78423	0-88318-932-1
No. 248	Computer-Aided Statistical Physics (Taipei, Taiwan, 1991)	91-78378	0-88318-942-9
No. 249	The Physics of Particle Accelerators (Upton, NY, 1989, 1990)	92-52843	0-88318-789-2
No. 250	Towards a Unified Picture of Nuclear Dynamics (Nikko, Japan, 1991)	92-70143	0-88318-951-8
No. 251	Superconductivity and its Applications (Buffalo, NY, 1991)	92-52726	1-56396-016-8
No. 252	Accelerator Instrumentation (Newport News, VA, 1991)	92-70356	0-88318-934-8
No. 253	High-Brightness Beams for Advanced Accelerator Applications (College Park, MD, 1991)	92-52705	0-88318-947-X
No. 254	Testing the AGN Paradigm (College Park, MD, 1991)	92-52780	1-56396-009-5
No. 255	Advanced Beam Dynamics Workshop on Effects of Errors in Accelerators, Their Diagnosis and Corrections (Corpus Christi, TX, 1991)	92-52842	1-56396-006-0
No. 256	Slow Dynamics in Condensed Matter (Fukuoka, Japan, 1991)	92-53120	0-88318-938-0
No. 257	Atomic Processes in Plasmas (Portland, ME, 1991)	91-08105	0-88318-939-9
No. 258	Synchrotron Radiation and Dynamic Phenomena (Grenoble, France, 1991)	92-53790	1-56396-008-7
No. 259	Future Directions in Nuclear Physics with 4π Gamma Detection Systems of the New Generation (Strasbourg, France, 1991)	92-53222	0-88318-952-6

No.	Title	Number	ISBN
No. 260	Computational Quantum Physics (Nashville, TN, 1991)	92-71777	0-88318-933-X
No. 261	Rare and Exclusive B&K Decays and Novel Flavor Factories (Santa Monica, CA, 1991)	92-71873	1-56396-055-9
No. 262	Molecular Electronics—Science and Technology (St. Thomas, Virgin Islands, 1991)	92-72210	1-56396-041-9
No. 263	Stress-Induced Phenomena in Metallization: First International Workshop (Ithaca, NY, 1991)	92-72292	1-56396-082-6
No. 264	Particle Acceleration in Cosmic Plasmas (Newark, DE, 1991)	92-73316	0-88318-948-8
No. 265	Gamma-Ray Bursts (Huntsville, AL, 1991)	92-73456	1-56396-018-4
No. 266	Group Theory in Physics (Cocoyoc, Morelos, Mexico, 1991)	92-73457	1-56396-101-6
No. 267	Electromechanical Coupling of the Solar Atmosphere (Capri, Italy, 1991)	92-82717	1-56396-110-5
No. 268	Photovoltaic Advanced Research & Development Project (Denver, CO, 1992)	92-74159	1-56396-056-7
No. 269	CEBAF 1992 Summer Workshop (Newport News, VA, 1992)	92-75403	1-56396-067-2
No. 270	Time Reversal—The Arthur Rich Memorial Symposium (Ann Arbor, MI, 1991)	92-83852	1-56396-105-9
No. 271	Tenth Symposium Space Nuclear Power and Propulsion (Vols. I–III) (Albuquerque, NM, 1993)	92-75162	1-56396-137-7 (set)
No. 272	Proceedings of the XXVI International Conference on High Energy Physics (Vols. I and II) (Dallas, TX, 1992)	93-70412	1-56396-127-X (set)
No. 273	Superconductivity and Its Applications (Buffalo, NY, 1992)	93-70502	1-56396-189-X
No. 274	VIth International Conference on the Physics of Highly Charged Ions (Manhattan, KS, 1992)	93-70577	1-56396-102-4
No. 275	Atomic Physics 13 (Munich, Germany, 1992)	93-70826	1-56396-057-5

No. 276	Very High Energy Cosmic-Ray Interactions: VIIth International Symposium (Ann Arbor, MI, 1992)	93-71342	1-56396-038-9
No. 277	The World at Risk: Natural Hazards and Climate Change (Cambridge, MA, 1992)	93-71333	1-56396-066-4
No. 278	Back to the Galaxy (College Park, MD, 1992)	93-71543	1-56396-227-6
No. 279	Advanced Accelerator Concepts (Port Jefferson, NY, 1992)	93-71773	1-56396-191-1
No. 280	Compton Gamma-Ray Observatory (St. Louis, MO, 1992)	93-71830	1-56396-104-0
No. 281	Accelerator Instrumentation Fourth Annual Workshop (Berkeley, CA, 1992)	93-072110	1-56396-190-3
No. 282	Quantum 1/f Noise & Other Low Frequency Fluctuations in Electronic Devices (St. Louis, MO, 1992)	93-072366	1-56396-252-7
No. 283	Earth and Space Science Information Systems (Pasadena, CA, 1992)	93-072360	1-56396-094-X
No. 284	US-Japan Workshop on Ion Temperature Gradient-Driven Turbulent Transport (Austin, TX, 1993)	93-72460	1-56396-221-7
No. 285	Noise in Physical Systems and 1/f Fluctuations (St. Louis, MO, 1993)	93-72575	1-56396-270-5
No. 286	Ordering Disorder: Prospect and Retrospect in Condensed Matter Physics: Proceedings of the Indo-U.S. Workshop (Hyderabad, India, 1993)	93-072549	1-56396-255-1
No. 287	Production and Neutralization of Negative Ions and Beams: Sixth International Symposium (Upton, NY, 1992)	93-72821	1-56396-103-2
No. 288	Laser Ablation: Mechanismas and Applications-II: Second International Conference (Knoxville, TN, 1993)	93-73040	1-56396-226-8
No. 289	Radio Frequency Power in Plasmas: Tenth Topical Conference (Boston, MA, 1993)	93-72964	1-56396-264-0
No. 290	Laser Spectroscopy: XIth International Conference (Hot Springs, VA, 1993)	93-73050	1-56396-262-4

No.	Title	LCCN	ISBN
No. 291	Prairie View Summer Science Academy (Prairie View, TX, 1992)	93-73081	1-56396-133-4
No. 292	Stability of Particle Motion in Storage Rings (Upton, NY, 1992)	93-73534	1-56396-225-X
No. 293	Polarized Ion Sources and Polarized Gas Targets (Madison, WI, 1993)	93-74102	1-56396-220-9
No. 294	High-Energy Solar Phenomena A New Era of Spacecraft Measurements (Waterville Valley, NH, 1993)	93-74147	1-56396-291-8
No. 295	The Physics of Electronic and Atomic Collisions: XVIII International Conference (Aarhus, Denmark, 1993)	93-74103	1-56396-290-X
No. 296	The Chaos Paradigm: Developments an Applications in Engineering and Science (Mystic, CT, 1993)	93-74146	1-56396-254-3
No. 297	Computational Accelerator Physics (Los Alamos, NM, 1993)	93-74205	1-56396-222-5
No. 298	Ultrafast Reaction Dynamics and Solvent Effects (Royaumont, France, 1993)	93-074354	1-56396-280-2
No. 299	Dense Z-Pinches: Third International Conference (London, 1993)	93-074569	1-56396-297-7
No. 300	Discovery of Weak Neutral Currents: The Weak Interaction Before and After (Santa Monica, CA, 1993)	94-70515	1-56396-306-X
No. 301	Eleventh Symposium Space Nuclear Power and Propulsion (3 Vols.) (Albuquerque, NM, 1994)	92-75162 156396-301-9	1-56396-305-1 (Set) (pbk. set)
No. 302	Lepton and Photon Interactions/ XVI International Symposium (Ithaca, NY, 1993)	94-70079	1-56396-106-7
No. 303	Slow Positron Beam Techniques for Solids and Surfaces Fifth International Workshop (Jackson Hole, WY 1992)	94-71036	1-56396-267-5
No. 304	The Second Compton Symposium (College Park, MD, 1993)	94-70742	1-56396-261-6
No. 305	Stress-Induced Phenomena in Metallization Second International Workshop (Austin, TX, 1993)	94-70650	1-56396-251-9

No. 306	12th NREL Photovoltaic Program Review (Denver, CO, 1993)	94-70748	1-56396-315-9
No. 307	Gamma-Ray Bursts Second Workshop (Huntsville, AL 1993)	94-71317	1-56396-336-1
No. 308	The Evolution of X-Ray Binaries (College Park, MD 1993)	94-76853	1-56396-329-9
No. 309	High-Pressure Science and Technology—1993 (Colorado Springs, CO 1993)	93-72821	1-56396-219-5 (Set)
No. 310	Analysis of Interplanetary Dust (Houston, TX 1993)	94-71292	1-56396-341-8
No. 311	Physics of High Energy Particles in Toroidal Systems (Irvine, CA 1993)	94-72098	1-56396-364-7
No. 312	Molecules and Grains in Space (Mont Sainte-Odile, France 1993)	94-72615	1-56396-355-8
No. 313	The Soft X-Ray Cosmos ROSAT Science Symposium (College Park, MD 1993)	94-72499	1-56396-327-2
No. 314	Advances in Plasma Physics Thomas H. Stix Symposium (Princeton, NJ 1992)	94-72721	1-56396-372-8
No. 315	Orbit Correction and Analysis in Circular Accelerators (Upton, NY 1993)	94-72257	1-56396-373-6
No. 316	Thirteenth International Conference on Thermoelectrics (Kansas City, Missouri 1994)	95-75634	1-56396-444-9
No. 317	Fifth Mexican School of Particles and Fields (Guanajuato, Mexico 1992)	94-72720	1-56396-378-7
No. 318	Laser Interaction and Related Plasma Phenomena 11th International Workshop (Monterey, CA 1993)	94-78097	1-56396-324-8
No. 319	Beam Instrumentation Workshop (Santa Fe, NM 1993)	94-78279	1-56396-389-2
No. 320	Basic Space Science (Lagos, Nigeria 1993)	94-79350	1-56396-328-0
No. 321	The First NREL Conference on Thermophotovoltaic Generation of Electricity (Copper Mountain, CO 1994)	94-72792	1-56396-353-1

No. 322	Atomic Processes in Plasmas Ninth APS Topical Conference (San Antonio, TX)	94-72923	1-56396-411-2
No. 323	Atomic Physics 14 Fourteenth International Conference on Atomic Physics (Boulder, CO 1994)	94-73219	1-56396-348-5
No. 324	Twelfth Symposium on Space Nuclear Power and Propulsion (Albuquerque, NM 1995)	94-73603	1-56396-427-9
No. 325	Conference on NASA Centers for Commercial Development of Space (Albuquerque, NM 1995)	94-73604	1-56396-431-7
No. 326	Accelerator Physics at the Superconducting Super Collider (Dallas, TX 1992-1993)	94-73609	1-56396-354-X
No. 327	Nuclei in the Cosmos III Third International Symposium on Nuclear Astrophysics (Assergi, Italy 1994)	95-75492	1-56396-436-8
No. 328	Spectral Line Shapes, Volume 8 12th ICSLS (Toronto, Canada 1994)	94-74309	1-56396-326-4
No. 329	Resonance Ionization Spectroscopy 1994 Seventh International Symposium (Bernkastel-Kues, Germany 1994)	95-75077	1-56396-437-6
No. 330	Computational Chemistry F.E.C.S. Conference (Nancy, France 1994)	95-75843	1-56396-457-0
No. 331	Non-Neutral Plasma Physics II (Berkeley, CA 1994)	95-79630	1-56396-441-4
No. 332	X-Ray Lasers 1994 Fourth International Colloquium (Williamsburg, VA 1994)	95-76067	1-56396-375-2
No. 333	Beam Instrumentation Workshop (Vancouver, B. C., Canada 1994)	95-79635	1-56396-352-3
No. 334	Few-Body Problems in Physics (Williamsburg, VA 1994)	95-76481	1-56396-325-6
No. 336	Dark Matter (College Park, MD 1994)	95-76538	1-56396-438-4
No. 337	Pulsed RF Sources for Linear Colliders (Montauk, NY 1994)	95-76814	1-56396-408-2

No. 338	Intersections Between Particle and Nuclear Physics 5th Conference (St. Petersburg, FL 1994)	95-77076	1-56396-335-3
No. 339	Polarization Phenomena in Nuclear Physics Eighth International Symposium (Bloomington, IN 1994)	95-77216	1-56396-482-1
No. 340	Strangeness in Hadronic Matter (Tucson, AZ 1995)	95-77477	1-56396-489-9
No. 341	Volatiles in the Earth and Solar System (Pasadena, CA 1994)	95-77911	1-56396-409-0
No. 342	CAM -94 Physics Meeting (Cacun, Mexico 1994)	95-77851	1-56396-491-0